A Flora of Cumbria

comprising the vice-counties of Westmorland with Furness (v.c.69),
Cumberland (v.c.70) and parts of North-west Yorkshire (v.c.65)
and North Lancashire (v.c.60)

by

Geoffrey Halliday

**with contributions by C.C. Haworth†, D.A. Ratcliffe and W. Pennington
and computational assistance by A.J.C. Malloch**

**Published by the Centre for North-West Regional Studies
University of Lancaster**

1997

This Flora is dedicated to the memory of those Recorders
who, sadly, did not live to see its realisation

Molly Birkett Muriel Coulson Margaret Gill

Chris Haworth Eileen Rhone Elizabeth Sterne Ralph Stokoe

Joan Williamson

**Oh! that we had a book of botany. All flowers now are gay and
deliciously sweet. The primrose still pre-eminent among the later
flowers of the spring. Foxgloves very tall, with their heads budding.**

***The Grasmere Journal*, May 16th, 1800, Dorothy Wordsworth**

[In March 1801 the Wordsworths acquired Withering's *An arrangement of British Plants
according to latest improvements of the Linnaean system.*]

Cover photograph: Ullswater from Gowbarrow
Copyright: © Geoffrey Halliday   ISBN 1-86220-020-3   Printed by Shanleys, Bolton, Greater Manchester

Contents

Maps of Cumbria

Water lilies and Brothers' Water

Introduction

Although a southerner, I felt even as a schoolboy on holiday in Westmorland that this must be God's own county. It was shortly after the last war that I came across a pristine copy of Wilson's *Flora of Westmorland* in Robert's bookshop at Kendal and this was sufficient to fire my enthusiasm for its flora. When Eventually I moved north in 1968 to settle in Westmorland, I became actively involved with the then Lake District Naturalists' Trust. This gave me an added excuse to travel widely throughout the county and over the whole area which was destined to become Cumbria.

The idea of up-dating Wilson's Flora and including Furness, so encompassing the whole of vice-county 69, crossed my mind intermittently but it became only too evident that Cumberland, vice-county 70, was even more in need of a Flora, Hodgson's having been published in 1898. Combining the two would have the obvious merit of treating the whole of the Lake District, for the first time since Baker's Flora exactly a century ago (1896). A further impetus was provided by the local government reorganisation of 1974 which gave Cumbria official recognition as a county. It merged Westmorland with Cumberland and added Furness, previously Lancashire north of the sands, as well as the old Sedbergh Rural District, part of vice-county 65, North-west Yorkshire, and formerly part of the West Riding of Yorkshire.

By 1974 the need for a new Flora was acute. Local authorities, the Lake District National Park and the Naturalists' Trust were all increasingly requiring information for planning and conservation purposes but if any was available it was all too often out of date, imprecise or subjective. It was evident that a new Flora would require the systematic collection of new records on a county-wide, grid-square basis. So the Flora of Cumbria Recording Survey came into being. This was based on the Division of Biological Sciences at Lancaster University, organised by myself and operated by a team of recorders. It quickly came to dominate my life and for much of the time I have been overwhelmed by a mass of notes, lists, queries and specimens from recorders and other contributors. I am all too conscious that not all have been dealt with adequately and that some important records will have been overlooked. For this I apologise, pleading only that time is short and that a curtain had eventually to descend after 23 years of recording. The recent review of local government induced panic among us that Cumbria would disappear as the Flora was published but to our intense relief we have been spared.

The reader will make his own judgement as to the value of the Survey. Personally I have found it immensely rewarding - as an exercise in extra-mural education, as a source of an ever-widening circle of friends with mutual interests, and as a continuing revelation of the extraordinary variety and beauty of the county.

Geoffrey Halliday
Burton-in-Kendal
July 1997

You dribble into lower Dunnerdale
Through wet woods and wood-soil and woodland flowers,
Tutson, the St John's-wort with a single yellow bead,
Marsh marigold, creeping jenny and daffodils;
Here from hazel islands in the late spring
The catkins fall and ride along the stream
Like little yellow weasels, and the soil is loosed
From bulbs of the white lily that smells of garlic,
And dippers rock up and down on rubber legs,
And long-tailed tits are flung through the air like darts;
From *To the River Duddon*, Norman Nicholson

Plate 1. Gorge of the River Duddon, April

6

Acknowledgments

It would have been quite impossible to produce this Flora without the computational assistance of my colleague A.J.C. Malloch. Andrew wrote all the necessary computer programs and with great patience ushered me belatedly into the computer era. Without a constant supply of updated tetrad print-outs and computer-generated maps our task would have been impossible. The high-quality computer-generated colour maps and the base maps used for the individual distribution maps have been painstakingly prepared by Nicky Higgitt. I should also like to thank Derek Ratcliffe for generously putting his accumulated Lake District and Cumberland records and knowledge at my disposal and, with Donald Pigott, reading and commenting on most of the draft systematic accounts. I am grateful also to him and Anne Pennington for their contributions to the introductory sections. The *Taraxacum* account was contributed by Chris Haworth and completed just before his tragic early death in 1989; it has since been revised and updated by Andrew Dudman.

I shall be forever indebted to the many who have contributed records for the Flora but particularly to the core of active recorders, listed below, for their persistence, endless patience, their tolerance of my many lapses and of the seemingly endless stream of requests to confirm records, to visit under-recorded areas, to check computer print-outs and to comment on draft accounts.

J. Adams	R.E.Groom	M. Smith
J. Atkins	†C.C. Haworth	C.F. & J. Steeden
M. Baecker	E. Marper	M.P.G. Tolfree
R.W.M. Corner	J. Parker	C.E. Wild
†M.G. Coulson	M. Porter	†J. Williamson
A.A. Dudman	†E. Rhone	C. Willink
J.C. Frankland	F.J. Roberts	K. & T. Wilson
†M.M. Gill	L. Robinson	F.L. Woodman
M. Gregory	H. Smith	

Invidious as it is to single out individual contributions, the persistent and sterling work done by Margaret Gregory and Jean Parker in the extreme north of the county and Colin Wild in the south deserve special mention. It is a matter of immense sadness that Molly Birkett, Muriel Coulson, Margaret Gill, Chris Haworth, Eileen Rhone, Elizabeth Sterne, Ralph Stokoe and Joan Williamson have not survived to see the fruits of their labours. Many others have contributed records and I should particularly like to thank J. Atkinson, I. Bonner, N. Botham, P. Buchanan, H. Caldwell, A. Cannell, M.J.Y. Foley, H. Frankland, A. Franks, K.A. Gunning, L. Henderson, M.M. Milne, I. Mortemore, J. Mounsey, K. Raistrick, T.C.G. Rich, C. Smith and A.E. Wilson. P. Bullard, J.S. Rodwell and P. Taylor have made helpful comments on the introductory sections.

We also owe an immense debt of gratitude to the many referees and specialists who have generously given of their time and patience including A.K. Bermani (*Festuca*), J. Bevan (*Hieracium*), J.C. Bowra (*Oenothera*), M.E. Bradshaw (*Alchemilla*), A.O. Chater, E.J. Clement (aliens), T. Cope (*Elytrigia*), J. Cullen (*Anthyllis*), M.G. Daker (*Fumaria*), R.W. David (Carex), T.T. Elkington (*Erophila*), I.K Ferguson (*Salicornia*), J. Fryer (*Cotoneaster*), G.G. Graham (*Rosa*), R.M. Harley (*Mentha*), B. Harold (*Potentilla*), A.C. Jermy, D.H. Kent (nomenclature), J.M. Mullin (*Chenopodium*), T.D. Pennington (*Epilobium*), C.D. Preston (*Potamogeton*), T.C.G. Rich, R.H. Roberts (*Polypodium*), N.K.B. Robson (*Hypericum, Ribes*), P.D. Sell (*Hieracium*), D.A. Simpson (*Elodea*), C.A. Stace (*Festuca*), N.F. Stewart (*Callitriche*), J. Voysey (forestry trees), S.M. Walters (*Alchemilla*), S. Webster (*Ranunculus*) and P.F. Yeo (*Aster, Geranium*). The following have given invaluable help and encouragement with the major critical groups: D.J. McCosh (*Hieracium*), R.D. Meikle (*Salix*), A. Newton (*Rubus*), A.J. Richards (*Taraxacum*) and A.J. Silverside (*Euphrasia*).

Special thanks are also due to C.D. Preston and J.M. Croft of the Biological Records Centre of the Institute of Terrestrial Ecology for endless patience in dealing with my requests and also to the almost universally co-operative Cumbrian landowners and farmers, with apologies to any who might have been inconvenienced by our recording activities.

Jim Adams has been a constant source of help dealing with the innumerable chores involved with the final editing and the preparation of the Index, and I am also grateful to K. Traynor and the staff of Shankleys our printers for their interest and close co-operation.

The following generously provided illustrative material: P. Arnold (painting of limestone pavement), E. Huckaby (computerised pollen diagrams), A. Sier (drawings of *Sorbus* leaves) and O.M. Stewart (drawings of *Calamagrostis*). Technical help has also been provided by R. Berry, P.W.H. Flint, C. Jeffery, S. Jenkins and J. Pritchard.

The following have contributed photographs:
R. Berry - plate 63
R.J. Cooper - 36, 111
V. Corbett - dust jacket, frontispiece, 11, 24, 32, 35, 48, 74
R.W.M. Corner - 14
B. Cross - 75, 77
A.A. Dudman - 51, 76
B. Fleming - 10, 18
M.J.Y. Foley - 84, 112
A.N. Gagg - 54, 100
G. Halliday - 6, 12, 13, 17, 19, 21, 26, 27, 29, 34, 37-41, 43, 44, 52, 55-60, 62, 64, 66-70, 79, 80, 82, 83, 85, 88, 92, 95-99, 104-109

J. Halliday - 49
C.C. Haworth - 91
R. Longstaff - 23
R. Mitchell - 2, 4, 7, 8, 15
J. Parker - 53, 86
D.A. Radcliffe - 1, 3, 5, 9, 16, 20, 22, 25, 28, 30, 31, 33, 45, 47, 50, 81, 87
K. Raistrick - 101- 103
L. Robinson - 46
J.S. Rodwell - 42
R. Stokoe - 110
I. Taylor - 65
P.R. Wallis - 113
P.F. Yeo - 78

We are grateful to David Higham Associates and A.H. Griffin for allowing reproduction of copyright material and we also acknowledge permission to use the following maps: Geological map of the Lake District (F. Moseley and the Yorkshire Geological Society), Average annual rainfall (The Meteorological Office, 1977), Sites of Special Scientific Interest in Cumbria (English Nature) and the Lake District (Cumbria Tourist Board, 1992), from which the base maps were prepared.

Publication would not have been possible but for the extremely generous financial support of the following bodies:

The Botanical Society of the British Isles
Cumbria Wildlife Trust
Glaxo Wellcome
Lancaster University
The Curwen Archives Trust
The Frieda Scott Charitable Trust
The Kirkby Archives Trust
The Tansley Fund of the *New Phytologist*.

Finally, I wish to thank the University and the Division of Biological Sciences for providing time, facilities and financial support for the project over the last 20 years.

G.H.

Plate 2. Whitbarrow, looking north, with the Scar in the foreground and Witherslack woods on the left

Geology and scenery
D.A. Ratcliffe

The two extremities of the lake [Derwent Water] afford most discordant propects: the Southern is a composition of all that is horrible; an immense chain open in the midst, whose entrance is divided by a rude conic hill, once topt with a castle, the habitation of the tyrant of the rocks.......But the opposite or northern view is in all respects a strong and beautiful contrast: *Skiddaw* shows its vast base, and bounding all that part of the vale, rises gently to a height that sinks the neighbouring hills; opens a pleasing front, smooth and verdant, smiling over the country like a gentle generous lord, while the *fells* of *Borrowdale* frown on it like a hardened tyrant.

A Tour of Scotland and Voyage to the Hebrides in 1772, Thomas Pennant

Cumbria has a complex geology (Fig. 2) and a greater variety of scenery than any other English county. A main contrast is between the central mass of older Ordovician and Silurian rocks forming the Lake District and the peripheral series of younger rocks of Carboniferous, Permian and Triassic age. These younger rocks form an encircling and partly coastal plain to the Lakeland fells, but the Carboniferous formations rise eastwards into the massive northernmost part of the Pennine Chain, and then again beyond the Tyne Gap as the south-western extremity of the Cheviot range. The orogenic processes that have given the Lake District its distinctive topography have been well described by Pearsall & Pennington (1973).

Successive and complex folding and uplifting of a great central dome, followed by extensive denudation of younger, superficial rocks and deep erosion along a developing radial drainage pattern, gradually formed the Lake District. The later folding and faulting of the Pennine uplift were also succeeded by erosion of the uppermost strata. Younger sedimentary rocks came to occupy most of the northern lowlands. The basic structure of the region was established by the end of the Tertiary period. The serial glaciations of the Quaternary then sculpted the final form of the fells, dales and lakes that we see today. Glacial action also overlaid most of the lower areas of country rock with great spreads of drift: fluvio-glacial sands, gravel and boulder clay, which somewhat complicate the relationship of plant distribution to solid geology. Morainic material also covers many mountain valley floors and slopes up to considerable altitudes. Coastal processes have led to the accumulation of marine sediments, especially flanking the larger estuaries; to the building of sand dunes and shingle beaches; and to the cutting of seacliffs. On waterlogged ground at all levels, from the coastal alluvial plains to the high mountain watersheds, deep deposits of peat have developed.

Geology has thus determined the range of physical features that become plant habitats and constitute scenery. It has given the range of slope angles, from cliffs to flat ground; lakes and watercourses of all sizes and forms; special features from bogs to gorges and fissured limestone pavements; and substrates of widely differing texture, from boulderfields and screes to clay soils and peats. It has also

given a wide range of local climate, because of the orographic effects of high land rising to 900 m above the coastal plains, with a range of aspects. And finally, it has established the range of soil chemistry which is such a crucial factor for plants.

Soils
In common with most of Britain, the prevalent soil-forming processes of the region have been towards the development of acidic, base-deficient substrates through leaching and podsolisation. The over-riding chemical soil factor is the availability of calcium, and this is largely dependent on the amount of calcium carbonate (calcite or lime) in the soil parent material. Markedly calcareous soils (pH 7.0 and above, exchangeable Ca > 300 mg/100 g) are well developed on the Carboniferous limestone formation, but occur in patches on other more localised exposures of lime-rich rocks, such as the Coniston limestone, beds of calcareous sandstone and shale, and igneous materials such as andesite. They include dark rendzinas and red-brown loams developed in situ over limestone bedrock and brown earths and clayey gleys formed from colluvium and boulder clay. Some contain free $CaCO_3$. These lime-rich soils support a distinctive calcicolous flora which includes many species not found on other geological formations, and typically containing a far larger number of species than communities of comparable situations on acidic substrates.

Typical free-draining podsols are localised and occur especially on the drier sandstone-derived soils of the Eden Valley and elsewhere, especially under heather-dominated vegetation. While such soils with surface horizons in the pH range 4.0-5.0 (Ca < 30 mg/100 g), are widespread on uncultivated ground, especially in the fell country, drainage impedance on the gentler slopes usually produces some degree of gleying and thus a range from gley podsols to peaty gleys. These support a variety of moist grasslands and heaths (especially with *Nardus stricta*, *Molinia caerulea*, *Juncus squarrosus* and *Calluna vulgaris*) and shallow marsh and bog (especially with *Juncus effusus*, *J. acutiflorus*, *Carex echinata*, *C. rostrata* and *Sphagnum*). Under native woodland of oak and on steeper hillsides are brown podsolic soils where marked profile development has been resisted by tree nutrient cycling or gravitational instability. Dry treeless slopes with such soils typically carry *Festuca ovina - Agrostis* grassland, or heaths of *Vaccinium myrtillus* and *Calluna*. In dry, rocky situations on the mountains the brown podsolic soils grade into rankers with raw humus overlying stones and a variable development of immature and leached mineral soil.

In the lowlands, the original soils have mostly been modified by some degree of cultivation or improvement, even where they still carry semi-natural vegetation. The changes involve both drainage activities and nutrient enhancement (especially nitrogen, potassium and phosphorus) from fertilisers. Brown earths with mull humus are typical of mixed broadleaved woodland on more base-rich parent materials, or where these are flushed from such substrates. Fertile, base-saturated, brown loams are associated with a variety of enclosed meadow grasslands, but tend to be clayey and calcareous on limestone (pH 6.0-7.0), and sandy and non-calcareous on sandstone (pH 5.0-6.0).

Alluvial soils are well developed on valley bottoms and, more generally, within the flood zone of streams. Peats are predominantly of the ombrogenous type, with marked acidity (pH 3.0-4.5) and base deficiency, both in the lowlands (raised bogs) and uplands (blanket bogs). The widespread poor fen peats are moderately acidic (pH 4.5-5.5), but rich fen peats with high pH (6.5-7.5) and base content occur in a few places, mainly on limestone. There is also a range of shallower peat horizons of intermediate type, associated with valley and flush mires and lateral surface water movement, often over deep mineral gleys.

The Ordovician formations

The oldest rocks of Cumbria are the Skiddaw Slates, which form most of the north Lakeland fells, including the group dominated by Skiddaw itself (931 m) and Blencathra (868 m), the Buttermere fells with the Grasmoor (851 m) and Robinson (737 m) groups, and then the lower fells (such as Melbreak, 512 m) and foothills between Lorton Vale and Egremont. A southern outcrop forms the massif of Black Combe north of Millom. The Skiddaw Slates are relatively soft and have been sculpted by erosion into great soaring smooth slopes, often steep and high, with extensive unstable screes, rising to graceful peaks and domes, and dissected by deeply cleft gills. Ice action has been especially marked around Blencathra, leaving narrow ridges and deep-cut corries at Bannerdale, Scales Tarn and Bowscale Tarn. The cliffs tend to be rather broken and sloping, but there are sheer faces of 120 m at Dove Crags, Grasmoor. A noteworthy feature of the Skiddaw Slate is that it is almost everywhere markedly acidic and deficient in lime, so that its flora is lacking in variety and nearly devoid of calcicoles. Perhaps because of the relatively high silt content of the derived soils, solifluction (frost-thaw movement) features of the type described by Hollingworth (1934) appear to be especially well developed on the higher fells of

this formation, although they are also prominent on the Helvellyn range (Hay 1937) and on the summit of Cross Fell (Plate 9).

Several masses of lava, taking their name from Eycott Hill east of Mungrisdale, date from late in the Skiddaw Slate period. They form a broad band around the northern flanks of the Skiddaw group from High Pike and Roughtengill to the western outlier of Binsey. Other occurrences include the northern end of the Helvellyn range, Ullswater side, Bampton and lower Swindale, and near Millom. Apart from the knobbly and undulating terrain of Eycott Hill itself, where hollows between ridges have given an interesting series of marshes, they do not contribute in any notable way to either topographic or floristic variety. Their soil-forming properties are, however, rather more favourable than those of the Slate in yielding exchangeable bases.

The most extensive of the Ordovician formations is the Borrowdale Volcanic Series, forming the central fells of Lakeland. These rocks vary from hard flinty tuffs (of stone-axe fame) and lavas (rhyolite and andesite) to friable breccias. Their variable hardness has given rise to the uneven and rugged character of the central fells, with irregular knobbly slopes, ridges and summits; while the crag-girt valley sides and coves, and great boulder fields and scree slopes reflect the past intensity of ice and frost action. There are many hanging valleys with lower waterfall ravines, such as the Watendlath valley and Lodore Falls (Plate 3).

Plate 3. The Lodore Falls, Borrowdale

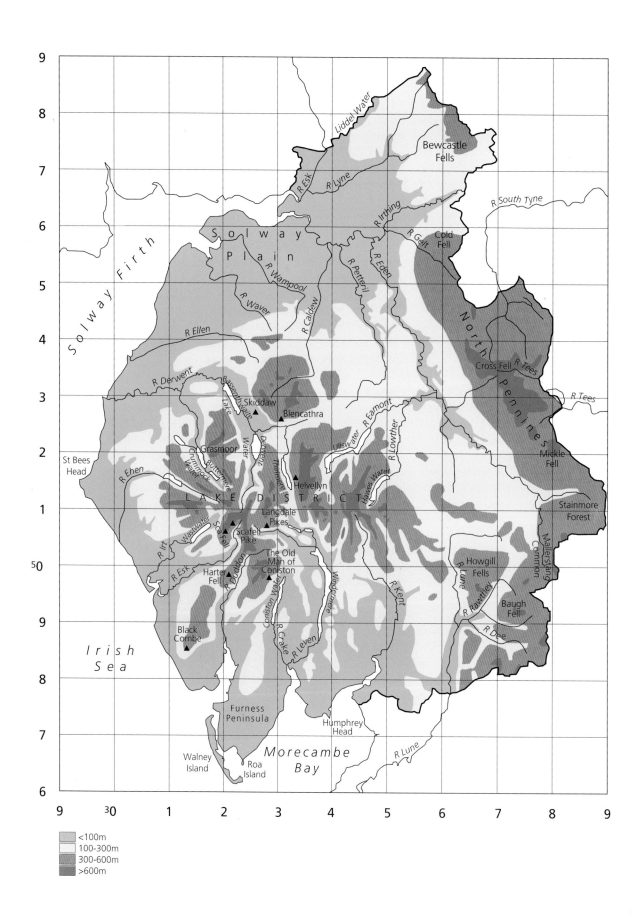

Fig. 1. Cumbria - Physical features

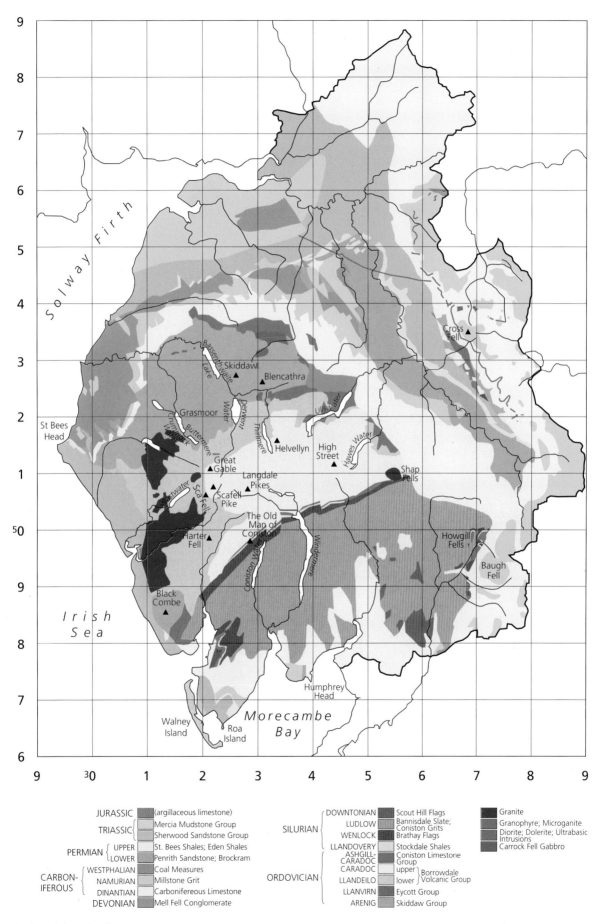

Fig. 2. Cumbria - Geology

These drain to the main dales of typically U-shaped form containing the large, ice-gouged lakes that are such a major feature and attraction. Corrie formation is most pronounced on the eastern ranges of Helvellyn (Plates 4, 12), Fairfield and High Street, while the Scafell range is notable for the deep chasms carved out along fault lines and dykes. The Lake District contains by far the largest and steepest inland cliffs in England, such as the Pillar Rock (200 m), Scafell Crag, Pavey Ark, Dow Crags and Dove Crag at Hartsop, and has accordingly become a major centre for rock climbing. The wild scene of Wast Water (Plate 5), in a deep trough below the huge gullied escarpment and talus fans of the Screes, and with an encircling ring of rocky mountains, shows the Borrowdale Volcanic terrain at its most characteristic.

Much of the Borrowdale Volcanic Series is non-calcareous and yields acidic substrates, so that the botanical variety of many central hills, such as Great Gable, is again extremely limited. There are, however, significant occurrences of lime-rich rock with an important bearing on the mountain flora. Andesites are calcareous in places, and both vesicular lavas and the friable rocks of shatter belts along faults have become charged with calcite. Where such rocks outcrop at high altitudes as cliffs and enrich surface drainage water in rills and flushes, notably on the Helvellyn and Fairfield ranges, they support an assemblage of calcicolous mountain plants. Most of these have a relict distribution in Britain and some hang on here in extremely small quantity. In places, fine-grained ashes have been secondarily altered by compression to produce slaty cleavage, and this rock has been quarried as the durable Lakeland greenslate especially at Honister and to the west of Ambleside. The Honister slate is quite calc-areous and botanically varied. Running along the southern boundary of the Borrowdale Volcanics from Broughton-in-Furness to Shap Wells is the narrow band of Coniston Limestone. This is mostly too thin and impure to have any pronounced botanical effects.

The Silurian formations

By contrast, relatively softer Silurian rocks form the southern part of the Lake District and extend east of Kendal into the outlying fell groups of the Howgills and the Middleton and Barbon Fells south of Sedbergh. The sequence of Stockdale Shales, Coniston Flags, Coniston Grits, Bannisdale slates and Kirkby Moor Flags is essentially a mixture of sedimentary grits and shales. The varying hardness of these rocks has given rise to the characteristic topography of southern Lakeland, with low, irregular and sometimes rocky hills and ridges with intervening valleys - a country of hillocks and hollows containing various tarns, as well as the large lakes of Windermere and Coniston Water. It is tamer scenically than the higher fell land to the north, well wooded in part and with relatively fertile valleys, but also very beautiful in its own way.

The Silurian reaches its highest point on the Calf (676 m) in the Howgill Fells which, with the Middleton and Barbon Fells, have the typically steep but smooth-sided form, with rounded summits and deep, waterworn gills that characterise the Skiddaw Slate hills. Sizeable cliffs are few but Cautley Crag and Spout (Plate 37), and Black Force are quite impressive features. Although generally non-calcareous, the Silurian formations are better than the Slates as soil parent materials, in yielding a more plentiful supply of calcium and thus substrates of modest fertility. The favourable terrain and soils have encouraged agriculture (and, hence, nutrient run-off) in the drainage basins of Esthwaite Water, Windermere and Coniston Water resulting in these becoming the most productive of the lakes.

Plate 4. Nethermost Cove, Striding Edge, Red Tarn and Swirrel Edge, Helvellyn

Plate 5. The Screes, Wast Water

The Devonian period

The Devonian was a period of mountain building, with much folding and faulting of rocks, and development of synclines and anticlines across the Lakeland area. This was accompanied by massive igneous intrusions in several areas. Intrusions of granite occur at Shap and over a much larger area in the west from the foot of Wast Water across lower Eskdale to Bootle. A mass of granophyre was intruded into the area either side of lower Ennerdale, and there were three different intrusions in the Skiddaw area: the gabbro and grano-phyre, diabase and felsite complex of Carrock Fell, triple outcrops of granite in Skiddaw Forest, and microgranite south of Threlkeld. The metamorphism of the surrounding rocks, mainly Borrowdale Volcanics and Skiddaw Slates, by these magma bodies involved extensive mineralisation, which led to metalliferous mining as an important human activity amongst the Lakeland fells.

These hard intrusive igneous rocks are mostly similar to the Borrowdale Volcanics in their effects on topography and scenery. Carrock Fell stands out as an angular, rugged mass against its rounded slate neighbours, but the hills formed by the other intrusions mostly blend in appearance with adjoining uplands on different formations. The major intrusions have little botanical significance, beyond adding yet more areas of essentially base-deficient parent rocks that give rise to acidic and infertile soils. Even the Carrock gabbro does little for soil quality despite being basic in the geological sense. Nor do the many mineralised areas and old mine workings have much effect on plants, except where the veins include calcite. Also of Devonian age is an area of conglomerate north of Ullswater, from which the tumulus-like hills of Great and Little Mell Fells near Troutbeck are formed.

The Carboniferous formations

The whole Lakeland area was once overlaid by Carboniferous rocks but extensive erosion has pared them down to a peripheral ring. Thick beds of limestone are the most important botanically, though in places their influence is moderated or negated by a covering of drift. Around the north end of Morecambe Bay and the Duddon estuary a broad but discontinuous belt of limestone extends from Millom and Dalton-in-Furness eastwards. It outcrops as small coastal cliffs in a few places, notably at Humphrey Head, and then is increasingly exposed as low, rocky hills at Arnside, east of Burton-in-Kendal, and to the west of Kendal. Their gleaming whiteness is itself a striking feature, though some are quite well wooded in part. In several places the limestone surfaces as bare, tabular formations, either flat or gently inclined, and deeply fissured with innumerable water-worn crevices. These are the famous limestone pavements, with clint and grike structure, at Middlebarrow, Hutton Roof Crags and Farleton Fell. Some of the limestone hills are more notable for scarp features, with crag and scree, at Whitbarrow Scar, Scout Scar, Cunswick Scar and Arnside Knott, and some have dip slope pavements as well (Plate 6).

These are some of the most important botanical localities in Cumbria, with a range of lime-rich habitats and calcicolous vegetation, including woodland and scrub, as well as varied rock communities. Northwards there is a break in the limestone belt, but this reappears massively north of Tebay, and curls north-westwards across the Shap Fells, past Penrith to Caldbeck. It then runs along the north edge of the Lake Fells to Cocker-mouth before curving southwest and becoming increasingly attenuated to end at Egremont. In the Orton area of Shap Fells, there are further important exposures of limestone pavement and low scarps, at a

Plate 6. Steeply dipping limestone pavement, The Rakes, Hutton Roof

higher elevation than the previous examples, though reaching only 300-400 m. At Sunbiggin there is a calcareous tarn and fens, but some of the adjoining moorland is covered with acidic drift carrying *Nardus* and *Calluna.* With increasing distance around the northern limestone ring, the bedrock is ever less exposed, and its influence is seen mainly in green pastures and floral roadside verges. There are minor scarps and pavement in Greystoke Park and an outlying fragment of pavement at the Clints near Blindcrake. Otherwise, the rock is revealed only in sundry quarries scattered along its length.

Running parallel to and outside the northern part of the limestone ring, where the foothills merge into the Solway Plain, is a band of Coal Measures with grits and shales. This broadens south of Maryport into the west Cumbrian coalfield, a low-lying but undulating tract between the north-west edge of Lakeland and the coast. It is a somewhat dismal area, where a former industrial age, founded on the local deposits of coal and haematite in limestone cavities, has passed by and left numerous signs of dereliction in the form of abandoned mines, quarries and waste tips. Such relics of an iron and steel industry are to be seen around Barrow-in-Furness and at Millom, where the huge excavation of the Hodbarrow mine has been flooded to form an artificial lake. Many of the disturbed soils of these old industrial areas are fairly calcareous and have acquired a moderate botanical interest. Between Harrington and Whitehaven, the railway runs below a line of crumbling cliffs and earthy banks in the Coal Measures.

Lying along the eastern border of Cumbria are the massive uplifts of Carboniferous strata forming the Pennines. At the south end and separated from the Howgill and Middleton Fells by the Dent Fault, lie Whernside, Rise Hill and Baugh Fell, isolated by the

deep valleys of Dentdale and Garsdale draining west to the Lune. Next are Wildboar and Swarth Fells facing High Seat across the Mallerstang valley near the source of the River Eden. These fell summits all lie at around 600-730 m. Limestone beds are little exposed in these hills and then only at moderate elevations (400-500 m): the main occurrences are as pavements on the west side of Wildboar Fell at Fell End Clouds, fragmentary pavements, potholes, cliffs and screes on Tailbridge Hill and Dukerdale head, and low-lying stream ravines between Mallerstang and Stainmore. Thick beds of gritstone and shale form the bulk of the ground and give mostly acidic soils and cliffs such as Mallerstang Edge. At the north end of this Askrigg Block of the Pennines the higher ground descends to the gently undulating or level moorlands of Stainmore Common where drainage impedance and, hence, development of blanket bog over deep peat, are extensive.

North of Stainmore lies the northernmost Alston Block of the Pennines, rising to Cross Fell (893 m) as the highest point of the whole range. There is a pronounced scarp and dip slope structure in the Cumbrian portion, with a long escarpment facing the 50 km length of the Eden valley between Brough and Castle Carrock, and gentle moorlands descending gradually eastwards into the broad valleys of the Rivers Tees and South Tyne. At the southern end, thick beds of limestone alternate with the grits and shales and are extensively exposed as pale grey scars and screes along the scarp slope. Tiers of these outcrops rise along the fell front between Helbeck and Long Fell, and the skeletal brown soils of the intervening slopes support extensive calcicolous grasslands. Other limestone bands appear here at successively higher levels, the final exposures girdling the entire summits of Little Fell (746 m) and Mickle Fell (776 m), and providing

Plate 7. High Cup Nick and Gill

important habitats for mountain plants. Only the western end of Mickle Fell lies in Cumbria.

The length of the scarp is penetrated at right angles by numerous deeply cut and steep-sided valleys, in some of which limestone crags and screes are again extensive, as in Scordale and Knock Ore Gill. It is here also that the great sheet of intrusive igneous rock known as the Great Whin Sill is exposed as dark, vertical and sub-columnar bands of cliff feeding screes below. The most impressive is the spectacular horse-shoe of crags at the head of High Cup Gill (Plate 7); its other outcrops include Black Doors in Ardale Head, the waterfall at Cash Force, north of Cross Fell, and, further north, the picturesque valley of the Black Burn under Cold Fell. The rock is a quartz dolerite, giving a mainly acidic substrate except where it receives seepage from limestone above. Along the foot of the scarp in its middle reach, a line of low, conical hills, exemplified by Knock and Dufton Pikes, marks the occurrence of the Cross Fell Inlier, an upthrust of the Ordovician Skiddaw slates and Eycott Hill lavas.

The greater part of the Cross Fell range is nevertheless built of acidic gritstones and shales, and beyond Knock Ore Gill the limestone exposures become steadily thinner and lower in elevation, with marked effects on botanical richness. At the northernmost extremity, fronting the Tyne Gap, they have almost disappeared, but the Coal Measures reappear here and have been worked on a small scale. The summit of Cross Fell (Plate 8) is a large plateau formed from a capping of hard gritstones which have shed a rim of block scree around the steep slopes immediately below. It contrasts with the slightly lower, rounded twin peaks of Little and Great Dun Fells in line to the south-east. Most of the main Pennine watershed is broad and plateau-like. Stretches of dry, stony grassland with grit-stone blocks alternate with peaty ground and from

Hartside to Tindale Fell extensive blanket bog occupies much of the high watershed. Much of the dip slopes are also buried under this deep mantle of peat, but the presence of limestone is revealed locally in drainage water which produces calcareous flushes and little marshes, as well as in occasional small outcrops. The blanket bogs are severely gullied or even, in places, subject to sheet erosion down to the bedrock.

The Carboniferous rocks of the Pennines are extensively mineralised, and lead has been widely worked, especially around Alston and Nenthead. The old mine tips form a substrate rich in lime as well and have become a distinctive plant habitat. The general abundance of lime in the glacial drift results in the wide occurrence of relatively fertile soils, and the hill farms extend well up the lower slopes of some of these Pennine hills.

The remaining upland area of Cumbria, the moorland district above Bewcastle and Gilsland north of the Tyne Gap, is also formed of Carboniferous rocks. These are predominantly sandstones and shales with thin or impure bands of limestone. The terrain is mostly gently undulating or even flat, with broad watersheds and plateaux, and the elevation rather low, reaching a maximum of only 518 m on Sighty Crag. The bedrock is little exposed, the only noteworthy crags being in the fine gorge of the Irthing extending for four km. above Gilsland. The sandstones and shales here are calcareous in places and give an interesting flora. Other minor outcrops on the moorlands include the distinctive summit tor of Christianbury Crags carved from hard, acidic gritstone. The thick layers of drift which overlie much of the area and form high, unstable banks to some of the streams, are in places quite strongly calcareous and clayey. Remote farms with good grassland and hay meadows were created on some of these better soils, and emergent seepage water produces

Plate 8. Cross Fell, looking south to Great and Little Dun Fells; Cow Green Reservoir is on the extreme left.

basic flushes and marshes scattered over the moors.

The prevailing vegetation is nevertheless blanket bog and wet acidic grassland or heath. The huge and featureless catchment of the Irthing and King Water known as Spadeadam Waste formerly had numerous areas of level bog with a spongy, *Sphagnum*-dominated surface. The best of these, Butterburn Flow, lies on a terrace above a loop of the upper Irthing, which here follows a deep and canal-like course cut in glacial drift up to seven metres deep. The headwaters of the Rivers Lyne and Liddel drain rather steeper valleys on the western side of this massif, but even here peaty and gleyed soils predominate and there is rather little dry heather moor. The wild and desolate character of this moorland district is best conveyed by its old name, The Waste of Cumberland.

The Permian and Triassic formations
Apart from superficial deposits, this is the last main geological series of the region. Collectively forming the New Red Sandstone, these rocks underlie the Solway Plain north of Maryport and the broad reach of the Eden valley as far as Kirkby Stephen. They also occupy the narrow coastal plain from Whitehaven to Millom, and then an outlier around Barrow-in-Furness. Most of the Permo-Triassic area lies at a low altitude, but it rises to a conspicuous ridge of minor hills at 200-280 m along the Eden valley. The hard flags of this ridge have been quarried for paving stones at Lazonby Fell and elsewhere, but much of the sandstone is fairly soft and there are thin shale beds locally. It is mainly a non-calcareous rock yielding acidic, sandy and permeable soils, but in places there are modest amounts of lime giving richer conditions. West of Carlisle is an area of marine Liassic strata belonging to the Jurassic, but this does not appear to produce any features of topographic or ecological note.

Some of the main rivers draining to the Solway have cut deeply into the New Red Sandstone to form lowland gorges which have some importance as botanical habitats. The finest of these is the Eden gorge in Baron Wood above Armathwaite, with 30 m cliffs, but other good examples occur on the Gelt near Brampton, the Irthing near Low Row, the Petteril near Wreay, and the Lyne between Kinkry Hill and Kirklinton. On the coast, these rocks form the only significant seacliffs in Cumbria, in the fine 100 m precipices fronting six kilometres of shore at St. Bees Head.

Most of the New Red Sandstone is otherwise occupied by unspectacular, low-lying farmland, within which the main islands of semi-natural habitat are the large raised bogs ('peat mosses'), with their deep peat, mostly developed over the marine warp of the Solway hinterland. Much of the formation is overlain by recent deposits, and east of Carlisle the undulating terrain of glacial sands has various hollows in which smaller peat

Plate 9. Solifluction stone polygons on Cross Fell

Plate 10. Walney Island and Barrow-in-Furness looking south

mosses of valley and basin mire type have developed. Another drumlin belt along the south-west coastal plain also holds a series of little bogs and fens.

Recent deposits

Included in this category are the main classes of coastal substrate. Extensive saltmarsh systems have developed on marine warp around the main estuaries of More-cambe Bay and the Solway, with lesser areas on the Duddon and Irt. The sediment tends to sand rather than silt and is mostly low in lime. Sand dunes have formed between Grune Point and Allonby, at Drigg, Eskmeals, Haverigg, Sandscale, and both ends of Walney Island (Plate 10). The sand is mostly non-calcareous, especially towards the north, but the Sandscale dunes (Plate 18) are notably lime-rich with good development of slacks. Most of the remaining part of the Cumbrian coast is bounded by shingle beaches, but around Walney and Foulney Islands these widen into extensive areas with vegetation.

> We scuffed through a scabbed and scruffy valley of ruddled rocks
> To Cumberland's southernmost point, a headland, half-blasted-away,
> Where the limestone met the tide. Here on the seaward side,
> Wave-action moulds the rocks, thumbs them like plasticine;
> Landward the crag splits vertical down to the old workings.
> We traversed the yard-wide col between quicksand and quarry, and there,
> In a cockle-shell dip in the limestone, matted with thrift and rock-rose,
> Was Sunday's flower, the Bloody Cranesbill, red as the ore
> It grew from, fragile as Venetian glass, pencilled with metal-thread
> Haematite-purple veins. ...
>
> from *The Bloody Cranesbill*, Norman Nicholson

Plate 11. *Geranium sanguineum*

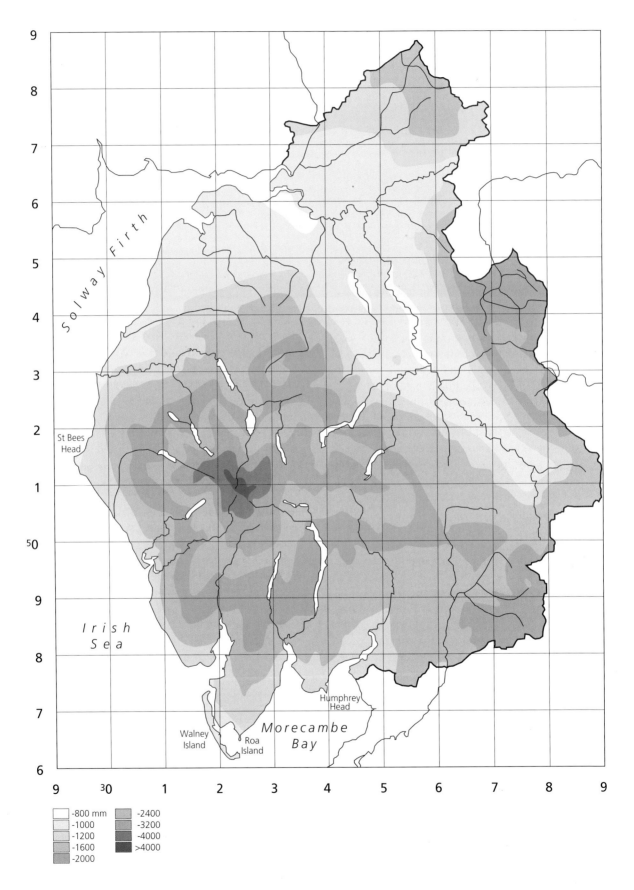

Fig. 3. Cumbria - Average annual rainfall, 1941 - 1970
(Crown copyright - reproduced with the sanction of the Controller of Her Majesty's Stationary Office)

D.A. Ratcliffe

The rain here comes down heartily, and is frequently succeeded by clear, bright weather, when every brook is vocal, and every torrent sonorous;...Vapours exhaling from the lakes and meadows after sunrise, in a hot season, or, in moist weather, brooding upon the heights, or descending towards the valleys with inaudible motion, give a visionary character to everything around them;.....Akin to these are fleecy clouds resting upon the hill-tops;...Such clouds, cleaving to their stations, or lifting up suddenly their glittering heads from behind rocky barriers, or hurrying out of sight with speed of the sharpest edge - will often tempt an inhabitant to congratulate himself on belonging to a country of mists and clouds and storms, and make him think of the blank sky of Egypt, and of the cerulean vacancy of Italy, as an unanimated and even a sad spectacle.

Description of the Scenery of the Lakes, William Wordsworth

Cumbria has a cool oceanic climate overall, but shows the greatest range of meteorological conditions of any English county. Its westerly location ensures Atlantic features, of equable temperatures and a tendency to damp, windy weather with less sunshine than southern England. This general pattern is exaggerated by topography, with steep gradients of increasing rainfall and snow, atmospheric humidity, cloud cover and windiness from the coastal lowlands to the high fells of central Lakeland, and an even sharper altitudinal decrease of temperature between the valleys and the summits of the mountains. Land use, and especially agriculture, reflect this broad climatic pattern. This is a region predominantly of pasture-lands, from productive grass leys in the lowlands to acidic rough grasslands on the hills. Arable farming is largely restricted to the driest and most fertile areas and is mainly of root crops, though cereals nowadays include fast-maturing barley as well as oats.

Precipitation (Fig. 3)

The driest, sunniest part of the county is the inner Solway, with a mean annual precipitation of 750 mm. Carlisle has 825 mm and most of the Solway Plain and the Eden Valley are fairly dry (850-950 mm), with rainfall rising to 1000 mm in foothill locations such as Caldbeck and Kirkby Stephen. The southern coastal belt from St Bees to Morecambe Bay is slightly wetter, with 1000-1250 mm. Kendal and Shap (1300 mm) and Keswick (1475 mm) are influenced by the mountains but the heart of Lakeland has a large area, including the heads of all the main valleys, with more than 2500 mm. The maximum is consistently reached around the central fells of Scafell, Bowfell and Great Gable, where the gauge at Sprinkling Tarn has averaged 4700 mm. The rainfall gradient is particularly steep along the Borrowdale valley, increasing from 1475 mm at Keswick to 3300 mm at Seathwaite in only 12 km, and a rise in elevation of only 50 m. While rainfall tends to increase with altitude, the relationship is not simple, and there is heavy rainfall in some of the valley heads in the lee of the highest hills. The high Pennines lie in the rain shadow of the Lake Fells and are not quite so wet, with 2065 mm at Moor House and 1200 mm at Alston. Rainfall on the Border moorlands

of the north-east is at most 1400 mm.

The distribution of rainfall as well as the total amount is important to soil development and vegetation. There is a greater frequency of damp, misty days with low cloud base along the hills and valleys close to the west coast than in eastern Cumbria, even when places with similar total rainfall are compared. Yet, although rain often falls in great deluges in the central fells, producing torrent streams with scoured courses, and flooding, especially in Borrowdale, there can be quite prolonged droughts of up to 30 days during anticyclonic weather. For this reason, there is not quite the oppressive degree of wetness, and prevalence of peat and moisture-loving plants, which characterise the western Scottish Highlands. The drying of grass and other plants around rocky outcrops in the lowlands is a characteristic feature of early summer drought. The great drought of summer 1976 gave rise to some catastrophic moorland fires, notably at Glasson Moss and around Devoke Water. When precipitation is measured as the number of days on which 1 mm or more falls ('wet days'), only the Solway Plain and the edge of Morecambe Bay have under 140, but even the wettest central fell area has only just over 200, compared with well over 220 in the high mountain systems between Argyll and west Sutherland.

Temperature

Climatological maps (Meteorological Office 1952) show that temperature, corrected to a theoretical sea level, decreases north-eastwards from an annual mean of 9.5° C around Morecambe Bay to under 9.0° north-east of the Eden Valley. Average maximum temperatures are 18.4 - 19.0° in a broad western coastal sector between Maryport and Ulverston, rising to 20.0 - 20.6° east of Shap. January average minima vary from 2.0-2.8° in this same coastal belt to 0.5 - 1.0° over the Pennines and Borders. The main overall gradient is thus from a western coastal belt with narrower temperature range between winter and summer, to an eastern inland zone with colder winters but warmer summers. In coastal localities, actual temperatures are the same as map temperatures, but in inland areas allowance has to be made for decreasing mean temperature with altitude, at the standard lapse rate of

1° C for every 150 m. These give mean temperatures on the high tops around 5° lower on average than in the adjoining valley floors and this is enhanced by the increasing wind chill effect at high altitudes; the day maxima fall even more rapidly. These effects have a pronounced influence on the length of the growing season (see below under Land Use). Since there is very little low ground in the Pennines, temperatures are lower in both winter and summer than the isotherm maps suggest. Examples of annual monthly temperature means are Millom (9 m) 9.4°, Keswick (76 m) 9.1°, Appleby (134 m) 7.8°, Moor House (560 m) 5.6° and Great Dun Fell (847 m) 3.4° (Manley 1952, 1973; Wilson 1938; Meteorological Office 1952).

The south-west coastal strip has the lowest incidence of frost and the least severe winter temperatures. The average annual number of days with air frost increases with both distance inland and altitude, from 42 at Sellafield to 84 in Appleby and 160 on Great Dun Fell (Manley 1973). Night temperature inversions during still winter weather may produce especially severe frosts in valley bottoms well inland, but only during the hardest winters do the larger lakes freeze over, and the lakes have an ameliorating effect on local temperatures. Prolonged and severe frosts sometimes cause winter browning of filmy ferns, heather, gorse and yew, but these native species usually recover later, in contrast to some of the less hardy garden plants. Manley remarks that the high-level climate is marked by the frequency with which temperatures oscillate about the freezing point. This is consistent with the frequent occurrence on the high ground of still active though small-scale frost-thaw (solifluction) features, such as the terraces, stone nets, soil hummocks, and striped screes described by Hollingworth (1934) and Hay (1937). The larger-scale stone nets and stripes well developed on the summit plateau of Cross Fell (Plate 9) probably owe their origin to a much colder period, closer to the Ice Age.

Snow

Frequency of snowfall and duration of cover depend on precipitation, but even more on the frequency of sub-zero temperatures. The obvious increase of snowfall frequency, amount and length of snowlie, with altitude has therefore also to be related to the underlying gradient of average number of days with snow falling on low ground, from less than 10 around St Bees Head to more than 25 in the extreme east of the Cumbrian Pennines. The effect of altitude is often shown by a sharply cut-off snow line after recent falls on the mountains. Snow cover on the higher fells under the oceanic Cumbrian climate is very variable. Some snow has usually fallen on the uplands by November, and in long hard winters there may be a more or less continuous cover over at least parts of the high ground until late March. Equally, in mild winters, series of Atlantic depressions may repeatedly strip periodic snow from the hills and give only an intermittent cover.

Plate 12. Contrasting snow-cover on Striding Edge, Helvellyn, looking east

Manley (1973) comments that most heavy snowfalls come from the east and north-east causing the Pennines to be conspicuously more snowy than the Lake District. He records an annual average of about 70 days with snow lying at Moor House and 105 days on Cross Fell. Atlantic depressions can result in heavy snowfall on the western Fells; this decreases eastwards to the Pennines where it lies lower and lasts longer. Late snowfalls, in March and April, have been a feature of recent decades, but have not noticeably contributed to the amount of late spring snowlie. Snow depth on the hills is determined by the form of the ground and exposure to wind, so that falls tend to blow off exposed ridges and accumulate in sheltered hollows. Late snow patches often linger into late May or even mid June just below the summit cap of Cross Fell, especially on the north side, and in high north- to east-facing gullies and hollows of the Scafell and Helvellyn ranges.

Wind

That part of the seaboard most exposed to the Irish Sea, between Silloth and Millom, is a notably windswept coast, the prevailing winds being south-westerly. Wind-shaping of trees is common here, but conditions soon become more sheltered inland where the ground is low-lying. Wind is very much a feature of the oceanic climate and it also increases markedly with altitude, being probably even more important than low temperature in producing an altitudinal tree limit on western British hills. During severe westerly gales there is tremendous wind blasting of the high tops. Manley (1973) notes that on Great Dun Fell the average windspeed is about force 5 (19 - 24 mph), and that an hourly wind of 99 mph and a gust of 134 mph have been recorded. It is likely that the high summits of the western Lakeland fells, closer to the Atlantic storm onslaughts, receive even more extreme winds. There is also another important but peculiarly local wind phenomenon experienced along the scarp slope of the Cross Fell range and adjoining Eden Valley - the notorious Helm Wind, described in detail by Wilson (1938). Under certain conditions, when the wind sets north-east to east, a relentless blast of cold air rolls down the fells to sweep a wide zone of the foothill country beyond, and to the accompaniment of distinctive cloud formations. Manley (1973) notes that similar winds occur at times among the Lake fells under such conditions, but that the phenomenon is less well defined due to the more irregular topography.

Sunshine and cloud cover

Manley (1973) comments that, despite the high rainfall of Lakeland, this area compares quite well with other English stations in similar latitudes, in the duration of bright sunshine. In Keswick this is greatest during May, when it amounts to 39% of the theoretically possible figure. Localities in deep valleys often experience a winter cut-off through the shadowing effect of surrounding mountains (Plate 13). Sunshine is clearly related to the influence of mountains in inducing cloud formation and rain, and so tends to decrease markedly with altitude. It increases towards the coast and is highest in the inner Solway. Average annual duration is only 840 hours on Dun Fell summit, compared with 1150 at Moor House, 1140 at Amble-side, 1270 at Keswick, 1425 at Silloth, 1515 at Sellafield and 1396 at Carlisle (Manley 1973). Hill fog is especially frequent at over 500 m, but its frequency at any particular elevation tends to increase westwards.

Associated features of vegetation and flora

These climate characteristics give Cumbria a number of distinctive vegetational features in comparison with other regions of Britain and they contribute to the remarkable variety of habitats within the county. Even on the drier and warmer coastal plains, the water balance is sufficient to have allowed the development of large raised bogs with deep acidic peat ('peat mosses'). The broad moorland watersheds of the Pennines and Border hills in the east and north have the topography which, under the cool and humid upland conditions, has favoured the formation of extensive blanket bogs. On the steeper Lakeland fells with narrower watersheds, blanket bog is more patchy, but where the abundant rainfall emerges on lower slopes and collects in hollows, there are frequent and sometimes extensive flush and valley marshes. The flora of acidic peatland, although a limited one, at best, is well represented in

Plate 13. Looking from Hartsop to the north-facing crags of Threshwaite Cove

Cumbria. Altitudinally widespread species include *Eriophorum vaginatum*, *E. angustifolium*, *Trichophorum cespitosum*, *Drosera rotundifolia* and *Narthecium ossifragum*. Those found mainly at lower levels include *Myrica gale*, *Andromeda polifolia* and *Drosera anglica*, while *Rubus chamaemorus* and *Carex magellanica* are upland.

The heavy and evenly distributed rainfall of the hill country is reflected more conspicuously in the abundance and variety of mosses and liverworts than in any element of the vascular flora. For some of these bryophytes, this is either the English headquarters or the only station. The profusion and luxuriance of ferns in the rocky woods and other upland habitats is nevertheless noteworthy, and in particular the frequency of *Hymenophyllum wilsonii* in Lakeland, especially the western half, is a strong botanical sign of the climatic wetness. This moss-like fern is confined to areas where precipitation exceeds 1150 mm, and it reaches its greatest abundance when it tops 2000 mm. Its absence or scarcity in the Pennines and Border moors may reflect the less extreme wetness of those districts, but probably also the lower frequency of suitable habitats.

Winter temperatures on the coast and at low levels inland in valleys facing the sea are high enough to allow the presence of southern, frost-sensitive plants, including ferns such as *Adiantum capillus-veneris*, *Asplenium obovatum*, *Dryopteris aemula* and *Hymenophyllum tunbrigense*. The absence of the last two species from central Lakeland, even Borrowdale, suggests that this area is distinctly colder in winter. Many species classed by Matthews (1955) as Continental Southern are at or near their northern British limits in Cumbria, suggesting that this region is close to the critical limits of summer warmth for this group. They include *Aster linosyris*, *Hippocrepis comosa*, *Hornungia petraea*, *Blackstonia perfoliata*, *Asperula cynanchica*, *Spiranthes spiralis*, *Ophrys apifera*, *Anacamptis pyramidalis*, *Acer campestre* and *Tamus communis*. The northern boundary of *Tamus* across the southern Lake District (Fig. 4) is particularly striking. *Tilia cordata*, which is near its northern limit, has been shown by Pigott & Huntley (1980) to produce viable fruit only in exceptionally warm years such as 1976, 1983 and 1984. Other notably southern species are *Ceterach officinarum*, *Umbilicus rupestris* and *Gentiana pneumonanthe*. There are many species, such as *Ranunculus sceleratus*, *Daucus carota* and *Malva sylvestris*, which are relatively widespread further south in England but in Cumbria are largely restricted to the coastal fringe and the dry and sunny Eden valley (Fig. 5). Oceanic and western species include *Hypericum elodes*, *Anagallis tenella*, *Scutellaria minor* and *Osmunda regalis*. Many of the above species are rare or scarce, but *Ulex gallii* is locally quite abundant. The occurrence of *Ulex minor* in the Carlisle area is a surprising outpost of a plant otherwise associated with sandy soils in southern England.

Conversely, the relatively cool conditions of the lowlands are reflected in the presence of Continental Northern species such as *Actaea spicata*, *Geranium sylvaticum*, *Cirsium heterophyllum*, *Circaea alpina*, *Meum athamanticum*, *Galium boreale*, *Andromeda polifolia* and *Carex ornithopoda*, which are absent from southern England. The Arctic Subarctic and coastal *Mertensia maritima* reaches its present southern limit in Britain in Cumbria.

The occurrence of a quite large area of fairly high mountains has allowed the survival of remnants of montane vegetation which migrated northwards and upwards during its post-glacial displacement by advancing scrub and forest. The original climatically determined altitudinal zonation of vegetation in the mountains has either been lost or profoundly modified by human activity. Most of Cumbria was once covered with woodland up to a level determined by a combination of wind severity and low temperature. Deforestation long ago removed the natural upper tree-line, and its potential present limits can be judged only by scattered trees on cliff faces, mainly of rowan, birch and occasional oak. It would evidently be around 450 - 520 m in the Lakeland fells (exceptionally to 580 m) and 550 - 600 m in the Pennines. The performance of the conifer plantation at Ashgill near Alston, and the experimental woodland at the nearby Moor House, suggest that in this easternmost part of Cumbria, trees and shrubs will grow successfully up to 550 m, the absolute limit being reached by larch at 660 m. The commercially viable limits of afforestation probably explain why the highest parts of the Bewcastle fells have remained unplanted. An upper zone of scrub, with juniper, birch, rowan and willows probably extended above the limits of the prevailing oak or mixed broadleaved woodland for at least another 100 m. Turner (1984) has presented evidence that during the postglacial period scrub occupied the summit plateau of Cross Fell up to 885 m, but in more recent times this ground has been covered by *Racomitrium lanuginosum* heath with abundant *Carex bigelowii*, representing the natural montane community so widespread on many summits in the Scottish Highlands.

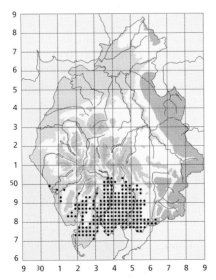

Fig. 4. *Tamus communis* at its northern limit on the west coast of Britain.

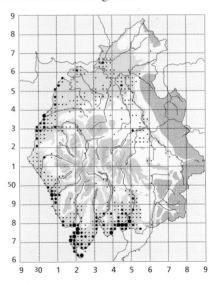

Fig. 5. Composite distribution of seven southern species: *Anagallis arvensis*, *Carduus tenuiflorus*, *Centaurium erythraea*, *Clematis vitalba*, *Cynoglossum officinale*, *Daucus carota* and *Pastinaca sativa*.

Plate 14. Late-snow-patch hollow with *Nardus stricta*, Great Dod, 850 m alt., May

Above the limits of woodland and scrub there were evidently montane heaths with *Calluna vulgaris* and other dwarf shrubs such as *Vaccinium myrtillus*, *V. vitis-idaea* and *Empetrum nigrum*. These have lost ground through centuries of heavy grazing by domestic stock and moor burning, which has virtually destroyed the dwarf *Calluna* mats at the upper limits of these heaths, and greatly fragmented the dwarf shrub heaths where these had replaced woodland. Extensive areas of botanically dull acidic grassland have become dominant on both the Lakeland fells and the scarp slopes of the Pennines: *Festuca - Agrostis* on drier slopes, often with heavy infestation by bracken below 450 m, *Nardus stricta* and *Juncus squarrosus* on poorly drained ground.

Although much modified and fragmented by subsequent land use, the high level communities above the potential tree line have retained a modest number of arctic-alpine and arctic or alpine species. There are many fewer than in the colder, loftier and more massive ranges of the Scottish Highlands, but rather more than in the slightly warmer region of Snowdonia to the south. In particular, communities owing their character to prolonged snow cover are rather poorly represented in Cumbria, though patches of *Nardus stricta* occur in high hollows on the Helvellyn range (Plate 14) and Cross Fell, where they appear to represent outliers of the characteristic snowbed community of the Scottish Highlands. The former pattern of late snow-influenced vegetation on the Cumbrian fells has probably been obscured by the changes caused by grazing and burning.

Many of the high mountain plants are true relics, with very small remnants hanging on in very few places. This is especially true of species which have become restricted to cliff faces through grazing and the competition of more aggressive plants. Only six clumps of *Dryas octopetala* are known, and tiny populations of

Saxifraga nivalis, *Salix lapponum*, *Cerastium alpinum*, *Carex atrata*, *Poa alpina*, *P. glauca* and *Phleum alpinum*. Their capacity for spread under present conditions may be very limited. Others such as *Alchemilla alpina*, *Saxifraga aizoides* and *Crypto-gramma crispa*, are less susceptible to grazing and have spread out on to deforested ground, becoming locally abundant, even at low levels. Some montane species previously present during the post-glacial period may have been lost from Cumbria altogether, especially if the tree-line was once higher than in recent times.

Although some species appear to be especially influenced by one or other of the main components of climate, plants react to climate as a whole, and distribution patterns should thus be compared with combinations of conditions. Many of the northern and upland species are adapted to the cool, humid, cloudy and windy hill climate, and in the mountainous parts of Cumbria they tend to replace those which are especially associated with the warm, dry, sunny and sheltered conditions of the lowlands. One of the most conspic-uous examples is the way in which *Geranium sylvat-icum* replaces *G. pratense* on roadside verges with proximity to the fell country. The distribution maps show that a whole group of plants characteristic of the English lowlands is conspicuously sparse or absent in central Lakeland. It includes species from a wide range of habitats, but especially the dry soils of arable land, roadside verges and disturbed or waste ground. Examples from a long list are *Viburnum opulus*, *Arum maculatum*, *Chaerophyllum temulum*, *Ononis repens*, *Vicia cracca*, *Fragaria vesca*, *Potentilla anserina*, *Glechoma hederacea*, *Anthriscus sylvestris*, *Lamium purpureum*, *Galeopsis tetrahit*, *Alliaria petiolata* and *Aegopodium podagraria* (Fig. 6).

This raises the further point that the effect of climate on plant distribution is at least partly through its indirect influence on soils and land use. Moreover,

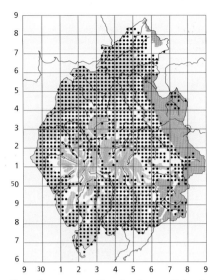 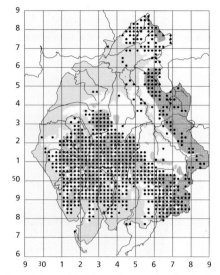

Fig. 6. Complementary distributions of the lowland *Aegopodium podagraria* (left) (though rare in the Lake District valleys) and the upland *Oreopteris limbosperma* (right)

in a county such as Cumbria, the mountain topography which so strongly modifies regional climate will itself generate particular habitats and land use. This complex interaction of ecological factors limits our ability to understand the distribution and composition of a regional flora as the direct expression of climate. Without detailed experimental studies, much of our interpretation of plant distribution as a response to climate must, indeed, rest upon inference, especially from map correlations between climate factors and patterns of species distribution limits or changing abundance (Fig. 6).

Land use

Arable farming occurs mainly on the lowland plains of the Solway, Eden valley and Morecambe Bay, where rainfall is least and sunshine and warmth greatest; though with modern farming methods, some arable crops are grown well into the hill country. Manley (1952, 1973) points out that, in an oceanic climate, the length of growing season (conventionally measured as the period with mean temperature >5.5° C) decreases rapidly with altitude, by about ten days for every increase of 75 m, giving the uplands late springs and early autumns. The rate of growth and yield of crops correspondingly falls. At Nenthead in the Pennines, for instance, the growing season at 460 m is nine weeks less than in the Cumbrian lowlands near sea level.

While the upper limits to farming are set by the climatic severity of higher altitudes, they also depend greatly on the quality of the soil and topography. Under the combination of heavy rainfall and cloudiness with hard acidic rocks and poor soils, the steep-sided Lakeland valleys mostly show an upper limit to enclosed farmland (pastures and hay meadows) at only 150 - 250 m in the west and 250 - 350 m in the east. On the drier north-east edge, this rises to just over 400 m at Matterdale. The limestone of the Pennines,

together with the less oceanic climate, gives a much higher farmland limit. As Wilson (1938) noted, there is a hay meadow at 560 m altitude at Moor House, where the rainfall is 1773 mm, and enclosed pastures at Nenthead reach 550 m (Manley 1952).

Above the limits of enclosed farmland are the open sheepwalks of the higher hills over which these animals range freely. The lambing season is much later among the hills than in the lowlands - late April to early May compared with February - March. While the sheep do not usually reach the high tops much before late May, they are fond of the summer growth on these upper levels, and are often found in considerable numbers there. In earlier times, mortality cut back sheep numbers to a level determined by the capacity of the ground and climate to sustain stock. Hard winters and backward springs could take a heavy toll of both ewes and lambs, and the timing of pasture growth was critical to summer performance.

The distribution of grouse moors has a connection with climate. In heavy rainfall areas, communities with heather (the staple food of the Red Grouse) often contain abundant grasses and also bracken which, under the combination of heavy grazing and repeated burning, have the competitive advantage and can rapidly replace *Calluna* and other dwarf shrubs. In drier, more eastern areas, heather more readily retains its dominance under grazing and burning. Topography is also important, and the gently contoured terrain suitable for driving grouse is chiefly found in the east. Former grouse moors in the Lake District are now virtually all abandoned, and the only managed grouse shoots now are in the Pennines, and perhaps some of the Shap fells, those on the Bewcastle fells having been replaced by afforestation.

Forestry is the other major land use in the uncultivated parts of Cumbria. The climate of the county is well suited to the growing of commercial exotic conifer crops, apart from the prevailing windiness which can cause premature wind-throw at the higher levels. Agricultural and forestry policy have determined that most post-1920 afforestation has been in the uplands. Public concern for scenic beauty has restricted afforestation in the Lake District National Park mainly to peripheral areas, while the extent of common land has severely limited its penetration of the Pennines. The main area of afforestation has been on the Border moorlands of Bewcastle and Gilsland, where a high proportion of the ground below 450 m has been planted, mainly with Sitka spruce, but some lodgepole pine and larch. Shelter belts and clumps, mainly of conifers, have been widely planted in the uplands, to give protection to stock from wind and rain. Many fell farmsteads also have sheltering groups of trees, notably of sycamores, which sometimes attain great size.

Pollution

In *The Flora of Westmorland*, Wilson (1938) wrote a perceptive section on atmospheric pollution which reads very 'fresh' today. He pointed out that even this non-industrial county was subject to fall-out of atmospheric acidity, especially during anti-cyclonic weather when south to easterly winds bring smoke haze from industrial areas in those directions. Wilson attributed the reduction in numbers and robustness of bark-loving lichens, mosses and liverworts in the county to this pollution effect. Nowadays, we know a great deal more about the whole subject of acid deposition and its adverse effects on plant life, especially corticolous lichens. While there has been a great reduction in old-fashioned smoke pollution, the wider dispersion of sulphur dioxide from the tall stacks of huge coal- and oil-fired power stations, and the increased output of nitrogen oxides, especially from road traffic, now pose a serious problem for most parts of Britain outside the Scottish Highlands. The large and conspicuous *Lobaria* and *Sticta* lichens are probably even scarcer and more depauperate in the region than

when Wilson wrote. There is no evidence of effects on vascular plants in Cumbria, though data for acidification of waters and soils have recently been presented for other regions of Britain similar in their prevalence of sensitive substrates. Acidification increases with rainfall and cloud cover, so that it is especially pronounced on the higher summits. Tyler (1988) states that rain falling through mountain-top cloud results in the summits receiving two to three times the weight of pollutants per unit area compared with the lower slopes and valleys, and Thompson & Baddeley (1991) have suggested that acid deposition is implicated in the striking recent decline of woolly fringe moss *Racomitrium lanuginosum* in the montane heaths of southern British mountains such as those of Cumbria.

The most threatening of all forms of atmospheric pollution, the release of 'greenhouse gases' and their promotion of global warming, has serious implications for the flora of Cumbria. Rising temperatures could, at least in theory, create increasingly unfavourable conditions for the tiny relict populations of several of the rarest mountain plants. Vigorous and competitive species such as bracken and gorse could extend to ever higher levels, while it might be possible for farmers to cultivate and foresters to plant trees at appreciably greater altitudes. The county might acquire new species through the northward migration of more southern plants, though the extent of artificial habitats must nowadays be a barrier to the free movement of native plants. The changes might favour some aliens and lead to an extension of their range. The great uncertainties about the precise form that climatic change could take in any one area make it unprofitable to speculate further on possible botanical consequences. The awesome potential for uncontrollable biological change and, on the wider scale, economic upheaval, is sufficient reason for urging that every effort be made to tackle the problem at source, by minimising the release of greenhouse gases.

The countryside reminded him of the Lake District "without that grow-nothing air of that soaking land".
Quotation from letter from William Morris in Ems (Germany), *William Morris*, F. McCarthy

A botanical tour of Cumbria
D.A. Ratcliffe & G. Halliday

This survey is intended to convey some idea of the main geographical subdivisions of the county and their characteristic plants and botanical highlights in the form of local and rare species. The subdivisions (Fig. 7) are not precisely delineated, for any such boundaries would be quite arbitrary, and the purpose is to give a general indication of floristic features rather than a detailed guide.

1. The Solway - Carlisle Plain
The Solway coast is flat, with the great saltmarshes of Skinburness, Newton Arlosh, Burgh and Rockcliffe lying east of Silloth. These saltings are heavily grazed by sheep and cattle, and consequently have a prevailing grassiness, with *Puccinellia maritima* as the chief pioneer halophyte, giving way at higher levels to dominance of *Festuca rubra*, *Agrostis stolonifera* and other species more typical of neutral grassland. This 'Solway turf' is removed on a long rotation and sold for high quality lawns. *Salicornia* species are rather localised, and the more robust saltmarsh plants, such as *Aster tripolium* and *Cochlearia officinalis*, grow mainly on the sides of creeks; *Limonium vulgare* is surprisingly scarce. Grazing-tolerant species such as *Armeria maritima*, *Glaux maritima*, *Juncus gerardii*, *Plantago maritima* and *Triglochin maritima* are abundant in the sward. *Apium graveolens* occurs near Anthorn, and around Bowness-on-Solway brackish marsh plants such as *Oenanthe lachenalii*, *Samolus valerandi*, *Centaurium littorale*, *C. pulchellum*, *Carex otrubae*, *C. distans*, *C. extensa* and *Blysmus rufus* appear, and there are species associated with base-rich conditions such as *Ononis spinosa*, *Pimpinella saxifraga* and *Blysmus compressus*. Salt-marsh pools and ditches feature *Ranunculus baudotii* and occasionally *Ruppia maritima* and *Potamogeton pectinatus*.

From Grune Point southwards the shores are of sand and shingle, backed by rather narrow ridges of sand dunes. The strand flora has a noteworthy amount of *Coincya monensis*, *Honckenya peploides* and *Leymus arenarius* and, locally, *Artemisia stelleriana*. The sand is non-calcareous and slacks are poorly represented, so that the dune flora is rather limited. There is however a spectacular abundance of *Ononis repens*, *Geranium sanguineum* and *Rosa pimpinellifolia* in places. The Silloth docks area was once a noted locality for introductions, associated with the unloading of grain cargoes and dumping of ballast. It still features aliens such as *Medicago sativa* subsp. *sativa*, *Muscari comosum*, *Salvia verticillata*, *Oenothera cambrica* and *Euphorbia waldsteinii*.

Immediately inland are large 'peat mosses', the raised bogs of the Solway hinterland, where deep layers of acidic peat have formed on the waterlogged ground over thousands of years. During the last century or so, marginal peat-cutting and reclamation to agriculture

Plate 15. Humphrey Head, looking north

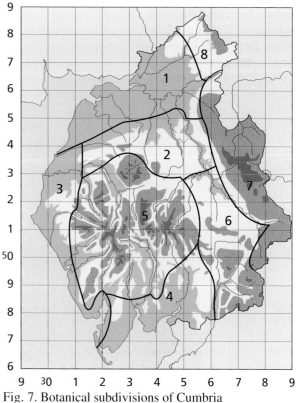

Fig. 7. Botanical subdivisions of Cumbria

have created artifical edges with the surrounding farmland which have tended to lower the water table and dry their surfaces. Repeated fires have further reduced the *Sphagnum* cover of the living bog and increased that of cotton-grasses and heather. Two of the largest, Solway Moss near Longtown, and Wedholme Flow near Kirkbride, have been extensively worked for commercial horticultural peat. Further inland, to the north-east of Carlisle, the mosses around Hethersgill are intermediate between raised and blanket bog. Bolton Fell is now largely cut-over but Walton and Broom Hill Mosses are still substantially intact. Scaleby and Moorthwaite Mosses were heavily cut long ago, but here and elsewhere *Sphagnum* communities have regenerated in the old cuttings. The peat mosses are the habitat of a distinctive acidic bog flora, with *Andromeda polifolia*, *Vaccinium oxycoccos*, *Narthecium ossifragum*, *Drosera rotundifolia*, *D. anglica*, *D. intermedia*, *Myrica gale*, *Rhynchospora alba* and *Empetrum nigrum*. *Sarracenia purpurea* on Wedholme Flow was presumably introduced.

This district once had expanses of richer fen along broad river valleys, as at Cardew Mires near Dalston and between Abbeytown and Allonby. These have been reduced by drainage and reclamation to a few small fragments, of which Biglands Bog on Bampton Beck is the most important, with an interesting mixture of fen, willow carr and acidic bog. The varied flora includes *Frangula alnus*, *Utricularia vulgaris*, *Stellaria palustris* and *Carex limosa*, as well as typical raised bog species. Salta Moss near Allonby is another mixed mire system with *Hippuris vulgaris*, *Lycopus europaeus*, *Scutellaria minor* and *Osmunda regalis*. The undulating glacial drift country east of Carlisle has numerous small pockets of bog in hollows and channels, such as Moorthwaite Moss, Faugh Moss, Black Dub and Unity Bog. *Vaccinium oxycoccos* carpets the ground in places and there are several

lowland stations for *Vaccinium uliginosum*, one with a remarkable quantity of this northern shrub (Plate 16). Scaleby Moss has a curious isolated colony of *Ledum palustre* subsp. *groenlandicum*.

In places, the edges of these bogs and the surfaces of the smaller examples are quite extensively colonised by birch, *Betula pubescens*, which typically develops a luxuriant field layer of *Dryopteris dilatata*. One open moss-edge near Bowness has an isolated population of *Trientalis europaea*. Some sites have a good deal of Scots pine which regenerates freely, and there are mixed mossland woods of birch, pine and oak with willow-dominated valley swamps in several places. These complexes of real wildwood are best represented at Orton Moss, Finglandrigg Woods and Cumwhitton Moss. *Vaccinium myrtillus* is locally dominant, and characteristic plants are *Dryopteris carthusiana*, *Pyrola minor*, *Luzula pilosa*, *Frangula alnus* and *Carex rostrata*.

Scots pine plantations in the Carlisle district have long been known as the habitat of *Goodyera repens*, but this orchid also grows in Cumwhitton Moss with other species typical of northern pinewoods, such as *Vaccinium vitis-idaea*, *Pyrola minor*, *Listera cordata* and certain mosses. The district has rather little woodland, but a few examples of oakwood occur on the predominantly acidic mineral soils. Oak is sometimes mixed with non-native species such as beech, and some oakwood has been largely replaced in recent years by conifers, as at Walton Wood near Brampton. The typical oakwood field layer has *Vaccinium myrtillus*, *Luzula sylvatica*, *Pteridium aquilinum*, *Blechnum spicant*, *Hyacinthoides non-scripta*, *Oxalis acetosella*, *Anemone nemorosa*, *Stellaria holostea*, *Holcus mollis* and *Deschampsia flexuosa*, often with an abundance of brambles.

Other fragments of woodland occur on more basic soils where there is lateral seepage or instability,

Plate 16. Luxuriant bushes of *Vaccinium uliginosum* in north Cumbria

often on the sides of little gills, and here there is a variety of trees, with a good deal of ash, wych elm, hazel and elder on the drier sites, and alder on stream alluvium or in swampy places. The more varied flora of these richer soils includes *Mercurialis perennis, Allium ursinum, Geum rivale, G. urbanum, Circaea lutetiana, Geranium robertianum, Sanicula europaea, Primula vulgaris, Stachys sylvatica, Galium odoratum, Veronica montana, Filipendula ulmaria, Carex sylvatica, C. laevigata, Brachypodium sylvaticum, Bromopsis ramosa* and *Melica uniflora.* Wet muddy ground has *Ranunculus repens, Caltha palustris, Chrysosplenium oppositifolium, Carex remota* and *C. paniculata.* More local plants of these basic woodland soils are *Paris quadrifolia, Neottia nidus-avis, Lathraea squamaria, Listera ovata* and *Epipactis helleborine.* Woods of both this mixed broadleaved type and the oak type are represented, to some extent at least, throughout Cumbria.

The rivers draining to the Carlisle basin - Eden, Lyne, Irthing, Liddel, Gelt and Petteril - have in places cut picturesque, craggy gills through the New Red Sandstone or Carboniferous series along their lower courses, and these mostly retain fringes or broader belts of woodland, often with both of the main types described above. Penton Linns on the Liddel is a good example. The rock is mostly acidic and the flora limited, but locally richer beds have basiphilous species such as *Polystichum aculeatum, Galium odoratum, Equisetum hyemale, Festuca altissima, Chrysosplenium alternifolium, Mycelis muralis* and *Myosotis sylvatica;* while clayey seepage soils have *Equisetum telmateia, Carex pendula, Crepis paludosa, Eupatorium cannabinum, Stellaria nemorum* and *Cardamine amara.* There is one clump of *Equisetum* x *trachyodon* by the Black Lyne, but a tiny outlying colony of *Hymenophyllum tunbrigense* on the Lyne appears to have gone. The wooded Gelt valley still holds two small colonies of

Cephalanthera longifolia. Valeriana pyrenaica occurs in wet woodland by the lower Esk and Liddel and the sandstone cliff at Rockcliffe by the Eden has *Anthriscus caucalis* and *Petroselinum crispum.* At Carlisle the Eden has *Potamogeton* x *cooperi* and *Eleocharis acicularis* while *Bunias orientalis* occurs on its banks and by the Petteril. South of Carlisle, the banks of the River Caldew below Dalston support small, precarious populations of *Artemisia absinthium, Dianthus deltoides, Clinopodium acinos* and *Trifolium striatum.*

The sandstone district is noteworthy for northern occurrences of heath characteristic of the southern England lowlands. Typical *Calluna vulgaris - Ulex gallii* heath occurs at Walby Moor, east of Carlisle, but in the area there are several authentic records of the southern *U. minor,* and a large population survives on heathland at Finglandrigg (Plate 17). Uncultivated ground, and old sand-pits and quarries have characteristic species of sandy soils, such as *Scleranthus annuus, Jasione montana, Ornithopus perpusillus, Filago minima* and *Gnaphalium sylvaticum,* and a sand-pit near Hayton has *Thymus pulegioides.* Roadside verges in the sandstone district are not usually floristically rich, but there are colourful displays of *Vicia cracca* and *V. sepium; Galium mollugo* is noticeably abundant on these non-calcareous soils and *Hieracium umbellatum* is a feature of this area. This is a largely agricultural district, of arable and high production grassland, with the 'weeds' of cultivation and disturbance, such as *Matricaria recutita* and *Chrysanthemum segetum,* well represented. *Anthriscus caucalis* survives on a few roadsides.

There are a few small tarns, such as Thurstonfield Lough, Monkhill Lough (now drained), Tarns Dub, Martin Tarn and Talkin Tarn. Talkin Tarn has *Potamogeton praelongus,* Thurstonfield *Eleocharis acicularis* and *Elatine hexandra,* and Monkhill *Glyceria maxima, Veronica catenata, Callitriche*

Plate 17. *Ulex minor,* Finglandrigg

obtusangula and *Salix triandra*. Flooded gravel pits at Longtown contain *Elatine*, *Eleogiton fluitans* and *Potamogeton pusillus*.

2. The middle Eden and Inglewood Forest to Cockermouth

This area includes the middle section of the Eden valley below Langwathby, the Petteril catchment above Wreay, the middle Caldew valley, and the northern foothills of the Lake District National Park extending past Caldbeck and Uldale to Cockermouth. The eastern part lies on the New Red Sandstone, but a band of Carboniferous limestone extends from Greystoke and Blencowe past Caldbeck to Bothel and Cockermouth. The gorge sections of the Eden at Wetheral, Cotehill, Armathwaite - Staffield, and Eden Lacy have fringing oak or broadleaved woodland in places. Noteworthy plants include *Gagea lutea*, *Campanula latifolia*, *Galium boreale*, *Juncus compressus*, *Scirpus sylvaticus* and *Carex ornithopoda*. The long course of the Eden itself has notable populations of *Scrophularia umbrosa*, *Senecio fluviatilis*, *Butomus umbellatus*, *Impatiens glandulifera*, *Zannichellia palustris* and *Ranunculus fluitans* and there are a few records of *Carex acuta*. The alluvial banks of the river at Langwathby were formerly a locality for *Orchis ustulata*. The heaths on the low Permian sandstone hills lying along the west side of the Eden are intermediate between lowland and upland heath with *Calluna*, *Vaccinium myrtillus* and *V. vitis-idaea*. These have been much reduced by agricultural reclamation and afforestation with the loss of *Radiola linoides*. One acidic valley mire has *Carex magellanica* and *Lycopodiella inundata*.

While there are minor occurrences of scar and pavement in Greystoke Park, and another small pavement area at Blindcrake, near Cockermouth, most of the limestone band is overlain by drift. The influence of the parent rock is especially evident on the broad roadside verges of the Greystoke - Sowerby Row area, and the banks of the old Penrith - Keswick railway around Newbiggin and Penruddock. Here, on calcareous clayey soils, there survive examples of herb-rich grasslands which were once widespread as meadow communities in the district. The flora includes *Primula veris*, *Campanula glomerata*, *Centaurea scabiosa*, *Serratula tinctoria*, *Geum rivale*, *Sanguisorba officinalis*, *Silaum silaus*, *Orchis mascula*, *Dactylorhiza fuchsii*, *Gymnadenia conopsea*, *Carex flacca* and *Briza media*, and, more rarely, *Coeloglossum viride*, *Platanthera chlorantha* and *P. bifolia*. Particularly characteristic of verges in the Eden valley is *Allium oleraceum*, and *Allium carinatum* and *Euphorbia esula* occur locally on roadsides north-east of Penrith. Open, sandy soils and banks provide habitats for *Cerastium arvense* and *Viola canina* respectively. On some verges with open, clayey soils, *Primula farinosa* grows in shorter, mossy grassland with *Valeriana dioica*, *Pinguicula vulgaris* and *Plantago maritima*. Johnby Moor formerly had an interesting grass-heath in which acid-loving or tolerant plants such as *Calluna*, *Genista anglica* and *Pyrola minor* were mixed with *Primula farinosa*, *Succisa pratensis*, *Silaum silaus* and other plants of basic soils. The ground is now largely planted with conifers.

The limestone area contains the only good example of a rich fen in northern Cumbria, at Newton Reigny Moss. Part of the fen is colonised by carr of willow thickets, mainly *Salix cinerea* and *S. pentandra* but also with *S. myrsinifolia*. Sedge swamp has *Carex rostrata*, *C. lasiocarpa*, *C. disticha*, *C. diandra* and formerly *Schoenus nigricans*. Other characteristic species are *Pyrola minor*, *Ranunculus lingua*, *Menyanthes trifoliata*, *Potentilla palustris*, *Galium palustre*, *Lychnis flos-cuculi*, *Caltha palustris* and *Eriophorum angustifolium*. *Cladium mariscus* and *Carex limosa* still survive but *Epipactis palustris*, *Utricularia vulgaris* and *Primula farinosa* have evidently vanished through war-time draining operations. Small patches of fen occur in Greystoke Park and in hollows on the crest of Eycott Hill, where *Carex limosa* is abundant as also is *C. pauciflora* in the more acid *Sphagnum*-dominated areas.

Some of the main river valleys and lesser gills have patches of oak or mixed broadleaved woodland. Limestone exposures at the Howk ravine, Caldbeck, in the Caldew valley above Sebergham, and in a gill on the north side of Catlands Fell have a limited calcicole woodland flora, with *Vicia sylvatica*, *Paris quadrifolia*, *Rubus saxatilis* and *Galium odoratum*. Two more westerly gills have *Equisetum hyemale*. The fine woodland complex at Park Wood, Brocklebank was spoiled by replacing most of the oaks with conifers. Fragments of limestone grassland survive at Catlands Fell, Aughertree Fell, the Clints at Blindcrake and perhaps elsewhere, but the pastures of the western limestone ground have mostly been botanically impoverished by 'improvement'. The foothill summits from Faulds Brow to Binsey are largely covered with heather moor or acidic grassland and have a very limited flora. Calcareous flushes on hill slopes west of Caldbeck contain *Blysmus compressus*.

3. The Western Coastal Belt

The coast on the Coal Measures between Maryport and Whitehaven is part of one of the two heavily industrialised districts of Cumbria, yet it has considerable botanical interest. At Maryport, waste ground adjoining the docks was quite recently found to support a large population of *Anacamptis pyramidalis*, along with *Orobanche minor* and the very rare *O. purpurea*. Along the shore and adjoining submaritime habitats south to Whitehaven are *Geranium sanguineum*, *Crambe maritima*, *Vicia lathyroides*, *Atriplex glabriuscula* and *Saxifraga granulata*. The broken cliffs and clayey banks around Harrington have *Lathyrus sylvestris*, *Equisetum telmateia*, *Eupatorium cannabinum* and *Dactylorhiza purpurella*. Inland, *Allium scorodoprasum* is conspicuous on railway banks from Bullgill southwards, and also grows on roadsides. *Senecio erucifolius*, *Crepis biennis* and *Serratula tinctoria* are other noteworthy species of this Coal Measures area. Disused quarries in the tail end of the Carboniferous limestone outcrop occur from Cockermouth to Egremont and provide the northernmost localities for *Ophrys apifera* on the west side of Britain.

The lofty sandstone cliffs of St Bees Head are mostly non-calcareous and lack some of the plants of basic seacliffs on the Scottish side of the Solway. The rocks have *Limonium recurvum*, *Crithmum maritimum* and *Asplenium marinum*, besides the common shore plants such as *Armeria maritima*, *Silene uniflora* and *Cochlearia officinalis*. The cliffs and railwaysides have *Lathyrus sylvestris* and the largest Cumbrian populations of *Vicia sylvatica* and *Polystichum setiferum*, and heathy communities have one of the few colonies of *Dryopteris aemula*. The sand and shingle shore from Allonby to Bootle has produced recent records for *Mertensia maritima*. A feature of the coast is the increasing amount of *Rosa rugosa* and *Lupinus arboreus*. The Drigg dunes are formed of rather acidic sand and are not rich botanically, though the flora includes *Anagallis minima*, *Radiola linoides*, *Calystegia soldanella*, *Euphorbia portlandica*, *E. paralias*, *Geranium sanguineum*, *Carlina vulgaris*, *Erigeron acer*, *Gentianella campestris* and *Ophioglossum azoricum*. The Eskmeals dunes across the Irt estuary have recently yielded interesting additions in *Pyrola rotundifolia*, *Corallorhiza trifida* and, briefly, *Lavatera arborea*. Between the two is salt-marsh, with *Atriplex portulacoides* and a small, declining population of *Seriphidium maritimum*. *Tamus communis* reaches its northern limit near the coast at Muncaster.

The coast between Drigg and Millom has most of the Cumbrian populations of *Hypochoeris glabra* and, near Silecroft, the only population of *Isolepis cernua*. At Haverigg is a dune fringe with some

botanical interest, though this is eclipsed in importance by the much larger calcareous dune system at Sandscale Haws across the Duddon estuary (Plate 18). Sandscale is the richest dune area in Cumbria with good slacks which support *Epipactis palustris*, *E. phyllanthes*, *E. leptochila* var. *dunensis*, *Anacamptis pyramidalis*, *Corallorhiza trifida*, *Ophrys apifera*, *Pyrola rotundifolia*, *Monotropa hypopitys*, *Centaurium littorale*, *Equisetum variegatum*, *Centaurium littorale*, *Selaginella selaginoides* and *Carex viridula* subsp. *viridula*. Inland of the slacks are wet heath and mire communities with *Hypericum elodes* and *Juncus subnodulosus*. The adjacent north end of Walney Island has a complex of salt-marsh, shingle, dunes, gravel pits and heath, with an extremely varied flora. It is the classic site for *Geranium sanguineum* var. *striatum*, and has many plants of Sandscale, besides others such as *Parentucellia viscosa*. The Duddon estuary is also fringed by extensive salt-marshes which are botanically similar to those of the Solway and Morecambe Bay.

The rather narrow plain between the coast and the Lakeland fells south of St Bees is covered with glacial drift and has a good many hollows variously holding small bogs or tarns. The best sites are Black Moss, Egremont; Silver Tarn, Nethertown; and Low Church Moss, Beckermet; Hallsenna Moor; Barfield Tarn; Whitbeck Moss; Silecroft and Kirksanton. Similar mires occur around Nether Wasdale just within the Lake District National Park. Snellings Mire and Sellafield Tarn have been lost to development. These wetland habitats mostly have base-poor water, and their flora includes *Anagallis tenella*, *Osmunda regalis*, *Carex curta*, *Drosera intermedia*, *Menyanthes trifoliata*, *Potentilla palustris*, *Vaccinium oxycoccos*, *Dactylorhiza maculata* and the south-western species *Scutellaria minor* and *Hypericum elodes*. Barfield Tarn has *Elatine hexandra* and Silecroft pond *Nymphoides peltata* and the invasive *Crassula helmsii*. Further

Plate 18. Sandscale dunes, Dalton-in-Furness

Plate 19. *Convallaria majalis*, Hutton Roof

Plate 20. *Viola rupestris*, Arnside

south at Hodbarrow are *Pilularia globulifera*, *Asplenium marinum* and, on the spoil from the old haematite mines, a number of calcicoles including *Inula conyzae* and *Epipactis atrorubens*. Other south-western species are *Asplenium obovatum*, on acid rocks and banks north of Millom, and *Umbilicus rupestris*, on banks and walls in several places. Raised bogs occur inland on either side of the Duddon estuary and have a similar flora to those of the Solway.

4. The Morecambe Bay - Kendal lowlands and foothills

The northern margin of Morecambe Bay is extensively fringed by salt-marshes floristically similar to those of the Solway. They are again heavily grazed by sheep and cattle, and subject to rotational turf removal. *Ruppia maritima*, *R. cirrhosa* and *Zostera angustifolia* occur in a few places. *Spartina anglica* has appeared on the estuarine flats in recent years and is locally dominant. In a few places the inner marsh shows transition to rich fen, with *Juncus subnodulosus*, *Schoenus nigricans* and *Thalictrum flavum*. At the western entrance to the Bay are the dunes and old gravel workings of South Walney and the shingle complex with Foulney Island. The varied flora includes *Glaucium flavum*, *Calystegia soldanella*, *Crambe maritima*, *Hyoscyamus niger*, *Echium vulgare*, *Carduus tenuiflorus*, *Hypochoeris glabra*, *Vulpia fasciculata* and, formerly, *Mertensia maritima*. *Stellaria pallida* occurs amongst the detritus on the gullery. The lime-stone cliff promontory of Humphrey Head (Plate 15) has a famous cluster of rarities in *Aster linosyris*, *Hypochoeris maculata*, *Veronica spicata* and *Helianthemum canum*, the first two at their northern British limits. Low coastal limestone outcrops elsewhere have *Adiantum capillus-veneris*. *Limonium britannicum* subsp. *celticum* occurs in the Kent and Leven estuaries.

In the hinterland of Morecambe Bay, the area around Barrow-in-Furness is of interest mainly for its wasteland and dock flora, with *Lepidium latifolium*, *Rumex scutatus*, *Diplotaxis tenuifolia*, *Senecio squalidus*, *Sisymbrium orientale*, *Hordeum murinum*, *Orobanche minor* and *Medicago sativa*. *Blackstonia perfoliata* grows on an old slag heap at Askam. Raised bogs occur at Rusland, Roudsea, Witherslack and Meathop, and feature *Andromeda polifolia*, *Vaccinium oxycoccos* and *Rhynchospora alba*. The large area of Foulshaw Moss was largely destroyed by afforestation, and a colony of *Gentiana pneumonanthe* with it, though this species still occurs at one site in this area. Drains around the Moss have *Lemna gibba*, *Potamogeton pusillus* and *Ceratophyllum demersum*.

The most important feature of the Morecambe Bay hinterland is the large extent of limestone between the Furness peninsula around Dalton to Hutton Roof in the east. There are several outstanding botanical areas here. Roudsea at the mouth of the Leven has a complex of habitats which include woodland on limestone and calcareous fen. East of the Kent estuary and bordering Lancashire is the Arnside area, with low, rocky limestone hills partly covered with woodland. Above Burton and Holme are the slightly higher hills of Hutton Roof Crags (Plate 6) and Farleton Fell, notable for their massive exposures of tabular limestone pavement, with exten-sive scrub and patchy woodland. South-west of Kendal are the scarps, screes and pavements of Scout, Cunswick and Whitbarrow Scars (Plate 2), also wooded in varying degree.

The limestone woodlands and scrub contain most of the species of the mixed broadleaved woodland described for the Solway district. Ash and hazel are locally dominant, but oak, wych elm and birch are often abundant. There are also trees and shrubs typical of calcareous soils in more southern woodland, including several at or near their native northern British limit - *Tilia cordata*, *Acer campestre*, *Sorbus torminalis*,

Plate 21. *Dryopteris submontana*, Hutton Roof

S. lancastriensis, Rhamnus cathartica, Euonymus europaeus, Cornus sanguinea, Ligustrum vulgare and *Viburnum opulus*. The pavement scrub has abundant juniper and yew. *Tamus communis* is widespread and *Clematis vitalba* occurs as an evident native. Herbaceous plants of the woodland and scrub include *Convallaria majalis* (Plate 19), *Ophrys insectifera, Hypericum montanum, Inula conyzae, Lithospermum officinale, Aquilegia vulgaris* and *Carex digitata*.

Dry grassland and rock habitats have a rich flora, with *Helianthemum nummularium, Scabiosa columbaria, Sesleria caerulea, Filipendula vulgaris, Geranium sanguineum, Thalictrum minus, Galium sterneri, Asperula cynanchica, Hippocrepis comosa, Minuartia verna, Potentilla neumanniana, Epipactis atrorubens, Polygonatum odoratum, Veronica spicata* and *Carex ornithopoda*. Vertical crevices (grikes) of the pavements have an abundance of ferns, including *Phyllitis scolopendrium, Polystichum aculeatum* and the nationally scarce *Dryopteris submontana* (Plate 21) and *Gymnocarpium robertianum*. Rarities include *Viola rupestris* (Plate 20) at Arnside, *Helianthemum canum* at Scout and Whitbarrow Scars, and *Polystichum lonchitis* in a single, low-level locality. Other notable plants of the Morecambe Bay limestone grasslands are *Ophrys apifera, Orchis morio, Spiranthes spiralis* and, at Arnside, *Carex ericetorum*.

The Roudsea fens have *Carex pseudocyperus, C. elongata, Calamagrostis epigejos* and *C. canescens*, and Roudsea Wood has *Carex flava* in its only British station. *Cladium mariscus* grows by Cunswick Tarn below the Scar. The calcareous marsh at Hale Moss near Burton features *Primula farinosa, Carex viridula* subsp. *viridula, C. elata* and the dominant *Schoenus nigricans*. The Lancaster - Kendal canal receives base-rich water and has *Ceratophyllum* species, *Potamogeton pusillus, Callitriche hermaphroditica, Lemna gibba* and *Azolla filiculoides*.

The rest of the Morecambe Bay lowlands and foothills lying on the Silurian formations are less rich botanically, but have a quite varied range of habitats. Numerous tarns and pockets of bog in the uplands between the Rivers Kent and Lune provide localities for aquatics such as *Ceratophyllum demersum, Utricularia vulgaris, Potamogeton gramineus* and *Sparganium natans*, and swamp species including *Carex magellanica, C. diandra, C. aquatilis* and, in one place, *Thelypteris palustris*. Of the rare Cumbrian plants, *Trientalis europaea* has an outlying locality near Sedbergh and *Dryopteris aemula* one in the west in the Kirkby Moor area.

5. The Lake District
This complex area is divided for convenience into the lakes and valley bottoms, the woods and lower ravines, and the fells.

The lakes and valley bottoms
The lakes and tarns form a graded series in water chemistry, from nutrient-poor (oligotrophic) to moderately nutrient-rich (mesotrophic). The western lakes, particularly Wast Water, Buttermere, Crummock Water, Ennerdale Water and many of the mountain tarns, lie at the oligotrophic end of the spectrum, and are characterised by stony shores with a lack of emergent vegetation. There is usually a submerged marginal growth of *Littorella uniflora* and *Isoetes lacustris*, with *Lobelia dortmanna* more locally and, in a few of the upland tarns, *Subularia aquatica*. *Myriophyllum alterniflorum* and the submerged form of *Juncus bulbosus* are often abundant. *Sparganium angustifolium* is typical of the upland tarns, some of which have open fringing stands of *Carex rostrata*.

It is in the richer waters of Derwent Water, Bassenthwaite Lake, Windermere and, especially, Esthwaite Water, where aquatic and fringing communities are more strongly developed. The pondweeds include *Potamogeton natans*, *P. perfoliatus*, *P. berchtoldii*, *P. gramineus*, *P. alpinus* and, more rarely, *P. pusillus* and *P. praelongus*. Coniston Water has *P.* x *zizii* and Esthwaite the very rare *Najas flexilis*. Succession from fringing reedswamp to willow and alder carr occurs in several places, with good examples at Esthwaite North Fen, which also features *Calamagrostis purpurea*, and at The Ings on the east side of Derwent Water. *Carex aquatilis* and *C. elongata* are uncommon but characteristic plants of carr and reedswamp respectively. *Lysimachia terrestris* is frequent in swamp vegetation around Windermere and *Cardamine raphanifolia* in the Ambleside - Grasmere area. The stony shores of several lakes have *Juncus filiformis*, and *Ranunculus* x *levenensis* occurs around Ullswater. *Elodea nuttallii* is abundant and in many places is replacing *E. canadensis*; *Crassula helmsii* has recently been found in Coniston and Derwent Water, and the latter also has *Isoetes echinospora*.

The woods and lower ravines

The native woods of Lakeland are more celebrated for their rich oceanic flora of mosses and liverworts than for their vascular plants. They are predominantly of sessile oak on acidic brown earths, but with a good deal of ash, wych elm and hazel on the richer soils. Downy birch is abundant and forms secondary woodland on some sites where oak has been cleared. Many of the woods are of the 'hanging' type, on steep slopes, often thickly littered with blocks and broken by outcrops. Some are traversed by waterfall ravines, as at Lodore (Plate 3) and Dalegarth, and many lower rocky gills carry fringing remnants of woodland on otherwise treeless slopes. The majority are subject above to grazing sheep with consequent suppression of the taller herbaceous plants. Ungrazed examples have lush *Luzula sylvatica* and *Vaccinium myrtillus*, but in the heavily grazed woods grasses and bracken tend to be the most conspicuous plants of the field layer. Juniper thickets are still extensive on the acidic soils of some lower hillsides, though retreating slowly through failure to regenerate, due in part to heavy grazing.

Borrowdale has the greatest extent of native woodland, but there are other good examples at Hallin Fell, Glencoyne and Low Wood in the Ullswater valley, Naddle Forest above Haweswater, Scales Wood at Buttermere and The Side woods in Ennerdale. There are also numerous oakwoods around Ambleside, the foot of the Langdales, and along the sides of Windermere and Coniston Water. There is a characteristic profusion of common ferns - *Dryopteris filix-mas*, *D. affinis*, *D. dilatata*, *Athyrium filix-femina*, *Oreopteris limbosperma* and *Blechnum spicant*, and also the more local *Gymnocarpium dryopteris* and *Phegopteris connectilis*. *Polypodium vulgare* is abundant as an epiphyte and *Melampyrum pratense* is usually common. *Dryopteris expansa* is occasional in rocky woodland. Many rocky woods and gills,

especially in the west, have an abundance of *Hymenophyllum wilsonii*. Its relative, *H. tunbrigense*, is much rarer and confined to the Eskdale area, where it is accompanied in a few places by *Dryopteris aemula*. The hill woods on acidic ground have a limited vascular flora similar to that of lowland oakwoods. On acidic rocks, especially in ravines, are scattered colonies of *Orthilia secunda*, but this also grows on treeless cliffs. *Circaea alpina* occurs locally, both in acid woodland and on now treeless boulder and scree-covered slopes.

Basic woodland soils are more productive, and the tree species include scattered occurrences of *Tilia cordata*, here at its northern limit. Bird cherry is frequent and yew locally so, especially on the Silurian rocks. Several colonies of *Sorbus rupicola* occur on basic crags around Keswick. One of the most notable field layer plants is *Impatiens noli-tangere*, growing wild in several valleys and forming considerable colonies. On wooded rocks there is usually *Polystichum aculeatum* but *P. setiferum* is surprisingly rare; other uncommon rock face plants include *Festuca altissima*, notably in Stock Ghyll, Ambleside, and *Melica nutans*.

The fells

Early deforestation was evidently followed by a great expansion of dwarf shrub heath, especially of heather, downwards from the higher levels where it was the natural vegetation. Centuries of heavy grazing by domestic stock, and repeated moor burning, then combined to cause the widespead replacement of dwarf shrub heath by acidic grassland at all levels, but often with much bracken up to 450 m. On many of the Borrowdale Volcanic hills heather communities are noticeably confined to cliff faces and block screes. In scattered places (Gillercombe, Kentmere, below Cam Spout, below Dove Crag at Hartsop, Wetherlam) huge detached blocks are crowned with luxuriant growths of heather and bilberry contrasting with the surrounding grassland - natural 'experiments' showing the effect of protection from sheep and fire. The Skiddaw Slate hills now have the main areas of heather moorland in Lakeland (Plate 22), which suggests that something in the chemistry of their soils has in general discouraged the levels of sheep stocking achieved on the Volcanics.

Some of the associated dwarf shrubs have declined even where heather has remained dominant. *Ulex gallii* has been greatly reduced on the slopes of Black Combe in recent years, and in several areas it is noticeably now a plant of crags and rocky ground. One of the chief casualties of fire is probably *Arctostaphylos uva-ursi*. The best remaining colony is on the west slope of Grasmoor where its scree habitat may have insulated it against recurrent fires. *Juniperus communis* subsp. *alpina* is also largely a rock face plant, except on the ridge of Whiteside above Lorton. *Empetrum nigrum* appears better able to cope with heavy grazing than some dwarf shrubs, forming numerous patches in some high level grasslands on the Skiddaw Slate. Yet

Plate 22. Skiddaw Forest from Carrock Fell - juniper scrub in foreground, heather dominated grouse moor in the distance to the right of the River Caldew and grasslands of the sheepwalks to the left.

E. nigrum subsp. *hermaphroditum* is confined to block screes and outcrops, and has declined in Riggindale since its discovery there in 1954, perhaps through heavy grazing. *Vaccinium vitis-idaea* is also relatively tolerant of fire, and abundant on some heavily burned situations, such as the lower slopes of Skiddaw.

The acidic *Festuca - Agrostis* and *Nardus* grasslands of the fells are botanically impoverished. *Potentilla erecta* and *Galium saxatile* are in great quantity, and, where seepage from basic substrates causes local enrichment, there are indicators such as *Thymus polytrichus, Linum catharticum, Euphrasia nemorosa, Potentilla sterilis, Prunella vulgaris* and, occasionally, *Carlina vulgaris. Alchemilla alpina* flourishes not only on the rock ledges but often in the short turf below, which suggests that it is somewhat tolerant of grazing, but most of the herbs of grazed situations are small species or chewed down rosettes. The clubmosses *Lycopodium clavatum, Huperzia selago* and *Diphasiastrum alpinum* are widespread in acidic grassland, often growing together, and at the head of Langdale small isolated southern populations of *L. annotinum* persist.

The almost entirely acidic Skiddaw Slate hills are generally botanically dull, yet in *Lychnis alpina* at Hobcarton Crag (Plate 23) they have the most famous Lakeland mountain plant. The presumed chemical peculiarity of the rock in the gullies where it grows has eluded positive identification, but the associated flora is that typical of other acidic rocks in the district. Even *Alchemilla alpina* is uncommon on the Slate, though there is a good colony at Dove Crags, Grasmoor. *Cryptogramma crispa* is luxuriant on the screes, while, on high stony summits, *Carex bigelowii* is relatively widespread and, above 650 m, *Salix herbacea* is locally plentiful. In the Skiddaw group itself, the more notable plants are mostly in acidic bogs or rather poor flushes:

Plate 23. *Lychnis alpina*, Hobcarton Crag

Carex pauciflora, Drosera longifolia, Hammarbya paludosa, Rubus chamaemorus, Andromeda polifolia, Saxifraga stellaris and *Myosotis stolonifera. Listera cordata* grows under heather on steep, shady slopes. *Saxifraga aizoides, S. hypnoides* and *Asplenium viride* are among the few calcicoles.

The main botanical interest of these grassy hills is either in wet habitats - bogs, flushes and rills - or on rocks. The undulating watershed of the Armboth Fells is a good area to see the bog flora, with *Andromeda polifolia, Vaccinium oxycoccos, Utricularia minor, Carex magellanica, C. pauciflora, C. limosa, C. curta* and, formerly, *C. lasiocarpa.* Calcareous flushes and little marshes at lower levels have *Saxifraga aizoides, Selaginella selaginoides, Parnassia palustris, Carex dioica, Eleocharis quinqueflora* and, rarely, *Erioph-*

orum latifolium. Local plants of rills are *Epilobium alsinifolium* and *Myosotis stolonifera*, and *Primula farinosa* is quite widespread in basic flushes among the eastern fells. *Lycopodiella inundata* survives at two localities, and other local plants of low-level heathland valley bog are *Drosera intermedia*, *Hammarbya paludosa*, *Hypericum elodes* and *Scutellaria minor*. Contrasting high-level plants of basic rills and flushes are *Juncus triglumis* and, rarely, *Sedum villosum*.

Calcareous cliffs occur most extensively above 600 m in the Helvellyn, Fairfield and High Street ranges, and these are the most productive for calcicole 'alpines'. Some are in extremely small quantity - *Saxifraga nivalis*, *Cerastium alpinum*, *Salix lapponum*, *Carex atrata*, *Poa alpina*, *P. glauca*, *Draba incana*, *Persicaria vivipara*, *Polystichum lonchitis*, *Dryas octopetala*, *Potentilla fruticosa* and *Ajuga pyramidalis*. Others are somewhat less rare: *Saxifraga oppositifolia*, *Silene acaulis*, *Potentilla crantzii*, *Hieracium subgracilentipes* and *Saussurea alpina*, while *Saxifraga hypnoides*, *Minuartia verna*, *Thalictrum alpinum*, *Sedum rosea*, *Alchemilla wichurae*, *Oxyria digyna* and *Asplenium viride* are relatively common, as also is *Euphrasia rivularis* in flushes and on wet rock ledges. The Scafell range is the next most productive area, mainly in deep ravines such as Piers Gill. *Dryas* survives at one site and there is a good colony of *Potentilla fruticosa* on Wasdale Screes. The westernmost fells are poorer, but Pillar has the *Potentilla*, *Hieracium subgracilentipes* and *Saussurea*. Honister Crag, around the old quarry workings, is a good locality for the more widespread calcicole mountain plants, along with the ascending lowland ones.

The basic cliffs also have a characteristic assemblage of lowland grassland or woodland plants, some of which ascend to a considerable elevation. They include northerners, such as *Geranium sylvaticum*, *Trollius europaeus*, *Galium boreale* and *Rubus saxatilis* but also local species of calcareous soils such as *Thalictrum minus*, *Arabis hirsuta*, *Aquilegia vulgaris*, *Crepis paludosa*, *Pimpinella saxifraga* and, rarely, *Vicia sylvatica*. Common plants of the ledge communities are *Heracleum sphondylium*, *Angelica sylvestris*, *Valeriana officinalis*, *Silene dioica*, *Alchemilla glabra* and *Geum rivale*. Maritime plants with isolated populations on the crags are *Silene uniflora*, *Armeria maritima* and *Plantago maritima*; the montane *Cochlearia pyrenaica* could also be included here. Hawkweeds are mainly on basic rocks, though the habitats of *Hieracium holosericeum* appear rather acidic. Borrowdale Volcanic crags formed of markedly non-calcareous rock have few species besides those mentioned for the Skiddaw Slate, though most localities for *Asplenium septentrionale* are on the former. The last known colony of *Woodsia ilvensis* is on rather acidic rocks, though some of its former stations were evidently more base-rich. Acidic screes and crags have an abundance of *Cryptogramma* and, locally, *Dryopteris oreades* and *D. expansa*.

6. The Lune valley to the Pennines

Between the eastern edge of the Lake District National Park and the Westmorland Pennines are the lower fells of Tebay and Shap, draining southwards to the Lune and the Asby - Orton limestone uplands draining north to the Eden. They include the Silurian massif of the Howgill Fells, steep but mainly smooth-sided hills covered largely by dull acidic grassland with dominant *Nardus*. Their few crags include Cautley Spout, with *Alchemilla alpina* and *Orthilia secunda*, Cautley Crag, with *Salix herbacea*, and Black Force. *Myosotis stolonifera* grows in the flushes. The Lune valley around Sedbergh and north to Tebay is a famous area for *Meum athamanticum* on roadsides and in pastures, in one place with *Vicia orobus*. From here south to Kirkby Lonsdale the main interest lies in the rocky banks and islands of the Lune and its tributary the Rawthey. The waterworn Silurian rocks support large populations of *Galium boreale* and the small, early-flowering *Solidago virgaurea* var. 'cambrica'. Associated species include *Viola canina*, *Scabiosa columbaria* and *Helictotrichon pratense*. The conglomerate riverside cliffs at Casterton harbour *Vicia sylvatica* and the gills draining Middleton Fell have small populations of *Tilia cordata*, *Hymenophyllum wilsonii* and *Festuca altissima*. Garsdale and Dentdale, with Baugh Fell, Crag Hill and the northern half of Whernside, are mainly acid ground. The summit of Crag Hill has a luxuriant cover of *Carex bigelowii*, and *Salix herbacea* occurs on the crag. There are exposures of limestone, including potholes on Ease Gill above Leck. Garsdale and Dentdale have wooded riversides and limestone waterfalls. Rarities of this area include *Alchemilla minima*, *Juncus compressus*, *Pseudorchis albida* and *Hammarbya paludosa*.

The most interesting area is the broad band of Carboniferous limestone which occupies the high land between Orton and Kirkby Stephen. Where drift-covered it has heather moor and *Nardus* grassland, but the limestone outcrops as extensive tabular, fissured pavements around Great Asby, Sunbiggin and Orton, with smaller areas on Crosby Ravensworth Fell. These have a characteristic grike flora with *Dryopteris submontana*, *Gymnocarpium robertianum*, *Phyllitis scolopendrium*, *Polystichum aculeatum*, *Cystopteris fragilis*, *Asplenium viride*, *Thalictrum minus*, *Mycelis muralis*, *Melica nutans* and, at one site, *Actaea spicata*. *Hornungia petraea* occurs here among shattered limestone at its northernmost British station and stony limestone grassland near Orton Scar has *Carex ericetorum*. There is an important calcareous wetland complex at Sunbiggin, with the Tarn and its fringing fen, and the adjoining Tarn Moor with wet pasture, marsh and incipient acidic bogs. The flora contains *Cladium mariscus*, *Schoenus nigricans*, *Eriophorum latifolium*, *Dactylorhiza purpurella*, *D. incarnata*, *Carex viridula* subsp. *brachyrrhyncha*, *C. elata*, *C. paniculata*, *C. diandra*, *Blysmus compressus*, *Parnassia palustris*, and *Primula farinosa* in glorious

Plate 24. *Dactylorhiza purpurella* and *Primula farinosa*

thera chlorantha, Orchis mascula, Listera ovata, Paris quadrifolia, Rubus saxatilis, Carex ornithopoda and *Melica nutans.* The banks of a disused railway line system at Waitby have dry limestone grassland with some of these species and *Serratula tinctoria, Filipendula vulgaris, Carlina vulgaris, Plantago maritima, Gentianella amarella, G. campestris, Rosa pimpinellifolia, Silaum silaus, Sanguisorba minor, S. officinalis* and *Trollius europaeus*; *Epipactis palustris* and *Primula farinosa* are conspicuous in the flushes.

The section of the Eden valley between Kirkby Stephen and Penrith falls within this subdivision. It has a few small fens and valley bogs of significance, such as Temple Sowerby, Udford and Cliburn Mosses, but the interesting Sandford Mire at Warcop has been drained. Temple Sowerby Moss has a fairly rich fen with *Carex diandra, Pyrola minor, Ranunculus lingua, Frangula alnus, Salix myrsinifolia* and *S. pentandra,* and formerly *Oenanthe fistulosa.* Cliburn Moss is more acidic with *Carex limosa, Andromeda polifolia, Vaccinium uliginosum, V. oxycoccos, Genista anglica* and *Listera cordata.* It lies within pinewood, and the adjoining Whinfell Forest has *Goodyera repens.* Flakebridge Wood near Appleby has *Calamagrostis canescens.* Other notable plants of the area include *Stellaria palustris* and *Carex riparia* at Great Ormside, *Potentilla argentea* near Udford and *Bryonia dioica* at Culgaith, with *Astragalus glycyphyllos* and *Vicia sylvatica* on railway banks nearby. *Groenlandia densa* is a feature of the Eden and its tributaries.

7. The Pennines

This is the most massive of the upland systems of Cumbria, and occupies the whole of the county's eastern edge, south of the Tyne Gap. The Pennines owe their botanical importance mainly to the extensive exposures of limestone and the calcareous soils derived from these. There is a strong and often sudden contrast with the acidic soils that nevertheless predominate. The Pennine dip slope, draining to the headwaters of the Rivers Tees and South Tyne, consists mostly of gently contoured moorlands. Here, and on the main broad watershed of the range, the extensive development of deep blanket bog (Plate 25) has obliterated the influence of the underlying limestone, except locally in drainage water emerging at the surface. The blanket bogs have undergone extensive erosion by gullying of the peat into 'haggs', and all stages in denudation down to the basal mineral soil may be seen. On the large plateau of Lodge Hags, east of Mallerstang Edge, there are both residual peat islands from an evidently once continuous cover (Plate 26), and stages in recolonisation of the bare debris to moss heath, grassland and even shallow bog.

While much of the River Tees and Tyne catchments have been managed as grouse moor and retain extensive areas of heather, they have been much affected by sheep grazing. The Eden Valley slopes have

profusion (Plate 24). It is feared that the large breeding colony of black-headed gulls at Sunbiggin Tarn has caused substantial nutrient enrichment damaging to the botanical interest of the surrounding fen.

The area was formerly important for its herb-rich meadows and limestone pastures, and though these are much reduced, there is still a rich grassland and roadside verge flora with *Geranium sylvaticum, Sanguisorba officinalis, Geum rivale, Alchemilla* species, *Cirsium heterophyllum* and, more rarely, *Galium boreale, Persicaria vivipara, Crepis mollis* and *C. praemorsa* in its only British locality. Between Orton and Crosby Ravensworth are moist pastures and flushes with *Bartsia alpina, Polygala amarella* and *Carex capillaris.* Grazed, upland grassland is usually dominated by *Sesleria caerulea,* and *Antennaria dioica* and *Linum perenne* occur in a few places. *Potentilla neumanniana, Hippocrepis comosa* and *Carex ornithopoda* are a feature of the limestone scars. The deep limestone valley of Smardale Gill has good ashwoods, and a varied flora with abundant *Pyrola minor, Geranium sanguineum, Helianthemum nummularium, Hippocrepis comosa, Ophrys insectifera, Epipactis helleborine, Coeloglossum viride, Neottia nidus-avis, Dactylorhiza fuchsii, Gymnadenia conopsea, Platan-*

Plate 25. Blanket bog, Moor House, looking east towards Teesdale

Plate 26. Erosion of blanket bog, isolated peat islands over basement sandstone, Lodge Hags, Mallerstang

been still more modified by heavy grazing and moor burning, to the point where heather and other dwarf shrubs have largely been replaced, on skeletal brown soils, by acidic *Festuca - Agrostis* grasslands, and by *Nardus stricta* or *Juncus squarrosus* on gley or peaty podsols. *Pteridium aquilinum* is abundant in the drier grasslands, though less luxuriant and dominant than on the Lakeland fells. Few upland species occur on the acidic soils, and even the clubmosses are scarce. On the blanket bogs of *Eriophorum vaginatum*, *E. angustifolium* and *Calluna*, there is an abundance of *Rubus chamaemorus*, which reaches extreme profusion on the broad ridge of Burnhope Seat on the Co. Durham border. *Vaccinium myrtillus* is generally abundant, and *V. vitis-idaea* and *Empetrum nigrum* occur widely on dry heaths and bogs alike. *Carex rostrata*, *C. curta*, *C. nigra* and *C. echinata* are common in acid flush

bogs, and *Carex magellanica* and *C. limosa* grow in spongy *Sphagnum* swamps in the Moor House area. *Listera cordata* occurs occasionally under the heather. Above 550 m *Carex bigelowii* occurs in some abundance, especially on the drier soils. *Cryptogramma crispa* is abundant on gritstone and Whin Sill screes; *Dryopteris dilatata* and, to a lesser extent, *Huperzia selago* are a characteristic feature of the block scree.

The Cross Fell range between Stainmore and Hartside is the most productive part of this subdivision for mountain plants since it has the greatest extent of limestone habitats. The best and highest exposures lie in the Westmorland part. In rills, flushes and springs with only slightly enriched water *Saxifraga stellaris* is widespread, and *Myosotis stolonifera* locally frequent. *Epilobium alsinifolium* is less common and *E. anagallidifolium* is confined to the higher levels. Perhaps the most notable plant of these slightly acidic springs is the grass *Alopecurus borealis*, found in several places from Cross Fell to Knock Fell. Where the water is more calcareous, mossy springs and flushes scattered over the higher fells of this massif are the British headquarters for the otherwise rare *Saxifraga hirculus*, and *Sedum villosum* is often in the more open flushes. Several species occurring abundantly on moist ground on the Teesdale sugar limestone are much less plentiful on the Cumbrian side of the Tees - *Thalictrum alpinum*, *Saxifraga aizoides*, *Kobresia simpliciuscula*, *Juncus alpinoarticulatus*, *J. triglumis* and *Equisetum variegatum*. North of Hartside, the flora is very limited, limestone being little exposed, but Cold Fell has one of the few populations of *Vaccinium uliginosum* in the Cumbrian Pennines. Natural lakes are scarce, the acid, peaty tarns above Dufton having little interest, but Tindale Tarn under Cold Fell has *Calamagrostis canescens*.

It is, however, the dry limestone habitats which are so conspicuously richer botanically than their acidic counterparts. Fine ashwoods occur at Helbeck on the slopes above Brough, and there are more fragmentary occurrences in the gills of that south-eastern area, such as Swindale, Augill, Argill, the River Belah, and Podgill and Stenkrith at Kirkby Stephen. The Mallerstang valley also has wooded limestone ravines, as at Aisgill and Hell Gill. Plants of these woods and gills include *Salix phylicifolia*, *Euonymus europaeus*, *Rhamnus cathartica*, *Convallaria majalis*, *Paris quadrifolia*, *Aquilegia vulgaris*, *Myosotis sylvatica*, *Vicia sylvatica*, *Campanula latifolia*, *Rubus saxatilis*, *Melica nutans*, *Equisetum hyemale* and, in single localities only, *Hordelymus europaeus* and *Cephalanthera longifolia*.

The lower level limestone grasslands and rocks have many widespread calcicoles such as *Helianthemum nummularium*, *Sanguisorba minor* and *Scabiosa columbaria*. Grazed swards typically contain a variety of grasses, but the most characteristic, all over the limestone, is *Sesleria caerulea*. On the scars, its dense, lush tufts are aggressive competitors with smaller plants. Rarer species of the lower rocks and screes include *Epipactis atrorubens*, *Carex ornithopoda*, *Dryopteris submontana* and *Gymnocarpium robertianum*. The southern *Hippocrepis comosa* extends north to Dufton Crag. *Tephroseris integrifolius* was formerly known from above Brough and may perhaps still persist.

At the higher levels, a few of the Upper Teesdale specialities of dry ground have spilled over into our area, but are rare: *Gentiana verna*, *Carex capillaris*, *Viola rupestris* and *Polystichum lonchitis*. *Potentilla fruticosa* has disappeared from the Maize Beck and presumed native *Polemonium caeruleum* from a pothole near Moor House - both evidently victims of grazing. Other more widespread upland calcicoles are *Draba incana*, *Potentilla crantzii*, *Galium boreale*, *Minuartia verna*, *Asplenium viride* (Plate 27), *Saxifraga hypnoides* and *Persicaria vivipara*. Close-grazed turf on high-lying limestone bands, notably of Little Fell, is the habitat of *Myosotis alpestris*, along with several of these species. *Sedum rosea* and *Thalictrum minus* are on several of the higher rocks but *Oxyria digyna*, frequent in the Lake District, is absent. Extreme rarities of these crags include *Saxifraga nivalis*, *Poa alpina* and *Phleum alpinum*. Limestone sinkholes on the summits provide isolated high-level stations for *Circaea alpina*, *Chrysosplenium alternifolium* and *Adoxa moschatellina*. Some lower-level plants of fertile soils, for example *Primula veris* and *Anthriscus sylvestris*, reach their greatest Cumbrian and even British elevation on the high exposures of Pennine limestone and the meadows and verges of the Alston area. Spoil from old lead mines is scattered widely over the moors, especially in the South Tyne catchment, and has a distinctive flora with *Minuartia*

Plate 27. *Asplenium viride* and *Minuartia verna*

Plate 28. *Armeria maritima* and *Minuartia verna* on lead mine waste, Alston Moor

Plate 29. Buttercup meadows, Mallerstang

verna, *Thlaspi caerulescens*, *Cochlearia pyrenaica*, *Viola lutea* and *Armeria maritima* (Plate 28).

The enclosed hay-meadows and pastures of the Pennine fell-foot country were once notable for their herbaceous communities with an abundance of northern species such as *Trollius europaeus*, *Geranium sylvaticum* and *Cirsium heterophyllum*. As in the Lake District, these have declined greatly through conversion to higher production leys, and their characteristic plants are now most often seen alongside roads, railways and rivers, or other ungrazed places. Even the colourful though botanically less interesting buttercup meadows (Plate 29) have declined during the last quarter century. Few if any species have disappeared - or not yet - but many have become much reduced during the last few decades. The once widespead

Primula farinosa and various orchids of these calcareous pastures in the marginal lands have also lost ground considerably. The meadows at the foot of the Cross Fell scarp were once a notable habitat of *Vicia orobus*, but this only just manages to survive in the area now. On the other hand *Crepis mollis* has recently been found in a number of new sites in the Alston area.

South of Stainmore there are good limestone pavements at Fell End Clouds, north-west of Wild Boar Fell, the highest locality in Britain for *Cardamine impatiens*. There are also crags, screes, fragmentary pavement and potholes to the east on Tailbridge Hill (Plate 30) and in nearby Dukerdale, but much of the ground is acidic. Notable plants include *Potentilla crantzii*, *Asplenium viride*, *Draba incana*, *Hippocrepis comosa*, *Saxifraga aizoides* and *Primula farinosa*.

Plate 30. Limestone sinkhole with *Geranium sylvaticum*, Tailbridge Hill

8. The Bewcastle - Gilsland moors

The northernmost corner of Cumbria is occupied by the south-western extremity of the Cheviot Hills. Bounded by Northumberland to the east and the Scottish border to the north-west, this formerly wild and little known area of moorland is now extensively afforested with conifers. The predominant vegetation of the unplanted ground ranges from typical Pennine blanket bog over deep peat to moist heaths and grasslands on peaty podsols and gley soils. On the flattest and wettest areas are expanses of spongy *Sphagnum* - dominated bog, the most important being Butterburn Flow in the upper Irthing valley. In places these flows grade into flush bogs where distinct lateral water movement is marked by the presence of sedges such as *Carex rostrata*, *C. echinata* and *C. curta*. Rare plants of the spongy bogs include *Drosera anglica*, *Carex lasiocarpa*, *C. magellanica*, *C. limosa*, and *C. pauciflora*. *Andromeda polifolia*, *Vaccinium oxycoccos* and *Rhynchospora alba* are widespread on the wetter moorland at lower levels, while *Rubus chamaemorus* occurs in profusion in the higher blanket bogs and is almost the only 'alpine' in this area. There are a few records of *Listera cordata* but it is *Trientalis europaea* which has conspicuously lost ground to the conifers. *Hymenophyllum wilsonii* and *Huperzia selago* grow on gritstone outcrops, *Lycopodium clavatum* is scattered,

mainly in old quarries and forestry rides, and there is a single, very isolated site for *Diphasiastrum alpinum*. *Antennaria dioica* has a few localities but is scarce. Among the species which have come in with the gravel forestry roads are *Draba muralis*, *Spergularia rubra* and *Bromopsis inermis*.

Where lime-rich drainage water emerges on the moorlands, there are flushes and marshes with a greater variety of species. They are the habitat of *Eriophorum latifolium*, *Eleocharis quinqueflora*, *Carex hostiana*, *C. dioica*, *Isolepis setacea* and *Parnassia palustris*. An isolated colony of *Saxifraga hypnoides* in a marsh on Christianbury Crags was eliminated by the conifers. The influence of calcareous rock is shown especially in the Irthing gorge above Gilsland (Plate 31), where the crags have provided a refuge for an interesting assemblage of plants. An isolated colony of *Saxifraga aizoides* grows here, far from the nearest occurrence, and there is an abundance of *Asplenium viride*, *Cystopteris fragilis* and *Polystichum aculeatum*. *Vicia sylvatica*, *Galium boreale* and *Melica nutans* are less frequent, while *Equisetum variegatum* appears to have died out. Native yew grows on the cliffs, and there is fringing woodland of the mixed broadleaved type. *Vaccinium uliginosum* was formerly present, and is still on the Northumberland side.

The basic soils of the farmland zone and roadside verges in this district were once similar to the Pennines in the occurrence of herb-rich meadow vegetation. While they are much reduced, an outstanding example remains, far up the Irthing valley at the farm of Butterburn. Here, on high banks of calcareous glacial drift above the river, little-modified communities have survived within a system of old hay meadows from which an annual cut is taken. The site, Gowk Bank, is a veritable oasis amidst barren sheep-grazed moorland. Scattered alders and willows, mainly *Salix cinerea* but also *S. aurita*, *S. pentandra* and *S. myrsinifolia*, suggest that woodland was once present on the bank, but the field communities appear to be natural remnants. *Geranium sylvaticum*, *Cirsium heterophyllum*, *Crepis paludosa*, *Geum rivale* and *Galium boreale* are abundant and there are good patches of *Trollius europaeus*. At least eight species of orchid are present, including *Dactylorhiza purpurella*, *D. incarnata* and *Coeloglossum viride*, and other herbs include *Sanguisorba officinalis*, *Trifolium medium*, and *Parnassia palustris*. The mown areas have a different type of grassland in which *Rhinanthus minor*, *Gentianella campestris* and *Euphrasia rostkoviana* subsp. *montana* are abundant.

The flora of the district is limited by the absence of some upland habitats. There is no ground high enough to have been above the former forest limit, and the scarcity of open rock habitats has therefore given montane plants little chance of survival within this woodland cover. The virtual absence of lakes and even tarns has severely restricted the occurrence of aquatic plants associated with standing water. The sluggish upper reach of the Irthing below Butterburn Flow is the nearest to lake conditions, and here are *Carex aquatilis*, *C. acuta*, *Potamogeton* x *nitens* and *Eleocharis aust-* *riaca*. Good examples of alderwood occur at Mollen Wood south of Bewcastle, but they are open to sheep grazing and so have a limited ground flora.

Plate 31. River Irthing in spate above Gilsland

And I remember with special joy the flowers of railway sidings and cuttings: colt's-foot along the banks in March and primroses in April......Soon after the primroses comes a time of sudden and transient whiteness along the railway banks, when the despised ribwort plantain hangs rings of white anthers around its dark-brown knob. If the stalks are twisted round the heads in a loop and tugged sharply, the heads will fly off as from a catapult. Then in summer there are tall, handsome flowers like the dog-daisy and yellow toadflax, and in the goods-yards, among the buffers and sleepers, tangles of wild carrot, crepises, cinquefoil, ivy-leaved toadflax on the walls, hogweed and many kinds of vetch. The greater bindweed climbs up stalks of sorrel like a spiral staircase, and curls around blocks of wood or pieces of iron left lying about, lolling its white bells on the rust. In the cinders of the track the least toadflax grows all along the west Cumberland coast. Perhaps the seeds are wafted by the draught from passing trains.

Cumberland and Westmorland, Norman Nicholson

Vegetational history
W. Pennington

Cumbria is remarkably well-endowed with geologically recent deposits which contain a fossil record of its flora and vegetation for the last 15000 years. This is the period which has elapsed since the last covering of ice began to disappear at the end of the most recent major glaciation of Britain, the Devensian. Fossiliferous deposits have accumulated in lakes of all sizes, from Windermere down to small kettlehole ponds and shallow depressions. Such deposits contain pollen and spores and also macrofossils, such as fruits and seeds. These remains can often be identified to species level and so provide a continuous record of the changing vegetation and flora at successive times in the past. The sequential changes in the composition of the pollen at any one site can be portrayed in a 'pollen diagram' and the various individual sampling horizons can be independently dated by radiocarbon. About 100 pollen diagrams are now available for sites in Cumbria.

We have very scanty information, from very few sites, about the vegetation history of Cumbria before the county was almost entirely covered by ice during the later stages of the Devensian glaciation. The icecap, generated on the central mountains, moved outwards and over-rode the Cumbrian lowlands, eventually joining the great mass of ice which occupied the Irish Sea basin. Its effects were mainly erosional in the mountains and depositional on the surrounding lowlands, where the land surface is largely formed of glacial deposits of gravels, sands and till. These are in general quite barren of fossils. Very little is known about the vegetation of any part of Britain during this period. Of the period before the Devensian, the Ipswichian interglacial, there are tantalising glimpses in reports of 'black muck' containing plant remains from beneath Devensian glacial deposits found in the shafts of the haematite mines in Low Furness, now long abandoned (Bolton 1862, Hodgson 1862, Kendall 1881). No samples were kept, so we shall never know whether Bolton's record of "leaves and fruit receptacle of beech" can be confirmed. The possibility remains that Cumbria was then within the native range of beech.

The wealth of data on Cumbria's flora and vegetation during the final stages of glaciation and the present temperate stage, i.e. the last *c.*15000 years, comes from detailed studies of infilled lakes and bogs (mosses), on the lowlands of the north (Walker 1966) and those around Morecambe Bay (Oldfield 1960, 1965, Smith 1958, 1959) and from investigation of lake sediments by the Freshwater Biological Association at Windermere (e.g. Pennington 1947, Pearsall & Pennington 1973). These studies were stimulated by the earlier work of Munn Rankin on the great raised bogs of the Morecambe Bay estuaries, especially Foulshaw with its corduroy road stratified into the peat, and by the first accounts of the Neolithic settlement site at Ehenside Tarn with its associated plant remains (Darbyshire 1874).

The Lateglacial Period
It is not known whether any plants were able to survive in Cumbria through the period of maximum glaciation. Very few possible habitats would escape being covered by ice. The highest summits, protruding through the ice sheet, were subject to intense erosion, with frost shattering their crags and a climate so severe that the existence of vegetation seems unlikely. It remains possible, however, that some upland slopes which now carry weathered soils did offer a refuge protected from the most severe erosive effects of ice. But no traces have been found of any plants associated with that long period, which began about 26000 years ago (BP). During the initial stages of retreat of the ice from Cumbria the landscape would appear to have been either unvegetated 'polar desert' or inhabited only by isolated plants which left no trace. During this period glacial landforms developed in the lowlands, including temporary glacier lakes of impounded meltwater in which stratified deposits were laid down, also kames, eskers and deltaic deposits.

At some time during the later stages of retreat, a great thickness of laminated clays and silts which contain no fossils was laid down in the major Cumbrian lakes including Windermere. These laminations result from the seasonal melt of glaciers and the sorting in the lakes of the glacial sediments. The laminations were termed varves and each pair, the varve couplet, represents the deposit of one year's winter plus summer layers. The barren varved clays of Windermere and the other major lakes represent a period after the ice had melted from the lowlands and the lake basins had filled with meltwater, when glaciers were still present on the mountains and pouring meltwater into the lakes. During this time it can be supposed that plants and animals were migrating northwards from the southerly parts of lowland England which had not been covered by ice.

The oldest sediments from Cumbrian lakes with plant remains are more than 14500 years old. The remains include oospores of the Characeae and leafless stems of the moss *Fontinalis*, both aquatic plants, and also the first pollen grains and unidentifiable cryptogam spores representing the pioneer plant communities of the land. The amounts and proportions of the pollen in these samples correspond with the values found today in surface sediments of some lakes of the middle alpine region in Norway, where the surrounding vegetation includes open communities of grasses and sedges, and snowbed communities dominated by *Salix herbacea*, together with a rich variety of herbs (Pennington 1970, 1977). The Cumbrian deposits include a saxifrage resembling

Saxifraga oppositifolia, Betula nana, Oxyria digyna, a sorrel (*cf. Rumex acetosa*) and several members of the Asteraceae, together with remains of the mosses *Polytrichum alpinum* and *Racomitrium* species. This assemblage, derived from the first plants to colonise the newly deglaciated land surfaces and the skeletal soils provided by the glacial deposits, represents primarily a pioneer vegetation, but one made up of cold-tolerant species including many which are now arctic-alpines. Species which are now rare mountain plants in Cumbria were then widespread in the lowlands, their distributions not restricted by either warm summers or competition from shrubs or taller herbs.

Soon after 14000 BP, this pioneer and species-rich herbaceous vegetation of the Cumbrian lowlands became much more luxuriant, increasing its annual pollen deposition by a factor of ten. This probably indicates a succession to closed communities of species-rich grasslands, into which were dispersing some taller shrubs including willows (there is a tentative record of *Salix phylicifolia* from Windermere) and a little juniper. The inference that temperatures must have been rising is supported by records of *Typha latifolia* (minimum summer mean temperature 13° C) in the upper part of this zone, together with *Helianthemum* sp. This richer herbaceous vegetation, characterised by an increased representation of species of *Galium* and *Thalictrum*, appears to represent full exploitation by the flora of conditions which made the lateglacial environment uniquely favourable to species tolerant of its temperatures, the as yet unleached base-rich soils, and the absence of competition for light by tall shrubs or trees.

From *c.*13000 BP there was, over most of England, Wales and Southern Scotland, a change to a milder climate which led to the deposition of more organic sediments and the spread of woody vegetation including tall shrubs, mainly juniper, and in many places sufficiently numerous trees of *Betula pubescens* to form open woodlands. The period from this date until 11000 BP, characterised in pollen diagrams by a 'woodland biozone' of juniper and birch, constitutes the Lateglacial Interstadial or 'Windermere Interstadial' of Coope & Pennington (1977). The percentage values for grasses, sedges, herbs and *Salix* in pollen diagrams decrease but are high enough to suggest that these plants of open vegetation maintained themselves through the interstadial, indicating that the woods were still very open. The widely synchronous rise of juniper at 13000 BP is interpreted as the response of this shrub to a general rise in temperatures at this date. This is shown very clearly by the faunal evidence which suggests that temperatures were at their highest during the juniper phase, and decreased after *c.*12000 BP (Atkinson, Coope & Briffa 1987). The very high rates of annual pollen deposition by juniper cannot be matched in any contemporary type of vegetation. The upper limit of luxuriantly flowering juniper appears to have lain at *c.*400 m in south Cumbria and for the

birchwoods *c.*100 m lower (Pennington 1970).

During the later part of this interstadial, changes in sediment composition in the upland lakes, associated with pollen changes, indicate a renewal of soil erosion and a change to more open vegetation at sites such as Blea Tarn above Langdale. In places there was by this time a development of acid soils with much *Empetrum*. Plants of open habitats such as species of *Rumex* and *Plantago* began to increase everywhere, and round Windermere the increases in *Huperzia selago*, *Cryptogramma* and *Selaginella* show the increasingly open nature of the lowland vegetation. This must have been a time of great floristic diversity. A notable record is that of *Koenigia islandica* from Windermere.

The interstadial ended with a pronounced fall in temperature at *c.*11000 BP which affected the whole of Europe, though vegetation changes were greatest in the north and west. During the following stadial an extensive sheet of ice built up on the western Scottish Highlands. In Cumbria, the boundaries of the glaciers which formed have been mapped in detail by Sissons (1980) from the distribution of fresh moraines and other features associated with recent glaciation. The associated sediments are poor in organic material but radiocarbon dates from Scaleby Moss show that the breakdown of established plant communities was apparent there before 10800 BP. Most of the corrie basins in Cumbria supported active small glaciers and from the larger corries glacier tongues descended to the floor of the main valleys. Beyond the ice cover there was severe periglacial erosion which must have destroyed all soil profiles and given rise to continuous disturbance of surface soils. The lakes which received glacial water contain varved clays which, in Windermere, consist of 400-500 couplets; indicating that the glaciers were active for about this number of years.

Those basins not receiving glacial water contain a clay with very little pollen. The plants which feature most in the pollen record are primarily those which tolerate continuous soil disturbance and are today well represented among ruderals and weeds of cultivation. *Artemisia* is the characteristic genus, with genera of the Caryophyllaceae, Brassicaceae and Asteraceae its regular accompanists. These assemblages occur at all altitudes from near sea level to Low Tarn, Wasdale at 480 m. At some sites there is an increase in *Salix herbacea* pollen suggesting a mosaic of snow-patch habitats and bare patches of eroding soil. There is no positive evidence for the presence of tree birches in Cumbria during this period and it seems very probable that trees disappeared altogether until around 10000 BP.

The Postglacial (Holocene) period

The opening of the Holocene is a geological horizon fixed at 10000 BP. Between *c.*10250 and 9750 BP at Scaleby Moss, juniper had once more become a dominant feature indicating rapidly rising temperatures. This juniper phase ended with the development of a closed canopy of birch woodland, which rapidly extended upwards to over 500 m, although at these levels it was probably more open, as at Red Tarn above Wrynose (Pennington 1964). Here, members of the lateglacial flora such as *Helianthemum* were able to persist for another thousand years, but subsequently pollen of herbaceous plants other than grasses disappeared, even from the sediments of Red Tarn, Helvellyn at 760 m.

The absence of fossil records of herbs at this altitude probably reflects unfavourable conditions for the preservation of their sparse pollen, rather than their complete disappearance from the summit and cliff vegetation. Species lists from the present vegetation of cliff ledges at high altitudes on Helvellyn (Ratcliffe 1960) contain many plants characteristic of the lateglacial vegetation, and it seems very likely that many of these, such as *Oxyria digyna, Rumex acetosa, Saxifraga oppositifolia, Filipendula ulmaria, Thalictrum minus* and *Salix herbacea*, have survived in that habitat from the lateglacial flora of more than 11000 years ago, persisting because the nature of the terrain kept the habitats open.

Both the record of fossil beetles and climatic models based on astronomical and other physical data suggest that temperatures rose rapidly in the early postglacial to levels approaching those of the present, so the subsequent spread into Britain of trees of the postglacial forest was not conditioned primarily by temperature. The birches spread early, from many foci, suggesting that they may have survived in particularly sheltered places, perhaps in the extreme south of Cumbria (Oldfield 1960, Smith 1958). Hazel, now thought to have entered Britain and spread from the north-west (Birks 1989), expanded early in Cumbria from *c*.9700 BP (Pennington 1991: fig.6) to equal and then surpass the birches. The oaks entered Britain from the south and reached Cumbria later, together with the elms (*ibid.*). The history of pine is difficult to decipher from its pollen record as there is much spatial variation in the amount of pollen and the timing of peak values. It does, however, seem clear that pine reached Cumbria from the south, rather late and after the expansion of hazel, elm and oak (Oldfield 1965, Pennington 1991). The peak values of pine were reached from *c*.8000 BP (*ibid.*). The suggestion is that a dry period from *c*.8000 - 7500 BP led to a rapid increase in pines as they spread over the dry peat surfaces of valley bogs (Pennington 1970). Although the pine maximum is present in all profiles from small lakes and bogs in mid- and south Cumbria, there is no record of any expansion at this time in north Cumbria, for example at Scaleby Moss (Walker 1966), showing that the county lay across a northern limit of spread of native pine. (The pine forests of northern Scotland had a different

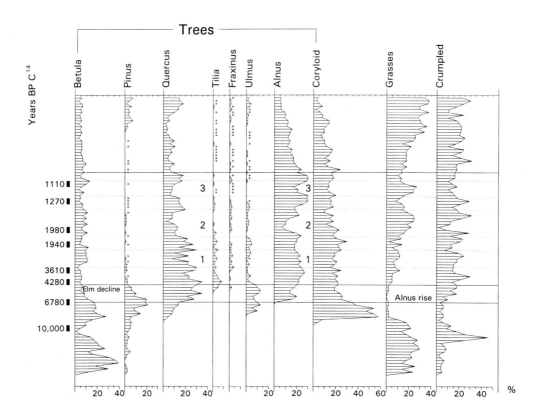

Fig. 8. Pollen diagram from Coniston Water. 1, 2 and 3 denote successive clearances of *Quercus* and *Alnus*, each followed by regeneration and a transient peak of *Betula*. There is total clearance from *c*.AD 1000, then partial return of oak and pine from the 18th century.

origin.) The early postglacial spread of these tree species is shown in the dated pollen diagram from Coniston Water (Fig. 8).

The rise of the alder pollen curve occurred in Cumbria from *c.*7500 BP, the time of the so-called 'Boreal-Atlantic transition'. The sediment stratigraphy in Cumbrian lakes supports the idea of Walker (1966) that this was a time of rising water tables, with consequent replacement of birch by alder in hollows and areas of impeded drainage. From *c.*7500 to 5000 BP the pollen deposited at most Cumbrian sites was dominated by oak, elm, hazel, alder and birch with very little non-tree pollen. These were the last truly native forests in the county.

The first appearance in pollen diagrams of ash and small-leaved lime, *c.*6000 BP, signalled the arrival in Cumbria of the last native tree genera. Ash of course continued its spread far beyond Cumbria, into Scotland, but the northern limit of lime lay across mid-Cumbria. This tree was frequent only in the south and south-east around Coniston, Windermere, the country south of Kendal, and on the Carboniferous limestone north and north-east of Morecambe Bay (Pigott & Huntley 1978, 1980, Pennington 1979, Birks 1982). Although often under-represented in pollen samples, ash became so generally recorded that it must have found some habitat in the native woodland, possibly in natural clearings resulting from the death of large trees. However, its main expansion came after the forest canopy was opened by Neolithic and later clearances.

By *c.*6000 BP, most of the native trees and shrubs were present, and it can be supposed that woodland communities similar to those of today had formed, and that others were in process of forming, as plant successions proceeded in response to soil development and minor changes within the prevailingly temperate climate. Comparison of about 60 pollen diagrams from the county suggests that there was spatial variation in forest composition along altitudinal and edaphic gradients. Oaks predominated over much of the lowlands and into the uplands, as shown by the presence of leaves in the sediments of Burnmoor Tarn (254 m, Fig. 9). Elm appears to have been commonest at intermediate altitudes, as at Blea Tarn (Langdale), Seathwaite and Angle Tarns (187 m, 537 m) (Pennington 1964, 1970, 1975). Birch remained commoner than in lowland England and, together with the remaining pines, appears to have formed a distinct uppermost forest type in places (Pennington 1975). Amounts of alder vary with the site; lake sediments may show local over-representation of this tree, but the large altitudinal range indicates a wide distribution. Hazel remained the most abundant pollen at many sites; apparently being associated successively with oak, elm and birch with increasing altitudes.

Birks (1982) provided a detailed analysis of the composition of lowland forest on Carboniferous limestone, from the sediments (dated *c.*6680-5150 BP) of a small hollow in Roudsea Wood, south Cumbria.

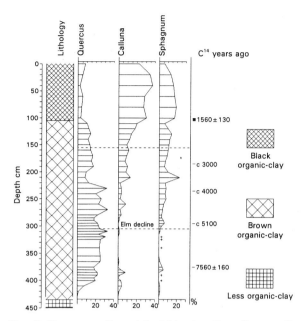

Fig. 9. Burnmoor Tarn (alt. 254 m) pollen diagram for *Quercus*, *Calluna* and *Sphagnum*.

The pollen here represents largely the vegetation of a sample patch of 20 to 30 m. radius and "suggests that a species-rich forest existed in mid-Flandrian times with approximately 10 to 30% *Quercus* in the canopy around the hollow, 5 to 15% *Betula*, 2 to 15% *Ulmus cf. U. glabra*, 5 to 25% *Tilia cordata*, 5 to 25% *Fraxinus excelsior*, 15 to 25% *Alnus glutinosa*, and 10 to 30% *Corylus avellana*". Other trees and shrubs included *Populus tremula*, *Taxus baccata*, *Sorbus aucuparia*, *Prunus padus* and *Ilex aquifolium*, together with five species which grow today at or near their northern British limits in south Cumbria - *Acer campestre*, *Euonymus europaeus*, *Rhamnus cathartica*, *Sorbus torminalis* and *Cornus sanguinea*.

It is now generally accepted that during these millennia men of Mesolithic hunter-gatherer cultures carried out extensive burning of vegetation, particularly in the uplands. The effects of this on British vegetation have been identified in pollen diagrams, and are often accompanied by charcoal, stratified into the peat or sediment. The Cumbrian coastal plain is rich in Mesolithic sites, with worked flints and dateable charcoal (Bonsall 1980) but there is as yet little evidence of a large enough Mesolithic presence in the Cumbrian hills to influence the vegetation. Occupation of sites on the south-west coast was probably associated with the small changes before 5000 BP in pollen diagrams from sites such as Barfield Tarn and the nearby 'Williamsons Moss' (Pennington 1970, 1975). These have been attributed to local and temporary small clearances of woodland or coastal scrub, which allowed the spread of *Plantago lanceolata* and other light-demanding herbs. There is as yet no evidence from Cumbria for any substantial disturbance of the forests by Mesolithic hunters, though in northeast England the creation and maintenance of forest clearings by men before 5000 BP is now accepted and recognised as a source of increasing floristic diversity. One upland site, Blea Tarn, Langdale, yielded evidence for a very minor opening of the forest before 5000 BP, in the area where there was subsequent activity associated with the axe factories.

The last 5000 years in Cumbria: vegetation and man
There can be no doubt that at *c.*5000 BP there was, in much of Britain, an environmental change which affected soils, erosion rates and vegetation. It is not yet clear whether this resulted from an early major impact of Neolithic farming or from a combination of this and some climatic change, which accelerated the deterioration and erosion of soils, bringing about or accelerating the formation of upland blanket peats. In all sediment profiles from Cumbrian lakes this change coincides with a marked fall in the percentage and absolute values for elm pollen, consistent with the widespread 'Elm Decline' of the British Isles and northwest Europe, just before 5000 BP. The between-site diversity in the changes in other pollen types at the time of the Elm Decline in Cumbria suggests the influence of some factor other than natural environmental change, and the pattern of variation is consistent with archaeological evidence of early Neolithic presence (Pennington 1965). At Barfield Tarn the Elm Decline coincides with signs of woodland clearance, as distinct from exploitation of the elms alone, and cultivation of cereals (Pennington 1970). Neolithic farmers were therefore active in Cumbria at the time of the Elm Decline. But the Elm Decline appears in all pollen diagrams, including many which show no other changes in vegetation such as might be expected from local Neolithic activity. The explanation now favoured is that of an almost synchronous, continent-wide attack on the elms by a pathogenic fungus.

On the coastal plain, Neolithic deforestation around Barfield Tarn, Ehenside Tarn (Walker 1966) and 'Williamsons Moss' followed the previous transient clearances and proved much more lasting in its effect, so that the amount and composition of the primary forest was never fully restored. The clear picture from Barfield (Pennington 1970: fig. 11) is of a parallel clearance of oak and elm with little change in birch, alder or hazel, which were presumably growing on the less-favoured and wetter soils. A sustained increase in the pollen of grasses and a wide variety of light-demanding herbs followed, together with cereal pollen indicative of cultivation. Cultivation was sufficiently intense to bring about a complete change in the sediment deposited in this small lake, which is surrounded by slopes of readily eroded drift. A great floristic increase is recorded for these coastal sites at this time, as new open habitats were created for a wide range of species which had been much restricted by the previous unbroken forest canopy.

Many small lakes, which recruit pollen mainly from the local vegetation within a few hundred metres of the site, show a characteristic pattern of vegetational change in which a 'primary' elm decline, affecting only the elm curve, is shortly followed by an episode of temporary reduction in all trees, with expansion of grasses, *Plantago lanceolata* and other herbs, and then of bracken. This 'Landnam' episode is explained as indicating nomadic farming in which temporary clearings of limited size, affecting all trees in the patch, were used for grazing livestock and often for cereal cultivation, and then abandoned. Such clearances of primary forest were widespread but not universal in Cumbria at and just after 5000 BP. At many lowland sites human pressure appears to have been relaxed after such episodes and the forests rapidly regenerated, but were more open than formerly (*cf.* the curve for grass pollen in Fig. 8) and showed a permanent reduction in the amount of elm. This could explain the associated increase in ash and, in the south of the county, lime, both of which had been present since *c.*6000 BP (Fig. 8).

In the uplands, vegetation changes at or just after 5000 BP are well recorded near sites of Early Neolithic activity, most notably in the hills round the Great Langdale axe-factory sites (Pennington 1970, 1975). Before 5500 BP the forest canopy existed up to at least 570 m (Angle Tarn), with an upper zone in which pine was still present. The Elm Decline took place between 5200 and 5100 BP at Blea Tarn and Angle Tarn and was shortly followed by a Landnam clearance. There is no sign of cereal cultivation at these altitudes, so the expansion of grasses and herbs indicates openings in the forest canopy and probably some grazing. Charcoal from the upland woods is associated with chipping floors (sites of axe manufacture) and was found by Walker (1965) at this horizon in the sediments of Langdale Combe. The charcoal from chipping floors, dated *c.*4500 BP, came from the wood of oak, hazel, alder, birch and probably rowan or hawthorn. A period of 400-500 years of human activity in the Langdale fells was associated with the transformation of upland woods, by clearance and burning, into heather moorland and grasslands. In other parts of the Cumbrian uplands the woodlands survived for 1000 - 2500 years, into Bronze Age and Iron Age times.

In all parts of the uplands the tendency towards soil acidification and paludification (Pearsall 1950) was already bringing about soil and vegetation changes independently of any action by man. Fig. 10 shows a sequence from a wet site, a level saddle above 500 m, in an area 2-3 km from the axe-factory sites, where at 5000 BP the only sign of man-made vegetational

Plate 32. The Langdale Pikes from Lingmoor

Seven
thousand years ago
trees grew
high as this tarn. The pikes
were stacks and skerries
spiking the green,
the tidal surge
of oak, birch, elm,
ebbing to ochre
and the wrackwood of backend.

Then
round the year Three
Thousand BC,
the proportion of elm pollen
preserved in the peat
declined from twenty
per cent to four.

Stone axes,
chipped clean from the crag-face,
ripped the hide of the fells.
Spade and plough
scriated the bared flesh
skewered down to the bone.
The rake flaked into fragments
and kettlehole tarns
were shovelled chock-full
of a rubble of rotten rocks.

from *The Elm Decline*, Norman Nicholson

change is the appearance of *Plantago lanceolata*. Here, by 5000 BP, mor humus had already begun to accumulate above the mineral soil; at this time the pollen deposited was mainly arboreal, but with *c*.30% heather and grasses. The disappearance of pine may have been caused by burning or paludification; that of the local elms is likely to have been due to disease. The decline of oak, alder and hazel through the next millennium, with a corresponding increase in heather, reflects upland clearance around the site. Birches remained sufficiently numerous to suggest that open birch woodland persisted on such damp organic soils; a layer of birch wood in the increasingly peaty deposit was dated at *c*.3890 BP. However, above this level there is no further trace of local trees and the peat seems to have accumulated more rapidly. This episode is similar to others in north-west Britain where peat formation began or accelerated *c*.4000 BP. It represents the spread of blanket bog (*Calluna, Sphagnum* etc.) over ground previously occupied by forest (Fig. 10).

The impact of Neolithic farming, together with the inferred change of climate towards increased oceanicity at the higher altitudes at *c*.4000 BP, resulted in increased spatial differentiation of vegetation, the pattern becoming increasingly related to that of human settlement and land use. The reduction in the upland woodland had proceeded to something approaching its present deforested state before there was any extensive clearance of the valley woods. As the areas of upland pastoral farming, rough grazings and dwarf-shrub heaths expanded at the expense of woodland, so many non-tree species must have been able to spread and establish new communities.

The vegetational history at intermediate altitudes can be seen in the pollen diagram from Burnmoor Tarn (254 m, Fig. 9). At the time of the Elm Decline, the small increases in grasses, heather and *Sphagnum* can be attributed to the general environmental change and there is no evidence of any man-made change until *c*.4000 BP. Then, the first of two major episodes in the destruction of oakwoods at this site is recorded. After *c*.1000 years of local Bronze Age occupation of the surrounding uplands, during which some oak persisted and there was a steady increase in heather, the remaining oaks disappeared during a Romano-British clearance phase which involved cultivation of cereals. The rapid input to the lake of acid, organic material, rich in heather pollen suggests that it was cultivation practices which led to rapidly increased soil erosion. At the same time, a spread of *Molinia - Myrica* communities over wetter parts of the catchment is suggested by the increases of grasses and 'Coryloid' pollen. The sequence of events may be related to the same climatic change which led to the renewed growth of the Morecambe Bay raised bogs, after a dry period with local cultivation ending *c*.400 AD (Smith 1959).

Raised bogs, carrying examples of several mire communities, are frequent in the north Cumbrian lowlands, and also at low altitudes round the Morecambe Bay estuaries where they developed from reedswamp over the marine clay deposited at the time of the maximum post-glacial sea-level; *c*.7500 years ago. Many investigations have shown changes with time in peat-forming vegetation and the degree of humification of the peat and these have been interpreted in terms of alternating periods of high and low precipitation combined with changes in bog hydrology. 'Recurrence surfaces', characteristed by sudden changes from dark, humified and slow-growing *Calluna* peat to light-coloured unhumified *Sphagnum* peat, have been studied and dated in many bogs, including several in the Morecambe Bay area (Smith 1959, Oldfield & Statham 1963) and Bolton Fell Moss near Carlisle (Barber 1981). It appears that some of these changes may indicate major changes in the regional climate from *c*.2500 BP. The general picture is towards an increasingly oceanic climate but interspersed with dry periods when the bog surfaces were not wet.

Pollen preserved in the raised bog peat also records changes in vegetation on the surrounding land, and has provided important data on the forest history of the early post-glacial period. Some of these bogs, such as Abbot, Moorthwaite and Scaleby Mosses in the north of Cumbria (Walker 1966) and Rusland Moss to the south of the Lake District (Dickinson 1973, 1975), have developed over the sediments of shallow lateglacial lakes. The general picture is of successive episodes of forest clearance and regeneration after 5000 BP and of the main regional clearance which was almost complete from *c*.1000 AD.

Some of the raised bogs retain an apparently undisturbed peat surface with the vegetation dominated by species of *Sphagnum*. Many however are now 'relict', with dry peat surfaces, some overgrown by birches and pines. The drainage and degeneration of these bogs has resulted from centuries of peat-cutting for fuel and attempts at drainage.

The long pollen diagram from Coniston Water (Fig. 8) illustrates the history of the surrounding valley woodland from Bronze Age time (*c*.3600 BP). The fluctuating but generally declining values for oak and alder indicate a series of clearance episodes followed by regeneration. The main decline in tree pollen began at *c*.1000 AD (with a peak of the holly curve which suggests that the formation of 'hollins', patches of holly maintained for winter feed, began at this time) and reached its minimum at about 1600 AD. Increases in pine from the eighteenth century, and in oak from the nineteenth, correspond with the documented planting of these trees in run-down woodland and wood pasture. One such oakwood is in fact called 'Grass Paddocks'. Birch and hazel decline from *c*.7000 BP from their early postglacial peaks, regenerating a little after each strong clearance of oak before declining with oak after 1000 BP. Like oak, alder also suffered the same three major cycles of clearance - Bronze Age, Romano-British and *c*.700-800 AD followed by regeneration before the final clearance beginning *c*.1000 AD. The parallel course of oak and alder clearances suggests that alder was possibly an important component of hillside woods, as in surviving woods in Martindale (Pearsall & Pennington 1973). The other trees, elm, ash and lime, declined in the early Romano-British centuries and again from *c*.700 AD. Ash did not here respond to clearance by expanding in secondary (opened-up) woodlands; today it is confined to the better soils of the existing woodlands. Elm is restricted even more severely to the flushed soils of the ravines (Pearsall 1950). Small-leaved lime appears to have reached its postglacial northern limits near the Coniston catchment; it began to retreat *c*.2000 years ago and is now present as scattered individuals which reproduce only vegetatively. These are mainly in sites within steep-sided gills which were never cleared (Pigott & Huntley 1978, 1980). This retreat is attributed to a decline in summer temperatures from levels at which fertile seed can be produced.

The vegetation which replaced the oak woods was at first, in some areas, a dwarf-shrub heath dominated by *Calluna*, but by mediaeval times the dominant vegetation was grassland (Figs. 8 & 10). This is consistent with a long-documented history of sheep farming in this region of mild winters where the hardy local breeds could sustain themselves on the hills. There is some genetic evidence which links the local Herdwick sheep with an Old Norse breed (Pearsall & Pennington 1973), suggesting that it may well have been introduced with the settlements and lowland clearance of *c*.1000 AD. This long history of heavy grazing by agile mountain sheep can explain much of the floristic poverty of the Lake District mountains. At the time of the settlement of the valleys and lower hills, the valley woods were cleared wherever farming was possible, first by Anglian invaders from Northumbria and then by those who spoke Old Norse. Records in pollen diagrams of this period show increasing amounts of cereals - barley, oats and rye, together with arable weeds and textile crops such as *Cannabis* (hemp, for ropes) and *Linum usitatissimum* (flax).

In this type of farming the woods which remained on land too difficult to clear, such as the interfluves between the steep-sided gills, were a valuable resource, supplying firewood and poles for

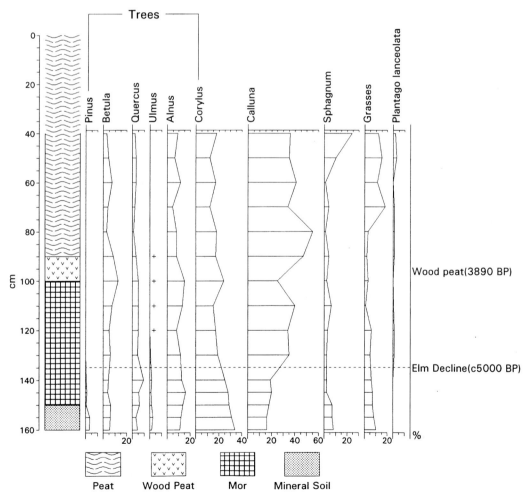

Fig. 10. Pollen diagram from Red Tarn Moss, Wrynose (alt. 503 m). The elm decline (E.D.) occurred *c*.5000 BP. *Sphagnum* peat developed *c*.4000 BP. *Betula*, which tolerated soil acidity, declines following paludification and peat formation.

fencing and a variety of other purposes. From Roman times local iron ores in south Cumbria have been smelted using locally produced charcoal, and it became profitable to manage woods, often on steep or rocky hillsides, as coppice. This was cut on a rotation of *c*.15 years, to supply both the charcoal-burners and a range of other woodland industries including bark for tanning. The pollen diagram in Fig. 11 summarises events in the Coniston area during the last 500 years. Oak was at its lowest from 1600 to 1800, coinciding with maximum pressure from subsistence farming and heavy demand for charcoal and wood. Its rise from *c*.1800 and the reappearance of pine from before that, indicates much planting of these trees. Hazel, abundant in the coppice woodland, declined from *c*.1900 as charcoal smelting ended and coppices were allowed to grow up, shading out the hazel. The grass curve illustrates the opening of the landscape which continued after the mediaeval period. This dating, and inferred history, are consistent with the position in Coniston Water sediments of a copper-rich layer, deposited by outwash from the 19th century mining in the catchment. Beech never reached Cumbria by natural spread during postglacial time. Together with the pine, which had probably died out as a native tree, it was widely planted in the county from the 18th century.

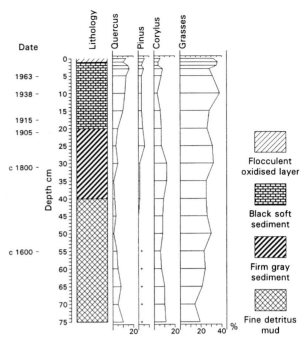

Fig. 11. The history of *Quercus*, *Pinus* and *Corylus* in the Coniston area as shown in the upper 70 cm of the core shown in Fig. 7.

The Irruption of Solway Moss

Solway Moss consists of sixteen hundred acres; lies some height above the cultivated tract, and seems to have been nothing but a collection of thin peaty mud: the surface itself was always so near the state of a quagmire, that in most places it was unsafe for any thing heavier than a sportsman to venture on, even in the driest summer.

The shell or crust that kept this liquid within bounds, nearest to the valley, was at first of sufficient strength to contain it: but by the imprudence of the peat-diggers, who were continually working on that side, at length became so weakened, as not longer to be able to resist the weight pressing on it.

Late in the night of 17th of November of the last year, a farmer, who lived nearest the moss, was alarmed with an unusual noise. The crust had at once given way, and the black deluge was rolling towards his house, when he was gone out with a lantern to see the cause of his fright: he saw the stream approach him; and first thought that it was his dunghill, that by some supernatural cause, had been set in motion; but soon discovering the danger, he gave notice to his neighbours with all expedition: but others received no other advice but what this Stygian tide gave them: some by its noise, many by its entrance into their houses, and I have been assured that some were surprised with it even in their beds: these passed a horrible night, remaining totally ignorant of their fate, and the cause of their calamity, till the morning, when their neighbours, with difficulty, got them out through the roof. About three hundred acres of moss were thus discharged, and above four hundred of land covered; the houses either overthrown or filled to their roofs; and the hedges overwhelmed; but providentially not a human life was lost; several cattle were suffocated; and those which were housed had very small change of escaping. The case of a cow is so singular as to deserve mention. She was the only one out of eight in the same cow-house, that was saved after having stood sixty hours up to the neck in mud and water: when she was relieved, she did not refuse to eat, but would not taste water: nor could even look without shewing manifest signs of horror.

The irruption burst from the place of its discharge, like a cataract of thick ink; and continued in a stream of the same appearance, intermixed with great fragments of peat, with their heathy surface; then flowed like a tide charged with pieces of wreck, filling the whole valley, running up every little opening, and on its retreat, leaving upon the shore tremendous heaps of turf, memorials of the height this dark torrent arrived at. The farther it flowed, the more room it had to expand, lessening in depth, till it mixed its stream with that of the Esk.

A Tour of Scotland and Voyage to the Hebrides in 1772, Thomas Pennant

Vegetation
D.A. Ratcliffe

This chapter provides a brief account of the present-day plant communities of Cumbria, in the sense of distinctive assemblages of plants defined by their dominant, constant and characteristic species. It is confined to semi-natural vegetation (that composed largely of native species though modified from its original state by human activity) and to the small number of types which can still be regarded as natural (virtually unmodified by human intervention). *British Plant Communities* (Rodwell 1991 *et seq.*) has provided a countrywide classification of contemporary vegetation types and, wherever possible, the codes used there are applied below to the equivalent Cumbrian community.

Woodlands and scrub

Even semi-natural woodland now exists in a highly fragmented state in Cumbria, and covers a far smaller total area than recent plantations of exotic conifers. The majority of extant woods have been re-planted on the sites of original woodland, and so have often retained much of the characteristic flora even if the structure bears obvious signs of planting and coppicing. In upland situations, grazing has modified the subsidiary layers and prevented regeneration of woody species. Other woods and probably most scrub areas have developed by recent tree and shrub invasion of open habitats (including felled woodland), and so are *seral,* lacking the stability of long-established woodland, though often possessing an interest and value of their own.

1. *Lowland oakwood* (W 10)

This is dominated by pedunculate or sessile oak, or mixtures of both, with some pendulous birch, and a variable tall shrub layer of hazel, rowan and holly. Field layer communities are variously dominated by brambles, honeysuckle, bracken, *Hyacinthoides non-scripta, Holcus mollis* and *Luzula sylvatica.* Other typical herbs are *Stellaria holostea, Anemone nemorosa, Oxalis acetosella, Silene dioica, Teucrium scorodonia, Viola riviniana* and *Luzula pilosa.* Parent rocks are usually base-poor and soils are rather acidic brown loams. Many examples adjoin enclosed farmland and are ungrazed as a result. Lowland oakwoods occur especially on New Red Sandstone in the north and the Silurian formation in the south.

2. *Upland oakwood* (W 11, W 16, W 17)

Sessile oak is usually dominant, with some downy birch. These woods are usually sheep-grazed, and typically lack a tall shrub understorey, though hazel, rowan and holly may be present. Under light grazing, bilberry may be dominant in the field layer, but more usually there is a grassland community of *Festuca ovina, Agrostis* spp., *Anthoxanthum odoratum* and *Deschampsia flexuosa,* or *D. cespitosa* and *Molinia*

caerulea on wetter ground. Grassland shows variable replacement by bracken, which forms completely dominant stands within many hill woods. The ground is often steep and rocky, with outcrops and block litter, and here other ferns may be abundant: *Dryopteris filix-mas* agg., *D. dilatata, Oreopteris limbosperma* and *Athyrium filix-femina.* These woods are widespread on lower hillsides in the Lake District and, under the high rainfall, the boulder-strewn examples have luxuriant bryophyte communities, with many rare species locally, and the moss-like *Hymenophyllum wilsonii.* The trees also have a notable abundance of epiphytic bryophytes and lichens, though atmospheric pollution has reduced their diversity. The sometimes skeletal soils are leached and acidic brown earths. Borrowdale has an especially fine range of examples.

3. *Mixed broadleaved woodland* (W 9, W 10)

This has a mixture of oak (either species), ash, wych elm and birch, with variable understorey of hazel and frequent bird cherry, hawthorn and rowan. Wych elm is now greatly reduced. Small-leaved lime probably belongs mainly to this woodland type in its scattered Cumbrian occurrences. Ungrazed lowland examples usually have field communities dominated by forbs such as *Merurialis perennis, Allium ursinum, Geranium robertianum, Primula vulgaris, Sanicula europaea, Geum rivale, G. urbanum, Circaea lutetiana, Ranunculus ficaria, Prunella vulgaris, Ajuga reptans, Fragaria vesca* and *Potentilla sterilis. Ranunculus repens* and *Chrysosplenium oppositifolium* are typical of wet, clayey seepages. Grazed upland woods have the same species, but greater amounts of grasses, such as *Brachypodium sylvaticum, Dactylis glomerata* and *Melica uniflora,* with *Deschampsia cespitosa* in damper places. Soils are fairly basic (pH 5.5 - 6.5) and parent rocks contain available lime. Patches of such woodland occur mixed with oakwood in many localities, both lowland and upland. They are frequent in the sandstone valleys of the north where more basic beds or calcareous drift have influenced soil development, on calcite-bearing beds of the Borrowdale Volcanic series in central Lakeland, and on the Silurian rocks in the south.

4. *Ashwood* (W 8) (Fig. 12)

Woodland dominated by ash is a more extreme form of the previous type. Hazel locally forms an understorey, and there are often tall shrubs listed under community type 11. The field communities have much in common with those of 3, but often contain a wider variety of species, for example *Myosotis sylvatica, Paris quadriolia, Convallaria majalis, Campanula latifolia, Rubus saxatilis* and *Aquilegia vulgaris.* Rocky examples are rich in lime-loving bryophytes. This type is almost confined to Carboniferous limestone in the Pennines

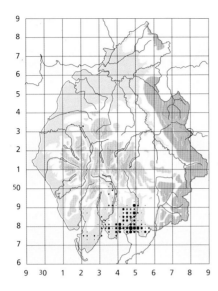

Fig. 12. Predictive distribution of lowland limestone ashwood (Type 4, W 8) based on the following indicator species: *Fraxinus excelsior*, *Acer campestre*, *Cornus sanguinea*, *Viburnum opulus*, *Rhamnus cathartica*, *Mercurialis perennis*, *Viola reichenbachiana*, *Clematis vitalba* and *Tamus communis*.

and the Kendal - Morecambe Bay area, and the soils are calcareous, clayey loams with pH 6.5 - 7.5. Good examples occur at Helbeck Woods, Smardale, Hutton Roof and Witherslack. In places on the Whitbarrow scarp, yew displaces ash as the dominant tree. The field communities of ashwood are well represented in the deep 'grikes' of the now mostly treeless, but sometimes scrubby, limestone pavements.

5. Birchwood (W 4, W 16, W 17)
Pure birchwood, usually of downy birch, is mostly a secondary, seral kind of woodland. In the lowlands, one type has developed by invasion of the drying peat of raised bogs: it is often dominated by luxuriant *Dryopteris dilatata*, over a ground cover of mosses, and with abundant *Molinia caerulea*. Clearance of both lowland and upland oakwoods, when not followed by re-planting, often results in development of birchwood (with either or both species), usually with field and ground communities similar to those previously present. There are numerous examples scattered over the county, and a long-established bryophyte-rich birchwood occurs at Watendlath.

6. Pinewood
Contemporary Scots pine in Cumbria is not regarded as native, and many woods are obvious plantations. Yet, where this tree has spread spontaneously onto drying peat mosses, it has produced stands with a more natural appearance. Field layers of bilberry are often present and *Dryopteris dilatata* is locally dominant. This is the habitat of *Goodyera repens*, and at Cumwhitton Moss other northern associates include cowberry, and the

mosses *Ptilium crista-castrensis* and *Dicranum polysetum*.

7. Alderwood (W 5-7)
Woods dominated by alder are extremely local, and found mostly on peaty soils associated with lowland fen, where they intergrade with the next type; or on periodically flooded alluvium beside streams. Bird cherry is often present. Ungrazed examples, with well-developed sedge and fen herb communities, occur around the edges of the larger lakes, for example The Ings, Derwent Water; Esthwaite North Fen and Elterwater. Other stands on spring seepages within oakwoods occur in the Lyne gorge below Kinkry Hill and Park Wood, Brocklebank, and have a local community with *Equisetum telmateia*, *Carex pendula* and *Eupatorium cannabinum*. Upland examples are usually sheep-grazed and open, with few distinctive associates. They are often stream-side fringes, as by Borrow Beck, but more extensive stands on wet gley soils occur at Mollen Wood near Bewcastle. The soils are usually somewhat base-enriched.

8. Willow carr (W 1, W 2, W 3)
Wet scrub dominated by *Salix cinerea* is frequent on or around lowland wet fens (e.g. Newton Reigny Moss, Braithwaite Moss, Silver Tarn), lake edges (Barfield Tarn, Derwent Water) and seasonally flooded woodland (Orton Moss, Finglandrigg Woods). Where the water is nutrient-poor, associates are those of 'poor fen' (29). Sites receiving more nutrient-rich water have other willows such as *S. pentandra*, *S. viminalis*, *S. fragilis*, *S. aurita*, *S. phylicifolia* and associates of intermediate (34) or rich (35) fen.

9. Juniper scrub (W 19)
In the Lake District, thickets of tall juniper occur on many dry lower fellsides, usually in association with acidic *Festuca - Agrostis* grassland and bracken, but sometimes on heather or bilberry ground. Scattered trees of yew may be present, but the field communities are usually much the same as those of type 2, with bilberry, *Oxalis acetosella*, *Deschampsia flexuosa* and mosses prominent. The soils are of the same types as under type 2. Juniper scrub is strikingly patchy in distribution, perhaps through burning and grazing effects, and in some areas this shrub survives mainly on cliffs. Good examples occur on Birk Fell above Ullswater, Wythburn and Little Langdale.

10. Hawthorn scrub (W21)
Crataegus monogyna occurs frequently in the tall shrub layer of many woods on the more basic soils, but some lower fellsides have open stands of hawthorn in similar situations to type 9, and also usually associated with dry acidic grassland and bracken on leached brown soils. Scattered crab apple and bird cherry trees are often present. A good example occurs in Glencoyne Park, Ullswater.

11. *Mixed limestone scrub* (W 21)
The rocky limestone hills of the Kendal - Morecambe Bay area show local occurrence of a scrub with calcicolous tall shrubs and small trees similar to that on the southern chalk. Ash, hazel, hawthorn, blackthorn and juniper are usually present, with variable frequency of spindle, purging buckthorn, guelder rose, dogwood, privet, small-leaved lime and *Sorbus lancastriensis*. The community is often developed over limestone pavement and contains a number of species associated with this habitat, such as *Sesleria caerulea*, *Rubus saxatilis* and *Melica nutans*. Older hedges in this part of Cumbria can also have rich mixtures of some of the above trees and shrubs.

12. *Gorse scrub* (W 23)
Common gorse often invades areas of lowland grassland, stretches of rough pasture around fell farms or waste ground and forms dense thickets with few other plants. It is well developed above high-tide level on the shore at Bowness-on-Solway, and inland on Caldbeck, Haltcliff and Brough Sowerby Commons.

Heaths and grasslands
Within the present climatic tree-limit (520-610 m), these communities have mostly been derived by the clearance of former forest and maintained by the grazing of domestic stock and repeated moor burning.

Dwarf shrub heaths
These are recognisably closest to the types present within the original woodlands, whereas the grasslands are mostly secondary derivatives of these, created by long continued management as grazing-range. Similarly, above the tree-line, the same factors have operated to convert the original montane dwarf shrub heaths to species-poor grassland.

13. *Lowland heather heath* (H 8)
Communities dominated by heather, with variable amounts of common and western gorse, and bracken are virtually identical with those typical of acidic heaths in southern England (H 8). Although southern species are mostly lacking, an example at Finglandrigg has an isolated colony of *Ulex minor*. *Genista anglica* occurs sparingly. The soils are mostly podsolised sands. This community usually covers only small areas, often around the drier edges of lowland and coastal peat mosses, and grades into wet heath and mire vegetation on moister, peaty ground, with the addition of *Erica tetralix*, *Drosera rotundifolia*, *Narthecium ossifragum*, *Juncus squarrosus*, *Trichophorum cespitosum* and *Sphagnum compactum* (M 16). Less constant but characteristic species here are *Anagallis tenella*, *Hypericum elodes* and *Scutellaria minor*. Good examples of dry to wet heath occur at Walby Moor, Finglandrigg Moss and Hallsenna Moor.

Heaths intermediate between lowland and upland types, with abundant bilberry (and, in one site,

cowberry), moss and lichen were formerly widespread on the low sandstone hills from King Harry's Common to Penrith Beacon, but have recently been extensively converted to farmland. A unique northern equivalent to southern 'chalk heath' on calcareous drift at Johnby Moor was largely destroyed by afforestation. It had an intimate mixture of calcifuge and calcicole species: *Calluna vulgaris*, *Genista anglica*, *Nardus stricta*, *Listera cordata*, *Pyrola minor*, *Primula farinosa*, *Valeriana dioica* and *Carex flacca*. Patches of a similar heath survive on Underbarrow and Whitbarrow Scars and on Farleton Fell where loess (wind-blown silt) was deposited over the limestone in the post-glacial period. A different example of limestone heath occurs on Sunbiggin Tarn Moor, where *Calluna* is mixed with *Anemone nemorosa*, *Sanguisorba minor*, *Galium verum*, *G. boreale*, *Lathyrus linifolius* and *Lotus corniculatus*.

14. *Upland heather heath* (H 10, H 12)
The most extensive type of dry heath on the fells is dominated by *Calluna vulgaris*, though *Erica cinerea* may take over during regeneration after fire. It occurs mostly on moderate to steep slopes up to 600 m on leached brown skeletal soils. Apart from a ground layer of mosses, there are typically few other plants apart from sparse growths of bilberry, *Deschampsia flexuosa*, *Agrostis capillaris* and *Festuca ovina* (H 12). The typical community is extensive where management has been for red grouse, as in Skiddaw Forest, Armboth Fells, Kirkby Moor and parts of the Pennines. On the Borrowdale Volcanic and Silurian rocks it is often restricted to cliff faces. On steep, shady slopes, *Sphagnum capillifolium* and *S. quinquefarium* are abundant in the ground layer, and *Listera cordata* and *Rubus chamaemorus* sometimes occur, a community rather similar to the damp *Calluna* heaths of the Scottish Highlands (H 21, H 22). A stand on the western slopes of Grasmoor has *Arctostaphylos uva-ursi* (Plate 33) and low juniper and resembles another type widespread in the Highlands (H 16).

Heather moorland has contracted in area during this century, and the widespread Highland montane variants, in which *Calluna* assumes a prostrate, mat-like form, and mosses or lichens become abundant (H 13 & 14), have been almost lost to Cumbria. Examples approaching this type and containing *Carex bigelowii* and *Empetrum nigrum* still occur on windswept spurs of Skiddaw at 600-730 m, and at similar elevations in the Pennines.

15. *Upland bilberry heath* (H 18)
The first stage in replacement of *Calluna* heath is dominance of bilberry, usually after particular conditions of burning followed by heavy grazing. Grasses become more abundant while cowberry and crowberry may be plentiful also. A high-level variant above the tree-line has more stunted growths of these dwarf shrubs, along with *Carex bigelowii*, *Diphas-*

Plate 33. *Arctostaphylos uva-ursi*, Grasmoor

iastrum alpinum, Huperzia selago, Polytrichum alpinum, Cladonia impexa and *C. arbuscula*. The crowberry in these high-level heaths is *Empetrum nigrum* and, although superficially similar, the community does not have the same relationship to prolonged snow-cover as the montane *E. nigrum* subsp. *hermaphroditum - Vaccinium* heaths of the Highlands (H 19).

Acidic grasslands
These are differentiated mainly by drainage conditions, and form a series from dry to wet. They are the most extensive of the semi-natural vegetation types of Cumbria, reflecting both the large extent of hill ground and its long history of intensive management for grazing animals.

16. *Upland* Festuca - Agrostis *grassland* (U 4)
The later stage of replacement of *Calluna* and *Vaccinium* heath under heavy grazing and repeated burning on the dry acidic soils of steep slopes is a close-grazed sward composed of a mixture of *Festuca ovina, Agrostis canina, A. capillaris, Deschampsia flexuosa* and *Anthoxanthum odoratum*, with *Galium saxatile* and *Potentilla erecta*. In places, grazed-down remnants of heather and bilberry persist to show the derivation of the community, which also shows all degrees of invasion by bracken on deeper soils, to the point of total replacement by dense, head-high stands of this fern (U 20). It is extremely widespread on the fells of both Lakeland and the Pennines. In Lakeland, the club-mosses *Lycopodium clavatum, Diphasiastrum alpinum* and *Huperzia selago* are locally plentiful, and in places on Borrowdale Volcanic ground *Alchemilla alpina* flourishes in the turf (grading to CG 11). High-level variants have evidently replaced former montane dwarf shrub heaths, but are usually species-poor, and lack arctic-alpines apart from *Carex bigelowii* and, occasionally, *Salix herbacea*. The summit of Brandreth

has one of the more species-rich examples of the montane type, while on Robinson it is associated with interesting frost-sorting patterns.

17. *Montane* Festuca - Racomitrium *grass-heath*
Some of the high Lakeland and Pennine summits at over 760 m formerly supported continuous carpets of the moss *Racomitrium lanuginosum*, with a sparse growth of vascular plants such as bilberry, cowberry, *Carex bigelowii* and *Salix herbacea*. This community, which was identical with the widespread Highland *Racomitrium* heath (U 10), was well represented on Grasmoor, Skiddaw, Helvellyn, the Scafell range and Cross Fell up to the 1950s. Since then it has been almost wholly replaced, except perhaps on Cross Fell, by a short turf mainly of *Festuca ovina, F. vivipara, Deschampsia flexuosa* and *Agrostis canina*, similar to the high level variant of type 16. The disappearance of the moss is evidently caused by the increased treading, manuring and grazing of sheep over the last few decades, although some ecologists believe enhanced nitrogen deposition is also responsible.

18. *Upland* Nardus *grassland* (U 5)
On less steep hill ground where acidic soils show some degree of gleying (from constant or periodic waterlogging), the previous grassland type gives way to dominance of mat grass, *Nardus stricta* (Plate 34), though the component species are mostly still present, or even abundant. This uniform, species-poor community has mostly been derived from the moister variants of earlier dwarf shrub heath by heavy grazing over a long period. It is one of the most extensive semi-natural communities of Cumbria, occurring in all upland areas and has little value as pasture. At higher levels the *Nardus* grasslands may represent the types associated with prolonged snow cover in the Highlands (U 7), but the spread of the secondary type has obscured

Plate 34. Upland *Nardus* grassland, Howgill Fells (Type 18, U 5)

their former natural distribution.

19. *Upland* Juncus squarrosus *grassland* (U 6)

The previous grassland type often contains *Juncus squarrosus*, but on still wetter and more strongly gleyed podsolic soils, this plant becomes dominant. Both this species and *Nardus* sometimes grow on eroded peat redistributed over lower slopes. Its vigorous competitive power reduces associates and it tends to be even more species-poor than *Nardus* grassland. The frequent presence of *Sphagnum* shows its relationships with blanket bog.

20. Molinia *grassland* (M 25)

Grasslands dominated by purple moor-grass, *Molinia caerulea*, occur on acidic waterlogged gley soils in a variety of situations, from the drained edges of lowland peat mosses to hillside seepages and valley mires at moderate elevations on the fells. Rodwell (1991) includes them with mires. They are characteristic of the south-western Lakeland fells (where they often contain an abundance of *Myrica gale*) and the Border moorlands, but are less frequent in the Pennines. Most examples are species-poor, but where they grade into bog the flora becomes more diverse. Very locally, richer *Molinia* grassland can also be seen around flushes and fens. It is, for example, co-dominant with *Schoenus* on Hale Moss.

21. Deschampsia cespitosa *grassland* (MG 9)

Tufted hair-grass, *Deschampsia cespitosa*, is often dominant on disturbed but uncultivated damp ground, such as that within or adjoining recently afforested areas. The soils are often mildly basic, and there may be a variety of rough pasture species, such as *Cirsium palustre*, *Centaurea nigra*, *Rumex acetosa*, *Carex binervis* and *Holcus lanatus*. It is often a somewhat short-lived type which develops by succession into

scrub or woodland.

Hay-meadow grassland

One of the most important botanical habitats in Cumbria is the traditional enclosed meadow, grazed by stock during autumn and winter but then kept ungrazed during spring and cut for hay in the summer. Originally derived from forest, these hay-meadows have either retained or acquired the forbs of woodland field communities on basic soils, in addition to the grass component for which they are especially managed. They belong to the class which Tansley termed neutral grasslands, occurring over deep brown loams of pH 5.5-7.0. The fertility of many of them was slowly increased under earlier management, but in ways that probably enhanced their floristic richness. Beginning with war-time intensification of use in 1940, modern farming methods (herbicides, ploughing, addition of NPK fertilisers, reseeding) have resulted in their systematic conversion into rye grass swards almost devoid of botanical interest. The effect has been to leave a thinly scattered remnant of the colourful hay-meadows which were such a conspicuous feature when Wilson (1938) published his *Flora of Westmorland*.

22. Arrhenatherum - Filipendula *herb-rich grassland* (MG 5)

The lowland types of these grasslands have almost disappeared, but examples remain along the broad roadside verges of the clayey limestone drift country between Greystoke and Sebergham. The grasses include *Arrhenatherum elatius*, *Dactylis glomerata*, *Holcus lanatus*, *Poa pratensis*, *Agrostis capillaris*, *Festuca rubra*, *Trisetum flavescens*, *Cynosurus cristatus*, *Deschampsia caespitosa* and *Briza media*. Characteristic forbs are *Geum rivale*, *Succisa pratensis*, *Alchemilla glabra*, *Sanguisorba officinalis*, *Filipendula ulmaria*, *Vicia cracca*, *V. sepium*, *Lathyrus pratensis*,

Rhinanthus minor, Stachys officinalis, S. sylvatica, Leontodon hispidus, Galium mollugo, Cruciata laevipes, Centaurea nigra, Geranium pratense, Leucanthemum vulgare, Serratula tinctoria, Silaum silaus, Angelica sylvestris, Valeriana officinalis, Trifolium pratense, Lotus corniculatus, Plantago lanceolata, Ranunculus acris, Achillea ptarmica, Orchis mascula, Dactylorhiza fuchsii, Listera ovata, Gymnadenia conopsea, Platanthera chlorantha, Coeloglossum viride, Carex panicea and *C. flacca.*

Variants on more acidic soils (such as over New Red Sandstone) have fewer species, and also more of others such as *Heracleum sphondylium, Senecio jacobaea, Tanacetum vulgare, Achillea millefolia, Hypochaeris radicata, Crepis capillaris, Genista tinctoria, Galium verum, Conopodium majus, Silene dioica, Rumex acetosa* and *Potentilla erecta.*

23. Geranium sylvaticum *meadows* (MG 3)

A few of the herb-rich hay-meadows have survived on the upland farms at 200-400 m, where they often have an abundance of *Geranium sylvaticum* (Plate 35) (replacing *G. pratense*), *Cirsium heterophyllum, Alchemilla* species, *Trollius europaeus, Crepis paludosa, Anthriscus sylvestris, Trifolium medium, Galium boreale, Athyrium filix-femina* and *Equisetum sylvaticum*. They were formerly widespread on the limestone around Kendal and Orton, and along the Pennine fell-foot, and examples remain here, as also along roadside and railway verges, but one of the best is at Gowk Bank on the upper Irthing. The colourful buttercup meadows of the Pennine farms (Plate 29), which have also declined, are a somewhat more modified stage and, apart from the profusion of *Ranunculus acris*, are rather poor in the distinctive herbs of the rich meadow communities. Flushed areas within these kinds of meadows often have an abundance of *Caltha palustris.*

24. *Herb-rich cliff ledges* (U 17)

Fragments of original herbaceous vegetation closely related to both the upland hay meadows and the field layer communities of ungrazed hill woods have survived on the bigger and more stable ledges of basic mountain cliffs. As well as many lowland species of type 23, such as *Geranium sylvaticum, Geum rivale* and *Silene dioica*, they typically have montane plants such as *Sedum rosea, Oxyria digyna, Thalictrum minus, Cochlearia pyrenaica* and *Alchemilla wichurae*. *Luzula sylvatica* and *Solidago virgaurea* are also usually present, emphasising the relationship with woodland field communities. *Saussurea alpina* is present in a few places. Good examples occur on Honister Crag, Helvellyn and High Street.

Calcicolous grasslands

On calcareous soils (pH 7.0 or above), especially shallow impoverished rendzinas and somewhat deeper, well-drained brown loams over Carboniferous limestone, unenclosed and grazed swards are composed of a species-rich mixture of grasses, sedges and small forbs.

25. *Lowland limestone grassland* (CG 9, CG 10)

This is the northern equivalent of the southern chalk grasslands, sharing many species in common: *Sanguisorba minor, Helianthemum nummularium, Thymus polytrichus, Scabiosa columbaria, Asperula cynanchica, Plantago media, Leontodon hispidus, Campanula rotundifolia, Sedum acre, Galium sterneri, Gentianella amarella, Linum catharticum, Polygala vulgaris, Carlina vulgaris, Ranunculus bulbosus, Hippocrepis comosa, Carex caryophyllea, C. flacca, Festuca ovina, F. rubra, Briza media, Koeleria cristata, Helictotrichon pratense, H. pubescens, Trisetum flavescens* and *Botrychium lunaria*. The distinctive grass of these northern calcicolous

Plate 35. *Geranium sylvaticum* hay-meadow, Shap (Type 23, MG 3)

grasslands is, however, *Sesleria caerulea*, which is often dominant. This is one of the richest Cumbrian botanical habitats, with a large flora and many rare species. Some Morecambe Bay stands are rich in orchids, including *Spiranthes spiralis*, *Ophrys morio* and *O. insectifera*. Moist examples often have *Primula farinosa*. The community is well represented on the extensive limestone exposures around Kendal and Morecambe Bay, on the Asby - Orton uplands, and all along the Pennines from Mallerstang to Geltsdale. Fragmentary outliers occur around Greystoke, Cockermouth, on the Border moorlands, and on the most lime-rich of the Borrowdale Volcanic rocks. Some of the more notable species are confined to the more southerly exposures.

26. *Montane limestone grassland* (CG 9 - 11)
With increasing elevation, the previous community changes gradually, as lowland and southern species drop out and are replaced by northern and montane types. At their upper limits, the Carboniferous limestone outcrops girdling certain Pennine summits at 500-750 m have a distinctive sward containing *Draba incana*, *Minuartia verna*, *Saxifraga hypnoides*, *Antennaria dioica*, *Cochlearia pyrenaica*, *Viola lutea* and *Persicaria vivipara*. Rare species of these high-level grasslands are *Myosotis alpestris*, *Gentiana verna*, *Viola rupestris*, *Potentilla crantzii* and *Carex capillaris*. Very small areas of a similar but less rich sward occur on some Lakeland fells, especially Helvellyn, where there is intermittent flushing from calcareous rocks, and these contain *Alchemilla alpina*. A related community with *Minuartia*, *Cochlearia* and *Viola lutea* has developed over old Pennine lead mine spoil containing calcite; it typically contains *Thlaspi caerulescens* and, occasionally, *Armeria maritima* (Plate 28).

Rock communities
The vegetation of loose rock debris, solid pavement and cliff faces is more difficult to classify into recognisable communities, but there are some distinctive plant assemblages.

27. *Acidic scree and block litter* (U 21)
Unstable talus slopes with small to medium-sized fragments are characterised by an abundance of clumps of *Cryptogramma crispa*, which acts as a pioneer around which other common plants from the surrounding grasslands and heaths can establish and form a stabilising patchwork of vegetation. The succession is often an 'arrested' one, though, and *Cryptogramma* may be abundant on screes which are still extending. Other ferns of the larger and more stable block litters include *Dryopteris* species: *D. filix-mas*, *D. affinis*, *D. oreades*, *D. dilatata* and, more locally, *D. expansa*. *Huperzia selago* is often present. Stable block litters often have dense carpets of the moss *Racomitrium lanuginosum*, and provide habitats for a large number of bryophytes.

28. *Acidic cliffs* (U 16)
The ledges and crevices have a rather limited flora, often with heather, bilberry, cowberry and other plants common in the widespread heaths and grasslands. *Luzula sylvatica*, *Solidago virgaurea*, *Succisa pratensis* and *Festuca vivipara* are also usually present. There are scattered localities for *Silene uniflora*, and *Carex bigelowii* and *Salix herbacea* may be quite luxuriant at the higher levels.

29. *Basic cliffs (other than limestone)* (U 15, OV 40)
Lime-rich crags provide the habitat for many of the more notable mountain plants, which tend to increase in number with altitude, and are best represented on the Helvellyn range. As well as the taller herb communities of large, moist ledges, there are distinctive groupings of smaller species on the steep, well-drained faces. They are not only strictly montane plants but include a number from low-level calcareous habitats. The list includes *Saxifraga aizoides*, *S. hypnoides*, *S. oppositifolia*, *Silene acaulis*, *Minuartia verna*, *Alchemilla alpina*, *Thalictrum alpinum*, *Galium boreale*, *Antennaria dioica*, *Sedum rosea*, *Carex flacca*, *Linum catharticum*, *Arabis hirsuta*, *Campanula rotundifolia*, *Euphrasia* species, *Asplenium viride* and *Cystopteris fragilis*. A number of other rarer species belong to such face and crevice communities.

30. *Limestone scree, pavement and scar*
These are among the most floristically rich of all the Cumbrian habitats. The woodland communities of the deep pavement crevices are referred to under type 4, but there is an additional element associated with the grikes of treeless pavements or the light shade of more exposed surfaces among open scrub, and shallower crevices of bare scree. The grike communities have general abundance of *Mercurialis perennis*, *Allium ursinum*, *Geranium lucidum*, *Phyllitis scolopendrium* and *Polystichum aculeatum*. The very local ferns *Dryopteris submontana* and *Gymnocarpium robertianum* are also plentiful in places. In more open situations *Sesleria caerulea* is profuse, and there are many other plants of limestone grassland, besides the rarer species listed in the Botanical tour.

Peatlands
These are separable into:
(a) bogs with deep, acidic peat where the nutrient supply is mainly atmospheric,
(b) fens with less acidic to alkaline peat where the nutrient supply is mainly from the ground water,
(c) hill marshes, flushes and springs, where humus and mineral particles are usually mixed, and nutrient supply is from ground water flowing near or at the surface.

Bogs
Acidic peat bogs are well represented under the oceanic

Cumbrian climate. They have developed in a range of topographic situations which cause impeded drainage, but their vegetation is differentiated primarily according to altitude, and secondarily as the result of modification, usually drying, through human activity.

31. *Lowland peat mosses* (M2, M3, M18, M20, M21)

These are mainly the domed raised bogs of the coastal plains, but include also certain valley and basin bogs with almost identical vegetation. In the most natural examples, now rather few, there is a gently undulating surface dominated by *Sphagnum*, and with a patchwork of communities differentiated by the height of the water table. The wetter ground of the hollows, seldom with open water, has *S. cuspidatum* and *S. pulchrum*, with open growth of *Eriophorum angustifolium*, *Rhyncho-spora alba* and *Drosera anglica* (M 2). The more general 'lawn' or low hummock has dominance of *S. magellanicum* and *S. papillosum*, with *Andromeda polifolia*, *Vaccinium oxycoccos*, *Erica tetralix*, *Narthecium ossifragum*, *Drosera rotundifolia*, *Eriophorum vaginatum* and *Trichophorum cespitosum*. The highest and driest hummocks have *S. capillifolium*, *Calluna vulgaris*, *Empetrum nigrum*, and more tussocky growths of *Eriophorum* and *Trichophorum* (M 18). *Myrica gale* is abundant on some sites.

Areas of bog most damaged by fire and drainage develop the tall hummock kind of vegetation, losing *Sphagnum* cover as the dwarf shrubs and tussocky cotton grass and deer sedge become dominant (M 20). Old peat cuttings often show redevelopment of earlier, active stages of bog growth, though the predominant bog-moss is usually *S. recurvum*; and this species and *Aulacomnium palustre* have also spread in valley and basin mires which receive fertiliser enrichment through internal drainage. More base-rich seepages within the acidic peat mosses, and more especially within the valley mires (M 21), are distinguished by growths of sedges (e.g. *Carex rostrata*, *C. lasiocarpa*, *C. curta*, *C. echinata* and *C. limosa*) and other typical fen plants.

These communities are well represented on the raised bogs of the Solway, the Duddon estuary and the Cumbrian side of Morecambe Bay. They are also characteristic of intermediate bogs farther inland, such as Bolton Fell (now largely cut-over) and Walton - Broomhill Mosses north-east of Carlisle. Acidic valley bogs occur at Hallsenna Moor, Tarn Moss (Troutbeck), Cliburn Moss, Wan Fell and Cumwhitton Moss, while Moorthwaite Moss is a basin bog.

32. *Blanket bogs* (M17-19)

Broad and poorly drained upland watersheds above 250 m are typically covered with deep peat mostly formed by the earlier growth of bog-mosses, but now retaining a very variable cover of these plants. The most typical community has a mixture of *Calluna vulgaris* and *Eriophorum vaginatum*, often accompanied by bilberry, cowberry, crowberry, *Eriophorum angustifolium*, *Trichophorum cespitosum*, *Erica tetralix*

and *Rubus chamaemorus*. The more undisturbed areas have abundant *Sphagnum papillosum*, *S. magellanicum* and *S. capillifolium*, and there may be pools with *S. cuspidatum*, *S. auriculatum* and *Uticularia minor*. *Carex magellanica* is a rare plant of wet hollows in blanket bogs, and *C. pauciflora* occurs locally.

Blanket bogs occur extensively along the watersheds of the Pennines and Border moors, and have a local profusion of *Rubus chamaemorus*. Butterburn Flow is the best undisturbed example, with a high cover of bog-moss. On the steeper slopes and narrower watersheds of the Lakeland fells, this type of vegetation is more local and contains a good deal of *Juncus squarrosus*. Some Pennine areas show extensive erosion by gullying of the peat, extending into more general denudation by sheet erosion in places. Some degraded areas have little plant life other than an open growth of the two cotton grasses.

Fens

Fens have developed in permanently waterlogged situations where the water has drained from mineral substrates and so contains nutrients from this source. They are found along lake margins, on wet flood-plains, and in channels and hollows of glacial drift. Their plant communities are differentiated mainly by the nutrient-richness of the water supply and the height of the water table. Those fringing open water bodies are often part of the 'hydrosere' sequence between floating aquatic vegetation and closed willow - alder carr woodland. Rodwell (1991) has distinguished the patchy dominance of certain species growing into the open water edge of the sequence as a range of swamp communities, but these have been treated here as part of the fen types into which they merge. Some of the vascular plants of fens have a wide tolerance of acidity and base-content of the ground water and so are not diagnostic of any particular type, for example *Carex rostrata*, *C. lasiocarpa*, *C. limosa*, *Eriophorum angustifolium*, *Phragmites australis* and *Menyanthes trifoliata*. The mosses forming the ground layer are usually the best indicators of the base-status.

33. *Poor fens* (M 4, M 5, S 9, S 10)

Sites influenced by nutrient-poor water typically have communities with *Carex rostrata*, *C. nigra*, *C. curta*, *C. echinata*, *C. lasiocarpa*, *Eriophorum angustifolium*, *Juncus acutiflorus*, *J. bulbosus*, *Galium palustre*, *Menyanthes trifoliata*, *Potentilla palustris*, *Eleocharis palustris*, *Ranunculus flammula*, *Vaccinium oxycoccos*, *Potamogeton polygonifolius*, *Narthecium ossifragum*, *Dactylorhiza maculata*, *Hydrocotyle vulgaris*, *Viola palustris*, *Molinia caerulea*, *Agrostis stolonifera*, *Dryopteris carthusiana* and *Equisetum fluviatile*. Bog mosses often form carpets of *Sphagnum recurvum*, *S. auriculatum*, *S. fimbriatum*, *S. squarrosum* and *S. palustre*. These poor fens usually cover only a few hectares, but good examples occur at Braithwaite Moss, Hallsenna Moor, Moorthwaite Tarn, Lazonby Fell and

Wan Fell. Poor fens may merge into willow carr or into acidic bog or heath.

34. *Intermediate fens* (M 8, M 9, M 22, S 1, S4, S 8, S 11, S 12, S 14, S 19)

These, the mesotrophic fens, receive rather more base-rich drainage water than the previous type. They may have some of the same sedges, especially *C. rostrata*, but additionally contain a wide range of other species such as *C. vesicaria*, *C. paniculata*, *C. disticha*, *C. acutiformis*, *C. elata*, *C. diandra*, *Sparganium erectum*, *Iris pseudacorus*, *Eleocharis palustris*, *Typha latifolia*, *Pedicularis palustris*, *Mentha aquatica*, *Myosotis secunda*, *Epilobium palustre*, *Caltha palustris*, *Lythrum salicaria*, *Scutellaria galericulata*, *Lycopus europaeus*, *Veronica scutellata*, *Galium palustre*, *Dactylorhiza purpurella* and *Equisetum palustre*. Grasses such as *Phragmites australis* and *Phalaris arundinacea* are variably present. *Sphagnum* cover tends to be less than in type 29 and species such as *S. warnstorfii*, *S. contortum*, *S. subsecundum* and *S. teres* are present. Other bryophytes include *Calliergon cordifolium*, *C. cuspidatum*, *C. giganteum* and *Marchantia polymorpha*. The drier edges of such fens often grade into tall herbaceous communities with *Filipendula ulmaria*, *Valeriana officinalis*, *Angelica sylvestris*, *Lotus pedunculatus*, *Solanum dulcamara*, *Oenanthe crocata*, *Eupatorium cannabinum* and *Epilobium hirsutum* (S 25, S 27, S 28). These in turn often merge into willow or alder carr.

Intermediate fen is a local type, usually of small extent, but good examples occur at Esthwaite North Fen, Blackdyke Bog (Matterdale), Biglands Bog, beside Derwent Water, and Cumwhitton Moss.

The moist alluvial banks of lowland rivers sometimes have tall herb communities resembling these 'fen meadows', with *Impatiens glandulifera*, *Carex acutiformis*, *Scirpus sylvaticus*, *Tanacetum vulgare* and *Senecio fluviatilis*.

35. *Rich fen* (M 10, M 13, S 1, S 2)

Where the ground water is strongly calcareous, the bog mosses disappear and the ground layer is often dominated by mosses which need lime-rich conditions. *Scorpidium scorpioides*, *Drepanocladus revolvens*, *Campylium stellatum*, *C. elodes*, *Ctenidium molluscum*, *Cratoneuron commutatum*, *C. filicinum*, *Calliergon giganteum*, *Fissidens adianthoides*, *Philonotis calcarea* and *Bryum pseudotriquetrum*. The typical vascular plants include *Cladium mariscus*, *Schoenus nigricans*, *Carex rostrata*, *C. lasiocarpa*, *C. elata*, *C. diandra*, *C. dioica*, *C. viridula* subsp. *brachyrrhyncha*, *C. pulicaris*, *C. hostiana*, *Eriophorum latifolium*, *Juncus articulatus*, *Eleocharis quinqueflora*, *Dactylorhiza incarnata*, *Epipactis palustris*, *Ranunculus lingua*, *Parnassia palustris*, *Primula farinosa*, *Valeriana dioica* and *Hippuris vulgaris*.

This is an extremely local community, almost confined to the Carboniferous limestone, and the best examples are at Sunbiggin Tarn, Cunswick Tarn and Newton Reigny Moss. Biglands Bog formerly had a remarkable system of 'brown moss' communities in hollows situated amongst the *Sphagnum* hummocks of acidic bog, but sewage seepage has destroyed these.

Hill marshes, flushes and springs

These are mainly upland types occurring where water draining from higher levels emerges at the ground surface or causes waterlogging immediately below. This involves substrate enrichment by the transport of nutrients from higher to lower levels, the process which Pearsall (1950) termed 'flushing'.

Upland marshes

Well-developed fens are usually on level ground, but in the uplands, where water draining laterally becomes concentrated near or at the ground surface, in hollows, channels or valley floors, there are little marshes with fen-like communities which are the recognisable equivalents of the previous three groups. Because of grazing, the vegetation tends to be shorter, lacking the taller herbs and species sensitive to cropping. Many overlie deep gley soils. On acidic substrates and in seepages among blanket bog are poor fen types, but occasionally the emergence of calcareous water in such situations produces communities approaching rich fen. Caudbeck Flow on the Bewcastle moors has a complex of bog and marsh reflecting wide variations in water chemistry, while Sunbiggin Tarn Moor also shows a range of communities, but in a different pattern.

36. *Rush-dominated marshes* (M6)

One of the most widespread types, found mainly below the tree-line and probably much influenced by sheep management, is usually dominated by *Juncus effusus*, with varying cover of *Sphagnum recurvum* and *Polytrichum commune*. It is poor in species, though *Epilobium palustre*, *Galium palustre*, *Stellaria uliginosa*, *Carex echinata* and *Holcus lanatus* are often present. In the Lake District *J. effusus* is often mixed with, or completely replaced by, *J. acutiflorus*, often with no change in the associated species, though sometimes with more base loving herbs and mosses where the latter rush is dominant.

37. *Acidic sedge marshes* (M 6)

The commonest type has a mixture of *Carex echinata*, *C. nigra* and *C. curta*, but there may also be an abundance of *C. rostrata* and, occasionally, *C. lasiocarpa* and *C. magellanica*. Bog mosses typically carpet the surface: *Sphagnum recurvum*, *S. cuspidatum*, *S. palustre*, *S. girgensohnii*, *S. russowii*, and other mosses include *Aulacomnium palustre* and *Calliergon stramineum*. *Erica tetralix*, *Vaccinium oxycoccos*, *Viola palustris*, *Narthecium ossifragum*, *Potentilla erecta*, *Molinia caerulea* and *Agrostis vinealis* are usually present. The community is widespread on all the more acidic upland massifs, but distinctive variants in western Lakeland have *Eleocharis multicaulis* and *Rhynchospora alba*.

38. *Basic sedge marshes* (M 10, M 38)

The hill marsh equivalents to intermediate and rich fens tend to be less well differentiated, and so are treated as a single group covering a range of conditions from moderately base-rich to strongly calcareous. Their characteristic sedges are *Carex viridula* subsp. *oedocarpa*, *C. panicea*, *C. nigra*, *C. hostiana*, *C. pulicaris* and *C. dioica*, accompanied by *Juncus articulatus*, *Eleocharis quinqueflora*, *Triglochin palustre*, *Leontodon autumnalis*, *Ranunculus acris*, *Plantago lanceolata*, *Pinguicula vulgaris*, *Parnassia palustris* and *Selaginella selaginoides*. *Primula farinosa* occurs locally and *Eriophorum latifolium* is occasionally present. Lime-loving mosses are usually present and reach a continuous cover on the most calcareous sites: they include the species listed under types 34 & 35. In the Pennines, these mossy spring-fed marshes are the habitat of *Saxifraga hirculus* and *Cinclidium stygium*. They are also well represented on the calcareous rocks of the Border moors and the Borrowdale Volcanics.

Flushes and springs

Where water wells up at the ground surface and either gives rise to rills or courses over it in more diffuse seepages, there are carpets of moss and liverworts or more open, muddy patches with a sparse plant cover.

39. Saxifraga stellaris - *moss springs* (M 32) (Fig. 13)

In springs fed by water poor in lime the bryophyte cushions have *Philonotis fontana*, *Dicranella squarrosa*, *Calliergon sarmentosum*, *Drepanocladus exann-

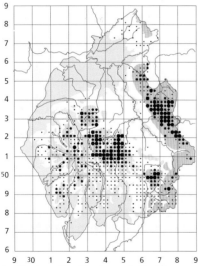

Fig. 13. Predictive distribution of *Saxifraga stellaris* - moss springs (Type 39, M 32) based on the following indicator species: *Saxifraga stellaris*, *Montia fontana*, *Chrysosplenium oppositifolium*, *Cardamine pratensis*, *Epilobium alsinifolium*, *Eriophorum angustifolium* and *Carex nigra*.

ulatus, *Sphagnum auriculatum*, *Jungermannia exsertifolia* and *Scapania undulata*. Forming an open growth of vascular plants in the cushions are *Saxifraga stellaris*, *Montia fontana*, *Chrysosplenium oppositifolium*, *Stellaria uliginosa*, *Cardamine pratensis*, *Caltha palustris*, *Cerastium fontanum*, *Deschampsia cespitosa* and *Agrostis stolonifera*. The spongy carpets often have *Epilobium alsinifolium* and *Myosotis stolonifera*, and these follow the rills down to low levels. Cold water springs high on the Cross Fell range are the habitat of *Alopecurus borealis* and *Epilobium anagallidifolium,* while those on slightly richer sites have *Sedum villosum* and *Saxifraga hypnoides*.

On the most acidic rocks of the Lake Fells is another spring community with dense grey carpets of the liverwort *Anthelia julacea*, in association with some of the previous species, especially *Sphagnum auriculatum* (M 31). Acidic, stony flushes with numerous cushions of *Campylopus atrovirens*, with *Eleocharis multicaulis*, appear to be the low level equivalent of this type in parts of western Lakeland.

40. Saxifraga aizoides - *sedge flushes* (M 11)

Where the drainage water is lime-rich, the spring-head is often marked by a large mound of the golden moss *Cratoneuron commutatum*, which may have sufficient associates to form a recognisable community with *Saxifraga aizoides*, *Cardamine pratensis*, *Festuca rubra* and various calcicolous bryophytes (M 37). The water often feeds a more diffuse flush with much bare stone, gravel and mud, in which these plants have a more patchy growth, along with *Carex viridula* subspp. *oedocarpa* and *brachyrrhyncha*, *C. panicea*, *C. flacca*, *C. hostiana*, *Juncus articulatus*, *J. bulbosus*, *Pinguicula vulgaris*, *Sagina nodosa* and *Eleocharis quinqueflora*. At high levels this is the habitat of *Juncus triglumis*, and *Saxifraga oppositifolia* and *Thalictrum alpinum* are often present. *Primula farinosa* occurs in places, and *Schoenus nigricans* on the Asby - Orton limestone. The community shares many species in common with type 38, and could be regarded as an open variant. The extensive systems of such flushes on the sugar limestone ground of upper Teesdale outside Cumbria have a still more distinctive assemblage of rare montane species.

Lakes and rivers

Open water vegetation is less easily grouped into distinctive communities, because of such practical difficulties as the listing of all species present, and the frequent vertical layering of the plant growth. As Rodwell (1995) has pointed out, communities tend to recognised according to single species dominants, though sometimes there are characteristic combinations. Floristic variation is mainly according to the nutrient content of the water, the organic and silt content of the substrate, altitude, and water movement.

Cumbria contains the largest natural lakes in England (though some of them have been profoundly modified for water supply), and their ecology has been closely studied over a long period. They show a range of variation in water nutrient content from the lower to the intermediate part of the range, though enrichment by inflow of fertilisers and sewage during recent years has elevated levels of nitrogen and phosphorus in some lakes. Even the best example of a lime-rich lake, Sunbiggin Tarn on the Carboniferous limestone, has suffered such eutrophication through the presence of a large breeding colony of black-headed gulls.

The rivers of Cumbria originate mainly as turbulent mountain streams, and remain relatively fast-flowing even along their lower courses. Some of them have more sluggish sections, and there are lowland becks and 'soughs' that are virtually stagnant, at least in places. There is a wide range in nutrient content of the water, from the low levels of dissolved minerals of many Lakeland gills, to the calcareous streams of the Pennines, the artificially enriched lowland rivers, and the seriously polluted sections near some of the industrial towns.

Although typical lakes and rivers obviously differ markedly in habitat features, there are examples which are difficult to assign to one or the other: broad and sluggish river sections, ox-bows and slow-moving or near-stagnant canals or ditches. Some plant communities also occur in both lakes and rivers, and the two will be treated here as a single continuous series of water bodies.

41. Littorella - Isoetes *swards* A 22, A 23)
These occupy the shallow, stony and sandy edges of almost all the lakes and tarns, forming low, submerged growths which are partly exposed when drought produces lowered water levels. The rosettes of *Littorella uniflora* and *Isoetes lacustris* are often mixed with those of *Lobelia dortmanna*. Other typical associates are *Myriophyllum alterniflorum*, *Ranunculus flammula*, *Juncus bulbosus* and *Eleocharis palustris*. There is sometimes a thin growth of *Carex rostrata,* and a few upland sites have *Subularia aquatica*.

42. *Water lily communities* (A 7)
In the deeper water of bays and other more sheltered places are floating rafts of *Nymphaea alba*, sometimes with *Nuphar lutea*. In shallower waters, the surface may sometimes have mats of *Potamogeton natans* and *P. polygonifolius*, with *P. obtusifolius*, *Sparganium natans* and *Juncus bulbosus* submerged below. Dense tangled growths of the submerged form of the *Juncus* are particularly characteristic of oligotrophic and peaty lakes and tarns. Emergent growths of *Schoenoplectus lacustris* are sometimes present. Beds of *Callitriche hamulata* are characteristic of the more oligotrophic lakes, tarns and especially the streams. This type is found in parts of the larger lakes, such as Derwent Water and Bassenthwaite Lake, and in some of the richer and peatier tarns, such as Overwater, Mockerkin, Braystones and Watendlath. It occurs occasionally in

canal-like sections of rivers, such as the Irthing above Butterburn. Rodwell (1995) regards floating mats of *Persicaria amphibia* as a distinct community (A 10), but they grade into both the present type and the next.

43. *Pondweed communities* (A 13)
Submerged growths of pondweeds, with *Potamogeton perfoliatus*, *P. gramineus*, *P. berchtoldii* and *P. obtusifolius*, occur in deepish (often to 2 m) standing waters with moderate nutrient levels and fine sediment. Floating mats of *P. natans* and *P. polygonifolius* may be present. Mixed with these are often *Myriophyllum alterniflorum* and *Nitella* spp. Particularly in the central and southern lakes, such as Windermere and Grasmere, vast submerged forests of *Elodea nuttallii* have severely suppressed the pondweeds. The classic locality for this pondweed community is Esthwaite Water where it also includes, very locally, *Najas flexilis*. In places it grades into type 42 through the addition of *Littorella* and *Isoetes*. Rodwell (1995) regards communities dominated by *M. alternifolium* in shallower and more turbulent water as a distinct type (A 14), while dominance of *P. natans* is also given separate status (A 9).

44. Ranunculus penicillatus *and* R. fluitans *communities* (A 17, A 18)
These communities are characteristic of moderate- to fast-flowing rivers. *Ranunculus fluitans* is particularly common in the River Eden and its tributaries where associated species include *Elodea canadensis*, *Myriophyllum spicatum*, *Potamogeton perfoliatus*, *P. gramineus* and *P. crispus*.

45. Callitriche *communities* (A 16, A 21)
The still or slow-moving waters of ponds, pools, canals or sluggish streams have growths of *Callitriche stagnalis*, occasionally also with *C. platycarpa* and, near the coast, *C. obtusangula*, *Myriophyllum spicatum*, *Potamogeton pectinatus*, *Ranunculus baudotii*, *Zannichellia palustris* and *Ruppia maritima*.

46. Lemna *communities* (A 2)
Floating mats of *Lemna minor* are widespread on the surface of lowland ponds, pools and ditches and, from the wide variety of other associates, can often be regarded as mixed with other aquatic communities.

Coastlands
Most of the main types of coastal habitat are well represented in Cumbria, though seacliffs are highly localised.

Salt-marshes
The extensive salt-marsh systems around the estuaries of Morecambe Bay, the Solway, Duddon and Irt have developed on a mainly sandy substrate. The successional sequence from colonisation of bare tidal sand flats to a consolidated sward submerged only by the highest

spring tides, is that typical of north-western Britain where the land has risen slowly in relation to sea level. The swards of many marshes are valued grazing range, and carry large numbers of cattle and sheep, which limit the growth of the taller herbs and favour stoloniferous and rhizomatous species.

47. Salicornia *and* Puccinellia *swards* (SM 8, SM 9, SM 10)

Salicornia europaea and *Puccinellia maritima* act as pioneers, invading the bare sand to form an undulating, open sward. This stage is absent where marshes are retreating through erosion.

48. Armeria *and* Cochlearia *communities* (SM 13)

A variety of salt-tolerant species invade and consolidate the sward: *Cochlearia officinalis, Aster tripolium, Plantago maritima, Triglochin maritima, Armeria maritima, Spergularia media, S. marina* and *Glaux maritima*. *Puccinellia* usually remains abundant but *Salicornia* disappears. *Suaeda maritima, Limonium vulgare* and, more rarely, *L. humile* (Plate 36) are locally common, and *Atriplex portulacoides* enters at this stage in the more southern marshes. On some marshes, hollows in the sand deepened by scouring to form salt-pans, are bare except for algae.

Plate 36. *Limonium humile*, North Walney (Type 48)

49. Festuca *and salt-marsh* Juncus *swards* (SM 16, SM 18, SM 19, SM 27)

With further upward growth of the marsh level and decreasing frequency of tidal inundation, the halophytes decline and are replaced by less salt-tolerant species, though *Juncus gerardii* often becomes abundant at this stage, and *Armeria* and *Glaux* sometimes remain so. Other grasses such as *Agrostis stolonifera* and, particularly, *Festuca rubra* invade, along with other herbs of grasslands on slightly base-rich soil, such as *Carex flacca, C. panicea, C. nigra, Leontodon autumnalis, Crepis capillaris, Ranunculus acris, Trifolium repens* and *Lotus corniculatus. Carex distans, C. extensa, Blysmus rufus, Juncus maritimus* and *Plantago coronopus* are more local constituents. This is the sward cut and sold as commercial salt-marsh turf: the cut-over areas regenerate fairly quickly by natural recolonisation.

50. *High marsh* (MG 11)

On some salt-marshes, type 49 represents the final stage in development, before the rising level of the land at the inner margins causes a change to a quite different habitat - usually enclosed farmland. In a few places, such as Burgh Marsh on the Solway, there is yet another inner zone of unenclosed and grazed marsh covered with a type of rather poor grassland found on some inland and half-improved pastures. Halophytes have largely disappeared and the ground is thickly invaded by clumps of *Juncus effusus*, while the sward contains *Holcus lanatus, Anthoxanthum odoratum, Poa annua, Potentilla anserina, P. erecta, Taraxacum officinale* agg. and *Bellis perennis*.

51. *Brackish marsh* (SM 20, S 4, S 21)

At the inner edge of some salt-marshes, fresh water seepages result in mildly saline conditions in ditches and hollows, and this has a distinctive assemblage of species. *Bolboschoenus maritimus* forms emergent growth in the wettest places, and other plants are *Juncus articulatus, Eleocharis palustris, Phragmites australis, Iris pseudacorus, Carex otrubae, C. hirta, Blysmus compressus, Oenanthe lachenalii, O. crocata, Samolus valerandi* and *Galium palustre*.

52. *Shingle beaches* (SD 1, SD 2)

Much of the west Cumbrian shore is lined by shingle beach, with varying proportions of sand. There is a characteristic flora, though hardly a community in the sense of consistent species assemblages within a space of a few m². The habitat usually remains very open, and lists such as the following are compiled by listing along 100 m or more of beach: *Eryngium maritimum, Honckenya peploides, Cakile maritima, Salsola kali, Crambe maritima, Silene uniflora, Glaucium flavum, Calystegia soldanella, Rumex crispus, Euphorbia paralias, Atriplex glabriuscula, A. laciniata, Coincya monensis* and *Elytrigia atherica*. More local species are *Euphorbia portlandica* and *Polygonum oxyspermum*, and the northern *Mertensia maritima* is very rare. At higher levels, the beaches have plants such as *Sedum anglicum*, *S. acre*, *Thymus polytrichus* and a range of other species found on fixed dunes.

Sand-dunes

The Cumbrian coast has several well-developed sand-dune systems, showing the sequence of sand stabilisation, from the loose material of the shore through the open sandhills to fixed dune and invasive scrub. Damp hollows among the more stable dunes have formed in only a few places, notably at Sandscale. Most Cumbrian dune systems are somewhat acidic, but those on either side of the Duddon estuary have a more calcareous sand supporting a richer flora. The grazing of rabbits is very important in maintaining the characteristic low herb swards of both dry dunes and damp slacks.

53. *Marram sandhills* (SD 4, SD 5, SD 6)

The outermost line of low foredunes has an open plant growth scarcely distinguishable in floristics from that of type 52, apart from the addition of *Elytrigia juncea*, but the taller and unstable sandhills behind have dense tussocky growths of *Ammophila arenaria*, with more local abundance of *Leymus arenarius* and *Elytrigia atherica*. Varying amounts of the shingle and foredune plants grow between the grass tussocks. These early stages of dune succession are little affected by the lime content of the sand.

54. *Fixed dune swards* (SD 7, SD 8)

Behind the unstable sandhills there is often a zone of stabilised sand on lower and less pronounced hummocks, carrying a closed community with *Ononis repens*, *Thymus polytrichus*, *Galium verum*, *Campanula rotundifolia*, *Lotus corniculatus*, *Carex arenaria*, *Pilosella officinarum*, *Achillea millefolium*, *Plantago lanceolata*, *Viola tricolor*, *Centaurium erythraea*, *Erodium cicutarium*, *Festuca rubra*, *F. ovina*, *Agrostis capillaris* and *Phleum arenarium*. Some localities have an abundance of *Geranium sanguineum*, *Carlina vulgaris*, *Vulpia fasciculata*, *Anthyllis vulneraria*, *Trifolium arvense* and *Daucus carota*.

The following species are indicative of the more calcareous dunes: *Ranunculus bulbosus*, *Linum catharticum*, *Sedum acre*, *Leontodon hispidus*, *Polygala vulgaris*, *Carex flacca*, *Koeleria macrantha*, *Trisetum flavescens* and, less commonly, *Anacamptis pyramidalis*, *Erigeron acer* and *Thalictrum minus*.

55. *Dune-slacks* (SD 14, SD 15, SD 16, SD 17, M 25)

On acidic dunes, such as those between Silloth and Allonby, waterlogged hollows have communities approaching those of poor fen (33). Where the ground is less wet, there is a damp meadow community with *Agrostis* species, *Molinia caerulea*, *Juncus effusus*, *J. conglomeratus*, *J. acutiflorus*, *Rumex acetosa*, *Potentilla anserina*, *P. erecta*, *Ranunculus repens*, *Hydrocotyle vulgaris*, *Anagallis tenella* and *Lotus pedunculatus*. *Salix repens* is abundant in some sites. Calcareous dune-slacks correspondingly approximate to intermediate or rich fen (34 & 35): characteristic plants include *Selaginella selaginoides*, *Dactylorhiza purpurella*, *D. incarnata* and *Parnassia palustris*. Those at Sandscale have many rarities such as *Equisetum variegatum*, *Corallorhiza trifida*, *Epipactis palustris* and *Pyrola rotundifolia*.

56. *Dune scrub and heath* (H 11)

The innermost dune areas with greatest stability have often developed dense growths of shrubs of varying stature. One of the most typical is *Rosa pimpinellifolia* which forms low, prickly mats, while the much taller alien, *R. rugosa*, produces dense thickets. There is often a mixed tangle of brambles, blackthorn, common gorse, broom and elder. Scrub with sea buckthorn is very local being best developed at Eskmeals (Plate 76). Some of the most acidic inner dunes have a *Calluna* heath with varying amounts of bracken, closely resembling the lowland heath of type 13.

Seacliffs

The six km long Triassic sandstone escarpment of St Bees Head is the only significant occurrence of seacliff in Cumbria. The rock is largely acidic, so that base-demanding species are few and confined to moist seepages. The lower and more sheltered limestone promontory of Humphrey Head has woodland and grassland showing rather little maritime influence.

57. *Open* Armeria - Silene uniflora *rock community* (MC 8)

The lower part of the St Bees cliffs most exposed to salt spray have the characteristic open growth of cushions and tufts of strongly salt-tolerant plants such as *Armeria maritima*, *Silene uniflora*, *Plantago maritima*, *P. coronopus*, *Festuca rubra*, *Tripleurospermum maritimum* and *Cochlearia officinalis*. *Crithmum maritimum* and *Limonium recurvum* are common in this community and *Asplenium marinum* is plentiful in shady crevices. At higher levels above the sea *Jasione montana* and *Sedum anglicum* become abundant.

58. Festuca rubra - Dactylis - Daucus *grassland* (MC 11)

Arable land encroaches closely to the cliff tops, but in places between the fields and the more abrupt cliff edge there is a narrow zone of a sub-maritime grassland of tussocky *Festuca rubra*, *Dactylis glomerata* and *Arrhenatherum elatius*, with forbs such as *Daucus carota*.

Where the escarpment is less abrupt, and long slopes run down to the sea, there are communities approximating to *Calluna* heath and to the mixed scrub and herb - fern layers of woodland, with elder, blackthorn, brambles and species listed in the field layer communities of woodland types 1-3.

Three centuries of botanical recording

The following historical account of botanical recording prior to the Survey does not, and could not, pretend to be exhaustive. Baker (1885), in his *Flora of the English Lake District* and Hodgson (1898), in his *Flora of Cumberland*, comment on the major early contributors. Their number has grown considerably since then and only the more prominent are included here.

The seventeenth and eighteenth centuries

Although William Nicolson, Bishop of Carlisle, is commonly regarded as the father of Cumbrian botany, it is at least arguable that this honour should go to his contemporary Thomas Lawson, the subject of a detailed biography by Whittaker (1981). In 1652, while an Anglican curate at Rampside, in Furness, Lawson welcomed George Fox to his pulpit and was immediately converted to Quakerism. He renounced his living and began a turbulent life as an evangelist, travelling around the country at large but living at Great Strickland where he established a school which he was forced to close on his excommunication in 1664. Then followed a brief period with the Fells, a Quaker family of Swarthmoor, near Ulverston, before his final return to Great Strickland where he succeeded in reopening his school and where he lived until his death in 1691.

Lawson's interest in botany was fired by reading John Ray's *Catalogus Angliae* (1670), which included records obtained by Ray during his visits to Westmorland. Lawson made careful records of his finds as he journeyed around Cumbria, as well as from further afield on his proselytizing excursions, and these he communicated to Ray from about 1680 until his death. About 150 of his records, mainly from the Great Strickland area, were sent in 1688 to Ray who subsequently included them in the second edition of his *Synopsis* (1696). These include *Scutellaria minor* in Longsleddale, *Dianthus deltoides* at Cliburn, *Senecio palustris* (cf. *S. fluviatilis*) at Burton-in-Kendal, *Ornithopus perpusillus* at Sandford Moss, *Orobanche rapumgenistae* near Skelsmergh, *Hottonia palustris* in the River Kent at Kendal, also *Allium oleraceum* at Kendal, and a white helleborine, presumably *Cephalanthera longifolia*, in woods by Askham Hall. He saw the white-flowered *Geranium sanguineum* var. *striatum* at Walney Island "thousands hereof I found in the Isle of Waney, and have sent roots to Edinburgh, York, London, Oxford, where they keep their distinction", and also *Mertensia maritima* there and near Whitehaven, although the latter had first been reported from Walney in 1640. Lawson also found *Lathyrus sylvestris* at Whitehaven. Whittaker, incidently, tells us that Ray very nearly accepted the living at Kirkby Lonsdale shortly before he resigned from the church following his refusal to accept the Act of Uniformity.

Nicolson is said to have been born in "characteristic haste" in the porch of the parish church at Great Orton, near Carlisle, where his father was the incumbent. He too went into the church becoming rector of Great Salkeld and Archdeacon of Carlisle. Like Lawson, he was keenly interested in field botany and the two men became acquainted in 1690, if not earlier. They frequently corresponded and Nicolson forwarded lists of their finds to Ray. In the latter part of Lawson's life they enjoyed several botanical excursions together finding, for example, *Astragalus glycyphyllos* near Great Strickland, *Radiola linoides* and *Rhychospora alba* at Cliburn, *Allium carinatum* at Mayburgh, Penrith, *Equisetum hyemale* and *Hippocrepis comosa* at Shap and *Vicia orobus* at Blencarn. As Whittaker observes, the conversations on these excursions, between the fiery Quaker and the archdeacon, charged with suppressing dissidents, are a fertile source of speculation. Nicolson was Bishop of Carlisle from 1703 until very shortly before his sudden death in 1726 and it seems likely that position somewhat curtailed his botanical activities. He was probably a less energetic and observant botanist than Lawson but he had a more methodical approach as is evident from his carefully compiled Catalogue of records mainly from within Cumbria and based largely on his own and Lawson's records. This Catalogue has been painstakingly edited and published by Whittaker (1981). Neither Lawson nor Nicolson appear to have spent much time in the hills; Nicolson's Catalogue, for example, includes only six montane plants, four first discovered by Lawson: *Diphasiastrum alpinum*, *Oxyria digyna*, *Alchemilla alpina* and *Saxifraga hypnoides* and two by Nicolson: *Salix herbacea* and *Sedum villosum*. The latter is particularly interesting as the record is from the pass between Haweswater and Longsleddale, near one of the very few present sites for this plant in the Lake District.

Ray visited Cumbria on four of his tours. In 1660 he found *Sedum rosea* "in excelsis montibus.....Westmorlandiae" (Ray 1670, p.264), *Primula farinosa* at Shap and *Impatiens noli-tangere* at Ambleside. He returned the following year finding *Saxifraga aizoides*, new to Britain and probably new to science, at Shap. In 1668 he discovered *Bartsia alpina*, new to Britain "near Orton by a stream running across the road to Crosby" (*ibid.*, p.86), *Galium boreale*, also at Orton, and *Meum athamanticum* "In Westmorland about two Miles from Sedberg in the Way to Orton" (*ibid.*, p.209). His last journey, in 1671, yielded *Alchemilla alpina* near Watermillock and *Lobelia dortmanna* in Ullswater. Somewhat later he reported the discovery by D. Newton of *Juncus filiformis* at Ambleside (Ray 1688 p.1305).

With Nicolson's death, the light which he, Lawson and Ray had kindled appears to have flickered, if not actually died, during the ensuing century. Surprisingly, although William Hudson was a native of Kendal he seems to have spent little time in his native county, his most important contributions being the discovery in 1762 of *Asplenium septentrionale* above Grasmere and of *Trientalis europaea* in Westmorland and Cumberland, but unfortunately he does not give a locality (Hudson 1778).

The nineteenth century
With the 1830s and 1840s we enter the golden age of exploration of the mountain flora of the British Isles, and the Lake District and Pennines certainly received their share of attention. We know that H.C. Watson visited the Kendal, Keswick and Shap areas in 1835 and made numerous records of mountain species, including the first record of *Thalictrum alpinum*, but there were few of major importance. These records were later extracted from his diaries and included by Baker in his own Flora.

Joseph Woods (1836) travelled through Cumberland during his northern journey of 1835. He was shown *Pyrola media* and *Orthilia secunda* below Walla Crag, Derwent Water, by Charles Wright of Keswick, and on the west coast he noted *Anagallis minima* at Ravenglass and *Asplenium marinum* at Whitehaven. William Borrer seems to have known the Lake District well, especially the Fairfield range and the fells around Ambleside (Borrer 1846). He found *Potamogeton praelongus* in Angle Tarn and Windermere, and spent much time in abortive searches for species such as *Moneses uniflora* and *Saxifraga cespitosa*, reported by Wright and William Hudson respectively. These are two of several montane species which have never been refound or satisfactorily authenticated. Others include Joseph Sidebotham's 1845 record of *Salix reticulata* above Brothers Water, Potter's record of *Lychnis alpina* on Coniston Old man, three reports of *Luzula spicata*, two of *Arabis petraea*, and single records of *Arctostaphylos alpinus*, *Gnaphalium supinum*, *Sagina saginoides*, and *Cerastium cerastoides*. Sidebotham, incidently, was taken by Wordsworth to see *Silene acaulis* above Rydal Mount, presumably on Fairfield, in 1847. He also found *Rumex pseudoalpinus* and *Peucedanum ostruthium* at Mungrisdale, where they have persisted until very recently. Although Borrer held Wright in high regard, he was frequently frustrated by his inability to refind many of Wright's records, even when accompanied by him, and the many records of the Keswick guide Mr Hutton, subsequently published by Hutchinson (1794), were considered most unreliable by N.J. Winch (1825), the Newcastle-upon-Tyne botanist, in the forward to his paper "Remarks on the flora of Cumberland". Hodgson was of much the same opinion. Miss Wright, Wright's daughter, collected numerous specimens from the Keswick area in the early 1840s.

A particularly interesting first record is that of *Ranunculus flammula* β *reptans*, reported by G.S. Gibson (1846) from Pooley Bridge. The identity of this plant, which still occurs around Ullswater, is problematical, collections having been variously ident-ified as the northern species *R. reptans* or its hybrid with *R. flammula*.

Not surprisingly the Backhouses, father and son, made a number of visits to Cumbria. In 1845 Backhouse junior climbed Helvellyn via Striding Edge (Backhouse 1846). Here he found *Hieracium holosericeum*, an *Hieracium* of "dwarf habit with large downy heads". He was much excited at finding *Cerastium alpinum* at the west end of the ridge, a species which he thought was new to England, although it had been found, together with *Saxifraga nivalis*, on Helvellyn at least 20 years earlier by N.J. Winch. In 1847 the Backhouses discovered the same *Hieracium* in Piers Gill, Scafell Pike and on Glaramara, where it grew close to the endemic *H. subgracilentipes*. In 1865, in a letter to J.D. Hooker, the father reported his son finding *Sagina saginoides*, *Carex atrata* and *Poa glauca* on Helvellyn. The *Sagina* has never been confirmed, but his record of the *Poa* predates by 40 years the previously accepted first English record - that of the Rev. A. Ley, also from Helvellyn and listed as *P. balfourii*.

The following year, 1866, Backhouse junior returned to admire *Lychnis alpina* "in fine profusion and luxuriance" on Hobcarton Crag, where it had been first reported by R. Mathews shortly before and where it still persists. Backhouse wrote ecstatically to Hooker of this visit: "The cliffs of Helvellyn and Fairfield were adorned for miles with cushions of *Silene acaulis*, multitudes of which were finely in bloom". How times have changed!, now, thanks to the sheep, the sight of two or three cushions is enough to raise the botanist's adrenalin level. In the same letter Backhouse expressed concern at over-collecting and commented "I had the great satisfaction of finding several fine *Woodsia ilvensis* upon crags where it could not, I believe, be touched without almost certain neck-breaking". Two years later, in 1869, he discovered *Potentilla fruticosa* and *Dryas octopetala* on Helvellyn, and *Ajuga pyramidalis* at what is still its only site in England and Wales. The same year he also discovered *Carex atrata* and *Saxifraga nivalis* on the High Street range, the latter has never been refound, and in 1872 he found *Salix lapponum* on Helvellyn, the only locality south of the Border.

On the Pennines, Backhouse visited High Cup Nick in 1843 and found *Saxifraga nivalis* at two sites in "tolerable abundance" and in "little danger of extermination", in marked contrast to its present extremely precarious state. In 1847 he saw *Saxifraga hirculus* in Knock Ore Gill, first found on the Cumbrian Pennines by J. Bell in 1840. It was growing with an unfamiliar *Epilobium* and a *Myosotis* "differing from the common form of *repens* [*secunda*] to which it approached the nearest". These are almost certainly references to *Epilobium anagallidifolium* and to *Myosotis stolonifera*, which was to remain unrecognised for a further 80 years. While walking over from Teesdale in 1844 he found *Tephroseris integrifolia* above Hillbeck, Brough, and *Equisetum pratense* in nearby Swindale. The latter has never been refound in Cumbria and the former is probably now extinct. In 1866 he told Hooker of his finds of *Viola rupestris* on Warcop Fell and of *Myosotis alpestris* on Little Fell, although he had first found the latter in 1852.

These were indeed men of stamina, charged with formidable reserves of energy and drive. In an account of an excursion in Weardale and Teesdale, the younger Backhouse (1844) made the terse comment "After a walk of twenty-seven miles, we arrived late in the eventing at the High Force Inn". In the Lake District he, and presumably his father, were lost for a while on Scafell Pike. In retracing their footsteps "the stormy wind....again burst upon us with great violence, accompanied by heavy rain; umbrellas became useless, one having been broken to pieces, leaving only the iron rod for a walking stick: soaked to the skin we pursued our dreary way amid the cloud....." (Backhouse 1847).

However, such tribulations are sometimes rewarded. In 1853 Professor J.H. Balfour with 44 pupils journeyed from Edinburgh to Carlisle by the Caledonian railway, and on to Ambleside for two days botanising in the Lake District. They rose at 4.30 a.m. and had botanised nearby Stock Gill before breakfast. After walking over Dunmail Raise they climbed Helvellyn from the west and reached the summit "in a thick mist, cold and wet. They waited patiently for a time, in the hope that some favourable change might tale place, but seeing no prospect of this, they proceeded to Swirrel Edge, with the view of descending to the rocks below. They had not however proceeded far down when the mist suddenly began to clear off, and the beautiful scenery around became visible. As the clearance continued to progress, there was a loud cry, 'to the summit! to the summit!'. Accordingly, they willingly retraced their steps. On reaching the top, they were gratified by one of the most splendid scenes they had ever witnessed: the thick mist gradually rose, and ever and anon displayed some interesting spot, some peaked or rounded summit, some lake or estuary; until at length the whole curtain rose, the sun shone in all its brilliancy, and the entire Lake District was illuminated. It is not easy to describe the ecstasy of the party as scene after scene opened up to their delighted eyes" (Balfour 1854). Shortly afterwards, Balfour was further rewarded with the first Lake District record of *Poa alpina*.

Of the pteridologists, F. Clowes, a Bowness doctor, Isaac and James Hudhart, and J. Coward were all extremely active during the 1840s and 1850s. Between them they were responsible for the first records of *Woodsia ilvensis* (in 1846), *Dryopteris aemula* (from near Windermere), *Polystichum lonchitis* (on Helvellyn and Fairfield), *Trichomanes speciosum* and the hybrid *Dryopteris carthusiana* x *D. filix-mas* (now extinct in the wild). *Dryopteris expansa* was collected by Wright at Keswick in 1859 but not then recognised as a species. The Rev. R. Rolleston of Ambleside discovered *Lycopodium annotinum* at its

only English locality at the head of Langdale in 1847 and Bolton King found *Cystopteris montana* on Helvellyn in 1880. Practically all the above species are now either extremely rare in Cumbria, or, in the case of the *Cystopteris*, extinct, no doubt a consequence of over-collecting.

Another profitable and lengthy father and son botanical partnership was that of H.J. and G.E. Martindale of Staveley, Kendal. They botanised assiduously from the mid-19th century into the 1930s and almost entirely in the lowlands. Between them they contributed the first British record of the North American loosestrife *Lysimachia terrestris* from Windermere in 1886 and found *Hottonia palustris* on Brigsteer Moss in 1840 and *Trientalis europaea* east of Kendal in 1909. Also active around the mid-19th century and in much the same area was Isaac Hindson of Kirkby Lonsdale whose unpublished manuscript of plants in south Westmorland and adjacent Lancashire and Yorkshire is in the Natural History Museum, London.

J.G. Baker was active over almost the same period as the Martindales. Despite the pressures of his work at Kew and his active botanising in Yorkshire and the north-east of England, he somehow managed to write his *Flora of the English Lake District*, which included a considerable number of his own records. This Flora is particularly important in its inclusion of the whole of Furness, many of the records being contributed and mostly published earlier by Miss E. Hodgson (1874) of Ulverston. This is the only compilation of Furness records in one volume. It was closely followed by a series of 19 short papers by S.L. Petty (1894-1898, 1902) giving an historical account of all Furness records. His own appear to be extremely few or non-existent.

William Hodgson spent most of his childhood at Raughton Head and his teaching career at Watermillock, with the exception of a brief interlude at Aspatria and Frizington and his retirement to Workington. His *Flora of Cumberland* was published towards the end of his life, in 1898. This was followed by two short papers of additional records (1899,1901), the last being written only three months before his death. In addition to innumerable records of his own, his Flora is important for the many records of his two contemporaries and friends, the Rev. R. Wood of Westward and W. Dickinson of Workington, and also of the Rev. H. Friend. Other contributors were J. Leitch, a Skinburness doctor, and Miss E.J. Glaister of nearby Silloth. Both took a particular interest in the ballast aliens generated at Silloth docks by the shipment of grain to Carr's mills and they collected there assiduously from the 1870s until the end of the century. Despite Hodgson's many contacts, he was apparently ignorant of the activities of W.H. Youdale, of Cockermouth, whose herbarium of about 300 sheets (Hofmann *et al.* 1990) was compiled in west Cumberland, from St Bees to Silloth, chiefly in the early 1890s. Hodgson and his friends botanised very largely in the lowlands. Although he seems to have climbed Helvellyn on a number of occasions his references are generally to having seen particular species on the Westmorland side.

In 1890 the Rev. H.E. Fox, a notable hawkweed enthusiast, visited Dollywaggon Pike and collected the endemic *Hieracium subgracilentipes* and the problematic *H. cumbriense* and *H. angustatum*. The Revd Augustin Ley visited the Lake District on several occasions. In 1905, accompanied by the Rev. W.R. Linton, he made several important hawkweed finds, including the first Lake District record of *H. leyi* (Ley & Linton 1906). The only other major hawkweed collector was H.W. Pugsley, who collected intermittently from 1920.

Charles Bailey visited Cumbria on three occasions during the 1890s; a solitary bush of *Rubus incurvatiformis* between Haweswater and Bampton could well be that which he recorded. Unfortunately his record of *R. sprengelii* from Ambleside has never been substantiated; it is the only one from Cumbria and near its northern limit in Britain. Linton also recorded numerous *Rubus* species (Ley & Linton 1906), including the first reference to a local and still unnamed endemic taxon belonging to the section *Rubus* and found in Langdale. There was further *Rubus* activity the following year when the Rev. W. Moyle Rogers visited the Lake District (Rogers 1907). His most interesting contribution was the discovery of *R. lacustris*, endemic to the Lake District. Its white, floriferous panicles are aptly described by him as "very handsome and starlike in sunshine" (Plate 70).

The twentieth century
The century opened with a lyrical account by F.A. Lees (1900) of an early summer stay at Kendal during which he found *Cephalanthera longifolia* at Cunswick. Hodgson (1901) published a further list of "waifs", derived from what was to prove his last visit to Silloth and Skinburness. These included *Ambrosia trifida*, *Silybum marianum* and *Medicago scutellata*. He also listed records, contributed by W. Thomson, which included *Salvia verbenaca*, *Silene dichotoma* and *Gypsophila pilosa* from the gravel bed of the River Eden opposite Grinsdale, Carlisle. This was probably also the site of Thomson's unlocalised record of *Anchusa ochroleuca*. His records of *Ulex gallii* from Kingsmoor and Todhills Moss probably refer to *U. minor*. These gravel beds proved also to be a favourite haunt of T.S. Johnstone who collected numerous aliens from there as well as contributing many records for the Carlisle area in general between 1900 and 1910. Another important recorder in Cumb-

erland at this time was The Hon. Marjorie Cross, of Thwaites, Millom, although she did not join the Botanical Exchange Club until 1936. In the 1920s there was H. Britten of Great Salkeld, who found *Herniaria glabra* at Lazonby, R.H. Williamson of Seascale and a friend of G.C. Druce, and the Rev. W.W. Mason of Melmerby, all frequent contributors of records to the annual *Reports of the Botanical Exchange Club*. Mason's annotated copy of Hodgson's Flora is now in the possession of L. Robinson, one of the Survey Recorders. In the north-east, George Bolam spent the last twenty years of his life before his death in 1934 at Alston. He was primarily an ornithologist but he knew the flora of Alston Moor well. He discovered several new sites for *Gentiana verna* and found *Dryas octopetala* on the limestone above Melmerby, although it has never been seen there since. In the Lake District, C.I. and N.Y. Sandwith discovered the Snowdonia and Lake District endemic *Euphrasia rivularis* on Honister Pass and G.W. Temperley, of Newcastle-upon-Tyne, found *Potentilla fruticosa* on Pillar on one of the many forays he and his wife made into the Lake District between 1890 and 1946. He used his knowledge of the Cumberland flora in collaborating with J.W.H. Harrison to write the account of "The flora of the three northern counties" [Durham, Northumberland and Cumberland] in *The three northern counties of England* (Headlam 1939). In the 1930s Miss J. Parkin of Wigton and her friend C.B. Sharp contributed many records although collecting few specimens. E. Blezard, although, like Bolam, primarily an ornithologist, collected mainly in the 1940s together with his wife, formerly Miss D. Stewart, whose mother, Mrs H. Stewart, collected during the 1920s and contributed the first Cumbrian record of *Cicerbita macrophylla*.

C.W. Muirhead was the first professional botanist since the time of Baker to accumulate Cumberland records systematically. She began collecting in the 1930s and continued almost up to her death in 1985. She seems to have visited most of the well-known sites in Cumbria but, living at Aglionby, near Carlisle, her records and collections are predominantly from Cumberland. Her collecting was interrupted by wartime service in the W.R.N.S. during which she extracted Cumberland records at the Natural History Museum, London. After the war she worked for a time at Tullie House Museum, Carlisle, and then in the herbarium of the Royal Botanic Garden, Edinburgh, while still returning frequently to Cumberland. She and J. Parkin led a ten day Botanical Society excursion to Cumberland in 1949 to collect records for a proposed county Flora. Notable finds included a single surviving specimen of *Epipactis palustris* at Newton Reigny bog

It had been hoped that some members of the party would be able to climb Pillar Mountain to see *Potentilla fruticosa* and *Saussurea alpina* in flower, both very rare plants in the county. That day, however, we were afflicted with a bus of uncertain age, which finally succumbed to its infirmities and could proceed no further, near Mockerkin Tarn.

BSBI Year Book (1951) p.47, C.W. Muirhead

and *Rumex scutatus* in Cockermouth. Her collections included a considerable number of specimens of critical groups, notably *Salix* and *Euphrasia*.

Another active mid-century botanist in the north of the county was Canon G.A.K. Hervey. His first post was as chaplain at Bryanston School, Dorset, during which he made a number of visits to the Lake District. In 1942 he moved north as Vicar of Gilsland, whence he made numerous botanical excursions across the Bewcastle Fells and moors and adjacent Northumberland as well as to the Lake District. In 1947 he followed in Nicolson's footsteps and moved to Great Salkeld where he spent the rest of his life. He was the founder of the Lake District Naturalists' Trust in 1962, a regular contibutor of records to *The Field Naturalist*, and part editor and author of the *Natural history of the Lake District* (1970), published after his death in 1967. He kept a natural history diary from the time of his arrival at Gilsland until 1960. His observations are chiefly ornithological and botanical and the plant records he transferred to two exercise books where they are arranged systematically. Unfortunately his entries are all too often tantalisingly brief. Notable records include *Pseudorchis albida* south of Bewcastle in 1943, *Orchis ustulata*, *Leucojum aestivum* and *Stellaria palustris* at Great Salkeld from 1947 to 1956, and *Anacamptis pyramidalis* at Melkinthorpe in 1951.

In the south of the county another prolific contributor of new and important records, particularly from about 1910 and into the 1920s, was W.H. Pearsall senior, who was then a schoolmaster at Dalton-in-Furness. Most of his records were from Furness, which was then botanically much under-recorded, and virtually all were from the lowlands. In 1925 he made the first and only reliable Cumbrian record of *Wahlenbergia hederacea* from the Winster valley, but by far his most significant discoveries, made in the company of his son and within a few days of each other in July 1914, were of *Hydrilla verticillata* and *Najas flexilis* in Esthwaite Water. These were respectively new to the British Isles and new to Britain. Sadly, the *Hydrilla* has long since disappeared, probably eliminated by pollution, and the *Najas* is now very rare. Pearsall was a close friend of Druce and he ultimately became secretary of the Botanical Exchange Club following his retirement to Kent in 1925. D. Lumb, also living in Dalton at that time, knew Pearsall and benefited from his experience. Their many records of aliens from Barrow-in-Furness and Askam feature prominently in the pages of the *Reports of the Botanical Exchange Club*. Lumb appears to have been quite catholic in his collecting, his 1912 submission including seedlings of *Acer* and *Rubus fruticosus* (*BEC Rpt* (1913) p.243)!

The most important pre-war figure in Westmorland botany was undoubtedly Albert Wilson (1862-1949) whose collecting spanned 60 years. In 1907 he published, in collaboration with J.A. Wheldon, *The Flora of West Lancashire* and in 1938 his own

Flora of Westmorland. Although he attended school in Kendal, he was never again resident in the county and his Flora relies heavily on numerous contributors such as J.F. Pickard and W.A. Sledge, both of Leeds, T.J. Foggitt of Thirsk, who found the first Lake District site for *Potentilla crantzii*, and the Sprotts of Appleby. Wilson's most important montane records include the first collections in Britain of *Myosotis stolonifera*, from the Howgill Fells and the Lake District, and the first Pennine record of *Phleum alpinum*.

In the 1950's, G. Wilson and the Barrow-in-Furness Field Club were actively recording throughout Furness and Wilson led the Botanical Society of the British Isles 1951 field meeting in Furness.

Without doubt the most impressive individual achievements during the post-war period are those of Derek Ratcliffe. He grew up in Carlisle, and by bicycle and on foot ranged widely throughout Cumberland, and more especially over the North Pennine and Lake District hills. His unequalled knowledge of the last was embodied in his stimulating paper on "The mountain flora of Lakeland" (Ratcliffe 1960). It was he who discovered *Empetrum nigrum* subsp. *hermaphroditum* in the Lake District and who found there a second site for *Dryas octopetala*.

Ratcliffe was able to confirm many old and notable Lake District records of species such as *Woodsia ilvensis*, *Polystichum lonchitis*, *Ajuga pyramidalis* and *Dryopteris aemula*. He also added numerous sites for *Draba incana*, *Orthilia secunda* and *Potentilla crantzii*. He discovered *Hymenophyllum tunbrigense* in Eskdale and in the Pennines he made the first Cumberland record of *Saxifraga hirculus* and, with A. Eddy, the first definite English record of *Alopecurus borealis*. He was assisted in his searches by his mentor E. Blezard, and in particular by J. Birkett, a quarryman of Little Langdale and a superb cragsman with a keen interest in plants. Not surprisingly a number of Birkett's finds continue to prove elusive. Birkett was also responsible for refinding *Vaccinium uliginosum*, *Carex atrata* and *Hieracium holosericeum* on the High Street range. In the same area in the 1960s H.J.B. Birks found *Saxifraga oppositifolia* and *Saussurea alpina*, neither of which have been seen since.

Finally, and overlapping with the first half of the Survey, there was Ralph Stokoe. An accountant, born at Wigton in 1921, he soon became a keen ornithologist and in 1962 published *The birds of the Lake Counties*. His interests subsequently widened. Whether his conversion to botany was gradual or dramatic is not known but by 1970 he was enthusiastically searching in Cumbria and elsewhere for orchids to photograph. His interests soon extended to pteridophytes: he discovered *Isoetes echinospora*, new to northern England, at three Lake District sites, and also several new sites for *Pilularia globulifera*. However, every spare moment of the last five or six years of his life, before his sudden early death in 1981, was devoted to recording meticulously the aquatic macrophytes in the lakes and tarns of

Cumbria, thereby amassing a body of data which was prepared for posthumous publication by E.M. Lind (Stokoe 1983). In addition to the *Isoetes*, his most notable finds include the second and third English records of *Nuphar lutea* x *N. pumila*, from Armboth Fell, and numerous Lake District sites for *Elatine hexandra*, previously known with certainty in the county only from Thurstonfield Lough on the Solway Plain.

The Sedbergh area, vice-county 65
The Sedbergh area, with Garsdale, Dentdale and the Rawthey valley, lies in vice-county 65, North-west Yorkshire, but being part of the old West Riding it was included by F.A. Lees in *The Flora of West Yorkshire* which he published in 1888. In this it comprises the northernmost part of his 'Lune' subdivision. Lees was at that time President of the Botanical Section of the Yorkshire Naturalists' Union (YNU) and the Sedbergh district was one of the more popular venues for Y.N.U. excursions. He included in his Flora specialities such as *Trientalis europaea* in a plantation in Garsdale, *Alchemilla alpina* from Dent Crag and Cautley Spout, its only two Yorkshire sites, and *Orthilia secunda* from Cautley Crag and near Black Force, with the exception of the *Trientalis* all finds attributed to the Revd G. Pindar in 1845, although Backhouse (1844) had already noted the *Alchemilla* at Cautley the previous year. *Trientalis* subsequently became extinct at its Garsdale station but was discovered in 1929 by R. Butterfield on Baugh Fell. In 1898 J. Handley, an active Y.N.U. member, published a list of plants from

the area. This included *Pseudorchis albida*, from by the Lune, and several adventives mostly associated with the railway. Unfortunately, few localities or recorders are cited and although he clearly includes plants from the Westmorland side of the Lune it is often not possible to tell which county a record is from. He continued with a short note in 1900 which included obvious errors such as *Thalictrum alpinum* on Cautley Crag and *Hypericum elodes* on the moors south-east of Sedbergh. A later note, in 1903, mentions *Orobanche rapum-genistae* and *Saxifraga hirculus* but the latter was based on a misidentification.

Wilson knew the district well, collecting several Hieracia and *Epimedium alpinum* from the banks of the Lune; in 1918 he found *Hammarbya paludosa* near Sedbergh and *Salix herbacea* on Cautley Crag and Dent Crag, and he provided briefing notes for two Y.N.U. excursions. Those for 1932 include a surprising reference to *Anagallis minima*, unfortunately without a locality. He was followed by C.A. Cheetham, W.A. Sledge, who found *Circaea alpina* by the upper River Rawthey, and G.A. Shaw. From 1937 to 1940, Cheetham published a succession of papers in the *Naturalist* detailing new records for the whole of Yorkshire which he and Lees, who died in 1920, had accumulated. These include a number of Sedbergh records, notably of *Rubi* recorded by W.M. Rogers in 1906, *Listera cordata* from below Combe Scar, several Hieracia records and a new site, discovered by Cheetham, for *Pseudorchis albida*. There have been few noteworthy finds since the war.

Plate 37. Cautley Crag and Spout in winter

Cumbria's changing flora

With its wide climatic variation and extraordinary range of geology and topography, Cumbria has an extremely rich flora for such a northern county. An analysis of the total native and alien flora of the county and the two major vice-counties is given in Table 1.

Hieracium, Taraxacum, Rubus and *Euphrasia* are excluded, Hodgson lists 1145 plants which are currently given specific rank. In addition to native species, this figure includes ones which are doubtfully native in Cumberland, as well as garden escapes and

	Species					Hybrids				
	Pre-Survey only		Survey		Survey	Pre-Survey only		Survey		Survey
	Native	Alien	Native	Alien	Total	Native	Alien	Native	Alien	Total
Westmld	51	47	868	288	1156	22	2	55	29	84
Furness	47	53	824	227	1051	5	1	51	23	74
v.c.69	52	68	935	331	1266	23	1	72	33	105
Cumbld	61	172	893	316	1209	16	5	79	37	116
Cumbria	32	172	945	428	1373	22	4	104	48	152

Table 1. Number of species,and hybrids recorded before and during the Survey
(excluding *Rubus*, *Hieracium*, and *Taraxacum*)
Aliens as a % of total Survey species: W 23%, F 20%, v.c.69 24%, C 24%, Cumbria 27%

		Native	Alien	Total
Wilson (1938)	Westmorland	876	83	959
	v.c.69			1010
Hodgson (1898)	Cumberland	946	250	1196

Table 2. Number of species recorded in Wilson's and Hodgson's Floras

The total for non-critical species, 1373, recorded since 1970, compares well with the 1525 recorded from Hampshire (Brewis *et al.* 1996) on the south coast, a county claimed to be floristically the richest in the British Isles and, like Cumbria, comprising two vice-counties. Cumbria is particularly well-endowed with members of the Cyperaceae, 71 species having been recorded during the Survey from Westmorland and Furness (v.c.69), although one became extinct during this period. The only other vice-counties with a similar total are Dorset and South Hampshire, with 65 and 66 extant species respectively. Of the 1373 species, 25% (233) of the native species and 66% (283) of the aliens occur in 20 or fewer (*c.*1%) of the 1780 tetrads. Listed below are those species which occur in more than 1700 tetrads. *Anthoxanthum odoratum* is absent from only one, on Rockcliffe Marsh.

Anthoxanthum odoratum	1779	Ranunculus acris	1732
Juncus effusus	1759	Plantago lanceolata	1722
Cerastium fontanum	1757	Holcus lanatus	1711
Rumex acetosa	1747	Urtica dioica	1709
Trifolium repens	1741	Potentilla erecta	1707
Ranunculus repens	1739	Cirsium palustre	1706

Table 1 can usefully be compared with the figures given in the early Floras (Table 2). Hodgson's Flora provides an opportunity to compare the major changes in the flora of Cumberland over the last century. If dubious records and microspecies of

other aliens. He lists 207 species in these last categories of which 86 are presumed garden-escapes or planted and 121 are best regarded as casuals; only 39 of these casuals have been recorded during the recent Survey. Most of the species which have disappeared are Hodgson's "waifs" of the ballast tips and dock areas of the then thriving west Cumberland ports, chiefly Workington, Silloth and Maryport, which were such a happy hunting ground for aliens from the 1870s well into the early decades of this century. This last figure could be substantially increased by the several species seen by J. Leitch and E.J. Glaister, but not listed by Hodgson, mainly in the Silloth area in the latter part of last century, and by T.G. Johnstone in the early years of the present century around Carlisle, particularly on the gravel beds of the River Eden. The picture is similar in south Cumbria where W.H. Pearsall and D. Lumb recorded numerous aliens from the then thriving steel-works and ports of Barrow and Askam-in-Furness.

Hodgson's total of 1196 species is remarkably similar to the 1209 seen during the Survey in Cumberland nearly a century later. 316 of the Survey species are aliens. These include established garden escapes and throw-outs. This massive increase in aliens (53%) is hardly surprising given the ever-increasing popularity of gardening and the wide variety of species now grown, as well as the deplorable habit of dumping surplus material, especially of excessively aggressive species, on roadside verges and in ditches where, not surprisingly, many persist. The similarity of the totals is surprising and indicates that the increase in number of aliens of garden origin and of 15 native species more than offsets the loss of the many 19th century casuals and 56 native species.

Were it possible to make a similar comparison over the century for Westmorland and Furness, it is likely that the result would be very similar.

Losses

No one can browse through Hodgson's Flora without being struck by the vast number of weeds from the ballast and industrial sites of the west coast ports. Many of these are natives of southern Europe but some are from even further afield, such as species of *Verbesina*, *Ximenesia* and *Bowlesia* from America. Most failed to survive the decline of these ports and have not been seen since. This is true, for example, of three species of *Trifolium* and four each of *Centaurea* and *Bromus*.

There has also been a substantial loss of native species in all three major areas of Cumbria. Appendix 2 lists 99 native species and 31 hybrids which were recorded for Westmorland, Furness or Cumberland in the earlier Floras but which were not seen in these areas during the Survey.

Major changes in agriculture, particularly since the last war, have had a major impact. Of the 31 native species which have been recorded in the past from Cumbria, but not seen during the Survey, 16 disappeared between 1930 and 1970. There has been a dramatic decline in traditional hay-meadows, which have been replaced by 'improved' grassland and leys either for hay, or especially in recent years, for silage. Such changes have resulted in the loss of *Orchis ustulata* and the increasing rarity in the dales of *Trollius europaeus* and *Cirsium heterophyllum*. *Primula farinosa* is now very rare in the limestone district between the Lake District and Wigton where Hodgson said it was abundant over several miles, and it has disappeared from Furness. The early Floras give 23 localitities in v.cs 69 and 70 for *Pseudorchis albida* but there are now only three; fortunately it still persists in a few sites in the former Yorkshire sector of Cumbria. During the same period, with the increasing use of herbicides, there was a corresponding decline in arable weeds resulting in the disappearance from Cumberland and Furness of *Agrostemma githago* and of *Scandix pecten-veneris*, *Valerianella dentata* and *V. rimosa* from Cumbria.

Moorland drainage and afforestation have also taken their toll. *Trientalis europaea*, once known from several sites in its stronghold on the Bewcastle Fells, is now very rare and *Lycopodiella inundata,* recorded from several areas in the Lake District, is now reduced to just three sites, one of which is outside the Lake District. Drainage and the drying out of wetland sites is no doubt also responsible for the disappearance of *Ceratophyllum submersum* and *Epipactis palustris* from Cumberland, of *Gentiana pneumonanthe* from its Westmorland and Cumberland sites and of *Carum*

verticillatum, *Wahlenbergia hederacea* and *Hottonia palustris* from Cumbria. Also lost from its two sites in Cumbria, though for no known reason, is *Astragalus danicus*.

The eutrophication of lakes and rivers, by fertiliser run-off and sewage effluent, has had a dramatic effect. It may well have been the cause of the extinction of *Hydrilla verticillata* from Esthwaite Water, its only British locality, and the apparent decrease there of *Najas flexilis* in its only English site. Most striking has been the increasing scarcity of *Potamogeton* species in the lakes.

There have probably been very few changes in status of the rarer montane species. Provided their cliff habitats are out of reach of the sheep, the major threats come from rock-falls and the inevitable fluctuations in numbers of small, isolated populations. *Pyrola media* has become extinct and rock falls appear to have been responsible for the loss in the Lake District of a site for *Polystichum lonchitis*, both sites of *Gymnocarpium robertianum*, and possibly the only sites there for *Thlaspi caerulescens* and *Phleum alpinum*. *Woodsia ilvensis* is now apparently reduced to a single population. Neither *Cerastium alpinum* nor *Lycopodium annotinum* have been seen in Cumberland this century and during the two decades prior to the Survey *Dryopteris submontana* vanished from its only Lake District site at Honister, probably as a result of a rock fall, and *Alchemilla monticola* from its only Cumbrian site on roadside verges on the border with Co. Durham. *Potentilla fruticosa*, lost from the Westmorland side of the River Tees and from Helvellyn, and *Dryas octopetala*, lost from the Cross Fell area, have, however, each been discovered at a new site in the Lake District. Grazing has probably eliminated the ragwort *Tephroseris integrifolia* from the fells above Brough, where it has not been seen since the 1920s. Fortunately collecting seems no longer to be a serious problem. It probably resulted in the disappearance last century of *Cystopteris montana* from Helvellyn, its only English locality, also of *Cyripedium calceolus* and the apparent restriction of sporophytes of *Trichomanes speciosum* to a single site.

Gains

Listed in Appendix 3 are 37 native species and 59 native hybrids which were recorded for the first time from Westmorland, Furness or Cumbria during the Survey. The gains, native and alien, fall into four general categories.

1. *Weeds and casuals*

These are the plants of roadsides, landscaped and reseeded ground, urban areas and waste sites. They include *Senecio squalidus*, first recorded in 1935 and now well-established around Carlisle, the west coast towns, and Barrow-in-Furness, and the North American willowherb *Epilobium ciliatum*, very widely distributed although unknown before c.1969. The extremely rare southern broomrape *Orobanche purpurea* was discovered at Maryport *c*.1983 and there are Survey records of *Crepis tectorum* and *C. setosa*, *Bromus inermis*, *Amaranthus retroflexus*, *A. albus*, *Kochia scoparia*, *Portulaca oleracea* and the decreasing southern weed *Legousia hybrida*, which was found in 1990 on a motorway bridge. The last is the only species restricted in Cumbria to v.c.60. *Lavatera arborea*, *Levisticum officinale*, *Artemisia stellerana* and *Lupinus arboreus* have all appeared on the west coast, but the former two failed to persist. Although known from only three pre-Survey sites, the North American grass *Hordeum jubatum* became very noticeable in the early years of the Survey along the saline verges of motorways and trunk roads although it appears to have declined subsequently.

2. *Aquatics*

Cumbria has shared in the spectacular national spread of the North American waterweeds *Elodea canadensis*, first seen in the county in 1874, and recently of *E. nuttallii* (1976) which seems to be replacing it, at least in several of the lakes. The figwort *Scrophularia umbrosa* (1924) is now well distributed along the whole length of the River Eden and even more at home throughout the county by rivers, lakes and canals is *Impatiens glandulifera* (1921). Gardeners and garden centres are no doubt responsible for the vegetative spread not only of the *Elodea* species but also the recent appearance of *Lagarosiphon major*, *Azolla filiculoides*, *Nymphoides peltata* and *Crassula helmsii*, perhaps also the appearance in inland fish-ponds of *Myriophyllum spicatum*. The *Crassula*, a native of New Zealand and Australia, is now an aggressive weed in southern England and it will need to be watched carefully in Cumbria following its discovery in Derwent Water and Coniston Water. The North American *Sarracenia purpurea* is well established on mossland by the Solway Firth and at a number of wetland sites between Windermere and Coniston, where it is known to have been planted. Another North American species is the elegant loosestrife *Lysimachia terrestris* (1884, Plate 38), now a characteristic feature of tarn and lake-side marsh communities around Windermere.

3. *Garden escapes*

The dumping of excess garden material and the movement of soil containing rootstocks and rhizomes is

Plate 38. *Lysimachia terrestris*, Windermere

Plate 39. *Mimulus guttatus* x *M. luteus*, River Eden, Mallerstang

increasingly noticeable throughout the lowlands. Especially conspicuous are *Heracleum mantegazzianum*, the dense thickets of *Fallopia japonica* (1930s) and, to a lesser extent, the related *F. sachalinensis* and *Persicaria wallichii*. The colonies of *Petasites albus*, *P. fragrans* and the eye-catching *P. japonicus* have probably originated in this way, although some roadside colonies are known to have been planted. Also conspicuous are *Lysimachia punctata*, the deep purple-flowered *Geranium* x *magnificum*, *G. endressii* and the pale blue *Cicerbita macrophylla* (1915). Other newcomers are *Tolmiea menziesii* and *Tellima grandiflora*. The American *Mimulus guttatus* and its hybrids (Plate 39) are eye-catching plants of lake-, river- and becksides and the showy, purple Pyrenean *Cardamine raphanifolia* is frequent in wet places in the Grasmere - Ambleside area. *Aconitum napellus* and *Polemonium caeruleum* are well distributed although not common. Bird-sown species of *Cotoneaster* are particularly frequent in the Arnside area and another species which has needed little in the way of help from man is the now almost ubiquitous New Zealand willowherb *Epilobium brunnescens* (c.1923). The attractive, early-flowering *Veronica filiformis* (1941) is now common in the lowlands in lawns and on roadside verges and this despite its inability to set seed. There have been a number of recent records of the mallow-like *Sidalcea*, several of the popular garden poppy *Papaver orientale* and a single remarkable occurrence on a limestone bank east of Sedbergh of the Pyrenean rock-garden plant *Ramonda myconi*, which, with *Alchemilla minima* and *Viola cornuta*, are the only species restricted in Cumbria to v.c.65. There is no mention in the early Floras, of the North American *Solidago gigantea* and *S. canadensis* which are spreading fast, nor, remarkably, of *Rhododendron ponticum*, now a pernicious weed in many of our woodlands. Other shrubs which appear to be spreading, although less spectacularly, are *Lonicera nitida*, *Syringa vulgaris* and *Buddleja davidii*.

As a result of 20th century commercial forestry, *Picea sitchensis* and *Pinus contorta* are now familar trees in the county and larch (*Larix decidua*) is so common that it comes as a surprise to find that Hodgson fails to mention it, although it has been present since the late 18th century. Also increasingly conspicuous are the roadside plantings of poplars, especially the balsam poplar *Populus trichocarpa*, and of *Cornus sericea*, *Viburnum lantana* and, occasionally, *Rosa rubiginosa*.

4. *Native species*

Excluding microspecies, 47 native species were not discovered in the county until this century. Pre-Survey records include *Epipactis leptochila* var. *dunensis* and *E. phyllanthes*, *Pyrola rotundifolia* and *Corallorhiza trfida*, all discovered on the dune-slacks of the south-west coast during the inter-war years. This coast also produced *Isolepis cernua* and *Hypochaeris glabra*.

Particularly interesting were the finds of *Carex elongata* in the Lake District and south Furness, and of *Viola rupestris* and *Carex ericetorum* in the Arnside area and of the latter on the Asby - Orton limestone. The same area yielded *Hutchinsia alpina* and, nearby, sites for another Teesdale rarity *Polygala amarella*. *Alopecurus borealis* was discovered in the Pennines, its British headquarters. The outlying Cumbrian sites of the otherwise southern *Ulex minor* near Carlisle were not discovered until the 1940s and more were added during the Survey. *Spartina anglica* arrived in Morecambe Bay in the 1940s and spread north during the Survey to Eskmeals but not, as yet, to the Cumbrian shore of the Solway. Only 15 native species have been added during the Survey but six are members of critical groups, for example *Alchemilla glaucescens*, otherwise restricted to the Craven area of Yorkshire, *Poa angustifolia*, but possibly introduced, *Salicornia fragilis* and *Stellaria pallida*. Other additions include *Isoetes echinospora* at three sites, the very rare *Eleocharis austriaca* in the upper River Irthing, *Ruppia cirrhosa* near Askam-in-Furness, one of its only two extant sites on the west coast of Britain, *Carex divulsa* near Barrow-in-Furness and *C. maritima* at Humphrey Head. The last two records are, respectively, the most northerly and southerly ones on the west coast of Britain. Unfortunately *C. maritima* failed to survive coastal defence operations. Probably the most important find of the Survey was of *Crepis praemorsa*, new to western Europe, in a hay-field in north Westmorland. Particularly encouraging were Survey finds of *Epipactis atrorubens* at Hodbarrow and *Carex ornithopoda* by the River Eden, both new to Cumberland.

Not surprisingly the recent systematic Survey has resulted in an appreciable increase in the number of known sites of species previously regarded as rare or very rare. These include *Saxifraga hirculus* in the Pennines, *Potentilla crantzii* both in the Lake District and the Pennines, the predominantly eastern *Eriophorum latifolium*, *Carex limosa* and *C. magellanica*. In some cases, as for example with *Hammarbya paludosa*, the number of 'new' sites seems to have kept pace with the loss of old ones. Prior to the Survey *Elatine hexandra* was known with certainty only from Thurstonfield Lough yet R. Stokoe's painstaking survey of Cumbrian lakes and tarns during the late 1970s produced records from eight of the lakes and 12 tarns. There has been a puzzling increase in the records of another waterweed, *Zannichellia palustris*. This is now widespread although Hodgson cites only five localities and Wilson two; none are from the Eden valley where it is now quite frequent.

As one would expect, increasing taxonomic knowledge has meant that the number of recorded microspecies and segregates is continually increasing. The bulk of the 40 Cumbrian species of brambles, 130 dandelions and 60 hawkweeds were described during this century. The same is true of three uncommon Lake District eyebrights: *Euphrasia frigida*, *E. rivularis* and

E. ostenfeldii (Plates 40 & 41).

In conclusion, 30 native species, excluding microspecies, which had previously been recorded from the county were not seen during the Survey but 15 have been added. However, six of these additions would probably not have received specific recognition in the early Floras. While this net decrease may be depressing, it is of far less consequence than the progressive loss of habitat and populations of a large proportion of the native flora. This aspect will be discussed further in the next section.

On Thursday morning Miss Hudson of Workington called. She said 'O! I love flowers! I sow flowers in the parks several miles from home, and my mother and I visit them, and watch them how they grow.' This may show that botanists may be often deceived when they find rare flowers growing far from houses. This was a very ordinary young woman, such as in any town of the north of England one may find a score.

The Grasmere Journal, June 19th, 1802,
Dorothy Wordsworth

Plate 40. *Euphrasia rivularis*, Haweswater

Plate 41. *Euphrasia ostenfeldii*, Eel Crags

Conservation
G. Halliday & D.A. Ratcliffe

- and there is an endless variety of brilliant flowers in the fields and meadows, which, if the agriculture of the country were more carefully attended to, would disappear.

Description of the Scenery of the Lakes, William Wordsworth

Cumbria is generally less intensively farmed than most of the predominantly arable shires in the south of the country and it might be thought that the threats to its wildlife were correspondingly less acute. However, the threats are many, varied and serious. This section will consider the major categories of threat but will end optimistically on a more positive note.

Forestry

Since the afforestation of the central Lake District was effectively suspended following the 1936 agreement between the Forestry Commission and the Friends of the Lake District, most plantings have taken place on the rolling moorlands on the eastern fringe of the Lake District and particularly in the extreme north and north-east of the county, at Kershope and on Spadeadam Waste, the Bewcastle Fells and Denton Fell. Following fencing, there is an increase in vigour and flowering of the moorland plants, and in bird numbers and species, noticeably the short-eared owl, but this is only transient. The tree canopy soon closes in and the only remaining vegetation of any consequence is in the rides. The planting has resulted in considerable habitat loss not only of drier moorland but, more importantly, of mires, with the loss of a number of sites for *Eriophorum latifolium, Andromeda polifolia, Carex lasiocarpa, C. pauciflora* and *C. magellanica*. The last survives in some of the rides, as also does *Trientalis europaea*, but it is very doubtful whether they are capable of surviving the traumatic effects of clear-felling in the near future. There has also been a substantial loss of heathland on the Lazonby Fells, with the apparent loss of *Radiola linoides* at its last inland site; basic grassland with *Primula farinosa* and *Trollius europaeus* has been lost on Greystoke Park, and 'limestone heath' nearby at Johnby. In the south, afforestation of Foulshaw Moss resulted in the disappearance of *Gentiana pneumonanthe*.

Just as damaging has been the progressive partial or complete conversion of lowland broadleaved woodland to conifers. Examples on the southern limestones are Middlebarrow Wood and Whitbarrow, and in the north-east Coombs Wood, Armathwaite; Park Wood, Brocklebank; and Walton Wood, Brampton. Particularly unfortunate is the recent 'coniferisation' of oakwoods on the west side of Windermere. Woodland felling in the Eden valley has resulted in the loss of a number of sites for *Gagea lutea*. On the other hand, species dependent on Scots Pine are at risk from clear-felling if there is no subsequent replanting. For at least 60 years coniferous plantations in north Cumberland featuring *Goodyera*

repens have been disappearing; the best Survey site, at Welton, was in fact clear-felled in 1995.

Farming

Any general programme of nature conservation can only be really effective if it has the co-operation and support of the farming community, yet it is an unfortunate fact that virtually everything a farmer seeks to do in the name of increased efficiency is likely to run counter to nature conservation interests.

Enforced wartime ploughing of heath and coarse grassland was not necessariuly detrimental in the long-term. An area near Hoff was allowed to recover after the war and supported *Galium boreale* and *Platanthera chlorantha* until about 1970 when it was ploughed up and reseeded. The superb hay-meadows in Ravenstonedale (Plate 42) were ploughed in the war and subsequently reseeded and managed traditionally, with no inorganic fertiliser other than occasional liming. The hay-meadow species reappeared and the meadows are now a Site of Special Scientific Interest (SSSI).

Plate 42. *Geranium sylvaticum* hay-meadow, *Cirsium heterophyllum* in foreground, Ravenstonedale

Plate 43. *Cirsium heterophyllum*

Plate 44. *Euphrasia rostkoviana* subsp. *rostkoviana*

Unfortunately, this is exceptional. In the lowlands the most serious consequence of post-war agriculture has been the loss of such species-rich hay-meadows for which the limestone country in particular was justly famous (Plate 35). This has resulted not only from the increased use of fertilisers, particularly nitrates, but also the switch, slow at first, from hay-making to silage. These changes have often been accompanied by improved drainage and re-seeding and the end product is a sward dominated by a few agricultural grass strains, except where excessive slurry has resulted in the dominance of dandelions and docks. The Lake District now has only 40 hectares of species-rich hay-meadows and the situation is nearly as serious along the Pennine foot and in the Sedbergh valleys. Grassland 'improvement' has resulted in the extinction of *Orchis ustulata* and the near extinction of *Pseudorchis albida*, *Linum perenne* and *Vicia orobus*. *Primula farinosa* has suffered a spectacular decline along with other conspicuous and attractive species such as *Cirsium heterophyllum* (Plate 43), *Trollius europaeus*, *Gymnadenia conopsea* and species of *Euphrasia* (Plate 44), together with, but to a lesser extent, similar indicators of unimproved swards such as *Sanguisorba officinalis*, *Geranium sylvaticum* and *Rhinanthus minor*. These once so familiar flowers are increasingly being restricted to an uncertain future by becksides and on roadside verges. Agricultural infilling of a quarry near Brigham resulted in the loss of one of the very few Cumberland colonies of *Ophrys apifera*. Occasional liming probably does little harm on the better soils but on the more acid soils there is a rapid impoverishment with the loss of *Succisa pratensis*, *Stachys officinalis*, *Genista tinctoria* and, locally, *Meum athamanticum*, which survives only on ground too rough or steep for tractors, and on field edges and roadside verges.

With far fewer cattle trampling the lower fell-sides, a much reduced farm labour force and dimin-

ishing heather cover, bracken has spread extensively, checked only locally by spraying. Kelly & Perry (1990) give the increase as 66% between 1940 and 1970.

Headage payments for sheep and a guaranteed price have ensured rapidly rising sheep numbers. They are thought, for example, to have inceased by about 50% in Langdale since the war and by 60% in Cumbria since 1977. The resulting overstocking has further inhibited tree regeneration, more importantly in the unfenced fellside woods which traditionally provide shelter, especially during the winter, increasingly restricting the more susceptible trees and shrubs to the more sheltered, as well as the less nutrient-depleted, gills. These woods, which are both a key feature in the Lake District landscape and a vital agricultural asset, will eventually disappear unless, at the very least, rotational fencing to allow regeneration can be negotiated or alternative shelter belts planted. The disappearance of the earlier, more continuous woodland leaving scattered hawthorn and the occasional holly and ash, is evident in most Lake District valleys, good examples being Kentmere, Dovedale and Haweswater. It has, however, been suggested that these trees became established at a time when grazing was less intense. Small lowland woods may also be at risk through progressive small-scale felling or grubbing out. Honey Bee Wood, near Kendal, for example, vanished during the period of the Survey.

However, the most obvious change resulting from increased sheep stocking has been the conspicuous loss, particularly outside the area of the Skiddaw Slates, of heather moorland to grassland. For example, at two sites near Ravenstonedale deep heather in the 1920s had gone by the 1940s. The grassland is initially characterised by *Festuca ovina* and *Agrostis capillaris* on the slightly more fertile and less peaty soils but increasingly with heavy grazing *Nardus stricta* becomes dominant on all except the wettest soils. Even the casual visitor speeding up the M6 motorway must

Plate 45. Grazed grassland and ungrazed *Calluna* moor, Faulds Brow, Caldbeck

have noticed the dull straw-colour of this grass blanketing the Howgill Fells (Plate 34). The loss of heather moorland has been dramatic on the wet Armboth Fells, between Borrowdale and Thirlmere, and along the road north of Orton Scar there is a striking contrast between fenced, lightly grazed heather moorland on one side and degraded limestone grass-land with sheep-resistant islands of *Carduus nutans* on the other. There is an equally dramatic example on Faulds Brow, above Caldbeck (Plate 45). Lowland *Calluna* heath has also largely disappeared, usually being converted to grassland, as at Baronwood Park, and there has been an associated decline in *Genista anglica*. Heavy grazing of grasslands has caused substantial loss of species such as *Viola lutea* and *Antennaria dioica,* and probably also rarities like *Gentiana verna* and *Myosotis alpestris* which are some-

times found uprooted. Heavy grazing of upland flushes can severely reduce the populations of such plants as *Primula farinosa* and *Saxifraga aizoides*: the latter was quite frequent in flushes above Crosby Ravensworth 25 years ago but it is now difficult to find a single plant. Shallow-rooted plants, such as the *Primula* and *Saxifraga hirculus,* are very vulnerable to uprooting. The effects of sheep-grazing are not restricted to the uplands. There are now virtually no ungrazed areas of salt-marsh around the coast. Broadleaved herbs, such as *Limonium* species, *Atriplex portulacoides* and *Aster tripolium*, have conspicuously declined and this is now being hastened by the spread of *Spartina anglica,* especially around Morecambe Bay.

The most noticeable effects of pollution resulting from the excessive application of inorganic fertilisers on agricultural land have already been referred to, but

Plate 46. Exclosure on summit of Great Dun Fell, 845 m, ungrazed since 1985

Plate 47. Sunbiggin Tarn, 1977

there is also the insidious progressive pollution of the lowland lakes, tarns and mires by agricultural run-off and sewage effluent. This is no doubt largely responsible for the impoverished submerged native macrophyte flora of the lakes and its replacement by often dense underwater forests of the two alien *Elodea* species. Particularly severely affected are Windermere, Blelham Tarn, Elterwater and Grasmere. Pollution by waterfowl can also be a serious problem, for example by Canada geese in the central lakes, especially Elterwater and Grasmere, and the vast numbers of black-headed gulls nesting around Sunbiggin Tarn (Plate 47) on the lime-stone uplands of north Westmorland. The tarn and the surrounding flushes and pools of varying acidity were superb before the area degenerated through the uncurbed breeding of the gulls, the pools becoming foetid and algal. The submerged macrophytes of

Tindale Tarn, in the extreme north-east, have declined, probably for a similar reason, and Bigland Bog has deteriorated through sewage effluent.

The loss of wetlands as a result of afforestation has already been mentioned, but there has also been extensive loss of upland wetland and rough grazing through drainage and conversion either to pasture or occasionally to poor quality arable land. This has happened, for example, over extensive areas of the uplands between Kendal and the Lune valley. Such conversion not only results in a major loss of flush habitats but inevitably affects the associated fauna, most noticeably the breeding birds such as lapwings. Progressive drainage of the low estuarine mosslands of the Duddon, Morecambe Bay and Solway has resulted in a lowering of the level of the pasture land and a progressive drying out of the residual islands of peat,

Plate 48. Limestone roadside verge with *Knautia arvensis*, north Westmorland

such as Meathop Moss, with the consequent accelerated invasion of the open moss by trees. Fire is an additional hazard causing deterioration of virtually all the raised bog surfaces which have escaped peat cutting. An almost ubiquitous feature throughout the lowlands has been the loss of ponds and marshy areas in fields. There must be few, if any, Survey recorders who have not had recent experience of this. In west Cumberland, Snellings Mire and Sellafield Tarn have been drained and what is left of Braystones Tarn is now in the middle of a caravan site. Infilling of a pond at Cleator Moor exterminated the only west Cumberland population of *Typha angustifolia*. In the north, Caldew Mires, at Dalston, and Wragmire Moss, near Cotehill, have been lost and Monkhill Lough was drained when the dam was destroyed, while in the Eden valley Sandford Mire has largely vanished and *Epipactis palustris* disappeared from Newton Reigny Moss when a drain was cut through it in the 1940s.

Development

Probably the single most damaging type of development, and that largely exempt from planning control, relates to road schemes. The field of *Meum* illustrated in Wilson's Flora became an M6 construction site. Although major road construction such as motorways can have disastrous local effects, it is the widespread minor 'improvements' combined with the excessive and ill-timed cutting of verges and hedges which are so destructive. The county's heritage of rich roadside verges, especially in the east (Plate 48), demands a range of treatments acceptable to both the Highways Authority and conservationists but it has to be said that the best intentions and agreements can be thwarted by the ill-informed and over-zealous action of the person actually carrying out maintenance. Much greater are the problems of protecting small populations of roadside rarities which may require very local management at variance with the agreed treatment for that general area. The farming community must share some of the blame for verge deterioration. Important wide and species-rich verges in the Lamonby - Johnby area have been 'improved' by being covered with soil, seeded and mown for hay. The habit of blanket spraying verges and roadside banks around farms is also deplorable. Occasionally an issue arises which pits amenity bodies against conservationists, creating not a few individual crises of conscience. One such was the A66 road improvement through the northern Lake District. A longer alternative, favoured by the amenity bodies but rejected, would have meant upgrading the old wide-verged drove road through Sebergham, along the north-east fringe of the National Park.

The damaging commercial extraction of peat continues by the Leven estuary in the south, on Wedholme Flow, south of the Solway, and north-east of Carlisle - first Solway Moss and then Bolton Fell, which has been all but stripped of its peat cover. The maintenance of lowland drains, dykes and streams by canalisation, dredging and the cutting back of marginal vegetation to facilitate run-off from low-lying areas means that marginal and submerged vegetation inevitably decline. Long stretches are 'cleaned out' at a time allowing little scope for recolonisation.

The few remaining miles of the Cumbrian section of the Lancaster - Kendal canal (Plate 49) has a particularly rich aquatic flora. This is currently threatened by an ambitious proposal to reopen it in its entirety to powered boats, a development which cannot conceivably do other than seriously affect its nature conservation value. Reed bed communities by Windermere have been lost or damaged by marina development and powered craft. Following the recent lengthy Public Inquiry into a proposed speed limit on the lake, the Inspector's recommendation that a limit be introduced was perversely overturned by the Secretary of State. Marginal vegetation is also particularly susceptible to tourist pressure, which is naturally greatest in popular access areas. The vegetation rapidly disappears and with it much of the shingle, producing the all too common scene of lakeside trees standing on exposed 'stilt' roots. It now requires considerable imagination to visualise *Baldellia ranunculoides* being plentiful in Windermere, as stated in Wilson's Flora.

Threats posed by reservoir construction in the valleys and estuaries over the last forty years have fortunately been either rejected or shelved, although,

Plate 49. The Lancaster - Kendal canal near Stainton

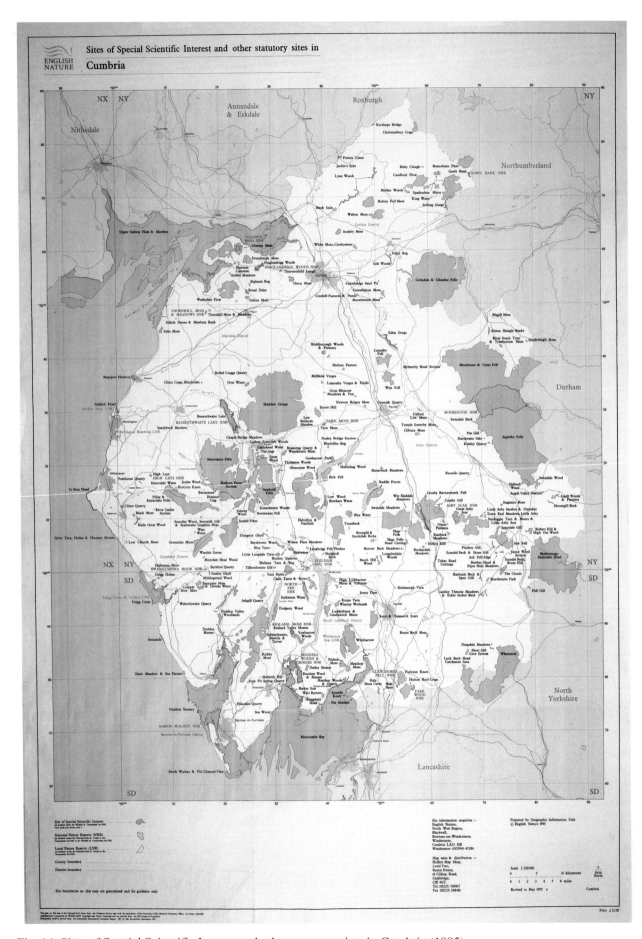

Fig. 14. Sites of Special Scientific Interest and other statutory sites in Cumbria (1995)

more recently, this has been largely as a consequence of the prolonged economic recession reducing the demand for water. However, plans for a barrage on the Duddon estuary and particularly the options of raising the Haweswater dam or flooding the Borrow Beck valley are still distinct possibilities, especially if the recent pattern of dry summers and low annual rainfall continues. The current appalling and environmentally reckless proposals for a Solway - Tyne canal and associated Solway, Duddon and Morecambe Bay barrages and a west Cumberland motorway would have absolutely devastating consequences for nature conservation.

Derelict land often supports a surprisingly varied and interesting colonist flora and its development presents something of a dilemma for the conservationist. The development of station sites at Keswick and Kendal led to the disappearance of *Vulpia myuros*, *Pimpinella major* and *Sisymbrium altissimum*. Fortunately the proposal for speculative housing on the *Orobanche purpurea* site at Maryport was rejected but a series of developments at Salthouse Pool, Barrow, resulted in the disappearance of populations of *Ophrys apifera*, *Cerastium arvense* and *Potentilla argentea*, the last two at their only site in Furness.

Land-based tourist threats include the excessive 'gardening' of cliffs by rock climbers, the present vogue for gill scrambling, the increasing use of 'off-road' vehicles of all descriptions and the heavy tourist pressure around popular sites, such as above Aira Force. It might be thought that a disused quarry in west Cumberland would be the obvious place to practise clay pigeon shooting; unfortunately it is the only west Cumberland site for *Eriophorum latifolium* (Plate 51). It is difficult now to believe that the wooded limestone pavements in the Arnside area were for long considered by planners to be ideal sites for caravans. A worrying development is the current construction of a holiday

complex in Whinfell Forest in the Eden valley, in close proximity to populations of *Goodyera repens* and *Vaccinium uliginosum*.

As mentioned in the section on Climate, the impacts of both acid rain and global warming are unpredictable and potentially extremely worrying.

Future prospects

The conservation picture, however, is not all gloomy. As regards agriculture, the Countryside Commission Farm Stewardship Schemes, now run by the Ministry of Agriculture, Fisheries and Food (MAFF), and the establishment of an Environmentally Sensitive Area (ESA) in the Sedbergh valleys and, more recently, one covering the Lake District, have considerable potential for more sympathetic farming. They can encourage, for example, the fencing off of riverside corridors, the preservation of small-scale features such as marshy areas and hay-meadows, especially by a reduction in the use of fertilisers, and also, hopefully, the long-sought-for reduction in sheep stocking rates. The Ravenstonedale meadows, referred to above, was the first site in north-west England to be included in the Stewardship Scheme. Both English Nature and the National Parks are able to influence management of SSSIs and other areas through their ability to effect management agreements under the provisions of the 1981 Wildlife and Countryside Act. Furthermore, the National Parks are notified of all applications for agricultural grant-aid and are therefore able to negotiate with the farmer concerning any which might have serious wildlife consequences. It is extremely unfortunate that this desirable, minimal safeguard of prior notification does not apply to the wider countryside outside the Parks, SSSIs and Areas of Outstanding Natural Beauty (AONBs). Even where a management agreement on an SSSI is in place there is no guarantee of renewal. Faced with such a situation, the Cumbria Wildlife Trust mounted a successful campaign to purchase Burns Beck Moss, Killington.

English Nature has more than 230 SSSIs and 20 National Nature Reserves in the county (Fig.14), probably the greatest density in the country. The NNRs include Bassenthwaite Lake, Esthwaite North Fen, North Walney, Roudsea Wood and Mosses by the Leven estuary, and Moor House in the high Pennines, the last being the fourth largest in the country and continuing to the south as the Appleby Fells SSSI. Since its inception in 1962, the Wildlife Trust has continued to acquire and extend its reserves. It now owns or leases nearly 40 (Fig.15), including 17 SSSIs. A rational long-term policy of acquisition is almost impossible to implement given the Burns Beck example just cited and the fortuitous nature of land left to the Trust. Further, the Lake District National Park also has substantial holdings, the broadleaved woods in the Borrowdale and Rusland valleys and the eastern coves of Helvellyn being particularly important, and within the Park a third of the land belongs to the National

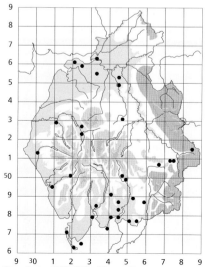

Fig. 15. Cumbria Wildlife Trust Reserves

Trust. Fortunately both bodies, and also the Yorkshire Dales National Park now employ an ecologist. The National Parks are actively involved in many aspects of conservation, for example in relation to farming grants and small woodland schemes. The Lake District Park established the Whitbarrow - Witherslack Local Nature Reserve and has recently taken over responsibility from the County Council for managing that at Drigg. It frequently works in close association with the Wildlife Trust. Actual purchase of land by the National Parks is, however, most likely to be directed towards areas of landscape importance, especially if there is actual or potential public access. For financial reasons such initiatives are practically impossible for other local authorities. With recurrent financial problems and faced with the high costs of reserve management, the Wildlife Trust has found a welcome ally in the Woodland Trust, whose reserves now include Sea Wood, Ulverston. Since the 1985 Wildlife and Countryside Amendment Act, English Nature has been able to delegate management of SSSIs and NNRs to bodies such as the Wildlife Trusts. As a result the Cumbria Trust now manages North Walney and, thanks to the by-laws obtained by English Nature, the horrendous problems of vandalism and motor-cycle scrambling on the dunes have now been largely overcome. Also, in recent years the even more important dune system at Sandscale (Plate 18), on the opposite side of the Walney Channel, has been acquired by the National Trust and is assiduously wardened and botanised by one of the Flora Survey recorders. In addition to participating on the Management Committees of Local Nature Reserves, the Wildlife Trust is also represented on the Ministry of Defence Conservation Committees at Spadeadam in the north-east and the Warcop range in the Eden valley. In both areas important nature conservation initiatives have been taken and conservation ranks second in priority only to military require-

ments. Both English Nature and the Wildlife Trust have a continual input into the local authority planning process and participate frequently in Public Inquiries.

The threat to limestone pavements (Plates 6 & 50) from the removal of surface waterworn limestone is still not generally appreciated, even by gardeners. Cumbria, the Craven area of Yorkshire and the Silverdale area of north Lancashire possess most of Britain's pavements, 97% of which have been damaged to varying degrees. Following victory in Planning Inquiries in south Westmorland and a sustained parliamentary campaign, a clause giving Local Authorities the duty to make Limestone Pavement Orders on the recommendation of the Countryside Commission or English Nature, was incorporated in the 1981 Wildlife and Countryside Act. To its credit, the Lake District National Park was the first authority in the country to make such an order. Practically all the pavements are now at last covered but not before one at Potrigg, Oddendale, had been destroyed, and with it a population of *Potentilla crantzii*. In 1996 Cumbria County Council succeeded in buying out an extant planning permission for the removal of pavement on Orton Scar, a long-standing grievance, but, unfortunately, permissions for deep quarrying, with consequent removal of surface stone, as at Sandside, cannot be rescinded without prohibitive compensation. The demand for stone from such sources and also the illegal removal from protected sites will not diminish without a major change in attitude on the part of the gardening community. The same is true regarding the demand for horticultural peat. Hopeful signs are an agreement not to extend peat removal on to the remaining third of Wedholme Flow and the recent successful outcome of the Public Inquiry on a proposal to extract peat from Black Snib, near Longtown.

A most encouraging feature over the last twenty years has been the marked shift in attitude towards

Plate 50. Limestone pavement above Great Asby

If it form the one landscape that we, the inconstant ones,
Are consistently homesick for, this is chiefly
Because it dissolves in water. Mark these rounded slopes
With their surface fragrance of thyme and, beneath,
A secret system of caves and conduits; hear the springs
That spurt out everywhere with a chuckle,
Each filling a private pool for its fish and carving
Its own little ravine whose cliffs entertain
The butterfly and the lizard;

          From *In praise of limestone*, W.H. Auden

conservation on the part of MAFF, from adamant refusal to contemplate expenditure of agricultural grant-aid on anything smacking of conservation, as was only too evident in the debates on the Wildlife and Countryside Bill, to the present active support for an increasing range of conservation activities within ESAs and responsibility for Stewardship Schemes. This change is also evident in the farming community. In the early 1970s the Chairman of the Lake District National Park Special Planning Board, a farmer, declared himself appalled at the suggestion that the Board might actually pay a farmer to continue to manage his species-rich hay-meadow in the traditional manner. Another attitude which needs to be confronted is that which sees any relaxation of farming pressure as retrograde, as damaging to a tidy, ordered landscape; that invasion of land by rushes or shrubs (usually emotively referred to as scrub) is automatically something to be opposed. However, present willingness, active or passive, can only be capitalised on if the information on the nature conservation importance of farmland is readily available. This is an impossible task for the very limited resources of a Wildlife Trust and the recent, belated appointment of a full-time Farming and Wildlife Advisory Officer is very welcome.

As regards forestry, there is now close liaison between English Nature and Forest Enterprise over the management of 'ancient and semi-natural woodland', such as Whitbarrow, and the problem of acid rain is increasingly leading to the removal of conifers along becksides and the resultant development of corridors of tall-herb communities. Forest Enterprise has the welcome policy of retaining the native broadleaved trees along gillsides although it has yet to be converted to the desirability of including more native trees in new plantings and in 'amenity screens' in place of the alien sycamore, beech and species of whitebeam. Also encouraging are the many small-scale replanting and 'rejuvenation' schemes within the National Park and the greater financial incentive for the latter. It is a great pity that such schemes are so fragmentary. How much stronger would the case for re-afforestation be if it had only been possible to carry out the large-scale replanting and regeneration of native species in a whole valley, such as Longsleddale, with its present patchy woods and scattered hawthorns.

Much of the foregoing has been written out of concern for our rich heritage of habitats, which is reflected in the fact that 11 of the possible U.K. Special Areas of Conservation under the European Community Habitats Directive lie wholly or partly in Cumbria. These are:

Asby limestone pavements
Borrowdale woodlands
Butterburn Flow
Hellvellyn and Fairfield
Moor House - Upper Teesdale
Morecambe Bay limestone pavements

North Pennine dales meadows
Ravenglass estuary
Roudsea Wood and Mosses
Solway Mosses
Wasdale Screes

Being well endowed with habitats, Cumbria is, not surprisingly, exceptionally fortunate in its flora. As mentioned in the preceding section, its 1373 species compare favourably with Hampshire. Cumbria has 24 extant species (Appendix 5) on the current list of British Red Data Book species, which are mostly those occurring in Britain in fewer than 15 10 km squares. Of these, 19 occur in Westmorland and Furness (v.c.69) and 10 in Cumberland (v.c.70). The following are protected by law:

Gentiana verna   Trichomanes speciosum
Lychnis alpina   Veronica spicata
Saxifraga hirculus   Woodsia ilvensis

A further four species are extinct: *Hydrilla verticillata*, *Alchemilla monticola*, *Cystopteris montana* and *Cypripedium calceolus*. Of the nationally Scarce Species listed in Table 1 of *Scarce Plants* (Stewart *et al.* 1994), 82 are native in v.c.69 (not 78 as stated there in Table 3). This figure is only exceeded by those rival vice-counties Dorset (92) and South Hampshire (89). A further 13 species are now extinct in v.c.69. Included in the latter are the losses during the Survey of *Carex maritima* and *Mertensia maritima*. This high proportion of extant species (86%) is only exceeded by five Scottish vice-counties (*Scarce Plants* Table 3). These are indeed facts to be proud of but, as stated at the beginning of this section, the protection of these habitats, their flora and associated animal life, requires constant vigilance and effort; this is not just a matter of local concern but, as the above figures demonstrate, we have a heavy national responsibility.

Plate 51. *Eriophorum latifolium* and clay pigeon shooting site in West Cumberland

The Flora Survey

For the purpose of the recording Survey it was decided to adopt the 2 x 2 km squares of the National Grid of which there are 1781 in Cumbria. Although with a county as large as Cumbria and one with so few resident botanists it seemed initially that the only practical course would be to adopt 5 x 5 km units, as has been the case in several montane Scottish counties. However, it was decided to be ambitious and adopt tetrad recording as in the adjacent vice-counties of West Lancashire and Durham. Recording commenced in earnest in 1974 although some earlier records were included, notably herbarium records from 1969 to 1973. This was the period during which R. Young was actively building up a Lake District herbarium at Brathay Field Studies Centre, Ambleside. However, the vast bulk of records were effectively accumulated over the period 1974-1996.

An informal group of recorders was quickly established and this grew to provide a nucleus of about 30 active members throughout the Survey. Record cards from the original Botanical Society of the British Isles (BSBI) *Atlas* scheme of the 1950s were used for field recording and sent in at the end of each season for transfer of the records to a set of master-cards. These cards had abbreviated species' names and corresponding code numbers. There was an annual Progress Report, an indoor autumn meeting to discuss progress and a series of summer field meetings to under-recorded tetrads. Gradually the challenging blank areas of unvisited tetrads diminished until eventually most had been visited several times and all at least twice. Specimens were collected of critical taxa and important finds, as far as conservation considerations would allow, and the bulk of these are incorporated in the herbarium at Lancaster University. To assist Recorders in assessing the importance of their finds, an interim check-list of *Flowering plants and ferns of Cumbria* was produced (Halliday 1978) and this has been followed by annual lists of amendments.

A.J.C. Malloch wrote a number of computer programs for the filing and accessing of the data, programs which were ultimately made more generally applicable and incorporated in his Vespan computer package. During 1981 and 1982 the data were laboriously entered into the University mainframe computer using punched cards but subsequent data entry was from the keyboard. When data was entered, a useful check was applied to ensure that the species' code was that of one known to occur in Cumbria. Once data had been entered by tetrad they could then be accessed as print-outs of species per tetrad, as a cumulative list of species for specified tetrads, as a list of tetrads for individual species or the latter could be combined with a digitised map of the county to produce species distribution maps. Co-incident distribution maps of species (Fig. 5) and habitat indicator species (Figs. 12 & 13) can also be generated. Lists of species in specified frequency classes can also be produced (p. 70). Of particular interest and use are lists of the rarer species. Some 515 species occur in fewer than 20 tetrads (*c.*1%). After eliminating the many casuals and garden-escapes, the resulting list of 233 native species is invaluable for any attempt to draw up a conservation strategy for species likely to be at risk in the county.

It was soon realised that the original timetable of 15 years over-optimistic and it was extended to 20. This extension had a number of very fortunate consequences. As recording progressed and individual expertise improved, emphasis tended to shift to certain critical groups and hybrids, for example *Erophila, Euphrasia, Polypodium, Spiraea* and *Salix*, to an extent that would otherwise have been quite impossible. The protracted gestation also meant that problems encountered during the preparation of the systematic accounts could sometimes be resolved during the ensuing season. However, a number of problems still remain, such as the relative frequency of *Brassica napus* and *B. rapa*, *Agrimonia eupatoria* and *A. procera*, and *Rorippa nasturtium-aquaticum, R. microphyllum* and their hybrid. The extended period saw the publication of Floras of Co. Durham and Northumberland. The appearance of *New Flora of the British Isles* (Stace 1991) meant that the Flora could adopt the same arrangement and nomenclature. In connection with the nationwide BSBI/Institute of Terrestrial Ecology Monitoring Scheme of selected tetrads in 1987 and 1988 and the preparation of *Scarce Plants in Britain* (Stewart *et al.* 1994), the

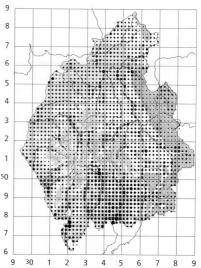

Fig. 16. Number of records per tetrad
-100, -200, -300, -400, -500, >500

Biological Records Centre distributed print-outs for each vice-county of all its records. This proved an immensely valuable source of records, mostly true but some certainly false!, which we would other-wise have missed.

The recording scheme has been characterised by close liaison at every stage. Recorders received maps at intervals of 'winter', 'spring' and common species and were exhorted to fill the gaps. Family accounts have been circulated for comment several times and quest-ions of format and the content of the final Flora have been discussed at the autumn end of season meetings.

The final total of records is 438,000 an average of 246 per tetrad (Fig. 16). This may seem low to some but a large proportion of Cumbrian tetrads lie in the uplands and a Pennine tetrad of blanket bog could well have as few as 70 species. At the other extreme the Brathay tetrad, with its variety of habitats, had 526, Arnside, thanks partly to its aliens, had 563 and

Askam, with its eagle-eyed recorder, 583. Before producing the final maps, recorders patiently and with the minimum of coercion completed a second complete check of tetrad print-outs against the master-cards.

In deference to the views of the Yorkshire Naturalists' Union, the Sedbergh area, in vice-county 65 (Fig. 17), was originally omitted from the Survey. This is the area which the late W.A. Sledge once referred to in a letter as "that part of Yorkshire annexed by Cumbria in 1974". The omission of such a large seg-ment, about 50 tetrads, in a Flora purporting to cover the whole of the county irked. Eventually the Y.N.U. agreed to drop their objections when it became increasingly evident that they were not themselves going to be recording there. A sustained onslaught on the area was carried out during the middle period of the Survey and in the event it has been covered no less thoroughly than any other comparable area in the county.

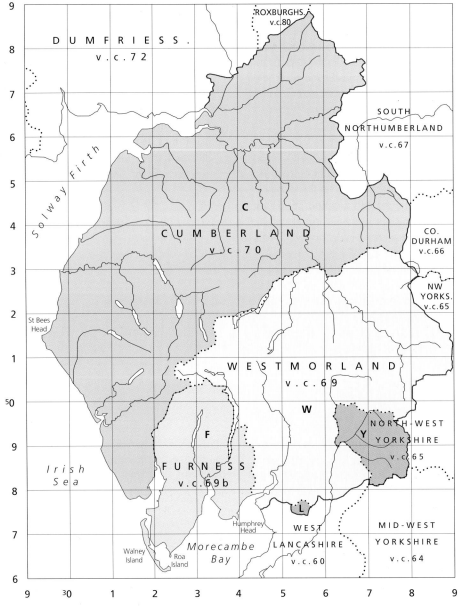

Fig. 17. The biological vice-counties of Cumbria and their abbreviations - L, W, F, C, Y
N.B. The bed of Windermere is in W. Rooted aquatics off the Furness shore are therefore in W, until the water level drops!

Scope and format of the Flora

The primary purpose of the Flora is to present as thorough an account as possible of the county's vascular flora as it existed during the 20 year period of the Survey. The county includes the whole of vice-counties Westmorland with Furness (v.c.69) and Cumberland (v.c.70), together with part of North-west Yorkshire (v.c.65) and a single parish of West Lancashire (v.c.60) (Fig. 17). Much as one would have liked to extend the scope to include bryophytes, the detailed scale of the recording and the scarcity of competent recorders rendered this impossible. This is definitely not an historical Flora: constraints of space and time made this impracticable. The search for old records is consequently very far from being exhaustive. The existing Floras are taken as starting points and references to Baker, Hodgson and Wilson throughout this Flora refer to their respective Floras unless indicated otherwise. In general, only in the case of species which have noticeably increased, decreased or become extinct are early records cited. On the other hand an attempt has been made to provide first records for those species which are new to Westmorland, Furness or Cumberland since the relevant Flora.

Nomenclature and sequence follow as far as possible that of *New Flora of the British Isles* (Stace 1991). The only synonyms are names used in *Excursion Flora of the British Isles* (Clapham, Tutin & Warburg 1981). The accounts of species not seen during the Survey appear in small print and those for which the early records are regarded as erroneous or very doubtful are placed in square brackets. For each species the last line gives the total number of tetrads and the vice-comital distribution. Where the records for a vice-county all pre-date the-Survey this is shown by the use of brackets. The accounts attempt to give the highest altitude to the nearest 5 m for all species occurring above 100 m. Altitudinal limits particularly interested Wilson who not only gives this information in his Flora but presented data for the whole of the British Isles in a posthumous paper (Wilson 1956). Not surprisingly most wide-ranging species reach their altitudinal limit in the Alston area. Distribution maps are generally given for all taxa, except for *Taraxacum*, occurring in more than seven tetrads. Major rarities and others which might be at risk are not mapped and only 10 km grid references are given in the text unless the localities are well known.

The Flora includes a substantial number of garden-escapes and planted species. Some will regard this as too generous but experience suggests that a liberal interpretation is the more valuable for future botanists. In general, the late D.A.Webb's simple criterion of 'well-established 100 m or more from a house or garden' has been the guiding principle. Inevitably there will have been appreciable differences of interpretation between individual Recorders. Non-native trees and shrubs in planted woodland and in urban and village situations have not been recorded. Neglected estates and other areas which have 'gone wild' constitute a grey area in this respect.

Which English name to use is a source of endless contention and there are many who object to the spate of highly artificial, relatively recently coined names with no general currency. The solution adopted here is to use the long-established names in the old Floras together with one or two local names where they exist.

The species' accounts are mostly fairly brief and, it is hoped, simple and self-explanatory. Abbreviations are minimal so avoiding the all too familiar clutter of irritating abbreviations characteristic of what the late Chris Haworth aptly described as "knitting pattern Floras".

Lack of time, man-power and expertise has meant that it has not been possible to include more than a very limited limited amount of ecological data in the species accounts or to complement these with a detailed study of the county's vegetation. On the other hand the Flora does aim to assist the reader with taxonomic comment where appropriate. This is particularly the case with such critical genera as *Rubus*, *Hieracium*, *Euphrasia*, *Sorbus* and *Cotoneaster* for which keys are provided. In the case of the first two, these should enable beginners to learn the local flora far more easily than by recourse to the standard works with their depressingly large number of 'irrelevant' species.

Coltsfoot laying claim to every new-dug clump of clay;
Pearlwort scraping up a living between bricks from a ha'porth of mortar;
Dandelions you daren't pick or you know what will happen;
Sour docks that make a first-rate poultice for nettle-stings;
And flat-foot plantain in the back street, gathering more dust than the dustmen.

Even the names are a folk-song:
Fat hen, rat's tail, cat's ear, old men's baccy and Stinking Billy
Ring a prettier chime for me than honeysuckle or jasmine,
And Sweet Cicely smells cleaner than Sweet William though she's barred from the garden.

From *Weeds*, Norman Nicholson

Sources

The main sources of records are listed below.

Principal herbaria and collectors

Botany School, Cambridge University (**CGE**) (*Hieracium*)

Carlisle City Council Museum, Carlisle (**CLE**) (E.J. Glaister, J. Leitch, T.S. Johnstone, C.W. Muirhead, D.A. Ratcliffe)

Institute of Environmental and Biological Sciences, Lancaster University (**LANC**) (P. Burton, R.W.M. Corner, A.A. Dudman, R.E. Groom, G. Halliday , M. Porter, C. Wild). This also houses the herbaria and diaries of R. Stokoe, on loan from The Cumbria Wildlife Trust, and the Lake District herbarium of the Brathay Hall Trust (**AMDE**) (R. Young)

Kendal Museum, Kendal (**KDL**) (J.A. & G.E. Martindale)

Natural History Museum, London (**BM**)

Plymouth University (**PLYP**) (C.W. Muirhead Memorial Herbarium)

Yorkshire Museum (**YRK**) (A. Wilson)

Literature

Baker, J. G. (1885). *A Flora of the English Lake District.*

Handley, J. (1898). *Catalogue of plants growing in the Sedbergh district.*

Hodgson, W. (1898). *Flora of Cumberland.*

Petty, S. L. (1894-1903). *Naturalist.*

Stokoe, R. (1983) *Aquatic macrophytes in the tarns and lakes of Cumbria.*

Wilson, A. (1938). *The flora of Westmorland.*

Proceedings of the Botanical Society of the British Isles.

Reports of the Botanical Exchange Club of the British Isles.

Watsonia

B.S.B.I. vice-county card indices for Westmorland, Furness and Cumberland

A.A. Dudman's *Taraxacum* data-base

D. McCosh's *Hieracium* data-base

C.W. Muirhead's card index of Cumberland records

A. Newton's *Rubus* data-base

Plate 52. Lunchtime for Survey recorders on the long seat, Hesket Newmarket

Abbreviations

Grid references
s and n

Grid references are usually given as four-figure references to tetrads, the co-ordinates refering to the bottom left-hand corner. A few tetrads in the extreme south and north of the county share the same reference and in these instances 's' or 'n' after the reference will distinguish them.

Recorders (abbreviations alphabetical by surname)

JA	J. Adams	KAG	K.A. Gunning	AJS	A.J. Silverside
JMA	J.M. Atkins	GH	G. Halliday	CS	C. Smith
MB	M. Baecker	CCH	C.C. Haworth	MS	M. Smith
NB	N. Botham	LH	L. Henderson	CFS & JS	C.F. & J. Steeden
PB	P. Burton	EEM	E.E. Marper	ES	E. Sterne
RWMC	R.W.M. Corner	MMM	M.M. Milne	RS	R. Stokoe
MC	M. Coulson	AN	A. Newton	PT	P. Tolfree
AAD	A.A. Dudman	JP	J. Parker	CEW	C.E. Wild
ESE	E.S. Edees	MP	M. S. Porter	JW	J. Williamson
JCF	J.C. Frankland	KR	K. Raistrick	CW	C. Willink
MMG	M.M. Gill	DAR	D.A. Ratcliffe	AW	A.E. Wilson
MG	M. Gregory	EHR	E.H. Rhone	TW & KW	T. & K. Wilson
REG	R.E. Groom	LR	L. Robinson	FLW	F.L. Woodman

Herbaria

AMDE	Brathay Hall, Ambleside	**KDL**	Kendal Museum
BM	Natural History Museum, London	**LANC**	Lancaster University
CGE	Cambridge University	**LDS**	Leeds University
CLE	City of Carlisle Museum	**LIV**	Merseyside Museums, Liverpool
DBY	City Museum and Art Gallery, Derby	**LTR**	Leicester University
E	Royal Botanic Garden, Edinburgh	**MANCH**	Manchester Museum
FBA	Institute of Freshwater Ecology, Windermere	**MNE**	Maidstone Museum and Art Gallery
		NMW	National Museum of Wales, Cardiff
HAMU	Hancock Museum, Newcastle upon-Tyne	**OXF**	Oxford University
HIWNT	Hampshire & Isle of Wight Wildlife Trust, Romsey	**PLYP**	Plymouth University
		UTLH	Loughborough University
K	Royal Botanic Gardens, Kew	**YRK**	Yorkshire Museum, York

Vice-counties

C	Cumberland, v.c.70
F	Furness, part of v.c.69 (formerly 69b)
L	part of West Lancashire, v.c.60
W	Westmorland, part of v.c.69
Y	part of North-west Yorkshire, v.c.65
W/Y	either W or Y
W?	either W or v.c.60 outside Cumbria

Status

*	species considered to be alien in Cumbria
[ ]	record erroneous or probably so
Small type	dubious or pre-Survey records only
R	Red Data Book Species
S	Nationally Scarce Species
†	Species extinct post 1974

Publications

Atlas	*Atlas of the British Flora* (Perring & Walters 1962)
Check-list	*Flowering plants and ferns of Cumbria* (Halliday 1978)
Critical Atlas	*Critical Supplement to the Atlas of the British Flora* (Perring & Sell 1968)
Fern Atlas	*Atlas of Ferns of the British Isles* (Jermy *et al.* 1978)
Scarce Plants	*Scarce Plants in Britain* (Stewart *et al.* 1994)
BRC	*Report of Botanical Record Club*
BEC Rpt	*Report of the Botanical Society and Exchange Club of the British Isles*
Handbook	the relevant BSBI Handbook

Organisations

BSBI   Botanical Society of the British Isles   N.H.S.   Natural History Society

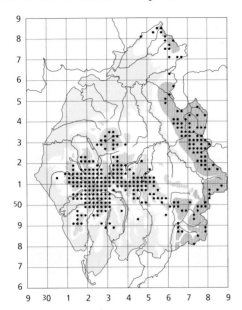

Map 1. *Huperzia selago*

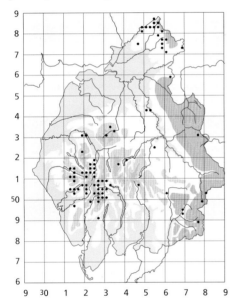

Map 2. *Lycopodium clavatum*

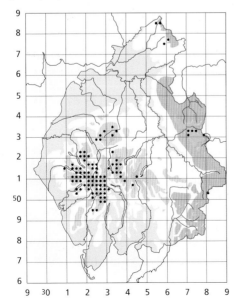

Map 3. *Diphasiastrum alpinum*

Plate 53. *Equisetum telmateia*

Plate 54. *Pilularia globulifera*

SYSTEMATIC ACCOUNTS

LYCOPODIOPSIDA, EQUISETOPSIDA, PTEROPSIDA
(PTERIDOPHYTES)

LYCOPODIACEAE

Huperzia selago (L.) Bernh. ex Schrank & C. Martius
Fir clubmoss
Common in the Lake District and widely distributed but less frequent in the Bewcastle Fells, Pennines and Howgill Fells. It is a species of rocky, grassy fellsides, but it also occurs in gullies and on rock ledges and extends to the highest summits. Tufts dislodged by sheep and walkers are a common sight in the Lake District. On the gentler ground of the Pennines the species tends to be confined to the crevices of block scree and it may have suffered in the past from heather-burning.

Hodgson gives an interesting low-level site for this, and also *Lycopodium clavatum*, by Wedholme Flow (2.5), but there are no subsequent records.
Alt. range: 185 m, Blea Tarn, Langdale (28.04) - 920 m, Scafell Pike (20.06)
Map 1   304   WFCY

Lycopodiella inundata (L.) Holub   Marsh clubmoss
Seen only at three sites, two in the Lake District, in the Wast Water and Scafell squares (1.0, *Watsonia* (1995) 20:419) 2.0) and the third an outlying site to the north-east in the Langwathby square (5.3, RWMC, 1983, *Watsonia* (1984) 15:125) where it was long thought to have become extinct. By contrast, Hodgson records it from Wast Water (1.0), Borrowdale (2.1) and Shoul-thwaite (3.2); Baker gives additional records from Little Langdale (2.0), Loughrigg (3.0), below Red Screes (3.0/4.0) and the foot of Long Sleddale (4.0/5.0), and Wilson lists it from near Codale Tarn (2.0, J.A. Martindale, 1889, KDL). The record in *Scarce Plants* for the Ambleside square (3.0) is an error.

This mossland species of bare, wet peat, has declined markedly in the country as a whole as well as in Cumbria. This is primarily the result of drainage but it has been accelerated by collecting. The two Cumbrian sites are now the only remaining ones in the north of England.
Alt. limit: 230 m, Wast Water (1.0) and Langwathby (5.3) squares
S 3   WC

Lycopodium clavatum L.   Stag's-horn clubmoss
Widespread though by no means frequent throughout the Lake District but somewhat commoner in the west. Elsewhere it is very scattered and chiefly in the east of the county. It occurs on rocky fellsides, in rough *Agrostis - Festuca* grassland but also under *Calluna* on a roadside verge near Sedbergh (68.92, 68.94), in a

railway cutting near Lazonby (50.42), by forestry rides in the Bewcastle Fells and on an old coal spoil-heap by the Northumbrian border (62.58). Its presence at these latter sites suggests that the species can be quite an effective coloniser. Nevertheless, like other clubmosses it appears to be declining. In the uplands this is probably a result of the ever increasing grazing pressure since the stems, although unpalatable, are easily uprooted, and in the lowlands it is the result of drainage.
Alt. range: 75 m, Wastwater (14.02) - 610 m, Scafell (20.04)
Map 2   82   WFCY

L. annotinum L.   Interrupted clubmoss
Known only from a few sites on the eastern flanks of Bowfell and Crinkle Crags at the head of Langdale (2.0), where it grows both in block scree and in grass-land. It formerly grew slightly further north as it was recorded from the Cumberland side of Rossett Gill by G. Stabler in 1889. There is a pre-1950 record in the *Fern Atlas* for the Grasmere square (3.0) but this must be an error.

These are the only extant sites in Britain south of the Scottish Highlands, where it is considered by Page (1988) to be a relic of former pinewoods.
Alt.: 300 m, Langdale (2.0)
S 2   W(C)

Diphasiastrum alpinum (L.) Holub   Alpine clubmoss
Fairly frequent on the western and northern fells of the Lake District, elsewhere confined to a single site near Stainmore (80.02) and a few in upper Teesdale above Moor House (7.3) and in the Bewcastle Fells (58.74n). It usually grows in short fellside grassland, sometimes amongst grazed *Vaccinium*. It is capable of withstanding considerable exposure as on Whiteside (32.16), at the north end of the Helvellyn range, where it forms small ridges in solifluction ground at 850 m. By contrast, it occurs on mine and quarry spoil heaps in the north-east, as at Moor House (74.32) and near Tindale Tarn (60.58). It has disappeared since the 1950s from Flinty Fell, Garrigill (76.42) and it has also been lost from the Hawkshead (3.9) and Kentmere (4.0) squares shown in the *Fern Atlas* as post-1950, and, like *Lycopodium clavatum*, from Holme Knott, Sedbergh (6.9, A. Wilson, 1878, YRK) and from Dentdale.
Alt. range: 75 m, Wast Water (14.02) - 915 m, Skiddaw (24.28) and Helvellyn (34.14)
Map 3   94   WFC(Y)

[D. alpinum x D. complanatum (L.) Holub

(*D. complanatum* subsp. *issleri* (Rouy) Jermy)

The *Fern Atlas* shows a pre-1950 record for this hybrid from the Langdale square (2.0). The basis for this is not known, but it could refer to Wilson's record of *Diphasiastrum complanatum* from Easedale (3.0), which is based on a specimen in W. Borrer's herbarium (K, *BEC Rpt* (1916) 4:222). Material collected from Wrynose Pass (26.02, J. & A. Harrington, 1980; LANC) has been identified by C.N. Page as the hybrid but by A.C. Jermy as *D. alpinum.* Jermy (*pers.comm.*) considers that all Cumbrian records of *D. complanatum* and the hybrid are referable to *D. alpinum.*

(W)]

SELAGINELLACEAE

Selaginella selaginoides (L.) P.Beauv.

Lesser clubmoss

A characteristic species of semi-open, base-rich upland flushes and particularly frequent from the eastern fells of the Lake District across to the Pennines. Rather surprisingly it is also quite frequent north of the Tyne Gap in the Bewcastle Fells. It usually grows with *Carex hostiana, C. dioica, Pinguicula vulgaris* and *Eleocharis quinqueflora.* Other habitats include irrigated rock ledges and steep slopes, commonly with *Saxifraga aizoides* and, more rarely, *Thalictrum alpinum.* Although essentially a montane species in Cumbria, it occurs near sea-level at Hale Moss, Burton-in-Kendal (50.76s) where it grows hidden amongst the tussocks of *Schoenus nigricans.* It occurs in dune-slacks on the coasts of Anglesey and formerly in Lancashire, but there are no records from such sites in Cumbria.

Alt. limit: 855 m, Cross Fell (68.34)

Map 4   427   WFCY

ISOETACEAE

Isoetes lacustris L.

Quillwort

Frequent in most of the lakes and tarns of the Lake District and extending southwestwards down to the coast. Like all aquatic macrophytes it is absent from the Haweswater and Thirlmere reservoirs which are subject to severe drawdown. It can grow in deeper water than *Littorella uniflora* and is more sensitive to wave action. It appears to thrive equally well in rocky tarns as in those with highly organic sediments. Viviparous plants were reported from Windermere by Manton (1950) and found in 1996 by A. Darwell (40.98).

Outside Cumbria, quillwort is restricted in England to three sites in Devon and Shropshire.

Alt. limit: 595 m, Sprinkling Tarn (22.08)

Map 5   109   WFC

I. echinospora Durieu

This species was collected from Cogra Moss, Lamplugh (08.18, AMDE, herb. R.S.) by RS in 1974 after a long search. The material was checked by A.C. Jermy and formed the basis of a short paper by Stokoe (1978). There are in his herbarium further specimens collected in 1980 from the south end of Derwent Water (26.18, *Watsonia* (1981) 13:327) and Baystone Bank Reservoir (16.84, 16.86, BM; LANC). At the first two sites it was growing with *Isoetes lacustris.* It was later found at the north end of Derwent Water by P. Thompson and F.J. Rumsey (26.22, 1992).

In his paper, Stokoe comments on the lighter green and more flexible and tapering leaves of this species compared with the darker, more rigid and acuminate ones of *Isoetes lacustris.* There is also the marked difference in the sculpturing of the megaspores, spiny in *I. echinospora,* warty in *I. lacustris.* The Cogra Moss population is remarkable in that it includes some

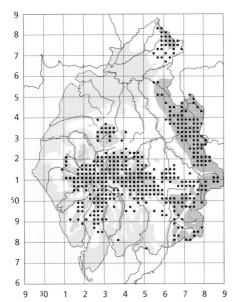

Map 4. *Selaginella selaginoides*

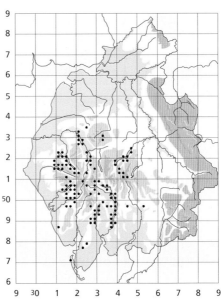

Map 5. *Isoetes lacustris*

autopolyploids (Rumsey *et al.* 1993).

The only other English records for this species are in the south-west, from Dorset to Cornwall. At the time of its discovery in Cumbria this was thought to be the only occurrence between North Wales and the Scottish Highlands, but it has since been found at several sites in Wigtownshire.

Alt. limit: 230 m, Cogra Moss (08.18)

S 4   C

EQUISETOPSIDA

EQUISETACEAE

Equisetum hyemale L.   Dutch rush
Rather rare, known only from a few sites in the north and west and an outlying one in the south near White Pike, Walna Scar (24.94, CFS & JS, 1985). The usual habitat is a wet, somewhat open and unstable wooded hillside; at higher altitudes it occurs in open flushes. It is probably decreasing having disappeared from Baker's site near Greenodd (3.8), by the River Lyne east of Longtown (44.68, DAR, 1957) and the Irthing above Brampton (58.64, DAR, 1957, 1962) and on Flinty Fell, Garrigill (74.42, DAR, 1964), also from the post-1950 Millom (1.8) and Castle Carrock (5.5) *Fern Atlas* squares and a site by the River Eden at Wetheral (4.5, T.S. Johnstone, 1905, CLE). C.N. Page considers that grazing is an important factor in the decline of this species in Scotland.

Alt. limit: 400 m, above Croglin (62.46) but formerly at 490 m, above Garrigill (7.4).

Map 6   20   WFC

E. hyemale x **E. variegatum**
E. x trachyodon A. Braun
This rare hybrid was discovered near Dodgsontown, Roadhead in 1982 (50.74n, GH, E, LANC, det. C.N. Page, *Watsonia* (1984) 15:125), a site at some appreciable distance from either parent. The few plants were growing on calcareous sandstone associated with calcicole mosses by a small stream draining off farmland and flowing down a steep wooded bank of the Black Lyne. The plants were very distinctive, the numerous stems being long and unbranched and quite procumbent. This was only the second English record, the first being from Northumberland. It has since been found in Cheshire.

Alt.: 110 m, Dodgsontown (50.74n)

1   C

E. variegatum Schleicher   Variegated horsetail
A rare plant of dune slacks in the south-west, of upland limestone flushes in mid-Westmorland and Teesdale. It is said by CW to have decreased considerably in mid-Westmorland during the Survey. It could not be found on the Cumberland side of the River Irthing near Gilsland (62.68) where it was seen by G.A. Swan in the late 1960s, nor has there been any confirmation of an old record from High Cup Nick (7.2, C.M. Rob, 1939, YRK). Baker cites C.C. Babington's interesting record from the shore near Holme Island, Grange-over-Sands (4.7s); Hodgson's Wasdale record may well refer to the Drigg dunes (06.96). It should be looked for at Penton Linns (4.7n), where it is known from rocks on the Scottish side of the river.

This species is easily overlooked, but it can also be confused during the growing season with small, unbranched stems of *Equisetum palustre*.

Alt. limit: 680 m, Great Dun Fell (70.32)

Map 7   S 16   WFC

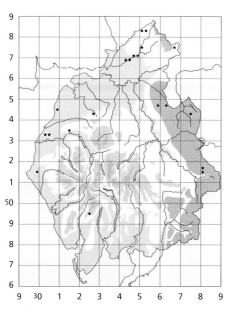

Map 6. *Equisetum hyemale*

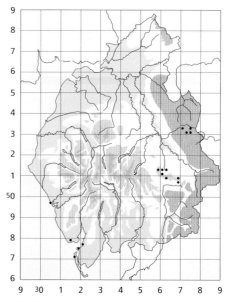

Map 7. *Equisetum variegatum*

E. fluviatile L.   Water horsetail

A common horsetail of lowland marshes and the sheltered margins of tarns, rivers and lakes, occurring usually in organic sediments at depths of up to 60 cm and often forming extensive pure stands. In the uplands it is not infrequent in *Sphagnum* pools.

Alt. limit: 750 m, Little Dun Fell (70.32)

Map 8   673   LWFCY

E. fluviatile x **E. arvense**

E. x litorale Kuhl. ex Rupr.

Scattered by ponds and in marshy ditches along the west and south-western coasts but with an isolated occurrence on the north shore of Ullswater (40.20, C.N. Page, 1964, E). This is the first Cumberland record. The first for Westmorland is that from Witherslack (42.82s, P. Jepson, 1977) and that for Furness from a sand-pit near Roosecote (22.68s, GH, 1978, E, LANC). These, and all the records from the south, have been checked by C.N. Page, and those from the west coast by CCH.

This rather uncommon hybrid is easily overlooked. It is best distinguished by its unbranched upper stem, branched middle section and the central hollow of the stem which is about half its width.

Alt. limit: 150 m, Glencoyne, Ullswater (40.20)

Map 9   19   WFC

E. arvense L.   Field horsetail

Very common everywhere in the lowlands and ecologically very catholic, occurring in hedgerows, marshes and derelict sites, railway sidings and sometimes as a nearly ineradicable garden weed. It is also widely distributed throughout the Pennines and the Bewcastle fells and this contrasts markedly with its behaviour in the Lake District. This horsetail is much more tolerant of wet conditions than is suggested in most Floras. It occurs, for example, in upland flushes, river shingle and lowland marshy meadows in habitats apparently more suited to *E. palustre* but probably always where the water is moving.

Alt. limit: 660 m, Tees Head (70.34)

Map 10   1508   LWFCY

E. pratense Ehrh.

The only authenticated record is from Swindale, Brough (7.2, BM) where it was collected by J. Backhouse in 1844. It has never been refound although Wilson's Kirkby Stephen record may refer to the same site. The *Fern Atlas* shows it as occurring in the upper Teesdale square (7.3) but this is probably based on a known erroneous identification from there.

The only extant sites for this northern horsetail in England and Wales are in north Yorkshire, Co. Durham and Northumberland.

S   (W)

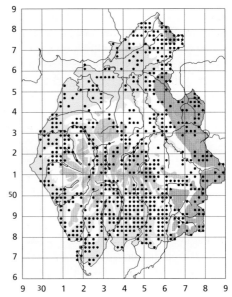

Map 8. *Equisetum fluviatile*

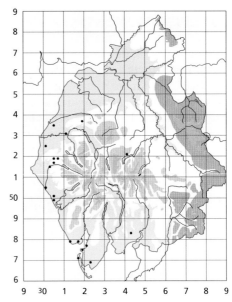

Map 9. *Equisetum fluviatile* x *E. arvense*

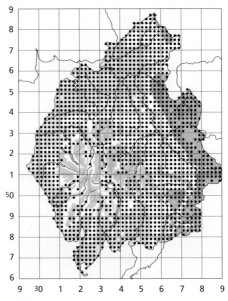

Map 10. *Equisetum arvense*

E. sylvaticum L.   Wood horsetail
Fairly frequent in the north but uncommon in the
south. This most attractive horsetail is characteristic of
damp, ungrazed peaty banks, roadside verges, riversides and railwaysides and marshy fields. In the
uplands it occurs also on rock ledges. It frequently
grows in light shade.
Alt. limit: 530 m, Force Burn, upper Teesdale (76.30)
Map 11   378   WFCY

E. palustre L.   Marsh horsetail
A common species of marshy fields and upland flushes
in the east, fairly frequent elsewhere but more
demanding in its soil requirements than *Equisetum
arvense*. Like the latter it is surprisingly uncommon in
the Lake District. It is frequent in the dune-slacks on
the south-west coast and a form with cones on the tips
of the upper branches, var. *polystachyum* Weigel,
occurs at Sandscale (18,74, REG, 1986, LANC).
Alt. limit: 750 m, Knock Fell (72.30)
Map 12   658   LWFCY

E. telmateia Ehrh.   Great horsetail
Not uncommon in the north-east and locally on the
west coast, elsewhere known only from a few scattered
sites near Morecambe Bay and in wooded gillsides
above Sedbergh (68.92) and Dent (70.82). In the north-
east it is characteristic of flushed, seepage areas in
steep wooded valleys, as along the Liddel Water and
the River Lyne. In the west it is locally abundant on the
clay railway embankments and near Morecambe Bay it
grows in roadside ditches, wet meadows and, at
Haverigg (16.78), on the sea-wall.
Alt. limit: 300 m, Deepdale, Dentdale (70.82)
Map 13, Plate 53   76   WFCY

OPHIOGLOSSACEAE

Ophioglossum vulgatum L.   Adder's tongue
An uncommon fern of short, damp, 'unimproved'
grassland and woodland rides and preferring base-rich
and calcareous soils. It is also a feature of dune-slacks
on the south-west coast and, with *Botrychium*, of old
lead mine workings high on the Pennines. Like that
species it is easily overlooked and is capricious in its
occurrence.
There are no recent records from the Lune valley
although Handley (1898) refers to it in "many of our
low-lying fields by the Lune".
Alt. limit: 660 m, Burnhope Seat (78.36)
Map 14   101   LWFCY

O. azoricum C. Presl
An extremely rare coastal species, particularly
characteristic of dune-slacks and found in Cumbria by
R. Walker in 1952 on the Ravenglass slacks where it
grows with *Ophioglossum vulgatum* (06.96, herb.
R. Walker, conf. C.N. Page; LANC).
The two species are very closely related and

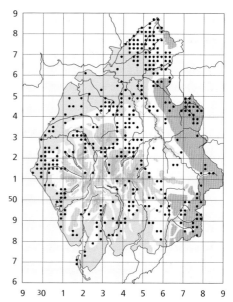

Map 11. *Equisetum sylvaticum*

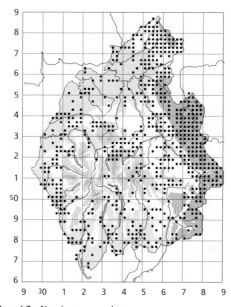

Map 12. *Equisetum palustre*

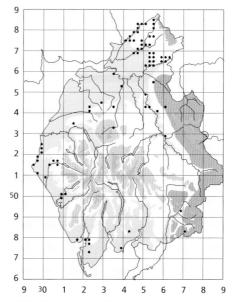

Map 13. *Equisetum telmateia*

Paul (1987) considers that *Ophioglossum azoricum* is best considered as the representing lower end of the variation of *O. vulgatum.* It has a sterile blade usually less than 3.5 cm and fewer than 15 sporangia. The only other sites in England are in the New Forest.

S 1   C

Botrychium lunaria (L.) Sw.   Moonwort
Scarce except in the northern Pennines where it is not infrequent in short limestone turf and on old lead mine workings. Elsewhere it occurs in short grassland, usually on the limestone or on sand dunes, but also very locally in the Lake District in well-drained *Agrostis - Festuca* grassland.
Alt. limit: 780 m, Great Dun Fell (70.30)
Map 15   138   LWFCY

OSMUNDACEAE

Osmunda regalis L.   Royal fern
Not uncommon in the coastal and valley mosslands in the south-west and extending into the southern Lake District and north to mosses on the Solway Plain. There is a single isolated locality in the Eden valley at Cliburn Moss (56.24). This distinctive western fern has in the past been extensively transplanted to gardens and its range has accordingly contracted so that it is now almost extinct in Westmorland. Even during the Survey it has gone from the Kirksanton area (12.78).
Alt. limit: 120 m, Cliburn Moss (56.24)
Map 16   81   WFC

ADIANTACEAE

Cryptogramma crispa (L.) R. Br.ex Hook. Parsley fern
Very common throughout the uplands. No other species is so characteristic of the Lake District non-calcareous dry-stone walls but its conspicuous tufts also play a major role in consolidating relatively fine scree. A number of the Eden valley occurrences are in sandstone quarries.

An upland species, it is virtually restricted in England to the north-west, and is probably commoner in Cumbria than anywhere else in Britain.
Alt. limit: 935 m, Helvellyn (34.14).
Map 17   530   WFCY

Adiantum capillus-veneris L.   Maidenhair fern
Known only from three tetrads along the shores of Morecambe Bay (3.7s, 4.7s), where it grows on shady, moist limestone rocks. One recent find appears to represent a recolonisation and it is interesting that at another site small fronds of the fern have appeared in some quantity among the limestone ballast of the coastal railway line. A specimen collected by W.H. Youdale from South Lodge, Cockermouth (1.3, 1892, LIV) must, like many of his records from there, be of garden origin.

The northernmost sites on the west coast for this

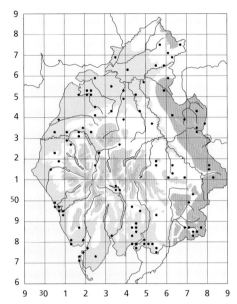

Map 14. *Ophioglossum vulgatum*

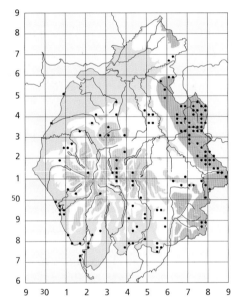

Map 15. *Botrychium lunaria*

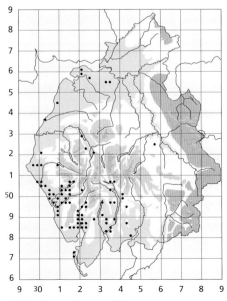

Map 16. *Osmunda regalis*

frost-sensitive south-western species are on the Isle of Man.

S 3   WF

MARSILEACEAE

Pilularia globulifera L.   Pillwort

Rare, occurring by the margins of a number of meso-trophic tarns in the south and south-west, also in one of the Derwent Water bays (24.22). The plant varies in abundance from year to year and sometimes may be sufficiently plentiful to form a sward along the margin of the tarn. On the other hand, subsequent attempts to refind it at Barkbooth (40.90) and some other early Survey sites have failed. Knowledge of its present status is due largely to the efforts of R. Stokoe, although his record from Stickle Tarn (2.0, Fryer (1991)) must be an error. He introduced the plant to Dubbs Moss (10.28) in l976, but it appears not to have survived. The anonymous and undated record in *Scarce Plants* for the Langwathby square (5.3) is best disregarded. Pillwort was formerly much commoner, particularly in Cumberland, Hodgson, for example, listing it from Ennerdale (0.1/1.1), Wythburn (3.1), St John's in the Vale (3.2) and Wigton (2.4) and G.A.K. Hervey recorded it from Loweswater (1.2) in 1928. It was also known to W.H. Pearsall from South Walney (1.6s).

Largely as a result of drainage, pillwort is now a declining species in Europe and the Cumbrian sites are the only extant ones in north-west England.

Alt. limit: 260 m, above Garnett Bridge (50.98)

Map 18, Plate 54   S 11   WFC

HYMENOPHYLLACEAE

Hymenophyllum tunbrigense (L.) Smith

Known from seven tetrads in mid-Eskdale (1.9, 1.0), where it grows on boulders in woods and ravines, at two sites with *Dryopteris aemula*. In at least two tetrads it is locally common. These sites, and ones in two other tetrads in the same area, were discovered by DAR in the 1950s (*Watsonia* (1979) 12:347), but it was reported as occurring in Eskdale by J. Parkin, probably in the 1930s. The only other definite record is from the extreme north where DAR found it in 1957 on a crag by the River Lyne in the wooded section below Shankhill Castle, near Waingatehead (44.68). It has never been refound and the rock may well have fallen.

Hodgson lists this fern from Ponsonby Hall, Calder Bridge (0.1), where it was said to be extinct, and Great Hall Gill, Wast Water (1.0), but he never saw the fern in Cumberland. The pre-1950 *Fern Atlas* record for Borrowdale (2.1), Petty's (1898 p.322) from Coniston (2.9/3.9) and the references by Wilson to its occurrence in v.c.69 are best treated as errors for *Hymenophyllum wilsonii*.

Alt. limit: 100 m, Eskdale (1.0)

7   [WF]C

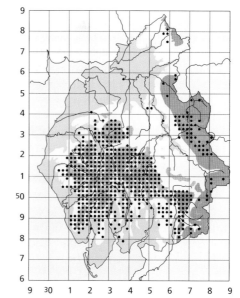

Map 17. *Cryptogramma crispa*

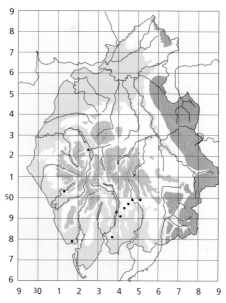

Map 18. *Pilularia globulifera*

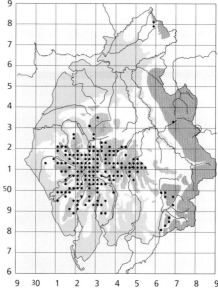

Map 19. *Hymenophyllum wilsonii*

H. wilsonii Hook.   Wilson's filmy fern
Fairly frequent in the Lake District, rather scarce in the tributary valleys of the River Lune and very rare elsewhere. In the Pennines it is restricted to a single site on block scree in Crowdundle Gill (68.32) and in the Bewcastle fells it occurs in deep crevices in two sandstone outcrops (58.78n, 58.80n). It descends almost to sea-level on the west coast. Its typical habitat is a shaded, usually wooded gill where it grows in moss communities on sheltered banks and near-vertical rocks, sometimes as an epiphyte, but not always so close to the water as to be directly affected by spray. Many populations appear moribund and look really healthy only after mild winters and during wet summers. There is little to suggest the species is decreasing in this its English headquarters.
Alt. limit: 670 m, Dollywaggon Pike (34.12)
Map 19   165   WFCY

Trichomanes speciosum Willd.   Killarney fern
Extremely rare, there being a single, relatively recent reliable record of sporophytes from Cumberland, a previous Cumberland population dying during the severe frosts of 1963. The only reference in the early Floras is the unlocalised one given by Wilson. The *Fern Atlas* has two undated dots, possibly misleadingly placed, in the Ambleside (3.0) and Keswick (2.2) squares. This is a fern of wet, shady riverside banks, caves and rocky crevices.

While on a visit to England in 1988, D. Farrar, of Iowa University, searched a number of ravines in the Lake District and succeeded in finding the delicate, filamentous gametophytes. With practice, these are fairly easily recognised. A.C. Jermy and others have continued this search and found gametophytes but no sporophytes in four areas in the central Lake District. This intriguing situation requires further investigation.

This species has a markedly south-western distribution in the British Isles. It has suffered much from collectors and it seems likely that in Britain sporophytes now occur at only a few sites in western Scotland, the Lake District, west Wales and Cornwall.
R 1   WC

POLYPODIACEAE

Polypodium vulgare L. sensu lato   Polypody
All three British segregates occur in Cumbria and their distributions mirror remarkably well the distributions in Britain as a whole.
Map 20   1538   LWFCY

P. vulgare L. sensu stricto
Common, both in the lowlands and in the fells. It occurs most typically on roadside banks and on dry stone walls, both limestone and siliceous, but it prefers a certain amount of shelter and humus accumulation. It can also be found as an epiphyte in sheltered woodland, growing usually on oak.
Alt. limit: 755 m, Knock Fell (70.30)
Map 21   899   LWFCY

P. vulgare L. x **P. interjectum** Shivas
P. x mantoniae Rothm. & U. Schneider
Rare, chiefly in the south, but no doubt commoner than the records suggest. The earliest records are for Westmorland from the north side of Rydal Water (3.0, F. Jackson, 1963, BM), for Furness the post-1950 one from the Duddon square (2.9) shown in the *Fern Atlas*, and for Cumberland from riverside rocks, Eden Lacy (56.38, RWMC, 1978, herb. R.W.M.C., *Watsonia* (1980) 13:132).

This hybrid is fairly easily distinguished from its parents by the intermediate number of annulus cells

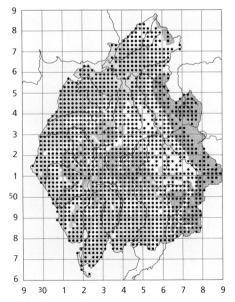

Map 20. *Polypodium vulgare* sensu lato

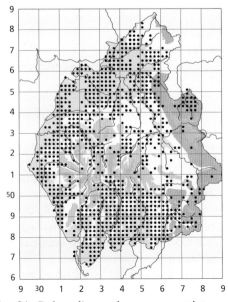

Map 21. *Polypodium vulgare* sensu stricto

and mostly abortive spores. All Survey records have been checked by R.H. Roberts.
Alt. limit: 215 m, Gowbarrow (40.20)

Map 22   21   WFC

P. interjectum Shivas

Frequent in the south and west and absent from large areas of the north and east. It occurs in similar habitats to *Polypodium vulgare* s.s. but it avoids banks and shows a marked preference for walls.

The first record is probably one from "Windermere" collected in 1876 (Anon., BM). The first for Cumberland was from Castle Crag, Borrowdale (2.1, F. Jackson, 1965)).
Alt. limit: 400 m, Hellbeck Wood, Brough (78.16)

Map 23   192   WFCY

P. cambricum L.

(*P. australe* Fée)

Very rare and virtually restricted to limestone rocks and walls around Morecambe Bay but with an outlying site on basic volcanic rock at Gowbarrow, Ullswater (40.20, FJR, 1982, LANC, *Watsonia* (1983) 14:419), the first Cumberland record. Newman's (1844) *History of British Ferns* lists it, as *Polypodium vulgare* var. *lacerum*, for both Westmorland and Furness. The earliest known specimen is that from Low Wood (J. Sidebotham, 1846, MANCH, *Proc. BSBI* (1968) 7:196). This could be on the narrow belt of Coniston limestone behind Low Wood, Ambleside (3.0), or, more likely, Low Wood, near Haverthwaite (3.8). The *Fern Atlas* dot for the 9.3 square on the west coast is an error, the square being invalid; that for the Arnside square (4.7s) refers to Lancashire.

The hybrid between this species and *P. interjectum* should be looked for as the latter probably occurs at all the *P. cambricum* sites.
Alt. limit: 215 m, Gowbarrow (40.20)

Map 24   9   WFC

DENNSTAEDIACEAE

Pteridium aquilinum (L.) Kuhn   Bracken

Very common, absent only from the more intensively farmed areas, the high summits of the Lake District, the limestone of mid-Westmorland and the peat moorlands on the eastern slopes of the Pennines. It is a species of the better-drained mosslands, of birch and oakwoods, also sometimes of hedgerows. However, it is most obvious as a pernicious weed of fell farms, clothing the lower slopes of the fells and suffering few competitors. In recent years it has extended its range considerably into many of the peripheral intake fields, partly as a result of farm amalgamations, with less labour available for cutting it, but also because of the decrease of cattle, which are very effective in trampling the stems. However, it is still cut for bedding, especially in Dentdale and the Rawthey valley. There has so far been rather little control by the use of systemic herb-

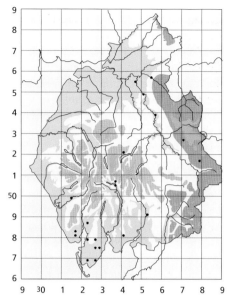

Map 22. *Polypodium vulgare* x *P. interjectum*

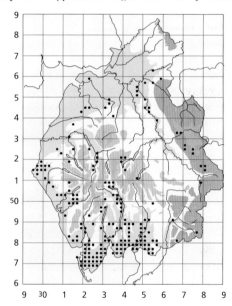

Map 23. *Polypodium interjectum*

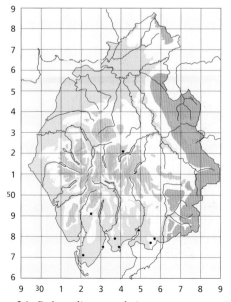

Map 24. *Polypodium cambricum*

icides. When applied incautiously these can be detrimental to many other plants as well as ferns.

Bracken occurs on well-drained soils of usually more than 25 cm depth. If less, then the rhizomes become susceptible to drought. Foresters regard good bracken growth as indicative of potential forestry land. Another limiting factor is frost, the emerging fronds being very sensitive. It is not strongly calcifuge but its frequency in many limestone areas perhaps reflects the prevalence of non-calcareous drift.

Page (1989) has recently drawn attention to the occurrence of three subspecies in Britain. Subsp. *aquilinum* is widespread; the only record of the western subsp. *atlanticum* C.N. Page is from White Moss Common, Rydal (34.06, PB, 1991, det. C.N. Page). It is characterised by the gradual progressive uncurling of the pairs of pinnae and the abundance, when young, of long, silvery hairs unmixed with red ones.
Alt. limit: 580 m, Striding Edge, Helvellyn (34.14) and Glenridding Common (34.16)
Map 25   1472   LWFCY

THELYPTERIDACEAE

Thelypteris palustris Schott   Marsh fern
(*T. thelypteroides* Michaux)
Extremely rare. The only Survey records are from moss woodland near Kendal (5.9) and Castle Carrock (5.5, FJR, 1978, LANC) and a wet wooded hollow near Arnside (4.7s, W.F. Grayson, 1994; LANC). Very few fronds were found at the former site but about 200 at the latter.

Baker describes it as very rare and lists only Irton woods, Gosforth (0.0), Roudsea Wood and peat mosses between Lakeside and Greenodd (3.8), and between Glencoyne and Blowick by Ullswater (4.1). Hodgson gives an additional site at Low Hall woods,

Egremont (0.1) and Wilson others from Heversham (4.8s) and between Appleby and Brough (7.1).

The *Fern Atlas* shows additional post-1950 records from the North Walney (1.7), Duddon (2.9), Cartmel (3.7s) and Hawkshead (3.9) squares. These and one from the Coniston area (3.0) were contributed by G. Wilson and the Barrow-in-Furness N.H.S.during the 1950s. In the absence of specimens, all except that for the Cartmel square, which is based on a Roudsea record, are best considered doubtful.
Alt. limit: 150 m, near Kendal (5.9)
                         S 3   W(F)C

Phegopteris connectilis (Michaux) Watt   Beech fern
This attractive fern is essentially an upland species being frequent in the Lake District and the Sedbergh area but less so in the Pennines. It occurs often in some profusion on rocky, moss-covered banks in wooded gills and at higher altitudes in more open situations such as rock crevices and ledges.
Alt. limit: 870 m, Helvellyn (34.14, Wilson)
Map 26   453   LWFCY

Oreopteris limbosperma (Bellardi ex All.) Holub
                    Lemon-scented mountain fern
A typical upland fern, being very common in the Lake District and extending through the Howgill Fells to the Yorkshire border. It is noticeably less frequent in the northern Pennines and the Bewcastle fells. There are, however, several localities virtually at sea-level, as, for example, on North Walney (16.72) and there are post-1950 *Fern Atlas* records for the Maryport (0.3), Aspatria (1.4) and Silloth (1.5) squares. It is a fern of gills, becksides, gullies and rock ledges, of pathside ditches and the sides of stone walls. It tolerates only light shade.

This attractive fern is often mistaken by visitors,

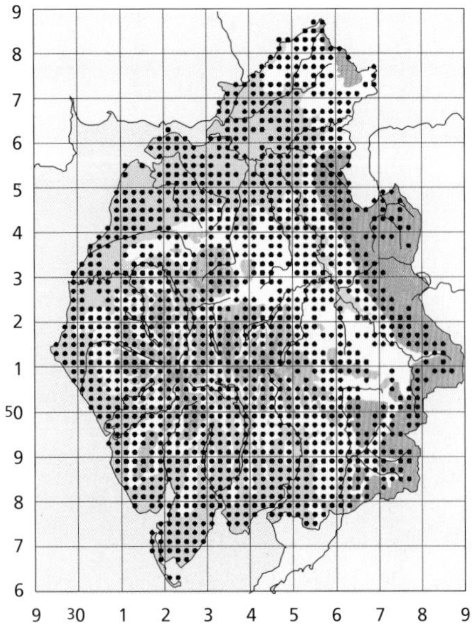

Map 25. *Pteridium aquilinum*

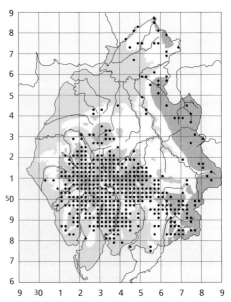

Map 26. *Phegopteris connectilis*

particularly from the south, for *Dryopteris filix-mas*, but it is easily distinguished by the tapering base of the frond, the pale, almost white scales of the young rachis, the lemon scent of the lightly bruised frond and the marginal sori.

Alt. limit: 610 m, Melmerby High Scar (64.38)

Map 27   737   WFCY

ASPLENIACEAE

Phyllitis scolopendrium (L.) Newman

Hart's-tongue fern

Locally common in the lowlands but preferring the richer soils. It is noticeably scarce in many of the Lake District valleys. It is particularly characteristic of shaded hedgerows, damp walls, limestone woodlands and the deep grikes of open or wooded limestone pavement, and on the Pennines it occurs in a number of the old mine shafts and adits at considerable altitude.

Alt. limit: 700 m, Great Dun Fell (72.32)

Map 28   657   LWFCY

Phyllitis scolopendrium x **Asplenium trichomanes**

x **Asplenophyllitis confluens** (T. Moore ex E. Lowe) Alston

The only British records of this hybrid are two 19th century ones: from Levens Park (4.8s, G. Stabler, 1865, K) and Whitby (Alston 1940).

(W)

Asplenium adiantum-nigrum L.   Black spleenwort

Rather infrequent over much of the county but becoming appreciably commoner in the south and especially on the west coast. It occurs chiefly on mortared walls but also on hedgebanks, in quarries and rocky gills, as in Sandbed Gill, Wanthwaite (32.22), and on limestone crags in the Pennines.

Alt. limit: 575 m, Moor House, Teesdale (74.32)

Map 29   342   WFCY

A. obovatum Viv.   Lanceolate spleenwort

(*A. billotii* F. Schultz)

Extremely rare. This fern is listed by Hodgson from the Cumberland side of the Duddon estuary. It still occurs there, below Black Combe (1.8), in two tetrads, in one on a roadside wall and the other on an outcrop of rotten rock. Both populations are small and vulnerable.

        This is a noticeably south-western fern in Britain. North of Cumbria it is known only from the Mull of Kintyre and Sutherland.

S 2   C

A. marinum L.   Sea spleenwort

Very rare. It is scattered along the west coast from Nethertown to Parton and is even locally abundant, as on the south side of St Bees (94.12). Further south it survives at Hodbarrow (18.78), on Piel Island in the Walney Channel (22.62s) and on limestone rocks by the Kent estuary east of Grange-over-Sands (42.78s), although it has apparently disappeared from Raven-

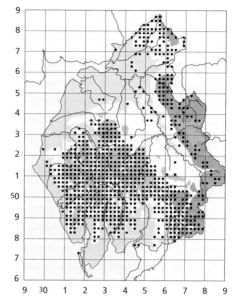

Map 27. *Oreopteris limbosperma*

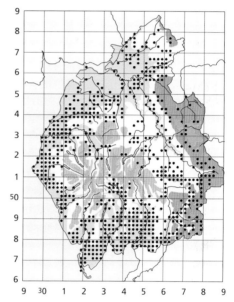

Map 28. *Phyllitis scolopendrium*

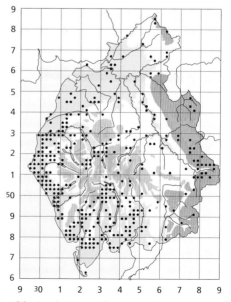

Map 29. *Asplenium adiantum-nigrum*

glass (0.9) and Humphrey Head (3.7s).
Map 30   7   WFC

[A. fontanum (L.) Bernh.
A fern of southern and central Europe listed by Baker from
Wythburn (3.1) but not surprisingly treated as an error.

(C)]

A. trichomanes L.   Maidenhair spleenwort
subsp. **quadrivalens** D Meyer
Common or even abundant, absent only from the
higher fells and scarce in the more intensive agricultural areas of the Solway Plain and the middle and
lower Eden valley, where stone walls tend to be
replaced by hedges. This is essentially a fern of mortared walls occurring usally with *Asplenium ruta-*
muraria and often also with *Cymbalaria muralis*. It
also occurs on limestone scars and pavements and, in
the Lake District, on many non-calcareous dry-stone
walls and crags.
Alt. limit: 730 m, Cross Fell (68.32)
Map 31 (of the species) 1138   LWFCY

subsp. **trichomanes**
A rather uncommon fern of northern and western
Britain which in Cumbria is known with certainty only
from a few sites in the Lake District, where it grows on
slate walls and on non-calcareous rocks. The only
Survey records are from roadside walls north of Lake-
side (36.88, PB, 1987, LANC, det. BM), Wolf Crags
(34.22, MC, 1979, det. A.C. Jermy), Stockley Bridge,
Borrowdale (22.10, FJR, 1980, herb. F.J.R.), Barf,
Thornthwaite (20.26, CCH, 1980), Beacon Tarn
(26.90, JA, 1983, det. A.C. Jermy) and the High Street
fells (42.10, 44.10, 46.10, C. Gransden, 1993, LANC).
There are earlier collections from Seascale (0.0, 1951,
BM) and Roudsea (3.7s/3.8, N.D. Simpson, 1951, BM)
and Walla Crag, Borrowdale (2.2, H.W. Pugsley 1903,
BM) and those in the *Fern Atlas* for the Dalton-in-
Furness (2.7), Langdale (2.0), Hawkshead (3.9),
Ambleside (3.0) and Helvellyn (3.1) squares.

In the absence of a chromosome count, this
subspecies can be distinguished by the less prostrate
fronds, the more rounded, non-overlapping pinnae and
the shorter guard cells of the stomata and smaller
spores (respectively less than 39μ and 31μ).
Alt. limit: 540 m, west of Pasture Beck, Hartsop
(42.10)
Map 32   8   WFC

A. trichomanes x **A. ruta-muraria**
A. x **clermontiae** Syme
The only British record, and that tentative, for this surprisingly rare hybrid is from Levens Park (4.8s/5.8s). A
specimen collected by W. Wilson from there was identified
by Stansfield (1919) as *A. trichomanes* x *A. trichomanes-*
ramosum but from the description Lovis (1975) considered it
more likely to be *A. trichomanes* x *A. ruta-muraria*. No
specimen is known.

(W)

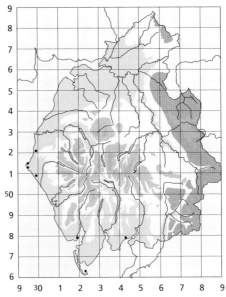

Map 30. *Asplenium marinum*

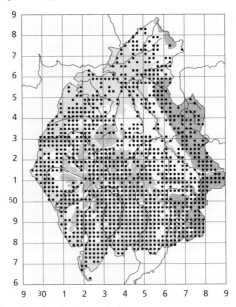

Map 31. *Asplenium trichomanes*

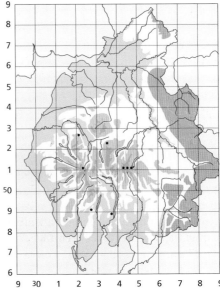

Map 32. *Asplenium trichomanes* subsp. *trichomanes*

A. trichomanes x **A. septentrionale**
A. x **alternifolium** Wulf.
Possibly now extinct having last been seen in the Borrowdale (2.1) and Keswick (2.2) squares by F. Jackson in *c.*1950. Wilson lists it from Langdale (2.0/3.0) and Hodgson from Borrowdale (2.1), Scafell (2.0), Skiddaw and Barf (2.2) and Helvellyn (3.1, J.S.P., 1877, BM). He also refers to another early record from a derelict conservatory at Underley Hall, Kirkby Lonsdale (6.8s)!

This very rare hybrid may now be restricted in Britain to Snowdonia.

(WC)

A. viride Hudson   Green spleenwort
(*A. trichomanes-ramosum* L.)
Common on the limestone in the east, not infrequent in the eastern fells of the Lake District but scarce in the south and west where it was not refound in five of the *Fern Atlas* post-1950 squares, including a near sea-level site at Humphrey Head (3.7s). This is a species of moist, sheltered rock crevices in basic rock.
Alt. limit: 820 m, Helvellyn (3.1, Wilson)
Map 33, Plate 27   214   WFCY

A. ruta-muraria L.   Wall rue
Common in most areas, but slightly less so than *Asplenium trichomanes* with which it very often grows. It is typically a species of mortared walls, but it is common on limestone scars and pavements, occasionally also on presumably basic rocks in the Lake District. It is less shade tolerant than *A. trichomanes*.
Alt. limit: 610 m, Melmerby High Scar (64.36)
Map 34   999   LWFCY

A. ruta-muraria x **A. septentrionale**
A. x **murbeckii** Dörfler
An extremely rare hybrid known in Britain only from two sites in Scotland, from which it has now disappeared, and one in Borrowdale (2.1; BM), where it was first collected by Miss Wright in the l9th century, rediscovered by F. Jackson (1959) and last seen by him in 1961.

(C)

A. septentrionale (L.) Hoffm.   Forked spleenwort
Very rare and elusive and largely confined to crags in the central fells of the Lake District and recorded from only six tetrads in the Wast Water (1.0), Langdale (2.0), Borrowdale (2.1), Keswick (2.2) and Martindale (4.1) squares.

Hodgson also lists it from Honister Crag and the Vale of Newlands (2.1) and above Crummock Water (1.1), and Baker from Patterdale and Red Tarn (3.1) and Red Screes (3.0). Wilson's additional records are from Grasmere (3.0) and Crosthwaite (4.9). Unlikely as this last site might appear to be, the fern was collected by W.H. Pearsall on a wall in the Rusland valley (3.8, 1913, BM) at an even lower altitude.

Between 1950 and 1960, R.J. Birkett, DAR and F. Jackson visited many of these sites, finding the fern

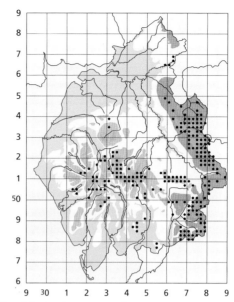

Map 33. *Asplenium viride*

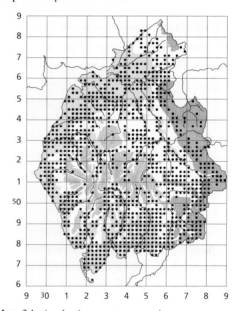

Map 34. *Asplenium ruta-muraria*

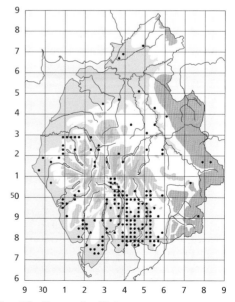

Map 35. *Ceterach officinarum*

at seven and also in a completely new area on the High Street Fells near the head of Mardale (4.1, DAR, 1958, *Proc. BSBI* (1963) 5:28).

Its preference for crevices on hard and often exposed, south-facing crags makes this fern less liable to chance extinction from rock falls than are several other montane species such as *Polystichum lonchitis*.
Alt. range: 180 m, Keswick (2.2) - 490 m, Wast Water (1.0)

7	W(F)C

Ceterach officinarum Willd.   Rustyback
Fairly frequent on the limestone in the south, becoming scarcer northwards and very rare in the east where it appears to be decreasing. The usual habitat is on limestone scars and walls, but it occasionally occurs on mortared walls of sandstone or slate. There is, for example, a particularly luxuriant colony by the A591 at Brockhole, Windermere (38.00).

This fern has a strongly south-western distribution in Britain and is generally considered to be frost sensitive.
Alt. limit: 400 m, North Stainmore (82.16)

Map 35	169	WFCY

WOODSIACEAE

*****Matteucia struthiopteris** (L.) Tod.
A handsome and widespread Continental species long naturalised in woodland near Witherslack Hall (42.86s, CFS & JS, 1970), on the steep eastern bank of the River Lune opposite Underley Hall (60.80s, GH, 1986), on the roadside near Matson Ground, Windermere (40.96, CEW, 1988) and in quantity at Thieves Hole quarry, Skelghyll Wood (38.02, M. Richards, 1989). In the first two sites it was growing with several other naturalised species.

4	W

*****Onoclea sensibilis** L.
A greenhouse fern and native of North America and eastern Asia which is well-established in a marshy field on the west side of the River Rothay below Brathay Hall (36.02, BM; AMDE, LANC). It has also been seen during the Survey in the same tetrad at Pull Garth, and nearby by the beck below Rydal Hall (36.06, GH, 1985, LANC), by Pelter Bridge (36.04, KAG, 1980s) and in Dove Nest wood, Ambleside (38.02, CEW, 1985, LANC). There is also a record from by a woodland pond north of Kirkby Lonsdale (60.78s).

It was first reported in Cumbria from marshy ground by Rydal Water (3.1, J.P.M. Brenan, 1945, *BEC Rpt* (1947) 13:75) and subsequently collected by K.M. Hollick in a field on the east side of the River Rothay, close to Ambleside (3.1, 1957).

The colonies are all small but, despite the comment in the *Fern Atlas*, it seems likely that at least some have become established naturally.

5	WF

Athyrium filix-femina (L.) Roth   Lady fern
A distinctive, delicate and very common fern. It prefers shaded damp habitats, such as by woodland streams, but it is also frequent in hedgerows, gills and moist mountain gullies.
Alt. limit: 825 m, Helvellyn (34.12)

Map 36	1506	LWFCY

Gymnocarpium dryopteris (L.) Newman   Oak fern
A calcifuge fern of shaded moss-rich woodlands, screes and rocks, chiefly in the uplands and particularly in the Lake District where it is relatively frequent. It also occurs at higher altitudes in gills and on rock ledges, sometimes associated with *Phegopteris connectilis*, and near sea-level at Arnside (44.76s).
Alt. limit: 825 m, Helvellyn (34.14, Hodgson)

Map 37	222	WFCY

G. robertianum (Hoffm.) Newman
               Limestone polypody
Not uncommon in fissures and grikes and in semi-stable scree on all the main Westmorland limestone outcrops and with an isolated occurrence on Scarrowmanwick Fell (58.46, DAR, 1956, *Watsonia* (1979) 12:348). Wilson's altitudinal limit of 610 m probably refers to the old record which he gives for Dollywaggon Pike (3.1), which is off the limestone, but DAR thinks this is probably referable to *Gymnocarpium dryopteris*. It has never been refound at the Scale Force, Buttermere (1.1) and Gilcrux (1.3) sites given by Hodgson. It has also disappeared during the Survey from a ravine on the east side of Hawes Water (48.10, RWMC, 1970), following a landslip, and from a roadside outcrop of Silurian slate in the Lune gorge (60.00, GH, 1970, LANC).

Cumbria and adjacent limestone areas in Yorkshire are the headquarters of this fern which has disappeared from many of its former sites in England and Wales and is extremely rare in Scotland.
Alt. limit: 335 m, above Hilton (76.22)

Map 38	34	WFC

Cystopteris fragilis (L.) Bernh.   Brittle bladder-fern
Common over much of the east and south, extending westwards to the gills and gullies of the eastern and central Lake District, elsewhere rather uncommon. It is typically a fern of shaded limestone outcrops and grikes, also of limestone and mortared walls and bridges in the lowlands. In the Lake District it is a valuable indicator of base-rich rock.
Alt. limit: 870 m, Helvellyn (34.14, Wilson (1956))

Map 39	622	WFCY

C. montana (Lam.) Desv.
A very attractive mountain fern of base-rich rock ledges in the western Scottish Highlands but long extinct in its only other British locality above Red Tarn, Helvellyn (3.1, OXF), where it was discovered by Bolton King in 1880. Its extinction was almost certainly the result of collecting. The

records cited by Baker and Wilson from Langdale (2.0) and Brothers Water (4.1) are probably errors, similarly material collected by Hayes purporting to come from Skiddaw (2.2, OXF).

R   (W)[(C)]

Woodsia ilvensis (L.) R.Br.
Reported by Baker and Hodgson from about a dozen sites in the central Lake District fells, from Wasdale to Mardale (1.0, 2.0, 3.0, 3.1, 4.0, 4.1) but said even then to be very rare. Only one new site has been found this century. These records have given the Lake District the distinction of being the British headquarters for this fern. Sadly, it is now known from only one of these localities, where, however, the population appears to be stable and luxuriant. This decline is usually blamed on the early fern collectors, but Rickard (1972) believes that the scarcity of localised herbarium specimens and the frequency of misidentification, particularly result-

ing from confusion with *Cystopteris fragilis*, indicate that probably several of the published records, such as that from Crosby Ravensworth church wall (6.1), cited by Baker, are erroneous. However, J. Backhouse's records from Helvellyn and Ill Bell are no doubt reliable.

The picture is similar elsewhere in Britain. Although it occurs from Cader Idris and Snowdonia to the Moffat hills and east central Scotland, only five populations apparently survive.

R 1   (W)C

[W. alpina (Bolton) Gray
In his paper on the genus *Woodsia* in Britain, Rickard (1972) refers to herbarium specimens at K one labelled " ? Helvellyn" (3.l) and another "Kinprite Co. Cumberland", which he thought was probably of foreign origin. There is also the record given by Baker, but unsubstantiated by any specimen, from Dove Crag, Fairfield (3.1). These records are

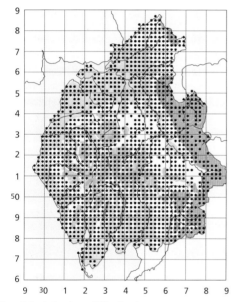

Map 36. *Athyrium filix-femina*

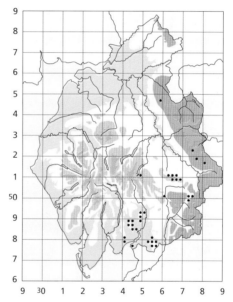

Map 38. *Gymnocarpium robertianum*

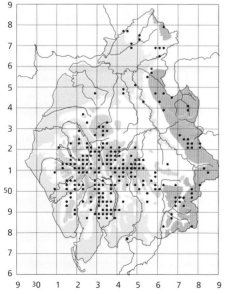

Map 37. *Gymnocarpium dryopteris*

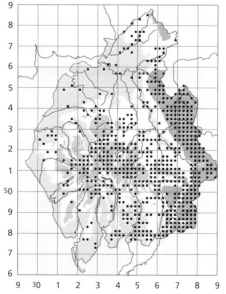

Map 39. *Cystopteris fragilis*

all best considered errors.

R   (WC)]

DRYOPTERIDACEAE

Polystichum setiferum (Forskål) Moore ex Woynar
                                    Soft shield-fern
Scarce or rare except in the extreme west and in south
Furness where it is a feature of roadside banks and
gills, growing both on the ground and in rock crevices.
It is particularly abundant on the sandstone cliffs at St
Bees (9.1). This is a south-western lowland species in
Britain and generally considered calcifuge, but there
are authenticated records in Cumbria from the grikes of
limestone pavement at Great Asby (66.14, CW, 1975,
LANC).

The map must, however, be interpreted with
some caution as it includes records from squares,
mainly in the east of the county, which are additional to
those shown in the *Fern Atlas* and which have not been
substantiated by specimens.
Alt. limit: 305 m, Walla Crag (26.20)
Map 40   85   WFC(W/Y)

P. setiferum x **P.aculeatum**
P. x bicknellii (Christ) Hahne
A vigorous hybrid discovered in 1988 by RWMC in the
gorge of the River Caldew south-west of Sebergham
(34.40, RWMC, LANC), the sandstone gorge of the
Liddel Water at Penton Linns (42.76n, RWMC, 1989,
LANC), by the River Lowther near Askham (50.22,
RWMC, 1989, LANC), and by the River Lyvennet near
Kings Meaburn (60.20, BM, LANC, *Watsonia* (1989)
17:464). The record from the Penton Linns square in
the *Fern Atlas* refers to material from the Scottish side
of the river.

All the specimens were characterised by abortive
spores and were checked by A.C. Jermy. It is certainly

under-recorded.
Alt. limit: 175 m, Askham (50.22)
                    4   WC

P. aculeatum (L.) Roth   Hard shield-fern
Well distributed over much of the county but scarcer in
the west and quite frequent, like *Cystopteris fragilis*, on
the limestone of the south and east. It too is very characteristic of shaded limestone crags and grikes, but it is
by no means restricted to obviously base-rich or calcareous rocks, occurring, usually in rock crevices, in
many wooded gills as, for example, in the central Lake
District and in the wooded cleughs in the extreme north
of the county. It occasionally grows in hedgerows.
There is an interesting concentration of records from
the belt of calcareous soils north of the Lake District
and west of Sebergham.
Alt. limit: 760 m, Knock Fell (70.30)
Map 41   428   LWFCY

P. lonchitis (L.) Roth   Holly fern
A very rare fern of base-rich rock ledges and crevices
and limestone pavement. There are Survey records
from only five tetrads in the Lake District (3.1, 4.1),
one in the south-east (5.7s) and three on the Pennine
limestone (7.9, 7.3, 6.3), the last being the only
Cumberland record (DAR, 1957, *Watsonia* (1979)
12:348).

It is listed by Baker from the east side of
Helvellyn and Fairfield (3.1), from the High Street
range above Blea Water (4.1), and from Harter Fell and
Nan Bield Pass above Small Water (4.0). He also
included records from Carrock Fell (3.3) and Swarth
Fell (7.9). Wilson's additional records are from Pennine
sites above Dufton (7.2) and the east side of Knock Fell
(7.3).

All the extant populations are small and vulnerable. This is particularly true of those in the Lake

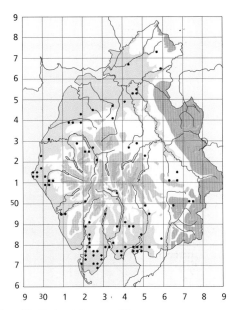

Map 40. *Polystichum setiferum*

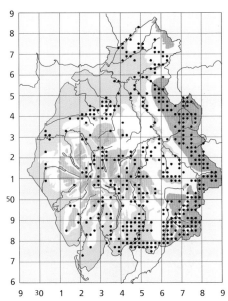

Map 41. *Polystichum aculeatum*

District where the plants are subject not only to collecting but also to the inherent instability of the habitat. A population at the south end of the Helvellyn range probably became extinct sometime during the 1960s.

Apart from Cumbria, the fern is restricted in Britain to the Scottish Highlands, the Moffat hills, upper Teesdale, the Craven area of Yorkshire and Snowdonia.

Alt. range: 275 m, south Westmorland (5.7s) - 825 m, Helvellyn (3.1)

5	WC

Dryopteris oreades Fomin   Mountain male-fern
Rather local but easily overlooked. It is a fern of screes, gullies and rock ledges in the northern half of the Lake District being commoner in the east where it descends to 150 m on roadside rocks and banks by Ullswater (44.20). It occurs locally in the Howgill Fells and on sandstone block scree along the Pennine scarp. There are additional records in the *Fern Atlas* from the Ennerdale (1.1) and Hawkshead (3.9) squares, but that for Arnside Knott (4.7s) is now regarded by the finder, A.McG. Stirling, as probably an error. There are no authenticated records from the limestone.

This species is generally smaller than *Dryopteris filix-mas*, with pale fronds and rather few, large pale scales. The fronds appear crisp, with the concave pinnae at an angle to the rachis. The tips of the pinnules tend to widen and their teeth are usually turned outwards. The relatively few sori are large and lead-coloured.

Alt. limit: 580 m, Piers Gill, Scafell Pike (20.08)

Map 42	79	W(F)CY

D. oreades x **D.filix-mas**
D. x **mantoniae** Fraser-Jenkins & Corley
Known only from four Lake District sites: on scree above Rydal Water (3.0, F. Jackson, 1963, BM), in

woodland north of Ferry House, Windermere (38.94, J.D. Lovis, 1975) and from the Matterdale (3.2) and Pooley Bridge (4.2) squares shown in the *Fern Atlas*.

These are the only English localities for this rare, sterile triploid hybrid.

Alt. limit: 305 m, above Rydal Water (3.0)

1	(W)F

D. filix-mas (L.) Schott   Male fern
Very common. It is less shade-demanding than *Athyrium filix-femina* and is perhaps most conspicuous as a fern of hedgerows, roadside ditches and verges where it tends to grow in the shelter of the stone walls. It is frequent in the fells growing in gullies and on rock ledges where it is sometimes confused with *Dryopteris oreades*.

Alt. limit: 820 m, Helvellyn (3.1) and Cross Fell (6.3; these Hodgson records could refer to *D. affinis*).

Map 43	1614	LWFCY

D. filix-mas x **D. affinis**
D. x **complexa** Fraser-Jenkins
(*D.* x *tavelii* auct.)
Known only from woodland north of Ferry House, Windermere (38.94, J.D. Lovis, 1975), near Brathay (34.02, 36.02, J.D. Lovis, 1975) and Johnny's Wood, Borrowdale (24.14, Brit. Pterid. Soc., 1980). A specimen tentatively identified by A.C. Jermy as this hybrid was collected at Roanhead, Dalton-in-Furness (20.74, PB, 1987, LANC).

In view of the frequency with which the two parent species occur together the hybrid is probably appreciably under-recorded.

5	FC

D. filix-mas x **D. carthusiana**
D. x **brathaica** Fraser-Jenkins & Reichst.
An extremely rare hybrid recorded only once in Britain. It was collected *c.*1854 by I. Huddart who gave a root to

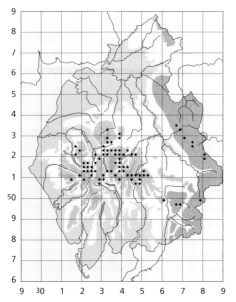

Map 42. *Dryopteris oreades*

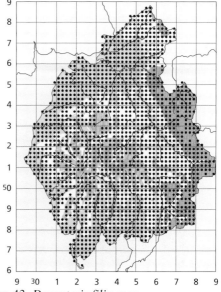

Map 43. *Dryopteris filix-mas*

F. Clowes. The hybrid is still in cultivation in the Oxford University Botanic Garden.

The specific name is misleading. Clowes' herbarium material is labelled "originally found near Bowness" (3.9/4.9, 1859, BM, K) and in subsequent papers (Clowes 1860 a & b) the locality is given as "at Windermere, Westmorland". Why, therefore, Fraser-Jenkins & Reichstein (1977) should have coined the epithet *brathaica* on the assumption that it came from Brathay, then in Lancashire, is a mystery.

The only other records are from France where it was found in 1860 and 1989.

(F)

D. affinis (Lowe) Fraser-Jenkins   Scaly male-fern
Frequent, and noticeably less common than *Dryopteris filix-mas*. It occurs in similar habitats, but it is, perhaps, more generally a woodland fern. In the uplands it is frequent on the fellsides and along becks where it is very conspicuous in the late spring with its expanding shuttlecock of yellowish fronds covered in tawny scales.
Alt. limit: 705 m, Little Dun Fell (70.32)
Map 44   865   LWFCY

The following three subspecies have been confirmed as occurring in the county. All the specimens have been identified by J. Camus or A.C. Jermy and many are in LANC. The paucity of records means that the maps can give only a very general indication of the distributions.

subsp. **affinis**
Apparently fairly widespread. It is distinguished by dense scales, at least on the lower part of the rachis, oblong pinnules, a black spot at the attachment of the pinnae and a rigid indusium.
Map 45   21   WFCY

subsp. **cambrensis** Fraser-Jenkins
The least common of the three subspecies and largely confined to the Lake District. It is characterised by fronds which taper markedly towards the base and with pinnules which are tapering rather than oblong.
Map 46   15   WFC

subsp. **borreri** (Newman) Fraser-Jenkins
The commonest subspecies. Its fronds have the fewest scales of the three; they are widest at the base, the pinnules are uneven in outline and the indusium is thinner, lifting at maturity.
Map 47   31   WFCY

D. aemula (Aiton) Kuntze   Hay-scented buckler-fern
Very rare, seen only on the cliff-top at St Bees (9.1, RS, 1971, herb. C.C.H.), in woodland in Eskdale, where it was first discovered in 1959 (1.9, 1.0, DAR, *Watsonia* (1979) 12:347), and on a roadside verge north-east of Ireleth (2.7, PB, 1993, LANC, conf. A.C. Jermy). At this last site it occurs with *Vaccinium myrtillus* and in Eskdale it grows on granite rocks and associated with *Hymenophyllum tunbrigense*.

The only sites in the early Floras for this rare fern are those given by Baker from St Bees (9.1),

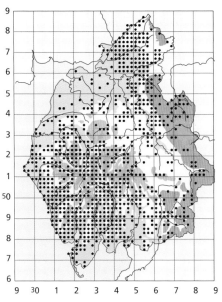

Map 44. *Dryopteris affinis*

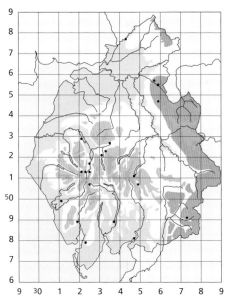

Map 45. *Dryopteris affinis* subsp. *affinis*

Coniston (2.9/3.9), Broughton-in-Furness and Kirkby Moor gills (2.8), south of Esthwaite (3.9) and "around Windermere", where it was collected by F. Clowes near Belle Grange (3.9, 1856, KDL). R.J. Birkett reported it from near Belle Grange, probably in the 1960s, but was unable subsequently to confirm the record.

The diagnostic characters include the convex upper surface of the pinnules, the dark, purplish base of the stipe and the disproportionately large pinnules on the lower sides of the lower pinnae.

Its rarity in the county and absence from south-west Scotland is rather surprising in view of its wide distribution in western Scotland, North Wales and south-west England.

Alt. limit: 220 m, north-east of Ireleth (24.78)

6 FC

D. submontana (Fraser-Jenkins & Jermy) Fraser-Jenkins Rigid buckler-fern
(*D. villarii* (Bellardi) Woynar ex Schinz & Thell.)
A typical and often frequent fern of the limestone pavements, scars and screes of the southern and mid-Westmorland limestones. It is quite rare on the Pennines and is not known north of Hilton Fell (76.22). It seems to require some degree of shelter but is intolerant of shade. During the Survey it was also seen on the end wall of a limestone building at Lupton (56.80s), although renovation subsequently removed it, and on the retaining wall of the old railway embankment at Heversham (50.82s). A single plant was reported from Honister Crag by Rowland (1938). This was cultivated by F. Jackson and the identity was confirmed by A.H.G. Alston. A second plant was later found by Jackson on a quarry wall but this was destroyed in a rock fall.

These Cumbrian localities, others in adjacent parts of Yorkshire and isolated sites in Wales and the north Midlands are the only extant ones in northern Europe for this central and southern European species.

The glaucous fronds and the abundance of pale scales make this fern readily recognisable.
Alt. limit: 460 m, above Helbeck (78.16)
Map 48, Plate 21 S 38 LWF(C)

[D. cristata (L.) A. Gray x D. carthusiana
D. x uliginosa (A. Braun ex Doell) Kuntze ex Druce
Hodgson's record of this now extremely rare hybrid is no doubt an error. There are no known specimens of the hybrid or of *Dryopteris cristata* from Cumbria.

(C)]

D. carthusiana (Villars) H.P.Fuchs
Narrow buckler-fern
Scattered throughout the lowlands but only really frequent on the mosslands by the Solway Firth, the Duddon estuary and around the head of Morecambe Bay. It is a fern of damp, peaty and relatively non-calcareous ground, often occurring amongst *Molinia* tussocks on mossland, marshy meadows and wet valley bottom woodlands. It is also frequently associated with

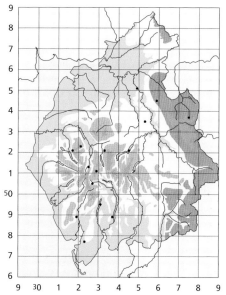
Map 46. *Dryopteris affinis* subsp. *cambrensis*

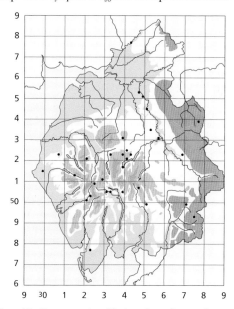
Map 47. *Dryopteris affinis* subsp. *borreri*

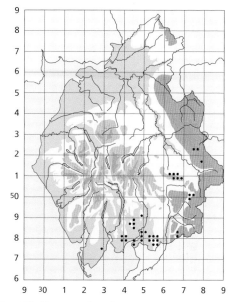
Map 48. *Dryopteris submontana*

Dryopteris dilatata, but is usually easily distinguished by its narrower, yellowish fronds and the pale, concolorous scales.

The statement that it is "frequent" in upper Teesdale (Eddy & Welch 1967) and the *Fern Atlas* dot for that square (7.3) must be based on misidentifications.

Alt. limit: 290 m, Long Cairn, Kershope (52.82n)

Map 49   259   WFCY

D. carthusiana x **D.dilatata**
D. x deweveri (J. Jansen) Wachter
A very rare hybrid found in a wooded flush near Brathay Hall (36.02) J.D. Lovis & R. Young, 1970, AMDE; BM) and in Cumberland at Culgaith (60.30, RWMC, 1982, BM, det. C. Fraser-Jenkins, *Watsonia* (1988) 17:183). There are pre-Survey records from Chapel Stile (3.0, G. Pincham, *c.*1879, BM) and more recently from near Matson Ground, Windermere (4.9, R. Kaye, undated, BM).

Wilson's records of *Lastrea spinulosa* var. *glandulosa* from Lyth Fell and Whitbarrow (4.8s) may refer to this hybrid which is certainly under-recorded.

Alt. limit: 145 m, Culgaith (60.30)

2   (W)FC

D. dilatata (Hoffm.) A.Gray   Broad buckler-fern
Extremely common in damp hedgerows and woodland. It is very shade tolerant, occurring in the less dense coniferous plantations and also, very characteristically, in birch woods over peat. It also grows in open mossland and extends high on the fells in block scree.

Alt. limit: 915 m, Helvellyn (30.14, Wilson)

Map 50   1630   LWFCY

D. dilatata x **D. expansa**
D. x ambroseae Fraser-Jenkins & Jermy
The only record of this hybrid south of the Scottish Highlands is one from Johnny's Wood, Borrowdale (24.14, M. Gibby & F. Jackson, 1972) where it was growing on scree.

1   C

D. expansa (C. Presl) Fraser-Jenkins & Jermy
Rare but no doubt overlooked. This critical species was seen during the Survey in seven Lake District tetrads, mostly in the central fells but there are outlying records from Wolf Crags, Matterdale (34.22, MC, 1978, BM) and in the south-west from Muncaster (10.98, AAD, 1990; 08.96, MMG, 1993, both LANC) and west of Wast Water (12.06, MMG, 1992, LANC). All these records have been confirmed by A.C. Jermy or M. Gibby. In the Pennines it occurs in Knock Ore Gill (70.30, K. Trewren, 1996, BM) and there is an early record from High Cup Nick (7.2, A.H.G. Alston 1939, BM). A.J. Richards has recorded it from there (74.26, 1975), Little Fell (78.22, 1970s) and limestone scree on Hillbeck Fell (80.18, 1976), but in the absence of verified specimens these records are omitted from the map. More recently, it was collected from the steep wooded bank of the Liddel Water at Penton Linns (42.76n, RWMC, 1989, LANC, det. A.C. Jermy).

There are earlier Lake District records, cited by Crabbe, Jermy & Walker (1970) from Holmrook (06.98, H.V. Corley, 1967, BM) and Keswick (2.2, C.H. Wright, 1859, K). The *Fern Atlas* omits the Holmrook record but includes post-1950 dots for the Ambleside (3.0), Martindale (4.1) and upper Teesdale (7.3) squares.

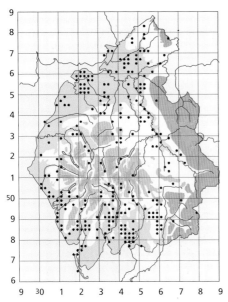

Map 49. *Dryopteris carthusiana*

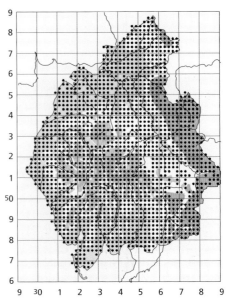

Map 50. *Dryopteris dilatata*

Although this is often a woodland species, several of the Cumbrian records are from sheltered crevices in block scree. In both habitats it is likely to be associated with *Dryopteris dilatata* rather than *D. carthusiana*. *D. expansa* is one of the diploid parents of the tetraploid *D. dilatata.*

Unfortunately there are few really satisfactory morphological characters to distinguish it. It has the pale, upright fronds of *D. dilatata,* but differs in having the pinnules almost dissected to their axis and with the segments clearly separated by a sinus. Apart from the chromosome number, the best diagnostic character is provided by the spore coat which is only sparsely furnished with short, blunt spines.

This fern is widespread in the Scottish Highlands and scattered in mountains to the south as far as west Wales.

Alt. limit: 915 m, Helvellyn (34.14)

Map 51 (post-1965 herbarium records)   13   WC

BLECHNACEAE

Blechnum spicant (L.) Roth   Hard fern
A strongly calcifuge fern, very common over much of the county but scarce on the Furness peninsula, the limestone and the more intensively farmed areas of the

Eden valley across to the Solway coast. It occurs chiefly on acid peat in woodland, both deciduous and coniferous, in hedgerows and the sides of ditches, and extending in the gills far up the fells.

Alt. limit: 760 m, Knock Fell (70.30)

Map 52   1167   LWFCY

AZOLLACEAE

*Azolla filiculoides Lam.   Water fern
This tropical American water-fern was discovered in 1985 in the Kendal canal at Stainton (52.84s, KR), where it was particularly abundant in 1991 and also at Crooklands (52.82s). It was later found further south at Farleton (52.80s, CEW, 1995). It is also known in the Lancashire section of the canal near Tewitfield locks and at several points north and south of Lancaster (Livermore & Livermore 1988). At Tewitfield it formed a conspicuous pink raft along the the edge of the canal and up to a metre wide, surviving severe frosts into early December. There is also a surprising record from a fellside pool above Rydal Water (34.06, PB, 1991, LANC), a most remarkable situation and possibly the highest and most northerly occurrence in Britain.

Alt. limit: 450 m, above Rydal Water (34.06)

        4   W

In the fellside woods the hard fern (*Blechnum spicant*) is abundant, its thick bush of barren fronds growing in damp places, with the fruiting fronds straight up the middle, like long combs with half the teeth knocked out.

Cumberland and Westmorland, Norman Nicholson

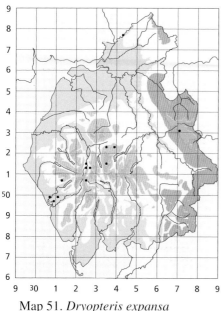

Map 51. *Dryopteris expansa*

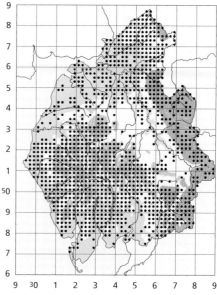

Map 52. *Blechnum spicant*

But this deformity, bad as it is, is not so obtrusive as the small patches and large tracts of larch-plantations that are over-running the hill sides...... It must be acknowledged that the larch, till it has outgrown the size of a shrub, shows, when looked at singly, some elegance in form and appearance, especially in spring, decorated, as it then is, by the pink tassels of its blossoms; but as a tree, it is less than any other pleasing: its branches (for boughs it has none) have no variety in the youth of the tree, and little dignity, even when it attains its full growth; *leaves* it cannot be said to have, consequently neither affords shade nor shelter. In spring the larch becomes green long before the native trees; and its green is so peculiar and vivid, that, finding nothing to harmonize with it, wherever it comes forth, a disagreeable speck is produced. In summer, when all other trees are in their pride, it is of a dingy,lifeless hue; in autumn of a spiritless unvaried yellow,and in winter it is still more lamentably distinguished from every other deciduous tree of the forest, for they seem only to sleep, but the larch appears absolutely dead.

Description of the Scenery of the Lakes, William Wordsworth

The planting is continuous, by the square mile.......What is seen is the rigid and monotonous ranks of spruce, dark green to blackish, goose-stepping on the fell side.......Bird life perishes,......And to the long unbroken mileage of drab, dead colour, with the sitka spruce king of this gloomy kingdom, you must add the curse of uniformity in growth. For conifers are patterned trees, branching evenly and subservient to arboreal geometry.......The sitka spruce and Norway spruce - the staple diet of waterlogged ground - Douglas fir, Corsican pine - how they repeat themselves: ages ago they learnt their lesson, and the 'damnable iteration' of them is an offence: they have the air of mass production, the efficiency of the machine.

Afforestation in the Lake District, H.H. Symonds

In any case, I do not think that conifer plantations invariably spoil the landscape. The dark colouring, the geometric lines and rectangles, give an oddness to the scene which I find not unpleasing. It is strange, it is sometimes harsh, but it is neither sentimental nor sham, and it is the sentimental and the sham which are the great dangers to the Lake District.

Cumberland and Westmorland, Norman Nicholson

The larch has certainly settled down, and there is not a lovelier green among the fells than the April green of the larch. It washes in a great gush up some of the ghylls, and well deserves to be defended by Faber, much less of a poet than Wordsworth, but, in this matter, surely clearer sighted.

"There is no other tree on earth but thou
Which brings the sky so near or makes it seem so blue"

Cumberland and Westmorland, Norman Nicholson

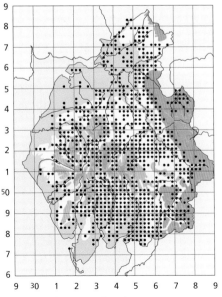

Map 53. *Larix decidua*

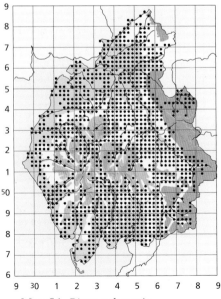

Map 54. *Pinus sylvestris*

SPERMATOPHYTINA

PINOPSIDA
(GYMNOSPERMAE)

PINACEAE

No attempt was made during the Survey to record the distributions of forestry conifers. Helpful comments on these species have been supplied by J. Voysey and P. Winchester.

*Pseudotsuga menziesii (Mirbel) Franco   Douglas fir
A widely planted and important timber tree although ecologically the most demanding of the commoner conifers requiring a deep, freely-drained soil and a sheltered situation. Most stands are pure and some of the most impressive are those by Thirlmere and in Grizedale Forest. The wood is particularly valuable for construction purposes.

*Tsuga heterophylla (Raf.) Sarg.   Western hemlock
Occasionally planted in the past but now rather rarely. Although a very elegant tree, very shade-tolerant and fast growing, the timber has few uses and the stems are susceptible to butt rot.

*Picea sitchensis (Bong.) Carrière   Sitka spruce
By far the most popular forestry conifer in Cumbria and in Britain as a whole, grown usually as pure, extremely dense and inpenetrable stands. Being fast-growing, profitable and tolerant of both exposure and peaty soils it is particularly well suited to upland afforestation, as, for example, in the Spadeadam and Kershope Forests. The wood is used both for pulp and construction.
Alt. limit: 510 m, Alston Moor (68.38), seedlings 600 m, Ruddy Gill (22.08)

*P. abies (L.) Karsten   Norway spruce
Frequent though less than in the past having been largely replaced by Sitka spruce. It is less tolerant of exposure and does best in somewhat flushed, sheltered situations. Most stands are pure but it is occasionally planted as a nurse for oak. There is a significant market for Christmas trees and the wood is used both for pulp and construction.

*Larix decidua Miller   European larch
Frequent, but absent in the more intensively agricultural lowlands and absent from the higher fells and peatlands. A native of central Europe, larch was introduced into Britain about 1620. It was widely planted both as an amenity tree, as, for example, along the Lancaster-Kendal canal, and by commercial forestry from the 18th century. It proved particularly popular as a nurse crop, usually for sycamore and beech, but its susceptibility to canker and insect attack has led to its extensive replacement by hybrid and Japanese larch.

Larch regenerates readily in deciduous woodland and in open grassland, as at the north end of Whitbarrow (4.8s). The wood is particularly useful for fencing. Surprisingly larch is not mentioned by either Hodgson or Baker.

It is widely regarded as a native tree, an illusion fostered by its prominent portrayal on the first edition of the Ordnance Survey Outdoor Leisure Maps of the Lake District.
Alt. limit: 660 m, south of Garrigill (76.38)
Map 53   870   LWFCY

*L. decidua x **L. kaempferi**   Hybrid larch
L. x marschlinsii Coaz
Occasionally planted when seedlings are available and producing better quality timber than Japanese larch.

*L. kaempferi (Lindley) Carriere   Japanese larch
Very common. This species has largely replaced European larch as a forestry tree being less susceptible to disease and more tolerant of exposure. There are fine stands above Thirlmere and in Grizedale Forest (3.9).

*Pinus sylvestris L.   Scots pine
Common and usually planted. It occurs on virtually all soil types, from lowland mosses to limestone pavement. It is often extensively self-sown, as on Meathop Moss by Morecambe Bay (44.80s). Although now native in Britain only in the Scottish Highlands it often looks thoroughly nativ as, for example, on the picturesque wooded rocky knolls between Ambleside and Rydal. The widespread plantings in England probably date from the 18th century following which it became commercially very popular but the plantings in Cumbria are mainly confined to the sandier soils around Carlisle and the Eden valley.
Alt. limit: 510 m, Alston Moor (68.38)
Map 54   1163   LWFCY

*P. nigra Arnold
subsp. **laricio** Maire   Corsican pine
Rarely planted being unsuited to a wet climate and exposure. There are, however, very prominent and inconcruous plantings at the southern end of Whitbarrow Scar (4.8s).

*P. contorta Douglas ex Loudon   Lodgepole pine
Not an important timber tree but being ecologically very tolerant it tends to be planted in the more exposed sites where it may be used as a nurse crop for Sitka spruce. It is particularly vulnerable to bark-stripping by deer. The wood is used mainly for chipboard.

CUPRESSACEAE

*__Chamaecyparis lawsoniana__ (A. Murray) Parl.

Lawson's cypress

Rarely planted and in little demand. The stems are particularly susceptible to forking and the timber is used for poles and fencing.

Alt. limit: 510 m, Kirkland Fell (66.32), but the record possibly refers to *Thuja plicata*.

*__Thuja plicata__ Donn ex D. Don   Western red cedar

Seldom planted but preferred to Lawson's cypress. Most of the plantings are on the limestone in the south, for example in Middlebarrow and Brigsteer Woods (4.7s, 4.8s), but the most conspicuous are those around the head of Thirlmere. When mature, the trees regenerate extremely freely.

__Juniperus communis__ L.   Juniper
subsp. **communis**

Widespread in the Lake District and south Westmorland and probably as frequent here as anywhere in England. Elsewhere in the county it is extremely rare. In the north Pennines, apart from planted bushes at Moorhouse (74.32), it is known only from two sites, both now reduced to a single bush. Ecologically juniper is extremely catholic, forming dense thickets on some of the Lake District fellsides, notably above Little Langdale, and appearing equally at home on the limestone plateau of Whitbarrow (4.8s). It also occurs, though rarely, on fen peat as at Hale Moss (50.76s).

Although long-lived, up to at least 200 years, the existing populations are likely to disappear as a consequence of grazing and, to a lesser extent, poor viability of the berries, unfavourable soil conditions and other establishment problems. Regeneration may also be encouraged by light burning. Juniper is a pioneer species but the prospects for the successful natural establishment of new populations appear bleak.

Morphologically, juniper is extremely variable. The Lake District bushes are mostly broad-topped, with ascending branches, although these may subsequently collapse with age. On Whitbarrow and Arnside Knott (4.7s) bushes such as these occur intermixed with a wide range of other forms including semi-prostrate and columnar ones.

Alt. limit: 685 m, Grasmoor (16.20)

Map 55 (of the species), Plates 55 & 56   335 LWFCY

The juniper may sometimes be seen in the lower dales as a tight dark flame of a tree, but more often it belongs to the desolate fellside. Here it does not grow tree-like, but straggles as a thick bush, from a distance looking like gorse when kissing is out of season. It is then a shrub of the damp slopes, where the mist hangs in it, bunched over the branches like muslin over currant bushes. Then it seems a bitter, sullen green. Yet when you look closer, its leaves have an under sheen of blue, and the spider-webs are laid all over them, gleaming with the wet. You forget its shapelessness then, because of the patterns of spine, and web and water.

Cumberland and Westmorland, Norman Nicholson

Plates 55 & 56. Ascending and columnar forms of juniper, above Whitbarrow Scar

subsp. **nana** (Willd.) Syme
(subsp. *alpina* Čelak.)
Scattered in the central Lake District, occurring chiefly on rock ledges. However, it is probably more widespread than is indicated by the map.

These prostrate plants are assumed to be all genetically dwarf. They are also characterised by shorter and only slightly divergent leaves.
Alt. range: *c*.500 m, Scar Crags (20.20) - 760 m, Horn Crag, Scafell (20.04)
Map 56, Plate 57   25   WFC

TAXACEAE

Taxus baccata L.   Yew
Very common in south Westmorland, elsewhere fairly frequent but usually planted. As a native tree it is very characteristic of the lowland limestone deciduous woodlands, often forming dense local subcommunities with few associated species. It is frequent also on the lower limestone scars and, by contrast, on the Silurian crags of the southern Lake District and in dense juniper scrub, both habitats affording some protection from sheep grazing.
Alt. limit: at least 300 m, Helbeck Wood, Brough (78.16) and above Barras (84.08)
Map 57   497   LWFCY

> Yews are particularly plentiful on the limestone: there is a wood of them near Beetham, dark and Gothic, like a crypt with red pillars and blue-green fan-vaulting.

Cumberland and Westmorland, Norman Nicholson

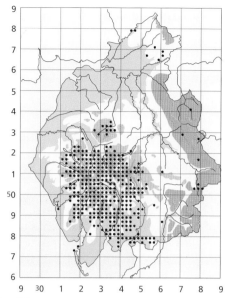

Map 55. *Juniperus communis*

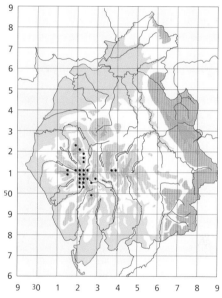

Map 56. *Juniperus communis* subsp. *nana*

Plate 57. *Juniperus communis* subsp. *nana*

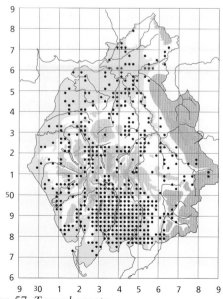

Map 57. *Taxus baccata*

MAGNOLIOPSIDA
(ANGIOSPERMAE)

MAGNOLIDAE
(DICOTYLEDONES)

ARISTOLOCHIACEAE

*Asarum europaeum L.   Asarabacca
The only Survey records are from Lindale (4.7s, KAG & FLW, 1970s), from where it has now disappeared, an estate on the north-east side of Bassenthwaite Lake (22.28, ES, 1974) where it is well-naturalised, and in the grounds of Dalton Hall, Burton-in-Kendal (54.74s), a site listed by Wilson. It persisted in a plantation at Dalton into the late 1970s when it was photographed by R. Entwistle.

Hodgson lists it from Gilgarran and Clifton Hall (0.2, E.J. Glaister, 1889, CLE), Troutbeck, Borrowdale (2.1), Ormathwaite (2.2) and Hutton Woods (4.3). There is also a record from an orchard at Armathwaite (5.4, ex herb. Winch, BM) and the churchyard at Isel (1.3, C.B. Sharpe, 1930s). Wilson mentions a site at Gill Head, Windermere (3.9, actually in Furness, which could be the same as "Lake Windermere", Miss Wright, 1843, BM), Kirkby Lonsdale (6.7s) and Dale Head, Martindale (4.1).

All the sites are near or in the grounds of houses and reflect the one-time medicinal importance of asarabacca.

               3   L(W)FC

NYMPHAEACEAE

Nymphaea alba L.   White water-lily
A common feature of tarns and sheltered lakeside bays, chiefly in the Lake District and in the south. It prefers peaty, mesotrophic water and is absent from the tarns of the limestone country between Asby and Orton, notably Sunbiggin Tarn (66.06). Its generally western tendency is typical of the species in northern Britain.
Alt. limit: 405 m, Dock Tarn, Watendlath (26.14) although formerly at 480 m in Angle Tarn, Patterdale (4.1).
Map 58   94   WFC

Nuphar lutea (L.) Smith   Yellow water-lily
Occurring in similar situations to *Nymphaea alba* with which it often grows. It also occurs occasionally in ditches and slow-flowing rivers. It is noticeably less common than *Nymphaea* and has a more pronounced south-western tendency.
Alt. limit: 395 m, head of Wythburn (30.10)
Map 59   90   WFC

N. lutea x **N. pumila** (Timm) DC.
N. x spenneriana Gaudin
This extremely rare hybrid was discovered in 1978 by RS in Blea Tarn, Armboth (28.14, herb. R.S., *Watsonia* (1979) 12:348). The only previous English record was from Northumberland. The following year he found a second colony, at nearby Dock Tarn (26.14, herb. R.S.). The mean carpel number in both populations lies between 11 and 11.5 and the material from Blea Tarn was verified by Y. Heslop-Harrison. With the exception of a site in Shropshire, *Nuphar pumila* is now restricted

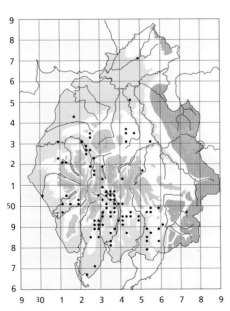

Map 58. *Nymphaea alba*

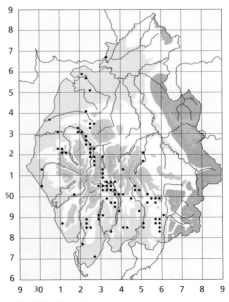

Map 59. *Nuphar lutea*

to Scotland.
Alt. range: 400 m, Dock Tarn (26.14) and 475 m, Blea Tarn (28.14)

2   C

CERATOPHYLLACEAE

Ceratophyllum demersum L.   Rigid hornwort
Very rare and now found chiefly in the south where it occurs in Wyndhammere (58.84s, RS, 1976, LANC), in the Lancaster - Kendal canal at Crooklands (52.82s) and Stainton (52.84s), where it appears to have almost ousted the more delicate *Ceratophyllum submersum*, filling the drains on Foulshaw Moss, Levens (46.80-84s), at Ormsgill reservoir, Barrow-in-Furness (18.70), and nearby at Roanhead (20.74). The only other record is from a recently constructed fish pond north-east of Penrith (52.34, KR, 1991; LANC, *Watsonia* (1992) 19:142) where it was probably introduced.

The earliest record is from Wray Castle boat-house on Windermere in 1933 (3.9, W.H. Pearsall, PLYP, *BEC Rpt* (1934) 10:542) and the following year Wilson found it on the Westmorland side of the ferry (YRK). Both records could be predated by M. Mason's from Little Musgrave (7.1), cited by Wilson, although in the absence of material and any other record from the Eden valley the record is perhaps best regarded as doubtful. The only previous Cumberland record is from Monkhill Lough, Carlisle (3.5, E. Blezard & C.W. Muirhead, 1950), which was drained shortly after-wards.
Alt. limit: 180 m, Wyndhammere (58.84s)
Map 60   9   WFC

C. submersum L.   Soft hornwort
Extremely rare and probably decreasing. It is now known only from the Lancaster - Kendal canal at Stainton (52.84s, RS, 1976; LANC), where it was first collected by M.R. Gilson in 1963 (FBA, *Proc. BSBI* (1964) 5:347) and where it is currently succumbing to competition from *Ceratophyllum demersum*. The only other record is from Monkhill Lough, Carlisle (3.5, E. Blezard & C.W. Muirhead, 1950, CLE, PLYP) where, as at Stainton, it was growing with *C. demersum*.

These sites are the most northerly ones in Britain.

1   W(C)

RANUNCULACEAE

Caltha palustris L. Marsh marigold; Kingcup
Very common throughout the county; rare or absent only on the highest fells of the Lake District and the more intensively farmed areas. It occurs in marshes, ditches and wet meadows, the margins of lakes and tarns, in wet woodlands and in flushes on the fells. Upland plants are smaller and more decumbent and are sometimes given subspecific rank as var. *radicans* (T.F. Forster) Hook.
Alt. limit: 915 m, Scafell (20.06, R.J. Birkett, pre-Survey)
Map 61   1370   WFCY

Trollius europaeus L.   Globeflower
Uncommon, mainly restricted to the uplands where it occurs in traditionally managed hay-meadows, on becksides and islands, in ungrazed areas of light wood-land, and on moist rock ledges and gullies where it is inaccessible to sheep. Although not plentiful, it is a characteristic feature, with *Galium boreale*, of rocky banks along the middle stretches of the River Lune and also of railwaysides in Mallerstang.

The globeflower has decreased markedly over the last 50 years. Wilson described it as "frequent and locally abundant" but this is now only true of meadows

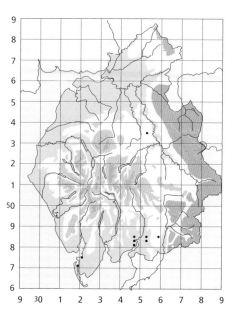

Map 60. *Ceratophyllum demersum*

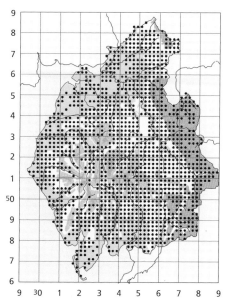

Map 61. *Caltha palustris*

in the Mallerstang area (7.9). This decline is partly due to its popularity as a garden plant but chiefly it results from increased sheep-grazing combined with the 'improvement' of upland hay meadows by drainage, artificial fertilisers and cutting earlier for silage. The Survey failed to refind it in a quarter of the *Atlas* squares.

Alt. limit: 900 m, Scafell (2.0, R.J. Birkett)

Map 62, Plate 58   218   WFCY

*Helleborus foetidus L.   Stinking hellebore

A garden escape naturalised in woodland in a few sites, mostly on the limestone around the Kent estuary. In the north of the county there are records from Southwaite (4.4, P. Hartley, 1947), Penrith (5.2, R.C.L. Howitt, 1957) and an undated, anonymous *Scarce Plants* record for the Wigton square (2.4). It still occurs at Sebergham (34.40, REG, 1984).

Alt. limit: 140 m, west of Kendal (50.90)

Map 63   S 9   WFC

H. viridis L.   Green hellebore

Scattered in limestone woodland around the Kent estuary. Elsewhere it is very rare, occurring usually on roadsides and probably a garden escape. Hodgson refers to a particularly fine site in Threapland Gill, Aspatria (1.4) devastated in the 19th century by the Maryport Iron Company. D. Hinchcliffe comments that it is sometimes grown by farms and has been used within living memory as a cure for cattle ailments. It still persists, for example, in the garden of a disused farmhouse above Crosby Ravensworth (62.10).

Alt. limit: 270 m, west of Shap (54.14) and above Crosby Ravensworth (62.10)

Map 64   21   WFC

*Eranthis hyemalis (L.)Salisb.   Winter aconite

Occasionally naturalised, sometimes abundantly so, in estate grounds, on roadside verges and, near Milnthorpe (48.80s), on a disused railway embankment. The only pre-Survey records of naturalised plants are from Abbey[town] (1.5, J. Leitch, 1876, CLE), the undated one for Thackwood Nook (4.4, J. Curwen, CLE), the *Atlas* one for the Sebergham square (3.4) and others for Great Salkeld (5.3, G.A.K. Hervey, 1946-57), Stainton (4.2, B.H. Thompson, 1944), the streamside at Rottington, St Bees (9.1, C.D. Pigott, 1944) and Aldingham (2.7, G. Wilson, 1960s).

Alt. limit: 180 m, Great Strickland (56.22)

Map 65   13   W(F)C

*Aconitum napellus L. and   Monk's-hood
A. napellus x **A. variegatum** L.
A. x cammerum L.

An occasional garden escape, occurring chiefly in the middle and lower Eden valley and south Westmorland. It occurs on roadsides but more usually in woodland, sometimes looking almost native as on the banks of the River Lowther and the Eden below Kirkby Stephen (76.10). It has long been cultivated and is particularly common in old farmhouse gardens.

It is not possible to make any definite statement about the occurrence, or relative frequency, of *Aconitum napellus* L. and its hybrid with *A. variegatum* L. Of the few recent herbarium specimens in LANC only one appears to be of the hybrid. One from Preston Patrick (56.84s, CEW, 1988) also agrees with the hybrid in having almost glabrous pedicels, but those of a plant from Garsdale Head (78.92, GH, 1987) are distinctly pubescent.

Plate 58. *Trollius europaeus*, Mallerstang

The first Furness record was from Old Park Wood, Holker (3.7s) where it was found by the Barrow-in-Furness N.H.S. in the early 1950s. Judging from the few records in the Floras (Wilson gives only one) the species appears to be increasing.

Alt. limit: 350 m, Garrigill (74.40)

Map 66   S 68   WFCY

*Consolida ajacis (L.) Schur   Larkspur
(*C. ambigua* auct.)
A Mediterranean species listed by Hodgson as a cornfield weed at Dean, near Cockermouth (1.2), Maryport (0.3) and Silloth (1.5, J. Leitch, 1883, CLE) and from the Duddon woods (1.8/1.9). It was also seen in 1872 at Grune Point (1.5, J. Leitch, CLE) and in 1932 at Thwaites, near Millom (1.8, M. Cross).

(C)

*C. regalis Gray
A widespread Continental species reported by Hodgson from Silloth (1.5, J. Leitch, 1889, 1890, CLE).

(C)

Actaea spicata L.   Baneberry
Known from two tetrads at Arnside (4.7s), where it grows on the edges of limestone woodland, by paths and sometimes on disturbed ground, but it is being gradually shaded out. An important unexpected record is that from a grike in limestone pavement above Sunbiggin (6.0, CW, 1976,1979).

Baker lists this from "a rocky wood on the limestone between Kendal and Arnside Knott" and he also gives two records which seem extremely unlikely: "mountainous pastures above Troutbeck" (4.0) and

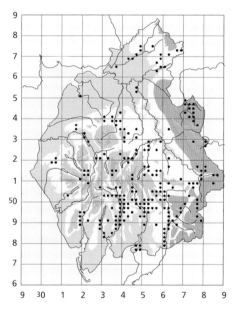

Map 62. *Trollius europaeus*

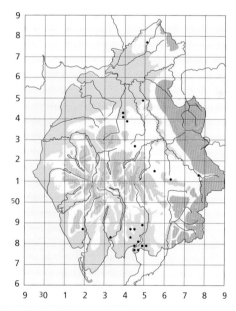

Map 64. *Helleborus viridis*

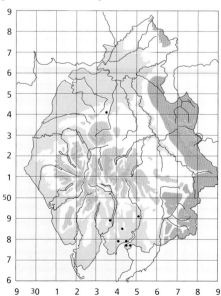

Map 63. *Helleborus foetidus*

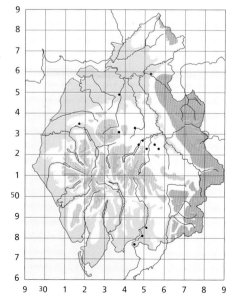

Map 65. *Eranthis hyemalis*

Sandwick, Ullswater (4.1/4.2). Hodgson cites only a "doubtfully native" record from the grounds of Gilgarran House, Distington (0.2, J. Curwen, undated, CLE) and Wilson accepts only the Kendal-Arnside record and one from near Arnside (J.A. Martindale, 1905, KDL). In the mid-1930s it was reported by J. Parkin in Herbert's Wood, Dovenby (1.3). It also occurs on the Lancashire side of Ease Gill (6.8s).

     In Britain this species is considered to be native only in northern England, from Morecambe Bay across to the Yorkshire coast.

Alt. limit: 345 m, Great Asby Scar (6.0)

               S 3   .W(C)

Anemone nemorosa L.   Wood anemone; Windflower
A common member of the spring flora of damp deciduous woodland throughout the county occurring on a wide range of soils. It grows also in shady hedgerows, in the grikes of open limestone pavement, in gills, on open fellsides - often under *Pteridium*, where it is probably indicative of former woodland, and high up the fells in gullies and on rock ledges.

Alt. limit: 870 m, Helvellyn (34.14)

Map 67   1070   LWFCY

*****Anemone blanda** Schott & Kotschy
A popular garden plant discovered in 1990 on a roadside verge far from houses between Staveley and Kentmere (44.00, CEW, LANC, *Watsonia* (1991) 18:420).

             1   W

*****A. ranunculoides** L.
Seen growing in silt by the River Eden at Wetheral (46.54, CS, 1978; 1993) and at Heversham (50.84s, CEW, 1987), where four plants were growing by a woodland path. The only previous record is Wilson's from a wood near Crackenthwaite Hall (6.2).

     This pretty, yellow-flowered anemone is not native in Britain but has a very wide distribution on the Continent.

             2   WC

Clematis vitalba L.   Old man's beard; Traveller's-joy
Locally common on the limestone around Morecambe Bay where it is a very conspicuous feature in late summer of hedgerows and woodland, particularly around Grange-over-Sands (4.8s), and there seems little reason to doubt that it is native there. Elsewhere it is rather rare and probably planted, although this seems unlikely on the Eskmeals (06.94) and Drigg (06.96) sand-dunes.

Alt. limit: 155 m, Hutton Roof Crags (54.76s)

Map 68   42   LWFC

Ranunculus acris L.   Meadow buttercup
Extremely common and absent only from the highest fells and moorland. In the lowlands it occurs chiefly in pastures and hay meadows, on roadside verges and by

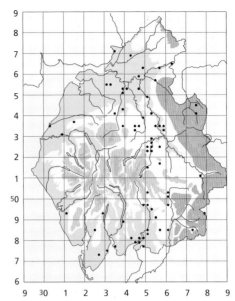

Map 66. *Aconitum napellus*

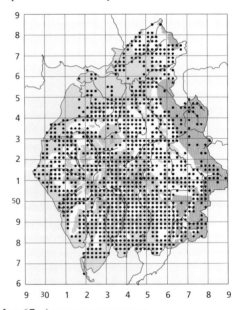

Map 67. *Anemone nemorosa*

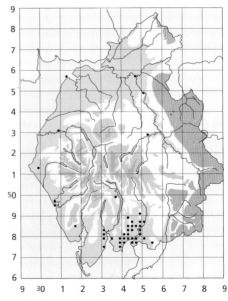

Map 68. *Clematis vitalba*

railwaysides. It ascends high up the fells in seepage areas and flushes, often in company with *Juncus effusus, Viola palustris* and *Stellaria uliginosa.*
Alt. limit: 870 m, Helvellyn (3.1, Wilson)
Map 69   1732   LWFCY

R. repens L.   Creeping buttercup
Like the last, this is one of the commonest plants in the county. It grows in similar habitats but generally prefers the wetter, heavier soils, extending into wet pastures and marsh communities. Its stoloniferous habit ensures that it is a particularly successful weed of grassland, gardens and waste sites.
Alt. limit: 850 m, Great Dun Fell (70.32) and Cross Fell (68.34)
Map 70   1739   LWFCY

R. bulbosus L.   Bulbous buttercup
A common and variable buttercup of well-drained lowland, mesotrophic grasslands, also along practically the entire coast and on the high limestone grasslands of the Pennines. On account of its rather brief flowering season it may be somewhat under-recorded.

    It is very variable morphologically, particularly in leaf-shape. This is to some extent correlated with habitat, although two collections (LANC) from grassland in east Westmorland have been determined by P.D. Sell as var. *dunensis* Druce.
Alt. limit: 580 m, Hartside (64.40), substantially higher than the 350 m given by Wilson. Hodgson's record from the summit of Cross Fell seems improbable.
Map 71   878   LWFCY

R. sardous Crantz
An extremely rare casual known chiefly from coastal areas and now all but extinct. The only Survey records are from a damp field between Siddick and Flimby (00.30, A. Mitchell, 1980; 1988, LANC), the coast near Bootle (06.88, MMG, 1983), North Walney (16.70, JMA, 1989) and the car-park at Askam-in-Furness (20.76, PB, 1993, LANC).

    Hodgson lists it from Workington (9.2), Drigg (0.9), Seascale (0.0), Maryport (0.3), Aspatria (1.4), Thackthwaite (2.4, E.J. Glaister, 1877, CLE) and Rockcliffe (3.6) and there are further records from a cornfield at Silloth (1.5, J. Leitch, 1882, CLE) and Bolton Wood (2.4, E.J. Glaister, 1877, CLE). It is recorded by Petty (1894 p.150) from near the Duddon (2.8) and on Walney Island.

    Rather surprisingly there are several post-1930 *Atlas* records, including ones for the mid-Walney (1.6s), Ravenglass (0.9), Gosforth (0.0), Maryport (0.3), Allonby (0.4), Wasdale (1.0) and Aspatria (1.4) squares.

                        4   FC

*R. parviflorus L.
The only record is that given by Hodgson for Silloth docks (1.5, J. Leitch, E.J. Glaister, 1878, CLE), although Druce

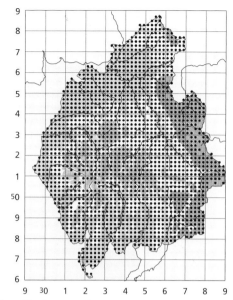

Map 69. *Ranunculua acris*

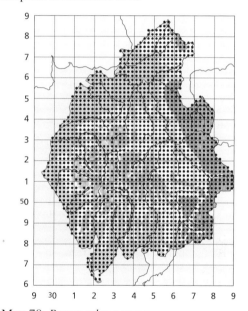

Map 70. *Ranunculus repens*

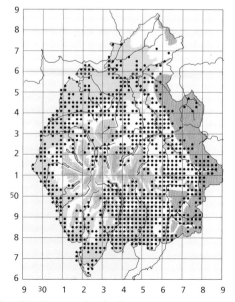

Map 71. *Ranunculus bulbosus*

(1932) lists it for v.c.69 and Wilson comments that it occurs in Furness. The Silloth record is the most northerly in Britain.

S   (FC)

R. arvensis L.   Corn buttercup

An extremely rare weed of cultivated ground and now almost extinct. It was seen only three times during the Survey: in gardens at Carlisle (36.56, REG, 1977) and Eskmeals (08.92, A. Warburton, 1975), where it survived for a few years, and at Haverigg (14.78, J.D. Williamson & A. Watson, 1977).

Hodgson records it from Nethertown (9.0), St Bees (9.1), Workington (9.2) and Dalston (3.5) and Wilson from Heathwaite, Windermere (4.9, M. Jackson, 1909, KDL). There is an 1899 record from Grune Point (1.5, E.J. Glaister), one from 1901 from by the River Eden at Grinsdale, Carlisle (3.5) and another from Thwaites, near Millom (1.8, M. Cross, 1932). In addition there is the pre-1930 *Atlas* record for the Grizebeck square (2.8), later ones for the lower Eskdale (1.9) and Kirkbride (2.5) squares and anonymous *Scarce Plants* post-1950 records for the Whitbarrow (4.8s) and Milnthorpe (5.8s) squares.

S 3   (WF)C

R. auricomus L.   Goldilocks

Common in a broad band to the west of the Eden valley and around its head, also in south Westmorland, but otherwise surprisingly local and practically absent from west Cumberland and the Furness peninsula. It requires some shade and occurs chiefly in hedgerows, woodland and roadside verges, usually on the better soils; there are also a number of records from lake shores and islands.

It flowers early and, like *Adoxa*, is rarely recorded after the end of May.

Alt. limit: 550 m, Moor House, Teesdale (74.32), where it was probably introduced unintentionally.

Map 72   283   WFCY

R. sceleratus L.   Celery-leaved buttercup

Not infrequent by muddy, brackish pools and the sides of dykes by the Solway Firth and around Morecambe Bay and the Duddon estuary. Elsewhere it is very rare although there is a cluster of records from the muddy margins of ponds in the middle Eden valley. The scarcity of inland records is characteristic of the species in north-west England and Scotland.

Alt. limit: 170 m, Askham (50.22)

Map 73   125   WFC

R. lingua L.   Greater spearwort

Rare, known now only from the margins of three eutrophic ponds and tarns on the Furness peninsula, from fens at Esthwaite (34.96), Newton Reigny (46.30) and Temple Sowerby (60.26) and a ditch at Eskmeals (08.92). This species has decreased particularly in Cumberland: Hodgson, for example, gives seven localities.

It is known to have been introduced into at least three recently created roadside ponds north of Ellonby (42.34, 42.36) and records from ponds at Holehird, Windermere (40.00, RS, 1978) and Sizergh Castle (48.86s) probably also represent plantings.

Alt. limit: 170 m, Newton Reigny (46.30)

13   WFC

R. flammula L.   Lesser spearwort

Very common in marshy ground - in ditches, by lakes and tarns, and in seepage areas and runnels on the fells. Only in areas of intensive agriculture, such as the Eden valley and the Solway Plain, is it at all uncommon. As Wilson remarks, it is morphologically very variable. Var. *tenuifolius* Wallr. (var. *radicans* Nolte), with nodal adventitious roots, is particularly characteristic of open, stony sites such as lake shores and has been much confused with the following hybrid with which it sometimes grows.

Alt. limit: 850 m, Cross Fell (68.34)

Map 74   1477   LWFCY

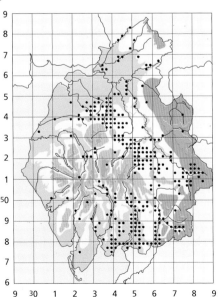

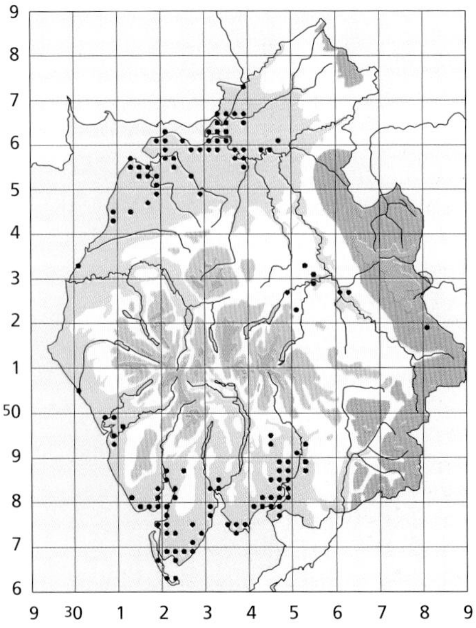

Map 72. *Ranunculus auricomus*

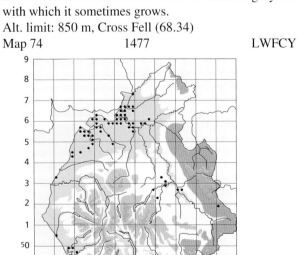

Map 73. *Ranunculus sceleratus*

R. flammula x **R. reptans** L.

R. x **levenensis** Druce ex Gornall

Very rare, occurring usually on mud and sand on stony lake shores, often intermixed with *Littorella*, along the length of Ullswater (38.l6 - 46.24), at the south end of Coniston (28.90), at Bowness, Windermere (38.94, GH, 1970s; LANC) and at the north end of Bassenthwaite (20.30, ES, 1987). Material from all these sites has been confirmed by R.J. Gornall. The only other confirmed record is from Thompson's Holme, Windermere (3.9, C. Waterfall, 1895, LIV). Its survival at Ullswater, particularly along the northern shore and at Pooley Bridge, is uncertain as the sites are under considerable tourist pressure.

For two centuries there have been frequent reports of a small spearwort occurring on lake shores in the Lake District, notably from Coniston, Windermere and Ullswater. These plants, which have narrow leaves and root at the nodes, have been variously referred to *Ranunculus reptans, R. flammula* var. *tenuifolius* (var. *radicans*) or var. *pseudoreptans* Syme. Druce was the first to suggest a hybrid origin for these plants and similar ones from Loch Leven, Fifeshire, and this view has recently been upheld by Gornall (1987), although, as mentioned under *R. flammula*, several collections supposedly of this taxon have proved to be *R. flammula* var. *tenuifolius*. Gornall has also reported hybrids from south-west Scotland. The hybrids are quite fertile and he speculates that hybridisation has occurred following the introduction of the occasional fruit of the non-British *R. reptans* by wildfowl migrating southwards from Iceland or Scandinavia.

Plate 59. *Ranunculus flammula* x *R. reptans*

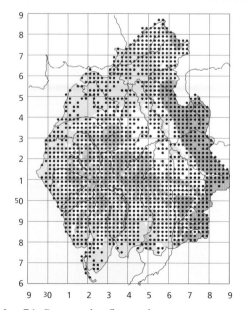

Map 74. *Ranunculus flammula*

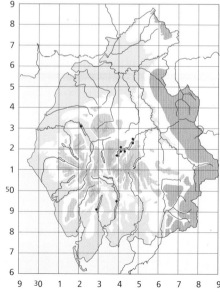

Map 75. *Ranunculus flammula* x *R. reptans*

The achenes of the hybrid have a proportionally shorter beak. However it is often possible to give a certain identification only after cultivation.

Alt. limit: 145 m, by Ullswater (3.1, 4.2)

Map 75, Plate 59   R 9   WFC

R. ficaria L.   Lesser celandine
Map 76   1361   LWFCY

subsp. **ficaria**
Very common in the lowlands, occurring in damp, often rather heavy soils, on roadside verges and banks, in deciduous woodland usually by streams and in seepage areas, and on the lower slopes of the fells in *Agrostis - Festuca* grassland.

This diploid subspecies is much under-recorded: its actual distribution is probably that of the species.

Alt. limit: 630 m, Dove Crag (36.10)

Map 77   582   LWFCY

subsp. **bulbilifer** Lambinon
(subsp. **bulbifer** Lawalrée)
Locally common in damp, shaded habitats, chiefly in the east. These are usually, but certainly not always, anthropogenic, for example gardens, where it can be a pernicious weed, and by woodland paths.

Its leaves are perhaps more distinctly lobed than those of subsp. *ficaria*, with which it sometimes grows. Large dense colonies usually belong to this sub-species. The plants die down early leaving a mass of axillary bulbils lying on the ground. Although most plants are probably tetraploid (Stace 1991), the only population counted during the Survey, from a garden at Burton-in-Kendal (52.76s), consisted mainly of triploids ($2n=24$) with only a few tetraploids ($2n=32$).

Alt. limit: at least 280 m, south and east of Alston (70.44, 72.46)

Map 78   358   LWFCY

R. hederaceus L.   Ivy-leaved crowfoot
Fairly frequent in shallow lowland ditches, muddy tracks and the margins of pools and often associated with *Montia fontana* and *Stellaria uliginosa*. It also grows high on the fells, particularly in the Pennines, amongst mosses in upland seepage areas and springs.

Alt. limit: 770 m, Little Dun Fell (70.32)

Map 79   391   WFCY

cf. **R. hederaceus** x **R. peltatus**
A curious plant with distinctive, acute-lobed leaves was collected by W.H. Pearsall in 1913 from a peaty drain close to the River Leven at Haverthwaite (5.8s, YRK). It has been tentatively identified by S.D. Webster as this hybrid. Material was distributed by Pearsall (*BEC Rpt* not traced). The cutting appended to the sheet indicates general agreement that *Ranunculus peltatus* was one of the parents and E.S. Marshall surmised that *R. omiophyllus* might be the other. Cook (in Stace 1975) gives no record for the *R. hederaceus* hybrid and only one for that involving *R. omiophyllus*.

(F)

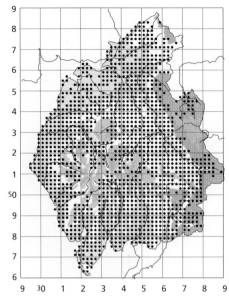

Map 76. *Ranunculus ficaria*

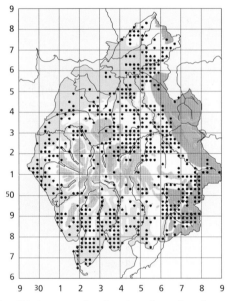

Map 77. *Ranunculus ficaria* subsp. *ficaria*

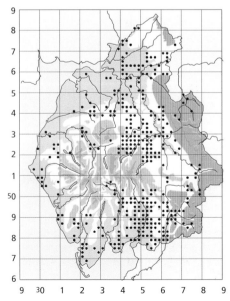

Map 78. *Ranunculus ficaria* subsp. *bulbilifer*

R. omiophyllus Ten.

In very similar habitats to *Ranunculus hederaceus*, but less frequent and with a somewhat more upland tendency. It comes into flower early and in mild winters may flower continuously.

Alt. limit: 780 m, Little Dun Fell (70.32, Wilson)

Map 80   389   WFCY

Fine-leaved water buttercups/crowfoots

Material of the following seven species can often be extremely difficult to identify. This is largely due to the high degree of environmental variation shown by the finely-divided submerged leaves but also to the difficulty of finding flowering material in many rivers. Specimens with abnormal terrestrial, early or late season leaves are particularly troublesome and often defy identification. Many of the Survey specimens were checked by N.T.H. Holmes and latterly by S.D. Webster. Because of the importance of accurate identification, all the species must be considered underrecorded.

R. baudotii Godr.

Scattered in ditches, pools and creeks around the entire coast but only frequent along the Solway Firth. There are two records from sites slightly inland: Longlands pond, Egremont (00.12) and Brickworks pond, Waverton (22.46), both contributed by RS. The records for the Aspatria (1.4), Bassenthwaite (2.3) and Shap (5.1) squares in *Scarce Plants* are based on misidentifications.

Map 81   28   WFC

R. trichophyllus Chaix.

Not infrequent in the limestone becks and tarns of north Westmorland. Elsewhere it is very scattered with no obvious pattern. Surprisingly one of the sites is at 310 m on Alston Moor (70.42, GH, 1993, LANC).

This species has often been confused with early-flowering and non-laminate leaved states of *Ranunculus aquatilis* and *R. peltatus*.

Alt. limit: 310 m, above Leadgate, Alston (70.42)

Map 82   32   WFC

R. aquatilis L.

A widely distributed but not particularly frequent plant of mesotrophic and eutrophic lowland tarns, ponds and ditches. It has also been recorded from the River Derwent (1.3), the middle Lune (60.84s) and Bassenthwaite Lake (22.26).

It is probably appreciably commoner than the records suggest since it is often confused with *Ranunculus peltatus* and this confusion has persisted during the Survey owing to the absence of the commoner *R. peltatus* from the recording card.

Alt. limit: 445 m, Small Water, above Haweswater (44.10)

Map 83   43   WFC

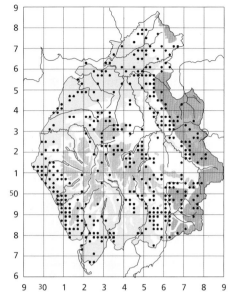

Map 79. *Ranunculus hederaceus*

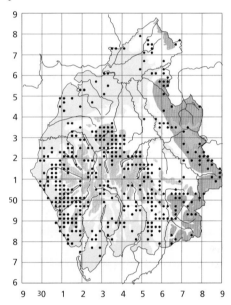

Map 80. *Ranunculus omiophyllus*

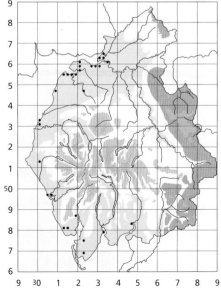

Map 81. *Ranunculus baudotii*

R. aquatilis x **R. fluitans**
R. trichophyllus x **R. fluitans**
R. x bachii Wirtgen
The only records for this rare sterile hybrid are from
the River Eden at Warwick Bridge and Great Salkeld
(respectively 46.56, 1983; 54.36, 1983, both GH,
LANC) and the River Kent north of Sedgwick (50.86s,
J.M. Lock, 1978, LANC). There is also a record probably of this hybrid from the Eden at Rickerby, Carlisle
(40.56, C.D. Preston, 1989, CGE, LANC). All the
collections were determined by Webster. Somewhat
surprising is the apparent absence of *Ranunculus
aquatilis* and *R. trichophyllus* from both rivers.

The only British records of this hybrid are from
northern England.

3   WC

R. peltatus Schrank
Widespread but scarce in ponds, tarns and lakes over
the entire trophic range but less common on the
limestone than *Ranunculus trichophyllus* and *R. aquatilis*. Several early Survey records of *R. penicillatus*
from the River Eden have been re-determined by
Webster as *R. peltatus*, and this appears to be the
predominant species in the lakes.

This species is more robust than *R. aquatilis*.
The best distinguishing characters are the larger petals
(12 mm or more) with pyriform, not circular, nectar
pits at the base.
Alt. limit: 500 m, Dogber Tarn, Musgrave Fell (78.18)
Map 84   118   WFC

R. peltatus x **R. fluitans**
R. x kelchoensis S. Webster
A robust and sterile hybrid reported by Webster (1990) from
the River Eden at Appleby (6.1/6.2, P.H. Raven, 1961, BM).
Its presence in the Eden is not surprising in view of the wide
distribution in the river of both parents.

(W)

R. penicillatus (Dumort.) Bab.
In her recent review of this species, Webster (1988)
concludes that it is represented in Britain by two
subspecies, both of which occur in Cumbria. Subsp.
penicillatus has laminate leaves when flowering and is
distinguished from *Ranunculus peltatus* by the summer
capillary leaves being usually appreciably longer than
the corresponding internodes. Flowering specimens of
subsp. *pseudofluitans* lack laminate leaves. Their
capillary leaves are shorter than those of *R. fluitans* and
at least 5-times forked compared with 4-times in that
species.

subsp. **penicillatus**
Recorded only from the River Eamont at Brougham
(52.28, GH, 1981, LANC) and the Haweswater Beck
(50.14, 1982, LANC, 50.16, 1994, both GH), the two
earlier records being confirmed by Webster. She has
redetermined as *R. peltatus* plants from the River Eden,
originally identified by N.T.H. Holmes as this taxon

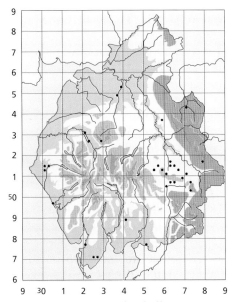

Map 82. *Ranunculus trichophyllus*

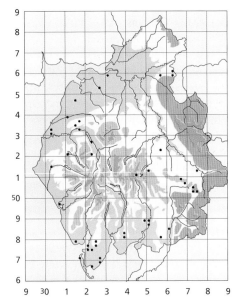

Map 83. *Ranunculus aquatilis*

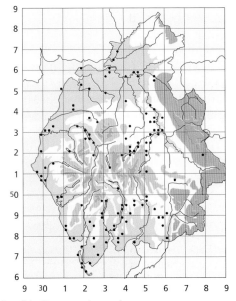

Map 84. *Ranunculus peltatus*

and her map (Webster 1988) and that of Preston & Croft (1997) require revision.

This is a western plant, restricted in England to Devon, Wales and Cumbria.

Alt. limit: 210 m, Haweswater Beck (50.14)

3   WC

subsp. **pseudofluitans** (Syme) S. Webster

This subspecies is considered by Webster to include two variants. Var. *pseudofluitans* has summer leaves usually exceeding the internodes and petioles up to 15 cm long whereas var. *vertumnus* C.D.K. Cook has leaves seldom exceeding the internodes and petioles less than 3 cm.

Var. *pseudofluitans* occurs in the River Kent (5.8s, 5.9), the Eamont at Brougham (52.28), the Ehen (0.0), the Petteril at Little Blencow (44.32), the Eden near Great Salkeld (56.34) and Warwick Bridge (46.56) and in Blackdike Beck, Matterdale (40.22). There is also an earlier record from the head of Ullswater (3.1, M.B. Gerrans, 1947, BM). Her map (Webster 1988) and that of Preston & Croft (1997) show additional records for the Cleator Moor (0.1) and Arnside (4.7s, v.c.60/69) squares.

Var. *vertumnus* is, in Cumbria, a distinctive plant with its short and rather rigid leaves. It is common in the Rivers Petteril (4.3, 4.4, 4.5) and Leith (5.2). It also occurs in the Kent (5.8s, 5.9) and Underbarrow Beck (46.92, GH, 1991, LANC) and there are isolated occurrences in the Caldew at Caldbeck (32.38, AAD, *c.*1987, LANC) and Smardale Beck (72.08, JCF *et al.*, 1991, LANC).

Subsp. *pseudofluitans* is generally distributed throughout England, Wales and the Scottish Borders.

Alt. limit: 265 m, Matterdale (40.22)

Maps 85 and 86   14   WC

R. fluitans Lam.

Very common in the River Eden and its major tributaries the Eamont and Lowther but surprisingly absent from the Caldew and the Irthing It occurs also in the Petteril. It is sometimes so abundant as to be a serious weed to the river authorities. A sterile double-flowered variant, probably of hybrid origin, is frequent in the middle stretch of the Eden. Elsewhere there is a single record from the River Bela at Milnthorpe (48.78s, GH, 1991, LANC).

Alt. limit: 145 m, Great Musgrave (76.12)

Map 87   79   WC

[**R. circinatus** Sibth.

This is listed with some doubt by Hodgson from Winscale, Gilgarran (0.2), Wythemoor (0.2) and Ullswater. There is a record from Thacka Beck, Penrith (5.3, J. Varty-Smith, 1917, *BEC Rpt* (1918) 5:94), presumably the same as J.A. Martindale's "Thacker Beck" referred to by Wilson. Wilson also lists it from the Lancaster-Kendal canal (5.8s). There is a pre-1930 *Atlas* record for the St Bees square (9.1) and post-1930 ones for the Ravenglass (0.9) and Skelton (4.3) squares.

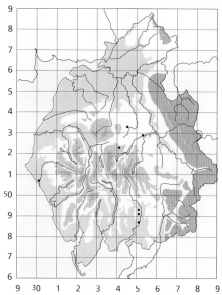

Map 85. *R. penicillatus* subsp. *pseudofluitans* var. *pseudofluitans*

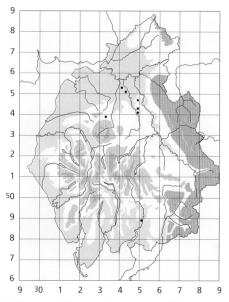

Map 86. *R. penicillatus* subsp. *pseudofluitans* var. *vertumnus*

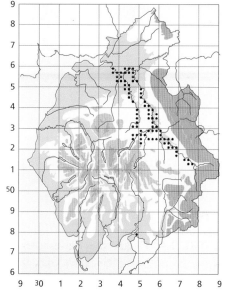

Map 87. *Ranunculus fluitans*

This species is predominantly a plant of mesotrophic and eutrophic waters in central and southern England. In the absence of authenticated material the above records are best considered doubtful, although apparently genuine material has recently been seen in the Lancashire section of the Lune valley below Kirkby Lonsdale.

Ranunculus circinatus is a distinctive plant, with the short leaf segments lying fan-like in one plane. It is probable that the early records from the Penrith area refer to *R. penicillatus* var. *vertumnus* which also has short and rather rigid leaves.

(WC)]

*Adonis annua L.   Pheasant's-eye
A rare and decreasing species of southern and south-eastern England recorded by Hodgson as a garden weed at Aspatria (1.4) and in a flax field at Flimby (0.3):

R   (C)

Myosurus minimus L.   Mousetail
Known only from the single late 19th century record cited by Hodgson from Wigton (2.4) and ones from Brough (7.1, G.A.K. Hervey & R.W. Robson, 1951-1956, CLE). The latter site was at the fairground and the plant may well have been introduced.

These records lie close to the northern limit in Britain of this rare and decreasing southern, arable weed.

S   (WC)

Aquilegia vulgaris L.   Columbine
Well scattered, occurring as a garden escape on roadside verges and railwaysides but probably as a native species in woodland, scrub and grassland in the limestone country around the Kent estuary and in the Sebergham area (3.4), also in gills and on rock ledges in the Lake District and probably elsewhere. In these last areas the flowers are of the deep, intense blue typical of the native plant.

Alt. limit: 470 m, Sandbed Gill (32.20)
Map 88   190   WFCY

Thalictrum flavum L.   Common meadow-rue
Rare and decreasing; known only from a few sites close to Morecambe Bay and by the Duddon estuary where it grows in ditches and open fen vegetation, associated with, for example, *Phragmites* and *Oenanthe crocata*.

In addition to Hodgson's records from Keswick (2.2), Dalston and Stainton (3.5), there are later ones from by the River Eden below Stainton (T.S. Johnstone, 1901, CLE) and Grinsdale, Carlisle (3.5, C.W. Muirhead, 1940) and also from Great Salkeld (5.3, H. Britten, 1904) and Buttermere (1.1, G.A.K. Hervey, 1929). Some of the early records, including that from Buttermere, suggest confusion with the following species. For example, Wilson gives Low Wood, presumably near Ambleside (3.0), for both species and many of the *Atlas* records must refer to *Thalictrum minus*.

The Cumbrian sites lie at the north-western limit of the species' native distribution in Britain.
Map 89   11   WF(C)

T. minus L.   Lesser meadow-rue
Fairly frequent in the Lake District, occurring by the stony and rocky shores of the major lakes and less commonly along the rivers, also in moist gullies and on rock ledges provided they are not too acidic. Curiously it is much more a feature of these montane sites than it is in Scotland. It also occurs in Cumbria on the contrasting habitats of the Duddon sand-dunes (1.7) and Grune Point (12.56), the limestone scars, screes and grassland and in the grikes of limestone pavement. Some outlying lowland sites as, for example, by the River Eden near Carlisle and in Whinfell Forest

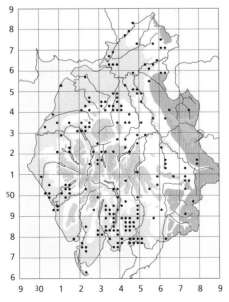

Map 88. *Aquilegia vulgaris*

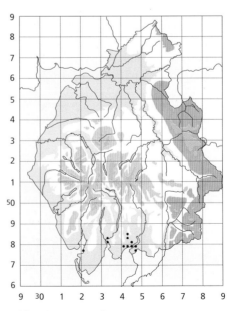

Map 89. *Thalictrum flavum*

(58.26), may represent garden escapes.

This species is probably commoner in Cumbria than anywhere else in Britain.

Alt. limit: 790 m, Hellvellyn (34.14)

Map 90   156   WFCY

T. alpinum L.   Alpine meadow-rue

An occasional and inconspicuous plant of the central and eastern fells of the Lake District occurring typically with *Selaginella selaginoides*, *Carex pulicaris* and *Sphagnum* spp. on irrigated ledges and slopes. It also occurs, though rarely, in upper Teesdale and on Alston Moor.

Alt. range: 455 - 870 m, Helvellyn (34.14)

Map 91   38   WC

BERBERIDACEAE

Berberis vulgaris L.   Barberry

A scarce shrub of old hedges throughout much of the Eden valley, the Carlisle area and south Westmorland but elsewhere distinctly rare. Handley (1898) recorded it from the roadside at Cautley, above Sedbergh (6.9), but there are no recent records from that area probably as a result of road-widening.

Whatever the general status of the species in Cumbria, the population on Wanthwaite Crags (32.22) looks very native.

Alt. limit: 395 m, Wanthwaite Crags (32.22)

Map 92   107   WFC(Y)

*B. darwinii Hook.

Seen in Red Hills Wood, Arnside (44.78s, MB, 1988). The only other record is that given by Petty (1894 p.151) who lists it from a hedge at Ulverston (2.7).

1   W(F)

*Berberis sp.

A non-flowering garden hybrid occurs extensively in woodland below the Long Meg mines at Great Salkeld (56.36, RWMC, 1991, LANC) and by a streamside south of Cartmel Fell (40.84s, CEW, 1995, LANC).

FC

*Epimedium alpinum L.   Barren-wort

Listed by Hodgson from the grounds of houses at Overwater (2.3) and Gilgarran, Distington (0.2) and, doubtfully, from Cockermouth (1.3, Miss Wright, c.1840, KDL), Borrowdale (2.1), Skiddaw (2.2), Helvellyn (3.1), Threlkeld and Saddleback (3.2) and Carrock (3.3). He clearly did not regard it as effectively naturalised.

There are also records from The Screes, Wastwater (0.1, C.H. Wright, 1837, CLE, 1839, BM), Santon Bridge (0.1, E. Forster, 1844, BM) and the banks of the River Lune near Sedbergh (6.9, A. Wilson, 1895, YRK). The only records this century are from a rocky bank at Lakeside, Newby Bridge (3.8, G. Caton, 1910, YRK), Wilson's record from Casterton Woods, Kirkby Lonsdale (6.8s) and Graythwaite, Newby Bridge (3.9, J. Lund, 1945, FBA).

(WFCY)

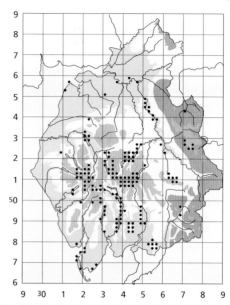

Map 90. *Thalictrum minus*

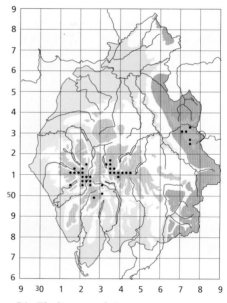

Map 91. *Thalictrum alpinum*

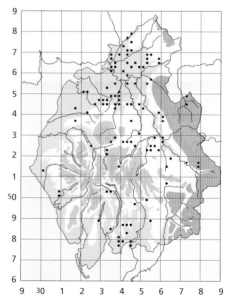

Map 92. *Berberis vulgaris*

Mahonia aquifolium (Pursh) Nutt.   Oregon grape
Rather rare, recorded mainly from estate woodlands in
the north and east, where it has been planted for ground
cover, although it also occurs as isolated plants. At
Rockliffe (34.60) it grows on the sandstone cliffs above
the River Eden. The first record for Cumberland is
from Great Salkeld (5.3, G.A.K. Hervey, 1957) and for
Westmorland from east of Skelwith Bridge (34.02,
R. Steele, 1973, AMDE).
Alt. limit: 180 m, Crosby Ravensworth (62.14)
Map 93   49   WFC

PAPAVERACEAE

Papaver orientale L.
Λ rare garden escape found at Skinburness in 1942
(1.5, C.W. Muirhead, CLE) and during the Survey in
seven scattered tetrads, occurring mainly on waste tips
and roadside verges but in rough coastal grassland at
Beckfoot, Wolsty (08.48). The first record for
Westmorland was from Oxenholme (52.90, CEW,
1988, *Watsonia* (1991) 18:420) and for Furness a
Survey record from Field Broughton (38.80, Anon.).
Map 94   16   WFC

P. somniferum L.   Opium poppy
An occasional garden escape, chiefly in the Eden
valley, on the Solway Plain and in the south. Like
Papaver dubium it occurs on waste ground, rubbish tips
and disturbed soil, also as a garden weed, although
probably initially sown. It is now appreciably
commoner than was indicated in the *Atlas.* Most of the
plants are double-flowered.
Alt. limit: 225 m, Little Strickland (54.20)
Map 95   93   WFCY

P. rhoeas L.   Common poppy
A local weed of arable fields in the Eden valley, and
also along the west coast and around the Kent estuary.
Elsewhere it is rare and it is practically absent from the
Lake District. It also occurs as a casual on roadside grit
piles.
      Although Hodgson regarded it as decreasing it
has probably changed little in frequency over the last
century.
Alt. limit: 250 m, Ravenstonedale (72.02)
Map 96   108   LWFCY

P. dubium L.   Long-headed poppy
subsp. **dubium**
Scattered and appreciably more widespread than
Papaver rhoeas, but still concentrated in the Eden
valley, the south and west. It is, perhaps, more a weed
of tips and waste ground than of arable fields. It also
occurs occasionally on the sand-dunes at Silloth (10.52)
and at times of low flow by the River Caldew and lower
Eden (3.5).
Alt. limit: at least 465 m, above Nenthead (78.44)
Map 97   377   WFCY

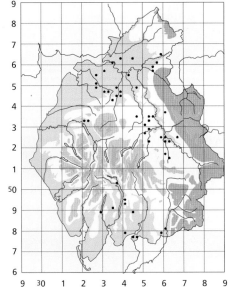

Map 93. *Mahonia aquifolium*

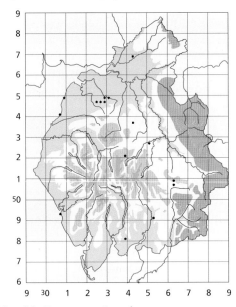

Map 94. *Papaver orientale*

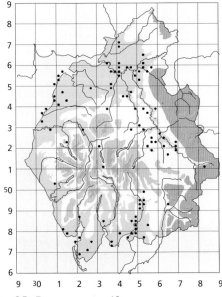

Map 95. *Papaver somniferum*

subsp. **lecoqii** (Lamotte) Syme
According to J.W. Kadereit (1988 *in litt.*), the yellow
latex is the only reliable character for distinguishing
this subspecies in Britain. Such plants were seen during
the Survey mainly in the south, around the Kent estuary
(4.7s, 4.8s), where it occurs sometimes as a garden
weed, and on disturbed ground on the Furness
peninsula. There is a solitary inland record from
Sandford in the Eden valley (72.16, A. Cannell, 1993).
The first Westmorland record is from Meathop (42.78s,
R. Young, 1971, AMDE). It is, however, likely to be
more frequent than these records suggest. The only
mention in the Floras is the unlocalised record given by
Petty (1894 p.152).
Alt. limit: *c.*150 m, Sandford (72.16)
Map 98   14   WF

*P. hybridum L.   Rough poppy
Listed by Hodgson from Brackenthwaite, Wigton (2.4), Blennerhasset (1.4) and Flimby (0.3), where a promising specimen had been prematurely "uprooted by juveniles". Handley
(1898) gives it for the Sedbergh area (6.9) and Wilson's only
record is from Cliburn (5.2). It was seen at Silloth (1.5, CLE)
by J. Leitch in 1879 and by J. Parkin in 1935.

    This is a decreasing species of southern and south-
eastern England and the Cumbrian records represent its
north-western limit.

                        S   (WCW/Y)

P. argemone L.   Prickly poppy
Seen only at Silloth (10.52, C.W. Muirhead, →1977),
Grange-over-Sands (40.76s, KAG, 1977), amongst
stubble at Newbiggin, Aldingham (26.68s, PB, 1989,
LANC), on the golf-course and roadside near Rampside
(24.66s, PB, 1993, LANC) and on disturbed ground at
Askam-in-Furness (20.76, PB, 1996).

    This poppy was once common in Cumberland as
a weed of light and sandy soils, especially near the
coast. Baker and Hodgson list numerous localities from
Middleton and Nethertown (9.0), in the south-west, to
Brampton (5.6), in the north-east, also an isolated one
from Roosecote (2.6s) in Furness. There are a number
of further records up to 1935 when it was seen at
Penrith (5.3, J.C. Varty-Smith) and Bullgill, near
Gilcrux (0.3, J. Parkin). G.A.K. Hervey recorded it at
Great Salkeld (5.3, 1956) and Wilson's only record is
from Kirkby Stephen (7.0). There are additional
undated *Atlas* dots for the Allonby (0.4), Aspatria (1.4)
and Longtown (4.6) squares.

    This dramatic decline is reflected nationally and
is presumably largely a consequence of improved agricultural practices.

              S 5   (W)FC

*Meconopsis cambrica (L.) Viguier   Welsh poppy
A frequent and locally abundant garden escape over
most of the county but particularly in the Lake District.
Most of the records are from roadsides and becksides,
although seldom far from houses where it can be a persistent and troublesome garden weed.

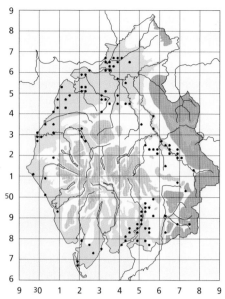

Map 96. *Papaver rhoeas*

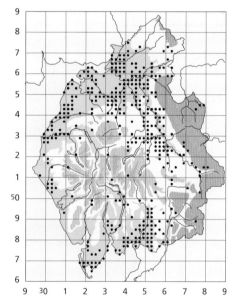

Map 97. *Papaver dubium* subsp. *dubium*

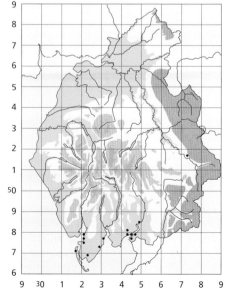

Map 98. *Papaver dubium* subsp. *lecoqii*

This conspicuous member of the early summer flora is so common that it is easy to forget that it is considered to be native only in south-west England and Wales. Significantly neither Nicolson nor Lawson mentions it. A century or more later Hodgson describes it as occurring "only in the centre of the lake district", although Petty (1894 p.153) quotes E. Hodgson's comment that it is present "near every hamlet in High Furness".
Alt. limit: at least 430 m, near Nenthead (78.42)
Map 99   S 712   LWFCY

Glaucium flavum Crantz   Yellow horned-poppy
A striking and sometimes frequent plant of the coast from Humphrey Head (38.74s) north to Mawbray (08.46). There is also an early record from Skinburness (1.5, C.W. Muirhead, 1932). It grows on the strand-line and shows a preference for shingle beaches where, unfortunately, it is often threatened by sea-defence schemes and erosion. It is a conspicuous feature of the slag banks at Ormsgill, Barrow-in-Furness (18.70) and Ulverston (30.76s).

This is a southern species which reaches its northern limit in Britain on the north side of the Solway Firth.
Map 100   41   FC

*G. corniculatum (L.) Rudolph
A southern European species listed by Hodgson from the Derwent Tinplate Works, Workington (9.2) and Silloth (1.5).
(C)

Chelidonium majus L.   Greater celandine
Frequent in the lowlands, occurring on roadsides and in hedgerows but nearly always close to houses, reflecting its earlier use for treating skin ailments.
Alt. limit: 270 m, near Alston (70.46).
Map 101   323   WFCY

*Eschscholtzia californica Cham.   Californian poppy
A garden escape seen only in the car park in the sand-dunes at Wolsty (08.50, REG, 1986).
1   C

*Hypecoum aequilobum Viv.
A North African species found at Silloth saltworks in 1890 (1.5, J. Leitch, CLE).
(C)

FUMARIACEAE

*Dicentra formosa (Andrews) Walp.   Bleeding heart
A garden escape recorded from a few scattered sites, mostly on roadside verges but also on the wooded Furness side of the river at Duddon Bridge (18.88, C. Cowan, 1970s; LANC). The only Westmorland record is from Fox How, Ambleside (36.04, B. Fisher, 1985).

It seems likely that the records of *Capnorchis canadensis* from the Vale of Newlands (2.1/2.2, H.P.

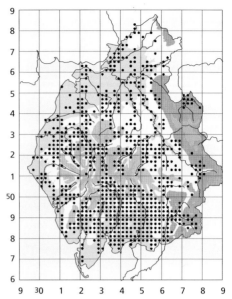

Map 99. *Meconopsis cambrica*

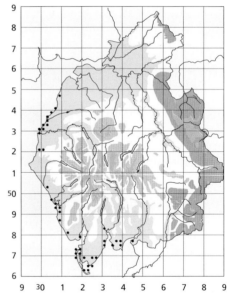

Map 100. *Glaucium flavum*

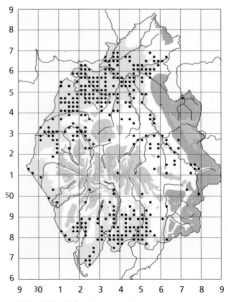

Map 101. *Chelidonium majus*

Bunell, 1924, *BEC Rpt* (1925) 7:555), Lamplugh (0.2, A. Templeman, 1922, *BEC Rpt* (1923) 6:716) and Millom (1.8, Mrs Mason,1923, *BEC Rpt* (1924) 7:166) refer to this species.
Alt. limit: 220 m, Newbiggin, Penrith (46.28)
Map 102   8   WFC

*Corydalis solida (L.) Clairv.
Noted by Hodgson as a garden escape in the Vale of Newlands (2.1/2.2), on one of the Derwent Water islands (2.2), by Wast Water (0.1), and Westward, Wigton (2.4), Cummersdale (3.5) and Castlesteads, Brampton (5.6). The first of these could be the same as the 1924 record listed above under *Dicentra formosa*. Baker gives additional records from Levens Park (5.8s), Kendal (5.9) and Ulverston (2.7) and Petty (1894 p.154) refers also to Furness Abbey (2.7). There are no subsequent records.

(WFC)

*Pseudofumaria lutea (L.) Borkh.   Yellow corydalis
(*Corydalis lutea* (L.) DC.)
Frequent in the lowlands, especially in the north. A colourful garden escape, it appears to have increased considerably during the last century although still virtually confined to walls and stonework near houses.
Alt. limit: 440 m, Nenthead (76.42)
Map 103   178   WFCY

Ceratocapnos claviculata (L.) Liden
(*Corydalis claviculata* (L.) DC.)   Climbing corydalis
Locally frequent, particularly in the Lake District. It is confined to siliceous and peaty soils where it grows in both deciduous and coniferous woodland. In the latter it tends to be restricted to the open phases, sometimes being extremely abundant for a while amongst the tangle of dead branches left after clear-felling. It also occurs on the fellsides under bracken. Its long flowering season can extend well into December.
Alt. limit: 400 m, above Coniston (28.96)
Map 104   272   WFC

Fumaria capreolata L.   White ramping-fumitory
Rare to occasional but locally frequent in the White-haven area (9.1). Like most of the genus it is a plant of rubbish tips, gardens, allotments, waste ground and arable fields. With its large pale flowers and reflexed fruiting pedicels it is usually easy to identify.
Alt. limit: 290 m, east of Ravenstonedale (74.02)
Map 105   76   WFC

F. bastardii Boreau
Rather uncommon, occurring in similar situations to *Fumaria capreolata* and like that species having a similar coastal tendency in Britain. There are add-itional untraced post-1950 *Scarce Plants* records for the Ravenglass (0.9), Allonby (0.4) and Cross Fell (6.3) squares and one from Carlisle (3.5, W.B. Waterfall, 1875, K).

        This species is best distinguished from *F. mur-alis*, which it superficially resembles, by the smaller

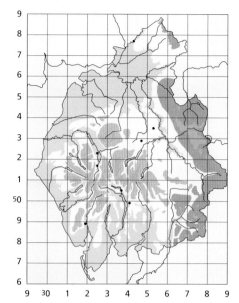

Map 102. *Dicentra formosa*

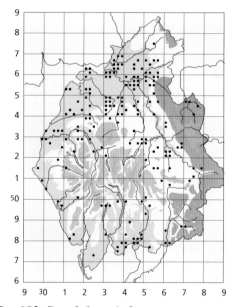

Map 103. *Pseudofumaria lutea*

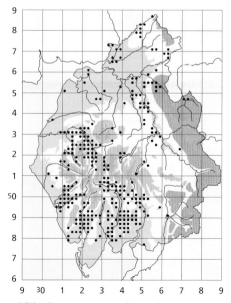

Map 104. *Ceratocapnos claviculata*

sepals (3 x 1 mm).
Alt. limit: at least 185 m, east of Threlkeld (34.24)
Map 106   S 28   WFC

F. muralis Sonder ex Koch
subsp. **boraei** (Jordan) Pugsley
                                    Common ramping fumitory
Quite common in the north, occasional to scarce elsewhere. This perhaps reflects the prevalence of arable
farming in the north. However, it is also a plant of
hedgerows and roadside banks, where it sometimes
survives the winter.
Alt. limit: at least 230 m, near Little Strickland (56.18)
Map 107   297   WFCY

F. purpurea Pugsl.   Purple ramping-fumitory
The only Survey collection identified as this species by
M.G. Daker is from the garden of Old Brathay, Amble-
side (36.02, T. Blackstock, 1978, LANC) where it had
been known for some years.
          This species has a puzzling history in Cumbria.
There are specimens in BM determined by H.W.
Pugsley from Walney island (*cf.*1.6s, Anon., undated),
Greenscoe Farm, Askam-in-Furness (2.7, D. Lumb,
1913, YRK), Eskmeals (0.9, A. Wallis, 1916), Elf Hall,
Broughton-in-Furness (1.8, W.H. Pearsall, 1926), Well
House, Bardsea (2.7, E. Hodgson & Mrs Jackson,
undated), Westward, Wigton (2.4, R. Wood, 1878),
Howtown, Ullswater (4.1, J.G. Baker, undated) and
Ambleside (3.0, Anon., 1849) and also one in HIWNT
from Threlkeld (3.2, A.W. Westrup, 1953).
          There are further specimens, not identified by
Pugsley, from Cark (3.7s, W.A. Sledge, 1932, BM) and
Thwaites, Millom (1.8, M. Cross, 1926, BM). Wilson
lists it from Arnside and Sandside (4.7s, J.A. & G.E.
Martindale, 1888, KDL) and near Staveley (4.9, J.A.
Martindale, 1912, KDL). The dots in *Scarce Plants* for
the Arnside (4.7s) and Burton-in-Kendal (5.7s) squares
refer to Lancashire; that for the Hawkshead square
(3.9) is an error for Ambleside (3.0).
          This dramatic decline may be partly illusory and
exaggerated by misidentifications but P.D. Sell (*verb.
comm.*) is of the opinion that the species is now almost
extinct in Britain.
          Fumaria purpurea is not easily distinguished
from *F. capreolata* on herbarium material. The former
has more oblong sepals and less strongly reflexed
pedicels but the most reliable character is its pinkish-
purple corolla.
                              1   (W)F(C)

F. officinalis L.   Common fumitory
A common lowland species of cultivated ground and
waste places but, like *Fumaria muralis*, most frequent
in the north and in the Eden valley. All the records
relate to subsp. *officinalis* although there is a single
Atlas record of the south-eastern subsp. *wirtgenii*
(Koch) Arcang. from the Brampton square (5.6). This
subspecies has smaller sepals (2 x 1 mm) and usually

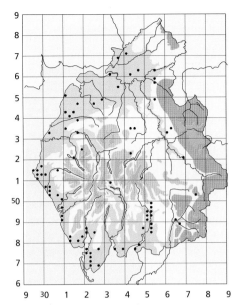

Map 105. *Fumaria capreolata*

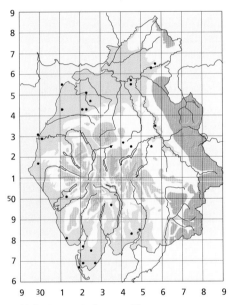

Map 106. *Fumaria bastardii*

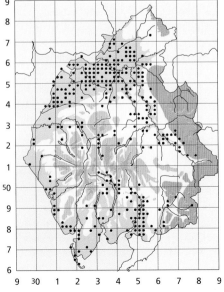

Map 107. *Fumaria muralis*

less than 20 flowers per raceme. Livermore & Livermore (1988) record both from the Lancaster District and surprisingly give subsp. *wirtgenii* as the more frequent.

Fumaria officinalis is characterised by finely dissected, glaucous leaves and small (-8 mm long), deep pink petals.

Alt. limit: 305 m, near Shap quarries (56.08)

Map 108   351   WFCY

F. densiflora DC.

A plant predominantly of south-eastern England and known in Cumbria only from the records cited by Baker, from Ambleside (3.0) and Hawkshead (3.9), and one from Silloth (1.5, E.J. Glaister, 1890, CLE, conf. P.D. Sell). Records unsupported by specimens are perhaps best regarded as doubtful.

([WF]C)

ULMACEAE

Ulmus glabra Hudson   Wych elm

Common, occurring around villages, as a hedgerow tree and as a fairly constant constituent of deciduous woodland, particularly with ash, on the richer soils, as on the limestones of south Westmorland and on the flushed gillsides of the Lake District, both on the volcanics and the Silurian slates.

During the Survey the county was badly affected by Dutch elm disease, though not to the same extent as southern England and the Midlands. The disease started slowly in the lowlands and at one time looked as if it might be contained there but it is now extending into the remote upland valleys and is ravaging the interesting high-level open ash - elm wood below Greencastle, on Alston Moor (68.38).

Alt. limit: 495 m, south of Garrigill (76.38)

Map 109   1218   LWFCY

U. glabra x **U. minor**

U. x vegeta (Loudon) Ley

The only records for this hybrid appear to be those made by G. Messenger in 1987 and all from Furness: street-plantings in Barrow (20.70, 20.72), a churchyard at Dalton (22.72), Holker Park (34.76s) and east of Old Park, Holker (34.78s). The distribution of both this hybrid and *Ulmus minor* clearly requires further study.

5   F

U. procera Salisb.   English elm

Fairly frequent in the north and west, elsewhere rather scarce. This is essentially a tree of hedgerows, lowland woods and parkland.It is thought to have been introduced into northern England only a few centuries ago. Although severely affected by Dutch elm disease, it seems likely that many trees will survive by suckering.

Alt. limit: *c.*185 m, Unthank, Hartside (60.40)

Map 110   145   WFC

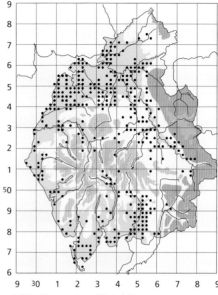

Map 108. *Fumaria officinalis*

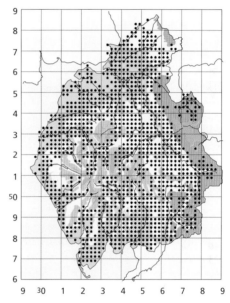

Map 109. *Ulmus glabra*

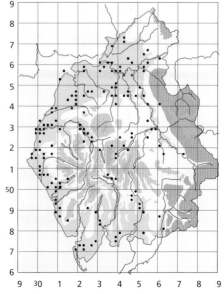

Map 110. *Ulmus procera*

U. minor Miller

G. Messenger has identified material of subsp. *angustifolia* (Weston) Stace (var. *cornubiensis* (Weston) Richens) and of its golden-leaved form from Holker Park (34.76s, 1987). He has also identified material from Kingstown, Carlisle (38.58, REG, 1983, LANC) and Hale Moss (50.76s, GH, 1991, LANC) as this subspecies and, tentatively, specimens collected from west of Pooley Bridge(44.24, MC, 1984) and Sandside, Milnthorpe (48.80s, GH, 1988, LANC). Baker's 19th century record of *Ulmus stricta* from Troutbeck (4.9) probably also belongs here. Messenger also noted the widespread planting of subsp. *sarniensis* (C. Schneider) Stace in Barrow-in-Furness and Vickerstown (1.6s, 2.7, 1987).

Although uncommon, this species is no doubt substantially under-recorded.

Alt. limit: *c.*150 m, Pooley Bridge (44.24)

5   WFC

CANNABACEAE

*Cannabis sativa L.

Known only from Silloth docks where it was discovered in 1883 (1.5, Anon., CLE).

(C)

*Humulus lupulus L.   Hop

Frequent in the lower Eden valley and the Solway Plain, also to the south of the Lake District, nearly always as a hedgerow plant around villages. W.F. Davidson has suggested that it may have been introduced locally through the feeding of brewery waste to cattle.

Hop has a very similar British distribution to *Parietaria* being essentially southern and, in the north,

fairly closely associated with man.

Alt. limit: *c.*125 m, Appleby (66.20) and Brampton (52.58)

Map 111   123   WFC(W/Y)

MORACEAE

*Ficus carica L.   Fig

A long-established plant was found in 1980 growing on the sea-wall of Cavendish Dock, Barrow-in-Furness (20.68s, GH; LANC), a number of plants were found later growing on an old mine tip south of Askam-in-Furness (20.76, PB, 1988) and there is a single tree on the east bank of the River Caldew in Carlisle (38.54, A. White, 1992).

3   FC

URTICACEAE

Urtica dioica L.   Stinging nettle

An abundant, ubiquitous and persistent weed of present and abandoned cultivated ground, roadside verges and pastures; it is also characteristic of wet alder woods. Its abundance in many churchyards and high on the fells around sheepfolds and abandoned farms testifies to its requirement of high soil phosphorus and nitrogen.

Alt. limit: 850 m, Great Dun Fell (70.32).

Map 112   1709   LWFCY

U. urens L.

A frequent lowland weed of arable land, allotments and gardens in the Eden valley and on the Solway coast, virtually absent from the Lake District but recurring around Morecambe Bay and the Duddon estuary.

Alt. limit: 470 m, Birkdale farm, Teesdale (80.26)

Map 113   161   WFC

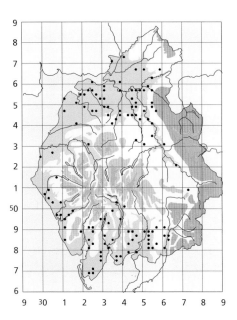

Map 111. *Humulus lupulus*

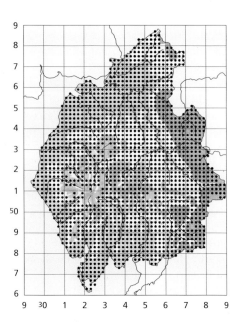

Map 112. *Urtica dioica*

Parietaria judaica L. Pellitory-of-the wall
A southern species and rather rare in Cumbria where it occurs on the warm coastal limestones by Morecambe Bay and inland on the stonework of abbeys, for example Furness (20.70), Wetheral (46.54), Abbeytown (16.50) and Cartmel Priory (38.78s), and castles, such as Kendal (52.92), Arnside Tower (44.76s) and Naworth (56.62), where it was noted by the Berwickshire Naturalist Club in 1880. It has, however, disappeared from Carlisle city walls (3.5), where it was said by Nicolson to be abundant, and also from Lanercost (54.62).

Its presence in such inland sites is a reminder of its old mediaeval usage for urinary disorders.
Alt. limit: 275 m, north of Shap (56.16)
Map 114 23 WFC

*Soleirolia soleirolii (Req.) Dandy

Mind-your-own-business
A very rare garden escape, chiefly coastal, the only inland records being from Caldbeck (32.38, REG, 1977), the first Cumberland record, and Burton-in-Kendal (52.74s). It is characteristic of bare open ground such as the edges of drives, between paving stones and the foot of walls. The first record for Furness was from Grange-over-Sands (40.76s, KAG & FLW, 1970s) and for Westmorland an ephemeral colony near Arnside (48.76s, MB, 1994).
Alt. limit: 160 m, Caldbeck (32.38)
Map 115 10 LWFC

JUGLANDACEAE

*Juglans regia L. Walnut
The only records of trees away from houses are single trees by a field south of Hall Dunnerdale (20.94, FLW, 1979), in Township Plantation, Whitbarrow (44.88s, I.R. Bonner, 1988), at High Newton (38.82, JA, 1980s) and west of Brampton (46.62, MG & JP, 1995). There is an avenue of 30 trees east of Old Park, Holker (34.78, JA, 1990) and a record of an old tree felled in the 1950s at Brackensgill, Dentdale (66.88s). There are no records in the Floras.

5 WFC(Y)

MYRICACEAE

Myrica gale L. Bog myrtle
Scattered, occurring in poor-fen communities in valley bottoms and, with *Molinia* and *Salix cinerea*, around lowland raised mosses, chiefly in the south-west of the county and by the Solway Firth. It has disappeared from a number of former sites, for example in west Cumberland, following drainage.

Bog myrtle has a pronounced north-western distribution in Britain.
Alt. limit: *c.*460 m, Armboth Fell (28.14)
Map 116 185 WFC

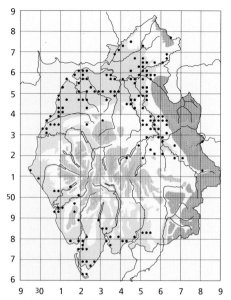

Map 113. *Urtica urens*

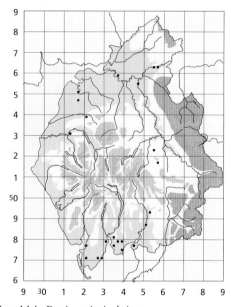

Map 114. *Parietaria judaica*

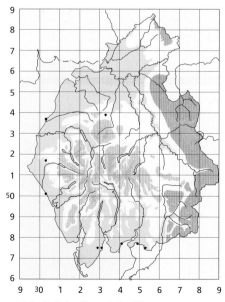

Map 115. *Soleirolia soleirolii*

FAGACEAE

*Fagus sylvatica L.   Beech
Very common in the lowlands in hedgerows, amenity woodland and parkland, as a 'cosmetic' screen for conifer plantations and as a valuable timber tree, sometimes in pure stands but often with larch as a nurse crop. It is grown chiefly on the drier, better-drained soils, especially in the southern limestone area, but like oak it is very susceptible to damage by grey squirrels. It is also prone to summer drought and to frost, the young leaves of trees in the upper Eden valley being extensively killed by frost in late May 1995.

     Although natural regeneration is locally very successful, most trees have probably been planted and the somewhat patchy distribution may be due in part to varying standards adopted by recorders.
Alt. limit: 650 m, south of Garrigill (76.38)
Map 117   1236   LWFCY

*Castanea sativa Miller   Sweet chestnut
Occasional in the lowlands, usually occurring in parkland but also planted by roadsides, more rarely in woodland. Wilson comments on the production of ripe fruit in the warm summer of 1933; seedlings have been noted in the Brathay (3.0) area and Barrowfield Wood (48.90).
Alt. limit: 300 m, near Tindale, on the Northumberland border (60.58)
Map 118   191   WFC

*Quercus cerris L.   Turkey oak
Rather rare. The earliest record appears to be that in the *Atlas* for the Great Strickland square (5.2). As is evident from the Survey map it is now fairly widespread, if sparse, as a planted tree, usually in private woodland or by roadsides, but occasionally in hedges and seen on coastal shingle at Aldingham (28.70). The first record from Furness is probably that from near Dalton (2.7, K. Jones, *c*.1970) and for Cumberland from Wreay woods (4.4/4.5, C.W. Muirhead, early 1970s). The trees are said to fruit freely but there are as yet no reports of established saplings.
Alt. limit: 145 m, Glenridding (38.16)
Map 119   51   WFC

*Q. cerris x **Q.suber** L   Lucombe oak
Q. x crenata Lam.
The only records of this infrequent hybrid are of a single, fairly old tree in a roadside field opposite the Swan Inn at Middleton, Barbon (62.86s, CEW, 1991, CGE; LANC), one by the edge of a wood at Grange-over-Sands (40.78s, G. Huse, 1980; LANC) and one in a field between Milnthorpe and Arnside (48.80, CEW, 1996, LANC). There is also a tree in the grounds of Sedgwick School (50.86s, CEW, 1995, LANC).
                   3   W

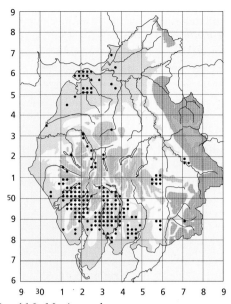

Map 116. *Myrica gale*

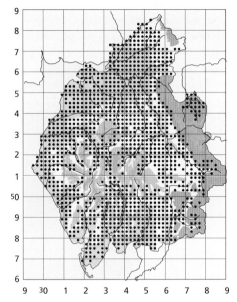

Map 117. *Fagus sylvatica*

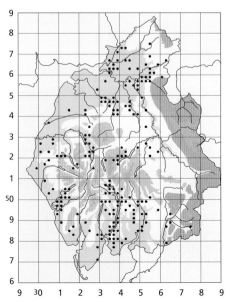

Map 118. *Castanea sativa*

*Q. ilex L. Holm oak

Known only from three tetrads in the Muncaster area (10.92, 10.94, 10.96) but perhaps overlooked. It was first recorded here in 1989 (10.94, MMG; LANC), but one of the two trees was shortly afterwards blown down.

3 C

Q. petraea (Mattushka) Liebl. and Sessile oak
Q. robur L. Pedunculate oak

The two native British oaks are notoriously difficult to distinguish, a fact which is usually attributed to extensive hybridisation particularly following the widespread planting of *Quercus robur* in northern Britain. Most investigators have confined themselves to population studies and refrain from attempting to identify individual trees. For this reason it was decided early in the Survey that the only practicable course was to record the aggregate.

Certainly pure trees of both species do occur, *Q. petraea*, like *Betula pubescens*, being widespread in the fells and along riversides and *Q. robur*, like *B. pendula*, favouring the richer soils and being the more commonly planted as a timber tree and in hedgerows. Significantly *Q. robur* also occurs on the south-facing Yew Crag above Ullswater (40.20), together with other southern species such as *Tilia cordata* and *Euonymus europaeus*.

Rather little oak is grown commercially at the present time, most planting being of amenity woodland. The wood is used for fencing and, if of high quality, for veneers. Like beech, oak is prone to damage by grey squirrels.

Oak woodland is the natural woodland of most of the Lake District but it is difficult to identify those areas of present oak woodland which actually represent native woodland. The isolated, high-level woods of *Q. petraea* above Keskadale and Birkrigg in the Newlands valley (20.18, 20.20), although usually considered to be such fragments, are now thought more likely to represent plantings. Over much of south Westmorland and Furness the oakwoods were coppiced over at least eight centuries for their bark, for tanning, and their wood, for charcoal burning. The present woods are mostly either old neglected coppice or planted woodland dating from the 17th century. A probable exception is Murthwaite Park, Ravenstonedale (70.96, 70.98) which is on stinted common land and unlikely to have been planted.

Oakwoods occur on a wide range of soils, commonly with ash on the richer soils and birch on the poorer. On the most acid soils the ground flora may consist of little other than *Holcus mollis* and *Deschampsia flexuosa*.

The distinctive knopper galls on the acorns, caused by the gall wasp *Andricus quercus-calicis*, have been reported from Levens Park (4.8s/5.8s) and Burton-in-Kendal (5.7s) and Kendal (5.9) and are likely to become more frequent as the insect spreads northwards. Alt. limit: 450 m, Keskadale (20.18) and at Tynehead (74.36), perhaps also on the south side of Bowscale Fell (34.30).

Map 120 (both species) 1324 LWFCY

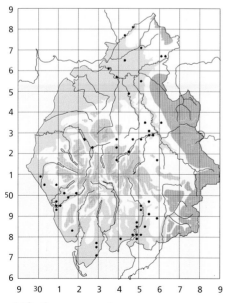

Map 119. *Quercus cerris*

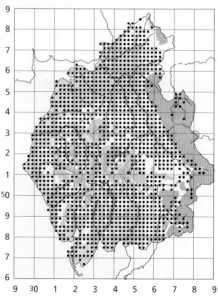

Map 120. *Quercus petraea* and *Q. robur*

*Q. rubra L.   Red oak
Known from a few scattered plantings. The first record
for Cumberland was from Coombs Wood, Armathwaite
(50.44, RWMC, 1989, LANC), for Westmorland from
woodland west of Hale (50.78s, CFS & JS, 1978) and
for Furness from the upper Duddon valley (22.00,
1993; 22.96, 22.98, 24.98, 1995, all CEW). There are
further records from Penruddock (40.28), Pooley
Bridge (46.24), the roadside near Netherby Hall
(40.72n), Great Strickland (56.22), a plantation at
Firbank (60.92) and Arnside (44.76s).
Alt. limit: 290 m, Penruddock (40.28)
                                        11   WFC

BETULACEAE

Betula pendula Roth   Silver birch
The Survey records have not distinguished between the
two birches but this is certainly the less common. It
occurs in hedgerows and is commonly planted by the
Highways Authority. It is most frequent in parts of west
Cumbria and on the limestones of south Westmorland,
being particularly abundant above the west side of
Whitbarrow (4.8s).

B. pubescens Ehrh.   Downy birch
Very common, usually as a minor constituent of valley-
side oakwoods and as a dominant in wet valley bottoms
and around relict lowland mosses, especially those
bordering Morecambe Bay and on the Solway Plain,
where it is otherwise rather patchy. In many of these
mossland sites, as, for example, on Meathop Moss
(44.80s), it is continually invading the drying surface of
the moss and can only be kept in check by persistent
uprooting.

    Birch also appears as a pioneer on abandoned
railway sites: sidings, cuttings, embankments, and on
areas of former heathland. It is perhaps less common
than the previous species as a hedgerow and roadside
tree.

    As in Co. Durham, most trees have puberulent
young twigs and appear referable to subsp. *tortuosa*
(Ledeb.) Nyman (subsp. *odorata* sensu E. Warb.).
Analysis of a number of populations in south Cumbria
(Williamson 1989) indicates some degree of hybrid-
isation with *Betula pendula*.
Alt. limit: 685 m, Hilton Fell (76.22)
Map 121 (both species), Plate 60   1402   LWFCY

Alnus glutinosa (L.)Gaertner   Alder
A common and characteristic tree of lake shores and of
waterlogged soils in valley bottoms where it is often
associated with *Salix cinerea*. With improved drainage
it is now all too often represented by a single row of
beckside trees. In a few places in the Lake District there
are examples of formerly more extensive alder woods
on the flushed and often flooded soils between the
valley-sides and the becks. Pearsall & Pennington
(1973) refer to such woodlands in Martindale and

Plate 60. *Betula pubescens*

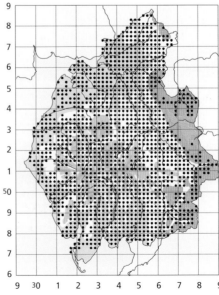

Map 121. *Betula pendula* and *B. pubescens*

another good example, though heavily grazed, is that on the south side of Borrow Beck, near the Lune gorge (60.00). Alder is also frequently included now in amenity planting on streamsides and soils subject to flooding.

Alt. limit: 470m, north-west of Garrigill (76.44)

Map 122   1281   LWFCY

*A. incana (L.) Moench   Grey alder

A native of eastern and central Europe and occasionally planted on roadside verges and by becks. There is a particularly good stand of old trees, mixed with *A. glutinosa*, on either side of the River Brathay east of Elterwater where it was first noted in 1948 (3.0, E. G[orham?], FBA; LANC). The first Westmorland and Cumberland records are from Arnside and Melmerby (4.7s, W.H. Pearsall, 1923; 6.3, W.W. Mason, 1922, *BEC Rpt* (1924) 7:211).

Alt. limit: 270 m, south of Alston (70.44)

Map 123   31   WCY

*A. cordata (Lois.) Duby

Seen only by the A6 near Shap quarry where a screen has been planted along the east side of the road (56.10, CEW, 1992, LANC).

Alt.: 305 m, Shap (56.10)

          1   W

*Carpinus betulus L.   Hornbeam

Not infrequent in hedgerows, more rarely in woodland, over a broad north-south band of the county. Although almost always planted, young self-sown trees have been seen in Wetheral Woods (46.54) and near Rash Bridge, Dentdale (64.90). A tree 30 m high at Hutton-in-the-Forest (46.34) is said to be one of the tallest in the British Isles.

Alt. limit: 380 m, south-east side of Great Mell Fell (40.24)

Map 124   147   WFCY

Corylus avellana L.   Hazel

Very common in hedgerows, deciduous woodland and open scrub and on the more low-lying limestone pavements. It prefers the richer soils and is particularly abundant on the Silurian rocks of the southern Lake District and the limestones around Morecambe Bay. In many of the woods the coppice, chiefly of hazel and oak, has been long neglected, but it is encouraging that at a number of sites, for example in the Rusland valley (3.9), the old coppicing regime is now being reintroduced by voluntary bodies and the National Park Authority. Regeneration by seed is very rare.

Alt. limit: 500 m, Wolf Crags, Matterdale (34.22)

Map 125   1315   LWFCY

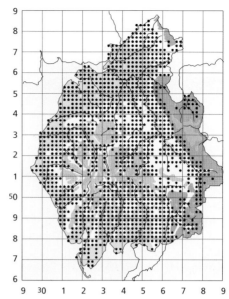

Map 122. *Alnus glutinosa*

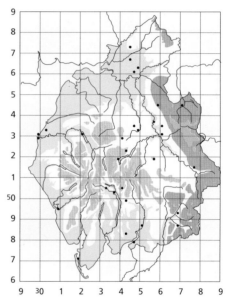

Map 123. *Alnus incana*

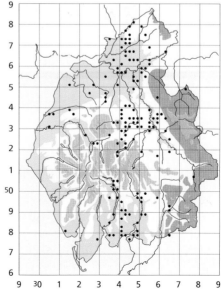

Map 124. *Carpinus betulus*

CHENOPODIACEAE

*Chenopodium capitatum (L.) Asch.
Known only from a collection from Silloth (1.5, CLE, det. J.M. Mullin) where it was seen by J. Leitch from 1890 to 1892.

(C)

*C. bonus-henricus L.   Good King Henry
Widespread and fairly frequent although scarce in the west and north. Its use as a vegetable has not entirely disappeared and until relatively recently it was sold in Penrith market. It is nearly always found close to buildings, particularly farmyards but it also occurs in other nitrogen-rich habitats such as rubbish dumps and on roadsides.
Alt. limit: 455 m, Helbeck Fell (80.18)
Map 126   326   WFCY

*C. glaucum L.
A southern species seen briefly during the Survey on the recently landscaped site at Eden Bridge, Lazonby (54.40, REG, 1984, det. J.M. Mullin, LANC, *Watsonia* (1988) 17:186). Mullin considers that Hodgson's specimen of 'C. botryoides' from Flimby (0.3, 1889, CLE) possibly belongs here.

1   C

C. rubrum L.   Red goosefoot
The only authenticated Survey records for this strongly nitrophilous species are from a brackish pool at South Walney (22.62s, GH, LANC, 1979) and the pond at Salt Cotes (18.52, GH, 1978, LANC, *Watsonia* (1979) 12:350), also as a casual in Carlisle (40.56, REG, 1986, LANC), on a manure heap at Milnthorpe (50.80s, CEW, 1988, LANC) and on a tip at the old Barrow-in-Furness steelworks site (18.70, PB, 1992).

This species is listed by Hodgson from Work-

ington (9.2) and Maryport (0.3). Baker refers to it as "being frequent on dunghills about Kirkby Lonsdale" (6.7s), a statement which Wilson regarded with suspicion. W.H. Pearsall reported it from Sandscale, Dalton-in-Furness (1.7, 1913, *BEC Rpt* (1914) 3:392) and there are post-1930 *Atlas* records from this square (G. Wilson, 1956), the Carlisle area, the Solway coast, and the Cartmel (3.7s), Arnside (4.7s, v.c.60/69) and Dent (7.8) squares.

5   WFC

[C. chenopodioides (L.) Aellen
(*C. botryodes* Smith)
Reported by Hodgson from Flimby (0.3) in 1889. This seems very unlikely as the species is otherwise known in Britain only from the south-east and East Anglian coasts. J.M. Mullin (*in litt.*) considers that Hodgson's specimen (CLE) is possibly *Chenopodium glaucum*.

(C)]

*C. polyspermum L.
Seen only in a garden at Talkin, Castle Carrock (54.56, MG, 1989→, conf. J.M. Mullin, LANC).

The only previous records for this southern and eastern species are Hodgson's from Workington (9.2), Maryport (0.3) and Penrith (5.3), Wilson's from Witherslack (4.8s) and one from Silloth in 1935 (1.5, J. Parkin).
Alt.: 150 m, Castle Carrock (54.56)

1   C

*C. vulvaria L.   Stinking goosefoot
A collection from a parking area at Eden Bridge, Lazonby (54.40, REG, 1984) has been tentatively identified as this species by J.M. Mullin.

The only other records for this very rare and decreasing species of southern and south-eastern England are those given by Hodgson from the Derwent Tinplate Works, Workington (9.2), Maryport (0.3) and

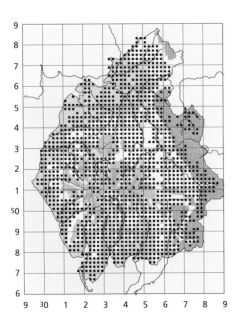

Map 125. *Corylus avellana*

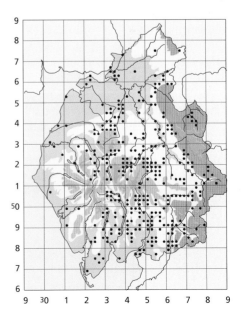

Map 126. *Chenopodium bonus-henricus*

the beach at Salt Cotes (1.5), and by Baker from Barrow-in-Furness (1.6s/2.6s).

R 1   (F)C

*C. hybridum L.
A rare southern species recorded in Cumbria only once, in 1951, from a Carlisle garden (3.5, D. Stewart, CLE, det. J.M. Mullin).

(C)

*C. urbicum L.
Found in 1995 growing on cattle-feed waste at Little Salkeld (56.38, RWMC, BM, LANC, conf. J.M. Mullin). This, and a record from Essex in the same week are the first British records for 30 years. The only other Cumbrian records are Hodgson's from the Maryport tips (0.3), others from Black Dyke, Silloth (1.5, E.J. Glaister, 1875), Silloth (1.5, W.H. Youdale, 1896, LIV; C.W. Muirhead, 1932) and Carlisle water-works (4.5, J. Parkin, 1935).

This is another decreasing species of southern and south-eastern England.

1   C

*C. murale L.
The only definite Survey records are from gardens at Great Salkeld (54.36, RWMC, 1982, det. J.M. Mullin, LANC, *Watsonia* (1988) 17:186) and Carlisle (36.56, REG, 1981), where it first appeared as a greenhouse weed before spreading into the garden.

This southern species was recorded by Hodgson from the Derwent Tinplate Works, Workington (9.2) and by J. Leitch from Silloth (1.5, both specimens 1889, conf. J.M. Mullin). There is a pre-1930 *Atlas* dot for the Arnside square, v.c.60/69 (4.7s).
Alt. limit: 110 m, Great Salkeld (54.36)

2   (W?)C

*C. opulifolium Schrader ex Koch & Ziz
A rare alien known only from Hodgson's records from the Derwent Tinplate Works, Workington (9.2) and Maryport (0.3).

(C)

C. album L.   Fat hen
Common in the lowlands in cultivated fields, especially potato fields, allotments, farms and rubbish dumps, also on the coast along the strand-line.
Alt. limit: at least 470 m, Nenthead (78.44)
Map 127   875   LWFCY

*Bassia scoparia (L.) A.J. Scott
(*Kochia scoparia* (L.) Schrader)
Like *Chenopodium glaucum*, this Asiatic annual appeared briefly at the Eden Bridge picnic site, Lazonby, in 1984 (54.40, REG, det. J.M. Mullin, LANC, *Watsonia* (1988) 17:186).

1   C

Atriplex prostrata Boucher ex DC.
A common plant along the entire coast, chiefly on the drift-line, on the upper parts of the salt-marshes and along the estuaries. It occasionally occurs inland on rubbish dumps, waste ground and roadside grit piles.
Alt. limit: 300 m, Hardendale Fell (56.12)
Map 128   170   WFCY

A. prostrata x **A. longipes**
A. x gustafssoniana Taschereau
The only record is from the Plumpton shore of the Leven estuary (30.78s, PB, 1995, LANC, det. A.O. Chater) but there is little doubt that it has been over-looked. The material is characterised by having bract-eoles and stalks of widely varying size.

1   F

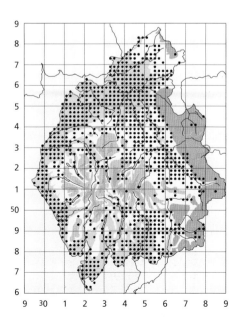

Map 127. *Chenopodium album*

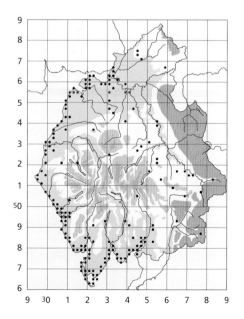

Map 128. *Atriplex prostrata*

A. glabriuscula Edmondston
Probably frequent on the shingle beaches of the west coast. The map is based on the examination of bracteole and seed characters in a number of fruiting specimens all of which proved to be this species. It may eventually prove to be as frequent as shown in the *Atlas* and it is clearly necessary now to review the coastal distribution of *Atriplex prostrata*. The only record shown on Taschereau's (1985) map of *A. glabriuscula* is for the Dalton-in-Furness square (2.7). The only inland record is from Ulpha, in the Duddon valley (18.92, AAD, 1996, LANC).
Map 129   45   FC

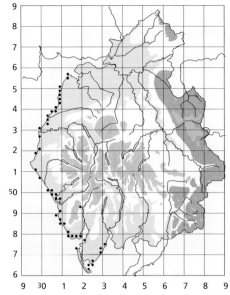

Map 129. *Atriplex glabriuscula*

A. longipes Drej.
This species was discovered in 1977 on the shore of the Kent estuary near Arnside station (46.78s, MB, LANC, *Watsonia* (1978) 12:170). The material was identified by P.M. Taschereau but was curiously overlooked in his papers (1985a,b) dealing with the taxonomy of the genus in Britain and with *Atriplex longipes* in particular.

This is a local plant which is scattered along the coasts of England and Wales and north to central Scotland. It has probably been much under-recorded.

Typical *A. longipes* differs from *A. prostrata* in having lower leaves which are narrower (c. twice as long as wide), with a cuneate leaf base and strong, patent or forwardly-directed basal lobes, and long-stalked axillary bracteoles.
                    S 1   W

A. littoralis L.
Fairly frequent along the Furness coast, elsewhere rather rare. It occurs in muddy, open ground in the upper parts of salt-marshes, usually along or near the drift line, and less commonly on sand and shingle beaches. There is also a record from waste ground by a farm building near Levens (46.84s, CEW, 1990, herb. C.E.W.). The first Westmorland record was from Arnside (44.78s, MB, 1980).

These Cumbrian sites are nearly at the species' northern limit on the west coast.
Map 130   33   WFC

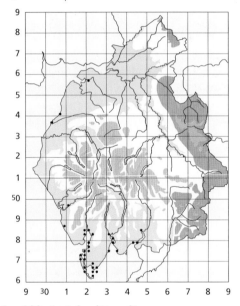

Map 130. *Atriplex littoralis*

A. patula L.   Common orache
Frequent in the lowlands as a weed of cultivated land, rubbish dumps, and occasionally of roadside grit piles. Ecologically it is very similar to *Chenopodium album* with which it is frequently confused early in the season. Alt. limit: at least 430 m, Garrigill (76.40).
Map 131   710   LWFCY

A. laciniata L.   Frosted orache
Scattered around the coast except for the far west, the inner parts of the Solway Firth and the Duddon estuary. It grows usually in sand or shingle along the strand-line and in the uppermost zone of the salt-marshes.
Map 132   33   WFC

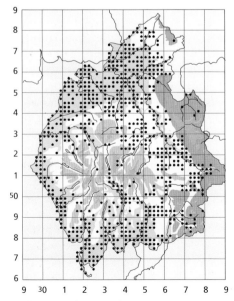

Map 131. *Atriplex patula*

A. portulacoides L.   Sea purslane
(*Halimione portulacoides* (L.) Aellen)
Common in the upper salt-marshes and by creeks along
Morecambe Bay and north to the Ravenglass estuary.
Despite careful searching there are no Survey records
further north although there are post-1930 *Atlas* dots
for the Workington (9.2) and Maryport (0.3) squares.

This species reaches its northern limit on the
west coast on the north side of the Solway Firth.
Map 133   46   WFC

Beta vulgaris L.   Sea beet
subsp. **maritima** (L.) Arcang.
Fairly common around the coast but scarce in the inner
parts of the Solway Firth and Morecambe Bay. It grows
scattered along shingle and sandy beaches, usually on
the drift line. It occurs also on sea walls and there are
two inland records, one from Thornthwaite, Keswick
(22.24, K.G. Messenger, 1978), when it appeared
following road improvements to the A66, and on a
farmer's dump in an old railway cutting near Temple
Sowerby (60.24, RWMC, 1988, BM, LANC).

Curiously, neither Baker nor Petty give any
Furness records, the first apparently being W.H.
Pearsall's from Roa Island (2.6s, 1913, *BEC Rpt* (1914)
3:392).
Alt. limit: 105 m, Temple Sowerby (60.24)
Map 134   57   WFC

Salicornia   Glassworts
Species of this genus are generally frequent in estuaries
and salt-marshes, occurring on bare mud mostly in the
middle and upper marsh. However, there are no records
from the inner part of the Solway Firth or to
substantiate the *Atlas* records for the Workington (9.2)
and Ulpha (1.9) squares. All the Survey records of
Salicornia ramosissima, S. fragilis and *S. dolichostachya* are based entirely on material submitted to I.K.
Ferguson, who has also provided diagnostic comments.
Most specimens can be referred to one of the four
species but in some populations it is very difficult to
draw the line between *S. europaea* and *S. ramosissima*,
and between *S. fragilis* and *S. dolichostachya*.

S. ramosissima J. Woods
Frequent, usually growing intermixed with *Salicornia
europaea*. The stems of both are glossy, with convex,
knobbly segments but those of *S. ramosissima* are
slenderer and more branched. They are generally
purplish-red and the short fertile spikes have less than
seven segments.

The first record was from Dunnerholme,
Askam-in-Furness (2.7, D. Lumb, 1913, DHM, *BEC
Rpt* (1914) 3:392), although the identification was
emphatically rejected by E.S. Marshall (*ibid.* p.493). In
Cumberland, specimens probably belonging to this
segregate were collected by C.W. Muirhead at Skinburness, Silloth and Grune Point (1.5, 1947, 1950,
1952, all PLYP) and the first Survey record was from

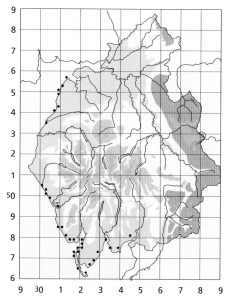

Map 132. *Atriplex laciniata*

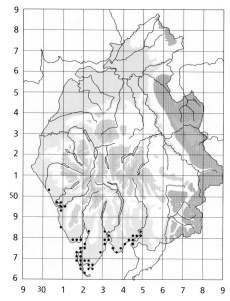

Map 133. *Atriplex portulacoides*

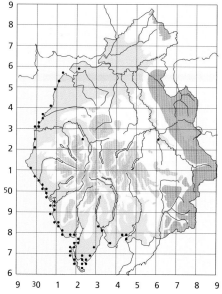

Map 134. *Beta vulgaris*

Newton Arlosh (18.56, REG, 1978, LANC, det. I.K. Ferguson, *Watsonia* (1979) 12:350).
Map 135   18   FC

S. europaea L. sensu stricto
The commonest species and probably the only one able to persist on grazed marshes. The shoots are fatter than in *Salicornia ramosissima*, usually green and have longer fertile spikes.

Although the map is of the aggregate species, none of the segregates has been found in tetrads lacking *S. europaea*.
Map 136   67   WFC

S. fragilis P. Ball & Tutin
Λ supposedly southern species but reported from an increasing number of sites along the west coast of Britain. The first records for Furness are from Roose-

cote and east of Rampside (respectively 20.68s, 22.64s, GH, 1978, *Watsonia* (1983) 14:422) and for Cumberland from the Moricambe marshes (1.5, GH, 1983).

This species differs from *Salicornia dolichostachya* chiefly in its yellow-green colour, oblong and truncate spike with cylindrical segments, and the smaller central floret which does not separate the two laterals.
Map 137   13   FC

S. dolichostachya Moss
Occasional and restricted to the open mud of the lowest part of the marshes. The first Survey records are from Newton Arlosh (18.56, REG, 1978, LANC, *Watsonia* (1979) 12:350) and South Walney (20.62s, GH, 1982, LANC, *Watsonia* (1983) 14:422). There are earlier specimens, almost certainly of this species, collected by T.S. Johnstone at Cardurnock (1.5, 1899, CLE) and by

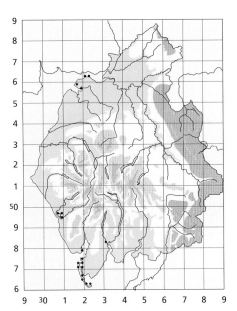

Map 135. *Salicornia ramosissima*

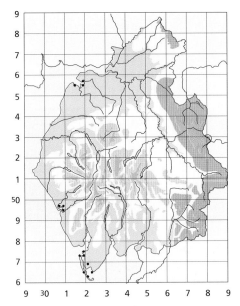

Map 137. *Salicornia fragilis*

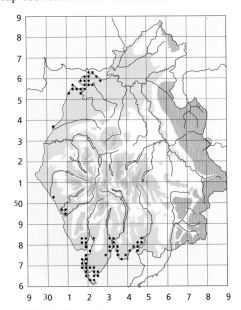

Map 136. *Salicornia europaea*

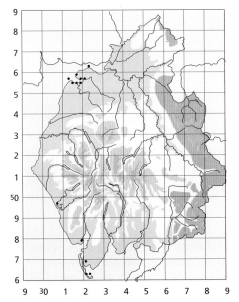

Map 138. *Salicornia dolichostachya*

C.W. Muirhead from Skinburness, Silloth and Grune Point (1.5, 1947, CLE; 1950; 1952, all PLYP) and Barrow-in-Furness (1.6/2.6s, 1952, PLYP).

This species is characterised by its dark green colour, long, tapering spikes, segments which bulge above and a central floret which separates the two laterals.

Map 138   13   FC

Suaeda maritima (L.) Dumort.   Annual sea-blite
Common in relatively protected situations along the entire coast, growing in bare mud below the strandline and in bare areas in the upper salt-marshes where it often forms a conspicuous pink zone in early autumn.

Map 139   66   WFC

Salsola kali L.   Saltwort
A frequent plant of the drift-line occurring usually on sand and shingle, seldom penetrating far into the estuaries. The first Westmorland record is of a brief occurrence at Far Arnside (44.76s, MB) in 1987.

Map 140   49   WFC

AMARANTHACEAE

*Amaranthus retroflexus L.
Seen only on a recently landscaped picnic area by Eden Bridge, Lazonby (54.40, REG, 1984, LANC), where it was growing with *Chenopodium glaucum* and *Bassia scoparia*, and on a disturbed roadside verge near Brunstock, Carlisle (42.58, PB, 1992, LANC). This North American casual was earlier recorded by Hodgson from tips at Flimby and Maryport (0.3), Parton and Workington (9.2) and Silloth (1.5, E.J. Glaister, 1878; J. Leitch, 1889, both CLE; Hodgson 1898 (1899)).

                         2   C

*A. albus L.
A North American weed found in a field south-west of Armathwaite in 1989 (48.44, JP & MS, LANC, det. J.M. Mullin), otherwise known only from Hodgson's records from the Derwent Tinplate Works, Workington (9.2), and later from Silloth (1.5, 1898 (1899)).

                         1   C

PORTULACAEAE

*Portulaca oleracea L.   Common purslane
An alien weed seen in a Penrith garden in 1983 and for a few years subsequently (52.28, AW, LANC, *Watsonia* (1985) 15:394).
Alt.: 135 m, Penrith (52.28)

                         1   C

*Claytonia perfoliata Donn ex Willd.   Spring beauty
(*Montia perfoliata* (Donn ex Willd.) Howell)
Very rare, known only as a garden weed in the north and north-east where it was seen at Warwick Hall

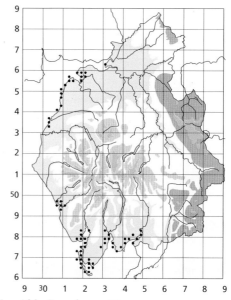
Map 139. *Suaeda maritima*

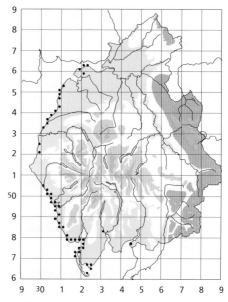
Map 140. *Salsola kali*

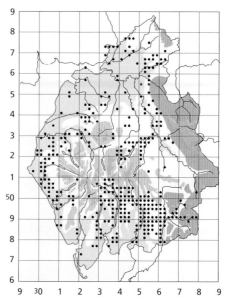
Map 141. *Claytonia sibirica*

(46.56, CS, 1978), between Penrith and Cliburn (54.26, AW, 1985), Baronwood and Little Salkeld (52.42, 56.36, LR, 1993), Penrith (50.30, RWMC, *c.*1970), Great Salkeld (54.34, RWMC, *c.*1970) and Talkin (54.56, MG, 1990).

This western North American species is listed by Baker and Hodgson from Coulderton (9.0), Calder Bridge and Seascale (0.0), Dub Beck, Frizington and Arlecdon (0.1), where it was known to be planted, Peel Wyke, Bassenthwaite (1.3/2.3) and Rogerscale, Lorton (1.2). There are later records from Salkeld Dykes (5.3, H. Britten, 1905), Thwaites, Millom (1.8, M. Cross, 1930), Scotby (4.5, *cf* C.W. Muirhead, 1949) and Nunwick Hall (5.3, G.A.K. Hervey, 1947-1950).

The first Furness records are from Broughton Mills and Hawkshead (respectively 2.8, 3.9, W.H. Pearsall, 1913, *BRC Rpt* (1914) 3:385) and Wilson's only record is from Windermere (4.9). There are, however, several additional post-1930 *Atlas* records from the Lorton (1.2), Cockermouth (1.3) and Great Salkeld (5.3) squares and, in the south, from the Millom (1.8), Grizebeck (2.8), Newby Bridge (3.9) and Sedbergh (6.9) squares. It has been suggested by CCH that Hodgson's records from the River Ehen and its tributaries (0.0, 0.1) might well refer to the abundant *Claytonia sibirica*, which significantly Hodgson fails to mention, and the same may well be true of some of the *Atlas* records.

Alt. limit: 150 m, Penrith (54.56)

7 (W)FC(W/Y)

*C. sibirica L. Pink purslane
(*Montia sibirica* (L.) Howell)
Fairly frequent in the lowlands in damp woodland, especially by streams and in shady gills, but avoiding the limestone.

This pretty introduction from North America appears still to be increasing. Previous Floras give only

about ten localities, the *Atlas* shows it in 23 squares and the Survey records are from more than 60.
Alt. limit: 425 m, Dockray (38.20)
Map 141 343 WFCY

Montia fontana L. Blinks
Common throughout the uplands in fellside flushes and bryophyte-rich springs where it frequently grows with *Chrysosplenium oppositifolium*, *Epilobium palustre*, *E. alsinifolium* and *Cochlearia pyrenaica*. At lower altitudes it is a characteristic plant of muddy ground on and alongside tracks and paths, often with *Callitriche stagnalis* and *Ranunculus hederaceus*, and also of the shores of tarns and lakes. It is, however, largely absent from the richer flushes on the limestone and from the more intensively farmed areas.

Seeds are essential for the recognition of the four subspecies. Rather surprisingly, three were found growing together on a limestone pavement on South Stainmore (88.10).
Alt. limit: 850 m, Great Dun Fell (70.32) and Cross Fell (68.34)
Map 142, Plate 61 1000 LWFCY

subsp. **fontana**
About two thirds of the seed samples collected during the Survey have proved to be this predominantly northern and western subspecies. This agrees with the observations of Stewart (1990) in Kirkcudbrightshire but contrasts with the situation in Co. Durham where Graham (1988) found subsp. *variabilis* to be the commonest. There are additional pre-Survey collections made by C.W. Muirhead in 1951 and determined by S.M. Walters from Seathwaite (2.0, CLE) and Coniston Old Man (2.9, CLE) and further records in the *Critical Atlas* from the Ambleside (3.0) and Spadeadam (6.7n) squares.

The seeds are smooth, black and shining.

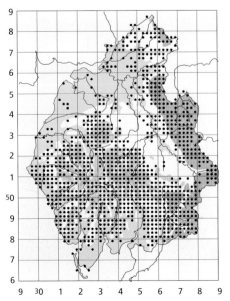

Map 142. *Montia fontana*

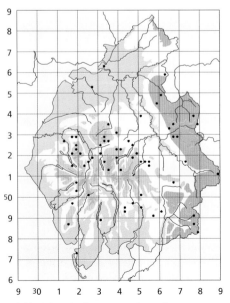

Map 143. *Montia fontana* subsp. *fontana*

Alt. limit: probably as for the species
Map 143   65   WFCY

subsp. **variabilis** Walters
A quarter of the seed collections are referable to this
subspecies. Collections, from near Whicham (14.84,
MMG, 1986, LANC), Brathay (36.02, T. Blackstock,
1977, LANC) and from Loweswater (12.22, CCH,
1978, det. S.M. Walters) have seeds intermediate
between this and subsp. *fontana*. Typical subsp. *variabilis* has seeds with 1-4 rows of narrow tubercles
around the margin.

    The *Critical Atlas* shows this subspecies as
occurring in a broad band from Northumberland south-
west to Wales; surprisingly it shows rather more Cumbrian records than for subsp. *fontana*.
Alt. limit: 450 m, South Stainmore (88.10) but probably higher.
Map 144   33   WFCY

subsp. **amporitana** Sennen
Apparently extremely rare. A southern and south-
western subspecies collected in Cumbria only at the
foot of Wast Water (14.02, CCH, 1978, det. S.M.
Walters) and upper Dentdale (76.84, CEW, 1990,
LANC, det. S.M. Walters), and mapped in the *Critical
Atlas* for the Gosforth (0.0), Dalton-in-Furness (2.7)
and Kirkby Lonsdale, v.c.60/69 (6.7s) squares.

    The seeds differ from those of subsp. *variabilis*
in usally having four or more rows of tubercles which
are wider, more acute and extend on to the face.
Alt. limit: at least 250 m, upper Dentdale (76.84)
                         2   (F)CY

subsp. **chondrosperma** (Fenzl) Walters
(subsp. *minor* Hayw.)
The most southern and early-flowering of the subspecies collected during the Survey from Drigg (*cf.*
06.96, CCH), Pooley Bridge (46.22, RWMC, 1989,

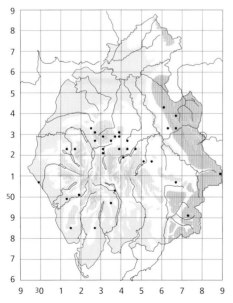

Map 144. *Montia fontana* subsp. *variabilis*

d) subsp. *chondrosperma*

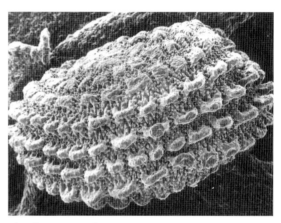

c) subsp. *amporitana*

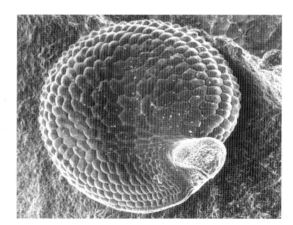

b) subsp. *variabilis*

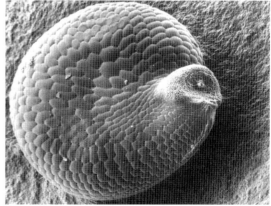

Plate 61. Seeds of *Montia fontana*: a) subsp.*fontana*

LANC) , Melmerby (62.36, RWMC, 1989, LANC, det. S.M. Walters) and South Stainmore (88.10, RWMC, 1993, herb. R.W.M.C.). There are also *Critical Atlas* records from the North Walney (1.7) and Gilsland, v.c.67/70 (6.6) squares, old collections in CLE determined by Walters from Black Dyke and Silloth (1.5, E.J. Glaister, 1876; J. Leitch, 1888) and Thurstonfield (3.5, P. Shepherd, 1884), and a record from Brough Hill (7.1, G.A.K. Hervey, 1953).

The seeds are characterised by broad, rounded tubercles over the face and around the margin.

Alt. limit: 450 m, South Stainmore (88.10)

4 WFC

CARYOPHYLLACEAE

Arenaria serpyllifolia L. Thyme-leaved sandwort
A lowland plant of well-drained, open ground. It is very much a feature of limestone outcrops, wall-tops, railway ballast, sandy areas in the Eden valley, open urban sites and the coastal sand-dunes, where it is especially frequent around rabbit burrows.

Map 145 544 WFCY

subsp. **serpyllifolia**
Fairly frequent but scarce or absent from much of the Solway Plain and north-east of Carlisle. There are no records of the sand-dune taxon subsp. *lloydii* (Jordan) Bonnier (subsp. *macrocarpa* F. Perring & Sell) characterised by a congested fruiting inflorescence.

Alt. limit: 610 m, Melmerby Fell (64.38)

Map as for the species WFCY

subsp. **leptoclados** (Reichb.) Nyman
Rare, authenticated specimens (mostly in LANC) were collected during the Survey from only seven tetrads, with most records being in the extreme south near Barrow-in-Furness (2.6s, 2.7) and the extreme north-east at Spadeadam (5.7n, 6.7n).

This subspecies was recorded by Baker and Hodgson from Humphrey Head (3.7s) and Talkin Tarn (5.5) respectively and by Wilson from Arnside (4.7s), Pooley Bridge (4.2) and Middleton (6.8s). There are later Cumberland records from Keswick (2.2, G.C. Druce, c.1902), Skirwith and Melmerby (6.3, W.W. Mason, 1926, *BEC Rpt* (1927) 8:109) and Lamplugh (0.2, A. Templeman, 1922, *BEC Rpt* (1923) 6:721). There are further records in the *Critical Atlas* for the Ambleside (3.0), Carlisle (4.5) and Arnside, v.c.60/69 (4.7s) squares based on material seen by F.H. Perring.

This is a critical taxon and records unsubstantiated by specimens should be treated as doubtful. It is best distinguished from subsp. *serpyllifolia*, with which it sometimes grows, by its slender pedicel (less than 0.5 mm diameter) and short (<3.0 mm), almost straight-sided capsule which is not readily dented, and slightly smaller seeds.

Map 146 11 WFC

*****A. balearica** L.
A garden escape naturalised on sandstone walls at Wetheral (46.54, 46.56) and Frizington (02.16, A. Riddell, 1980s), formerly at St Bees (96.10, AAD, 1993, 96.12, CCH, 1970s) and on limestone walls at Kents Bank (38.74s, FLW, 1970s), probably the basis of the undated *Atlas* dot. It was first recorded by C.D. Pigott at St Bees and Hensingham in 1943 (9.1, *BEC Rpt* (1946) 12:708) and from a bridge near Acre Walls, Frizington (0.1, Anon.) at about the same time.

6 FC

Moehringia trinervia (L.) Clairv.
Three-nerved sandwort
A fairly common and readily overlooked lowland species of hedgerows and woodland, usually on the lighter and better soils.

Alt. limit: c.300 m, Barras (84.10)

Map 147 693 LWFCY

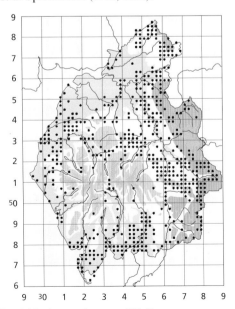

Map 145. *Arenaria serpyllifolia*

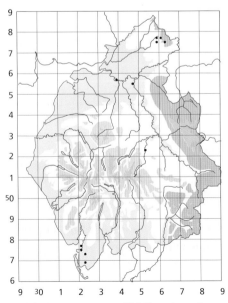

Map 146. *Arenaria serpyllifolia* subsp. *leptoclados*

Honckenya peploides (L.) Ehrh   Sea sandwort
Locally frequent to abundant along the west coast and
Morecambe Bay where it is a plant of the drift line on
shingle and sandy beaches. With its vigorous rhizomes
and its ability to withstand burial it plays an important
role in accretion and stabilisation.
Map 148   88   WFC

Minuartia verna (L.) Hiern   Spring sandwort
A common and characteristic species of open, short
limestone grassland and of ledges and crevices on limestone scars in north and south Westmorland and the
north Pennines. In the old lead and zinc mining areas
of the Pennines it is particularly prominent in the open
vegetation of old mine tips. Although some of the populations are resistant to toxic concentrations of these
metals, the toxicity is very substantially reduced if there
is any appreciable amount of limestone in the tip. Tips
with *Thymus polytrichus*, for example, are unlikely to
be toxic. The most toxic ones occur in the Nenthead
area where *Minuartia verna* and *Thlaspi coerulescens*
may be the only two flowering plants. In the Lake
District it occurs in a number of sheltered base-rich
gullies in the Helvellyn, Fairfield and High Street
ranges. In contrast to the situation on the east side of
the Pennines, the only definite riparian record is from
an island in the River Eden at Great Salkeld (5.3,
G.A.K. Hervey, 1953), although such a site may be the
source of the *Atlas* record for St John's in the Vale (3.2,
Anon., 1840, BM).

There are other *Atlas* records for the North
Walney (1.6) and Dalton-in-Furness (2.7) squares.
These are probably errors, perhaps resulting from the
misidentification of *Sagina nodosa*. The *Scarce Plants*
record for the Broughton-in-Furness square (2.8, W.M.
Hind, 1876, BM), apparently off the limestone, is based
on a misinterpretation of a specimen from Blawith
Grange [Grange-over-Sands] (4.7s, W.M. Hind, 1876,
BM).
Alt. limit: 855 m, Helvellyn (34.14)
Map 149, Plate 27   S 124   LWFCY

Stellaria nemorum L.   Wood stitchwort
Practically confined to the north and north-east, where
it is very characteristic of the rich valley woodlands of
the lower Eden, Caldew, Lyne and Liddel, and a
smaller area in south-east Westmorland and around
Sedbergh. It grows, often in some quantity, in shady,
damp, flushed or occasionally flooded ground by
wooded rivers and gills.
Alt. limit: 200 m, Knock (68.26)
Map 150   217   WCY

S. media (L.) Villars   Common chickweed
Extremely common in damp open ground, by roadsides
and on waste ground, and especially in nitrogen-
enriched sites such as farmyards, gardens, tips and
arable fields. A specimen of var. *apetala* was collected
at Rampside in 1984 (22.66s, GH, LANC, det. A.O.

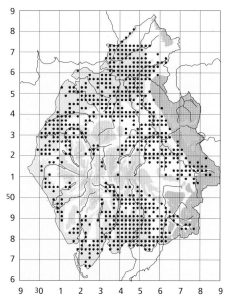

Map 147. *Moehringia trinervia*

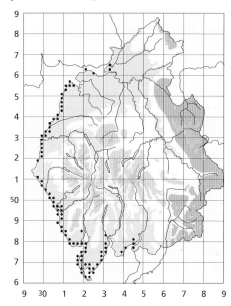

Map 148. *Honckenya peploides*

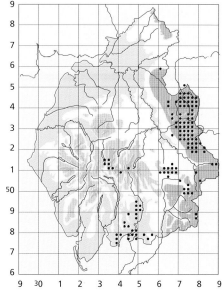

Map 149. *Minuartia verna*

Chater).

Alt. limit: 850 m, Great Dun Fell (70.32)

Map 151   1607   LWFCY

S. pallida (Dumort.) Piré   Lesser chickweed
Known only from the south-west where it occurs on
sand-dunes and in gulleries, also on bare slag at Roanhead (20.74) and on limestone at Hodbarrow (18.78).
The first definite record for Furness is from South
Walney (20.60s, A.O. Chater & GH, 1978, *Watsonia*
(1980) 13:133) and for Cumberland from Ravenglass
Local Nature Reserve (06.94, RS & J. Rose, 1979,
ibid.). All records are based on material in LANC.

The only reference in the Floras to this member
of the *Stellaria media* aggregate is Hodgson's record of
var. *'boraeana'* from Flimby (0.3). This is probably the
basis of the pre-1930 *Atlas* dot but it possibly refers to
apetalous plants of *S. media*.

Map 152   12   FC

S. neglecta Weihe
Apparently extremely rare. Numerous robust specimens
of '*Stellaria media*' were examined during the Survey
but only two possessed the requisite combination of
large (-1.4 mm), dark, strongly papillose seeds and
flowers with 10 stamens characteristic of *Stellaria
neglecta*. These were from lanesides in the extreme
south: east of Milnthorpe (50.80s, GH, 1977, LANC,
det. P.M. Benoit) and near Lupton (54.80s, GH, 1979,
LANC, det. A.O. Chater).

Baker refers to it as "a common Lakeland
variety", Wilson gives only a single record, from Kings
Meaburn (6.2), and there are two post-1930 *Atlas*
records for the Pooley Bridge (4.2) and Langwathby
(5.3) squares. In view of the confusion between this
species and *S. media*, all records unsupported by herbarium material are best regarded as doubtful.

2   W[(FC)]

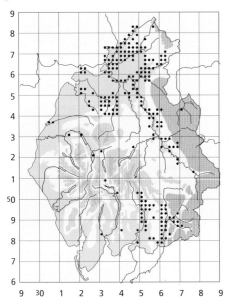

Map 150. *Stellaria nemorum*

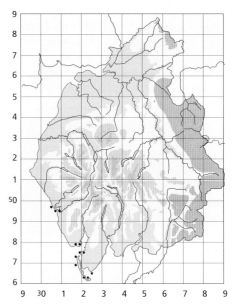

Map 152. *Stellaria pallida*

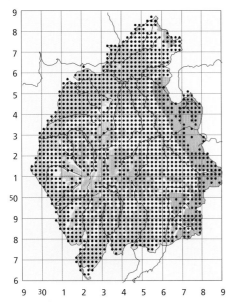

Map 151. *Stellaria media*

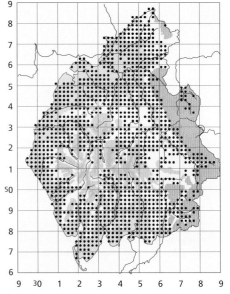

Map 153. *Stellaria holostea*

S. holostea L.   Greater stitchwort
Very common in the lowlands where it is a typical spring species of hedgerows, banks and woodland margins.
Alt. limit: 460 m, above Hayeswater (42.12)
Map 153   1187   LWFCY

S. palustris Retz.   Marsh stitchwort
Extremely rare and decreasing. The only recent authenticated records are from a marshy field near Great Ormside (70.16, CFS & JS, 1984, LANC), where it was growing with *Carex acutiformis*, and amongst *C. rostrata* in tall fen vegetation at Biglands bog (24.52, D. Shaw, 1986, LANC); vegetative material almost certainly of this species has also been collected at Thurstonfield Lough (3.5, D. Shaw). The populations are small and that at Biglands could well become shaded out.

Although Baker lists it from the Muncaster woods, Ravenglass (0.9) and near Whitehaven (9.1), Hodgson makes no reference to this species. There are, however, more plausible Cumberland records from the Solway Plain at Red Moss (1.5, E.J. Glaister, 1876), Monkshill Lough (3.5, P. Shepherd, 1884), and Abbeytown (1.4/1.5, J. Hodgson, 1899; J. Parkin,1935), the latter probably the site on Southerfield Moss by the old railway where it was collected by C.W. Muirhead in 1949 (1.4, CLE). It was also recorded from Great Salkeld (5.3, H. Britten, 1900-1908; G.A.K. Hervey, 1948), and Edenhall (5.3, A. Wallis & C.E. Salmon, 1919, BM), where it was growing very luxuriantly by the pond. Wilson's only record is from Temple Sowerby moss (6.2) where it was first seen in 1894.

The *Atlas* shows additional post-1930 records for the Workington (9.2), Distington (0.2), Millom (1.8), Langdale (2.0), Newby Bridge (3.8), Hawkshead (3.9), Ambleside (3.0) and Arnside (4.7s) squares. In view of the relative paucity of earlier records and the frequency with which it is misrecorded for *Stellaria uliginosa* and *S. graminea*, most if not all of these *Atlas* records should be considered very doubtful. Handley's (1898) unlocalised record for the Sedbergh area should be regarded similarly.
Alt. limit: 135 m, Great Ormside (70.16)
                      2   W[(F)]C[(W/Y)]

S. graminea L.   Lesser stichwort
A species with a similar distribution and frequency to *Stellaria holostea* but ascending higher. Although less tolerant of shade it occurs in rather poor soils, both dry and damp, and it is characteristic of the coarse vegetation of neglected pastures. It flowers later than *S. holostea*, continuing well into August.
Alt. limit: 600 m, Tynehead Fell (76.36)
Map 154   1420   LWFCY

S. uliginosa Murray   Bog stitchwort
(*S. alsine* Grimm)
Very common in wet places everywhere, occurring in ditches, by becks and ponds, in marshy fields and fens, in flushes on the lower slopes and high up the fells as a common constituent of *Juncus effusus* soakways.
Alt.limit: 885 m, Helvellyn (30.14)
Map 155   1548   LWFCY

[Cerastium cerastoides (L.) Britton
This species, not otherwise known in Britain south of the Scottish Highlands, is stated by Druce (1932) to occur at 445 m on Cross Fell (6.3). The basis for this statement is unknown but DAR thinks the finder subsequently withdrew the record. No supporting herbarium material has been found. The record was probably the result of confusion with large-flowered plants of *Cerastium fontanum*.

(C)]

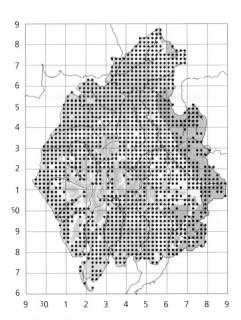

Map 154. *Stellaria graminea*

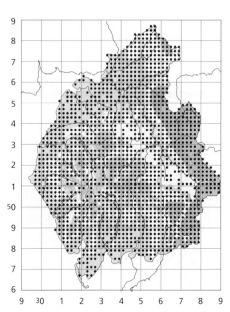

Map 155. *Stellaria uliginosa*

Cerastium arvense L.   Field mouse-ear chickweed
Seen only five times during the Survey. It was growing
in waste ground by the Cavendish dock at Barrow-in-
Furness (20.68s, GH, LANC, 1983), in an old sand pit
near Plumpton (52.36, RWMC, 1990, LANC, *Watsonia*
(1991) 18:422), on a sandy bank in Whinfell Forest
(56.26, RWMC & MP, 1993), on a riverside bank in St
John's in the Vale (30.20, CFS & JS, 1991, LANC) and
on Grune Point (12.56, A. Cannell, 1992). Why this
species should be so scarce in the sandy coastal areas of
south-west Cumbria is puzzling.

Hodgson considered it an uncommon species of
sandy fields and waste places and listed from Whitehaven (9.1), from between Maryport and Cockermouth
(0.3), in quarries at Penrith Beacon and near Edenhall
(5.3). It was seen early this century at Dalston (3.5,
T.S. Johnstone, 1905) and Great Salkeld (5.3,
H. Britten, 1900-1908) and in 1935 at Skinburness
(1.5, J. Parkin). It was reported by G.A.K. Hervey from
Udford (5.3, 1953), Great Salkeld (1953-1958) and
Eamont Bridge (5.2, 1955). Wilson's only records are
from Brackenber Moor, Appleby (7.1) and Melkinthorpe (5.2).

In 1950 it was collected at Ulpha, Meathop
(4.7s, herb. P. Reiss, KDL) and there are post-1930
Atlas records from the Distington (0.2), Maryport (0.3),
Brampton (5.6), Langdale (2.0) and Ambleside (3.0)
squares. This is surprising and the last two records in
particular are best regarded as very dubious, possibly
the result of misidentification of large-flowered
Cerastium fontanum.
Alt. limit: 185 m, Whinfell (56.26)
5   (W)FC

*****C. tomentosum** L.   Snow-in-summer
An occasional but increasing garden escape found on
roadside verges and banks, usually near houses. A
particularly fine colony grows on the limestone cutting

at the northern end of the Kendal by-pass (48.94).
There are no published records prior to the *Atlas* which
shows it as occurring in only two squares, Gosforth
(0.0) and Skelton (4.3), compared with the present
number of 28.
Alt. limit: at least 430 m, Nenthead (76.42)
Map 156   54   WFC

C. alpinum L   Alpine mouse-ear chickweed
Known only from the Helvellyn square (3.1) where it
occurs in four tetrads, growing on rock ledges, crevices
and scree. It was not refound at Browncove Crags, on
the west side of Helvellyn, where it had been seen by
DAR from 1954 to 1965, nor on Dove Crag (3.1, H.E.
Fox, OXF) where it was collected in 1862 and
rediscovered by R.J. Birkett in 1961. This last site is the
basis of the misplaced *Atlas* dot for the Ambleside
square (3.0).

This arctic-montane species was listed by Baker
from Helvellyn, Fairfield and Deepdale (3.1), also from
Langdale. This last record may refer to Great End
where the plant was collected by J. Ball in 1838 (2.0,
E). The pre-1930 *Atlas* record for the Ambleside square
(3.0) is an error.

These sites, together with those in Snowdonia
and Dumfriesshire, are the only ones in Britain south of
the Scottish Highlands.
Alt. range: 760-870 m, Helvellyn (3.0)
S 4   W(C)

C. fontanum Baumg.   Common mouse-ear chickweed
Extremely common in the lowlands in meadows and
pastures and as a weed of waste ground, gardens and
lawns. In the uplands it occurs in damp *Agrostis -
Festuca* grassland, also in flushes and particularly in
Juncus effusus dominated soakaways. It is the commonest roadside plant at 750 m on Great Dun Fell
(70.30).

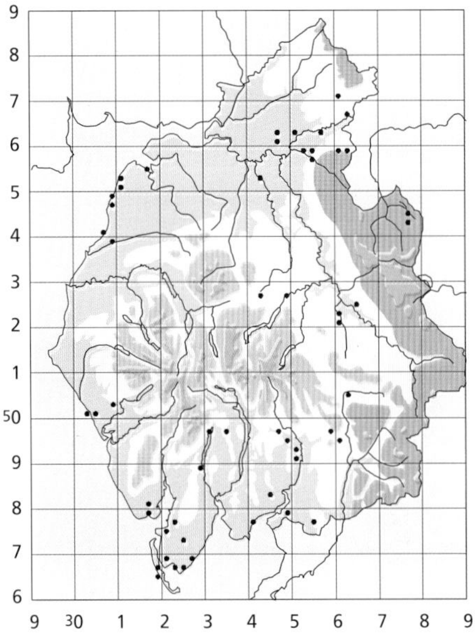

Map 156. *Cerastium tomentosum*

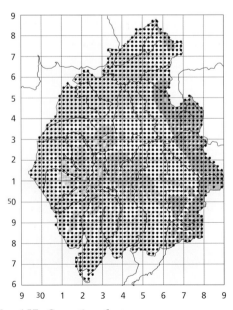

Map 157. *Cerastium fontanum*

Alt. limit: 890 m, Cross Fell (68.34), although Baker gives it as 925 m on Helvellyn (3.1).

Map 157   1757   LWFCY

C. glomeratum Thuill.   Sticky mouse-ear chickweed
A common plant of lowland meadows, pastures, field gateways, verges and waste places. With its erect stem, pale leaves and clustered flowers it is a much more conspicuous species than the last. It appears to be spreading, being seen during the Survey in every square, more than twice the number shown in the *Atlas*.
Alt. limit: 520 m, south of Alston (68.38)

Map 158   1234   LWFCY

C. diffusum Pers.
Present in dry, sandy or gravelly places around most of the coast, rare inland where it occurs chiefly as a casual, particularly on the ballast of railway lines.
Alt. limit: 150 m, Penrith (50.28)

Map 159   77   WFC

C. semidecandrum L.
A species with a similar distribution and ecology to the last but appreciably commoner and not infrequent inland on limestone scars and open turf. The first Furness record was from Sandscale (1.7, D. Lumb, *BEC Rpt* (1915) 4:127).
Alt. limit: 340 m, Knipe Scar (52.18)

Map 160   72   WFC(W/Y)

Myosoton aquaticum (L.) Moench   Water chickweed
Seen in a garden at Arnside (46.78s, MB, 1978, LANC), where it persisted for six years, on a trackside near Salta Moss (08.44, MM, 1985, LANC, *Watsonia* (1986) 16:185), where it still occurs, and by a dyke at Little Strickland (56.18, CEW, 1992, LANC). It is a plant easily confused by visiting botanists with *Stellaria nemorum*.

The only previous records are a passing reference by Hodgson to a drawing of a specimen said to have been collected at Waverton (2.4), a collection from "Holwell" (R. Wood, 1882, CLE), an unlocalised reference to its occurrence in the Sedbergh area (Handley 1898) and an old record, cited by Wilson, from between Chapel Stile and Little Langdale (3.0).

3   WC(W/Y)

Moenchia erecta (L.) Gaertner, Mayer & Scherb.
The only records are those given by Baker from St Bees (9.1) and Coulderton (9.0), one in Wilson, attributed to Baker, from Grange-over-Sands (4.7s) and a collection from Silloth (1.5, E.J. Glaister, 1876, CLE).

These sites are at the northern limit of this species' range on the west coast of Britain.

(FC)

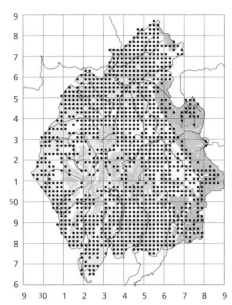

Map 158. *Cerastium glomeratum*

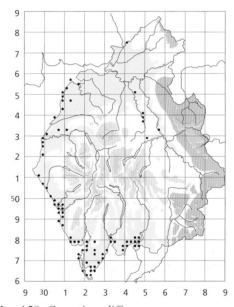

Map 159. *Cerastium diffusum*

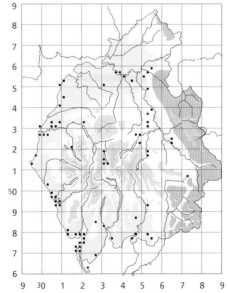

Map 160. *Cerastium semidecandrum*

Sagina nodosa (L.) Fenzl   Knotted pearlwort
Frequent in the limestone areas of south Westmorland
and the north Pennines, also in the Bewcastle Fells, but
otherwise rather local and coastal. In the fells it occurs
in open, flushed areas, growing around tussocks of
sedges or mosses or in irrigated gravel, also on damp
tracksides. On the coast it grows on the upper margin
of salt-marshes and particularly in dune-slacks.
D. Lumb (*BEC Rpt* (1913) 3:236) even recorded it as
growing plentifully in the "rivet holes of flattened
boiler plates" at Askam-in-Furness (2.7) ironworks!
Alt. limit: 740 m, Knock Fell (70.30)
Map 161   390   WFCY

[S. subulata (Sw.) C. Presl
Hodgson gives an 1854 record from "Workington rabbit-
warren" (9.2). He comments that the warren had since ceased
to exist and he appears to have reservations about the record.
    Sagina subulata has a scattered but fairly continuous
distribution in western Britain. It occurs across the Solway
Firth in Kirkcudbrightshire where Stewart (1990) describes it
as "fairly common". It could yet be found in Cumbria.
                                    (C)]

[S. saginoides (L.) Karsten
Listed by J. Backhouse Snr in a letter to Sir Joseph Hooker in
1865 (K) as having just been seen by his son on the Helvellyn
- Fairfield range. It has never been refound and in the
absence of herbarium material the record is best treated as an
error. This is the only British record from outside the Scottish
Highlands.
                                    (C)]

S. procumbens L.   Procumbent pearlwort
Very common. In the lowlands it behaves as a most
persistent weed occurring in gardens and lawns, footpaths and pavements. Its natural habitat is damp, open
ground by rivers and becks, particularly in short grassland where poached by cattle, and in the fells it is characteristic of the wetter *Festuca - Agrostis* grassland.
Alt. limit: 850 m, Cross Fell (68.34)
Map 162   1653   LWFCY

S. apetala Ard.   Annual pearlwort
A lowland plant of dry, open ground such as gravel
paths, pavements, forestry tracks, wall-tops and railway
ballast.

subsp. **apetala**
(*S. ciliata* Fries)
Scattered in the north, elsewhere rare or absent but no
doubt to some extent under-recorded. All the records
are based on herbarium material. The only Flora ones
are Baker's and Wilson's from Kents Bank (3.7s). The
first records for Cumberland appear to be from Silloth
(1.5, R.A. Boyd, 1949, *BSBI Yr Bk* (1951) p.48),
Abbeytown (1.5, C.W. Muirhead, 1950), Baron Wood,
Armathwaite (5.4, DAR, 1956, LANC) and a post-
1930 *Atlas* dot for the Ravenglass square (0.9). The
only Westmorland record is from Soulby (74.10, TW,
1988).

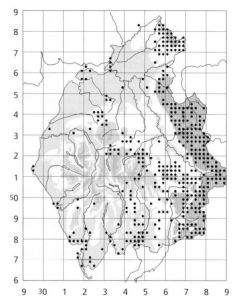

Map 161. *Sagina nodosa*

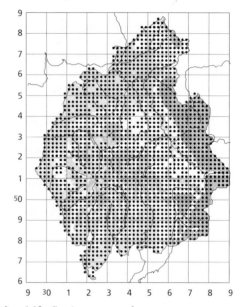

Map 162. *Sagina procumbens*

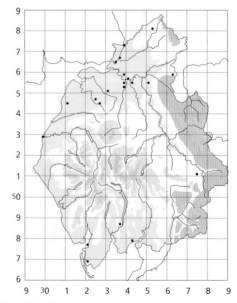

Map 163. *Sagina apetala* subsp. *apetala*

Alt. limit: 200 m, Midgeholme (62.58)

Map 163   21   WFC

subsp. **erecta** F. Herm.
Fairly frequent and widespread. Rather surprisingly it occurs in salt-marsh turf by the Solway Firth.
Alt. limit: 310 m, Spadeadam (62.74)

Map 164   220   WFCY

S. maritima G. Don   Sea pearlwort
Scattered but locally frequent around much of the coast occurring in open sandy ground, in the higher parts of the salt-marshes and on sea-walls. By the Kent estuary (4.7s) it grows on the limestone embankment of the railway viaduct. There are no Furness records prior to that in the *Atlas* for the Newby Bridge square (3.8).

Map 165   26   WFC

Scleranthus annuus L.   Annual knawel
A rather rare species of sandy ground, fields and sand-pits occurring chiefly in the Wetheral - Hayton area east of Carlisle (4.5/5.5) and around Silloth (1.5), also on dry Silurian outcrops to the south of the Lake District. This species appears to have decreased considerably since both Baker and Hodgson described it as "not uncommon". Since the 1950s it has disappeared from Whins Pond, Penrith (5.3) and Eamont Bridge (5.2) where it was recorded by G.A.K. Hervey.
Alt. limit: 170 m, Brackenber Moor, Appleby (70.18), although Wilson gives a record from 380 m on the hause between Fusedale and Martindale (4.1).

Map 166   24   WFC

Herniaria glabra L.   Smooth rupturewort
Found in 1964 by M. Mason on the ballast of the old railway line at Hartley, near Kirkby Stephen (78.08, *Proc. BSBI* (1966) 6:238). A few plants still survive. in a nearby quarry. The only other record for this East Anglian and Lincolnshire species is H. Britten's from a gravelly field near Great Salkeld (5.3, CLE) in 1905 (*North Western Naturalist* (1926) 1:214; *BEC Rpt* (1927) 8:129) where for at least 21 years it persisted and spread.
Alt.: 230 m, Hartley (78.08)

                 R 1   W(C)

Spergula arvensis L.   Corn spurrey
A frequent lowland calcifuge weed of arable fields, disturbed ground and gardens. It occasionally occurs in some quantity around reservoirs at times of draw-down.

       The proportion of individuals in a population with non-papillose, distinctly winged seeds, var. *sativa* (Boenn.) Mert., has been shown by New (1958) to increase in a north-northwesterly direction across Britain. Her only Westmorland sample showed 79% non-papillose seeds, appreciably higher than a recent one of 60% from Killington Reservoir (58.90).
Alt. limit: 300 m, Gawthrop, Dentdale (68.86s)

Map 167   633   WFCY

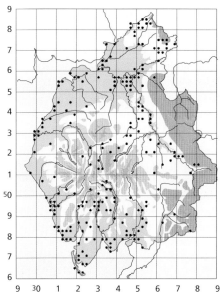

Map 164. *Sagina apetala* subsp. *erecta*

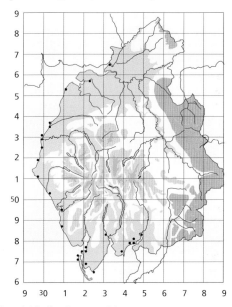

Map 165. *Sagina maritima*

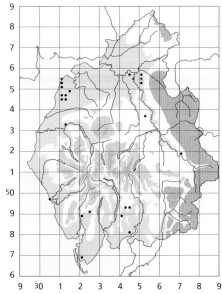

Map 166. *Scleranthus annuus*

Spergularia media (L.) C. Presl   Greater sea-spurrey
Common in the salt-marshes occurring chiefly in the
upper marsh, being scarce both as a pioneer species and
as a member of the mature marsh. Common associates
include *Plantago maritima, Glaux maritima* and *Puccinellia maritima.*
Map 168   62   WFC

S. marina (L.) Griseb.   Lesser sea-spurrey
Commoner than the last species and usually prefering
drier situations. It is typical of the more mature marsh
and it also occurs on sea-walls, by coastal paths and at
the foot of sea-cliffs along most of the west coast. There
are also inland records from a roadside verge between
Kendal and Sedbergh (60.92, CEW, 1988, LANC), a
pile of sand in Witherslack churchyard (42.84s, CEW,
1990, LANC) and on dumped refuse in Broughton
Crags quarry (08.30, CCH, 1989).
Alt. limit: 230 m, between Sedbergh and Kendal
(60.92)
Map 169   90   WFC

S. rubra (L.) J.S. & C. Presl   Sand spurrey
Scattered in open, sandy ground around most of the
coast and inland in a few places, notably in the Eden
valley and south of Thursby (3.4). It has recently
appeared on forestry tracks in the remote north-east
corner of the county.
Alt. limit: c.380 m, north of Christianbury Crag
(56.84n)
Map 170   68   WFC

*Lychnis coronaria (L.) Murray
A handsome garden escape naturalised in scrub on the
sandstone cliff above the River Eden at Rockcliffe
(34.60, REG, 1979, *Watsonia* (1988) 17:185), in grassland at Grune point (12.56, REG, 1981), by a field edge
at Silloth (10.52, A. Cannell, 1992) and in a coalyard
at Arnside (46.78s, MB, 1993, *Watsonia* (1995)
20:421).
                           4   WC

L. flos-cuculi L.   Ragged robin
Common in ditches, fens and wet meadows, also on wet
roadside verges. It is often associated with *Filipendula
ulmaria, Achillea ptarmica* and, particularly in the
uplands, *Juncus acutiflorus.* Sadly this attractive plant
has disappeared from many sites following drainage.
Alt.limit: 550 m, Moor House, Teesdale (74.32) and
Helbeck Wood, Brough (78.16)
Map 171   1009   WFCY

L. alpina L.   Alpine catchfly
Known for more than a century from Hobcarton Crag
(1.2) above the Whinlatter Pass where it was first
recorded by R. Matthews in 1844. The population, of
about 60 plants, grows on a crumbling rock spur and
there has been frequent speculation that British plants
may have a requirement for certain heavy metals, part-

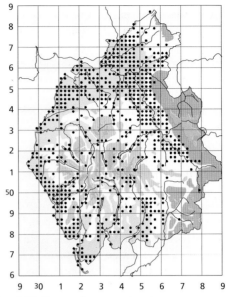

Map 167. *Spergula arvensis*

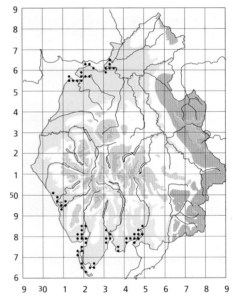

Map 168. *Spergularia media*

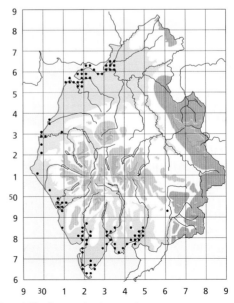

Map 169. *Spergularia marina*

icularly in view of its occurrence on serpentine rock in the Clova mountains of Angus, the only other British locality for this arctic-montane species. However, Proctor & Johnston (1977) state that although there are pyritic outcrops the soil, unlike that in Clova, is not particularly rich in heavy metals.

The record from Coniston Old Man (2.9), cited by Baker and shown in the *Atlas*, is thought by DAR to be the result of erroneous labelling of herbarium material.

Alt. range: 610-700 m, Hobcarton Crag (1.2)

Plate 23　　　　　　R 1　　　　　　[(F)]C

Agrostemma githago L.　　　　　　Corncockle

Seen only once during the Survey, in a roadside lay-by at Milnthorpe where several vigorous plants were seen in 1994 (48.80s, CEW, LANC). Baker gives no localities for this species only commenting that it "is frequent in cultivated fields", a statement which contrasts with Hodgson's observation only ten years later that it was "becoming every year more rare from the gradual abandonment of cereal cropping". His only records are from Bennet Head Banks, Ullswater (*cf*.4.2), Keswick (2.2), Ainstable (5.4) and Carlisle (3.5/4.5). There are later Cumberland records from Great Salkeld (5.3, H. Britten, 1904), Penrith (5.3, M. Cross, 1907), Cumwhinton (4.5, E. Blezard, 1914), "Bridgemill" (J. Parkin, 1931), Brigham quarry (0.3, J. Parkin, 1932) and Silloth (1.5, T.S. Johnstone, 1899), where it was last seen by C.B. Sharpe in 1939.

There are no *Atlas* records for the Sedbergh area although Handley (1898) notes that "it is not uncommon". Wilson's records are all pre-1930 - from Laverock Bridge, Kendal (5.9), Levens (4.8s) and near Appleby (6.2). The two post-1930 *Atlas* records, from the mid-Walney Island (1.6) and Roanhead (1.7) squares, both date from the mid-1950s. The last record as an agricultural weed was in 1960 when it was seen at Dyke Nook Farm, near Brough (7.1, M.P. Lord, *Field Nat*uralist (1960) 5(4):42).

　　　　　　R 1　　　　　　W(FCY)

*Silene nutans L.　　　　　　Nottingham catchfly

A rare casual listed by Hodgson only from Silloth (1.5, E.J. Glaister, 1882, CLE). The record from Moorland Close and Dean, Workington (0.2) was considered by both him and Baker to refer to *Silene gallica*.

　　　　　　S　　　　　　(C)

S. vulgaris Garcke　　　　　　Bladder campion

Fairly frequent on lowland roadside verges and railway-sides, chiefly in coarse grassland and scrub, and favouring the lighter, drier soils and the limestone. It is virtually absent from the Lake District.

Alt. limit: *c*.250 m, Hardendale (56.12)

Map 172　　　　　　197　　　　　　WFCY

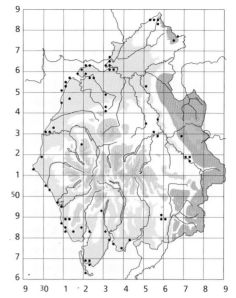

Map 170. *Spergularia rubra*

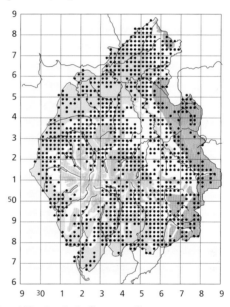

Map 171. *Lychnis flos-cuculi*

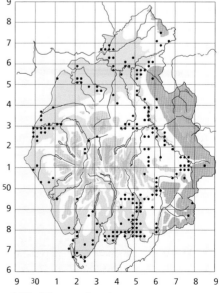

Map 172. *Silene vulgaris*

S. uniflora Roth   Sea campion

(*S. maritima* With.)

Very common around the entire coast, occurring on
sand and shingle beaches, along the drift line, on sea-
walls, cliffs and waste ground. Inland it grows on often
markedly base-poor screes and cliff-ledges in the Lake
District, chiefly in the Fairfield and Langdale fells but
it is surprisingly scarce on the Helvellyn range. There
are also recent records of ephemeral populations on
roadside grit piles east of Kendal (58.94) and in the
Lune gorge (60.98), on Shap summit (54.06) and above
Garrigill, and an early one from river gravel by the
Eden at Great Salkeld (5.3, H. Britten, 1907).

Alt. limit: 775 m, Fairfield (34.10)

Map 173   130   WFC

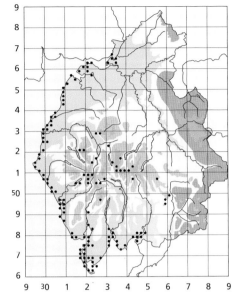

Map 173. *Silene uniflora*

*****S. csereii** Baumg.

A south-east European species found by D. Lumb at Askam-
in-Furness ironworks in 1915 (2.7, YRK, *BEC Rpt* (1916)
9:545).

                                                      (F)

S. acaulis (L.)Jacq.   Moss campion

A very local arctic-montane species found in eight Lake
District tetrads, mostly in the Helvellyn square (3.1) but
with single sites in the Scafell (2.0) and High Street
(4.1) squares. Most of the sites are moist, usually north-
or east-facing base-rich cliffs, both on ledges and on
steep open slabs.

There are no recent records from Langdale (2.0)
and St Sunday Crag (3.1), listed by Baker, or from the
Ambleside square (3.0), indicated as a post-1930 record
in the *Atlas*. Baker also refers to a site above Rydal
Mount, probably Fairfield (3.l), where it was shown to
J. Sidebottom by Wordsworth in 1847.

South of the Scottish Highlands, moss campion
occurs only in the Lake District, in Snowdonia and on
Cader Idris.

Alt. range: 455 m, Scafell square (2.0) - 790 m,
Helvellyn square (3.1)

Plate 62   8   WC

*****S. noctiflora** L.   Night-flowering catchfly

Seen only in two arable fields near Brough (78.14, LH,
Watsonia (1977) 11:393) in 1975 and 1976. Hodgson
records it from the Derwent Tinplate Works, Workington (9.2), Maryport (0.3) and Floshgate, Ullswater
(4.2), surprisingly overlooking an 1877 record from
Silloth (1.5, E.J. Glaister, CLE). The only other Cumberland records are H. Britten's from Great Salkeld (5.3,
1905) and W.W. Mason's from Melmerby (6.3, 1917).
This southern and eastern species was later seen by the
docks at Barrow-in-Furness (2.6s, Barrow N.H.S,
1950s).

Alt. limit: 190 m, Brough (78.14)

                    1   W(FC)

Plate 62. *Silene acaulis*

S. latifolia Poiret   White campion
(*S. alba* (Miller) E.H.L. Krause)
Fairly frequent in the lowlands in the northern part of the county but quite scarce in the south, except on the Furness peninsula. It is a plant of generally more open habitats than *Silene dioica*, occurring in hedgerows, waste places and field margins. It is surely worth perpetuating the bizarre Cockermouth belief, as recorded by Hodgson, that if a child picks this or red campion, his mother or father will shortly die, hence the popular names "mudder dees" and "fadder dees" respectively.
Alt. limit: *c.*370 m, Nenthead (76.44)
Map 174   319   WFCY

S. latifolia x **S. dioica**
S. x **hampeana** Meusel & K. Werner
Rather rare except in the Eden and Petteril valleys. This hybrid is not always easy to recognise as the flowers of *Silene dioica* vary appreciably in the intensity of their colour and may even be white. There appear to be no published records prior to the *Critical Atlas* which shows a single dot for the Ravenstonedale square (7.0).
Alt. limit: 300 m, Ravenstonedale (74.04)
Map 175   89   WFC

S. dioica (L.) Clairv.   Red campion
An abundant plant of hedgerows, copses and woodland margins extending from sea-level (there is a remarkably fine population in an old Walney Island cemetery) to high up into the sheltered gullies of the Lake District. In the lowlands it is rather scarce in the intensively farmed parts of the Solway Plain and surprisingly local in the upper Eden valley.
Alt. limit: 845 m, Great Dun Fell (70.32)
Map 176   1189   LWFCY

*S. inaperta L.
A species of the western Mediterranean known only from Hodgson's records from the Derwent Tinplate Works, Workington (9.2).
               (C)

*S. dichotoma Ehrh.
An eastern European species listed by Hodgson (pp.47, 380) as growing with *Vaccaria hispanica* at the Derwent Tinplate Works, Workington (9.2). There are also records from Silloth docks (1.5, J. Leitch, 1883, CLE, 1891).
               (C)

*S. gallica L.   Small-flowered catchfly
Listed by Hodgson from the Derwent Tinplate Works, Workington (9.2) and Silloth (1.5, E.J. Glaister, 1877, CLE), the latter find being of the larger-flowered var. *gallica*. He also mentions finds of var. *anglica* (L.) Mert.& Koch at Portinscale (2.2) and as a garden weed by Ullswater (4.2). H. Britten recorded both varieties from Great Salkeld (5.3) early this century. Var. *gallica* was also seen near Thwaites, Millom (1.8, M. Cross, 1932). The only other record of this southern and decreasing species is Wilson's of var. *anglica* from a sandy bank in Windermere (4.9).

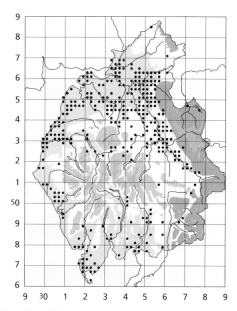

Map 174. *Silene latifolia*

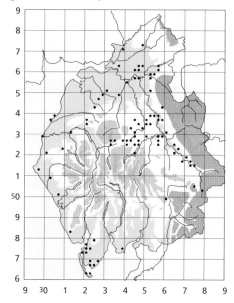

Map 175. *Silene latifolia* x *S. dioica*

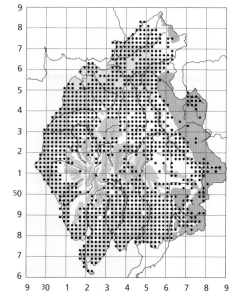

Map 176. *Silene dioica*

Baker cites records of *Silene nutans* from Moorland Close and Dean, Workington (0.2), but he regarded these as probable errors for *S. gallica*.

S   (WC)

*S. conica L.
Hodgson gives two localities for this south-eastern species, the Derwent Tinplate Works, Workington (9.2) and Silloth (1.5, J. Leitch, 1889, CLE). The Atlas record for the Arnside square (4.7s) is from v.c.60.

S   (C)

Saponaria officinalis L.   Soapwort
Very scattered and chiefly in the west, occurring on roadside verges and in hedgerows, nearly always close to houses. Most of the records are of double-flowered plants and white-flowered ones are frequent.
Alt. limit: 160 m, Great Musgrave (76.12)
Map 177   59   WFC

*Vaccaria hispanica (Miller) Rauschert
(*V. pyramidata* Medikus)
The only Survey record, and the first for Westmorland, is from a cornfield near Brough (78.14, LH, 1971, *Watsonia* (1977) 11:393).

This Continental species was noted by Hodgson from a garden at Whitehaven (9.1), the Derwent Tinplate Works, Workington (9.2), Risehow and Maryport (0.3, H. & J. Adair, 1894, BM) and Silloth (1.5, J. Leitch, 1889, BM, CLE). There are later records from gravel beds by the River Eden at Etterby, Carlisle (3.5, T.S. Johnstone, 1901, CLE), Melmerby (6.3, W.W. Mason, 1918) and Silloth (C.W. Muirhead, 1939, CLE). It was recorded in 1913 by W.H. Pearsall from Barrow-in-Furness docks and by D. Lumb from Dalton-in-Furness (respectively 1.6/2.6s, 2.7, *BEC Rpt* (1914) 3:402-403) and by M. Cobbe from Walney Island in 1924 (*BRC Rpt* (1925) 7:562).
Alt.: 190 m, Brough (78.14)

1   W(C)

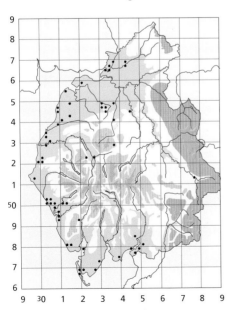

Map 177. *Saponaria officinalis*

*Petrorhagia nanteuilii (Burnat) P. Ball & Heyw.
A rare British casual collected by C.W. Muirhead at Silloth in 1931 (1.5, PLYP).

(C)

*Gypsophila pilosa Hudson
An Asiatic species found on the gravel beds of the River Eden at Grinsdale, Carlisle in 1901 (3.5, T.S. Johnstone, CLE).

(C)

[Dianthus gratianopolitanus Villars   Cheddar pink
Listed by Baker from limestone rocks in Furness, although he considered the record to be erroneous.

R   (F)]

*Dianthus plumarius L.   Common pink
An eastern central European species known from the railway near the viaduct over the Leven estuary (32.78s) where it persisted from at least the early 1950s until *c*.1970.

1†   F

D. deltoides L.   Maiden pink
Seen by L.A. Cowcill at High Nibthwaite (28.88, 1951, CLE), growing on roadside rocks and surviving until 1974. The only other Survey records are of a single plant on the eroding sandstone cliff by the River Eden at Rockcliffe (34.60, REG, 1987, LANC), and of small populations on the sand-dunes at Silloth (10.52, MP, 1993) and by the River Caldew between Cummersdale and Dalston (38.52, REG, 1978, LANC), where it was growing amongst gorse in heavily grazed turf overlying sand and shingle. Erosion has largely destroyed this last site and in 1988 only one plant was seen, and that some distance from the river.

Hodgson gives records from near the foot of Skiddaw (2.2) and Heathfield, near Braystones (0.0, G.J. Carr, 1887, CLE). C.H. Wright collected it near Crosthwaite, Keswick (2.2) in 1837; Baker refers to "common pastures in Furness" and Lawson's Westmorland record from Common Holme Bridge, Cliburn (5.2) dating from 1688. *Scarce Plants* shows pre-1930 unlocalised dots for the Aspatria (1.4) and Kirkoswald square (5.4) squares.

(W)FC

*D. barbatus L.   Sweet William
A garden escape recorded during the Survey from Brampton, Appleby (68.22, ES & AW, 1988), sand-dunes at Askam-in-Furness (20.76, PB, 1989), a quarry west of Kendal (50.92, CEW, 1997 LANC) and a riverside bank at Garrigill (74.40, RWMC & LR, 1989).
Alt. limit: 330 m, Garrigill (74.40)

4   WFC

*D. armeria L.   Deptford pink
A southern and decreasing species listed by Baker and Hodgson from Orton (6.0), Nunnery, Kirkoswald (5.4), Hensingham (9.1) and Bromfield (1.4) and seen on North Walney (1.7) in 1951 by E.J. Gibbons and 1954 by G. Wilson.

                                                    (WFC)

POLYGONACEAE

*Persicaria campanulata (Hook. f.) Ronse Decraene
(*Polygonum campanulatum* Hook. f.)
An attractive Himalayan species extensively naturalised in wet woodland in the Witherslack woods, Grange-over-Sands (42.86s, E. Hodgson, 1965, LANC). It has also been seen during the Survey on a laneside near Frizington (02.18, A. Riddell, 1989, LANC), north of Gosforth (06.04, MMG, 1992, LANC) and in Knowe-field Wood, Carlisle (40.56, REG, 1989, LANC, *Watsonia* (1991) 18:428). Conolly (1977) records it also from a plantation at Ravenglass (0.9, C.R. Lanc-aster, 1960, BM) but there are no recent records from there.

This species is characterised by rather thick leaves which are dark green above and finely pubescent below, and by the bright pink perianth segments.

                            4   WC

*P. wallichii Greuter & Burdet   Himalayan knotweed
(*Polygonum polystachyum* Wallich ex Meissner)
Rather rare, mainly in the south and west but with outlying sites at Carlisle (40.56) and, surprisingly, in the upper Eden valley (76.04). Like *Fallopia japonica*, this is an aggressive garden escape which is clearly very much at home in the Ambleside area. The most impressive stands are around the White Moss Common car park by Rydal Water (34.06), at High Close above Grasmere (32.04) and in Brankenwall Plantation,

Ravenglass (08.96).

The first record for Westmorland is from Bowness (4.9, R.K. Brummitt, 1957, LIV), for Cumb-erland from Eskdale (1.9, V. Gordon, 1930s) and for Furness from Arrad Foot (30.80, JA, 1981). There are no Survey records for the Matterdale (4.2) square shown on Conolly's (1977) maps.

Conolly has examined material in AMDE and LANC from the Grasmere - Windermere area. The specimens are all short-styled, except for a specimen from the Holehird estate (40.00), and most are var. *pubescens* Meissner which is softly pubescent on the main and primary veins as well as on the undersurface and margins of the leaves. It was first collected in 1969 (40.98, R. Young, AMDE) from a roadside near Wind-ermere; the population in the White Moss Common car park is a mixture of the variety and typical plants. When in full flower, in late summer, the large clumps are extremely fragrant.
Alt. limit: 200 m, Brough (78.14)
Map 179   18   WFC

*P. weyrichii (F. Schmidt ex Maxim.) Ronse Decraene
(*Polygonum weyrichii* F. Schmidt ex Maxim.)
The first and so far the only report of this Asiatic species naturalised in Britain was in 1978 when it was seen in the vicinity of Wasdale Hall (14.04, CCH, LANC, *Watsonia* (1980) 13:138) growing vigorously by the roadside, in acid grassland and in woodland. The date, 1970, cited by Lousley & Kent (1981) is an error. There is a recent record from the Holehird estate, Windermere (40.00, A. Boucher, 1996, LTR), where it was presumably planted.

In habit, this species resembles *Persicaria camp-anulata* but it is usually taller, the leaf pubescence is appreciably less fine and the perianth is pale green.

                            1   C

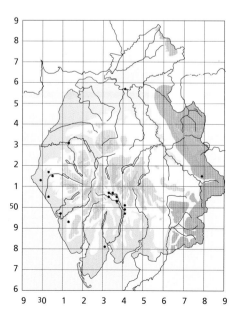

Map 179. *Persicaria wallichii*

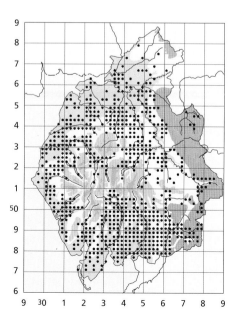

Map 180. *Persicaria bistorta*

P. bistorta (L.) Samp.   Common bistort; Easter ledges
(*Polygonum bistorta* L.)
A common plant of damp roadside verges and hay-
meadows over most of the lowlands but less so in the
north and east and on the Furness peninsula. It extends
far into the dales where it is a conspicuous and
welcome sight along the lanes in early summer.

It was long used, and perhaps still is, as a major
ingredient in the traditional herb pudding Easter
ledges.
Alt. limit: at least 430 m, above Garrigill (74.38)
Map 180   729   WFCY

*P. amplexicaulis (D. Don) Ronse Decraene
(*Polygonum amplexicaule* D. Don)
A handsome species and native of the Himalayas,
found in 1982 naturalised on the banks of the River
Eamont near Dalemain (46.26, W.F. Davidson, LANC,
Watsonia (1983) 14:425).
Alt.: 145 m, Dalemain (46.26)
                                    1   C

P. vivipara (L.) Ronse Decraene   Alpine bistort
(*Polygonum viviparum* L.)
Occasional in damp, upland limestone grassland
around the upper Eden valley and along the Pennines to
Alston, becoming much rarer westwards and known in
the Lake District from moist, ungrazed ledges only
from the Helvellyn area (3.1) and Hanging Knotts
(24.06), although there are *Atlas* records from the
Borrowdale (2.1) and Ambleside (3.0) squares. The
Hanging Knotts site is probably the "north-east face of
Bowfell" cited by Wilson although the recent record is
definitely in Cumberland. Perhaps the most surprising
occurrence was at 825 m on Raise (34.16) where it
grows in very short turf with *Antennaria dioica* and
Salix herbacea. The first definite Cumberland record
was from Comb Crags on the west side of Helvellyn
(3.1, G. Moule, 1950s), but attempts to refind it there
have failed.

Plants in short, grazed grassland often remain
vegetative and are easily overlooked.
Alt. range: 250 m, east of Crosby Ravensworth (64.14)
- 870 m, Helvellyn (34.14)
Map 181   62   WCY

P. amphibia (L.) Gray   Amphibious bistort
(*Polygonum amphibium* L.)
A frequent species in the coastal lowlands and along
the River Eden but elsewhere rather uncommon. It is a
plant of ditches, wet fields, pond margins, sheltered
bays, as on Windermere, and of calm river backwaters.
The erect and distinctly hairy terrestrial form is the
commoner, at least in the north; it flowers infrequently
and can occur in quite dry situations.
Alt. limit: 570 m, Blind Tarn (26.96)
Map 182   319   WFCY

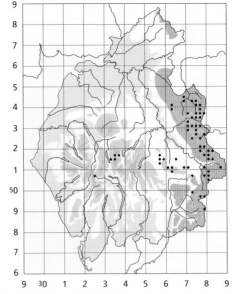

Map 181. *Persicaria vivipara*

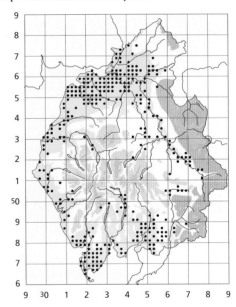

Map 182. *Persicaria amphibia*

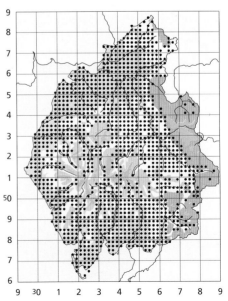

Map 183. *Persicaria maculosa*

P. maculosa Gray   Redshank
(*Polygonum persicaria* L.)
An extremely common weed of rubbish dumps, waste and cultivated ground, often appearing in abundance following disturbance and the ploughing of old pasture. It occurs also by roadsides and pavements and by ponds.
Alt. limit: at least 450 m, east of Garrigill (76.40)
Map 183   1265   LWFCY

P. lapathifolia (L.) Gray   Pale persicaria
(*Polygonum lapathifolium* L., *Polygonum nodosum* Pers.)
A weed of allotments, arable fields, rubbish dumps and disturbed ground, almost invariably associated with the previous species but noticeably less common and probably having a higher requirement for soil nitrogen and moisture. White-flowered plants of *Persicaria maculosa* are sometimes mistaken for this species.
Alt. limit: at least 440 m, Nenthead (78.42)
Map 184   525   WFCY

P. hydropiper (L.) Spach   Water pepper
(*Polygonum hydropiper* L.)
Common in the lowlands in ditches and by streams and rivers, ponds, tarns and lakes, sometimes forming extensive swards in dried-up ponds and by drawn-down reservoirs. It occurs also in damp arable fields and in wet woodland rides.
Alt. limit: 260 m, Watendlath (26.16), although Wilson (1956) gives it as 410 m.
Map 185   884   WFCY

[P. laxiflora (Weihe) Opiz
(*Polygonum mite* Schrank)
Listed somewhat doubtfully by Hodgson from St Bees (9.1). This probably refers to *Persicaria hydropiper* with which it is often confused. There is also an unlocalised record from the Sedbergh area (Handley 1898).

(W/YC)]

P. minor (Hudson) Opiz   Small water-pepper
(*Polygonum minus* Hudson)
A local species of tarn margins and lake shores, mainly in the south, although there are interesting old records from the north and east, for example from by the River Eden at Stainton, Carlisle (3.5, T.S. Johnstone, 1901, CLE) and Great Salkeld (5.3, H. Britten, 1904; G.A.K. Hervey, 1948).

It usually grows quite prostrate in open, stony ground, often associated with *Persicaria hydropiper* and its occurrence often in some quantity by the Wet Sleddale (54.10), Tarnhouse, Lupton (56.82s) and Killington (58.90) reservoirs indicates a surprising tolerance of draw-down.

This species is probably declining nationally. It is therefore encouraging that the Survey has trebled the number of *Atlas* squares, Cumbria now ranking as the species' British headquarters. The first Furness record is that from Finsthwaite High Dam (3.9, W.H. Pearsall, 1913, *BEC Rpt* (1914) 3:494).
Alt. limit: 315 m, Skeggles Water (46.02)
Map 186   54   WFC

Mr Read: All that I had in mind, Dr Newbold, when I put the question to you was to whether it [*Persicaria minor*] is to be found distributed in the north, the east, the midlands and the south. In the presence of those black spots that are to be found in those areas I have just nominated, are they not the north, the east, the midlands, the south?

Dr Newbold: Yes, generally distributed.

Mr Read: Very well. May I come back to the second, at which I have to confess I still flounder. It is a very local distribution. That is to say in where is it very local? Are you saying in Cumbria or in the Lake District you only find it at Wastwater?

Dr Newbold: No, I am not saying that. When I used the phrase "local" I am talking about in a national sense, so in Britain it is very local, or in England and Wales it is very local.

Mr Read: I am taking assistance from my learned junior. Mr Bartlett suggests to me that I ask you whether you mean it is only to be found in a few localities.

The Inspector: All over England.

Mr Read: All over England. Precisely.

Dr Newbold: Yes. All right. As you wish.

Mr Read: In a few localities all over England. Is that what you mean?

Dr Newbold: No. I meant what I said, but I accept your description.

Mr Glidewell: I am under the distinct impression, if I can try to help, that what is intended is that these plants are not unlike county surveyors. Generally speaking, one can find one in each county, although when you go to the county you have a bit of a job to find a county surveyor.

Mr Read: I see. Do you mean that, Dr Newbold?

The Inspector: A rare plant indeed.

Mr Read: I think really what I want to know is to what extent in the Lake District the Small Water Pepper is to be found elsewhere than at Wastwater.

Day 43 of the Wastwater/Ennerdale Public Inquiry, April 16th, 1980

*Fagopyrum esculentum Moench   Buckwheat
The only Survey records are of casual occurrences east
of Carlisle (40.54, ER, 1970s), in a roadside field at
Seathwaite, Borrowdale (22.12, D. Earl, 1984, LANC),
at Bolton (64.22, CW, 1981) and a brickworks at
Askam-in-Furness (20.76, PB, 1988, LANC).

Hodgson recorded it from ballast at Maryport
(0.3) and Silloth (1.5, E.J. Glaister, CLE; J. Leitch,
CLE) and there are later records from the railway
at Port Carlisle (2.6n, T.S. Johnstone, 1899, CLE),
Dalston (3.5, G.A. Field, 1934, CLE), a railway bank at
Carlisle (3.5/4.5, R. Martindale, 1951, CLE) and Melkinthorpe (5.2, G.A.K. Hervey, 1957). In the south, J.A.
Martindale collected it from a cornfield near Staveley
(4.9, undated, KDL). There is also a post-1930 *Atlas*
record from the Arnside square (4.7s, v.c.60/69). Some
of these records probably indicate only brief persistence
following cultivation.
Alt. limit: 125 m, Seathwaite (22.12)
                    4   (W)FC

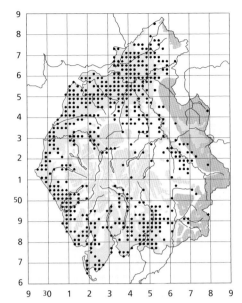

Map 184. *Persicaria lapathifolia*

*F. tataricum (L.) Gaertner
The only record is from Dalton-in-Furness (2.7, D. Lumb,
BEC Rpt (1914) 3:403) where it was seen in 1913.

                                        (F)

*Polygonum maritimum L.   Sea knotweed
Seen at Silloth docks (1.5) by Hodgson and others from 1890
to 1896, and in 1897 on ballast at Maryport (0.3, Hodgson,
(1899). This rare coastal plant has been recorded from
several counties in the south of England but is now known
only from Cornwall and Sussex.
                    R   (C)

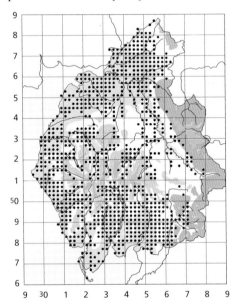

Map 185. *Persicaria hydropiper*

P. oxyspermum Meyer & Bunge   Ray's knotweed
Scattered along the drift-line of the entire coast from
the outer part of Morecambe Bay north to Grune Point
(12.56), rare or absent from the inner parts of the
Solway Firth and Morecambe Bay although seen in the
Longtown square (3.5, DAR) in the 1950s. The only
Westmorland record is from Far Arnside (44.76s, MB,
1982) from where it was washed away in 1988.

This species is as frequent in Cumbria as
anywhere in Britain.
Map 187   50   WFC

P. arenastrum Boreau
A common species of trampled ground, field entrances
and dry stony sites. The first definite record is that from
Silloth in 1923 (1.5, G.C. Druce & R.H. Williamson,
CLE, *BEC Rpt* (1924) 7:209).
Alt. limit: at least 450 m, Birkdale, Teesdale (80.26)
Map 188   346   LWFCY

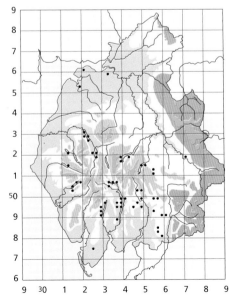

P. aviculare L. sensu stricto   Knotweed
Very common as a weed of cultivated ground, waste
places, roadsides and pavements and bare trampled
ground.

Although the maps suggest that this segregate is

Map 186. *Persicaria minor*

appreciably less common than *Polygonum arenastrum*, this is probably due to the natural bias to record segregates other than the typical one.

Alt. limit: 670 m, Great Dun Fell (70.30)

Maps 189 and 190 (agg.) 427, 1467 (agg.) LWFCY

*Fallopia japonica (Houtt.) Ronse Decraene

(*Reynoutria japonica* Houtt.)   Japanese knotweed
An aggressive garden plant, probably never now intentionally grown but almost impossible to eradicate and commonly dumped surreptitiously by roadsides and waste ground with the result that it is now widespread and all too evident over much of the county. Its sparseness in the east may be related to its sensitivity to late frosts. It is very noticeable from the map how it follows the A591 through the Lake District from Ambleside to Bassenthwaite Lake.

    The remarkable spread of this species has been detailed by Conolly (1977). It was first noted in Cumb-

erland *c.*1940 at Cotehill (4.5) by JA who remembers using the stems for pea-shooters. The first Furness record dates from *c.*1950 (4.7s, Grange-over-Sands, GH) and Conolly's maps indicate that it was first noted in Westmorland during the following decade. There are no Survey records from the Allonby (0.4) and Lorton (1.2) squares shown in her maps.

Alt. limit: 230 m, Reagill (60.16)

Map 191   341   LWFCY

*F. japonica x **F. sachalinensis**

F. x bohemica (Chrtek & Chrtkova) J. Bailey
An extremely rare hybrid found at Beckhead, Witherslack (44.84s, MB & T.C.G. Rich, LANC, *Watsonia* (1995) 20:287) in 1991 and identified by A.P. Conolly and J. Bailey. It was also found, presumably planted, on the Holehird estate, Windermere (40.00, A. Boucher, 1996, LTR). The hairs on the underside of the leaf are appreciably shorter, 3 or 4 cells long, than the long

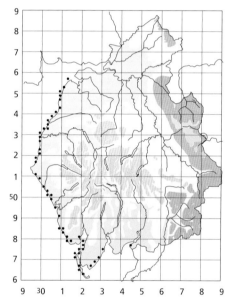

Map 187. *Polygonum oxyspermum*

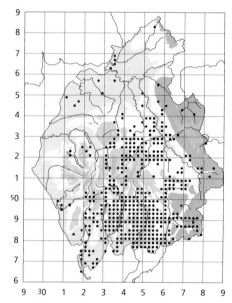

Map 189. *Polygonum aviculare* sensu stricto

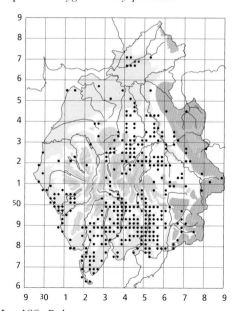

Map 188. *Polygonum arenastrum*

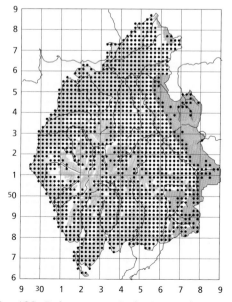

Map 190. *Polygonum aviculare* sensu lato

flexuous ones of *Fallopia sachalinensis* (9-14 cells).

1   W

*F. sachalinensis (F. Schmidt ex Maxim.) Ronse
Decraene   Giant knotweed
(*Reynoutria sachalinensis* (F. Schmidt ex Maxim.)
Nakai)
A rather rare garden escape largely confined to the
Ambleside (3.0) area. Most of the records are from
roadsides or waste ground but at Aldingham (28.70) it
grows on the old shingle beach. It has disappeared
during the Survey from Tendley Hill quarry (08.28) as
a result of infilling.

A.P. Conolly (*in litt.* 1990) gives the first record
as from the then Yorkshire part of the Barbon square
(6.8s) during the 1950s. It was first found in Westmorland in 1961 at Ambleside (36.04, R.K. Brummitt,
LIV), in Cumberland in the Silloth (1.5) and possibly
the Pooley Bridge (4.2) squares in the 1960s and in
Furness in 1970 at Grizedale (32.94, R. Young).
Alt. limit: *c.*250 m, Watermillock (42.22)
Map 192   32   WFC(Y)

*F. baldschuanica (Regel) Holub   Russian vine
There are two records of well-established colonies of
this common garden climber, from near St Bees (96.12,
98.10, CCH, 1978, herb. C.C.H., *Watsonia* (1979)
12:354) and from the roadside between Kendal and
Cunswick (50.92, R. Young, 1970, AMDE).
Alt. limit: 135 m, west of Kendal (50.92)

3   WC

F. convolvulus (L.) Á. Löve   Black bindweed
Common as a weed of cultivated ground, rubbish
dumps and waste tips, chiefly in the Eden and Irthing
valleys and the Solway Plain and elsewhere mainly
coastal, sometimes occurring actually along the drift-
line as on Foulney Island (24.64s).
Alt. limit: at least 440 m, Nenthead (78.42)
Map 193   334   WFCY

[F. dumetorum (L.) Holub
Reported by Hodgson from Woodrow, Wigton (2.4) and later
from Scotby, Carlisle (4.5, "D.S.", 1908). Wilson gives a
single record from Appleby (6.2). In the absence of
specimens these records are best regarded as doubtful since
this southern species is often confused with *Fallopia*
convolvulus, and even during the Survey, by a recorder from
southern England, with *Tamus communis*!

(WC)]

*Rheum x hybridum Murray   Rhubarb
Well-established clumps have been seen during the
Survey at several sites, mostly in the north, the first
being from roadsides at Penrith (52.32, 1990; 52.34,
1991) and between Penrith and Langwathby (56.32,
1991, all RWMC). It occurs on roadsides, on dumps of
garden rubbish, in the gardens of remote, long derelict
cottages and by the railway in Mallerstang (76.02, JCF,

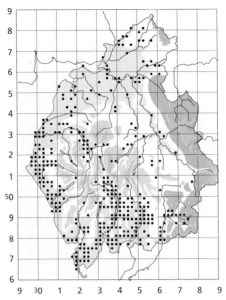

Map 191. *Fallopia japonica*

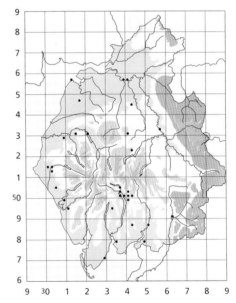

Map 192. *Fallopia sachalinensis*

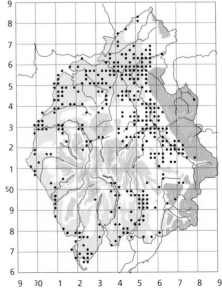

Map 193. *Fallopia convolvulus*

1980s; 76.96), the first Westmorland record.

Alt. limit: 340 m, by the railway, Aisgill (76.96), although it persists in the old garden of Moor House (74.32) at 550 m.

Map 194   16   WFC

Rumex acetosella L.   Sheep's sorrel

(*R. tenuifolius* (Wallr.) Á. Löve)

Common, absent only from the highest fells, some of the limestone areas and upland blanket peat. It is a calcifuge occurring in open sandy ground, on heaths and dry hillsides and on dry field banks.

Plants with very narrow leaves and referable to var. *tenuifolius* Wallr. have been recorded from coastal sands at Drigg (06.94, CCH, 1982, herb. C.C.H., *Watsonia* (1983) 14:425), Askam-in-Furness (20.76, PB, 1988, LANC), Sandscale (18.74, PB, 1990, LANC), Beckfoot (10.48, D.A. Simpson, 1980s, LANC) and Silloth (10.52, EEM, 1992, LANC), also inland from a trackside east of Mosedale (36.32, I. Mortemore, 1982, LANC), above Holmescales quarry, Old Hutton (54.86s, GH, 1982, *Watsonia* (1984) 15:132) and a trackside near Brougham, Penrith (56.28, RWMC, 1985, LANC). The only pre-Survey record is from Grune Point (1.5, C.W. Muirhead, 1938, 1953, both PLYP).

Alt. limit: 845 m, Great Dun Fell (70.32)

Map 195   1450   LWFCY

*****R. scutatus** L.   French sorrel

Found during the Survey at Cockermouth (12.30, MS, 1982, LANC), where it was first noted by G.H. Day in 1949 (CLE), and in the south by a roadside hedge at Haverthwaite (34.84, JA, 1978, LANC), on a fallen wall at Bouth (32.84, JA, 1981), by Ramsden Dock, Barrow-in-Furness (20.68s, B. Fisher, 1978) and on a wall at Allithwaite (38.76s, FLW, 1989).

Hodgson gives records from Harrington (9.2), Keswick (2.2) and Tirril, Penrith (4.2), and it was collected at Brownrigg, Abbeytown in 1877 (1.5, E.J. Glaister, CLE). Baker refers to its cultivation as a pot-herb and mentions a further occurrence at Allithwaite (3.7s), Wilson's only record is from Bowness (4.9). It was reported from Ulverston (2.1, CLE) by G. Wilson in 1948.

This is a montane species of southern and central Europe.

        6   (W)FC

R. acetosa L.   Common sorrel

Extremely common, probably absent only from the highest summits and blanket peat. This is essentially a plant of damp grassland, meadows, marshy fields and roadside verges, and in the fells it is characteristic of *Juncus effusus* flushes and areas of ungrazed grassland alongside becks.

Alt. limit: 945 m, Helvellyn (34.14)

Map 196   1747   LWFCY

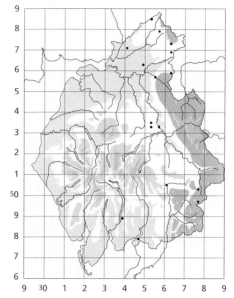

Map 194. *Rheum* x *hybridum*

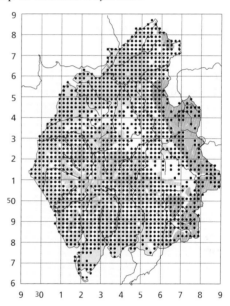

Map 195. *Rumex acetosella*

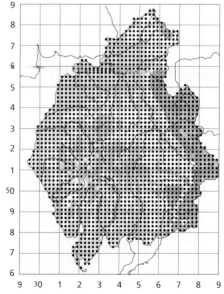

Map 196. *Rumex acetosa*

*R. pseudoalpinus Hoefft   Monk's rhubarb
(*R. alpinus* L.)
Very rare, apparently now restricted to the area around Mungrisdale and Caldbeck, Matterdale, Martindale and Monk Coniston (30.98, FLW, 1982). It is pleasing to be able to report that it is "still at its old localities, Ulcat Row and Mossdike Farm, Mungrisdale" (40.22, 36.30, J.E. Lousley, *BEC Rpt* (1938) 11:501). However, there are no recent confirmations of the rather surprising *Atlas* records from the Broughton-in-Furness (2.8), Cleator Moor (0.1), Distington (0.2), Ennerdale (1.1) and Borrowdale (2.1) squares, nor of an earlier record from Eskdale (1.0, M. Cobbe, *BEC Rpt* (1925) 7:595).

Most of the recent records are from roadsides near farms. This dock has a long history as a herbal remedy and is still sought after by herbalists. The leaves are reputed to have been used for wrapping around butter.
Alt. limit: 375 m, Dowthwaite Head, Dockray (36.20)
Map 197   9   WFC

[R. aquaticus L.
Baker cites a record from Shap (5.1) and it was found by T.S. Johnstone (Thomson 1902) at Rockcliffe (3.6). These records must be errors since the dock is known in Britain only from the shores of Loch Lomond.

          R   (W)]

R. longifolius DC.
Scattered in the east from Mallerstang north to Naworth, Brampton (56.62). There is also an *Atlas* record for the Spadeadam square (6.7n), slightly further north. It grows in rather undistinguished damp grassland, often by becks and roadsides and sometimes in abundance. During the 1980s it was, for a while, an impressive feature of disturbed ground by the A66 on Stainmore (8.1) and the A685 around Kirkby Stephen

(7.1)
The main centres of this species in Britain are the eastern Scottish Highlands, the Yorkshire Dales and east Cumbria.
Alt. limit: 520 m, Hartside (64.42)
Map 198   34   WCY

R. longifolius x **R. obtusifolius**
R. x **hybridus** Kindb.
Extremely rare but it is easily overlooked and should be looked for throughout the range of *Rumex longifolius*. All the sites are in the east and mostly on roadsides. It was first found near Culgaith (60.28, 1982, LANC, *Watsonia* (1983) 14:425) by RWMC, who also recorded it from Inglewood Bank (54.34, 1982, LANC), Carleton, Penrith (52.30, 1988), Flusco Pike, Newbiggin (46.28, 1994), south-east of Garrigill (74.40, 1997, LANC) and north of Alston (72.48, 1994). There is also a record from Garsdale (74.88, GH, 1986, LANC). Most of the specimens have been confirmed by D.H. Kent.
Alt. limit: 430 m, Garrigill (74.40)
          7   CY

R. hydrolapathum Hudson   Water dock
Seen only twice during the Survey, at Mere Tarn (26.70) and Urswick Tarn (26.74), although also recorded from the pond at Holehird, Troutbeck (40.00) where it had no doubt been planted. Hodgson listed it from Moss Pools, near Wreay (4.4) and Black Dyke, Silloth (1.5) and there is a later record from Plumpton (5.3, J. Parkin, 1935). Wilson's only record is from Arnside (4.7s).

The disappearance of this southern dock from most of its early sites is no doubt the result of improved lowland drainage.
          2   (W)F(C)

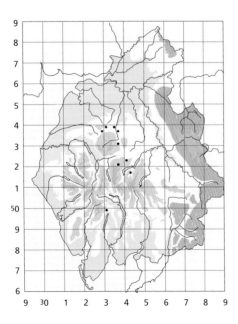

Map 197. *Rumex pseudoalpinus*

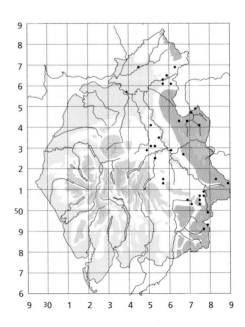

Map 198. *Rumex longifolius*

R. crispus L.   Curled dock
A very common dock of roadsides, rough grassland and waste ground, also of river gravels and the shore where it is particularly characteristic of the drift-line. There are numerous Survey records of coastal plants with three equally prominent perianth tubercles and referable to subsp. *littoralis* (J. Hardy) Akeroyd.
Alt. limit: 845 m, Great Dun Fell (70.32)
Map 199   1285   LWFCY

R. crispus x **R. obtusifolius**
R. x **pratensis** Mert.
Probably not infrequent but only recorded during the Survey from a number of sites around Penrith, chiefly by RWMC, but with outliers on the west coast at Whitehaven (98.16) and in the north at Penton Linns (42.76n) and Castle Carrock (54.54). The first records for Cumberland, all confirmed by D.H. Kent, are from Whitehaven (98.16, CCH, 1982, herb. C.C.H., *Watsonia* (1984) 15:133), Bennet Head, Watermillock (44.22, MC, 1982, herb. M.C.) and the old railway south-west of Penrith (50.28, RWMC, 1982, LANC). The only previous records are those from Roose at the end of last century (2.6s, Henry (1897)) and that given by Baker from Bowness (3.9).
Alt. limit: 245 m, Bennet Head, Watermillock (44.22)
Map 200   11   (WF)C

R. conglomeratus Murray
Generally if rather sparsely distributed but probably over-recorded in error for *Rumex sanguineus*. Although it occurs in similar habitats to that species, it seems to prefer damper, unshaded situations such as pond margins and heavy, wet, riverside soils.
Alt. limit: *c*.420 m, Swindale, Brough (80.18)
Map 201   233   WFCY

R. sanguineus L.
Common to very common in the lowlands in the east of the county but less so in the west and south-west, although this may well be a consequence of the over-recording there of *Rumex conglomeratus*. This dock prefers light shade, occurring by woodland margins and paths, in scrub and in ditches.
        There are apparently no records of the species in Furness prior to those in the *Atlas*. Most apparently refer to var. *viridis* (Sibth.) Koch; the only definite record of the strongly red-veined var. *sanguineus* are from around Burgh-by-Sands (3.5, REG).
Alt. limit: at least 300 m, south of Alston (70.42)
Map 202   793   LWFCY

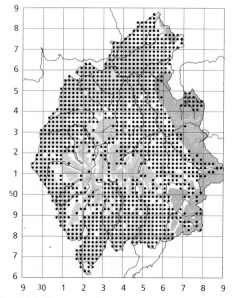

Map 199. *Rumex crispus*

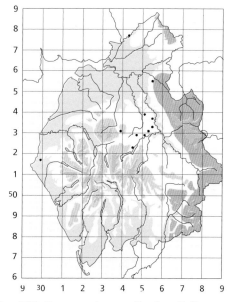

Map 200. *Rumex crispus* x *R. obtusifolius*

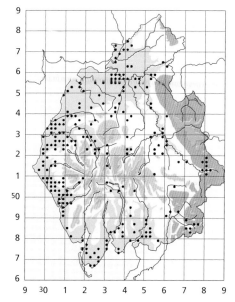

Map 201. *Rumex conglomeratus*

[R. pulcher L.   Fiddle dock
According to Baker a specimen collected at "Sir John Lowther's Newtown" (5.2) was identified as this species by C.C. Babington. As he remarks, the identification must have been at fault. The northern limit in Britain for this southern species is on the north coast of Wales.

(W)]

R. obtusifolius L.   Broad-leaved dock
Very common, absent only from the upland peats and the higher fells. It generally prefers damper soils and a higher nitrogen level than *Rumex crispus* and it thrives in damp meadows and pastures, roadside verges and ditches and cultivated ground. It has increased markedly in haymeadows and silage fields with the recent intensive use of slurry.
Alt. limit: 845 m, Great Dun Fell (70.32)
Map 203   1535   LWFCY

R. maritimus L.   Golden dock
Listed by Hodgson from ballast tips at Maryport (0.3) in 1896 and Wilson refers to it having been seen by W.H. Pearsall at North Scale, Walney Island (1.6s). The only other records are from the 1950s when it was seen by G. Wilson and the Barrow-in-Furness N.H.S. near Newbiggin (2.6s) and Bardsea (3.7s). It has been recently seen just outside the county at Leighton Moss (4.7s, Livermore & Livermore (1987)).

This species reaches its northern limit on the west coast of Kintyre.

(FC)

Oxyria digyna (L.) Hill   Mountain sorrel
Frequent on ungrazed rock ledges and particularly on the sides of gills and wet shady gullies in the central, western and eastern Lake District. There are records dating from the 1950s and made by DAR, but unconfirmed during the Survey, from Lingmell (18.08), Harter Fell (20.00) and Ashness Gill (26.18); it was also seen by R.J. Birkett on Great Gable (20.10, 1967).

South of the Scottish Highlands, this arctic-montane species is known in Britain only from the Moffat hills, the Lake District and North Wales.
Alt. range: 105 m, Wast Water Screes (14.04) - 885 m, Helvellyn (34.14)
Map 204   69   WFC

PLUMBAGINACEAE

Limonium vulgare Miller   Sea lavender
Fairly frequent on the upper levels of salt-marshes; sometimes very abundant and colourful in late summer, as on the landward side of North Walney (16.72), but generally becoming sparser and even disappearing as sheep-grazing extends and becomes more intensive. It has now, for example, virtually disappeared from the Leven estuary where it occurred, with *Limonium humile,* in the 1950s. There are no reliable records north of the Ravenglass estuary (0.9).
Map 205   35   WFC

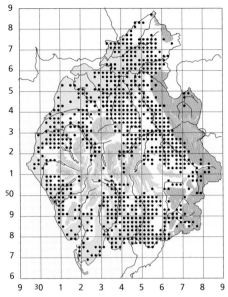
Map 202. *Rumex sanguineus*

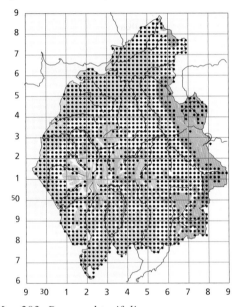
Map 203. *Rumex obtusifolius*

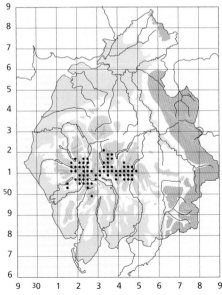
Map 204. *Oxyria digyna*

L. humile Miller

Occurring in very similar habitats to *Limonium vulgare* but rather less frequent and generally found a little lower on the marsh. It appears to be the only species on the Solway marshes.

Plants may sometimes be difficult to distinguish from *L. vulgare*. Dawson & Ingrouille (1995) examined mixed populations on South Walney and the Kent estuary and found intermediate individuals which they interpreted as hybrids.

Map 206, Plate 36   S 27   FC

L. britannicum Ingrouille
subsp. **celticum** Ingrouille

A few plants of this member of the *Limonium binervosum* complex were discovered in 1988 on the railway embankment stonework at Arnside (44.78s, C. Webb; LANC, *Watsonia* (1993) 19:281) and in 1995 in a similar habitat at Plumpton (30.78s, PB, LANC) on the Leven estuary. The only previous record is from the east side of the estuary near Frith Hall where a population of very small plants, less than 6 cm high, was found by W.H. Pearsall in 1916 growing in tidal mud overlying limestone (3.7s, BM, YRK, *BEC Rpt*

(1917) 4:577,). All three collections were identified by M.J. Ingrouille. Baker's record from "a saltmarsh between Tridley and Greenodd", on the west side of the estuary, probably refers to the same taxon. There are *Atlas* records of *L. binervosum* from this square and also from South Walney (2.6s) contributed by the Barrow-in-Furness N.H.S. during the 1950s.

This subspecies is known only from the coasts of Anglesey, Cheshire and Lancashire.

            S 2   WF

L. recurvum Salmon
subsp. **humile** (Girard) Ingrouille

This segregate of the *Limonium binervosum* complex is known in Cumbria only from St Bees where it grows on the sandstone cliffs on the south side of the headland (94.12) and also on the coastal railway embankment south of the village (96.10, M.J. Ingrouille, 1980 (Ingrouille & Stace 1986)). This segregate is characterised by a distinctly tuberculate scape and dense spikes with 7 or 8 flowers/cm. Ingrouille has assigned the Cumbrian and Galloway plants to var. *humile* and those from Co. Donegal to var. *recurvum*.

            R 4   C

...no one plucks the rose,
Whose proffered beauty in safe shelter blows
'Mid a trim garden's summer luxuries,
With joy like his who climbs, on hands and knees,
For some rare plant, yon Headland of St. Bees.
From *Suggested in a steamboat off Saint Bees' Heads, on the coast of Cumberland*, William Wordsworth

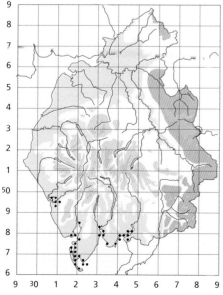

Map 205. *Limonium vulgare*

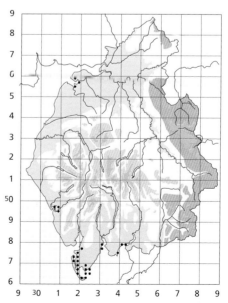

Map 206. *Limonium humile*

Armeria maritima (Miller) Willd.   Thrift; Sea pink
Plentiful around the entire coast, on cliffs, rocks and
sea-walls, in cliff-top grassland and on the upper parts
of salt-marshes, where it is more resistant than most
salt-marsh species to trampling and grazing. By
contrast, it occurs locally in the north-east on Alston
Moor, where it occurs exclusively around old lead-
workings. There are no recent records for the Teesdale
square (7.3) shown in the *Atlas*. Just as remarkable is
the species' ability to thrive on rock ledges high up on
Pillar (16.10, 16.12), Steeple (14.10), Scafell Pike
(20.06) and Helvellyn (32.14, 34.14). There are no
Survey records from Dove Crags, Grasmoor (1.2) or the
fells between the Vale of Newlands and Borrowdale
(2.1), both cited by Hodgson, from Great Gable (2.1,
C.W. Muirhead, 1940) or from "a moor near Appleby "
referred to by Wilson. During the Survey it was
discovered well-established on Cartmel racecourse
(36.78s).
Alt. limit: 875 m, Scafell and Scafell Pike (20.06)
Map 207   149   WFC

PAEONIACEAE

*Paeonia officinalis L.   Peony
A garden escape found in 1990 by CEW by the fell road
above Casterton (62.78s, LANC) and in wood-land by
the River Kent near Sedgwick (50.86s, LANC,
Watsonia (1991) 18:420), and subsequently on a roadside verge west of Dalton-in-Furness (20.74, PB,
LANC, 1993) and near Arnside (46.78s, MB, 1995).
                    4   W

ELATINACEAE

Elatine hexandra (Lapierre) DC.
Rather rare but seen during the Survey in eight of the
lakes and 12 tarns, and occurring in both nutrient-rich
and nutrient-poor water. Most of the records were
contributed by RS who found it by the shores of
Ennerdale Water in 1976 (08.14, LANC, *Watsonia*
(1977) 11:392) and at five further sites during the
following year, mostly in the Lake District. Also in
1977, J.D. Allenby collected it in Grasmere (34.06,
AMDE), the first definite Westmorland record).

By contrast, there are only two literature records
for this nationally scarce and declining aquatic plant.
Wilson lists it, with understandable reservation, from
near Arnside (4.7s) and it was reported by R.W.
Butcher from Thurstonfield Lough (3.5, 1937, *BEC Rpt*
(1938) 11:469).

There is no reason to suppose that this species
has suddenly and dramatically expanded its range
within the county. It is very easily overlooked and
impossible to spot should the water be disturbed. It
grows usually in shallow water, rooted in sand or
gravel and during dry summers it may be exposed but
this is not necessary for flowering as this occurs freely
underwater. There is no evidence that it behaves in any
of its Cumbrian sites as an annual, contrary to the
statements made in many Floras.

The Lake District can fairly claim to be the
British headquarters of this species.
Alt. limit: 200 m, Tewet Tarn (30.22)
Map 208   S 35   WC

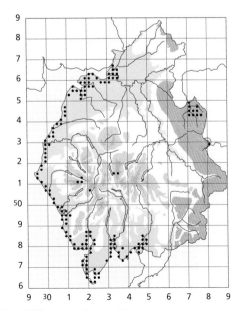

Map 207. *Armeria maritima*

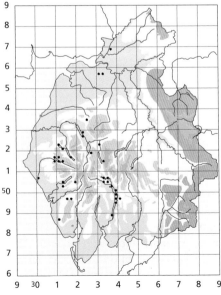

Map 208. *Elatine hexandra*

CLUSIACEAE
(GUTTIFERAE)

*Hypericum calycinum L.   Rose of Sharon
Known only at a few sites, all but four in west Cumberland. Most of the records are from roadside banks and hedgerows and usually close to houses. Most represent garden-escapes, but it may occasionally have been planted for ground cover, as perhaps on the riverbank near Gosforth (08.02).
Alt. limit: 140 m, Castle Carrock (54.54)
Map 209   12   WC

H. androsaemum L.   Tutsan
Occasional in the south, scattered and rare in the north and east. It occurs in damp deciduous woodland, particularly on the southern limestone but also in several of the wooded Lake District gills. Many of the outlying records are probably of plants bird-sown from gardens.
Alt. limit: 310 m, Ill Gill, Kirkfell (20.08)
Map 210   172   WFC

*H. androsaemum x **H. hircinum**
H. x **inodorum** Miller
A garden escape found in Carlisle by the River Caldew (38.56, REG, 1995, LANC). It is otherwise known only from the 19th century record cited by Wilson from between Bowness and the ferry (3.9/4.9) and records from Dalton-in-Furness (2.7, D. Lumb, 1913, *BEC Rpt* (1914) 3:403) and near Sebergham (3.4, E. Lomax, 1886, BM, CLE).
                              1   (WF)C

*H. hircinum L.
A Mediterranean species reported as naturalised at Grange-over-Sands (4.7s, A.W. Westrup,1953) and in the Rampside square (2.6s, Barrow-in-Furness N.H.S., 1950s).
      This species and the preceding hybrid are often confused and in the absence of material records should be regarded with some doubt.
                                                    (F)

H. perforatum L.   Common St John's wort
Common in scrub, rough grassland, quarries, roadside verges and railwaysides but avoiding the more acidic soils. It is a lowland species which in the Lake District is more or less restricted to the corridor of the A591 trunk road.
Alt. limit: *c.* 340 m, above Hilton (76.22)
Map 211   809   LWFCY

H. perforatum x **H. maculatum**
subsp. **obtusiusculum**
H. x **desetangsii** Lamotte
An uncommon, variable and fertile hybrid. 35 specimens (LANC) collected, uncritically, as *Hypericum maculatum* have been examined by N.K.B. Robson and

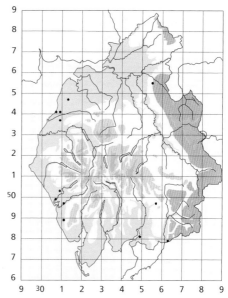

Map 209. *Hypericum calycinum*

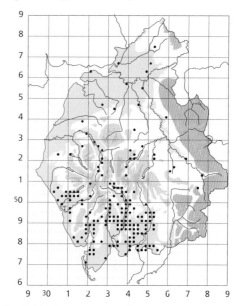

Map 210. *Hypericum androsaemum*

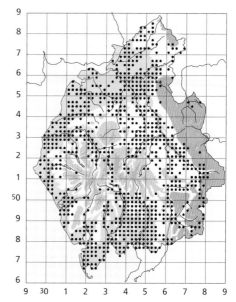

Map 211. *Hypericum perforatum*

half have been identified as the hybrid. The collection has recently been examined by A. Hartley using the hybrid index devised by Crackles (1990). This showed a normal distribution centred on plants backcrossed to *H. maculatum* and only one plant with the expected *H. maculatum* score. On the other hand, a few specimens showed considerable superficial similarity to *H. perforatum*, for example those from the new Greenodd by-pass (30.80, 30.82, PB, 1987, LANC), differing only in their denticulate sepals. Interestingly, there are no records of *H. maculatum* from this area.

Most of the records are from roadsides; all the specimens have been checked by Robson and most are in LANC. The first record for Cumberland is from east of Dalston (38.50, REG, 1979, herb. R.E.G.); for Westmorland Heltondale (50.20, GH, 1979, LANC, *Watsonia* (1981) 13:329) and for Furness the records from Greenodd.
Alt. limit: 230 m, Heltondale (50.20)
See Map 212   WFC

H. maculatum Crantz   Imperforate St John's wort
Not infrequent in the upper Lune valley and above Tebay on the north Westmorland limestones but generally considerably less frequent than *Hypericum perforatum*. As mentioned above, there is a complete transition to the hybrid, *H.* x *desetangsii*, and pure specimens of *H. maculatum* may well be rather rare. It usually occurs in rather damper habitats and in taller vegetation as, for example, by the Great Langdale Beck (30.04), by the River Rothay near Ambleside (36.04) and the shore of Ullswater at Pooley Bridge (46.24).

All but one of the specimens submitted to N.K.B. Robson are of the much commoner subsp. *obtusiusculum* (Tourlet) Hayek, distinguished from the northern subsp. *maculatum* by its wider branching and denticulate sepals. The single pre-Survey Furness record of the latter in the *Check-list* cannot now be traced and is probably an error. However, a specimen collected in 1990 by the A685 in the Lune gorge (60.00, CEW, LANC, *Watsonia* (1991) 18:421) has been identified by Robson as this subspecies. This and a record from south Lancashire are the only ones from northern England.
Alt. limit: *c.*350 m, east of Alston (76.46)
Map 212   94   WFCY

H. tetrapterum Fries   Square-stemmed St John's wort
Frequent in the lowlands by roadside ditches and riversides and in wet meadows and fens. It is absent from most of the Lake District.
Alt. limit: *c.*300 m, above Alston (72.46) and Brough (78.16, 80.16)
Map 213   578   LWFCY

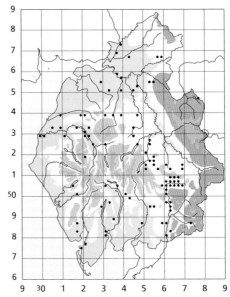

Map 212. *Hypericum maculatum* and *H.* x *desetangsii*

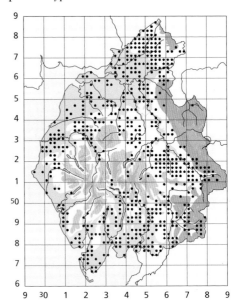

Map 213. *Hypericum tetrapterum*

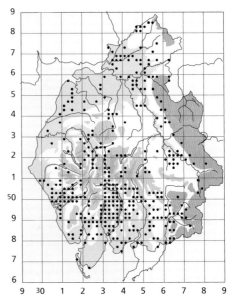

Map 214. *Hypericum humifusum*

H. humifusum L.   Trailing St John's wort
Occasional to fairly frequent, absent from the high
ground and scarce in the more intensively farmed areas
of the Solway Plain and the Petteril valley. It is a plant
of open ground, both on and off the limestone, despite
Wilson's statement to the contrary. It is characteristic of
damp, open or cleared woodland and woodland rides as
well as of dry, heathy ground and sandy hedgebanks. It
can be a persistent garden weed.
Alt.limit: 250 m, Cautley Spout (68.96)
Map 214   366   WFCY

H. pulchrum L.   Slender St John's wort
A common plant of coarse grassland, heaths, wood-
land margins, hedgerows, upland gills and rock ledges,
particularly on rather dry, siliceous soils. It is, for
example, characteristic of sunny hedgebanks in the
Carlisle area together with *Hieracium umbellatum* and
Jasione montana.
Alt. limit: 530 m, Force Burn, upper Teesdale (76.30)
Map 215   1060   WFCY

H. hirsutum L.   Hairy St John's wort
Frequent in south Westmorland and from the limestone uplands of north Westmorland northwestwards
towards the Solway Firth, elsewhere rather rare.

        This is a plant of woodland margins and road-
sides, chiefly on base-rich soils where it is occasionally
misidentified as *Hypericum montanum*.
Alt. limit: 450 m, Garrigill (76.36)
Map 216   274   WFCY

H. montanum L.   Mountain St John's wort
The only Survey records are from the limestones
around Morecambe Bay and the Asby - Orton area
where it occurs in open woodland, limestone pavement
and on riverside rocks.

        The few records from outside these areas have
not been substantiated and probably refer to the
previous species. Hodgson, for example, cites Workington (9.2), Rose Bridge, Raughton Head (3.4) and
Egremont (0.1). Baker certainly had his doubts about
the last of these. The *Atlas* records for the Scafell (2.0)
and Ambleside (3.0) squares and those in *Scarce Plants*
for the Tebay (6.0) and Alston (7.4) squares all date
from the 1950s and are anonymous field records; they
are best treated as suspect. Handley (1898) describes it
as scarce in the Sedbergh area but there are no
subsequent records from there.
Alt. limit: 330 m, above Great Asby (64.10)
Map 217   17   WF[(CW/Y)]

H. elodes L.   Marsh St John's wort
Rather rare and with a particularly pronounced south-
western distribution in Cumbria, as in Britain
generally. It extends inland as far as Crook (42.92,
42.94) in the south and the head of Derwent Water
(24.18) in the north. Sadly, it seems to have disappeared from the *Atlas* sites on South Walney (1.6s,

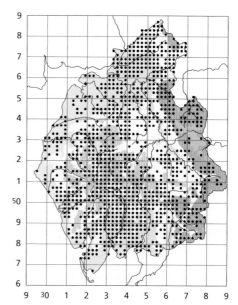
Map 215. *Hypericum pulchrum*

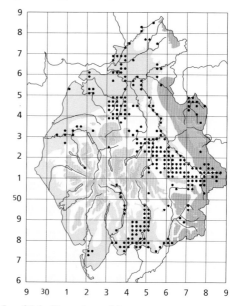
Map 216. *Hypericum hirsutum*

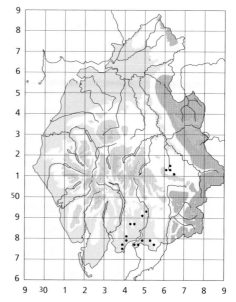
Map 217. *Hypericum montanum*

2.6s) and in the Cartmel (3.7s) and Newby Bridge (3.8) squares. It is a plant of peaty, base-poor shallow water by tarns, lakes and rivulets. Its distribution and habitats in Cumbria are mirrored by the south-western dragonfly *Orthetrum coerulescens*.

Alt. limit: 175 m, west of Crook (42.94)

Map 218   42   WFC

*H. forrestii (Chitt.) N. Robson

A few bushes of this widely cultivated south-east Asian species occur on the disturbed limestone scree below Ravens Scar, Whitbarrow (44.84s, GH, 1988, LANC, det. N.K.B. Robson, *Watsonia* (1989) 17:466).

1   W

TILIACEAE

*Tilia platyphyllos Scop.   Large-leaved lime

Apparently very rare and almost always planted. The records (all LANC, det. C.D. Pigott) are mostly of single trees: Chapel Island in the Leven estuary (30.74, T.C.G. Rich & E. Nic Lughadha, 1990), on waste railway land at Cark (36.76, JA, 1996), by a roadside plantation at High Ireby (22.36, REG & MB, 1980), by Naworth Castle car park (56.62, MG, 1990), by the River Eamont at Pooley Bridge (46.24, RWMC, 1995), by the River Eden at Eden Lacy (56.36, 56.38, RWMC & C.D. Pigott, 1996), south of Great Salkeld (54.34, RWMC, 1996), and at one site with *Tilia cordata* on the roadside between Langwathby and Skirwith (58.32, RWMC, 1995). There is natural regeneration at Eden Lacy and also in Barrowfield Wood (48.92, C.D. Pigott, 1968), where it is known to have been planted *c.*1910 (C.D. Pigott *verb.comm.*). Hodgson refers to an avenue of *T. platyphyllos* at Naworth. This has been felled and replaced by common lime. He mentions also records from the west side of Derwent Water (2.2) and near Gaitsgill, Dalston (3.4). The only record from Furness is that given by Baker from below Gummers How,

Windermere (3.8) and Wilson's only localised record is from Levens Park (4.8s/5.8s).

Alt. limit: 130 m, south of Great Salkeld (54.32)

S 9   WFC

*T. platyphyllos x **T. cordata**   Common lime

T. x **europaea** L.

Widely distributed in lowland hedges and in estate grounds, around towns and villages, and occasionally in broad-leaved woodland on the better soils.

Probably all the individuals have been planted or are descended from planted trees. Virtually all trees appear to belong to 'clonal group A' ('*pallida*'), one of the two clonal groups recognised by Pigott (1992a). This is characterised by dense epicormal branching.

Alt. limit: 415 m, north-west of Nenthead (76.46)

Map 219   465   LWFCY

T. cordata Miller   Small-leaved lime

Frequent on the limestone around Morecambe Bay and also in the southern Lake District where it reaches its northern limit in Britain as a native tree. However, it occurs further north at Wetheral (46.52, EHR, 1970s) on a vegetated sandstone cliff and steep slope, a very native-looking site. There has been no confirmation of an undated record from Walla Crag, Derwent Water (2.2, R.C. Steele & G.H. Peterken) and the record is best treated as doubtful. Roadside trees on Langwathby Moor (58.32, RWMC & AAD, 1991, LANC) and by the River Eden at Eden Lacy (56.36, RWMC, 1996), and in a hedge below Great Salkeld (54.36, GH, 1983, LANC) have no doubt been planted.

This is typically a tree of wooded limestone scars and of steep Lake District gills, both on the slates and the Borrowdale volcanics, and where the soils are less impoverished, the grazing is reduced and man's forestry activities are least. It is usually associated with other more nutrient-demanding trees such as ash and elm. The individual trees can fairly be described as potent-

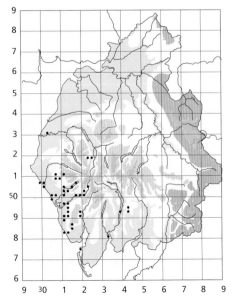

Map 218. *Hypericum elodes*

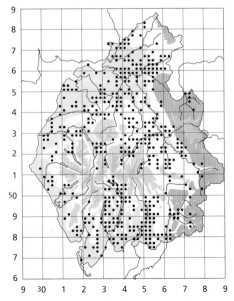

Map 219. *Tilia platyphyllos* x *T. cordata*

tially immortal. The massive bases have been dated by C.D. Pigott as up to 2,300 years old, a period long enough for soil erosion in the gills to expose 2.5-3 m of the original root. Pigott has, in fact, demonstrated that most of these old 'bases' are of root tissue rather than shoot. When trees fall, they sprout freely from the base and, if the original trunk remains partly rooted, shoots grow up from the crown and may eventually root and form individual trees. In this way an extraordinary zig-zag of trunks may develop down the gillside.

Pigott & Huntley (1978, 1980) have shown that Cumbrian populations of *Tilia cordata* only produce viable seed in exceptionally warm years and that they must be considered relicts from earlier, warmer periods, possibly even from the middle post-glacial warm period which lasted until about 3,000 BC.

Alt. limit: noted by R.J. Birkett at *c*.600 m at sites in the southern Lake District.

Map 220, Plate 63   112   WFC

MALVACEAE

Malva moschata L.   Musk mallow
Scattered but locally frequent south of Penrith and in south Westmorland. It is almost exclusively a plant of dry pastures and roadside verges, both on and off the limestone. It seems likely that in many areas the populations are of garden origin.
Alt. limit: 260 m, Spadeadam (62.72n)
Map 221   136   LWFCY

M. sylvestris L.   Common mallow
A southern species and, outside the Eden valley, with a strong coastal tendency. It grows usually on waste ground in towns and villages and in dry situations near the shore. Allen (1984) comments on its close assoc-iation with farmyards and cottages in the Isle of Man.
Alt. limit: at least 200 m, Kirkby Stephen (78.06)
Map 222   135   WFCY

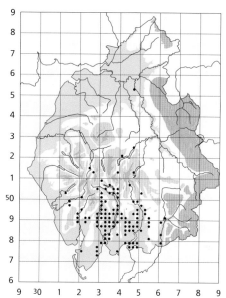

Map 220. *Tilia cordata*, excluding plantings

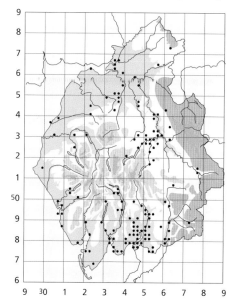

Map 221. *Malva moschata*

Plate 63. Base of *Tilia cordata*, *c*.1000 years old, exposed by erosion of the gillside.

***M. nicaeensis** All.

A southern species like the last and known only from a grassy bank at Silloth (10.52, NB & MM, LANC, *Watsonia* (1986) 16:185) where it was found in 1985 and where it still persists.

1   C

***M. parviflora** L.

A Mediterranean species listed by Hodgson from Silloth (10.52, J. Leitch, 1889, BM, CLE), where it was later collected by C.W. Muirhead (*c.* 1940, CLE).

(C)

***M. pusilla** Smith

An uncommon Continental species listed by Hodgson from Silloth (10.52, J. Leitch, 1889, CLE) where it was growing with *Malva parviflora* and *M. neglecta*.

(C)

M. neglecta Wallr.   Dwarf mallow

Very rare and probably decreasing. It was seen during the Survey only in a garden at South Walney (20.62s, GH, 1979, 1986, LANC), at Silloth (10.52, MM, 1986, LANC), by the roadside at Underbarrow (46.92, CEW, 1990, LANC) and on waste ground at Great Salkeld (54.36, CEW, 1992, LANC), where it was later seen as a garden weed.

By contrast, there are numerous early records. Hodgson lists it from Workington (9.2), Flimby (0.3), Allonby (0.4), Cockermouth (1.3), Silloth (1.5, E.J. Glaister, 1878, CLE), Dalston and Stanwix, Carlisle (3.5), Great Salkeld and Lazonby (5.3), Armathwaite (5.4) and Wetheral (4.5). There are further Cumberland records from Dundraw (2.4, R. Wood, *c.*1878), Skirwith, Ousby and Melmerby (6.3, 1913, 1914, 1919) and Langwathby (5.3, 1916), all recorded by W.W.

Mason, and Abbeytown (1.5, C.W. Muirhead, 1949). Baker gives three localities in Furness: Cark, Kents Bank and Allithwaite (3.7s) and Wilson collected it near Grange-over-Sands (4.7s, YRK) in 1892. His only Westmorland record is from Brigsteer (4.8s).

In the 1950s it was recorded at North Walney (1.7, Grange-over-Sands N.H.S., 1959) and Sandscale (2.7, K. Jones, 1957) and there were further sightings at Great Salkeld (G.A.K. Hervey, 1950) and Wetheral (M.I. Tetley, 1951, CLE). There is also an *Atlas* record for the Arnside square (4.7s, v.c.60/69).

7   WFC

***Lavatera arborea** L.   Tree mallow

A few plants were discovered by A.B. Warburton on shingle banks at the Eskmeals gunnery range (06.92, *Watsonia* (1977) 11:393) in 1976 and they persisted until about 1982.

This south-western species is known from several sites on the Isle of Man. It reaches its northern limit on the west coast on Ailsa Craig.

1   C

***Althaea officinalis** L.   Marsh mallow

A southern species found in 1989 in a ditch at Rampside (22.66s, PB, LANC, *Watsonia* (1991) 18:422). The only previous records are Hodgson's from Rockcliffe (3.6) and E.J. Glaister's from Skinburness (1.5, 1876, CLE).

S 1   F(C)

***A. hirsuta** L.   Rough marsh mallow

Recorded by Hodgson from Silloth (1.5, E.J. Glaister, 1887, CLE), probably the northernmost record in Britain.

(C)

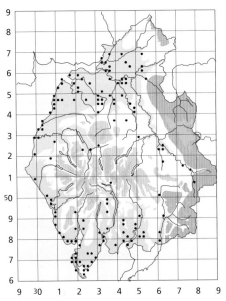

Map 222. *Malva sylvestris*

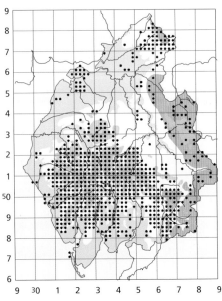

Map 223. *Drosera rotundifolia*

*Alcea rosea L.   Hollyhock
A single plant of this popular garden species was found in 1991 on the banks of the Hodbarrow lagoon, Millom (16.78, B. Fisher).

                    1   C

*Sidalcea malviflora (DC.) A. Gray ex Benth.
A garden escape recorded in 1992 from near Distington (02.24, AAD, LANC), by the River Kent in Kendal (50.92, CEW, LANC), on the roadside between Helton and Brampton (50.20, RWMC, LANC), and at Workington (00.26, D. Wildridge, 1996). All three collections are noticeably different, but E.J. Clement was unable to assign them to any particular cultivar or hybrid.

                    4   WC

*Abutilon theophrasti Medikus
A plant of south-east Europe found in 1975 by the roadside during the construction of the Keswick by-pass (26.24, G. Wilson, LANC).

                    1   C

SARRACENIACEAE

*Sarracenia purpurea L.   Pitcher-plant
There appear to be no published records of this striking North American plant prior to the Survey. It is now known from four sites, at all of which it appears to be well established. These are on Wedholme Flow (20.50, 20.52, EHR, 1978), by the Solway Firth, and in the southern Lake District by the lake at Grasmere (32.06, T. Blackstock, 1978), on Nor Moss, Claife (36.98, M. Hutcheson, 1979) and in a wet field at Wray (36.00, R. Young, 1969, AMDE), where it is obviously spreading. The species has no doubt been introduced at all these sites. The earliest record is of plantings by

M. McGill on Claife Heights in 1963.
Alt. limit: 150 m, Nor Moss, Claife (36.98)

                    5   WFC

DROSERACEAE

Drosera rotundifolia L.   Sundew
Common in suitable habitats particularly in the uplands where it grows in wet, rather open, peaty soils, often subject to some flushing. In the lowlands it is chiefly a plant of the relict raised bogs, growing on and around *Sphagnum* tussocks although it grows by the beach at Askam-in-Furness (20.76).
Alt. limit: 600 m, Yad Moss (76.36)
Map 223   701   WFCY

D. rotundifolia x **D. anglica**
D. x **obovata** Mert.& Koch
This hybrid is very difficult to identify with certainty other than by its sterility. There are convincing herbarium specimens from the Dunnerdale Fells (2.9, J.C. Cooper, 1961, CLE), near Low Wray (36.00, T. Blackstock, 1977, AMDE), and from Solway Moss (3.6, [N.A.?] Dalzell, 1837, E). It is also thought to occur in the north-east on Butterburn Flow (6.7n).

                    1   F(C)

D. anglica Hudson   Great sundew
(*D. longifolia* L.)
Rather rare. It persists in several places on the peat-mosses of the Solway Plain and on Butterburn Flow (6.7n) in the north-east, but it is otherwise restricted to a few *Sphagnum*-rich valley bogs and hillside mires in the Caldbeck area (3.3) and in the Furness fells. It is no longer to be found in the post-1930 *Atlas* Longtown (3.6), Brampton (5.6) and Langwathby (5.3) squares although it was seen at Newton Reigny Moss (5.3) by

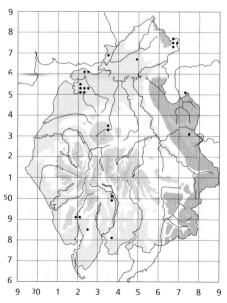

Map 224. *Drosera anglica*

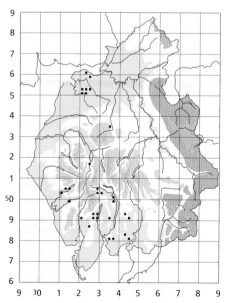

Map 225. *Drosera intermedia*

DAR in 1966. Hodgson gives additional 19th century sites by Crummock Water (1.1) and in Borrowdale (2.1) and Wilson old records from Foulshaw and Meathop Mosses (4.8s) and, presumably his own, from near Blea Tarn (2.0).
Alt. limit: 280 m, Butterburn Flow (66.74n)
Map 224   25   (W)FC

D. intermedia Hayne   Oblong-leaved sundew
A local plant of raised bogs on the Solway Plain and of valley bogs in the Lake District, particularly in the Furness Fells. It grows amongst *Sphagnum* around the margins of bog pools and not infrequently as floating rosettes. Wilson collected it from Terrybank Tarn, west of Barbon (5.8s, 1935, YRK) but it has since all but disappeared from the south-cast.
Alt. limit: 250 m, Carrock (34.34) and above Eskdale (14.98)
Map 225   31   WFC

CISTACEAE

Helianthemum nummularium (L.) Miller
                                        Common rockrose
Common on the limestone in short, species-rich turf and in the crevices and ledges of the scars. The few sites off the limestone include the base-rich ledges of Wanthwaite Crags (32.22) and Glenridding Common (34.16).

There are no recent records of the plant with red-tipped petals said by Baker to occur on Whitbarrow (4.8s).

The rockrose is the food plant for the larvae of the Northern Brown Argus butterfly (*Aricia arta-xerxes*), which is restricted in north-west England to the limestone hills around the head of Morecambe Bay

and the head of the Eden valley.
Alt. limit: at least 560 m, Knock Ore Gill (70.30, Rawes 1981)
Map 226   174   WFCY

H. canum (L.) Hornem.   Hoary rockrose
Very rare but locally abundant on very thin soil at the upper edge of scars and in rock crevices on the lime-stone hills of Cunswick and Scout Scars (4.9), Whitbarrow Scar (4.8s) and the coast at Humphrey Head (3.7s). With its small, densely hairy grey leaves and smaller, paler flowers, it is readily distinguished from the last species. Seen from below, the dark, leafless, cup-shaped bases of the tufts, seemingly cemented against the vertical cliffs, are particularly striking.

Elsewhere in Britain this rockrose is known only from the coasts of north and south Wales and Cronkley Fell in upper Teesdale.
Alt. limit: 200 m, Cunswick and Scout Scars (48.90, 48.92)
Map 227   S 8   WF

VIOLACEAE

Viola odorata L.   Sweet violet
An infrequent plant of damp woodland edges, hedge-rows and verges, usually on the better soils and seldom far from houses. Although generally considered to be native, at least in the south of the county, it is likely that most records are of garden escapes.

White-flowered plants appear to be the commoner.
Alt. limit: *c.*190 m, Kershope Forest (48.78n) and Unthank, Hartside (60.40)
Map 228   202   WFCY

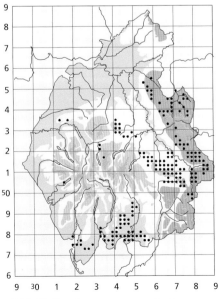

Map 226. *Helianthemum nummularium*

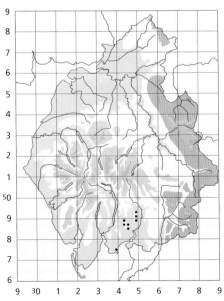

Map 227. *Helianthemum canum*

V. odorata x **V. hirta**

V. x scabra F. Braun

Known only from below a limestone scar near Kings Meaburn (60.20, RWMC, 1989; LANC, conf. D.M. Moore), where it was growing in open woodland with both parents.

Alt.: 150 m, Kings Meaburn (60.20)

1	W

V. hirta L.   Hairy violet

Fairly frequent on the limestone, occurring on scars, in grassland and in scrub, chiefly in the south and in north Westmorland. Common associates include *Sesleria caerulea, Galium sterneri* and *Helianthemum nummularium.* Records from the Carlisle area are suspect and have been omitted, but it was seen by G.A.K. Hervey on the railway embankment at Great Salkeld (5.3) from 1947 to 1957 and during the Survey on flushed riverside sandstone by the River Eden above Armathwaite (52.42).

These are the most northerly records on the west coast of Britain.

Alt. limit: 610 m, Long Fell, above Brough (76.18)

Map 229   93   LWFC(W/Y)

V. rupestris F.W. Schmidt   Teesdale violet

Extremely rare, known only from shallow, open limestone turf on Arnside Knott (4.7s) and Long Fell, Brough (7.1).

The history and ecology in Cumbria of this rare species were reviewed by Valentine & Harvey (1961). It appears to have been first collected, although not recognised as such, by T.F. Cuthwick from Arnside in 1927. Although J. Backhouse reported it from Long Fell, above Brough (7.1), in 1881, all his herbarium material (E) consists of *Viola riviniana.*

Valentine and Hervey visited Arnside Knott and Long Fell in 1960 and were successful in rediscovering

the plant at both sites. The population of 100 to 200 plants at the former site was entirely white-flowered and growing in semi-open south-facing limestone grass-land above the scree and immediately below woodland. The turf consisted largely of *Festuca ovina* and *Sesleria caerulea* and the violet was associated with such local rarities as *Asperula cynanchica* and *Carex ericetorum.* At the large Long Fell site above the Pennine scarp the plants were very scattered, extremely small, mostly sterile and growing in rather closed *Festuca - Sesleria - Thymus* turf and associated with *Draba incana.*

Unfortunately during the intervening years the Arnside population has been seriously diminished, first by excessive trampling by tourists and, following fencing, by the spread of *Sesleria caerulea.*

The only other British localities are the well-known one on the sugar limestone of Widdybank Fell, on the Co. Durham side of upper Teesdale, and the more recently discovered sites on Ingleborough, but Valentine & Harvey suggest that the violet could well occur on limestone elsewhere in Cumbria. It is worth noting that the journals of G. Bolam refer to the finding of a single plant above Tynehead (7.3).

Important characters distinguishing it from *V. riviniana* are the rounded, smaller leaves and the short dense pubescence of the stems, petals and capsules.

Viola rupestris has a wide distribution through-out central, eastern and northern Europe.

Alt. limit: 600 m, Long Fell, above Brough (76.18)

Plate 20   R 2   W

V. riviniana Reichb.   Common dog-violet

Extremely common and very catholic in its soil requirements, shunning only the most acid habitats and blanket peat. It occurs most commonly in hedgerows, short grassland, deciduous woodland, screes, gullies

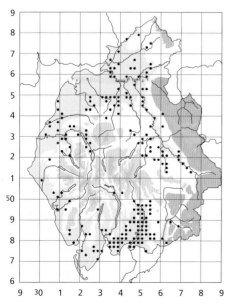

Map 228. *Viola odorata*

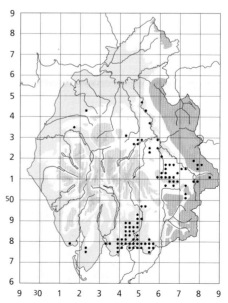

Map 229. *Viola hirta*

and on rock ledges.

Alt. limit: 770 m, Knock Fell (72.30)

Map 230   1656   LWFCY

V. riviniana x **V. canina**

V. x intersita G. Beck

This hybrid, very rarely found in northern Britain, was first collected by RWMC at two sites near Plumpton, north of Penrith (50.36, 52.36, LANC, conf. D.M. Moore, *Watsonia* (1986) 16:184) in 1985 and later from several others in the same general area as well as from the sand-dunes at Eskmeals on the west coast (08.92, GH, 1986, LANC). All are areas of open, sandy soil or roadside verges. In 1988 RWMC discovered it on riverside rocks on the Westmorland side of the River Lune near Sedbergh (62.92, LANC, *Watsonia* (1989) 17:466) and in 1989 it was found by LR at two sites near Melmerby (58.38, 60.38, LANC), probably a rediscovery of W.W. Mason's 1915 record from there of *Viola canina* x *V. lactea* (*BEC Rpt* (1930) 9:105). *V. canina* was absent from most of these sites. There seems no reason why the hybrid should not be widely distributed on the west coast dunes.

The hybrid is characterised by an intermediate leaf-shape, large flowers and, particularly noticeable, the absence of developing capsules later in the season.

Alt. limit: 220 m, Renwick (58.42)

Map 231   14   WC

V. reichenbachiana Jordan ex Boreau

                                    Pale wood-violet

Fairly frequent in deciduous woodland on the better soils in the south of the county but rapidly becoming rare northwards where it just reaches the band of base-rich soils to the south of Wigton. Its northern limit appears to be by the River Caldew near Sebergham (34.40, 1986, RWMC, LANC), although a specimen from by the River Eden at Low House Wood, Holm-wrangle (50.48, FJR, 1991, LANC) was determined by D.M. Moore as probably this species. All Survey records from the north of the county unsupported by specimens have been omitted from the map. They are probably the result of confusion with *Viola riviniana*.

This violet flowers early, about three weeks before *V. riviniana*. Hybrids probably occur but none has been recorded. There are only a few sporadic records of this southern species north of Cumbria.

Alt. limit: 245 m (Wilson)

Map 232   107   LWFCY

V. canina L.   Dog violet

Locally common on the coastal sand-dunes but extremely rare inland. Many of the inland Survey records, those in the *Atlas* and Rawes' (1981) record from Knock Ore Gill (7.3) are probably errors; those shown on the map have all been carefully checked.

These inland sites have been recently studied by Corner (1989). Most are rocky lake shores, as at Ullswater, or riverside rocks as along the River Lune

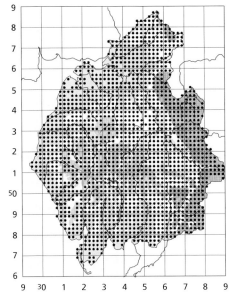

Map 230. *Viola riviniana*

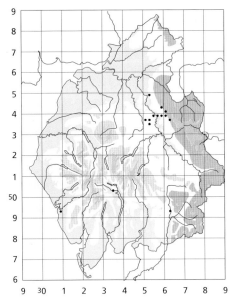

Map 231. *Viola riviniana* x *V. canina*

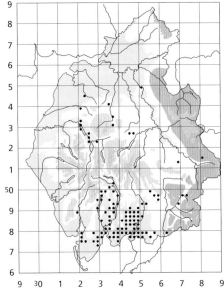

Map 232. *Viola reichenbachiana*

near Sedbergh (6.9) and the Eden near Armathwaite (52.40). There has since been confirmation of an old record from rocks by the River Brathay (36.02, K.M. Hollick) and the discovery of a lakeside population on Coniston Water (28.90). However it has also been found on sandy banks north of Penrith (5.3) and by the old railway line in the upper Eden valley near Sandford (70.18, 72.16). There is an authenticated 1950 record from the margin of a bog pool by the old railway between Abbeytown and Bromfield (1.4, C.W. Muirhead, CLE, det. D.H. Valentine), but this area has now been drained.

Alt. limit: 200 m, West Brownrigg (50.36)

Map 233   37   WFCY

V. palustris L.   Marsh violet

Common in slightly flushed but acidic marshy communities over most of the county. In the uplands it is a feature of the *Juncus effusus, Epilobium palustre, Stellaria uliginosa* and *Sphagnum recurvum* flushes, and in the lowlands of base-poor marshes and riverside alder and willow carr, again usually associated with *Sphagnum.*

      Wilson includes a record of subsp. *juressii* (Link ex Wein) P. Fourn. from Cliburn Moss (5.2). This is a south-western plant here at its very isolated north-eastern limit. It is characterised by hairy petioles and a bract half-way up the peduncle. There are no Survey records.

Alt. limit: 850 m, Cross Fell (68.34)

Map 234   1043   WFCY

*V. cornuta L.

A garden escape and native of the Pyrenees discovered in 1985 below a roadside hedge east of Sedbergh (66.92, CFS & JS, 1985; LANC).

Alt. 120 m, Sedbergh (66.92)

                              1   Y

V. lutea Hudson   Mountain pansy

Frequent in the north Pennines, occurring in *Festuca - Agrostis* grassland, usually around 300m and on the deeper, less calcareous soils. It is also a feature of the old mine-workings. It is curiously absent from the limestone uplands of north Westmorland and of those around Morecambe Bay. It occurs in an arc around the north of the Lake District and in the Lake District itself it is restricted to the north-east and north, being particularly frequent on Matterdale Common (38.22). The only Furness record is that given by Petty (1894 p.289) from Colwith (3.0).

      Plants with purple flowers are slightly commoner and mixed populations are frequent and eye-catching.

Alt. limit: 855 m, Helvellyn (34.14)

Map 235   181   W(F)CY

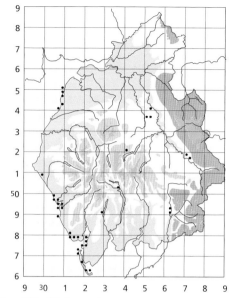

Map 233. *Viola canina*

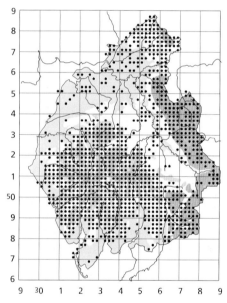

Map 234. *Viola palustris*

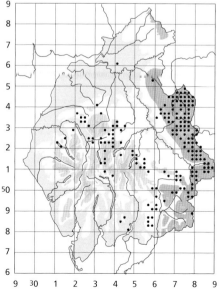

Map 235. *Viola lutea*

*V. lutea x V. tricolor x V. altaica Ker Gawler
V. x wittrockiana Gams ex Kappert   Garden pansy
Recorded in 1993 growing on tipped soil in an old quarry at Oakshaw Hill, near Longtown (40.70n, MG & JP).

                              1   C

V. tricolor L.   Heart's-ease; Wild pansy
Well distributed but frequent only in the Eden valley, especially to the north and south of Carlisle, and on the west coast sand-dunes. The dune plants are perennial and usually referred to the yellow-flowered subsp. *curtisii* (E. Forster) Syme although most of the Cumbrian dune plants are blue. For this reason similar plants occurring on the Isle of Man are treated by Allen (1984) as a maritime form of subsp. *tricolor*.

Inland, *Viola tricolor* is mainly a weed of cultivated ground, including gardens, although it also occurs in short, dry and somewhat acidic grassland where the plants are probably perennial. There is also a record from sandstone cliffs by the River Eden at Rockcliffe (34.60).
Alt. limit: *c*.300 m, Leadgate, Alston (70.42)
Map 236   282   WFCW/Y

V. tricolor x V. arvensis
V. x contempta Jordan
This uncommon hybrid was collected by D. Lumb in 1913 at Foxfield, on the Duddon estuary (20.84, DHM).

                                            (F)

V. arvensis Murray   Field pansy
A fairly frequent plant of disturbed and cultivated ground, particularly allotments and arable fields, chiefly in the Eden valley and on the Solway Plain. There are also a few records from open sandy pastures. It probably prefers more basic soils than does *Viola*

tricolor, but mixed populations do occasionally occur and may well include hybrids.
Alt. limit: *c*.300 m, Alston (72.44)
Map 237   234   WFC(W/Y)

TAMARICACEAE

*Tamarix gallica L.   Tamarisk
(*T. anglica* Webb)
A few bushes of this southern coastal shrub still occur by the shore at Barker Scar (32.78s) on the Leven estuary where they were first recorded by the Grange-over-Sands N.H.S. in 1964. They were presumably planted.

                              1   F

CUCURBITACEAE

*Bryonia dioica Jacq.   White bryony
Extremely rare. The only Survey records are from a roadside hedge near Culgaith (58.30, AW, *c*.1980) and in the village itself (60.28, RWMC, 1981; LANC), where it has been long-established in a hedge near the post office and also in the churchyard. It is probably not native at this locality which lies far to the west of the species' native range in Britain. These two sites are very close to Woodside, on the Westmorland side of the River Eden, and the only site for *Bryonia* given by Wilson (5.2, A. & W.A.P. Sprott; A. Wilson, 1935, YRK).

Baker's statement that "bryony is frequent in hedges and woods at Kirkby Lonsdale" and Hodgson's that it is "not seen north of Bootle" must refer to black bryony, *Tamus communis*. This is likely to be true also of all but one of the *Atlas* records for *Bryonia*, namely the Millom (1.8), Dalton-in-Furness (2.7), Seathwaite (2.9), Cark (3.7s) and Newby Bridge (3.8) squares.

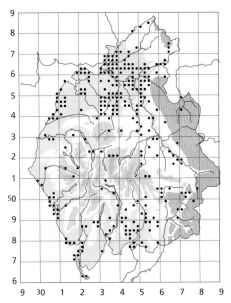

Map 236. *Viola tricolor*

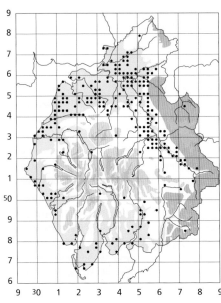

Map 237. *Viola arvensis*

That for the Melkinthorpe square (5.2) presumably refers to the Woodside record. Wilson's record of *Tamus* from Melkinthorpe Wood probably refers to *Bryonia*. The only definite early record from Cumberland is from a hedge at Kelsickhouse, Abbeytown (1.4, Anon., 1941) where it was introduced.

5   (W)C

SALICACEAE

*Populus alba L.   White poplar
Occasionally planted as an amenity tree in hedgerows, estates, private woodland, villages and towns, and growing best in wet, heavy soil.
Alt. limit: *c.*160 m, Great Strickland (56.22)
Map 238   104   WFC

*P. alba x **P. tremula**   Grey poplar
P. x canescens (Aiton) Smith
Apparently restricted to the middle Eden valley and the Solway Plain but with outlying records from Arnside (46.78s) and Coniston (30.96, PB, 1991, herb.PB), the only Furness record. The first Westmorland record is from near Brougham (52.26, RWMC, 1983, LANC, *Watsonia* (1984) 15:133).

     The only pre-Survey records are the *Atlas* dot for Dentdale (7.8) and a specimen from Newbiggin, Carlisle (4.5, J.J. Taylor, 1949, CLE). Some of the trees are fine specimens, usually taller and faster growing than *Populus alba* with which it has probably been confused in the past.
Alt. limit: 170 m, Catterlen, Penrith (48.34)
Map 239   16   WFC(Y)

P. tremula L.   Aspen
Fairly frequent, particularly in the Eden valley and Solway Plain, but rather scattered towards the south and west coasts. It occurs in hedges and woodland, wooded gills and riversides, where it grows particularly well, and occasionally high on Lakeland crags. Some of these populations are clonal and only of one sex.

     It is also sometimes planted commercially, presumably for veneer and matchwood.

     A roadside specimen from near Orton (62.06, GH, 1971, LANC), with leaves only 3 cm wide, could well be f. *microphylla* A. Brown ex Schneid. Var. *villosa* (Lang) Wesm. was said by H. Britten to be common in hedges at Great Salkeld at the beginning of the century.
Alt. limit: 600 m, Pavey Ark, Langdale (28.08)
Map 240   433   WFCY

P. nigra L.   Black poplar
subsp. **betulifolia** (Pursh) W. Wettst.
A single tree was reported in 1994 by J. Kerr (*Cumberland & Westmorland Herald*, July 16th 1994) growing on the bank of the River Eden west of Culgaith (58.28). The identity of material from there and the riverside at Langwathby and Holmwrangle (56.32, 50.48, RWMC,

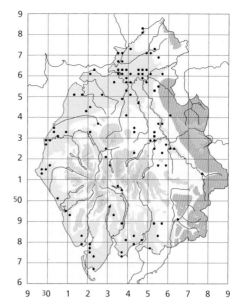

Map 238. *Populus alba*

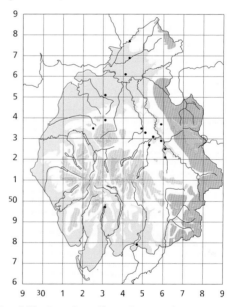

Map 239. *Populus alba* x *P. tremula*

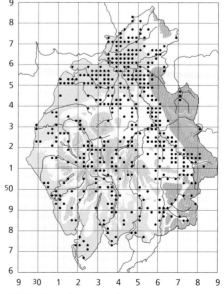

Map 240. *Populus tremula*

1995, LANC) has been confirmed by J. White and D. Hobson. These are the northernmost native records in Britain.

3   C

*var. **italica** Muenchh.   Lombardy poplar
This striking and unmistakable poplar has rather surprisingly been reported only from Naworth Park (56.62), Houghton (40.58) and east of Wray (44.48, 46.48) in the north, and in the south, from the Kent estuary near Arnside (46.78s).

WC

*P. **nigra** x P. **deltoides** Marshall   Hybrid black poplar
P. x **canadensis** Moench
Widely distributed and conspicuous, occurring as isolated specimens or in small groups, usually by roadsides or woodland margins and always planted. Probably all the literature records for *Populus nigra* belong here.

All specimens for which the sex has been recorded are male. Most of the records probably refer to var. *serotina* (Hartig) Rehder, characterised chiefly by its broad crown. The only certain record of var. *robusta* is from the Westmorland side of the River Eamont at Pooley Bridge (46.22, GH, 1991, LANC, det. R.D. Meikle).
Alt. limit: 260 m, south of Shap (56.12)
Map 241   128   WFC

[P. balsamifera Duroi non L.
Hodgson's two records, from Dearham (0.3) and Water-millock (4.2), may refer to this species, similarly W.W. Mason's from Melmerby (6.3, *BEC Rpt* (1919) 5:399), but in the absence of specimens their identity is best considered doubtful.

(C)]

*P. **balsamifera** L. x P. **deltoides** Marshall
P. x **jackii** Sarg.   Balm-of-Gilead
(*P.* x *gileadensis* Roul., *P.* x *candicans* auct.
)
Apparently very rare and confused in the past with the now appreciably commoner *Populus trichocarpa*. The only Survey specimens, all confirmed by R.D. Meikle, are from the roadside south-west of Scaleby (42.60, MG, 1991, LANC), the Cumberland side of the River Eamont at Pooley Bridge (46.24, MC, 1985), Mill-wood, Dalton-in-Furness (20.72, PB, 1990, LANC) and Arnside (46.78s, MB, 1990). It is recorded by Hodgson, as a variety of *Populus balsamifera*, from Stockdale-wath (3.4).

There is no general agreement regarding the parentage of this tree. Jobling (1990) states that it is readily distinguished from *P. balsamifera* by its spreading branches, pubescent young shoots and petioles, and broader, more strongly cordate leaves.

4   WFC

*P. **trichocarpa** Torrey & Gray ex Hook.
Western balsam poplar
Scattered in the north and south, absent from much of the west and east and occurring usually as roadside plantings and small plantations. It appears to be increasingly popular with the Highways Authority. The clone planted along the A590 between Levens Bridge and Lindale (4.8s) is particularly fast-growing and large-leaved.

Although not grown in Britain until 1892, this is by far the commonest balsam poplar in the county. The first certain record is from by the old railway near Brampton (52.60, C.W. Muirhead, 1951, CLE; MG, 1990, LANC). The first Furness collection was in 1984 from by the River Crake at Nibthwaite (28.86, J. & A. Harrington, 1984) and from Westmorland by the A590 near Town End, Witherslack (44.82s, GH, 1988). Being always planted and often near houses, this poplar

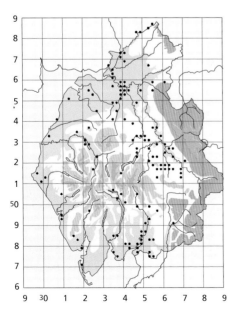

Map 241. *Populus nigra* x *P. deltoides*

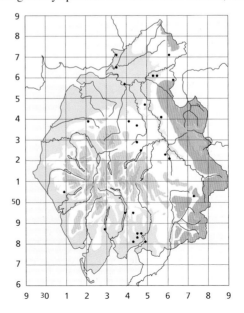

Map 242. *Populus trichocarpa*

is probably much under-recorded.
Alt. limit: 225 m, Spadeadam (60.70n)
Map 242   26   WFC

*P. trichocarpa x **P. balsamifera** L.   Balsam spire
Although stated by Stace (1991) to be one of the
commonest planted poplars, the only Survey record of
this striking fastigiate clone is of a group of trees on the
Cumberland side of the River Eamont at Pooley Bridge
(46.24, GH, 1991, LANC, det. A.O. Chater).
                          1   C

*P. trichocarpa x **P. angulata** Aiton
P. x **generosa** Henry
The only records, all determined by R.D. Meikle, are
from the banks of the River Eden at Carlisle (40.56,
REG, 1978, LANC) and near Colby, Appleby (66.20,
RWMC, 1988, LANC), and from waste ground in
Kendal (52.90, CEW, 1989, LANC). According to
Jobling (1990) this hybrid is becoming increasingly
rare due to its susceptibility to canker. The leaves are
less leathery and less markedly white beneath than
those of *Populus trichocarpa*.
Alt. limit: 125 m, Colby (66.20)
                          3   WC

*P. simonii Carr.
Nine small trees of this Asiatic poplar were found by
REG on the benks of the River Eden at Carlisle (38.56,
1995, LANC, det. E.J. Clement).
                          1   C

Salix
With 17 species and 20 hybrids Cumbria has an
exceptionally rich willow flora. The following account
is largely based on extensive collections during the
Survey and a study of these and earlier herbarium
collections by R.D. Meikle.

S. pentandra L.   Bay willow
Common in the north and east, scarce in the south-west
and curiously absent from most of the Lake District.
This attractive willow is characteristic of becksides and
the margins of tarns and lakes, and more generally,
with *Salix cinerea*, in areas of carr. It also occurs in
hedges and in scrub on the wider and wetter roadside
verges.
Alt. limit: *c*.400 m, south of Garrigill (74.38)
Map 243   507   WFCY

[S. pentandra x S. fragilis
S. x meyeriana Rostkov ex Willd.
Although listed by Wilson and Meikle (1975) and recorded in
1949 by P. Taylor (*BSBI Yr Bk* (1959) p.49, det. R. Melville)
from Newton Reigny (4.3) and Newsham (3.3), all records
are now considered by Meikle to refer to the following
hybrid.
                          (WC)]

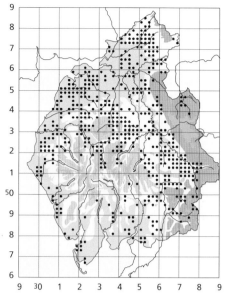

Map 243. *Salix pentandra*

Plate 64. *Salix pentandra* x *S. alba* by the River Eden

*S. pentandra x S.alba

S. x ehrhartiana Smith

A frequent willow widely planted throughout the Eden valley and extending on to the Solway Plain. Although this willow is characteristic of the river, its tributaries, pondsides and ditches, several of the records are from roadsides and hedges.

Meikle has examined herbarium and fresh specimens from Cumbria over a number of years. Initially he was inclined to think that both this hybrid and *Salix pentandra* x *S. fragilis* (*S.* x *meyeriana*) were present, the latter having been reported by Sprott (1936) as the common male hybrid willow of the upper Eden, a statement later fully supported by Wilson. However, C.A. Sinker was of the opinion that the trees by the Eden were not the same as the female trees of *S.* x *meyeriana* in Shropshire and this is also the conclusion reached by Meikle. The first Cumberland record is probably that from Newsham cross roads, Greystoke (3.3, P. Taylor, 1949) determined by R. Melville as *S.* x *meyeriana*; no specimens have been located.

This is a delightful willow, especially when the male catkins are out, fully justifying Wilson's remark "highly ornamental".

Alt. limit: 175 m, Brough Sowerby (80.12)

Map 244, Plate 64　　　49　　　　　WC

S. fragilis L.　　　　　　　　Crack willow

A common lowland willow throughout the north and east but, like *Salix pentandra*, scarce in the Lake District and the south-west. Most of the records are from riversides, wet woodland and by field ditches; it also occurs by the Ravenglass estuary (0.9). It is generally assumed to be native in northern England but there is no doubt that it has been very extensively planted.

A majority of the 33 specimens examined by Meikle proved to be the female var. *russelliana* (Smith) Koch. There were only two collections of var. *fragilis*,

from Kirkby Thore (60.24, GH, 1982, LANC) and Stangerthwaite, Killington (62.88s, GH, 1981, LANC), ten of var. *decipiens* (Hoffm.) Koch and four of var. *furcata* Ser. ex Gaudin.

Alt. limit: 275 m, south of Alston (70.44)

Map 245　　　　457　　　　　WFCY

S. fragilis x S. alba

S. x rubens Schrank

Very rare, the only records of this hybrid are an unlocalised collection by C.W. Muirhead in 1952 (PLYP) and Survey collections from by the River Esk near Kirkandrews (38.72n, REG, 1979, herb. R.E.G.), a streamside near Anthorn (20.58, REG, 1985, LANC) and Mirehouse, Whitehaven (98.14, CCH, 1979, herb. C.C.H.). There are also specimens of *Salix alba* var. *vitellina* x *S. fragilis* (*S.* x *basfordiana* Scaling ex J. Salter) from near St Bees (98.08, CCH, 1979, herb. C.C.H.), St Herbert's Island, Derwent Water (24.20, GH, 1992, LANC) and Newtown, Carlisle (38.56, REG, 1978, herb. R.E.G.). All the material has been determined by Meikle who also tentatively identified as the hybrid a specimen from the Cumberland side of the river at Pooley Bridge (46.24, MC, 1981, herb. M.C.).

　　　　6　　　　　　　　　C

S. alba L.　　　　　　　White willow

Not infrequent in the Eden valley and around the coast, elsewhere scarce. There seems little reason to doubt the statements in the literature that it is always planted, occurring as it does almost invariably by roadsides, near houses and in estate woodland. A distinctive and beautiful tree, it is currently in fashion with the Highways Authority for planting in association with road-improvement schemes.

The only Survey records of var. *vitellina* (L.) Stokes are from by the River Mite above Ravenglass station (08.96, GH, 1991, LANC), where it was first noted in 1928 (Anon., CLE), and by the roadside in

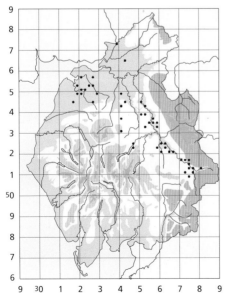

Map 244. *Salix pentandra* x *S. alba*

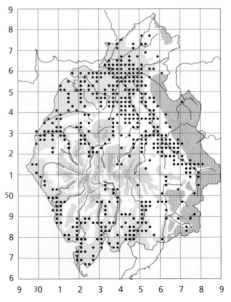

Map 245. *Salix fragilis*

Kearstwick, Kirkby Lonsdale (60.78s, GH, 1995). Meikle comments that the Ravenglass material is atypical in that the twigs lack the usual lustrous orange colour. Hodgson lists the variety from by the River Ellen (0.3/1.4), Wilson from Hoff, Appleby (6.1) and H. Britten recorded it from Great Salkeld (5.3, 1900-1908).

Alt. limit: *c*.350 m, south of Alston (72.44)

Map 246   117   WFCY

*S. amygdaloides Anderss.

A North American willow collected at Wetheral in 1950 (4.5, C.W. Muirhead, PLYP, det. Meikle). Unfortunately no further information is known about the location or habitat.

(C)

S. triandra L.   Almond willow

Fairly frequent in the upper Eden valley, by the main river and its tributaries and in hedges and willow thickets. Outside this area there are only scattered records from the north side of the Lake District, Monkhill Lough, Carlisle (32.56, 32.58), where it is locally dominant and curiously has a second flowering season in late summer, and above Bewcastle, in the extreme north (56.74n). The only record from Furness is that given by Petty (1898 p.37) for Humphrey Head (3.7s) but this seems very doubtful.

Alt. limit: 175 m, Great Asby (66.14)

Map 247   35   W[(F)]C

S. triandra x **S. viminalis**

S. x mollissima Hoffm. ex Elwert

Known only from a 1950 collection of C.W. Muirhead's from near Ravenglass station (0.9, PLYP, det. R. Melville) and Survey ones from Temple Sowerby Moss (60.26, RWMC, 1985, LANC, *Watsonia* (1988) 17:190) and by the River Eden above Langwathby bridge (56.32, RWMC, 1982, LANC), the last two being determined by Meikle.

2   WC

S. purpurea L.   Purple willow

Fairly frequent in the east but rather uncommon elsewhere. It occurs in willow carr, by river and streamsides and, rarely, in hedges. Like so many willows this species is commonest in the Eden valley, although frequent also along the Irthing and the Lune, where it is very much a feature of gravel islands, its branches often festooned with plant and plastic flood debris.

This is a delicate willow, easily recognised by its pale, almost glaucous, sub-opposite leaves.

Alt. limit: *c*.270 m, Mallerstang (78.98) and Greenhead Gill, Grasmere (34.08)

Map 248   164   WFCY

S. purpurea x **S. viminalis**

S. x rubra Hudson

Scattered throughout the Eden valley, usually by the

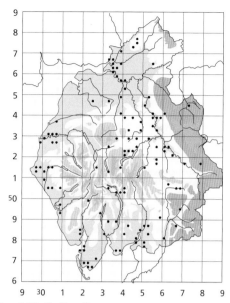

Map 246. *Salix alba*

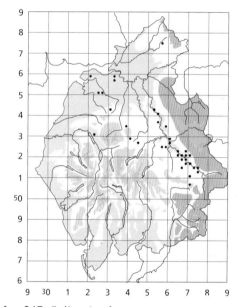

Map 247. *Salix triandra*

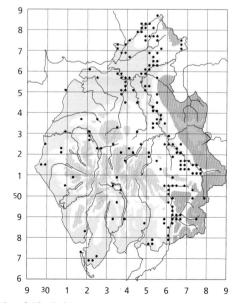

Map 248. *Salix purpurea*

riverside, but with scattered localities in west Cumberland and an isolated one in the south near Sedbergh (66.88s, MB, LANC). All records have been checked by Meikle.

This hybrid could well be commoner in the west since Hodgson gives several sites from Egremont northwards, including by the River Ellen where he describes it as abundant from Aspatria to Maryport (0.3, 1.3, 1.4). C.W. Muirhead also collected it by the Ellen in 1950 and additionally at Monkhill Lough, Carlisle (3.5) and Lessonhall, Wigton (2.5), all identified by R. Melville (PLYP).

This hybrid is easily recognised by the leaves, which are silky when young and tend to be opposite.
Alt. limit: 380 m, Dowthwaite Head (36.20)
Map 249   19   WCY

*S. daphnoides Villars

Very rare and probably always planted. It was first recorded in Cumberland in 1978 by the River Eden at Carlisle (40.56, REG, LANC, *Watsonia* (1980) 13:139) and in Westmorland in a roadside hedge at Lambrigg, Grayrigg (58.96, MB & R. Dalton, 1981, LANC, *Watsonia* (1984) 15:133).

All the specimens have been seen by Meikle and most show the characteristic dense glaucous bloom on the young stems.
Alt. limit: 210 m, Lambrigg (58.96)
Map 250   7   WC

S. viminalis L.   Osier

Common in the north and east, especially in the Eden valley, scattered in the south and almost absent from the Lake District. This is one of the most distinctive of willows and probably always planted, occurring not only in the usual river and streamside habitats but also by farm ponds and in amenity plantings. It is particularly popular at the present time with the Highway Authority.
Alt. limit: 380 m, between Watermillock and Matterdale (42.22)
Map 251   400   WFCY

S. viminalis x S. caprea

S. x sericans Tausch ex A. Kerner

A fairly common and usually readily recognised hybrid of roadside hedges and streamsides in the lowlands. As Meikle (1984) says, this is frequent as a spontaneous hybrid as well as being commonly planted. Probably many of the literature records of *Salix* x *smithiana* belong here rather than to the appreciably less common *S. viminalis* x *S. cinerea*.

This hybrid is characterised by its softly hairy and large, rather broad, oblong leaves, and the twigs lack the underlying striae of *S. viminalis* x *S. cinerea*. It is very vigorous and has considerable potential for biomass production.
Alt. limit: 460 m, head of Gildersdale (66.44)
Map 252   80   WFCY

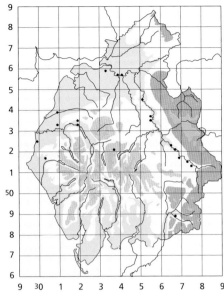

Map 249. *Salix purpurea* x *S. viminalis*

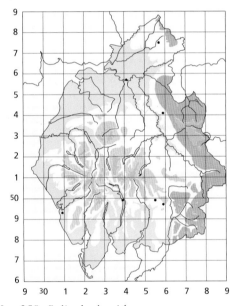

Map 250. *Salix daphnoides*

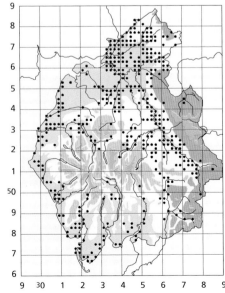

Map 251. *Salix viminalis*

S. viminalis x **S. caprea** x **S. cinerea**
S. x **calodendron** Wimmer
(*S. dasyclados* Wimmer)
Recorded only from a laneside on Millom Marsh (18.80, GH, 1981, LANC, det. Meikle, *Watsonia* (1982) 14:193).

This willow is characterised by large leaves, 10 cm or more long and about three times as long as wide, and prominent stipules. The trees are female and probably always planted.

1   F

S. viminalis x **S. caprea** x **S. aurita**
S. x **stipularis** Smith
Known only from Paddaburn (62.76n, CCH, 1977, LANC) and an old collection from Black Dyke, Silloth (1.5, E.J. Glaister, CLE, det. Meikle) where it was collected in 1877.

1   C

S. viminalis x **S. cinerea**
S. x **smithiana** Willd.
Rather rare, occurring in similar habitats to *Salix viminalis* x *S. caprea* but noticeably less common and probably usually planted. The leaves are narrower and less pubescent below and the twigs have the underlying striae of *S. cinerea*. However, as Meikle (1984) states, there are some specimens which cannot confidently be assigned to either hybrid.

As mentioned above, the early records of *S.* x *smithiana* no doubt refer to both these hybrids.
Alt. limit: 200 m, near Crosby Garrett (70.10)
Map 253   30   WFC

S. viminalis x **S. aurita**
S. x **fruticosa** Doell
Known only from the two records given by Wilson: Bolton

(6.2) and Ormside road, Appleby (6.1).

(W)

S. viminalis x **S. repens**
S. x **friesiana** Andersson
A single bush of this rare hybrid was discovered in the Sandscale car park in 1992 (18.74, 20.74, PB, LANC, *Watsonia* (1995) 20:288). The only other British records are from east Sutherland and in Lancashire on the Ainsdale dunes and the Fylde.

2   F

*S. udensis Trautv. & Meyer
Two clumps of this eastern Asiatic species were found in 1985 growing by the roadside on Viol Moor, Kirkoswald (58.40, RWMC, BM, LANC, det. R.D. Meikle, *Watsonia* (1988) 17:191). Although occasionally cultivated, this is apparently only the second reported occurrence in Britain of established plants away from habitation.

The trees are about 5 m high and quite distinctive with their narrowly elliptic, subentire, long, pendent leaves. These are pale, almost glaucous below with a strong, deep orange-brown mid-vein and a thin pubescence of short hairs.
Alt.: 170 m, Viol Moor, Kirkoswald (58.40)

1   C

S. caprea L.   Goat willow
Common in the lowlands occurring chiefly in hedgerows and scrub, occasionally in deciduous woodland and not infrequent on recently disturbed or abandoned sites. As Wilson observes it prefers rather drier sites than either *Salix aurita* or *S. cinerea*; it is also appreciably taller than either. The map suggests that it is generally more lowland than is indicated by Wilson's reference to rocky gills and his upper limit of 505 m.

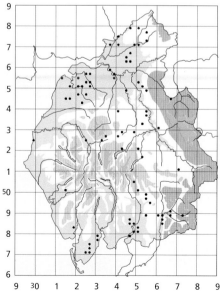

Map 252. *Salix viminalis* x *S. caprea*

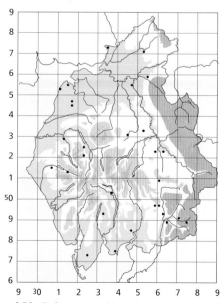

Map 253. *Salix viminalis* x *S. cinerea*

Material collected from some trees in a gorge in upper Eskdale (22.04, GH, 1991, LANC) and one by Thirlmere (30.16, RWMC, 1994, LANC) have been identified by Meikle as subsp. *sphacelata* (Smith) Macreight, which is distinguished chiefly by its persistent leaf pubescence The former is the first English record. Two other collections, from becksides at the western and eastern ends of Hard Knott Pass (20.00, 22.00, GH, 1992, LANC) were considered by Meikle to approach this subspecies.
Alt. limit: 440 m, Tynehead (76.34)
Map 254   1204   LWFCY

S. caprea x S. cinerea
S. x reichartii A. Kerner
Recorded chiefly by RWMC from several localities, particularly on disturbed ground, in the Penrith area and the north-east but otherwise apparently rather scarce. This is a distorted picture since Recorders, having noted both the parents in a tetrad, not unnaturally tend to be content to pass over the inconvenient intermediates. The 14 Survey records, all determined by Meikle, are no doubt the tip of the proverbial iceberg. The only record of this hybrid in the Floras is that given by Wilson from Cliburn Moss (5.2). The first Cumberland record was from Penrith (50.30, RWMC, 1982, LANC).
Alt. limit: 450 m, Cash Burn, Alston Moor (70.38)
Map 255   14   WCY

S. caprea x S. aurita
S. x capreola J. Kerner ex Andersson
Recorded during the Survey only from Mirehouse, Whitehaven (98.14, CCH, 1979, herb. C.C.H., det. R.D. Meikle). It was earlier reported in Cumberland by W.W. Mason from Melmerby (6.3, 1913) and Newton Reigny (4.3, 1917). The only Westmorland record is that given by Wilson from near Milburn (6.2).
1   (W)C

S. caprea x S. myrsinifolia
S. x latifolia James Forbes
A rare northern hybrid collected in Cumbria during the Survey in a disused railway cutting near Penton (42.76n, RWMC, 1988, LANC) and there are earlier collections from Greystoke Park (4.3, P. Taylor, 1948, PLYP, *BSBI Yr Bk* (1951) p.49) and near Ravenglass station (0.9, C.W. Muirhead, 1950, PLYP). The last two were deter-mined by R. Melville and all three by Meikle, who (1975) also lists the hybrid for West-morland.
1   (W)C

[S. caprea x S. phylicifolia
Material tentatively identified as this hybrid by R. Melville and Meikle was collected at Red Moss, Abbeytown in 1950 (1.5, C.W. Muirhead, PLYP). According to Meikle (1975) there are no definite records south of the Scottish Highlands.
(C)]

S. cinerea L.   Common sallow; Grey willow
subsp. oleifolia Macreight
Very common almost everywhere below about 300 m. It occurs frequently in old hedges and on the broad verges of the old drove roads, such as the B5305 through Sebergham, but it is most characteristic as the domin-ant component of willow carr in fens, around tarns and sheltered bays of the lakes, in poorly drained valley bottoms and around lowland mosses, often associated with *Molinia* and alder.

Although very variable, there are no records of the south-eastern subsp. *cinerea* which, interestingly, has been reported by Allen (1984) from a site on the Isle of Man.
Alt. limit: 845 m, Great Dun Fell (70.32)
Map 256   1459   LWFCY

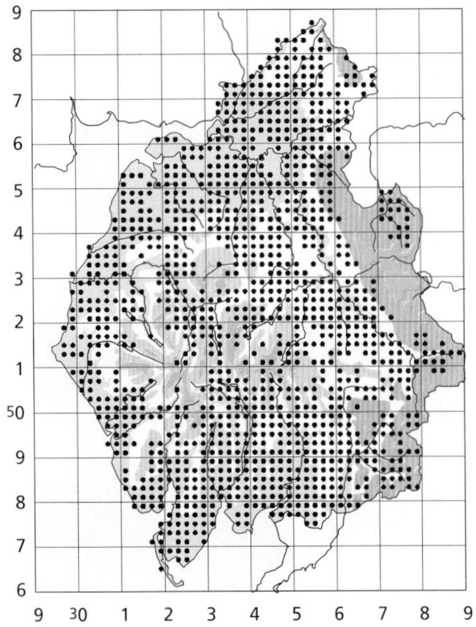

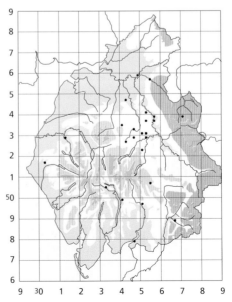

Map 254. *Salix caprea*

Map 255. *Salix caprea* x *S. cinerea*

S. cinerea x **S. aurita**
S. x **multinervis** Doell
Rare and very scattered, chiefly in the north and east.
Most of the records are from river and lanesides; all
have been determined by Meikle.

The first Cumberland record is that of W.W.
Mason's from Melmerby (6.3, 1913), for Westmorland
Wilson's from Cliburn (5.2) and "Looking Flat" and for
Furness W.H. Pearsall's from Ireleth (2.7, 1925, YRK,
det. R.D. Meikle, *BEC Rpt* (1926) 7:1065).
Alt. limit: 320 m, northern Howgill Fells (64.00)
Map 257   17   WFCY

S. cinerea x **S. myrsinifolia**
S. x **strepida** Forbes non Schleicher
The only Survey records, all identified by Meikle, are
from Smardale, Crosby Garrett (72.08, E.F. Green-
wood, 1973, LIV), the roadside south of Greystoke
(44.28, MC, 1984, herb. M.C.), Flusco Lodge quarry,
Stainton (46.28, RWMC, 1993, LANC), a laneside near
Ireby (22.38, REG, 1994, LANC) and by the Liddel
Water above Riddings (40.74n, GH, 1991, LANC).

Prior to the Survey it was listed by Wilson from
Ormside Road, Appleby (6.1) and there are three
collections from Cumberland: Newsham cross roads
(3.3, C.W. Muirhead & P. Taylor, 1949, CLE, PLYP,
BSBI Yr Bk (1951) p.49), Thurstonfield (3.5, C.W.
Muirhead, 1949, PLYP) and Newton Reigny moss (4.3,
C.W. Muirhead, 1949, PLYP). These three collections
were identified by both R. Melville and Meikle.
Alt. limit: *c.*195 m, Smardale (72.08)
                           5   WC

S. cinerea x **S. phylicifolia**
S. x **laurina** Smith
Probably not uncommon in the eastern half of the Lake
District, the north-east of the county and Garsdale in
the south-east. There are isolated occurrences well
outside the range of *Salix phylicifolia* at Workington
(00.28, CCH, 1979, herb. C.C.H.) and in Furness north
of Holker (34.78s, GH, 1986, LANC). All the records
have been determined by Meikle.

The only pre-Survey records are from Newton
Reigny (4.3, W.W. Mason, 1917) and that given by
Wilson from Bolton (6.2).
Alt. limit: 330 m, Garrigill (72.40)
Map 258   21   WFCY

S. aurita L.   Eared willow
Fairly common along the west coast and in the north
and east, elsewhere rather scattered. It occurs in hedgerows, scrub and woodland margins and by becksides on
the lower fells.
Alt. limit: 450 m, Cauldron Snout, Teesdale (80.28)
Map 259   477   WFCY

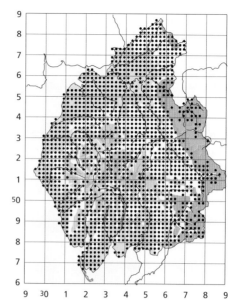

Map 256. *Salix cinerea*

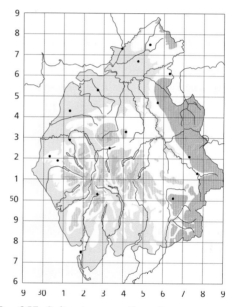

Map 257. *Salix cinerea* x *S. aurita*

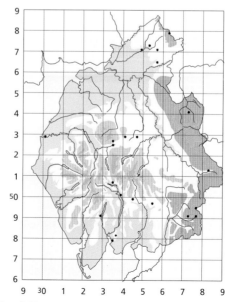

Map 258. *Salix cinerea* x *S. phylicifolia*

S. aurita x **S. myrsinifolia**
S. x coriacea James Forbes
Seen during the Survey by the River Lyvennet at Crossrigg Hall, Cliburn (60.24, RWMC, 1984, herb. R.W.M.C.), near Bowscale moss (36.30, RWMC, 1984, LANC), on Newton Reigny moss (46.32, RWMC, 1985, LANC), in Greystoke Park (40.32, RWMC, 1992, herb. R.W.M.C.), on the old railway line at Penruddock (42.26, MC, 1991, herb. M.C.) and on the roadside to the west (40.26, RWMC, 1989, LANC).

The only pre-Survey records are those given by Wilson from Helm, near Appleby (7.1) and two Cumberland ones: Roadhead, Bewcastle (5.7n, M.P.H. Kertland, 1952) and by the River Ellen (1.3/1.4, C.W. Muirhead, 1957, PLYP). All identifications except Wilson's have been checked by Meikle.
Alt. limit: 270 m, Penruddock (40.26)
6   WC

S. aurita x **S. phylicifolia**
S. x ludificans F.B. White
Collected from a laneside in upper Garsdale (76.90, GH, 1986, LANC) and a beckside west of Lowthwaite (40.22, GH, 1966, LANC). It was earlier listed by Wilson from Ormside road, Appleby (6.1) and collected in a limestone pothole on Knock Fell (7.3, A. Eddy, 1959, LANC); the herbarium specimens were determined by Meikle.

The only English records for this northern hybrid are from Cumbria and the Craven area of Yorkshire.
Alt. limit: *c.*760 m, Knock Fell (7.3, 1959)
2   (W)CY

S. aurita x **S. repens**
S. x ambigua Ehrh.
Seen during the Survey at two sites in Greystoke Forest (40.32, RWMC, 1990, 1992, LANC, det. R.D. Meikle), at one of which it was growing with *Primula farinosa* near the road between Lamonby and Johnby. This hybrid is otherwise known only from the two Westmorland sites given by Wilson: between Common Holme Bridge and Newby Head (5.2) and Asby Road, Appleby (6.1). The basis for the pre-1962 Cumberland record in the *Check-List* cannot now be traced and may be erroneous.
Alt. limit: 350 m, Greystoke Forest (40.32)
1   (W)C

S. myrsinifolia Salisb.   Dark-leaved willow
(*S. nigricans* Smith)
Rather rare and apparently restricted to the north and east where it occurs in fens and by becksides and roadsides. There are, however, reliable pre-Survey records from the Lake District: Sourmilk Gill, Easedale (3.0, A.E. Ellis, 1939, LANC, erroneously mapped in *Scarce Plants* as 0.0) and a streamside near Kendal (5.9, E. Hodgson, 1966, LANC). These and all the Survey records have been verified by Meikle. There is also Baker's record from Witherslack (4.8s), Wilson's from Cliburn (5.2), High Cup Gill (7.2), Orton (6.0) and Mallerstang (7.0), and one from Garsdale (7.8, Lees (1888)).

This species is not always readily distinguished from *Salix phylicifolia* and hybrid derivatives. The most reliable characters are the persistent stipules and dull twigs and leaves, the latter blackening on drying.

It is largely restricted to northern Britain, being most frequent in the east: in Co. Durham and Northumberland, the Southern Uplands and the eastern central Scottish Highlands.
Alt. limit: at least 470 m, Alston Moor (78.38)
Map 260   S 22   WC(Y)

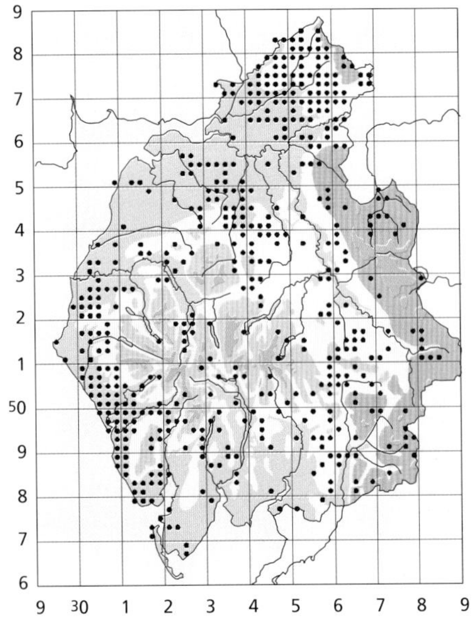

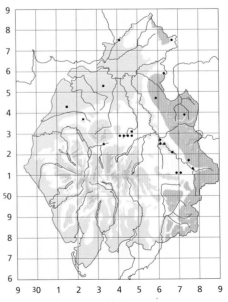

Map 259. *Salix aurita*

Map 260. *Salix myrsinifolia*

S. myrsinifolia x S. phylicifolia
S. x tetrapla Walker

This hybrid was collected during the Survey from by the River Eamont at Brougham Castle (52.28, GH, 1982, LANC), Butterburn, Spadeadam (68.72n, GH, 1988) and by the River Lune at Gaisgill bridge (64.04, R.C.L. Howitt, 1977, AMDE) and the Liddel Water above Riddings (40.74n, GH, 1991, LANC).

Meikle has identified material from all these sites and considers that the following three collections (all GH, LANC) probably also belong here: lane between Moorhouse and the A66, Warcop (74.16, 1982), by the River Eden, near Great Musgrave (76.12, 1982) and Ingmoor Fen, Sunbiggin (66.08, 1980), also material from Newton Reigny Moss (46.30, RWMC, 1992, LANC) and an old collection from Longstone Fell, Sedbergh (6.9, C.G. Trapnell, 1924, YRK).

Wilson gives a record from Warcop bridge (7.1) and there are other early collections, verified by Meikle, from between Brough and Appleby (7.1, J.E. Woodhead, 1937, LANC), the laneside south-west of Brampton (5.5, C.W. Muirhead, 1950, PLYP) and Roadhead, Bewcastle (5.7n, M.P.H. Kertland, 1952, PLYP). There is also a record from Seascale (0.0, D. Lumb and G.C. Druce, 1915, *BEC Rpt* (1916) 4:281) which is surprising in view of the distribution of the parent species.

Hybridisation between these two species can produce extensively introgressed populations. This problem does not apparently occur in Cumbria on the scale with which it is encountered in Scotland or even in Northumberland and Co. Durham, presumably because Cumbria lies towards the southern limit of *Salix myrsinifolia*.

Alt. limit: 245 m, Butterburn (68.72n)

4	WC

S. phylicifolia L.   Tea-leaved willow

A northern willow, like *Salix myrsinifolia*, but in Cumbria appreciably more frequent, occurring over much of the east and extending into the eastern Lake District. There are no recent records from the west although there are *Atlas* ones for four squares in the south-west of the Lake District.

This willow is particularly characteristic of river and becksides, but there are a few records from high crags in the Lake District.

Alt. limit: 685 m, Catstycam (34.14) and Blea Water Crag, Haweswater (44.10), although Wilson gives it as 760 m on Dollywaggon Pike (3.1).

Map 261	131	WCY

S. repens L.   Creeping willow

Rather uncommon but locally frequent as, for example, on the Bewcastle and Gilsland moors in the north and in the Crosby Ravensworth - Great Asby limestone area of north Westmorland. It grows in short vegetation, in somewhat open ground and over a wide range of pH, occurring on hillsides, roadside verges, in lowland fens and mosses but probably nowhere more luxuriantly than in the dune-slacks around the Duddon estuary. Some of the plants from there have the silvery leaves characteristic of var. *argentea* (Sw.) Wimm. & Grab. Similar plants occur in other populations, for example, on Newton Reigny Moss (46.30) and Greystoke Moor (40.28, RWMC, 1994, LANC).

Alt. limit: *c*.420 m, upper Swindale, Brough (80.18)

Map 262	164	WFCY

S. lapponum L.   Downy willow

This attractive woolly willow still occurs on rock ledges in three of the eastern coves of Helvellyn (3.1) where it

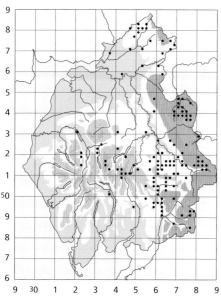

Map 261. *Salix phylicifolia*

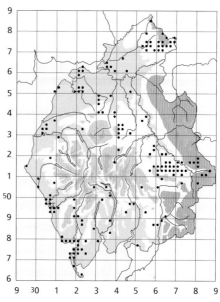

Map 262. *Salix repens*

was first found by J. Backhouse in 1872 (E).

Although quite widespread in the Scottish Highlands, this northern willow is known elsewhere in Britain only from Helvellyn, the Moffat Hills and Galloway.

Alt. range: 810-870 m, Helvellyn (3.1)

S 2   W

S. herbacea L.   Dwarf willow; Least willow
Not infrequent in the Lake District, around the summits and on ridges and stony plateaux, and with outlying sites on Cautley Crag (68.96) and Great Comb, Dentdale (68.82s). It became extinct on Cross Fell (6.3) possibly early this century. It has probably declined in the Lake District, yet in a few places, notably on the summit of St Sunday Crag (36.12), it still occurs in some abundance. Not surprisingly summit plants are usually extremely stunted in comparison with those growing on sheltered ledges. The only Furness site is on Dow Crag (26.96, R.J. Birkett, 1959; LANC, *Proc. BSBI* (1963) 5:34). There are no Survey records to confirm the earlier records from Dale Head (2.1, J.H. Vine-Hall, 1938) and Eel Crags (2.1, M. Johnstone, 1952).

Outside the Lake District and the Scottish Highlands this willow is now known in Britain only from the Cheviots, Ingleborough and a few sites in southern Scotland and Wales.

Alt. range: 640 m, Cautley Crag (68.96) - 945 m, Scafell Pike (20.06)

Map 264   48   WFCY

[S. reticulata L.
Baker refers to two doubtful records from the vicinity of Brothers Water (3.1/4.1) and Greystoke (4.3). In the absence of specimens these must be considered errors. This willow is restricted in Britain to the Scottish Highlands.

R   (W)]

BRASSICACEAE
(*CRUCIFERAE*)

*Sisymbrium irio L.   London rocket
Recorded by Hodgson from Cockermouth Castle (1.3), Silloth (1.5) and High Head Castle, Ivegill (4.4). The only other records are from Silloth docks (N.W. Simpson, 1949) and the valley of the River Roe (3.4, J. Curwen, *c.*1890s, CLE), although this may refer to High Head Castle.

R   (C)

*S. altissimum L.   Tall rocket
A very rare but persistent urban alien known from Silloth docks and the immediate neighbourhood (1.4, 1.5), the Carlisle area (3.5, 3.6, 4.5) where it occurs on the city walls locally intermixed with *Sisymbrium orientale*, and from the railway stations at Coniston (28.96), now disused, and Kendal (50.92), the latter being the only Westmorland record.

There are additional *Atlas* records for the Workington (9.2), Sedbergh (6.9) and Barrow-in-Furness (1.6s, 2.6s) squares, where it was first recorded by W.H. Pearsall in 1913 (*BEC Rpt* (1914) 3:402), and others from Melmerby (6.3, W.W. Mason, 1917, 1924, *BEC Rpt* (1925) 7:556) and Thwaites, Millom (1.8, M. Cross, 1936-1949).

Map 265   10   WFC(W/Y)

*S. orientale L.   Eastern rocket
A common plant of waste ground on Walney Island and around Barrow-in-Furness, also along the west coast from Workington to Silloth. Elsewhere it occurs only as an occasional casual.

The only pre-Survey records are from Workington (9.2, W. Hodgson, 1894-1897), Silloth (1.5, T.S. Johnstone 1909; C.W. Muirhead 1932-1943), Melmerby, (6.3, W.W. Mason, 1924, *BEC Rpt* (1925) 7:556),

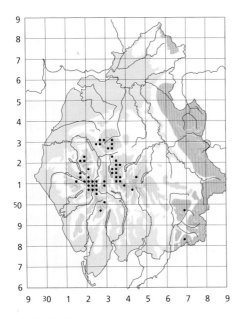

Map 264. *Salix herbacea*

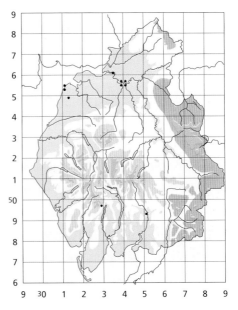

Map 265. *Sisymbrium altissimum*

Thwaites, Millom (1.8, M. Cross, 1937-1949) and the docks at Barrow-in-Furness (1.6s/2.6s, W.H. Pearsall, *BEC Rpt* (1914) 3:402). The first Westmorland record was of a brief occurrence at Burton-in-Kendal (52.76s, GH, 1973, LANC, *Watsonia* (1974) 10:175).
Alt. limit: 140 m, Long Marton, (66.24)
Map 266   32   WFC

S. officinale (L.) Scop.   Hedge mustard
Frequent in the lowlands, chiefly in towns and villages, where it occurs on waste ground, along roadsides and in open spaces; it also occurs commonly by farms.
Alt. limit: 315 m, north-west of Garrigill (72.42)
Map 267   672   LWFCY

***Descurainia sophia** (L.) Webb ex Prantl   Flixweed
A very rare casual occurring around the Kent estuary, on the west coast at Sellafield (02.02),.by the Solway Firth and around Carlisle, rarely further inland.

Hodgson recorded it from the Derwent Tinplate Works, Workington (9.2), Maryport (0.3) and Silloth docks (1.5, E.J. Glaister, 1877, CLE), and also from Penrith (5.3). It was seen by W.W. Mason at Bitterlees, Silloth in 1921 and by C.W. Muirhead (CLE; PLYP) in some abundance at Silloth in 1937 and 1939. There are no subsequent records from that area or from the South Walney (1.6s, G. Wilson, 1950s) and Rampside (2.6s) squares, for which there are post-1930 *Atlas* dots. These latter are the only Furness records.
Alt. limit: 260 m, west of Wan Fell (50.36)
Map 268   9   W(F)C

Alliaria petiolata (M. Bieb.) Cavara & Grande
                Hedge garlic; Jack-by-the-hedge
A common spring-flowering plant of roadsides, banks, hedgerows and waste ground throughout the lowlands.
Alt. limit: 535 m, south of Garsdale Head (78.88)
Map 269   1080   LWFCY

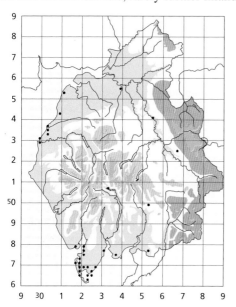

Map 266. *Sisymbrium orientale*

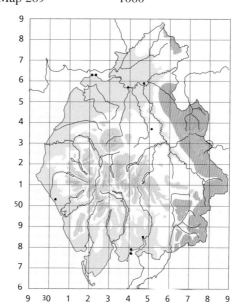

Map 268. *Descurainia sophia*

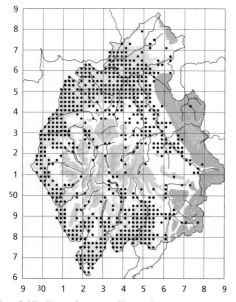

Map 267. *Sisymbrium officinale*

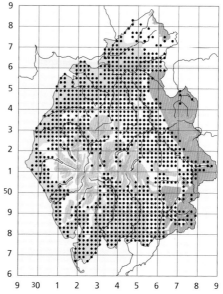

Map 269. *Alliaria petiolata*

Arabidopsis thaliana (L.) Heynh.   Thale cress
Generally frequent but less so in west Cumberland and on the Solway Plain. This is an inconspicuous weed of gardens and urban sites but it is also frequently found on wall-tops, in rather dry open ground on hedgerow banks and occasionally on rock ledges in the fells.
Alt. limit: 550 m, Kentmere (44.08)
Map 270   740   LWFCY

*Bunias orientalis L.
Well established in rough grassland by the River Eden at Carlisle (40.56, REG, 1978, LANC, *Watsonia* (1979) 12:349; 38.56, EHR, 1978) and later found nearby by the River Petteril (40.54, A. Frank & PT, 1984) a few years later.

    The British distribution of this species is centered on the Thames valley and the south-east, but it has a wide distribution as a casual.
                                                  3   C

*Myagrum perfoliatum L.
Recorded only from Silloth (1.5, T.S. Johnstone, CLE, *BSBI News* (1991) 58:38) in 1901 and as a garden weed at Dalton-in-Furness (2.7, D. Lumb, YRK, *BEC Rpt* (1918) 5:210) in 1917.
                                                  (FC)

*Erysimum cheiranthoides L.   Treacle mustard
A very rare casual. Of the ten Survey records, only that from Arnside (46.76s) coincides with any of the earlier ones.

    It is listed by Hodgson from Workington (9.2) and Abbeytown (1.5) and there are other Cumberland records from Silloth (1.5, J. Leitch, 1884, J. Parkin 1939), Etterby, Carlisle (3.5, J. Leitch, 1873) and Braithwaite marshes (2.2, G.A.K. Hervey, 1955). The six *Atlas* dots include the first records for Furness: Walney Island (1.6s, G. Wilson, 1954) and the adjacent mainland (2.6s), and for Westmorland possibly the

Arnside square (4.7s, v.c.60/69).
Alt. limit: 150 m, Appleby by-pass (70.18)
Map 271   10   WFC

*E. cheiri (L.) Crantz   Wallflower
(*Cheiranthus cheiri* L.)
Rather scarce but perhaps recorders may be inclined to disregard records from urban sites. It occurs as a garden escape on walls and waste ground, on the walls of Calder Abbey (04.06. CCH, 1978), Furness Abbey (20.70, Anon., 1970s) and Millom Castle (16.80, RS, 1980), also on limestone rocks at Castlehead, Grange-over-Sands (42.78s, FLW, 1974) and on the sandstone cliffs at Rockcliffe (34.60, EHR, 1970s).

    There are additional records in the Floras from the walls of several castles, from the coast between Silverdale and Arnside (4.7s, v.c.60/69) and on walls high up in the Pennines at Alston (7.4). It was also recorded from the walls of Lanercost Priory (5.6) where it persisted until at least 1945.
                                                  5   WFC

*Hesperis matronalis L.   Dame's violet
An increasingly common and colourful garden-escape in the north of the county and the Eden valley but still rather infrequent in the south. Although many of the records are from roadsides and not far from houses, it is abundant around a gravel island in the River Lune above Kirkby Lonsdale (60.78s), and it is also frequent along the Eden and in damp woodland in the north. The first Furness records are those shown in the *Atlas*.
Alt. limit: at least 280 m, south of Alston (70.44)
Map 272   250   WFCY

*Malcolmia maritima (L.) R. Br.
An eastern Mediterranean plant found by A. Cannell on disturbed ground at Skinburness (12.56) in 1991 but otherwise recorded only from Silloth in 1939 by C.W.

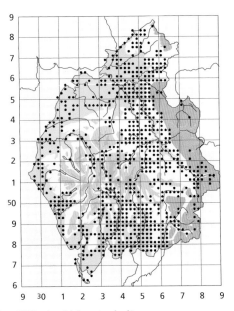

Map 270. *Arabidopsis thaliana*

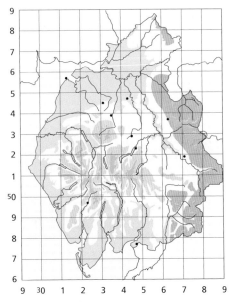

Map 271. *Erysimum cheiranthoides*

Muirhead (1.5, CLE, *BSBI News* (1991) 58:38).

C

*M. africana (L.) R. Br.
A southern European species found in 1990 at Askam-in-Furness (20.76, PB, LANC, *BSBI News* (1991) 58:38) where it had earlier been collected by G. Wilson (1951, CLE).

1 F

*M. crenulata Boiss.
Found at Silloth by J. Leitch in 1890 (1.5, CLE, *BSBI News* (1991) 58:38). This is the only substantiated British record for this casual.

(C)

*Matthiola longipetala (Vent.) DC. Night-scented stock
A few plants of this southern European species were found on the foreshore at Mawbray Bank, Allonby in 1986 (08.48, REG, LANC). The only previous record is from gravel beds in the River Eden at Grinsdale, Carlisle (3.5, T.S. Johnstone, 1902, CLE, *BSBI News* (1991) 58:38).

1 C

*Chorispora syriaca Boiss.
A extremely rare alien collected at Silloth in 1889 and 1890 (1.5, J. Leitch, CLE).

(C)

Barbarea vulgaris R. Br.　　　　　　Winter cress
A frequent and conspicuous spring-flowering ruderal over most of the lowlands, occurring in waste places and car parks, on dumps, railwaysides and recently disturbed roadside verges, occasionally also in arable fields and quite frequently along stream and riversides.
Alt. limit: *c.*380 m, north-east of Alston (72.48)
Map 273　　　　　494　　　　　WFCY

[B. stricta Andrz.
The only records for this rare casual are two undated ones in the *Atlas* for the South Walney (1.6) and Rampside (2.6s) squares. No specimens are known and the identifications may well be erroneous.

(F)]

*B. intermedia Boreau
Rather rare, particularly in the south, but certainly more widespread than the Floras and the *Atlas* suggest. It occurs in similar habitats to *Barbarea vulgaris*. The first Furness record is that of D. Lumb's from the railway at Askam-in-Furness (2.7, 1914, BM).
　　　Although often confused with the last and the following species, it is characterised by distinctly pinnate upper stem leaves, with a narrow terminal lobe, pale yellow flowers and relatively short, fat fruits.
Alt. limit: *c.*340 m, north of Alston (70.48)
Map 274　　　　　46　　　　　WFC

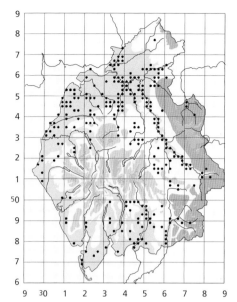

Map 272. *Hesperis matronalis*

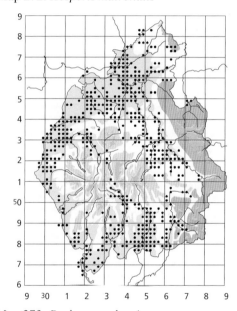

Map 273. *Barbarea vulgaris*

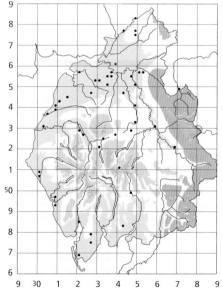

Map 274. *Barbarea intermedia*

*B. verna (Miller) Asch.

Rare and well scattered in the lowlands, occurring in similar habitats to *Barbarea vulgaris*. There is a single strand-line record from Rampside (22.66s) and two from gardens but the latter may be derived from previous cultivation as 'winter cress'.

This species is distinguished by the conspicuously pinnate upper stem-leaf, the large, deep-yellow flowers, and the long (> 3.5 cm), narrow, arcuate fruits.
Alt. limit: 280 m, Ravenstonedale (74.02)
Map 275   19   WFC

Rorippa nasturtium-aquaticum (L.) Hayek sensu lato
(*Nasturtium officinale* R. Br.)   Water-cress
Map 276   895   LWFCY

R. nasturtium-aquaticum sensu stricto
Common in the lowlands occurring in ditches, rivers, ponds and tarns but preferring running and nutrient-

rich or calcareous water.

None of the three segregates is identifiable in the absence of ripe fruit but this is probably the commonest, although it is likely to have been recorded sometimes in error for the aggregate. It is interesting, however, that Allen (1984) states that in the Isle of Man the commonest of the three is the hybrid while, in Co. Durham, Graham (1988) says it is *Rorippa microphylla*.
Alt. limit: 550 m, Moor House (74.32)
Map 277   168   LWFCY

R. nasturtium-aquaticum x **R. microphylla**
R. x sterilis Airy Shaw
Scattered, mainly in the Eden valley and in west Cumberland but no doubt under-recorded. There are no published records prior to the *Critical Atlas*.

T.C.G. Rich has drawn attention to the difficulty of distinguishing this sterile hybrid from plants of the

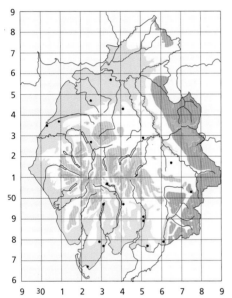

Map 275. *Barbarea verna*

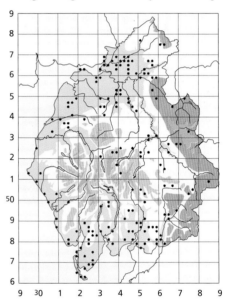

Map 277. *Rorippa nasturtium-aquaticum* sensu-stricto

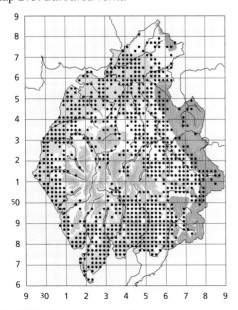

Map 276. *Rorippa nasturtium-aquaticum* sensu lato

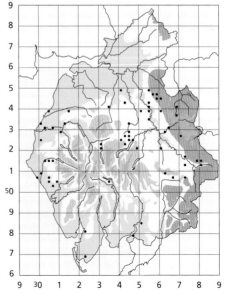

Map 278. *Rorippa nasturtium-aquaticum* x
R. microphylla

parent species with poorly-formed pods.
Alt. limit: 290 m, Renwick (60.44)
Map 278   61   WFC

R. microphylla (Boenn.) N. Hylander
(*Nasturtium microphyllum* (Boenn.) Reichb.)
Scattered in the lowlands in similar habitats to
Rorippa nasturtium-aquaticum. It is less common but
it is certainly under-recorded. There appear to be no
published records prior to the *Critical Atlas*.

It can be reliably distinguished only by its
narrower pods and more finely pitted seeds, with 12-18
depressions across their width.
Alt. limit: *c*.260 m, Croglin (68.36)
Map 279   77   WFC

R. palustris (L.) Besser
Fairly frequent on the Solway Plain and over most of
the Eden valley and quite common in the south of the
county. It is most frequently found in open habitats by
the cattle-poached margins of mesotrophic becks, tarns
and pools, sometimes occurring in abundance when the
latter dry out. It is, for example, particularly noticeable
in the draw-down zone at the head of Haweswater in
times of drought. There is a single record of its occurrence as a garden weed. Judging from the paucity of
records in the Floras, this species has increased
considerably this century. There are no published
records for Furness prior to the *Atlas*.

The seeds may well be widely dispersed by birds.
The association, for example, of an abundance of this
species around a small tarn east of Kendal with a large
colony of nesting black-headed gulls can hardly be
coincidental.

There have been no reports of the related but
much rarer *Rorippa islandica* (Oeder) Borbás which
curiously is the only one of the pair represented on the
Isle of Man.
Alt. limit: at least 270 m, north of Sunbiggin Tarn
(66.08)
Map 280   176   LWFCY

R. sylvestris (L.) Besser
Common in the middle and lower Eden valley,
otherwise occasional or rare. It is typically a plant of
muddy, silty banks of dykes, ditches and rivers, and
also of river shingle, but there are several records of it
occurring as a casual. It is, for example, abundant as a
nursery weed near Crooklands, Milnthorpe (52.82s).
There are no Survey records for the Walney Island and
Rampside squares (1.6s, 2.6s, G. Wilson, 1950s).
Alt. limit: 270 m, near Ravenstonedale (74.02)
Map 281   132   WFC(Y)

*R. amphibia (L.) Besser
Reported by W.H. Pearsall from the docks at Barrow-in-
Furness in 1913 (1.6s/2.6s) *BEC Rpt* (1914) 3:383).

(F)

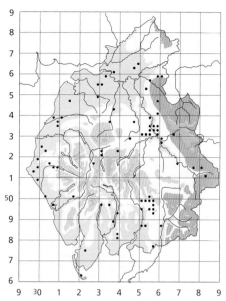

Map 279. *Rorippa microphylla*

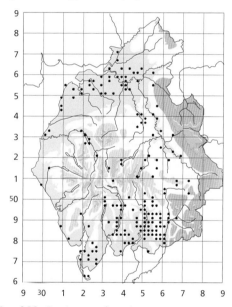

Map 280. *Rorippa palustris*

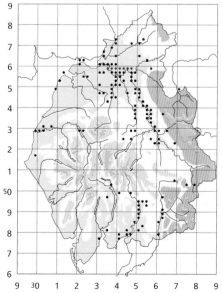

Map 281. *Rorippa sylvestris*

*Armoracia rusticana P. Gaertner, Meyer & Scherb.

Horse-radish

An occasional garden throw-out and, like *Lunaria annua*, seldom seen far from houses. It usually occurs on roadsides and waste ground, although able to persist and spread in denser vegetation than that species. It appears to be increasing: Hodgson, for example, gives only three localities and Wilson one.

Alt. limit: *c.*300 m, south of Alston (70.44)

Map 282   99   WFCY

*Cardamine bulbifera (L.) Crantz   Coral-wort

A rare plant of south-east England once recorded, no doubt as a garden escape, from Bowness (4.9, W. Clitheroe, 1940, YRK).

S   (W)

*C. trifolia L.

Recorded in 1972 from a plantation east of the River Lune near Underley Hall, Kirkby Lonsdale (60.80s, J.A. Wood) where it was originally found by T.J. Foggitt in 1926 (BM). Subsequent searches of the wood have been unsuccessful.

This central European mountain plant was first recorded in 1903 as well-naturalised in woods at Gill Foot, Egremont (0.1, J. Adair, BM, K, LANC, YRK, *BEC Rpt* (1904) 2:9). It was later collected at Windermere by F. Long in 1916 (BM) and noted as naturalised in the grounds of Bonny Hall, Alston (7.4, G. Bolam, 1924).

1†   W(C)

C. amara L.   Large bitter-cress

Frequent in the lowlands in the north and south, elsewhere rather scattered. It grows usually in flushes in wet woodland and on shady riversides, occasionally in marshy fields. Of particular interest is its presence, first noted by RWMC (Corner 1990), in seepage areas and around springs in the uplands, especially in the Pennines but also in the eastern Lake District and the Howgill Fells. Common associates in these bryophyte-rich communities are *Cochlearia officinalis, Epilobium alsinifolium, Myosotis stolonifera* and *Saxifraga stellaris*. The plants rarely flower in such sites, but they are distinguishable from *Cardamine pratensis*, with which they commonly grow, by the larger terminal leaflet and the generally more undulate leaflets. There appears to be no previous report in Britain of the occurrence of this species in upland flushes although there is an old record at 455 m from the Ochil Hills, Perthshire, where RWMC has recently refound it at 640 m.

Alt. limit: 540 m, Kirkdale (66.32)

Map 283   511   LWFCY

*C. raphanifolia Pourret

Almost restricted to the Ambleside - Grasmere (3.0) area where it is thoroughly naturalised in marshes, by rivers and along lake margins. Its large mauve flowers

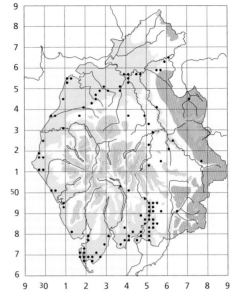

Map 282. *Armoracia rusticana*

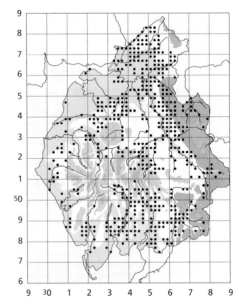

Map 283. *Cardamine amara*

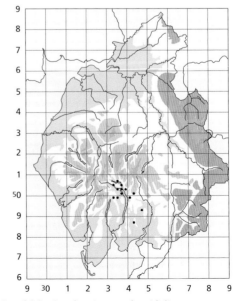

Map 284. *Cardamine raphanifolia*

render it very conspicuous in late spring. The Wither-slack (42.86s) record is of plants naturalised in a woodland estate.

The earliest records of this southern European mountain plant are from Loughrigg (3.0, H. Taylor, 1932, K) and Waterhead (3.0, B. Allen, 1932, BM).
Map 284   13   WF

C. pratensis L.

Cuckoo flower; Lady's smock; Milkmaid
Very common, absent only from the more intensively farmed areas and the western fells of the Lake District. In the lowlands it is typically a plant of damp fields and roadside verges, of marshy ground and ditches. In the uplands it occurs in *Juncus effusus* flushes and at higher altitudes it is a characteristic member of bryo-phyte-rich springs where it seldom flowers.
Alt. limit: 850 m, Cross Fell (68.34)
Map 285   1658   LWFCY

C. impatiens L.

Rare, known only from a cluster of sites by the Kent estuary and isolated ones on a wall at Keswick (26.22), by a field path at Gowbarrow (40.20) and on the lime-stone pavement of Fell End Clouds, Ravenstonedale (72.98, 72.00, 74.00). The estuary sites are quite varied: by the sides of tracks and roads, in a hedge-bank and actually on the foreshore, but all are on the limestone. The only records additional to those in the Floras are from Silloth (10.52), where it was seen by J. Parkin in 1935, and the Kendal square (5.9, E. Hyde, 1954), and *Scarce Plants* post-1930 records for the Barbon (6.8s) and Sedbergh (6.9) squares.
Alt. limit: *c.*430 m, Fell End Clouds (72.98)
Map 286   S 12   WCY

C. flexuosa With.

Wavy bitter-cress
Very common and probably under-recorded. It prefers damper habitats than the following species: marshes, wet meadows, hedgerows, riversides, ditches and shelt-ered mountain gullies. It also occurs as a garden weed and is then particularly likely to be confused with *Cardamine hirsuta.*
Alt. limit: 760 m, Knock Fell (70.30)
Map 287   1528   LWFCY

C. hirsuta L.

Hairy bitter-cress
Very common in the lowlands and preferring drier habitats than the last species with which it is often confused. The most reliable means of identification is the prevalence of flowers with four stamens. It occurs most frequently as a garden weed but it can also be found on wall tops, open sandy ground and dry limestone scars and pavements.
Alt. limit: 610 m, east of Cross Fell (70.34) and above Nenthead (78.44). Wilson gives it as 745 m on Great Dun Fell (7.3), substantially higher than his figure for *Cardamine flexuosa.*
Map 288   1070   LWFCY

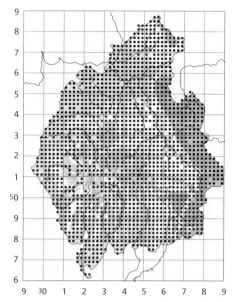

Map 285. *Cardamine pratensis*

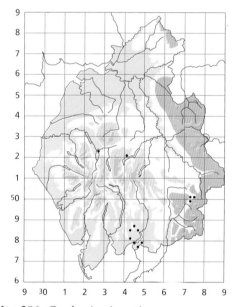

Map 286. *Cardamine impatiens*

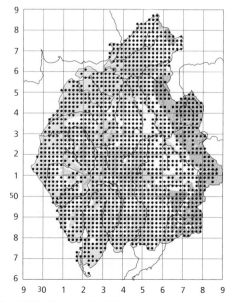

Map 287. *Cardamine flexuosa*

[Arabis petraea (L.) Lam.
(*Cardaminopsis petraea* (L.) Hiit.)
Baker lists, without comment, records from the Wast Water
Screes (1.0) and Scafell (2.0). These are best regarded as
errors in the absence of herbarium material or subsequent
confirmation. This species is known in Britain only from
Snowdonia and the Scottish Highlands.

S   (C)]

Arabis glabra (L.) Bernh.   Tower mustard
An extremely rare casual and apparently never persisting. The only Survey records are from Middlebarrow quarry, Arnside (46.76s, MB, 1974), Brigsteer
(48.88s, FLW, 1974), Jenkyn's Crag, Ambleside (38.02,
L. Holmes, 1977) and a roadside bank at Cliburn
(58.24, W.F. Davidson, 1983, LANC).

It is listed by Hodgson from Stainburn,
Workington (0.2), near Cockermouth and Isel Bridge
(1.3) and Silloth (1.5). There are later Cumberland
records from between Edenhall and Great Salkeld (5.3,
B. Lovejoy, 1901) and Great Salkeld in 1910 and an
undated and dubious one from "gills and walls about
Alston" (7.4, G. Bolam). Wilson lists it from by the
River Lyvennet, Cliburn (6.2, A. Sprott & A. Wilson,
1936, YRK), Temple Sowerby and Appleby (6.2,
D. Oliver, 1886, K).
Alt. limit: 150 m, Ambleside (38.02)

S 4   W(C)

*A. caucasica Willd. ex Schldl.   Garden Arabis
A very rare garden escape, occurring on walls, roadsides and quarries. The first record for Cumberland of
naturalised plants is from Alston (70.46, REG, 1978),
for Westmorland from a limestone quarry at Levens
(48.86s, M. Andrew, 1984, LANC, *Watsonia* (1985)
15:393) and for Furness from Spark Bridge (30.84, JA,
1996).
Alt. limit: 300 m, Alston (70.46, 72.46)
Map 289   9   WFC

A. hirsuta (L.) Scop.   Hairy rock-cress
Frequent on the limestone in the south and east, elsewhere infrequent to scarce. It is a plant of limestone
screes, pavement and scars, also of base-rich rock
ledges and gullies in the Lake District and, very
noticeably from the map, along the middle section of
the River Lune (6.9) where it grows on Silurian rock
outcrops. In the lowlands its usual habitats are
mortared walls and bridges.
Alt. limit: 755 m, Knock Ore Gill (70.30, Wilson)
Map 290   309   LWFCY

*Aubrieta deltoidea (L.) DC.   Aubretia
This popular garden plant was recorded as particularly
well-established and colourful on the ruins of Pen-
dragon Castle, Mallerstang (78.02, JCF, 1970s) and
Brougham Castle (52.28, GH, 1981) and on the nearby
bridge over the River Eamont. A single flourishing
plant was seen in a disused quarry near Hallbankgate
(58.56, MG, 1988) and there is a record from the

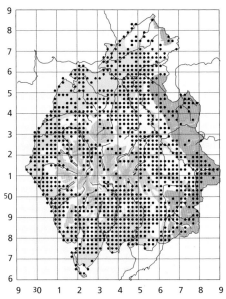

Map 288. *Cardamine hirsuta*

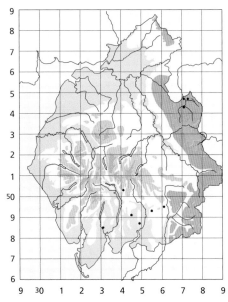

Map 289. *Arabis caucasica*

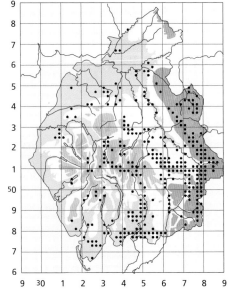

Map 290. *Arabis hirsuta*

Barrow-in-Furness steelworks (18.70, PB, 1989).

4   WFC

*Lunaria annua L.   Honesty
An occasional garden escape and chiefly in the south. It is seldom seen far from houses or other than by roadsides and on waste tips. The only pre-Survey record is an undated one from Beaumont, Carlisle (3.5). The first ones for Westmorland and Furness appear to be from Ambleside (36.02, R. Young, 1975, AMDE) and Rusland (34.88, JA, 1977) respectively.
Alt. limit: 250 m, Reagill (60.16)
Map 291   128   WFCY

*Alyssum alyssoides (L.) L.   Small alison
Known only from Silloth saltworks (1.5, J. Leitch, 1889, CLE), by the River Eden at Etterby, Carlisle (3.5, T.S. Johnstone, 1901, CLE) and at Alston (7.4, G. Bolam, 1919).

This casual species has now all but disappeared from most of England.

R   (C)

[A. serpyllifolia Desv.
A south-west European species listed by Hodgson from Abbey Holm, Abbeytown (1.5). The genus is taxonomically difficult and in the absence of a specimen the record is best considered dubious.

(C)]

*Berteroa incana (L.) DC.
Listed by Hodgson from the Derwent Tinplate Works, Workington (9.2), Maryport and Flimby (0.3) and Silloth (1.5, J. Leitch, 1889, BM, CLE), and by Handley (1898) from near Tebay (6.0). In 1913 it was found by W.H. Pearsall at Barrow-in-Furness docks (1.6s/2.6s, BM, *BEC Rpt* (1914) 3:402) and there are additional Cumberland records from by the River Eden at Great Salkeld (5.3, H. Britten, 1900-1908) and from Silloth where it was seen by C.W. Muirhead from 1934 (PLYP) until 1938.

This species is widespread in Europe although native only in the centre and east.

(WFC)

*Lobularia maritima (L.) Desv.   Sweet alison
Like *Lepidium sativum*, an ephemeral escape from cultivation. The only pre-Survey records are Baker's from Grange-over-Sands (4.7s) and one from Carlisle waterworks (4.5) where it was seen by J. Parkin in 1935. The ten Survey records indicate its increasing popularity as a garden plant. The first Westmorland record was from Knock (68.26, CW *et al.*, 1980).
Alt. limit: 200 m, Knock (68.26)
Map 292   10   WFC

Draba incana L.   Hoary whitlow-grass
Common on the limestone of the northern Pennines, less so further south and now increasingly rare on the Asby - Orton pavements. It occurs also on sandstone south-east of Nateby (80.02), in a few places on basic rocks in the eastern Lake District and there are outlying sites on Coniston Old Man (26.96, 26.98).

On sheltered, damp, limestone ledges in the Pennines the plants can be remarkably luxuriant - up to 40 cm high in a gill on Stainmore.
Alt. range: *c.*300 m, Longsleddale (48.06) - 750 m, Green Castle, Great Dun Fell (70.30)
Map 293   76   WFCY

Draba muralis L.   Wall whitlow-grass
Rather rare. Most sites are in the lowlands where it grows on dry banks, on wall-tops and as a garden weed, but there are several from the uplands: on a wall in Alston (70.46), from which it has recently disappeared, as a recent and locally abundant colonist of forestry tracks at Kershope and Spadeadam (4.7n, 6.7n), and on a small, lightly wooded limestone scar by the upper River Belah on Stainmore (84.08, JMA, 1987; LANC) where it was growing with *Arabidopsis thaliana* and *Erophila verna*.

There are Cumberland records early this century from a nursery garden at Stanwix, Carlisle (3.5/3.4, T.S. Johnstone, 1902, CLE), Wigton (2.4, J. Parkin,

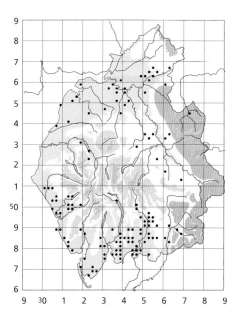

Map 291. *Lunaria annua*

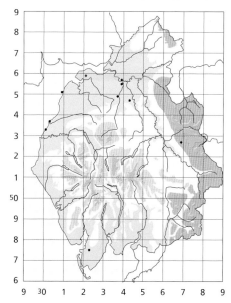

Map 292. *Lobularia maritima*

1931) and Penrith (5.2/5.3, N.M. Stalker, 1949; CLE) and further *Scarce Plants* records from the Brampton (5.6) and Arnside (4.7s, v.c.60/69) and Kershopefoot (4.8n, v.c.70/80) squares. Wilson records it only from limestone rocks by the Augill Head waterfall, near Brough (8.1, 1937, YRK), but it has not been refound there, and Handley gives records from Garsdale and Dent dating from 1666! The only Furness records are from Torver (28.94, T.G. Tutin, 1973, LTR) and a garden at Allithwaite (38.76s, FLW, 1970s).

Whether or not all the lowland sites represent non-native, casual occurrences, the species appears quite native on the Pennine limestone. The native distribution of this species is largely confined to the Mendips, Cotswolds, the Peak District and the Craven area of Yorkshire.

Alt. limit: 375 m, South Stainmore (84.08)

Map 294   S 12   WFC(Y)

Erophila verna (L.) DC. sensu lato   Whitlow-grass
Well distributed in the lowlands but, like *Arabidopsis,* thinning out on the Solway Plain and parts of the west coast. It occurs characteristically on wall-tops, pathsides, sand-dunes, waste ground, railway tracks and rock ledges, also on shallow open ground and in quarries. The maps of the two segregate species are based on material collected during the Survey and identified by T.T. Elkington.

Alt. limit: 770 m, Great Dun Fell (70.30)

Map 295   563   LWFCY

E. verna (L.) DC. sensu stricto
This is clearly the commoner of the two segregates in Cumbria and its distribution is no doubt essentially that of the complex. The records from sand-dunes by the Duddon estuary (1.7) include some of var. *praecox* (Steven) Diklic. (subsp. *spathulata* (Láng) Vollm.), characterised by fruits only twice as long as wide.

Map 296   34   WFCY

E. glabrescens Jordan
Apparently infrequent but easily overlooked. This segregate is distinguished by the sparse pubescence of the leaves and the lower part of the flowering stem, and by the longer petioles.

Map 297   11   WFCY

Cochlearia anglica L. and **C. officinalis** L.
English and Common scurvy-grass
Very common around the entire coast but especially in the estuaries where it is a conspicuous feature in late spring. The typical habitat is on the upper parts of the salt-marshes, also by creeks and among rocks.

During the course of the Survey it became apparent that *Cochlearia anglica* and *C. officinalis* are rarely distinguishable. In 1987 T.C.G. Rich investigated the Cumbrian and Lancashire Morecambe Bay populations, as well as others around the Irish Sea, and concluded that it was impossible to find convincing

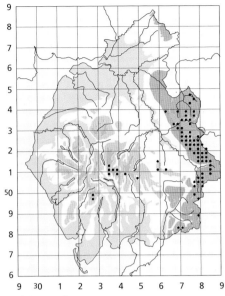

Map 293. *Draba incana*

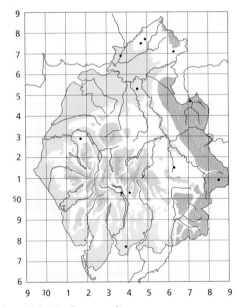

Map 294. *Draba muralis*

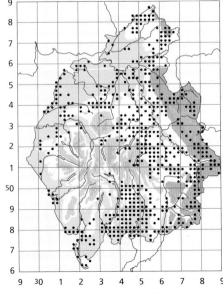

Map 295. *Erophila verna* sensu lato

specimens of either. He subsequently studied material from the west coast and the Solway, also from North Wales, and he thinks it likely that the plants may be of hybrid origin or even belong to another unnamed member of the *C. officinalis* complex. However, subsequent collections from mainly non-estuarine sites along the west coast between Walney Island (1.6s) and Allonby (0.4) and examined by D.H. Dalby are almost all typical *C. officinalis*.
Map 298 (incl. *C. pyrenaica*)   391   WFC

C. pyrenaica DC.
subsp. **pyrenaica**
Frequent in the fells although rare in the northern and north-western parts of the Lake District. It is a plant of wet gullies, bryophyte flushes and seepage areas, often growing with *Montia fontana*, *Saxifraga stellaris* and *Chrysosplenium oppositifolium*. Such populations have given rise to downstream colonies, often at some considerable distance as, for example, along the middle section of the River Lune (6.9), but nowhere have they reached the coast.

These montane plants have the typical, glossy, orbicular basal leaves of *Cochlearia officinalis*. The two species cannot be reliably distinguished morphogically, although *C. pyrenaica* is diploid (2n=12) and the other tetraploid. The only two chromosome counts from Cumbria (Gill *et al.* 1978) are from Helvellyn (3.1) and Great Dun Fell (7.3). The situation is further complicated by the existence of tetraploid montane plants elsewhere in Britain which are also almost indistinguishable morphologically. These are treated by Rich (1991) as subsp. *alpina* (Bab.) Dalby.
Alt. limit: 870 m, Helvellyn (34.14, Wilson)
See Map 298   WFCY

C. danica L.   Danish scurvy-grass
A generally early-flowering but easily overlooked winter annual of open habitats along most of the coast occurring in sand-dunes, on sea-walls, in the drift-line and, in Morecambe Bay, on the stonework of the Arnside viaduct. This last record was the only one for Westmorland (46.78s, MB, 1978; LANC) until the plant was discovered in 1990 and in subsequent years by the edges of trunk roads and particularly along the edges of the central reservation of the M6. The species is rapidly colonising this motorway habitat in Britain and is readily spotted both in flower and fruit. It was also seen in 1996 in Burton-in-Kendal (52.76s).
Alt. limit: 250 m, west of Firbank Fell (58.92)
Map 299   86   WFC

*Camelina sativa (L.) Crantz
Listed by Hodgson from Whitehaven (9.1), Workington (9.2), Stainburn (0.2), Silloth (1.5, J. Leitch, 1882, CLE), Thackthwaite and Watermillock (4.2), and collected from Stoneraise, Westward in 1868 (2.4, J. Wood, BM). The only records this century are from Melmerby (6.3, W.W. Mason, 1918, *BEC Rpt* (1919) 5:368), Blaithwaite, Wigton (2.4,

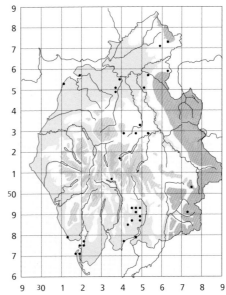

Map 296. *Erophila verna* sensu stricto

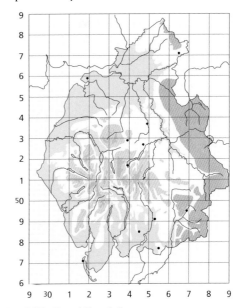

Map 297. *Erophila glabrescens*

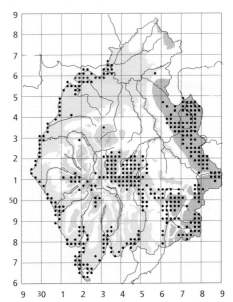

Map 298. *Cochlearia officinalis*, *C. anglica* and *C. pyrenaica*

J. Parkin, 1935) and Silloth (C.W. Muirhead, 1939).

(C)

*Neslia paniculata (L.) Desv.

A central and southern European species reported by Hodgson from Maryport (0.3) and Silloth (1.5, J. Leitch, 1889, CLE), where it was later seen by C.W. Muirhead in 1939 (CLE). It was also reported in 1913 by D. Lumb from Dalton-in-Furness (2.7, *BEC Rpt* (1914) 3:403).

The Silloth plants are of *Neslia paniculata* sensu stricto (*BSBI News* (1991) 58:37).

(FC)

Capsella bursa-pastoris (L.) Medicus

Shepherd's purse

A generally common weed of gardens, arable fields, waste places, disturbed ground and farmyards.

Alt. limit: 780 m, Knock Fell (70.30)

Map 300   1396   LWFCY

Hornungia petraea (L.) Reichenb.

Although listed by Watson (1883) for v.c.69, the only authenticated and localised records are from amongst the shattered fragments of limestone pavement and low scars above Raisbeck, Orton (6.0, *Proc. BSBI* (1963) 5:29), where it occurs, sparingly, at two sites within one tetrad. It was originally found here by H. Milne-Redhead in 1961. This locality is somewhat isolated from the Craven area, the headquarters of this species in northern England.

Despite its altitude, the populations flower very early, sometimes among the melting snow and once observed on New Year's Day.

Alt.: 395 m, above Raisbeck (6.0)

S 1   W

*Bivonaea lutea (Biv.) DC.

A native of Sicily collected by Hodgson at Silloth (1.5) in 1890.

(C)

Teesdalia nudicaulis (L.) R. Br.

A rather rare winter and spring annual largely restricted to the Lake District and the coast of southwest Cumberland, especially the dunes of the Ravenglass estuary (0.9). It occurs in rather well-drained, open sandy ground, by paths and among scree and boulders. It has proved impossible to refind the Longsleddale (4.0, J.A. Martindale, 1888, KDL) and Kentmere (4.0) sites listed by Wilson, or Martindale's 1889 Langdale Pikes locality (2.0, KDL) or to confirm nearly half of the post-1930 *Atlas* records. Its presence in the Lake District is interesting in view of the widespread belief that this is a lowland species. As with many annuals, there is considerable annual variation in the size of the populations.

Alt.` limit: 455 m, Wasdale Screes (16.04), although Eddy & Welch (1967) mentions a record early this century from the source of the River Tees (7.3), which would be about 700 m, and there is a post-1930 *Atlas* record for this square. These last records seem very dubious.

Map 301   46   WC

Thlaspi arvense L.   Field penny-cress

A casual of waste sites, roadsides and disturbed ground throughout the lowlands and only occasionally occurring as an arable weed. It sometimes appears in abundance on recently disturbed or tipped soil and during the latter part of the Survey it was briefly conspicuous along long sections of the verges and central reservation of the M6 following extensive roadworks.

Alt. limit: 310 m, by the M6, Shap summit (58.12)

Map 302   127   WFC

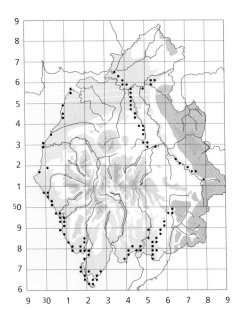

Map. 299. *Cochlearia danica*

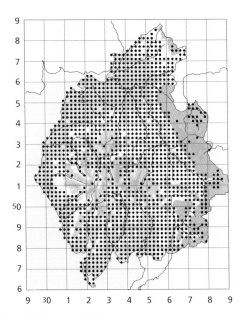

Map 300. *Capsella bursa-pastoris*

T. caerulescens J.S. & C. Presl   Alpine penny-cress
(*T. alpestre* L.)

Very common on old lead-mine workings on Alston Moor and continuing southwards, though less frequently, as far as Brough (78.16). In these habitats it is commonly associated with *Minuartia verna*, but it is more restricted than that species to metalliferous soils. Populations certainly occur on non-metalliferous soils, for example limestone outcrops and scree as at Helbeck Scar (80.16), the whinstone of High Cup Nick (72.24, 74.24) and the old sites at Ardale Head (6.3, E. Blezard, 1927, CLE), which may be limestone or whinstone, and in the Lake District, where it was collected last century in Fusedale, Haweswater (4.1) by W. Foggitt and where a small population was known on the eastern cliffs of Helvellyn (34.14, DAR) from 1871 until 1959. Baker gives two puzzling and unlikely records: H.C. Watson's from Skiddaw (2.2) and his own from the roadside between Kendal and Ambleside (4.9).

    This species is largely confined in Britain to the mining areas of the north Pennines, Peak District, north and west Wales and the Mendips.

Alt. range: 275 m, Alston (70.46) - 750 m, Knock Ore Gill (70.30), although formerly at 855 m on Helvellyn.(3.1).

Map 303   S 25   WC

*Iberis amara L.   Wild candytuft

Listed by Hodgson from Bassenthwaite Lake station yard (1.3), Aspatria (1.4) and near Abbeytown (1.5, E.J. Glaister, 1876, CLE) and, in addition, by Baker from Kendal (5.9) and "Jacklands Tarn", Ulverston, and by Petty (1894 p.247) from Coniston (3.9). The only record this century is from Carlisle waterworks (4.5) where it was seen by J. Parkin in 1935.

    This species is native in Britain only on the chalk of south-east England.

                        S   (WFC)

*I. umbellata L.   Garden candytuft

A popular garden plant from the Mediterranean found in some quantity on the exposed shingle of the River Caldew above Cummersdale (38.50, REG, 1979), in an old quarry at Oakshaw Hill, Netherby (40.70n, MG & JP, 1993) and on disturbed ground at Skinburness (12.56, A. Cannell, 1991). The only previous record was from Silloth (1.5, A.W. Trethewy, 1919, *BEC Rpt* (1920) 5:641).

                        3   C

*Lepidium sativum L.   Garden cress

A very rare casual recorded from gardens at Thurstonfield (30.56, REG, 1977) and at Calgarth Park, Windermere (40.98, M. Gladding, 1979, LANC), the first Westmorland record, and from waste ground at Melkinthorpe (54.24, AW, 1986, LANC) and Penrith (50.28, RWMC, 1986, LANC). It was also recorded from Barrow-in-Furness (18.68s, 20.68s, PB, *c.*1957), where it was seen in 1913 by W.H. Pearsall (*BEC Rpt* (1914) 3:402) and again during the 1950s. It is a

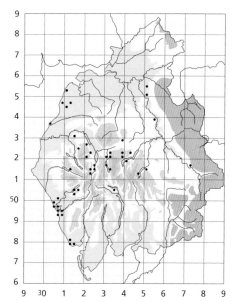
Map 301. *Teesdalia nudicaulis*

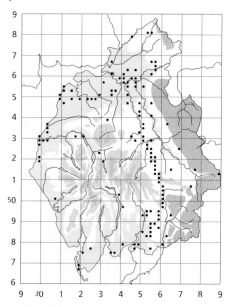
Map 302. *Thlaspi arvense*

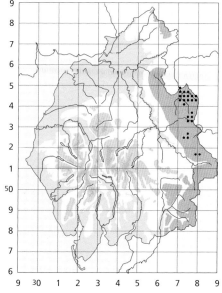
Map 303. *Thlaspi caerulescens*

frequent constituent of bird-seed.

It was recorded by Hodgson from Workington (9.2), Maryport (0.3), Silloth (1.5), Dalston (3.5) and by the River Caldew at Hawksdale (3.4), also by T.S. Johnstone from the gravel beds of the River Eden at Grinsdale, Carlisle (3.5, 1902, CLE).
Alt. limit: 140 m, Melkinthorpe (54.24)

	6	WFC

*L. campestre (L.) R. Br.   Field pepperwort
Extremely rare, seen only four times during the Survey: in gardens at Burton-in-Kendal (52.76s, GH, 1977, LANC) and Kendal (50.92, TW, 1986, LANC), on gravel at Cumwhinton station (44.52, FJR, 1970s) and on a roadside verge south of Great Corby (48.52, MS, 1970s, herb. M.S.). In view of the paucity of records in the Floras, the 14 post-1930 *Atlas* records are surprising.

	4	W(F)C

L. heterophyllum Benth.   Smith's pepperwort
Rather uncommon but well distributed throughout the lowlands. It occurs in rather dry heathy situations such as roadside and field banks and dry open grassland, especially near the sea. It also occurs on the ballast of disused railway lines as near Kirkby Lonsdale (62.78s), on the old slag bank at Ulverston (30.76s) and by gravel forestry tracks in Kershope Forest (48.76n, 48.78n). It is probably declining as its habitats disappear.
Alt. limit: 400 m, Kershope Forest (58.86n)

Map 304	126	WFC

*L. virginicum L.
A North American species collected at Silloth by J. Leitch in 1883 (1.5, CLE) and reported by T.S. Johnstone in 1901. A piece probably of this species was later collected there by C.W. Muirhead (1939, CLE, *BSBI News* (1991) 58:38). There is a single Furness record from Askam in 1914 (2.7, D. Lumb, YRK, det. T.C.G. Rich, *BEC Rpt* (1915) 4:121, under *Lepidium neglectum*).

		(C)

*L. ruderale L.
Found only on a disturbed roadside verge at Brunstock, Carlisle and on the Rockcliffe Marsh gullery (respectively 42.58, PB; 30.62, D. Hawker, both records 1992, LANC, *Watsonia* (1993) 19:283).

The only previous records for this southern species are Hodgson's from the Derwent Tinplate Works, Workington (9.2), Seaton and Maryport (0.3) and Silloth (1.5, E.J. Glaister, 1887, CLE), and later ones from Wetheral (4.5, H. Britten 1900-1908), and that from Windermere (4.9) cited by Wilson.

	2	(W)C

*L. perfoliatum L.
Recorded from the sand-dunes at Silloth in 1890 (1.5, J. Leitch, CLE) and also from 1935 to 1940 (C.W. Muirhead, CLE, PLYP, *BSBI News* (1991) 58:38), and from Penrith (5.3, A. Wallis, 1916, BM).

		(C)

*L. latifolium L.   Dittander
Seen only around Barrow-in-Furness where are robust colonies of this rare casual near Cavendish Dock (20.68s, 22.68s), a small one nearby at Roose station (22.68s) and another at Rampside (22.66s). It had earlier been reported from Rampside, and also North Walney, by G. Wilson in the 1950s. The only other early records are Hodgson's from Silloth (1.5, E.J. Glaister, 1878 ,CLE) and near Little Salkeld (5.3).

The presumed native distribution of this species in Britain appears to lie along the coasts of south-east England and the Severn estuary.

S 3		F(C)

*L. densiflorum Schrader
A widely naturalised North American species seen by A.W. Trethewy at Silloth in 1919 (1.5, *BEC Rpt* (1920) 5:640) and by A.B. Cobbe on Walney Island in 1924 (*BEC Rpt* (1925) 7:558).

		(FC)

*L. draba L.   Hoary cress
(*Cardaria draba* (L.) Desv.)
subsp. **draba**
Not infrequent by the shore along the Solway coast and on Walney Island. Most of the records are from open grassland, waste sites and shingle. The few 'inland' records are mostly near the coast although there is a record from Langwathby station in the Eden valley (56.32). It occurs as a persistent garden weed at Burton-in-Kendal (56.76s).

Map 305	25	WFC

subsp. **chalepense** (L.) Thell.
A rare casual recorded in 1952 from the railwayside between Durranhill and Scotby (4.5, R. Martindale, BM, OXF, PLYP, *Proc. BSBI* (1955) 1:577). It differs in having broadly ellipsoid fruits instead of the broadly triangular ones of subsp. *draba*.

		(C)

*Coronopus squamatus (Forskål) Asch.   Swine-cress
A very rare plant of waste ground occurring chiefly near the west coast. It is now rarer than *Coronopus didymus* although apparently commoner at the end of the last century, being recorded by Hodgson from Parton and Workington (9.2), Flimby and Maryport (0.3), Allonby (0.4) and Skinburness (1.5). Baker also listed it from Seaton (0.3) and Kendal (5.9) and there are additional Cumberland records from Bowness-on-

Solway (2.6n, P. Shepherd, 1884), an undated one from Thwaites, Millom (1.8, M. Cross) and Blaithwaite, Wigton (2.4, J. Parkin, 1935). The first Furness record dates from the 1950s when it was seen by G. Wilson in the Rampside (2.6s) and Dalton-in-Furness (2.7) squares.

Alt. limit: 150 m, Howtown (44.18)

Map 306   7   WFC

*C. didymus (L.) Smith   Lesser swine-cress

Occasional to locally abundant in the Whitehaven - Workington and Barrow-in-Furness areas, elsewhere rather rare. There are also a few inland casual occurrences. It is a plant of open, disturbed and often trampled ground.

Alt. limit: 180 m, Crosby Ravensworth (62.14)

Map 307   46   WFC

Subularia aquatica L.   Awlwort

Rare; restricted in England to the Lake District where it occurs in a few tarns and lakes, mainly in the west and centre. It usually grows in about 15 cm of water and rooted in sand or silt, often on exposed shores. It can, however, flower and fruit successfully in water of about 30 cm or more and J.M. Lock noted it at depths down to 2.5 m in Brothers Water (40.12). It is also able to withstand brief periods of exposure. *Lobelia dortmanna* and *Littorella uniflora* are common associates and occasionally also *Elatine hexandra*.

Unfortunately it appears to be declining, having been seen during the Survey in only four of the eight post-1930 *Atlas* squares. It was, however, seen at one new square at Brothers Water (40.12) and refound in a pre-1930 square at High Nook Tarn, Loweswater Fell (12.18). There is a surprising outlying locality on the

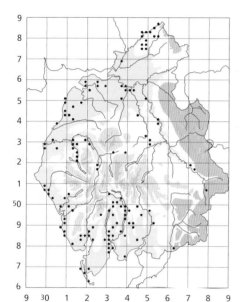

Map 304. *Lepidium heterophyllum*

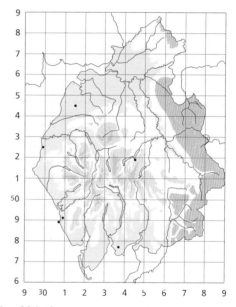

Map 306. *Coronopus squamatus*

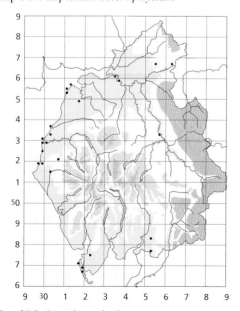

Map 305. *Lepidium draba*

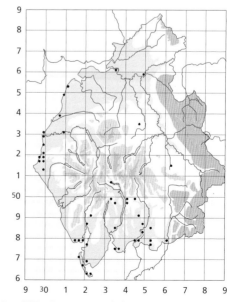

Map 307. *Coronopus didymus*

Solway Plain at Thurstonfield Lough (3.5, R.W. Butcher, 1937, *BEC Rpt* (1938) 11:466), but it has never been refound there, nor in the Ulpha (1.9, Mrs A.H. Chandler, 1958), and Keswick (2.2, B.F.C. Sennitt, 1957) squares. The same is true of Wilson's Easedale Tarn (3.0) and Red Tarn, Helvellyn (3.1) sites, of H.E. Fox's at Angle Tarn, Patterdale (4.1, 1887, E) and Angle Tarn, Bowfell (2.0, 1888, BM), and also of DAR's at Bowscale Tarn (3.3, 1950s).

Alt. limit: 600 m, Sprinkling Tarn (22.08), although it earlier grew at 710 m in Red Tarn prior to the tarn being dammed.

Map 308   11   WC

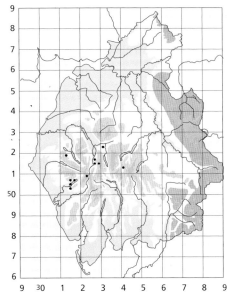

Map 308. *Subularia aquatica*

*Conringia austriaca (Jacq.) Sweet
A rare British casual recorded in 1953 from the roadside between Appleby and Dufton (6.2, M. Cross, det. K, *Proc. BSBI* (1954) 1:163).

(W)

*C. orientalis (L.) Dumort.
Listed by Hodgson from the usual alien hunting grounds at Workington (9.2), Maryport (0.3) and Silloth (1.5, E.J. Glaister, 1877, CLE), and by Wilson from near Maulds Meaburn (6.1). There are later Cumberland records from the River Eden at Etterby (3.5, T.S. Johnstone, 1902), Brampton (5.6, H. Britten, 1900-1908), Gamblesby (6.3, W.W. Mason, 1918), Thwaites, Millom (1.8, M. Cross, 1929) and Edenhall, Penrith (5.3, T.R. Gambier-Parry,1923, *BEC Rpt* (1924) 7:168). The last record of this central and eastern European species was from Silloth in 1950 (R.C.L. Howitt, CLE).

(WC)

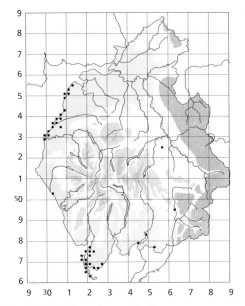

Map 309. *Diplotaxis tenuifolia*

*Moricandia arvensis (L.) DC.
A Mediterranean casual found at Silloth in 1887 and 1919 (l.5, J. Leitch, CLE; A.W. Trethewy, CLE, *BEC Rpt* (1920) 5:640, *BSBI News* (1991) 58:38). There are only about six British records for this species.

(C)

*Diplotaxis tenuifolia (L.) DC.   Perennial wall-rocket
A fairly common species of waste ground in the Barrow-in-Furness area, at Sellafield (02.02) and along the coast from Workington(98.28) to Skinburness 12.54), its northern limit on the west coast of Britain. It also occurs inland as a rare casual, for example by the roadside near Firbank, Sedbergh (62.94), growing on a grit pile which had probably originated from the west coast, and there is an old record from Gilsland (6.6, G.A.K. Hervey, 1944, v.c.67/70).

Alt.: limit: 190 m, Firbank, Sedbergh (62.94), but there is a surprising *Atlas* record for the Alston square (7.4).

Map 309   39   WFC

*D. muralis (L.) DC.   Annual wall-rocket
Not uncommon on spoil heaps and disturbed and waste ground by the sea in the Barrow-in-Furness area but mostly occurring in the county only as a casual. It reaches its northern limit on the west coast near Wolsty (08.50). The first Westmorland record may be that shown in the *Atlas* for the Sedbergh square (6.9),

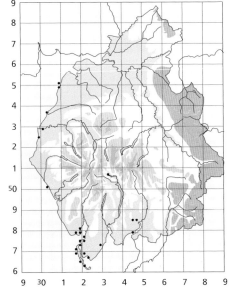

Map 310. *Diplotaxis muralis*

although this may refer to v.c.65.
Map 310   24   WFC(W/Y)

*Brassica oleracea L.   Wild cabbage
According to Wilson wild cabbage was discovered on limestone rocks at Arnside point (4.7s) in 1775 and it was last seen there in 1877. The *Scarce Plants* record for this square is from Lancashire. This is the only apparently native record. The most northerly extant native populations on the west coast of Britain are in north Wales and the Isle of Arran. The only Survey record is from Barrow-in-Furness (18.68s, PB, 1993, LANC). This and early records from the shore at Whitehaven (9.1, Anon., 1910, LANC), Haverigg (1.7, M. Cross, 1930s), Ravenglass (0.9, G.A.K. Hervey, 1948) and near Abbeytown (1.5, J. Parkin, 1935), also the *Scarce Plants* record from the Sebergham square (3.4, Anon., 1950), no doubt represent escapes from cultivation. Similar recent occurrences have probably been disregarded by recorders.
                        S 1   (W)F(C)

*B. napus L. and ***B. rapa** L.   Rape and Turnip
Occasional to locally frequent, occurring chiefly in south Westmorland and the Eden valley. These are casual species of waste ground and roadside verges, especially the intermittently disturbed verges and central reservations of trunk roads and the M6 motorway where they can be a conspicuous feature in late spring.

During the early part of the Survey it soon became apparent that the relative position of the buds and open flowers was proving a difficult diagnostic character. It was thought that most if not all of the records were of *Brassica rapa* and all the records were consequently pooled under this species. However, in 1991 T.C.G. Rich recorded *B. napus* from a number of

sites in south Westmorland and drew attention to the importance of petal size provided it is measured on fresh material. In 1991 T. Alexander examined several populations in south Westmorland and north Lancashire and concluded that overall *B. rapa* was slightly commoner. Her measurements of petal length gave means of 10 mm and 15.5 mm respectively for *B. rapa* and *B. napus*, although there is a slight overlap. The latter species also has rather paler flowers, less glaucous leaves and more divided lower leaves. The maps are based on post-1989 records only.

The records of *B. napus* presumably refer to the increasingly cultivated, autumn-sown var. *oleifera* (DC.) Metzger, which is grown to some extent in the south of the county but chiefly in the Eden valley. Records of *B. rapa*, which is not currently grown in the county, refer to the wild subsp. *campestris* (L.) Clapham. Presumably spring-flowering plants germinate in the autumn and late summer plants in the spring.

Hodgson gives few records for *B. rapa* but regarded *B. napus* as "probably the commonest of the brassica tribe...found in every part of the county more or less abundantly". Wilson did not distinguish between the two saying only that "the three forms, turnip, swede and rape all occur". The *Atlas* maps must be regarded as unreliable.
Alt. limit: c.420 m, by the A66, Stainmore (88.12)
Maps (post-1989 records) 311, 312   WFCY
Tetrads: 434 (all records), 46 *B. napus*, 33 *B. rapa*

*B. juncea (L.) Czernj.   Chinese mustard
Found in 1979 on a freshly disturbed roadside verge east of Smithfield (44.64, CS & MS, LANC, *Watsonia* (1980) 13:132). It was earlier seen at Silloth and a mill-yard at Whitehaven (respectively 1.5, J. Leitch, 1887, CLE; 9.2, R.H. Williamson, 1924, CLE, *BSBI News*

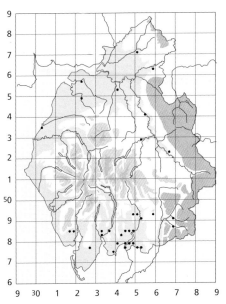

Map 311. *Brassica napus*

Map 312. *Brassica rapa*

58:37), by the shore at Greenodd and the railway at Askam-in-Furness in 1912 and 1914 (respectively 3.8, 2.7, D. Lumb, BM, *BEC Rpt* (1913) 3:231;(1915) 4:120 under *cf. B. erucastrum* and *B. balearica* respectively).

This species occurs sporadically in Britain but is widely distributed as a weed on the Continent.

1   (F)C

*B. elongata Ehrh.

An uncommon British casual found on the Ulverston slag bank (30.76s, GH, LANC, *Watsonia* (1985) 15:392). The only previous records were made by W.H. Pearsall in 1913 from near Cavendish Dock, Barrow-in-Furness and "Lindal Bank" (2.6s, 2.7, 1913, BM, *BEC Rpt* (1914) 3:307, 402, 451).

1   F

*B. nigra (L.) Koch   Black mustard

Seen during the Survey only at Silloth (10.52), where it was last seen in 1990 by A. Cannell, and briefly on old railway land at Carlisle (38.56, REG, 1979). The species has declined markedly in the county and could well become extinct.

Hodgson recorded it from the Derwent Tinplate Works, Workington (9.2), from ballast at Maryport (0.3) and Silloth (J. Leitch, 1889, CLE), turnip fields at Thirlmere (3.1) and by the River Eden at Stainton, Carlisle (3.5). It was seen at Great Salkeld (5.3, 1900-1908) by H. Britten, at Thwaites, near Millom (1.8, 1908) by M. Cross and again at Silloth by J. Parkin in 1935. Apart from a pre-1930 dot for the Burton-in-Kendal square (5.7s), there are later *Atlas* records for the South Walney (1.6s), Rampside (2.6s), Millom (1.8), Distington (0.2), Silloth (1.5), Carlisle (3.5) and Great Salkeld (5.3) squares.

2   (WF)C

Sinapis arvensis L.   Charlock

A frequent lowland weed of arable fields, farmyards and disturbed ground. Like *Brassica rapa* and *B. napus* it very often occurs following roadworks, especially on the motorway, and is readily recognised by its paler flowers and by flowering about a month later.

Alt. limit: at least 430 m, near Nenthead (78.44)

Map 313   540   WFCY

*S. alba L.   White mustard

A rare weed of arable land and waste ground chiefly in the north-east. It was recorded by Hodgson from St Bees (9.1), Maryport (0.3), Silloth (1.5) and Westward (2.4). There are later Cumberland records from Thwaites, near Millom (1.8, M. Cross, 1907), Lang-wathby (5.3, H. Britten,1900-1908) and again at Silloth in 1935 where it was seen by J. Parkin. In Furness, Petty (1894 p.246) recorded it as a garden escape at Grange-over-Sands (4.7s). There are post-1930 *Atlas* dots for the South Walney (1.6s), Rampside (2.6s), Millom (1.8), Workington (9.2), Bowness-on-Solway (2.6n, v.c.70/72) and Warcop (7.1) squares.

Alt. limit: 190 m, near Bewcastle (54.72n)

Map 314   10   (W)FC

*Eruca vesicaria (L.) Cav.
subsp. **vesicaria**

A plant of Spain and North Africa collected by J. Leitch at Silloth in 1887 (1.5, CLE, *BSBI News* (1991) 58:38). This is the first and only British record of this subspecies.

(C)

*Erucastrum gallicum (Willd.) O. Schulz

A single plant was discovered in 1984 on a recently landscaped area by Eden Bridge, Lazonby (54.40, REG, LANC, *Watsonia* (1985) 15:392). This was followed by records from waste ground by the M6 near New Hutton

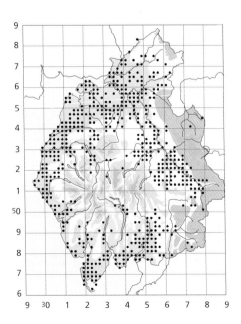

Map 313. *Sinapis arvensis*

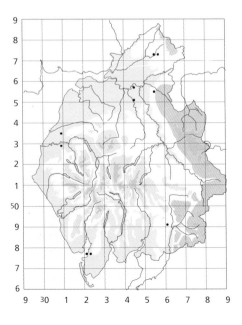

Map 314. *Sinapis alba*

(58.90, CEW, LANC, *Watsonia* (1988) 17:184), a nearby roadside verge (58.86s, CEW, 1989, LANC), a trackside at Spadeadam (60.74n, MG & JP, 1991, LANC, *Watsonia* (1992) 19:142) and a roadside verge near Houghton (42.58, PB, 1992, LANC, *Watsonia* (1991) 18:420).

This is a species of central and south-west Europe which is naturalised in Britain chiefly in the south and south-east.

Alt. limit: 340 m, Spadeadam (60.74n)

5   WC

Coincya monensis (L.) Greuter & Burdet
subsp. **monensis**   Isle of Man cabbage
(*Rhynchosinapis monensis* (L.) Dandy)
Frequent along the coast from Bowness-on-Solway to Workington and then more scattered south to Walney Island. It is a plant of the strand-line and open ground close to the sea. The first certain Furness record was from Askam-in-Furness (20.76, D. Lumb, 1912, *BEC Rpt* (1913) 3:231), where it is still common on the old slag bank behind the salt-marsh.

Apart from an isolated occurrence in South Wales, this British endemic is now restricted to the Firth of Clyde, the Isle of Man, the Solway Firth and the coast southwards to Merseyside. The Cumbrian and Lancashire populations seem to be maintaining themselves well, or even, locally, increasing, but elsewhere the subspecies appears to be declining.

Map 315   S 32   FC

*subsp. **cheiranthos** (Villars) Aedo, Leadley & Monoz-Garm.
(subsp. *recurvata* (All.) Leadlay
The first definite record is from Workington docks (98.28, CCH, LANC, *Watsonia* (1979) 12:349) where it was discovered in 1978 but is probably now extinct. It is possible that W.W. Mason's 1919 record of *Brassica erucastrum* from Melmerby (6.3) refers to this species.

This subspecies occurs mainly around the Bristol Channel. Workington is its northernmost recent site.

1   C

***Hirschfeldia incana** (L.) Lagr.-Fossat   Hoary mustard
Seen by B. Fisher in 1979 at Ramsden Dock, Barrow-in-Furness (20.68s, LANC, *Watsonia* (1980) 13:132). The only other record for this southern European crucifer is from Thwaites, Millom (1.8, M. Cross) where it was seen in 1932.

1   F(C)

The dunes stretch from Walney Island to St. Bees Head, and from Maryport northward along the edge of the Solway. Here, tipped over the ground, are great splashes of colour in summer, the soft colours of women's dresses, pinks, mauves, pale yellows and orange. Lady's bedstraw is yellow and frilly, and never seems to get dirty. Yellow stonecrop is also pale, but a more acid colour, pressed closer to the dune. The Isle of Man cabbage, one of the rarer brassicas, grows here and there, lemon-coloured and rather tender-looking. You can slip out and gather it when the train stops at Seascale Station. Then there are fat dandelions when the sand has more soil in it, and hawkweeds by the million that I cannot recognise.

Cumberland and Westmorland, Norman Nicholson

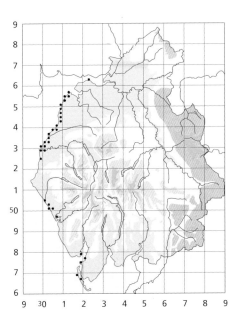

Map 315. *Coincya monensis* subsp. *monensis*

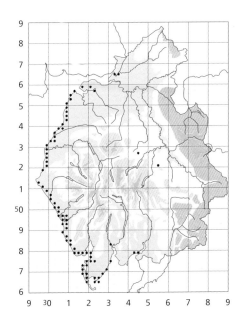

Map 316. *Cakile maritima*

Cakile maritima Scop.　　　　　　　Sea rocket
A common strand-line species along the entire coast
but less frequent in the inner parts of Morecambe Bay
and the Solway Firth. There are inland records, all
from roadside grit-piles, from near Dalton-in-Furness
(22.74), Dacre, Penrith (44.26) and near Thrimby Hall,
Little Strickland (54.20, RWMC, 1978, LANC), the
last being the only Westmorland record.
Alt. limit: 260 m, Dacrebank (44.26)
Map 316　　　　　　78　　　　　　WFC

*Rapistrum rugosum (L.) Bergeret
An uncommon casual seen on waste ground in Carlisle
near the River Petteril (40.54, A. Franks, M. Dickinson & PT, 1984, LANC, *BSBI News* (1985) 40:16)
and, in the south, in a copse at Heversham (48.82s),
MB & A. Cannell, 1986, LANC, *Watsonia* (1988)
17:184). The latter is the first Westmorland record.
　　　Hodgson gives records from Workington (9.2),
Maryport (0.3), Silloth (1.5) and Penrith (5.3). It was
seen in 1901 by the River Eden at Carlisle (3.5, T.S.
Johnstone, CLE) and Silloth, where it was also collected by N.M. Stalker in 1949 (CLE). There are additional post-1930 *Atlas* dots for the South Walney (1.6s)
and lower Eskdale (1.9) squares.
　　　　　　　　　　　　2　　　　　W(F)C

*R. perenne (L.) All.
A rare southern casual recorded in 1913 from Dalton-in-
Furness (2.7, Anon., NMW) and in 1951 from the docks at
Barrow (1.6s/2.6s, J.G. Dony, BM).
　　　　　　　　　　　　　　　　(F)

Crambe maritima L.　　　　　　　Sea kale
Frequent along the coast from Bardsea (30.74s) in
Morecambe Bay to Skinburness (12.54) on the Solway
Firth. It has long disappeared from Arnside Point

(4.7s) where Wilson noted it in 1934.
　　　This is a characteristic and conspicuous plant of
shingle beaches. It is probably commoner in Cumbria
than anywhere else along the west coast of Britain and
it appears to be increasing.
Map 317　　　　　　45　　　　　　(W)FC

*C. cordifolia Steven
A single plant was found in 1990 in coastal turf at East
Cote, Silloth (10.54, REG, LANC). This is the first
record north of the Midlands for this rare Russian
casual.
　　　　　　　　　　　　1　　　　　　　C

Raphanus raphanistrum L.
*subsp. raphanistrum　　　　　Wild radish
An uncommon lowland weed of disturbed ground,
roadside verges and arable land. It is probably less
common than in the past. As mentioned by previous
Flora writers, both the pale yellow- and white-flowered
forms occur, the latter being the commoner, but there
are also single records of the deep yellow and purple
forms from Furness.
Alt. limit: 200 m, Crosby Ravensworth (62.14)
Map 318　　　　　　64　　　　　WFC(W/Y)

subsp. maritimus (Smith) Thell.　　　Sea radish
Frequent along the coast from Morecambe Bay north
to Grune Point, Silloth (12.56), occurring usually
along the strand-line. This subspecies appears to be
spreading, the only published records being Hodgson's
from Workington (9.2), Petty's (1894 p.248) from
Rampside (2.6s), one from Askam-in-Furness (2.7,
D. Lumb, 1912, *BEC Rpt* (1913) 3:231) and the *Atlas*
ones from the North Walney (1.7) and Silloth squares.
Map 319　　　　　　50　　　　　　FC

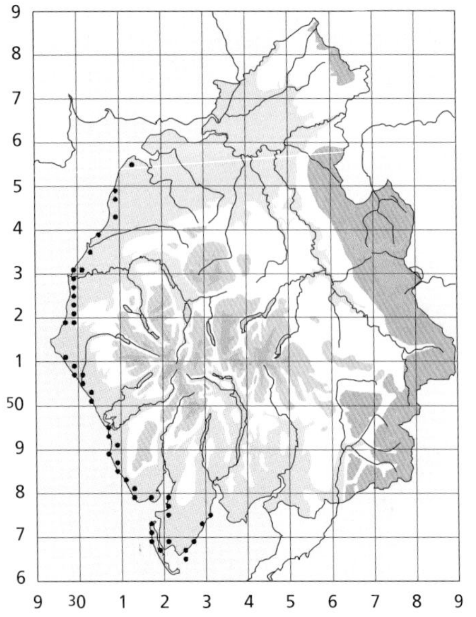

Map 317. *Crambe maritima*

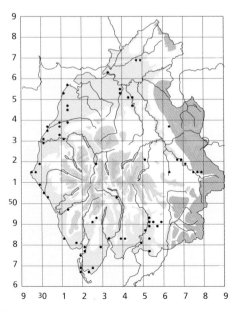

Map 318. *Raphanus raphanistrum* subsp. *raphanistrum*

*Raphanus sativus L.   Radish
Recorded in *c*.1993 from rough ground south of
Askam-in-Furness (20.76, PB, LANC).

                                             1   F

RESEDACEAE

Reseda luteola L.   Dyer's rocket; Weld
Widely distributed in the lowlands as a casual on waste
land, in limestone quarries and on roadside grit-piles
but persistent only in the urban areas of Carlisle, west
Cumberland and Barrow-in-Furness. Handley (1898)
makes the interesting comment that in the Sedbergh
area it is "well-established, probably owing to its use in
our woollen mills".
Alt. limit: 400 m, north of Shap Summit (54.06)
Map 320   171   WFC(Y)

*R. alba L.
Listed by Hodgson as a garden escape on the ballast hills at
Maryport (0.3) in 1891.

                                                        (C)

R. lutea L.   Wild mignonette
A species with a similar distribution and ecology to
Reseda luteola although rather less frequent, particularly around Carlisle. Both are characteristic of the
west Cumberland coast and it is from there that the
roadside grit-piles and presumably their associated
flora, which includes these two species, originate. The
grit is said to have been brought from Hamburg.
Reseda lutea is particularly associated in the west with
ash ballast and railway lines and is one of the
commonest plants at Maryport docks.

      This species appears to have spread considerably
during the last few decades, Hodgson recording it only
from the west coast and Wilson from Kendal (5.9).
Alt. limit: 440 m, by the A66 on Stainmore (88.12)
Map 321   129   WFCY

EMPETRACEAE

Empetrum nigrum L.   Crowberry
subsp. **nigrum**
Widespread and locally common in the uplands
occurring on crags and in moorland and mires, very
commonly associated with *Vaccinium myrtillus* and
Calluna. Like *Vaccinium*, it is more resistant to moor-
burning and grazing than *Calluna*. In the lowlands it is
rather rare although locally common as on the Morecambe Bay and Solway mosses, usually occurring under
pine or birch rather than on the open moss. It also
grows on sea-cliffs at St Bees Head (94.14).
Alt. limit: 905 m, Skiddaw (26.28). Wilson gives
915 m on Helvellyn (34.14), but it no longer occurs
around the summit.
Map 322   459   WFCY

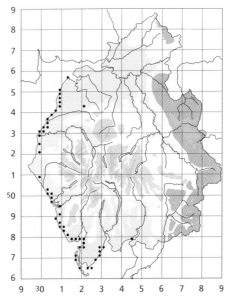

Map 319. *Raphanus raphanistrum* subsp. *maritimus*

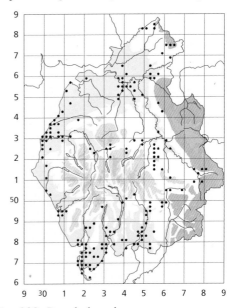

Map 320. *Reseda luteola*

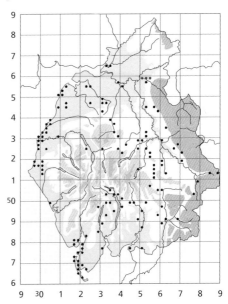

Map 321. *Reseda lutea*

subsp. **hermaphroditum** (Hagerup) Böcher
Present in the four Lake District sites discovered in
the1950s by DAR and refound, except on Tarn Crag, by
RWMC (LANC): Riggindale, Haweswater (44.10),
Dove Crag, Fairfield (36.10), Tarn Crag, Grisedale
(34.12) and Hanging Knotts, Bowfell (24.06) (the first
two finds being in *Proc. BSBI* (1963) 5:34). It was also
seen by RWMC on Great End (22.08). The usual
habitat is block scree but at Tarn Crag it occurs on the
cliff itself.

The identification of this tetraploid subspecies is
rendered difficult by the occasional occurrence of
hermaphrodite diploid plants and of plants with stam-
inodes. In the absence of specimens, the Pennine
records from Knock Fell (72.30, 74.28, 74.30, A.J.
Richards, 1970s) and Cross Fell (68.34, REG, 1973)
are best regarded as unconfirmed.

The Riggindale and Hanging Knotts specimens
do, however, agree with the tetraploid in leaf length: 3x

as long as broad (4x in subsp. *nigrum*) and in habitat,
subsp. *hermaphroditum* generally occurring at higher
altitudes and in rockier situations. It is said also to have
larger seeds: 1.9 mm (*cf.*1.6 mm) and pollen tetrads:
>34 μm diameter (<34 μm) and less red stems.

South of the Scottish Border, this subspecies is
restricted to Cumbria, North Wales and, possibly, the
Cheviots (Swan 1993).
Alt. range: 320 m, Riggindale (44.10) - 720 m, Hanging Knotts (24.06)

4   WC

ERICACEAE

*Ledum palustre L.   Labrador tea
The only Survey records are from Scaleby Moss
(42.62n, LANC). It was reported by H. Ruddick (PLYP;
LANC) in 1948 to have been established at Scaleby for
many years and despite frequent fires and disturbance

Plate 65. *Ledum palustre* subsp. *groenlandicum*, Scaleby Moss

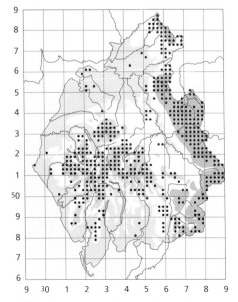

Map 322. *Empetrum nigrum* subsp. *nigrum*

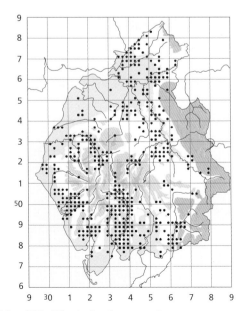

Map 323. *Rhododendron ponticum*

to the moss about 20 plants still survive. Wilson refers to "two strong bushes" at Foulshaw Moss in the south where he collected it in 1934 (4.8s, YRK), but these have now disappeared following afforestation. Both populations are, or were, some distance from houses and intentional introduction seems unlikely. The plants have the broad leaves of subsp. *groenlandicum* (Oeder) Hultén. This is a native of Greenland and North America and is not generally regarded as native in north-west Europe, although H. McAllister (*in litt.*) states that the British chromosome number is $2n=52$ whereas that of the American and Greenland plants is $2n=26$.

The nearest sites to the Cumbrian one are on the north side of the Solway Firth and a peat moss south of Lancaster.

Plate 65   R 1   (W)C

*Rhododendron ponticum L.   Rhododendron
Widespread and locally abundant on lowland mosses and on usually peaty soils in the hills, both in the open and under trees, usually oak and birch, and forming a dense, often impenetrable, shrub layer. Wilson omits any mention of it and its present wide distribution in the county must have been largely attained during the last half century, although it is said to have been extensively planted along shooting drives in lower Dentdale in the 1890s. It is a most aggressive colonist and its spread by seed has been abetted by 'amenity' planting. Its dense shade effectively eliminates the previous ground flora, but it is only recently that it has become generally recognised as a major pest rather than an attractive newcomer. Unfortunately it is extraordinarily difficult to eradicate.
Alt. limit: 600 m, Eel Crags, Newlands (22.16)
Map 323   423   LWFCY

*R. luteum Sweet
Well-naturalised on the fellside east of Brantwood, Coniston (30.94, 32.94, GH, 1994).
Alt. limit: 190 m, above Brantwood (32.94)
                    2   F

*Rhododendron sp.
Unidentified bushes are well established by Blea Tarn, Langdale (28.04, CFS & JS, 1980s) and Beacon Tarn, Blawith (26.90, B. Fisher, 1990).
                    2   WF

[Loiseleuria procumbens (L.) Desv.   Trailing azalea
The 19th century record from St Bees lighthouse (9.1) was regarded by Hodgson as an introduction. It may well have been a misidentification since persistent reports of it from the Lake District during the Survey all proved to be *Empetrum nigrum*.
                                 (C)]

*Kalmia angustifolia L.
An eastern North American shrub found in 1979 well-established on Ellerside Moss, Holker (34.78s, I.R. Bonner, *Watsonia* (1982) 14:193).
                  1   F

Andromeda polifolia L.
       Bog Rosemary; Marsh Andromeda
Locally frequent on the coastal raised mosses by the Solway Firth, the Duddon estuary and Morecambe Bay. It is less frequent inland, but it still persists on the remaining unimproved peatlands between Kendal and the Lune valley and, in the north, it is widespread up the Irthing valley and towards the Northumbrian border. However, a substantial part of the population on the Gilsland moors has been lost through afforestation. This last area, together with the Solway mosses, can fairly be described as the species' headquarters in Britain. There are no Survey records for the Keswick (2.2, Anon., undated) and Hawkshead (3.9, G. Wilson, 1952) squares. The *Scarce Plants* record for the Dentdale square (7.8) is an error.

Common associated species include *Vaccinium oxycoccus, Narthecium ossifragum, Erica tetralix, Sphagnum papillosum, S. magellanicum* and *S. capillifolium*.
Alt. limit: 490 m, Armboth Fells (28.16) and Skiddaw Forest (26.30)
Map 324   121   WFC

*Gaultheria shallon Pursh   Shallon
A North American shrub found by Long Cairn, Kershope Forest (52.82n, EHR; LANC), where it is long established, in estate woodland at Underley, Kirby Lonsdale (60.78s, 60.80s, GH, 1975), in woodland near Eskdale Green (14.00, MMG, 1990, LANC, *Watsonia* (1991) 18:428), at Muncaster (10.96, MMG, 1992) and on the west side of Derwent Water (24.20, GH, 1978).
Alt. limit: 290 m, Kershope Forest (52.82n)
                  6   WC

*G. mucronata (L. f.) Hook. & Arn.
(*Pernettya mucronata* (L. f.) Gaudich. ex Sprengel)
A North American garden-escape found on Belle Isle, Windermere (38.96, C.D. Pigott, 1970), by the roadside at Dale Bottom, Keswick (28.20, ES, 1984) and on Penrith Beacon (52.30, RWMC, 1984, LANC; all records in *Watsonia* (1985) 15:398), and also by a path near Ambleside (38.04, R. Steele, 1973, AMDE).
Alt. limit: 275 m, Penrith (52.30)
                  4   WC

Arctostaphylos uva-ursi (L.) Sprengel   Bearberry
Extremely rare. The only definite records this century are from the Grasmoor fells, where it occurs in some abundance on screes and crags on Grasmoor (16.20, DAR, 1956, CLE) and on nearby Robinson (18.16, DAR, 1977), also in one area of crags on the Wast Water Screes (14.02, DAR, 1956) and on Goat Crag, Thirlmere (28.18, L. Reinecke, 1988; LANC). The habitats are mostly rather unpromising steep crags with

scattered *Calluna* and *Erica cinerea*, and, on Grasmoor, stunted *Juniperus*, although *Lathyrus linifolius*, *Galium boreale* and *Rubus saxatilis* are associates on Goat Crag.

Baker lists earlier records from Corney Fell (1.9) and Dalehead, Martindale (4.1) and Hodgson and Wilson ones from the "Higher Ennerdale Fells" (1.1), Pavey Ark (2.0) and the Westmorland side of Cauldron Snout, Teesdale (8.2, J.A. Martindale & G. Stabler, 1868, KDL). It was also noted by J. Parkin on Hobcarton Crag, Whinlatter (1.2, 1935). Some of these early records may refer to *Vaccinium vitis-idaea*, with which it is often confused, but DAR is of the opinion that *Arctostaphylos* has been much reduced in northern England by moor-burning.

The only other extant sites in England and Wales are in Teesdale, Northumberland and Derbyshire.
Alt. range: 380 m, Goat Crag (28.18) - 580 m, Grasmoor (16.20)
Plate 33   4   (W)C

[A. alpinus (L.) Sprengel
A species of the north-western Scottish Highlands listed by Baker from Scafell (2.0), but he regarded the record as an obvious error.

(C)]

Calluna vulgaris (L.) Hull   Heather; Ling
Very common except in the lowlands where it continues to decrease as a result of agricultural reclamation and afforestation. It occurs often as a dominant species on dry, podsolic heaths and wet mosses in the lowlands. In the fells it is a feature of rock outcrops and of semi-stable scree, especially on the Skiddaw slates. In the Pennines it is a constant feature of upland blanket peat, often dominating the low rounded spurs between the

sikes. The grouse are highly dependent on heather and the cover is maintained by rotational burning. Regular burning is also carried out by many sheep farmers.

Heather moorland is sensitive to overgrazing and 'improvement' and in many places in the Lake District and Pennines the heather is rapidly giving way to grassland. The effect of excluding sheep is very noticeable along the M6 near Killington (58.92) where heather is rapidly invading in many places. Heather also occurs in a number of places on the limestone, usually where there is sufficient depth of non-calcareous drift but also on herb-rich, calcareous drift as on Tarn Moor, Sunbiggin (66.06).
Alt. limit: 860 m, Skiddaw (26.28), 55 m below Baker's height on Skiddaw
Map 325   1375   LWFCY

[Erica ciliaris L.   Dorset heath
Listed by Hodgson from "Seaton brick-works" (0.3) where it was presumably introduced.

(C)]

E. tetralix L.   Cross-leaved heath
A common species over most of the county except in the more intensively agricultural areas and, surprisingly, rather uncommon in the Alston area and on the moors between Hartside and Geltsdale. It is typically a plant of *Sphagnum - Eriophorum* peatlands occurring in raised mosses, valley mires and upland blanket peat.
Alt. limit: 610 m, east of Cross Fell (72.34) and Bellbeaver Rigg (76.34)
Map 326   808   WFCY

E. cinerea L.   Bell heather
Fairly common on ungrazed acid moorland in the fells, especially in the Lake District, and on sea cliffs and dune-heaths along the west coast. It prefers drier

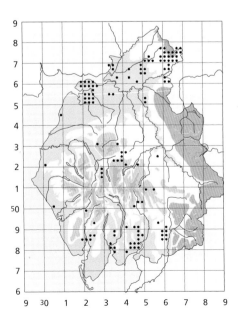

Map 324. *Andromeda polifolia*

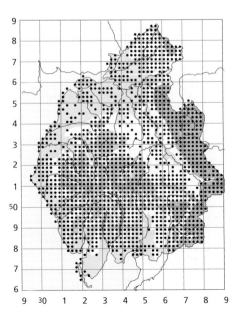

Map 325. *Calluna vulgaris*

situations than *Erica tetralix* and characteristically occurs in the fells on rock outcrops, particularly along gillsides. This may be a consequence of its sensitivity to grazing. Where non-calcareous outcrops are rare, as in much of the Pennines and the Bewcastle Fells, it can sometimes be quite elusive. It is often favoured initially by burning although subsequently largely replaced by *Calluna*.

Alt. limit: *c.*620 m, Great Rundale (72.26)
Map 327, Plate 66   612   WFCY

*E. herbacea L.
A very rare garden escape known in Cumbria only from Arnside (44.78s, MB, 1994, LANC) where a well-established colony occurs on a coastal limestone cliff.
                          1   W

*E. vagans L.   Cornish heath
Known only from Siddick (98.30, MP, 1995, LANC) where there are a few plants in rough grassland between the A596 and the sea.
                          1   C

Vaccinium oxycoccos L.   Cranberry
A frequent mossland species, mainly in the lowlands and conspicuously concentrated around the periphery of the Lake District. Like *Andromeda*, with which it often occurs, it grows on the Solway mosses and on the moors in the north-east, but it also occurs sparingly on the Pennine blanket peat. It usually occurs growing over *Sphagnum* hummocks and sometimes with the relatively mesotrophic *S. recurvum*. It can also grow beneath pine and birch where these are not too dense.

Plate 66. *Erica cinerea* and *Juniperus communis* subsp. *nana*, Scar Crags

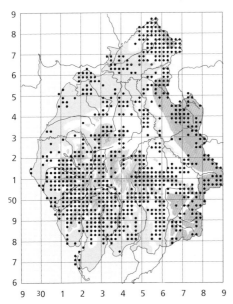

Map 326. *Erica tetralix*

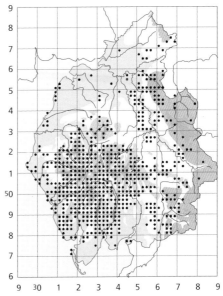

Map 327. *Erica cinerea*

In the lowland sites it usually flowers and fruits profusely.

Alt. limit: 700 m, Burnhope Seat (78.36)

Map 328   524   WFCY

*****V. macrocarpon** Aiton   American cranberry

This large-fruited North American cranberry was found in 1943 growing in *Sphagnum* by a small tarn between Nethertown and Braystones (9.0, C.D. Pigott; BM, CLE, *BEC Rpt* (1946) 12:736) where it was last seen in 1945.

(C)

V. vitis-idaea L.   Cowberry

Not infrequent in the Lake District and along the whole of the Pennines but also with a number of low-lying sites. It occurs, for example, near sea-level on Meathop Moss (44.80s) by Morecambe Bay, and on Wedholme Flow (2.5) by the Solway Firth and at 60 m in the wooded gorge of the Liddel Water at Penton Linns (42.76n). It has disappeared from its former sites on Stribers Moss, Holker and on Rusland Moss (3.8) but it is still present at several sites in the Eden valley, usually under pine or birch.

It usually occurs intermixed with *Vaccinium myrtillus*, both in the open and under trees, deciduous and coniferous, but it is much more sensitive than that species to grazing. In consequence it flowers and fruits well only when growing on rock ledges, usually on gillsides.

Alt. limit: 915 m, Scafell (20.06) and Helvellyn (34.14, both Wilson (1956))

Map 329   298   WFCY

V. uliginosum L.   Northern bilberry

Local and virtually restricted to the north-east. There are a few records from blanket peat on the Pennines, for example a single shoot on Knock Fell (74.30) and several colonies on Cold Fell (60.54), but most are from the Geltsdale and Eden lowlands where it occurs, for example, amongst *Calluna* and *Empetrum* on Cliburn Moss (56.24), with *Vaccinium myrtillus, Narthecium ossifragum* and *Erica tetralix* in a bog near Bailey Water, Bewcastle (52.76n) and amongst *Molinia* and *Calluna* on a hillside above the River Gelt (56.54). Most of these lowland populations are small, seldom flower and appear vulnerable to agricultural 'improvement' schemes. The individual plants are, however, often quite robust, more so than is usually the case in Scotland. There is a particularly fine population, with stems to 60 cm, by the River Gelt near Brampton (52.58n). The only Lake District site is on the High Street range (44.12), a site originally discovered more than a century ago by J. Backhouse.

This bilberry has declined considerably during this century in northern England, and south of the Scottish Border it is now restricted to Northumberland and Cumbria. It has disappeared from its old Lake District sites on Great Mell Fell (3.2), Whinfell Common (5.0) and Grisedale Pike, Coledale (1.2, G. Ward, 1944), and from near Gamblesby (6.3, H.H.,

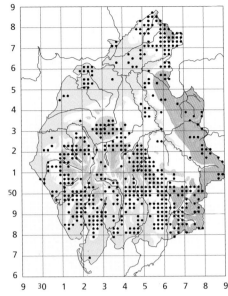

Map 328. *Vaccinium oxycoccos*

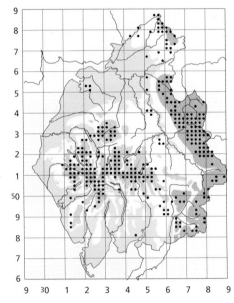

Map 329. *Vaccinium vitis-idaea*

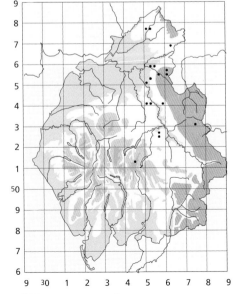

Map 330. *Vaccinium uliginosum*

1781, BM). Baker's records for Reake Mosses, Holker (3.7s) and Hodgson's for Lamplugh (0.2) are probably errors. It is nevertheless encouraging that the Survey records nearly treble the post-1930 *Atlas* squares and three squares appear to be new.

Alt. limit: 825 m, High Street range (44.12)

Map 330, Plate 16   17   W[(F)]C

V. myrtillus L.   Bilberry; Blaeberry

Very common throughout the uplands, on lowland heaths and mosses and even locally, like *Calluna*, on the limestone. It is essentially a fellside shrub, often growing under oak but, like *Vaccinium vitis-idaea*, frequently grazed out, surviving and fruiting in the shelter of the gills. It is, however, more resistant to heavy grazing than *Calluna*. With increasing altitude the plants become very dwarfed and a short wind-pruned 'turf' of bilberry is a common feature of many Lake District ridges. It also occurs in lowland pine, birch and oak woods on peaty and acidic soils.

Alt. limit: 965 m, Scafell Pike (20.06)

Map 331   1267   LWFCY

PYROLACEAE

Pyrola minor L.   Common wintergreen

Scattered throughout the Eden valley, the lowlands south of Carlisle and extending up to the North-umberland border. Elsewhere it is very rare with isolated occurrences at Carleton (02.08), south of Cockermouth (10.26) and in upper Dentdale (76.84). Most of the records are from rather open, damp coniferous plantations, as in Orton Woods, Great Orton (32.54), where it is quite abundant, and also, less commonly, under birch. At a number of such sites it is associated with *Goodyera repens* and it has been suggested that some populations may have been introduced when the plantations were established. There is another flourishing population on the disused

railway line in Smardale, Crosby Garrett (72.06).

It has disappeared over the last century from its Lake District sites, although there is an unconfirmed report from Far Easedale (3.0, M. Davies-Shiel, 1970). In the south-east of the county it has gone from Combe Scar, Dent (6.8s), the Lune gorge (6.9), Wilson's locality at Lily Mere, Killington (6.9), and Uldale, Sedbergh (7.9, Handley (1898)).

Alt. limit: 320 m, upper Dentdale (76.84)

Map 332   39   WCY

P. media Sw.

Listed by Baker from Brayton Woods (1.4), Kirklinton moors (4.6), Walla Crag, Derwent Water (2.2, Miss Wright, 1843, BM) and Stock Gill (3.0, F.A. Lees, *c*.1860s). Lees (1870) is quite definite that he collected it in Stock Gill although on subsequent visits he could find only *Pyrola minor* and a specimen of the latter collected in 1869 is in BM. A specimen is said to have been sent to H. Britten from Calthwaite (4.4) in 1907 and there are also records from Kingmoor Wood, Carlisle (3.5, P. Shepherd, 1884) and Dalston Wood (3.5, J. Parkin, 1935). In view of the frequency with which this species is confused with *P. minor* all records unsubstantiated by herbarium specimens are best disregarded.

This species is characteristic of coniferous woodland and *Calluna - Arctostaphylos* heath in north-east Scotland; the only extant English sites are in the north-east and on the Shropshire-Worcester border.

S   (WC)

P. rotundifolia L.   Round-leaved wintergreen
subsp. **rotundifolia**

The only records for this subspecies are from High Borrans, Windermere (4.0, herb. R. Meinertzhagen, *c*.1950, BM), perhaps the site of the present reservoir, and Sandford bog, Warcop (7.1, R.W. Robson, 1947, CLE, PLYP). The first seems somewhat unlikely and the specimen may be wrongly labelled but there is no reason to doubt the authenticity of the Sandford record.

(W)

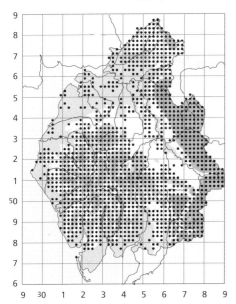

Map 331. *Vaccinium myrtillus*

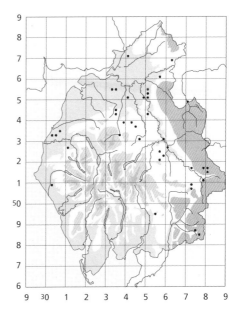

Map 332. *Pyrola minor*

subsp. **maritima** (Kenyon) E.Warb.
Rare; known only from grazed dune-slacks along the Duddon and Ravenglass estuaries. Most of the populations are small, but an exception is the flourishing one at Sandscale (18.74) where it grows intermixed with *Salix repens, Parnassia palustris, Hydrocotyle vulgaris* and *Anagallis tenella*. It was first recorded from here in 1913 (D. Lumb, *BEC Rpt* (1914) 3:389). The first Cumberland record was from Drigg (06.94, C.W. Muirhead, undated). These are the most northerly British records.

This subspecies is characterised by orbicular leaves, obtuse sepals and a short (4-6 mm) style.

Map 333   S 8   FC

Orthilia secunda (L.) House
Rare, occuring only in a few ravines and on cliffs in the Lake District, Cautley Spout, Sedbergh (68.96) and the basalt cliffs of High Cup Nick (74.24), where it has not been seen since 1971. Most of the sites are on acid rock, out of reach of sheep and with an undistinguished flora. There are additional records, contributed by DAR during the 1950s, from both sides of the Thirlmere valley (2.1, 3.1). Of the locations given in the early Floras, there are no recent records from Walla Crag, Derwent Water (2.2) or near Black Force, in the Howgill Fells (6.9, Handley (1898)). It seems likely that Hodgson's record from Westward, Wigton (2.4) is an error for *Pyrola minor*.

These sites, and one each in Wales and North-umberland and a few in southern Scotland, are the only extant ones south of the Scottish Highlands.
Alt. range: 250 m, Thirlmere (30.14, DAR, 1950) - 555 m, High Cup Nick (74.24)

Map 334   12   WCY

[**Moneses uniflora** (L.) A. Gray
A species of northern Scotland reported in the 19th century from a wood near Bardsea (3.7s), a record rightly regarded by Baker as an error. A specimen collected in 1820 from "Westmorland" (LIV) may have been mislabelled.

(F)]

MONOTROPACEAE

Monotropa hypopitys L.   Yellow bird's-nest
Very rare, seen only at three sites: Grange-over Sands (40.78s, R.J. Elliott, *c*.1970, KAG, 1980), the Sandscale dune-slacks (18.74), Arnside Park (42.76s, C.D. Pigott, 1973) and in Arnside (44.78s, MB, 1977), where the population has now been lost as a result of recent building operations. The Grange-over-Sands and Arnside Park populations occur under *Corylus* in deciduous limestone woodland, that at Arnside being of *Quercus, Fraxinus* and *Tilia cordata*. At Sandscale it apparently parasitises *Salix repens*. All the populations are small and capricious in their appearance from year to year.

It is listed by Baker from Barrowfield Wood, Kendal (4.9) and by Wilson from Whitbarrow (4.8s). There are also previous records from Grange-over-Sands (4.8s, W. Duckworth, 1907) and Sandscale (1.7, D. Lumb; W.H. Pearsall, 1913, both *BEC Rpt* (1913) 3:389) and Underlaid Wood, Arnside (4.7s, D. Walker, 1940). That cited by Baker from "Holme area", below Kirkby Lonsdale, is actually in Lancashire.

Plants with fewer flowers, a glabrous ovary and a shorter style are referred to subsp. *hypophegea* (Wallr.) Holmboe. The *Monotropa* records in the *Critical Atlas* for the Sandscale (1.7) and Arnside (4.7s) squares are both included in this segregate. Unfortunately there has been no recent critical study of the Cumbrian plants, which are the northernmost in western Britain.

6   WF

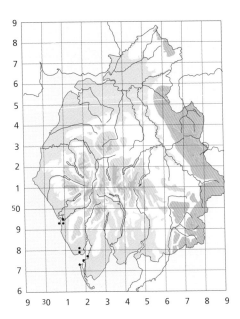

Map 333. *Pyrola rotundifolia* subsp. *maritima*

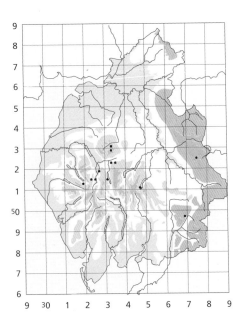

Map 334. *Orthilia secunda*

PRIMULACEAE

Primula vulgaris Hudson — Primrose

Generally common but scarce or lacking on the higher ground and in the more intensively farmed areas. It is typically a plant of deciduous coppice woodland, preferring the moister and often heavier soils. It is also frequent in hedgerows, along roadside and motorway verges and on railway banks, often in association with *Primula veris*. In the fells it occurs in ungrazed habitats, such as rock ledges and crevices by becks and gullies.

There are populations in a wood near Edenhall Grange (54.32) which include plants ranging in colour from light to deep pink and ones at Roadhead (50.74n), Spadeadam (62.70n), Bampton (48.18) and by the River Lune near Sedbergh (62.92) which include pink flowers, but none shows any other evidence to suggest hybridisation with garden cultivars.

Alt. limit: at least 500 m, Musgrave Fell, Brough (78.18)

Map 335   1120   LWFCY

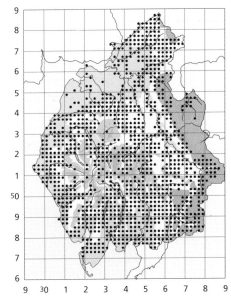

Map 335. *Primula vulgaris*

P. vulgaris x **P. veris** — False oxlip

P. x polyantha Miller

(*P.* x *tommasinii* Gren. & Godron)

Not infrequent in the Eden valley and south Westmorland, usually in scrub and woodland margins. A complex population which includes pink-flowered and also scapeless hybrids has been reported from near Dalton-in-Furness (22.74, PB, 1986). Such plants, and also deep red 'polyanthus' plants from Tebay (64.04), the M6 at Penrith (50.28) and similar pale yellow ones near Natland (50.88s) and Langwathby (56.32), probably result from hybridisation between *Primula veris* or *P. vulgaris* and garden 'polyanthus' cultivars, which in turn have been derived by hybridisation between these two species.

Alt. limit: *c*.260 m, Helbeck Wood, Brough (78.16)

Map 336   87   WFC

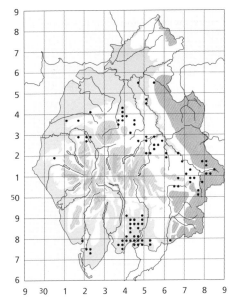

Map 336. *Primula vulgaris* x *P. veris*

P. veris L. — Cowslip

Very common in south Westmorland, north-west Cumberland south of the Solway Plain and throughout the Eden valley, otherwise scattered or rare. It is virtually absent from the Lake District and surprisingly very scarce in the extreme south-east, despite the numerous limestone exposures in Dentdale and Garsdale. Although typically a plant of base-rich or calcareous grassland, it is now often restricted to motorway verges, railway banks and scrub as a result of grassland 'improvement' and grazing. The conspicuous and large population along the A590 on the Silurian slate cutting between Newby Bridge and Haverthwaite (3.8) is largely derived from specimens transplanted from a threatened site near Barrow-in-Furness.

Alt. limit: 845 m, Great Dun Fell (70.32)

Map 337   540   LWFCY

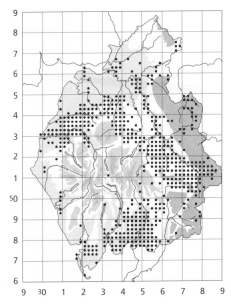

Map 337. *Primula veris*

P. farinosa L.   Bird's-eye primrose
Locally still fairly common in open, base-rich or calc-areous flushes in north Westmorland extending north-west to Bothel (18.38) and into the Lake District as far as Loughrigg (34.04). It thins out noticeably towards the Yorkshire and Lancashire borders. There are a few old records from Furness but it disappeared from its last site in Tilberthwaite Gill (30.00) about 1970. There is some evidence that it formerly occurred in the north-east of the county: Hardy (1881 p.284) says it "was said to bloom profusely in marshes near Brampton" and there is an 1864 specimen from Gilsland, v.c. 67/70 (6.6, R. Elliot) in the Hawick Museum.

    Like *Primula veris*, this most attractive primrose has suffered considerably as a result of grassland 'improvement' and increased grazing, also uprooting by gardeners. Hodgson refers to it as abundant "over several miles" between Tallentire and Ireby (1.3, 2.3), an area from which it has largely vanished since the war; similarly it is now very scarce in the Sedbergh area although Handley (1898) described it as "in all our valleys". There are also no recent records from Fing-land Rigg (2.5, J.H. Vine-Hall, 1950), where it may have been introduced, Cliburn Moss (5.2, H. Britten, 1923, CLE) or Newton Reigny moss (46.30), where it was buried in the 1960s beneath builders' rubble, or from the Wigton square (2.4, W. Scott. 1958). The *Scarce Plants* record for the Langwathby square (5.3) is anonymous and undated and best regarded as dubious. The thriving introduced population at Rough Syke, Moor House (74.32) is omitted from the map.

    The British distribution of bird's-eye primrose is centred on the Pennines, from Lancashire to Cumbria. It has disappeared from the Lothians and most of its sites to the east of the Pennines.
Alt. limit: 472 m, Tailbridge Hill (80.04))
Map 338, Plate 24   S 165   W(F)CY

***Hottonia palustris** L.   Water violet
An attractive aquatic plant of eastern and southern England listed by Wilson from the dykes of Brigsteer Moss (4.8s), the last record from there being in 1840 (G.E. Martindale, KDL), by Baker from the River Kent, which seems unlikely, and the mill-pond at Bardsea (3.7s), and by Petty (1897 p.89) from Urswick Tarn (2.7).

    These were the northernmost records for this species in western Britain.

                                    (WF)

Lysimachia nemorum L.   Yellow pimpernel
Frequent to common although scarce on the Solway Plain and in parts of the Eden valley. It occurs in damp habitats in deciduous woodland, by ponds and ditches and, very characteristically, by becksides in the fells.
Alt. limit: at least 690 m, Ousby Fell (68.36)
Map 339   1007   LWFCY

L. nummularia L.   Creeping jenny
Scarce but well distributed, occurring in similar habitats to *Lysimachia nemorum* but more obviously lowland and several of the populations almost certainly represent garden escapes. Nevertheless, it seems likely that, as in Co. Durham (Graham 1988), this south-eastern species could be native.
Alt. limit: *c.*250 m, above Casterton (64.78s)
Map 340   99   WFCY

L. vulgaris L.   Yellow loosestrife
Occasional in ungrazed tall-herb communities by tarns and rivers and in sheltered bays by the lakes, often forming dense stands with *Lythrum salicaria* and *Filipendula ulmaria*.
Alt. limit: 200 m, Shap (56.12)
Map 341   254   WFCY

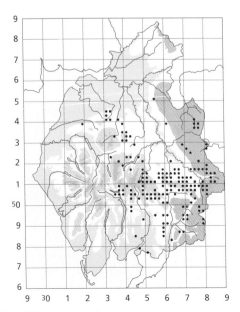

Map 338. *Primula farinosa*

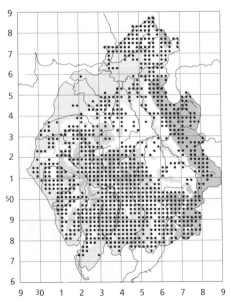

Map 339. *Lysimachia nemorum*

*L. ciliata L.

A North American species collected in 1953 by K.M. Hollick near Belle Grange, on the west side of Windermere (38.98, DBY; LANC), where it still grows by the laneside. It was earlier noted by Baker and Hodgson as naturalised since at least 1820 on the roadside near Monkhouse Hill, Sebergham (3.4, W. Borrer, 1844, BM) and in a slate quarry on Warnell Fells (3.4). It was seen at the former site on many occasions up to 1930 (M. Horsfall, CLE) and there are eight collections in the BM alone.

    This species can be distinguished from the much commoner *Lysimachia punctata* by its glabrous leaves and usually solitary flowers.

1   F(C)

*L. punctata L.   Dotted loosestrife

A native of southern and central Europe and a popular, if invasive, garden plant which is becoming increasingly established from garden throw-outs along roadside verges and riverbanks, especially in the south. The first record for Westmorland is from Clappersgate, Ambleside (36.02, K.M. Hollick, 1947), for Furness, the *Atlas* record for the Hawkshead square (3.9) and for Cumberland an unlocalised collection from 1934 (C.E. Sharp, CLE) and one from by the River Ehen between Ennerdale Bridge and Cleator Moor (04.14, CCH, 1978, *Watsonia* (1979) 12:355).

Alt. limit: 430 m, west of Nenthead (76.42)

Map 342   131   LWFCY

*L. terrestris (L.) Britton, Sterns & Pogg.

Locally frequent in reed beds and low-lying lakeside meadows by Windermere and along the River Leven, also west of the lake by Esthwaite Water (36.94), Boretree Tarn and Rusland Pool (34.86), and at High Dam, Finsthwaite (36.88). It grows with *Iris pseudacorus*, *Lysimachia vulgaris* and *Lythrum salicaria*, and by Rusland pool amongst *Carex paniculata*.

    The first record dates from 1883 when it was shown to G.E. Martindale by L. Jackson on the west side of Windermere (3.9, BM, *Watson BC Rpt* (1887) 1886-1887:5), a specimen being subsequently identified by J.G. Baker. There is an *Atlas* but no Survey record for the Windermere square (4.9).

    In North America three types of plants occur producing only flowers, flowers and bulbils or only bulbils. Only the latter two types occur in Cumbria. Some populations consist only of bulbiferous plants but in most flowering plants predominate. However, no capsules develop and flowering is followed by the rapid development of the elongate, purplish, ribbed and constricted bulbils, which are solitary in the axils of the lower leaves, each subtended by a pair of small leaves.

    The above localities were the only ones known in Britain for this eastern North American loosestrife until it was discovered in Lancashire near Fleetwood by

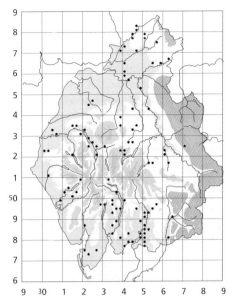

Map 340. *Lysimachia nummularia*

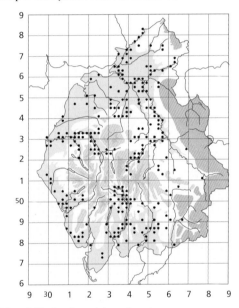

Map 341. *Lysimachia vulgaris*

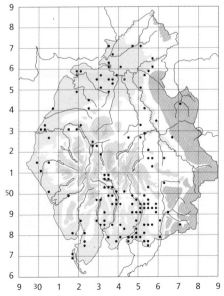

Map 342. *Lysimachia punctata*

CFS & JS in 1985.
Alt. limit: 205 m, Boretree Tarn (34.86)
Map 343, Plate 38   14   WF

*L. thyrsiflora L.
A northern European species reported by Baker from Sellafield Tarn (0.0) where it was presumably introduced and
which no longer exists. It is thought to be native in north-east
Yorkshire and in central Scotland.

                         S   (C)

Trientalis europaea L.   Chickweed wintergreen
A northern species known in Cumbria from only four
sites. It occurs sparingly by a forestry ride north of
Bewcastle (5.7n, LANC), where it was seen by DAR in
1958 and shown to G. Maughan in 1972. More than
100 plants were discovered to the west in woodland
rides near Catlowdy (4.7n, D.R. Briggs & T.R.
Harwood, 1987). The only other extant sites are on the
Solway Plain in an open area of birchwood near
Kirkbride (2.5, DAR, 1959, CLE; 1986) and in south
Westmorland on Firbank Fell (6.9, TW & KW, 1985,
Watsonia (1986) 16:190), where more than 20 plants
were discovered growing under *Calluna* at J.A.
Martindale's 1909 site (BM, KDL). It formerly occurred elsewhere in the Bewcastle Fells. G.A.K. Hervey,
for example, found it during the 1940s on the Gilsland
moors near Wiley Sike (6.7n) and EHR on moorland
north of Blackpool Gate (5.7n, c.1960s). It grew mostly
amongst *Molinia caerulea* and *Trichophorum cespitosum*. Although it has survived afforestation at one
site, there must be doubt about its ability to withstand
clear-felling.

It has never been refound at three other early
sites. Hodgson (p.381, 1899) records it from a hillside

above Boot in Eskdale (1.0, E. Pearson, 1897), and
Baker refers to a report from the "hills of north-west
Furness", but this is probably an error. Lees (1888)
mentions an old Yorkshire record from a heathy larch
plantation in Garsdale, later identified by Cheetham
(1939) as Widdale Carr. This is probably the "Baugh
Fell" site discovered by R. Butterfield on a Yorkshire
Naturalists' Union excursion to Garsdale Head (7.9, not
7.8 as shown in the *Atlas*, *Naturalist* (1929) 1929:348).
Unfortunately, this site was not precisely noted and the
plant has never been refound.
Alt. limit: 305 m, Firbank Fell (6.9)
Plate 67   4   W[(F)]C(Y)

Anagallis tenella (L.) L.   Bog pimpernel
Frequent in the south-west and the Lake District growing in short damp and rather open turf around flushes
and along becksides, also on the coast in dune-slacks
where it is often abundant and flowers profusely.
Despite the indication by Duncan & Robson (1977
p.88) that this species is widespread in the Pennines,
there are neither *Atlas* nor Survey records from there.
Alt. limit: 320 m, Glencoynedale (36.18, Hodgson)
Map 344   226   WFCY

A. arvensis (L.) L.   Scarlet pimpernel
subsp. **arvensis**
Fairly frequent. This southern lowland plant has a pronounced coastal distribution in Cumbria. The map
reflects well the dry climate of the area around Carlisle
and the head of the Solway Firth. Most of the records
are from cultivated ground: gardens, allotments and
arable fields, and waste ground, but it also occurs on
the coastal sand-dunes.

The only record of the blue-flowered f. *azurea*

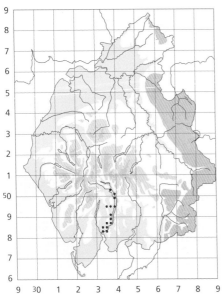

Map 343. *Lysimachia terrestris*

Plate 67. *Trientalis europaea*

N. Hylander is from Shannon House, Workington (9.2, E.J. Glaister, 1889, CLE), a locality cited by Hodgson under *Anagallis caerulea*.
Alt. limit: 180 m, Tebay (60.04)
Map 345   241   LWFCY

subsp. **foemina** (Miller) Schinz & Thell.
(subsp. *caerulea* Hartman)
The only records of this blue-flowered taxon are specimens from Silloth (1.5, J. Leitch, 1889, 1890, CLE; H. Stewart, 1930, CLE), the Etterby gravel beds of the River Eden, Carlisle (3.5, T.S. Johnstone, 1901, CLE) and a record from The Green, Millom (1.8, W.H. Pearsall, 1933, *BEC Rpt* (1934) 10:533). In the absence of specimens it is impossible to know which of the following records of *Anagallis caerulea* listed by Baker and Hodgson belong to this subspecies or to subsp. *arvensis* f. *azurea*: Roosebeck (2.6s), Whitehaven (9.1), Maryport and Camerton Hall (0.3), also a later one from Thwaites, Millom (1.8, M. Cross, 1948) and Handley's (1898) from Birks Mill, Sedbergh (6.9). Baker also refers to blue- and dull purple-flowered plants from Colton (3.8), the latter presumably belonging to the rare var. *purpurascens* Thurston & Vigurs.

Subsp. *caerulea* differs from subsp. *arvensis* f. *azurea* in having shorter flowering pedicels and petals with fewer or no marginal glandular hairs.

[(F)](C)[(Y)]

A. minima (L.) E.H. Krause   Chaffweed
Known only from dune-slacks at Ravenglass (04.96, 06.96), Eskmeals dunes (08.92, MMG, 1993, LANC), close to the shore of the Kent estuary near Ulpha (44.80s, P. Jepson, 1981, *Watsonia* (1983) 14:426) and by a spring on the south-west side of Arnside Knott (44.76s, C.Webb, 1987).

The only other records for this diminutive and probably under-recorded species are Baker's from Ravenglass (0.9) and coastal meadows at Newton-in-Cartmel (3.7s), Wilson's own from Terrybank Tarn (5.8s, 1935, YRK; conceivably his Sedbergh site (*Naturalist* (1932) 1932: suppl. p.vii)) and further Cumberland ones from between Armathwaite and Cotehill (4.4, W.W. Mason, 1914), Eskdale and Wasdale (1.9, 1.0, M. Cobbe, 1924; CLE, *BEC Rpt* (1925) 7:583) and Thurstonfield Lough (3.5, T.S. Johnstone, 1899, CLE), where it was last seen by C.W. Muirhead (CLE) in 1949.

5   W(F)C(Y?)

Glaux maritima L.   Sea milkwort
Common in salt-marshes, occurring over practically the whole zonation although absent from the lowest pioneer zone and scarce in the upper mature marsh. It is well able to withstand grazing and it is a common constituent of *Puccinellia - Festuca rubra* cropped turf. Together with *Plantago maritima* and *Juncus gerardii*, it has persisted on the summit of Hartside (64.40, RWMC, 1994) on soil originating from the west coast.
Alt.limit: 570 m, Hartside (64.40)
Map 346   111   WFC

Samolus valerandi L.   Brookweed
Scattered around the coast, usually in flushed open areas at the upper limit of the salt-marsh or on or below irrigated rocks, also in dune-slacks. There are a few records from a little way inland, for example near Beetham (48.78s) in seepage areas over limestone and also by Urswick Tarn (26.74). Wilson refers to such a site on the top of Whitbarrow (4.8s) but it no longer occurs there, nor at his Hale Moss site (5.7s).
Map 347   41   WFC

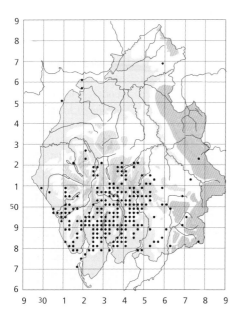

Map 344. *Anagallis tenella*

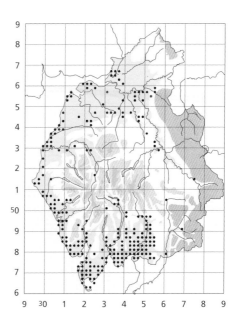

Map 345. *Anagallis arvensis*

HYDRANGEACEAE

*Philadelphus coronarius L.   Mock orange
An occasional garden escape found in 1987 by a roadside plantation near Barbon (62.80s, CEW, LANC, *Watsonia* (1988) 17:189), also near Windermere (38.98, CEW, 1990, LANC), at Arnside (46.78s, AF, 1991), by a lane west of Coniston (28.96, PB, 1993) and, in the north, in Wreay Wood (44.48, MG, 1991, LANC) and south of Alston (70.44, REG, 1993).
Alt. limit: 285 m, Alston (70.44)

     5   WFC

*Deutzia scabra Thunb.
Recorded from a wood at Arnside (44.78s, MB, 1988; LANC), woodland north of Great Salkeld (56.36, RWMC, 1991, LANC) and at Ulpha (20.92, CEW, 1995). The two collections were determined by E.J. Clement.

     3   WC

GROSSULARIACEAE

*Escallonia macrantha Hook. & Arn.
Known only from the coast at Siddick (98.30, MP) where several bushes were discovered in 1994 well established on waste land.

     1   C

Ribes rubrum L.   Red currant
A lowland species, common throughout the Eden valley but otherwise rather scattered. It occurs in hedgerows, scrub and woodland, often in damp places by rivers and becks, also near houses. It is probably widely bird-sown and perhaps also distributed by water.
Alt. limit: 370 m, Shap Fells (54.06)
Map 348   434   WFCY

*R. spicatum Robson   Downy currant
Seen at seven sites, all but one in the north. All the Survey records and pre-Survey herbarium records have been verified by N.K.B. Robson. It has been found by the Kershope Burn and in a nearby lane (48.82n, REG, 1977, herb. R.E.G.), in damp scrub near Liddel Lodge (46.78n, REG, 1990, LANC), in a roadside hedge north east of Catlowdy (46.76n, REG, 1988), on a stream-bank near Kirkbride (24.56, REG, 1985, LANC), on the rocky and sandy banks of the River Eden near Holmwrangle (50.48, FJR, 1988, LANC) and Arma-thwaite (50.46, FJR, 1991, LANC) and by the River Rawthey just east of Sedbergh (66.90, MB & R. Dalton, 1984; LANC), the last probably being Wilson's 1922 Sedbergh site (YRK). None of the sites is on limestone, its habitat in the Ingleborough area, and it seems prob-able that the species is not native in Cumbria (Roberts 1994). *Ribes spicatum* is practically confined to northern England, from Yorkshire northwards, and the eastern Highlands of Scotland.

This species is listed by Baker from roadside hedges near Keswick (2.2) and north-east of Kendal (5.9). There are later records from near Egremont and Ennerdale (0.1, H. & J. Adair, 1895, 1904, BM, YRK, *BEC Rpt* (1905) 3:22), Wetheral (4.5, C.W. Muirhead, 1946, CLE) and Haybergill, near Warcop (7.1, Anon., 1938, BM). Wilson's records are from Docker Brow, Kendal (5.9), near Casterton (6.7s), Low Borrow Bridge (6.0) and his own from Flakebridge Wood (6.2/7.2, 1934, YRK). *Scarce Plants* has additional pre-1930 records for the Distington (0.2), Hawkshead (3.9), Arnside (4.7s, v.c.60/69) and Gilsland (6.6, v.c.67/70) squares. That for the Burton-in-Kendal square (5.7s) is from Lancashire. It is difficult to know what credence to give these early records in the absence of supporting specimens since this species has been much confused with *R. rubrum*.

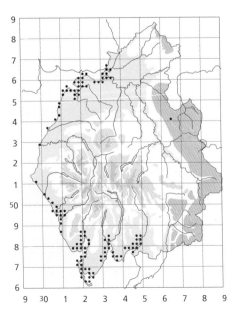

Map 346. *Glaux maritima*

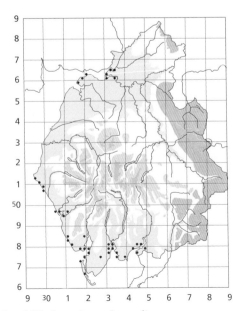

Map 347. *Samolus valerandi*

This species is conspicuous in flower with its erect, yellowish racemes. The most reliable distinguishing characters, best seen when fresh, are the contiguous anther lobes and cup-shaped receptacle which lacks the raised ring between the anthers and style which is present in the flatter receptacle of *R. rubrum*.
Alt. limit: 200 m, Liddel Lodge (46.78n) and Catlowdy (46.76n)
Map 349   S 7   (WF)CY

R. nigrum L.   Black currant
Quite widespread and in similar habitats to *Ribes rubrum* but less common and seldom far from houses.
Alt. limit: at least 330 m, Garrigill (74.40)
Map 350   243   WFCY

*R. sanguineum Pursh   Flowering currant
Occasional in the south towards Morecambe Bay, rare or absent elsewhere. The bushes originate either as garden throw-outs or by bird-dispersal from gardens.

The first record is from the Dalton-in-Furness square (2.7) where it was noted by K. Jones in 1956. It is not possible to identify from the Survey records the first ones for Westmorland and Cumberland.
Alt. limit: 300 m, Helbeck Wood, Brough (78.16)
Map 351   58   WFCY

R. alpinum L.   Mountain currant
Largely restricted to the south-east where it occurs in roadside hedges, in limestone woodland and on rocky becksides where it may well be native. There is an outlying record from north-west of Renwick (58.44) and earlier ones from a hedge at Unthank, Carlisle (3.4, E. Blezard, 1940, CLE; PLYP) and Cumwhitton (5.5, F.H. Day, 1920, *Naturalist* (1920) 1920:239), and *Scarce Plants* records for the Wast Water (1.0), Carrock Fell (3.3), Carlisle east (3.5), Kirkby Thore (6.2), Sedbergh (6.9), Mallerstang (7.8) and Raven-

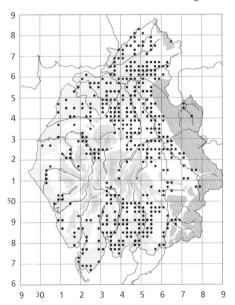

Map 348. *Ribes rubrum*

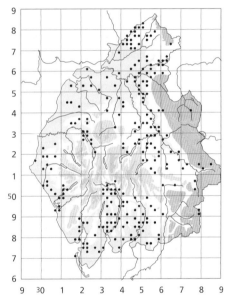

Map 350. *Ribes nigrum*

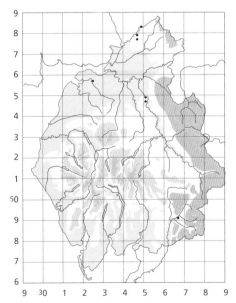

Map 349. *Ribes spicatum*

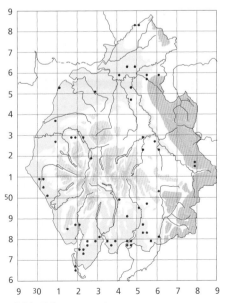

Map 351. *Ribes sanguineum*

stonedale (7.9) squares in none of which was it seen during the Survey.

Alt. limit: 230 m, Longsleddale (48.04) and east of Grayrigg (58.96)

Map 352   S 10   W(F)C(Y)

R. uva-crispa L.   Gooseberry

A common plant of lowland hedges and woodland, also by stone walls. It is one of the earliest shrubs to come into leaf. Most Floras assume that the bushes are bird-sown from gardens, but Allen (1984) says that in the Isle of Man it was widely planted in hedgerows.

Alt. limit: 410 m, Leadgate (68.42)

Map 353   1022   LWFCY

CRASSULACEAE

*Crassula helmsii (Kirk) Cockayne

First seen in 1986 in a garden pond at Bampton Grange (50.16, D. McClintock). It was found in a pond at Sile-croft in 1988 (12.80, R. Jerrans) and in 1992 on the north-east side of Derwent Water (26.20, 26.22, GH, LANC, det. C.D. Preston, *Watsonia* (1993) 19:283), growing in 70 cm of water with *Lobelia dortmanna*. It was later found by J. Hughes close to the last site in short lakeside turf as well as washed up in quantity on the stony shore. In 1995 it was found washed up at the north end of Coniston (30.96, PB, LANC).

     This invasive antipodean species has spread rapidly over the last 20 years but is fortunately still rare and local in the north of Britain.

Alt. limit: 175 m, Bampton Grange (50.16)

                            4   WFC

Umbilicus rupestris (Salisb.) Dandy   Navelwort

Occasional and mainly in the south-west, occurring on stone walls, usually near villages and farms, and only rarely on rock outcrops, as at Strands (12.04) and Waterhead (36.02). There are no recent records from the north, although there is a post-1930 *Atlas* dot for the Silloth square (1.5) and a 1942 collection from Dalston (3.4/3.5, Mrs Inglis, CLE), and the species seems to be on the decline.

Alt. limit: *c.*150 m, Crook (44.96)

Map 354   43   WFC

*Sempervivum tectorum L.   Houseleek

Rare but widely distributed, growing usually on walls and roofs and rarely seen away from buildings. Its recording, therefore, has been unsystematic and it has not been mapped. One unexpected find was on a lime-stone road cutting near Penruddock (42.26). Hodgson says the leaves were "much in use among the peasantry as a cooling application to malignant sores" and during this century it has been used as a cure for warts.

                        17   WFC

Sedum rosea (L.) Scop.   Roseroot

Present and conspicuous in most of the basic gullies in

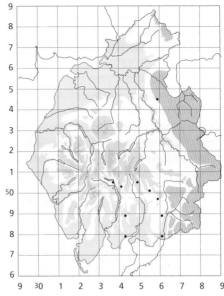

Map 352. *Ribes alpinum*

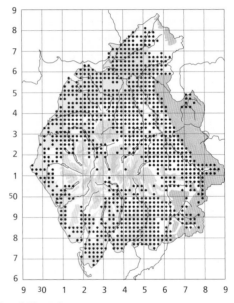

Map 353. *Ribes uva-crispa*

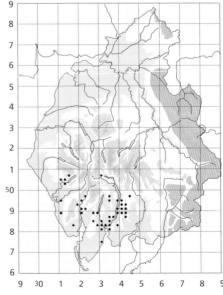

Map 354. *Umbilicus rupestris*

the Lake District growing on rock slabs, in crevices and on moist ledges. By contrast it is surprisingly local in the Pennines, occurring only between Murton Fell (74.24) and Cross Fell (66.34) and growing on both limestone and whinstone. There is an interesting outlying record in the Lune gorge (58.00), not far from that cited by Handley (1898) from near Black Force (6.9). The *Atlas* records for the Broughton-in-Furness (2.8) and Holker (3.7s) squares must be errors or refer to garden escapes and should be disregarded.

Alt. range: 305 m, Wast Water Screes (14.04) - 885 m, Helvellyn (34.14)

Map 355   75   WFC(Y)

S. telephium L.   Orpine; Livelong

In contrast to the last this is essentially a lowland species. It is locally frequent on the Solway Plain between Aspatria (1.4) and Silloth (1.5), and further south along the coast as at St Bees Head (9.1), and also in the south-east of the county. It occurs usually by roadside hedges and walls, where it is particularly prominent when flowering and where it may be native. It grows also on limestone pavement and on some of the lower crags in the Lake District.

Alt. limit: 320 m, above Coniston (28.98)

Map 356   273   WFC(Y)

*S. spurium M. Bieb.

An occasional garden escape but apparently increasing. There are no records in the Floras and the *Atlas* shows only three in the north of England, all post-1930. Two of these are from Cumbria, from the Lorton (1.2) and Hawkshead (3.9) squares. It occurs on walls, rocks and roadside banks.

Alt. limit: 270 m, Greystoke Moor (40.28)

Map 357   28   WFCY

*S. rupestre L.

(*S. reflexum* L.)

A rather rare garden escape but chiefly in the south and west, growing usually on walls but occasionally on rock faces. In a lyrical aside Hodgson comments that it is "grown in flower-pots and sedulously trained through and over the rounds of miniature ladders by cottage damsels", hence presumably the old Lakeland name 'Love-in-a-chain'.

Alt. limit: 260 m, Haweswater (48.12)

Map 358   35   WFCY

*S. forsterianum Smith   Rock stonecrop

The only Survey records are from a wall below Bank End, Torver (26.92, JA, 1980), roadside walls near Askam-in-Furness (20.76, PB, 1994) and Pull Wyke (36.02, CFS & JS, 1985; 34.02, 36.00), the latter probably the same site as reported by K.M. Hollick in 1950, on an old slag tip at Roanhead, Dalton-in-Furness (20.74, PB, 1992, LANC), and on a semi-bare roadside embankment west of Sedbergh (60.92, GH, 1995, LANC).

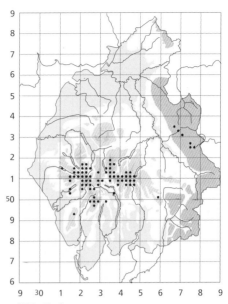

Map 355. *Sedum rosea*

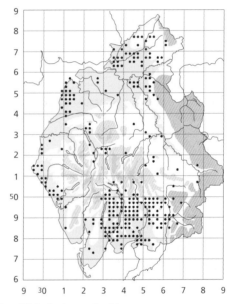

Map 356. *Sedum telephium*

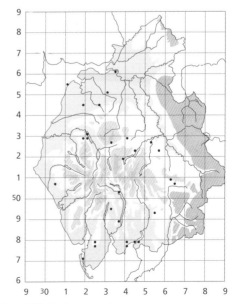

Map 357. *Sedum spurium*

Baker gives records from Borrowdale (2.1), Broughton-in-Furness (2.8) and Sawrey and the Windermere ferry (3.9); Hodgson from Ennerdale Lake shore (0.1/1.1) and Greenup Gill, Borrowdale (2.1), erroneously plotted as the Helvellyn square (3.1) in *Scarce Plants*, and Wilson from the roadside near Grasmere (3.0) and near Underbarrow (4.9). In 1954 it was seen by G. Wilson in the Grizebeck square (2.8). There are also anonymous *Scarce Plants* records from the 1950s for the Cark (3.7s) and Newby Bridge (3.8) squares and a mis-plotted one for the Helvellyn square (3.1).

This species is native only in Wales and south-west England.

It is often confused with *Sedum rupestre*, but its generally leafless lower stems and the dense terminal rosettes of the sterile shoots are very characteristic.

S 6   WF(C)

S. acre L.   Biting stonecrop
Very common on the limestone and on open grassland, sand and shingle by the coast. Elsewhere it is largely restricted to mortared walls and roofs, especially around farms, although present in open, sandy ground by the River Caldew at Dalston (36.50, 38.52). It is a very characteristic plant of sand-dunes, of fine limestone debris and of open, occasionally irrigated areas on the surface of limestone pavements.
Alt. limit: 845 m, Great Dun Fell (70.32)
Map 359   761   LWFCY

*S. sexangulare L.
Seen only on a laneside on Kendal Fell (50.92, TW, 1986), perhaps the same locality where it was noted some 30 years previously by J.N. Frankland. The only other records for this central European species are those listed by Baker from Workington (9.2), Watermillock (4.2) and Grange-over-Sands (4.7s).
Alt.: 170 m, Kendal Fell (50.92)
                           1   W(FC)

*S. album L.   White stonecrop
A fairly frequent and locally abundant garden escape, long naturalised on wall-tops and roofs, occasionally on rocks far from houses and, less commonly, on roadsides.
Alt. limit: 570 m, Hartside (64.40)
Map 360   278   WFCY

S. anglicum Hudson   English stonecrop
Fairly frequent, especially in the south-west, rare in the east. This is a very characteristic plant of open ground on acidic outcrops, especially on the Silurian slates of the southern Lake District where it usually grows with *Rumex acetosella*. It is also very common along the south-west coast and it is generally much commoner than is suggested in the *Atlas*.
Alt. limit: 550 m, Coniston Old Man (26.96)
Map 361   305   WFCY

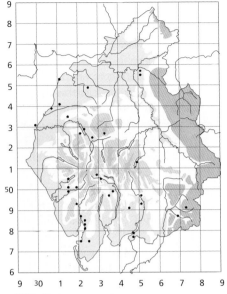

Map 358. *Sedum rupestre*

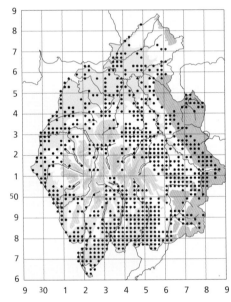

Map 359. *Sedum acre*

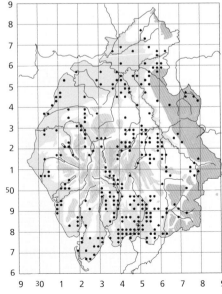

Map 360. *Sedum album*

S. villosum L.   Hairy stonecrop
Frequent in the north Pennines, especially on the eastern slopes of the watershed. There is only one Survey record from south of Stainmore, although recorded in the *Atlas* from the Mallerstang (7.9) and Dentdale (7.8) squares. This eastern species is very rare in the Lake District. The cluster of records from the head of Kentmere (42.08, E.G. Hall, 1985; 44.08, CFS & JS, 1990) and Harter Fell (46.08, RWMC, 1987) are of particular interest since it was from the head of Longsleddale (5.1) that Lawson recorded it three centuries earlier. The only other Lake District sites are Wolf Crags (34.22, E.G. Hall, 1959, RWMC, 1986) and two sites in nearby Mosedale (32.20, 34.20, RWMC, 1990). The usual habitat is rather open, stony flushes although it also occurs amongst mosses in small sykes.

The record given by Baker from "north of Furness" is very suspect as is also an herbarium specimen from Loweswater (1.2, J. Leitch, 1876, CLE). Leitch did not apparently collect in the Lake District and the specimen is probably mislabelled.
Alt. range: 460 m, Hartside (66.42) - 850 m, Cross Fell (68.34)
Map 362   S 51   W[(F)]C

SAXIFRAGACEAE
(PARNASSIACEAE)

*Astilbe cf. x **arendsii** (Morren & Decne) A. Gray
Material apparently referable to this garden species was collected during the Survey from by the River Caldew at Sebergham (34.40, RWMC, 1985, LANC), in woodland at Rigmaden, near Barbon (60.84s, CEW, 1987, LANC), on a laneside verge north of Grasmere (32.08, CFS & JS, 1994, LANC) and on the banks of the River Irthing above Gilsland (62.66, GH, 1994).
Alt. limit: 145 m, above Gilsland (62.66)
4   WC

*Rodgersia aesculifolia Batalin
Naturalised by the pool in Witherslack woods (42.84s, R. Young, 1971, AMDE) where it still occurs.
1   W

*Darmera peltata (Torrey ex Benth.) Voss ex Post & Kuntze
Found in 1988 well naturalised by the River Eden downstream from Nunnery Walks, Staffield (52.42, RWMC, LANC, *Watsonia* (1989) 17:472) and the following year persisting as a garden escape in a shady laneside east of Stainton (52.84s, CEW, 1989; LANC, *Watsonia* (1992) 19:147).
2   WC

Saxifraga hirculus L.   Yellow marsh saxifrage
A rare plant of high-level bryophyte-rich flushes on the northern Pennines. It is an inconspicuous species, the flowering stems being often eaten. Several of the Westmorland sites were discovered in the middle of the last century, but it was not recorded from Cumberland until 1958 when it was found on Cross Fell (70.38, DAR, CLE, *Watsonia* (1979) 12:353). The majority of the extant sites in Britain for this saxifrage are in Cumbria.
Alt. range: 550 m, south of Cash Burn, Alston Moor (70.38) - 730 m, Knock Fell (70.28)
Map 363, Plate 68   R 14   WC

*S. cymbalaria L.
The only records for this small but attractive yellow-flowered annual are of a few plants as a garden weed and on walls at Great Corby (46.54, CS, 1977; MS, 1984) and Warwick (44.56, CS, *c.*1980), at Winderwath, Culgaith (58.28, AW, 1986), and in Carlisle (40.54, EEM, 1994, LANC), where it appeared in flower pots which may have originated from Wetheral.
4   WC

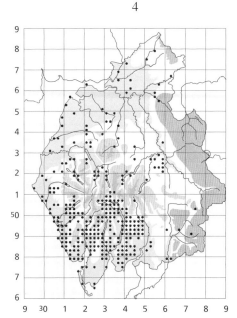

Map 361. *Sedum anglicum*

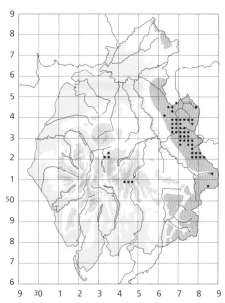

Map 362. *Sedum villosum*

S. nivalis L.   Alpine saxifrage
Extremely rare, seen only on the whinstone at High
Cup Nick (7.2), where it was originally discovered by
J. Backhouse in 1843 (E), and in a gully on the east
side of Helvellyn (3.1). The plants are small, to 13 cm,
and both populations are extremely small, fewer than
six plants, and very vulnerable to rock-falls. There
are additional 19th century sites listed by Baker on
Scafell (2.0), the Thirlmere side of Helvellyn (3.1) and
Backhouse's in the High Street range (4.1, 1869, E). At
neither of these has it been seen this century.

Together with a few sites in Snowdonia and the
Moffat hills, these are the only ones in Britain south of
the Scottish Highlands.
Alt.: 550 m, High Cup Nick (7.2) and 855 m, Helvellyn
(3.1)

                    S 2   W(C)

S. stellaris L.   Starry saxifrage
One of the more widespread mountain plants although
never really common. It grows in irrigated gravel
flushes but is especially characteristic of bryophyte-rich
high-level springs where it grows with *Epilobium
alsinifolium*, *Montia fontana* and *Cochlearia pyrenaica*. Like the *Cochlearia*, it may be carried far downstream.
Alt. range: 130 m, Lune gorge (62.98) - 915 m,
Helvellyn (34.14)
Map 364   266   WFCY

*S. umbrosa L. x **S. spathularis** Brot.   London pride
S. x **urbium** D. Webb
An occasional but usually very persistent garden throw-
out of waste ground, roadsides, banks and quarries,
occasionally in very natural surroundings as on a limestone waterfall of the River Dee above Dent (72.86).
Wilson's two records of *Saxifraga umbrosa*, one of
which dates from 1744, and also Winch's (1825) from
Ormathwaite (2.2) no doubt belong here. It is not poss-

ible from the Survey records to identify the first definite
Furness record.
Alt. limit: *c.*300 m, Garsdale Head (78.90)
Map 365   57   WFCY

*S. spathularis Brot. x **S. hirsuta** L.
S. x **polita** (Haw.) Link
Long-naturalised on old walls and pathsides by the
river above Gilsland (62.66, C.W. Muirhead, 1947,
PLYP; MG, 1991) and found during the Survey in an
old quarry at Arnside (44.78s, K. Hearn, 1972, AMDE;
both LANC and det. R.J. Gornall).
Alt. limit: 140 m, Gilsland (62.66)
                    2   WC

*S. hirsuta L.   Kidney saxifrage
A garden escape found in 1975 with other well-
established throw-outs in a roadside ditch at Rather
Heath, Crook (48.94, GH, LANC, *Watsonia* (1979)
12:353). It is also known from between Force Forge
and Satterthwaite (32.90, PB, 1987, LANC), Thwaite
Flat, Dalton-in-Furness (20.74, PB, 1991, LANC), in a
hedge near Grizebeck (24.84, JA, 1996), by Bull Pot,

Plate 68. *Saxifraga hirculus*

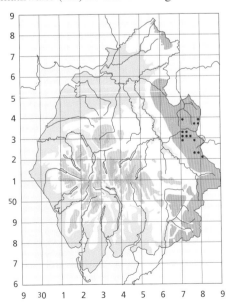

Map 363. *Saxifraga hirculus*

Casterton (66.80s, GH, 1988, LANC) and by a beckside at Heads Nook, Wetheral (48.54, FJR, 1985, LANC). There is a also an earlier specimen collected in woodland at Kendal (5.9, E. Hodgson, 1960s, LANC). All the material was determined by D.A. Webb or R.J. Gornall.
Alt. limit: 300 m, Bull Pot (66.80s)
6   WF

S. oppositifolia L.   Purple saxifrage
Scarce but quite widespread on sheltered base-rich rock ledges and gullies and, more rarely, in grassland, across the central Lake District from the Wast Water Screes to the High Street fells. It is easily overlooked since, as Hodgson observed, it flowers "before tourist pedestrians have begun to scale the lofty heights on which it grows".
There are additional post-1930 *Atlas* records for the Duddon (2.9) and Borrowdale (2.1) squares.
Alt. range: 245 m, Wast Water Screes (14.04) - 885 m, Helvellyn (34.14)
Map 366   17   WC

S. aizoides L.   Yellow mountain saxifrage
A common species of Lake District gills, conspicuous in mid-summer with its festoons of yellow flowers over dripping rocks. It also occurs in gravel flushes and, like *Saxifraga stellaris*, it may be carried down to quite low altitudes. Its rarity in the high Pennines is surprising since it is widespread at lower altitudes around Widdybank and Cronkley Fells (v.cs.65,66). Particularly interesting is the isolated record from the Irthing gorge, Gilsland (62.68).
Most of the localities for this species south of the Scottish Highlands are in Cumbria.
Alt. range: 75 m, Wast Water Screes (16.06) - 855 m, Helvellyn (34.14)
Map 367   204   WFCY

S. granulata L.   Meadow saxifrage
Scattered throughout the Eden valley and the coastal areas of west Cumberland becoming locally frequent along the Solway coast and also in the Sebergham area (3.4). In the south it occurs in base-rich meadows and by shady river banks, as by the Kent and Lune and their tributaries, growing usually in bare silt below flood level. Although not generally considered a woodland plant, it is characteristic of damp, species-rich woodland in the north, where it is often associated with *Ranunculus auricomus*, and in the north-west it occurs in dune turf by the Solway Firth (0.4).
Alt. limit: 580 m, north-east of Helbeck Fell (82.20)
Map 368   175   WFCY

S. hypnoides L.   Mossy saxifrage
A characteristic and frequent species of limestone ledges, pavements and rocky turf in the northern Pennines and the east. Off the limestone, it occurs quite widely though less frequently in the Lake District,

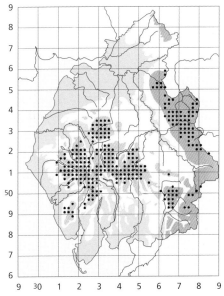

Map 364. *Saxifraga stellaris*

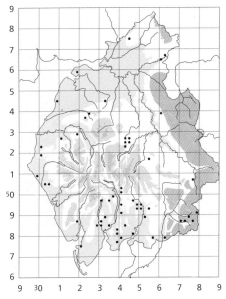

Map 365. *Saxifraga umbrosa* x *S. spathularis*

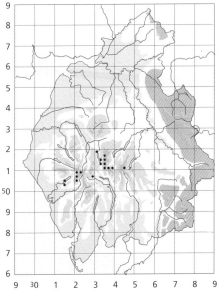

Map 366. *Saxifraga oppositifolia*

particularly in the eastern fells where, on damp ledges and in shaded gullies, it can be remarkably luxuriant. There are records of garden escapes from a streambed near Penruddock (42.28) and near Carwinley (40.72n).

Chromosome counts were made by Parker (1986) on plants from three Westmorland sites. These, and a later one from the Wast Water Screes, all proved to be that of the northern tetraploid race (2n=52).

Early records of *Saxifraga platypetala*, *S. muscoides* and *S. cespitosa* probably belong here.
Alt. range: *c.*270 m, Penruddock (introduced) (42.28) - 855 m, Helvellyn (34.14)
Map 369   203   WFCY

*S. x **arendsii** group
Recorded in 1992 from Roanhead (20.74, PB, LANC) where a large population of this popular garden plant was growing on an area of old mine workings.
                1   F

S. tridactylites L.   Rue-leaved saxifrage;
                       Three-fingered saxifrage
A very common winter annual in the limestone country but elsewhere scarce or absent. It is a very characteristic feature of limestone and mortared walls, also of fine limestone debris and rock ledges, and often associated with *Erophila verna*, *Sedum acre* and bryophytes on occasionally irrigated areas of limestone pavement.
Alt. limit: at least 425 m, Hartside (66.42). Baker's record of 730 m on Swirrel Edge, Helvellyn (3.1) seems unlikely as there are no Survey records for this square.
Map 370   346   LWFCY

*Tolmiea menziesii** (Pursh) Torrey & A. Gray
                       Pick-a-back plant
A rare garden escape first recorded from a roadside ditch at Rather Heath (48.94, GH, 1975, LANC, *Watsonia* (1979) 12:354). The first Furness record was from a roadside near Mansriggs, Ulverston (28.80, JA,

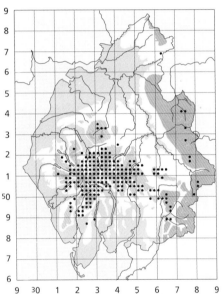

Map 367. *Saxifraga aizoides*

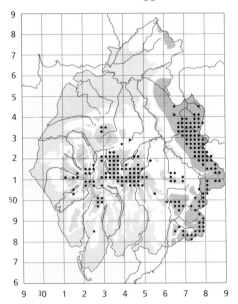

Map 369. *Saxifraga hypnoides*

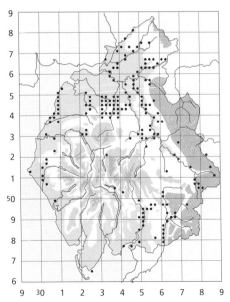

Map 368. *Saxifraga granulata*

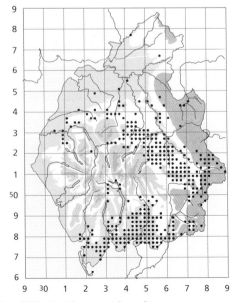

Map 370. *Saxifraga tridactylites*

1979, LANC, *Watsonia* (1980) 13:137), and for Cumberland from woodland on the west side of the river above Duddon Bridge (18.88, GH, 1985, LANC).
Alt. limit: 190 m, Stenkrith, Kirkby Stephen (76.06)
Map 371   12   WFCY

*Tellima grandiflora (Pursh) Douglas ex Lindley
Rare and mostly in the south. This garden escape has been known on a roadside verge near Kings Meaburn (60.20, Eden Field Club) since at least 1974. It was first noted in Furness on a waste dump at Blawith, Grange-over-Sands (40.78s, J.M. Lock, LANC, *Watsonia* (1979) 12:353) and in Cumberland by the River Esk near Carwinley (38.72n, REG, 1979).
Alt. limit: 150 m, Kings Meaburn (60.20)
Map 372   14   WFC

Chrysosplenium oppositifolium L.
                    Opposite-leaved golden-saxifrage
Common, absent only from the intensively farmed lowlands of the Solway Plain and the Eden valley and from areas of blanket peat. It grows, often in profusion, in a wide range of moist habitats: hedgebanks, wooded seepage areas and river- and streamsides, shady cliffs and mountain gullies, also high up in bryophyte-rich springs and seepage areas.
Alt. limit: 870 m, Helvellyn (32.14, Wilson)
Map 373   1222   LWFCY

C. alternifolium L.   Alternate-leaved golden-saxifrage
A mainly eastern species only really frequent in Cumbria in the Eden valley and in the north, and with a few scattered occurrences in the south. It is quite frequent on the limestone of the northern Pennines where it occurs in grikes and sink-holes. In the lowlands it is typically a plant of flushed and seepage areas in wet woodland and wooded riversides, where it is sometimes associated with *Saxifraga granulata*.

Surprisingly, the *Atlas* shows it as occurring in three squares in the north and west of the Lake District: Ennerdale (1.1), Borrowdale (2.1) and Keswick (2.2) for which there are no Survey records.
Alt. limit: 795 m, Cross Fell (68.34)
Map 374   178   WFCY

Parnassia palustris L.   Grass of Parnassus
A much-loved and locally frequent high-summer plant of upland flushes, also of marshy fields and, rarely, of dune-slacks as at Sandscale (18.74), where it occurs in profusion. It appears to be decreasing in the lowlands as a result of drainage. There are a few records from mountain gullies, Glaramara (24.10) and pre-Survey ones of DAR, where it flowers somewhat earlier.

The Sandscale plants, unlike those of the south Lancashire dunes, do not appear to belong to the shorter-stemmed and larger-flowered coastal var. *condensata* Travis & Wheldon.
Alt. limit: 730 m, Fairfield (36.10, DAR, pre-Survey)
Map 375   368   WFCY

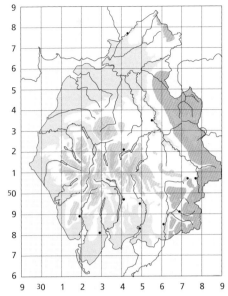

Map 371. *Tolmiea menziesii*

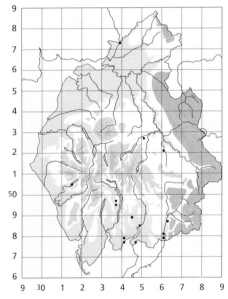

Map 372. *Tellima grandiflora*

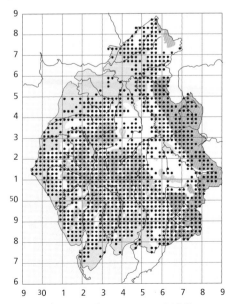

Map 373. *Chrysosplenium oppositifolium*

ROSACEAE

Spiraea

It was only gradually during the course of the Survey that the taxonomic complexity of this popular garden genus became apparent. Indeed, encouraged by the recording card, it was initially assumed that only *Spiraea salicifolia* was present. However, it is now evident that this species is in fact absent and that four other species and three hybrids are represented in the county. The following account is based entirely on material seen by A.J. Silverside.

[S. salicifolia L.
Although previous Flora writers give numerous records, none can be substantiated and it seems very probable that the pure species has never occurred in Cumbria.

(WFC)]

*S. salicifolia L. x **S. alba** Duroi
S. x **rosalba** Dippel
Known from lanesides and hedgerows in the north: Gaitsgill (38.46, REG, 1989, herb. R.E.G.), Smithfield (42.64, MG & JP, 1995), Park House (44.72n, REG, 1982, LANC), Carwinley (40.72n, MG & JP, 1994, LANC), Penton (42.76n, RWMC, 1994, LANC) and above Kershope Bridge (50.82n, REG, 1988, LANC). It also occurs in the Hawkshead area (34.96, R. Young, 1970, AMDE; 34.98, CEW, FLW, 1989, LANC) where it was collected by Wilson (YRK) in 1879, and John Dalton's 1796 record from here (Petty 1895, p.137, under *Spiraea salicifolia*) probably refers to this hybrid. The earliest Cumberland specimen, from "Springfield" (R. Wood, 1884, CLE), is the common nothovar. *rubella* (Dippel) Silverside. Interestingly, the material from Carwinley, Park House and Kershope Bridge are of nothovar. *rosalba*, previously known with certainty only from Cardiganshire.

This hybrid differs from *S. salicifolia* in having rose-pink flowers and broader, less regularly toothed leaves.
Alt. limit: 140 m, above Kershope Bridge (50.82n)

8   FC

*S. salicifolia L. x **S. douglasii** Hook.
S. x **pseudosalicifolia** Silverside
Rare and apparently largely absent from the west and the south-east. The first Survey record for Westmorland was from the north-west corner of Grasmere (32.06, GH, 1984, LANC, *Watsonia* (1986) 16:187), perhaps the 1831 site referred to by Wilson under *Spiraea salicifolia*. The first certain record for Cumberland was from Seascale in 1923 (0.0, R.H. Williamson, CLE) and for Furness from the roadside east of Newby Bridge (38.88, REG & TW, 1988; LANC).

The leaves of this hybrid are rather regularly toothed and pubescent below.
Alt. limit: 435 m, Nenthead (78.42)
Map 376   19   WFC

*S. alba Duroi
Seen only in field hedges near Maryport (04.34, AAD, 1991; LANC) and Aikshaw, near Westnewton (12.46, EEM, 1992, LANC). The latter is no doubt the site where it was collected by E.J. Glaister in 1876 (CLE). All the specimens are of var. *latifolia* (Ait.) Dippel.

2   C

*S. alba x **S. douglasii**
S. x **billardii** Hérincq.
Extremely rare, known only from the side of a plantation at Priestfield, near Burgh-by-Sands (34.60, REG, 1990, LANC), east of Cockermouth (14.30, AAD, 1993, LANC) and by a pond east of Langwathby (58.32, RWMC, 1990, LANC). An earlier record from Bampton (50.18, *Watsonia* (1983) 14:423) is an error for *Spiraea salicifolia* x *S. douglasii*.

These two hybrids can be difficult to distinguish but the leaves of *S.* x *billardii* are broader and, usually,

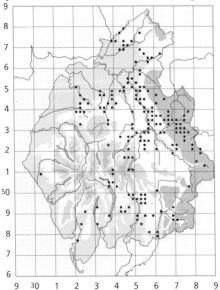

Map 374. *Chrysosplenium alternifolium*

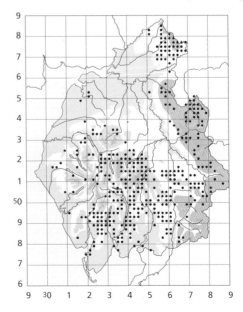

Map 375. *Parnassia palustris*

more coarsely and irregularly toothed.
Alt. limit: 115 m, east of Langwathby (58.32)

3   C

*S. douglasii Hook.
Rare. It was first recorded from Westmorland from a plantation near Wyndhammere, Killington (58.84s, 1976, GH, LANC, *Watsonia* (1981) 13:331) and from Cumberland from near Park House in the Lyne valley (44.72n, REG, 1982, LANC).

This distinctive North American shrub has leaves which are tomentose beneath and sparsely and coarsely toothed towards their tips; the panicle is dense and the flowers deep pink.
Alt. limit: 185 m, Wyndhammere (58.84s)
Map 377   11   WC

*S. japonica L. f.
A garden shrub discovered in 1991 by an old flooded mine near Askam in Furness (20.74, PB, LANC).

1   F

*S. media Schmidt
A few bushes of this south-east European shrub were found in 1989 in the disused Catcrag limestone quarry, Whitbarrow (44.88s, CEW, LANC, *Watsonia* (1991) 18:423) and subsequently one was seen by a lane at Low Meathop (42.78s, JA, 1995, LANC).

2   W

*S. x **arguta** Zabel
Seen on a roadside bank near the jetty at Pooley Bridge (46.24, CEW, 1997, LANC), presumably the site given by Baker and Hodgson for *S. hypericifolia* L.

C

*Aruncus dioicus (Walter) Fern.
Extremely rare. It was first found naturalised by a quarry near Grizebeck (22.84, JA, 1981, LANC,

Watsonia (1982) 14:190). A large clump was later seen on the roadside between Newby Bridge and Borrow Bridge (40.88s, GH, 1988), and there are also records from Ulverston (28.78, PB, 1993, LANC) and the north side of Ullswater (40.20, FJR, 1991, LANC, *Watsonia* (1992) 19:145).
Alt. limit: 200 m, north-east of Newby Bridge (40.88s)

4   FC

*Kerria japonica (L.) DC.
Known only from by the River Irt between Strands and Santon Bridge (10.02, 10.04, MMG, 1992).

2   C

Filipendula vulgaris Moench   Dropwort
Frequent on the limestone of south Westmorland, less so on that of the Asby - Orton area. Elsewhere it is known only from a roadside between Aspatria and Waverton (16.44, M.V. Shorrock, 1978, LANC), by the coast road north of Maryport (04.38, NB, 1991, LANC) and under bracken on a verge near High Newton (40.82s, PB, 1992). These are presumably accidental introductions as are probably all records from the west and north given by Hodgson and the *Atlas*.

This south-eastern species is characteristic of limestone grassland and scrub and often associated with *Helictotrichon pratense* and *Koeleria macrantha*; it also occurs in vegetated runnels and shallow grikes in limestone pavement.
Alt. limit: 260 m, above Crosby Ravensworth (62.12)
Map 378   30   LWFC

F. ulmaria (L.) Maxim.   Meadowsweet
Very common, absent only from the highest fells and areas of blanket peat. It is most familiar as a conspicuous high-summer species of rather damp uncut roadside verges, together with *Chamerion angustifolium* and *Valeriana officinalis*, but it is also common in the tall-herb vegetation of fens and in sheltered

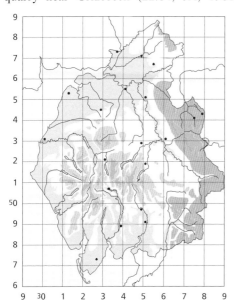

Map 376. *Spiraea salicifolia* x *S. douglasii*

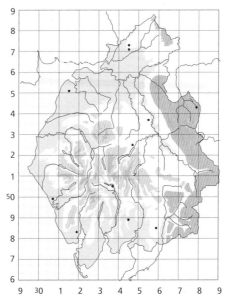

Map 377. *Spiraea douglasii*

damp gullies in the fells.
Alt. limit: 730 m, High Street (44.10)
Map 379 1543 LWFCY

Rubus chamaemorus L. Cloudberry
A common and characteristic plant of blanket bog on
the Pennines and the Bewcastle Fells. In the latter it
descends to about 300 m but in the Pennines it seldom
occurs below 450 m. In the Lake District it is very rare
and despite frequent searches it has not been refound
on the north side of Raise (34.16) where it was seen by
DAR in 1959.

It usually flowers and fruits only rarely but on
the Geltsdale and Bewcastle Fells it does so very freely.

Wilson quotes the first record from the 1597
edition of Gerard's Herbal: "Knotbcrries do live upon
snowie hills and mountains...they grow upon Stane-
more between Yorkshire and Westmorland.".
Alt. range: 240 m, Butterburn (68.72n) - 840 m, Cross

Fell (68.34), but formerly at 855 m on Raise (34.16).
Map 380 120 WCY

R. saxatilis L. Stone bramble
Scattered on base-rich and usually shaded rock-ledges
and crevices, chiefly on the Westmorland limestones
but also in the better Lake District gills and, in the
extreme north-east, on sandstone riverside outcrops.
Alt. limit: 790 m, Helvellyn (34.14.)
Map 381 133 WFCY

R. idaeus L. Raspberry
Very common throughout most of the lowlands,
frequently forming, with *Chamerion angustifolium* and
Urtica dioica, a dense mass of vegetation on thc uncut
and damper verges, often against a stone wall. It grows
very well at high altitudes, ascending far up some Lake
District gullies.

Male plants from Penruddock (42.26, RWMC,

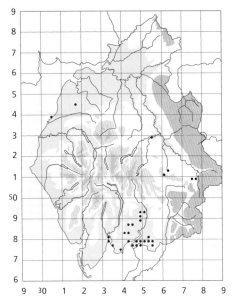

Map 378. *Filipendula vulgaris*

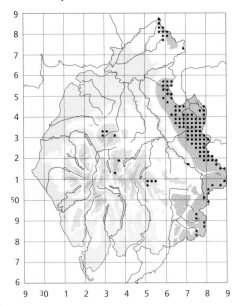

Map 380. *Rubus chamaemorus*

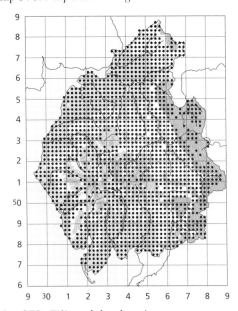

Map 379. *Filipendula ulmaria*

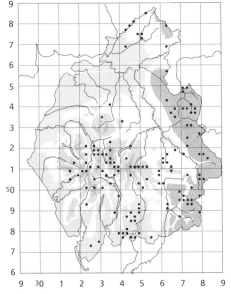

Map 381. *Rubus saxatilis*

1989, LANC), with trifoliate leaves and orbicular leaflets, appear to be f. *anomalus* J. Arrh. Yellow-fruited plants occur between Greystoke and Berrier (42.28).

Alt. limit: 855 m, Fairfield (36.10)

Map 382   1243   LWFC

*R. spectabilis Pursh

Known from a number of sites in the Ambleside - Grasmere area (3.0) and six outlying sites. It prefers shady, damp situations and in at least some sites it was originally planted, as in the grounds of Netherhall, Maryport (04.36) and High Close, Elterwater (32.04), where it is abundantly naturalised.

This attractive North American species appears not to have been recorded until 1958 when G.A.K. Hervey noted it in the Stockgill - Sweden Bridge area of Ambleside, although he had earlier seen it at Gilsland (6.6, v.c.67/70) in 1943. The first certain Cumberland record is from Keswick (26.22, G. Wilson, 1974).

Alt. limit: 250 m, High Ireby (22.36)

Map 383   10   WC

*R. loganobaccus L. Bailey   Loganberry

Recorded in 1993 from the roadside between Grange-over-Sands and Meathop (42.78s, PB, 1993).

1   F

Rubus subgenus **Rubus**

R. caesius L. and *R. fruticosus* L. sensu lato

This account and the accompanying maps are based chiefly on records made by GH from 1980 and contributions from A. Newton (AN, from 1972), E.S. Edees (ESE, 1954-1965) and C.W. Muirhead (1943-1972). The time span is therefore appreciably longer than the 20 years of the recording scheme, although of the 42 species recorded only *Rubus bertramii* and *R. amplificatus* have not been seen during the Survey. A number of recorders have also assisted with the collection of material, notably JA, AAD and MG, and the initial stimulus leading to the recent activity was provided by a collection from south Cumbria by A.J. Sherwood in 1980 and 1981. A significant result of these efforts is a large, modern, comprehensive collection of Cumbrian specimens in LANC. Almost all the Survey records, except those of the commoner species, have been checked by A. Newton. He has also seen most of C.W. Muirhead's collection (CLE, PLYP).

Lack of space and uncertainties regarding identification and synonyms mean that references in this account to pre-1940 literature records and first vice-county records are largely omitted. Early herbarium records, accepted by ESE and AN, are cited where these supplement recent records. Few of the records in the three county Floras were made by the authors and the only early papers which make any significant contribution to our knowledge of Cumbrian brambles, particularly of the Lake District, are those of Ley & Linton (1906) and Rogers (1907).

Despite the daunting nature of attempting to record brambles on a tetrad basis, probably the first such vice-county survey, the often well defined and varied distributions suggest that there are probably few areas which are so under-recorded as to render the maps misleading, and further that distortions due to over-collecting along roadsides is of little significance. In fact woodlands and especially heaths, the characteristic habitats in southern England, are relatively unimportant in Cumbria where brambles are very much a feature of roadside verges, the edges of stone walls, waste ground and neglected pasture. An exception is the section Rubus, the species of which occur chiefly around lowland mosses and by becksides, particularly in the Lake District.

Brambles are essentially lowland, rarely being found over 250 m, and no single species is characteristic of higher altitudes. In the south-east of the county *R. robiae* ascends to 305 m in Garsdale and the coastal *R. ulmifolius* to 270 m. north-east of Sedbergh.

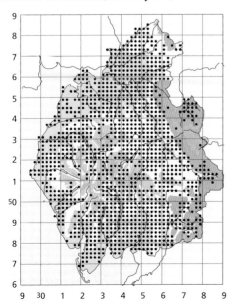

Map 382. *Rubus idaeus*

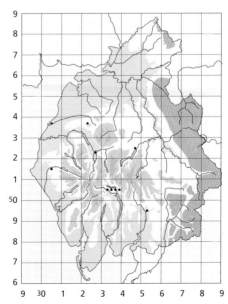

Map 383. *Rubus spectabilis*

Plate 69. *Rubus cumbrensis*

Plate 70. *Rubus lacustris*

In the east, *R. dasyphyllus* and *R. latifolius* reach 270 m at Alston and *R. adenanthoides* 475 m at Nenthead, while the garden *R. armeniacus* grows at 210 m at the M6 service station north of Tebay. Members of section Corylifolii are noticeably absent from the Lake District, a fact which makes brambling there distinctly pleasurable! Yet these are the only brambles present in the upland limestone area between Asby and Orton.

Although 40 species of brambles may seem quite a rich flora, and it is by northern standards, it is nevertheless paltry compared with that of southern English counties. This relative paucity does however ease the task for beginners and it is hoped that the following key and the brief diagnostic notes will assist them in becoming acquainted with the species. Care is of course necessary. Material from extremes of shade or exposure should be avoided. Well-developed mature leaves and stem from new vegetative shoots and a terminal flowering panicle of a shoot of the previous year are usually both necessary. It is also important to note the colour of newly opened flowers. A particular problem is the frequency with which species intermingle making it only too easy unwittingly to collect vegetative and flowering material of different species.

This account follows the sequence and nomenclature of Edees & Newton (1988), the species being arranged alphabetically within the relevant groups. It must be stressed that within subsection Hiemales the series represent artificial divisions of a spectrum of variation and they are not therefore as precise as one would like. The following names used in earlier Floras differ from those used by Edees & Newton.

R. affinis (= mostly *R. nemoralis*)
R. bellardii (=?)
R. carpinifolius (= mostly *R. polyanthemus* and *R. errabundus*)
R. corylifolius var. *conjungens* (= *R. latifolius*) var. *sublustris* (= *R. pruinosus*)
R. discolor (= *R. ulmifolius*)
R. drejeri (= *R. anisacanthos*)
R. dumetorum var. *diversifolius* (= *R. tuberculatus*)
R. fissus (= *R. scissus*)
R. fuscus (=?)
R. hirtifolius var. *danicus* (=?)
R. hystrix (= *R. hylocharis*)
R. incurvatus (= *R. incurvatiformis*)
R. koeleri subsp. *dasyphyllus* (= *R. dasyphyllus*) var. *infestus* (= *R. infestus*) var. *pallidus* (= *R. dasyphyllus*)
R. leucostachys (= *R. vestitus*)
R. maasii (= *R. polyanthemus*)
R. macrophyllus (=?)
R. mercicus var. *bracteatus* (= *R. cumbrensis*)
R. opacus (unnamed Langdale plant)
R. pyramidalis (= *R. incurvatiformis*)
R. radula var. *echinatoides* (= *R. echinatoides*)
R. rhamnifolius (= *R. errabundus* and *R. furnarius*) var. *cordifolius* (=?)
R. rogersii (= *R. fissus*)
R. rosaceus (= *R. hylocharis*)
R. rudis (= *R. echinatoides*)
R. rusticanus (= *R. ulmifolius*)
R. salteri (= *R. errabundus*)
R. scheutzii (= *R. errabundus*)
R. selmeri (= *R. nemoralis*)
R. suberectus (= *R. nessensis*)
R. umbrosus (= *R. polyanthemus*)
R. villicaulis (= *R. nemoralis*)

The following species occur in neighbouring vice-counties but as yet have not been found in Cumbria.

R. rubritinctus and *R. sprengelii* (v.c.60, W. Lancs.), although Baker cites an unsubstantiated record from Storrs, Windermere (3.9).
R. hebridensis and *R. lanaticaulis* (v.c.71, Isle of Man)
R. leptothyrsos (v.c.72, Dumfriess.)
R. scotticus (v.c.73, Kirkcudbrights.)

KEY
Stems refer to first-year vegetative shoots.

1. Stems more or less erect, not tip-rooting
   Section **Rubus** subsection **Rubus**
2. Prickles more than 10 per 5 cm, slender, not confined to stem angles ....5. *scissus*
2. Prickles usually fewer than 10 per 5 cm, more or less confined to the stem angles

3. Petals distinctly pink ....4. *plicatus*
3. Petals white or very pale pink
   4. Lflets 5; ripe fruits black ....4. *plicatus*
   4. Lflets sometimes 6 or 7; ripe fruits black or dark red
      5. Lflets somewhat unevenly serrate; stamens reflexed after anthesis; ripe fruits dark red ....3. *nessensis*
      5. Lflets finely serrate; stamens not reflexed after anthesis; ripe fruits black ....2. *fissus*
1. Stems arching, often tip-rooting
   6. Stems slender; lflets 3; fruits pruinose
      Section **Caesii** ....42. *caesius*
   6. Stems not slender; lflets (3-)5; fruit not pruinose
      7. Basal lflets sessile, or almost so   Sect. **Corylifolii**
         8. Stems with very few if any stalked glands, prickles equal
            9. Terminal lflet rhomboidal; petals pink ....37. *eboracensis*
            9. Terminal lflet broadly ovate or rounded, large; petals white
               10. Stem angled, not shining ....38. *latifolius*
               10. Stem rounded, often shining 39. *pruinosus*
         8. Stems with stalked glands, prickles unequal
            11. Lflets large, broadly ovate; petals *c.*17 mm long ....40. *tuberculatus*
            11. Lflets small, rounded-cordate; petals *c.*10 mm long ....41. *warrenii*
      7. Basal lflets shortly but distinctly stalked
         Section **Rubus** subsection **Hiemales**
         12. Panicle without conspicuous stalked glands
            13. Lvs chalky white beneath; petals pink
               14. Lvs coriaceous, terminal lflet < 10 cm long; petals *c.*10 mm ....23. *ulmifolius*
               14. Lvs not coriaceous, terminal lflet > 10 cm long; petals *c.*18 mm` ....22. *armeniacus*
            13. Lvs not chalky white beneath; petals white or pink
               15. Fls white, rarely pink in bud
                  16. Panicle as broad as long, branches patent, with many fine, pale prickles ....10. *lindleianus*
                  16. Panicle distinctly longer than broad, branches ascending
                     17. Panicle with stout, recurved prickles
                        18. Lvs felted beneath .....18. *lindebergii*
                        18. Lvs thin, not felted beneath ....9. *lacustris*
                     17. Panicle with mostly straight, declining prickles
                        19. Lvs appearing green beneath, not obviously hairy ....13. *silurum*
                        19. Lvs hairy beneath
                           20. Lvs distinctly softly felted beneath ....17. *incurvatiformis*
                           20. Lvs hairy beneath, not softly felted
                              21. Terminal lflet large, irregularly serrate to subentire, the base cuneate or truncate

22. Lflets rounded, subentire ....6. *cumbrensis*
22. Lflets obovate, cuspidate, irregularly serrate
...........14. *amplificatus*
21.Terminal lflet small, neatly serrate; cordate
...........16. *furnarius*
15. Fls pale to deep pink
23. Lflets deeply incised ....8. *laciniatus*
23. Lflets not deeply incised
24. Anthers conspicuously hairy
25. Terminal lflet large, rounded 7. *errabundus*
25. Terminal lflet rhomboidal, cuspidate
...........21. *rhombifolius*
24. Anthers not conspicuously hairy
26. Panicle with strongly recurved prickles
...........19. *nemoralis*
26. Panicle with straight, usually declining prickles
27. Lflets usually large and rounded, those of the panicle often concave and felted below; panicle narrowly pyramidal
...........20. *polyanthemus*
27. Lflets flat, not felted beneath; panicle lax
28. Prickles long and narrow; young carpels hairy ....15. *elegantispinosus*
28. Prickles broad-based; young carpels glabrous
29. Terminal lflet rounded-truncate, finely serrate,hairs on underside along veins crisped; sepals reflexed ....12. *robiae*
29. Terminal lflet ovate, somewhat irregularly serrate, hairs on underside along veins straight and shining; sepals patent ....11. *pyramidalis*
12. Panicle with conspicuous stalked glands
30. Fls white
31. Prickles uniform, not grading into pricklets
...........24. *vestitus*
31. Prickles grading into pricklets
32. Stems rough with numerous gland-tipped acicles
33. Panicle large, with long patent branches
...........33. *pallidus*
33. Panicle narrow, dense ....27. *newboldii*
32. Stems almost smooth, with few gland-tipped acicles
34. Lflets large, rounded, the basal often almost sessile ....29. *anisacanthos*
34. Lflets small-medium, obovate, the basal never sessile ....32. *echinatoides*
30. Fls pink
35. Prickles uniform, not merging into pricklets
36. Stems smooth, lacking acicles and conspicuous stalked glands
37. Stems distinctly hairy; lflets felted beneath
...........24. *vestitus*
37. Stems glabrous or almost so; lflets not felted beneath
38. Lflets medium-sized; anthers glabrous
...........26. *wirralensis*

38. Lflets large; anthers hairy........25. *mucronulatus*
36. Stems rough with acicles, stalked glands present
39. Stems glabrous ....28. *raduloides*
39. Stems hairy
40. Lflets often 3, the lower often lobed; panicle dense ....31. *adenanthoides*
40. Lflets 5, entire; panicle not dense .....4. *radula*
35. Prickles unequal, merging into pricklets
41. Terminal lflet large, rounded-cordate, the lowest often sessile ....29. *anisacanthos*
41. Terminal lflet not rounded, lflets all stalked
42. Panicle with stout, strongly recurved prickles
...........30. *infestus*
42. Panicle with straight, declining prickles
43. Lflets often 3, the lower often lobed; panicle large and dense ....31. *adenanthoides*
43. Lflets usually 5; panicle narrow and few-flowered or large and lax
44. Panicle large, with long patent branches; petals 15-20 mm, narrow ....36. *hylocharis*
44. Panicle few-fld, narrow; petals *c.*12 mm
...........35. *dasyphyllus*

Section **Rubus**
Subsection **Rubus**

Stems suberect, not tip-rooting. Flowering in Cumbria in June.

This is a difficult subsection with a significant overlap of characters between the species.

1. **R. bertramii** G. Braun

Known only from the main island on Rydal Water (34.06, Cookson, 1858). The specimen is said to be in CGE but it cannot be located. Attempts to refind the branble at Rydal have failed, the only species on this small island being *Rubus plicatus*.

This bramble has a very scattered distribution in Britain being most frequent in Wales and reaching its northern limit in Cumbria.

The stems have rather strong, pale, falcate prickles. The leaflets are smaller, darker and more sharply serrate than those of most Cumbrian species of this subsection and the terminal leaflet is more longly acuminate; the petals are white and the stamens exceed the styles.

(W)

2. **R. fissus** Lindley

A very rare bramble with only four Survey records, all in the southern Lake District: by the River Esk near Eskdale Green and in mossland south of Duddon Bridge (14.98, 18.86, 1985), moss woodland near Bouth (32.84, 1995) and by Sour Milk Beck, Easedale (32.08, 1995, all GH, LANC). There are earlier records from between Gosforth and Wast Water (1.0, ESE, 1965, herb. E.S.E.), and the Hawkshead square (3.9, ESE) and early collections from Skelwith Bridge and Red Bank, Grasmere (3.0, W.M. Rogers, 1905, LIV), nearby Clappersgate (3.0, W.P. Linton, 1905, LIV) and near Great Salkeld (5.3, H. Britten, 1909, BM).

This species is rare in England and Wales and

commonest from central Scotland northwards.

Rubus fissus most closely resembles *R. nessensis* differing chiefly in the finer toothing of the leaves, the longer, stronger prickles, the black berries and the stamens which remain erect after flowering.

4   WFC

3. **R. nessensis** W. Hall

A very rare plant of Lake District valleys where it grows along becksides. There is a single isolated occurrence in the Eden valley in Flakebridge Wood, Appleby (68.22, AN & GH, 1984, LANC). There are additional records from the shore of Coniston (3.9, D.L. Nash, 1976) and the Ambleside square (3.0, ESE) and an early collection from Keswick (2.2, G.C. Druce, 1919, OXF).

This species has a wide distribution in Britain but is commonest in central Scotland and south-east England.

This is not an easy species to recognise but it is usually possible to find some leaves on the more vigorous canes with more than five leaflets. The leaflets are irregularly serrate, the long stamens are reflexed after flowering and the fruits are unusual in being dark red.

Alt. limit: 150 m, Flakebridge Wood (68.22)

Map 384, Plate 71a 10   WFC

Unnamed Langdale bramble

This plant was seen during the Survey at four sites: Skelwith Bridge (34.02, AN, 1976), the riverside in Great Langdale and nearby (28.06, 30.04, GH, 1985, LANC), by the Easedale Beck, Grasmere (32.08, GH, 1985, LANC) and the roadside between Bampton and Haweswater (50.16, GH, 1993, LANC). It was first collected in 1905 by Ley & Linton (1905) (LIV) from Great and Little Langdale and erroneously included in *Rubus opacus*. There are no subsequent records from Little Langdale.

Although extremely local and unnamed, this is

quite a distinctive plant with its strong, broad-based prickles and subobtuse leaflets varying from broadly ovate to oblong and even obovate. An unusual feature in this subsection is that the lowest leaflets are noticeably stalked (1-2 mm). The petals are white and the stamens at most equal the styles.

5   W/FW

4. **R. plicatus** Weihe & Nees

Scattered in the south and with a strong cluster of sites on the Solway Plain. It occurs by becksides, roadsides, hedgerows, in damp, unimproved pasture and in mossland. There are additional records from Bowness Common (1.6n, AN, 1976) and the Silloth (1.5, ESE, herb. E.S.E.), Longtown (3.6, AN) and Newby Bridge (3.8, AN) squares and early collections from Haweswater (4.1, W. Borrer, 1845, CGE), near Ulverston (2.7, ex herb. W. Robertson, undated, HAMU) and Arnside Moss (4.7s, A. Wilson, 1934, YRK).

This species is widespread and locally frequent throughout Britain but less so than *Rubus scissus*.

With its generally small, long-acuminate and plicate leaves, *R. plicatus* most closely resembles *R. bertramii*, but it is distinguished by its longer, slender and curved prickles and flowers with generally pink petals and short stamens.

Alt. limit: 150 m, Flakebridge Wood (68.22)

Map 385, Plate 71b 19   WFCY

5. **R. scissus** W.C.R. Watson

Widely distributed in the Lake District and elsewhere in the south, but otherwise rare apart from an area north-east of Carlisle. It grows in damp places in upland woods, alongside becks and on lowland mosses, as well as by roadsides and in hedges. There are further records from the Crook square (4.9, ESE) and Bowness Common (1.6n, 2.6n, AN, 1976) and an early specimen from Dufton (6.2, D. Oliver, 1887, SLBI). The record given by Edees & Newton for the Distington square

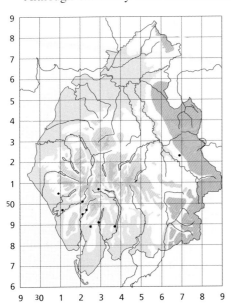

Map 384. *Rubus nessensis*

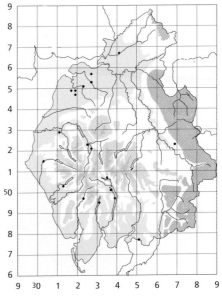

Map 385. *Rubus plicatus*

(0.2) is an error for the Keswick square (2.2).

This bramble has a wide British distribution and is the commonest member of the subsection.

It is usually smaller than the other species. The prickles are rather numerous, pale and very slender; the lowest pair of leaflets are typically sessile, and the ripe fruits are dark red, like those of *Rubus nessensis*. Shade plants can be difficult to identify.

Alt. limit: 180 m, Grizedale Forest (32.92)

Map 386, Plate 71c   33   WFCY

Subsection **Hiemales**

Stems arching, tip-rooting. Flowering in Cumbria from mid-July.

Series **Sylvatici**

Stems lacking glandular hairs; prickles on stem angles only; leaflets not soft or white-felted beneath.

6. **R. cumbrensis** Newton

Fairly common in the south and south-east, especially on hedges and along stone walls in the Ambleside - Troutbeck area (3.0) and on the north side of Ullswater (4.2). There are also records from the Cross Fell square (6.3, AN) and an early one from the north-east side of Derwent Water (2.2, W.M. Rogers, 1906, BM). It was discovered in south-west Cumberland during the Survey and it may prove to be more widely distributed in the west than the map suggests. Edees & Newton's record for the Arnside square (4.7s) is from v.c.60.

Outside Cumbria this attractive bramble is known only from north Lancashire and the Isle of Man.

Although described only relatively recently, this bramble was first collected in Cumbria more than a century ago. The leaves resemble those of *Rubus polyanthemus* and *R. errabundus* in having large, rounded, shiny, rather floppy concave leaflets. However, those of *R. cumbrensis* are of a lighter green. When fresh, the deep red base of the prickles contrasts strikingly with the usually greenish stem. The large panicles are white-flowered, with well-developed lower branches and often broadly ovate leaves. The general impression, especially when in fruit, is of a very heavy, leafy panicle.

Alt. limit: 160 m, below Forest Hall (54.00)

Map 387, Plate 69   58   LWFC

7. **R. errabundus** W.C.R. Watson

Very common in the north-west, becoming abundant along the coast south to Bootle (08.86); elsewhere rather scattered and very rare in the south-east. There is also a record for the Brampton square (5.6, ESE, 1954, herb. E.S.E.). Edees & Newton's record for the Burton-in-Kendal square (5.7s) refers to Lancashire.

Although having a wide English distribution, this is essentially a bramble of the northern Irish Sea coasts and central Scotland.

The leaves resemble those of *Rubus cumbrensis* and *R. polyanthemus* in having large, roundish, con-

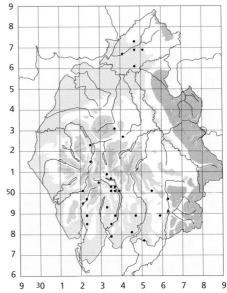

Map 386. *Rubus scissus*

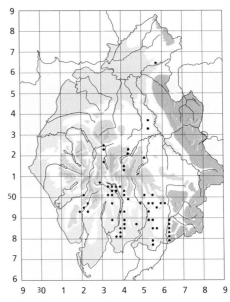

Map 387. *Rubus cumbrensis*

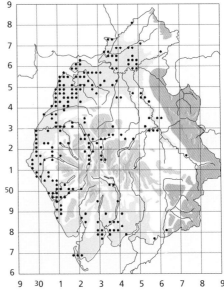

Map 388. *Rubus errabundus*

a) Rubus nessensis

d) Rubus lindleianus

b) Rubus plicatus

e) Rubus robiae

c) Rubus scissus

f) Rubus furnarius

Plate 71

cave leaflets. It is, however, easily distinguished from the former by its pink flowers and from the latter by its hairy and usually pink anthers.

Alt. limit: 150 m, Sandford (72.16) and Birdoswald (60.66)

Map 388   215   WFC

8. **R. laciniatus** Willd.

A garden escape known only from roadsides at Holker (34.76s, KAG & FLW, 1970s) and Skelwith Bridge (34.02, KAG, 1983; LANC), an old record from a railway siding at Scotby (44.54, R. Graham, 1940, CLE) and an untraced Edees & Newton record for the lower Lyne square (4.6).

This is a quite unmistakable bramble with its strongly recurved prickles, deeply pinnately-lobed and serrate leaflets, sepals with conspicuous leafy tips and pink petals.

                          3   WFC

9. **R. lacustris** Rogers

Thought for long to be endemic to a limited area of the central Lake District, this bramble is now known to have an appreciably wider range, although still apparently restricted to Cumbria. In 1965 it was found by ESE at the foot of Coniston Water and in the mid-1980s by GH at several sites from Eskdale to Longsleddale and the Winster valley to Uldale (24.36), also at a very isolated locality north-west of Kirkoswald (52.42).

The obovate leaflets, pale, strongly recurved prickles on the rachis and white flowers suggest *Rubus lindebergii*, but the leaflets are thin, pale green and almost glabrous below, and the panicle is very open with numerous very beautiful large flowers, with relatively narrow, long petals (-1.4 mm).

Alt. limit: 260 m, by the A6, Borrow Beck (54.02)

Map 389, Plate 70   35   WFC

10. **R. lindleianus** Lees

Frequent in the south, thinning out to the north and rare in the east. It has also been recorded from the head of Ullswater (3.1, AN, 1976) and there is an early collection from Grange-in-Borrowdale (2.1, C. Bailey, 1894, NMW). Edees & Newton's record for the Scafell square (2.0) cannot be traced and seems unlikely.

Rubus lindleianus is widespread over much of England and Wales, but it becomes increasingly western in northern England and Scotland.

The leaflets are coarsely toothed and plicate, as in *R. nemoralis*, but they are of a lighter green and appreciably narrower, with undulate margins; the panicles are large and distinctive, white-flowered, about as long as wide, with patent branches densely armed with subulate pale prickles and with a number of simple leaves. The wide panicles projecting above the hedgerows make this bramble very conspicuous.

Alt. limit: *c*.150 m, Lowgill (62.96)

Map 390, Plate 71d   144   LWFCY

11. **R. pyramidalis** Kaltenb.

Extremely rare, found in 1983 in a mossland birchwood in the lower Duddon valley (22.84, GH & AN, LANC) and later in a roadside hedge near Santon Bridge (10.02, GH, 1985, LANC).

This southern bramble has a general resemblance to *Rubus polyanthemus*, but the leaflets are distinctly acuminate, the veins on the underside of the leaflets are densely covered with long, straight hairs, the panicle is sparsely glandular and the pink flowers are somewhat smaller.

                          2   FC

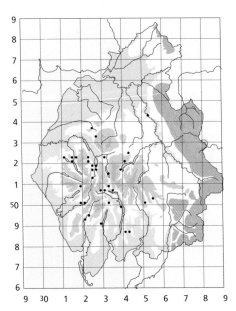

Map 389. *Rubus lacustris*

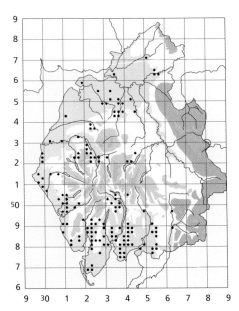

Map 390. *Rubus lindleianus*

12. R. robiae (W.C.R. Watson) Newton
Occasional in parts of the south-east and in the Lowther area (5.2), but elsewhere widely scattered and rare. There are additional records from Bowness Common (1.6n, AN, 1976) and the Cartmel square (3.7s, Anon.).

The distribution of this species extends from the Solway Firth south to Hertfordshire.

The leaflets are finely serrate and softly hairy on both surfaces, the terminal being rounded with a truncate or even cordate base.
Alt. limit: 305 m, Garsdale Head (78.90)
Map 391, Plate 71e   23   LWFCY

13. R. silurum (Ley) Ley
A single large bush was discovered in 1993 on a lane-side bank near Middleton in the Lune valley (62.88s, GH, LANC).

As the name suggests, this is essentially a Welsh species which reaches its northern limit in Britain in Cumbria.

Diagnostic characters include the green, glabrescent leaves, the straight and rather sparse prickles and the sometimes large, ovate, more or less entire upper panicle leaf.
                          1   W

Series **Rhamnifolii**
Like Sylvatici but leaflets felted beneath.

14. R. amplificatus Lees
Known only from a small area on the Solway Firth between East Coteand Skinburness Marsh where it was collected by C. Bailey in 1896 (1.5, MANCH). Several attempts to rediscover it during the Survey have failed.

This bramble has a wide distribution in England and Wales but is very local in Scotland.

Its most distinctive feature is the general downiness, the hairs of both the leaves and panicles being conspicuously long and straight. The terminal leaflet is narrowly, obovate, strongly acuminate and irregularly serrate. The panicle is narrow and the flowers dull white.
                                                (C)

[R. cardiophyllus Lef. & P.J. Mueller
Shown in Edees & Newton's map as occurring in the Brampton square (5.6). Since AN has a field record of *Rubus furnarius* in that square it seems likely that their map record is an error for that species, the two species having been much confused in the past.
                                                C]

15. R. elegantispinosus (A. Schum.) H.E. Weber
Seen only at two sites: in an overgrown garden at Alston (70.46, GH, 1993, LANC) and in a hedge opposite gardens at Arrad Foot (30.80, GH, 1993, LANC).

This bramble, said to be of horticultural origin, is scarce and very scattered in England although not infrequent across central Scotland.

Diagnostic characters include the dark purple stems and prickles, the latter being long, slender and narrow-based, and the broadly-pyramidal, leafy panicles.
                          2   FC

16. R. furnarius W.C. Barton & Riddelsd.
A predominantly eastern species and especially characteristic of the middle Eden valley and Pennine slopes. In the western Lake District it grows on ungrazed rock ledges in the lower sections of gills such as the Tarn Beck above Troutal in the Duddon valley (22.98, GH, 1988, LANC). It has also been recorded from Bowness and Windermere (4.9, F.J.A. Hort, 1849, CGE) and Edees & Newton give additional records for the Hawkshead (3.9), Sedbergh (6.9) and Dentdale (7.8) squares. That for the Burton-in-Kendal square (5.7s) refers to Lancashire.

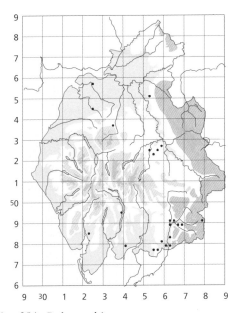

Map 391. *Rubus robiae*

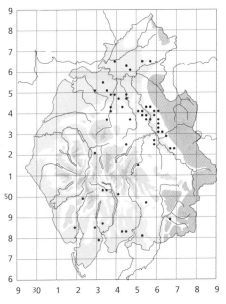

Map 392. *Rubus furnarius*

This bramble is known only from north-west England where it occurs from west Yorkshire to the Solway Firth.

It is characterised by the strongly angled stem with long, strong, pale-tipped prickles almost equalling the width of the stem. The terminal leaflets are rather small, long-stalked, broadly ovate-acuminate, finely toothed, white-felted beneath and with a cordate base. The panicle is narrow with purplish or pale, slightly deflexed prickles and medium-sized white flowers. It differs from the more southern *Rubus cardiophyllus* in its longer prickles and smaller, more finely serrate terminal leaflet.

Alt. limit: 275 m, south of Alston (70.42)

Map 392, Plate 71f   55   WFCY

17. **R. incurvatiformis** Edees

An eastern bramble and really frequent only in the south-east, and around Ullswater and Kirkoswald (5.3, 5.4).

It is known only from Wales and north-west England and is probably commonest along the slopes of the Pennines.

This is a very robust species with strong, long and broad-based prickles; the leaflets are typically large, broadly ovate, acuminate, somewhat cordate and usually softly felted beneath; the panicle is large, white-flowered and very prickly. *Rubus incurvatiformis* has been much confused in the past with *R. incurvatus*, from which it differs chiefly in its laxer, white-flowered panicles.

Alt. limit: 200 m, Garsdale (72.90)

Map 393, Plate 72a   58   WFCY

[R. incurvatus Bab.

Edees & Newton's map shows this species as occurring in the Hawkshead square (3.9). The source of the record cannot now be traced and it seems very likely that it refers to *Rubus incurvatiformis*.

(F)]

18. **R. lindebergii** P.J. Mueller

Widely distributed but most frequent in the Lake District and especially in the east of the county. There are also records for the Sebergham (3.4, ESE), Crook (4.9, GH), Kentmere (4.0, Anon.), Gilsland (6.6, AN, 1976) and Dentdale (7.8, GH) squares. Edees & Newton's map records for the Arnside (4.7s) and Burton-in-Kendal (5.7s) squares refer to Lancashire, and their Geltsdale (6.5) and Winton Fell (8.0) records refer respectively to v.cs 67 and 65.

In Britain, *Rubus lindebergii* ranges from the Severn - Wash line north to central Scotland.

This bramble is usually readily identifiable by its small, narrow, obovate acuminate leaflets which are grey-white felted beneath and the white-flowered panicle with numerous strong, pale, sharply recurved prickles. It is, in fact, a very prickly bramble.

Alt. limit: 250 m, above Melmerby (62.38)

Map 394, Plate 72b   107   WFCY

19. **R. nemoralis** P.J. Mueller

Generally common and one of the commonest brambles in the north, although like most species absent from the Asby - Orton limestone area. There is an early record from Brampton (5.1, C. Bailey, 1896, MANCH). The basis of Edees & Newton's dot for the Tebay square (6.0) cannot be traced. It is a plant of generally poor soils and when occurring on mossland it can be confused with species of section Rubus when not in flower.

This species is generally distributed throughout Britain but surprisingly scarce east of the Pennines.

Distinguishing features include the dark green, thick, plicate leaflets, which are irregularly and rather strongly toothed, the dark, strongly recurved, stout prickles, the pink, notched petals and the distinctive thimble-shaped fruits.

Alt. limit: at least 260 m, north of Alston (70.46)

Map 395   366   LWFCY

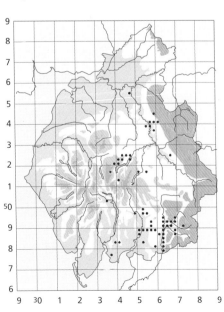

Map 393. *Rubus incurvatiformis*

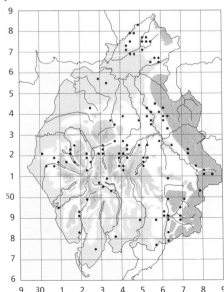

Map 394. *Rubus lindebergii*

20. **R. polyanthemus** Lindeb.

Frequent to locally common, as it is throughout most of lowland Britain. There are additional records from the Blindcrake (1.3, AN), Cartmel (3.7s, AN), Whitbarrow (4.8s, AN) and Arnside (4.7s, Anon., v.c.60/69) squares.

Like *Rubus cumbrensis* and *R. errabundus*, this species has typically rather large, obovate, convex leaflets, although those of the lateral shoots are more ovate and appreciably smaller. The leaves are dark green, as in *R. errabundus*, and when only sparingly hairy beneath are virtually indistinguishable from that species. However, the panicle is narrowly pyramidal with some small simple leaves which are usually ovate and noticeably pale and felted beneath. The flowers are slightly smaller, opening pink but often bleaching white; the anthers are yellow and glabrous.

Alt. limit: 210 m, Unthank (60.40)

Map 396   153   LWFCY

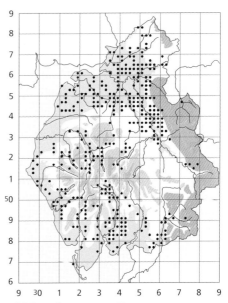

Map 395. *Rubus nemoralis*

21. **R. rhombifolius** Weihe ex Boenn.

Collected during the Survey at a number of places in the Bewcastle area (5.7n, LANC) and at isolated sites west of Carlisle airport (46.60, GH, 1991, LANC), Kingmoor, Carlisle (38.58, C.W. Muirhead, undated), around Windermere (36.88, H. Caldwell, 1983; 36.98, GH, 1991, both LANC), Hartsop (40.12, GH, 1991, LANC) and straddling the Lancashire border south of Hutton Roof (56.76s, GH, 1994, LANC).

This species has a scattered distribution through England and Wales, thinning out northwards although with a local concentration of records in north Cumbria and mid Scotland.

The most important features are the narrowly rhomboidal, long-acuminate, thin leaflets, the fairly strong, reddish, recurved prickles, the sparse, glandular hairs of the panicle, the pink flowers and slightly hairy anthers.

Alt. limit: 170 m, Roadhead (52.74n)

Map 397   11   WFC

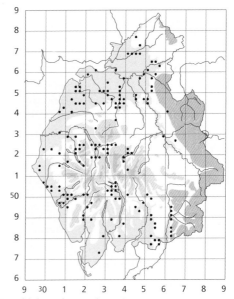

Map 396. *Rubus polyanthemus*

Subsection **Discolores**

Stems lacking glandular hairs, prickles confined to the angles. Leaflets chalky-white-felted beneath, glabrous or subglabrous above, often leathery.

22. **R. armeniacus** Focke

(R. procerus auct.)   Himalayan Giant

This, the common garden blackberry, is naturalised in a few places in lowland Furness, also at Skelwith Bridge (34.02, AN, 1978), Tebay M6 service station (60.06, GH, 1985, LANC) and on the roadside west of Renwick (56.44, GH, 1993, LANC).

This is a much more robust plant than *Rubus ulmifolius*. The stem is strongly angled, but less pruinose, and the prickles are large, slightly deflexed and broad-based. The leaflets are appreciably larger, often broadly ovate-acuminate, of a lighter green, white-felted beneath and not leathery. The large panicle

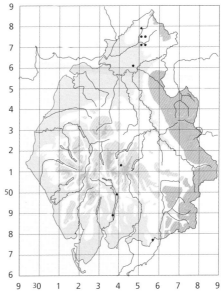

Map 397. *Rubus rhombifolius*

is densely hairy.

Alt. limit: 210 m, north of Tebay (60.06) and west of Renwick (56.44)

Map 398   9   WFC

23. **R. ulmifolius** Schott.

Frequent around Morecambe Bay and scattered along the west coast as far as the Seaton - Camerton area (3.3). A large, vegetative clump was seen at a railway site at Carlisle (38.54). It is rarely found any distance from the coast and it is by far the commonest bramble on Walney Island. However, a large bush, no doubt bird-sown, was seen by the roadside north-east of Sedbergh (72.98) at 270 m.

This southern species reaches its northern limit in western Britain on the Clyde coast.

It is easily recognised by its pruinose stems with strong, reflexed prickles, the rather small, oblong and leathery leaflets, dark green and glossy above and white-felted beneath, the late-flowering panicles of deep pink flowers and the small, firm berries.

Alt. limit: 270 m, Cautley, Sedbergh (72.98)

Map 399   132   LWFCY

R. ulmifolius x **R. vestitus**

Putative hybrids between the sexual *Rubus ulmifolius* and *R. vestitus* have been seen at a number of places around Morecambe Bay and along the south-west coast of Cumberland.

WFC

Series **Vestiti**

Stems hairy, with few glandular hairs and prickles confined to the angles. Leaflets often felted beneath. Rachis with stalked glands.

24. **R. vestitus** Weihe

A frequent to abundant species in the south and west, appearing equally at home both on and off the lime-stone. Rather surprisingly it extends far into Borrow-dale and the Langdales and there is a very isolated site on the Pennine slopes at Renwick (60.42, GH, 1993, LANC). There is also a record for the Carrock Fell square (3.3, ESE). Edees & Newton's record for the Wigton square (2.4) cannot be traced.

This bramble is frequent over most of England but scarce in the west and rare in Scotland.

This is usually a relatively easy species to recognise: the stems are dull purple, softly pubescent and have strong, rather slender prickles; the leaflets are ovate to round, dull green and softly white-felted beneath. The flowers of the Cumbrian plants are dull white, tinged with pink, (var. *albiflorus* Boul.) but bushes with handsome, deep pink flowers occur by the roadside at the White Moss Common car parks (34.06).

Alt. limit: 150 m, Longsleddale (50.00)

Map 400   200   LWFC

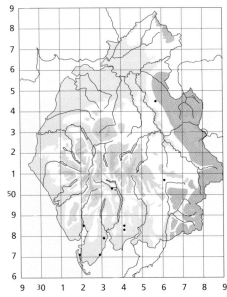

Map 398. *Rubus armeniacus*

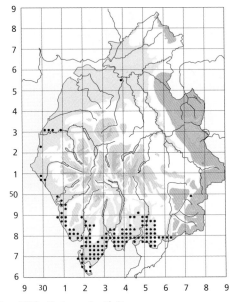

Map 399. *Rubus ulmifolius*

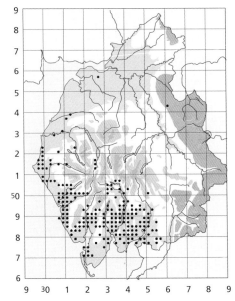

Map 400. *Rubus vestitus*

Series **Mucronati**
Like Vestiti but leaflets thinly hairy beneath, not felted.

25. **R. mucronulatus** Boreau
Extremely rare and apparently restricted to the north-east where it was found in 1976 by AN at Banks, near Brampton (56.64) and in the Castle Carrock square (5.5). It was later collected by the roadside near Stapleton (50.70n, GH, 1983, LANC).

This is a northern bramble being commonest in central and eastern Scotland.

Its most distinctive feature is the large, broad, almost orbicular leaflets which are distinctly mucronate. The narrow panicle has a thick, downy, glandular main stem and noticeably thinner, ascending lateral shoots. The flowers are pink and the anthers hairy.
Alt.limit: 120 m, Stapleton (50.70n)

2   C

26. **R. wirralensis** Newton
Rather scarce, chiefly in the Lake District and south Westmorland. There is also an old record from between Grange-over-Sands and Cartmel (3.7s, C. Bailey, 1893, MANCH) and untraced Edees & Newton ones for the Barbon (6.8s) and Sebergham (3.4, AN) squares.

Its distribution extends north to central Scotland and south to Devon.

Distinguishing characters are the almost smooth, somewhat glandular but only sparsely hairy crimson stems, the broadly obovate and mucronate terminal leaflets with glandular stalks, and the long compound panicles with usually deep pink flowers with rather narrow petals.
Alt. limit: 120 m, Cumwhitton (52.50)
Map 401, Plate 72c   23   LWFC

Series **Micantes**
Stems with prominent unequal acicles, pricklets and stalked glands, prickles subequal.

27. **R. newbouldii** Bab.
Known only from four sites in the south. It was first recorded from a hedgerow near Spark Bridge in 1974 (30.84, AN) and subsequently seen by roadsides near Haverthwaite (34.84, JA, 1984, LANC) and north of Holker (34.78s, GH, 1986, LANC) and by the Lancaster - Kendal canal at Crooklands (52.82s, GH, 1984, LANC).

This bramble occurs in a narrow belt from Lincolnshire north-west to Morecambe Bay.

Its most distinctive features are the large, oblong-acuminate leaflets which are very shallowly toothed, the teeth being patent or even slightly deflexed. The petals are dull white.

4   WF

28. **R. raduloides** (Rogers) Sudre
Known only from three sites. It occurs in the grounds of the High Close Youth Hostel, Grasmere (32.04, GH, LANC), where a few very vigorous bushes were found in 1985. This is probably the same area where it was originally reported by Rogers (1906). It was also collected in the north at Nether Welton (36.46, GH, 1993, LANC) and in the Eden valley at Long Marton (66.24, GH, 1992, LANC). The Edees & Newton record for the Crook square (4.9, A. Sherwood, 1980, LANC) is based on a misidentified specimen of *Rubus echinatoides*.

There is only one other site for this bramble in northern England, which is surprising in view of its wide, if patchy, British distribution.

The stems are glabrous with very unequal

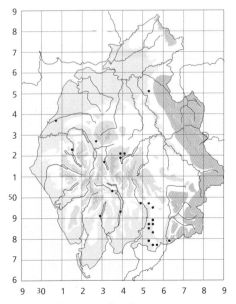

Map 401. *Rubus wirralensis*

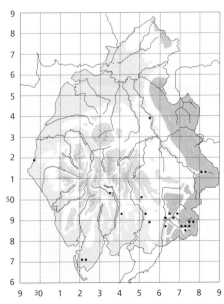

Map 402. *Rubus anisacanthos*

a) Rubus incurvatiformis

d) Rubus anisacanthos

b) Rubus lindebergii

e) Rubus infestus

c) Rubus wirralensis

f) Rubus adenanthoides

Plate 72 (a, b and f lower surface)

prickles and short pricklets. The leaflets are broadly ovate and the panicles much branched. The flowers are pale pink.

Alt. limit: 175 m, High Close, Grasmere (32.04)

3   WC

Series **Anisacanthi**

Stem hairy or glabrescent; prickles, pricklets and stalked glands very variable in quantity, prickles sub-equal or grading into pricklets.

29. **R. anisacanthos** G.Braun

Generally uncommon and predominantly in the south-east, being commonest in the Sedbergh area. There are additional records from near Pooley Bridge (4.2, AN, 1976) and Esthwaite (3.9, AN, 1978), a collection from Killington (5.8s, AN, 1972, herb. A.N.) and Edees & Newton records from the Broughton-in-Furness (2.8, Anon.), Brampton (5.6, AN, 1976) and Kirkby Thore (6.2, AN, 1984) squares.

Its British distribution ranges from the Midlands to southern Scotland.

Useful diagnostic characters are the large, ovate, almost rounded terminal leaflets, cordate at the base, the almost sessile lower leaflets, resembling those of the Corylifolii, the somewhat hairy, angled stem and the dull white flowers. In their colour, shape and soft texture, the leaves resemble those of *Rubus vestitus*. Glandular hairs are almost restricted to the panicle.

Alt. limit: 230 m, upper Dentdale (76.86)

Map 402, Plate 72d   22   WFCY

30. **R. infestus** Weihe ex Boenn.

Scarce except in the middle Eden valley and the north and north-east. There are also records for the Scafell square (2.0, ESE) and from St Bees Head (9.1, C.W. Muirhead, 1948, PLYP). Several of the records are

from heathy areas, for example the Flimby site in the west (02.32) and Brough Sowerby Common (80.12) in the east, where it is particularly abundant.

This is essentially an upland species in Britain, occurring from the Pennines to central Scotland.

No other Cumbrian bramble has a densely glandular panicle furnished with strong, recurved prickles. Other features are the strongly angled stem with robust, unequal prickles, sepals which are patent in fruit and small, pink-cupped flowers.

Alt. limit: 190 m, east of Kirkoswald (58.40)

Map 403, Plate 72e   79   LWCY

Series **Radulae**

Stems subglabrous, rough with numerous short acicles or pricklets and stalked glands, prickles subequal, confined to the angles.

31. **R. adenanthoides** Newton

Common or locally common in south-west Cumberland and Furness, elsewhere rather rare. There is also an Edees & Newton record from the Scafell square (2.0, Anon.). Their records for the Arnside (4.7s) and Burton-in-Kendal (5.7s) squares refer to Lancashire. The paucity of records from the east is surprising in view of its relative frequency across the Pennines in Co. Durham and north Yorkshire.

Its British distribution extends from the Thames valley to central Scotland.

The stem is sparsely hairy, rough with short, often gland-tipped pricklets and the prickles are pale brown, slender and somewhat deflexed. The leaves are often ternate, with the basal pair of leaflets crowded and often lobed and the terminal elliptic-rhomboidal. The panicle has numerous pale, subulate, patent prickles, the branches are congested above making it top-heavy in fruit; the flowers are pink.

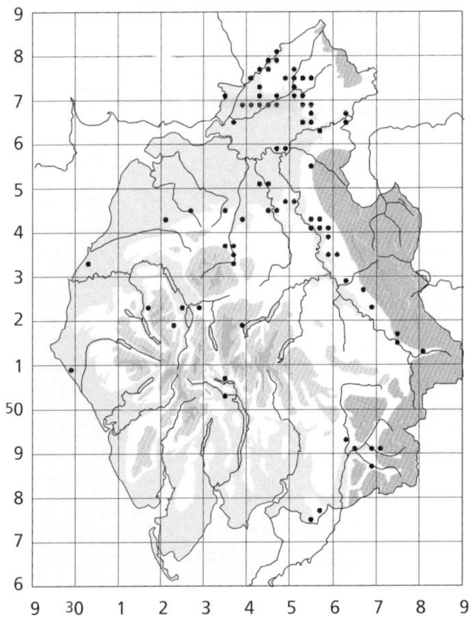

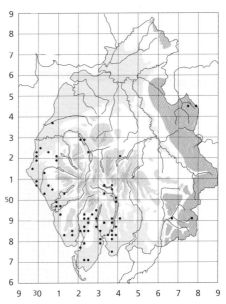

Map 403. *Rubus infestus*

Map 404. *Rubus adenanthoides*

Alt. limit: 475 m, above Nenthead (78.44)
Map 404, Plate 72f   67   WFCY

32. **R. echinatoides** (Rogers) Dallman

The distribution of this species is unfortunately unclear. Recent records are all from a limited area south-east of Windermere. There is a record from Skelwith Bridge (34.02, ESE, 1965, herb. E.S.E.), but only a piece of vegetative stem was collected and all attempts to rediscover the bush have failed; there are also field records from west of Ennerdale (1.1, AN, 1976), the head of Ullswater (3.1, AN, 1976), and a 1908 specimen from Keswick (2.2, R.S. Standen, OXF). The Survey plants are very distinctive with obovate-acuminate leaflets and usually a very broad, strongly lobed leaf, or terminal leaflet, on the rachis. AN (*verb. comm.*) regards such plants as a local variant. The leaves do not match those of ESE's specimen.

    The species has a fairly widespread but scattered British distribution with a general eastern tendency.
Alt. limit: 200 m, north-east of Newby Bridge (38.86)
Map 405, Plate 73a,b   WFC?

33. **R. pallidus** Weihe

Known only from four sites to the south and west of Windermere. It was discovered in 1983 by H. Caldwell on the roadside between Esthwaite Water and Graythwaite Hall (36.94, LANC) and subsequently on the roadside near Graythwaite Hall (36.92, GH, 1985, LANC), in woodland below High Dam, Finsthwaite (36.88, GH, 1986, LANC) and amongst rhododendrons at Fell Foot, Newby Bridge (38.86, D. Earl, 1994). These sites and others in north Yorkshire represent the species' northern limit in Britain.

    Characteristic features include the ovate and long-acuminate terminal leaflets, the rather broad panicle and the white flowers. The styles are red and

the sepals are erect in fruit.

                            4   F

34. **R. radula** Weihe ex Boenn.

Locally frequent in the middle Eden valley with isolated occurrences to the north and north-west. The only records from the south are from High Wray (36.98, H. Caldwell, 1989, LANC) and Grange-over-Sands (38.76s, GH, 1986, LANC), where it was earlier collected by C. Bailey in 1893 (MANCH). Bailey also collected a specimen the same year from Hampsfield Fell, Grange-over-Sands (3.7s/4.7s). It has not been possible to trace Edees & Newton's record for the Keswick square (2.2).

    This species has a wide but generally eastern distribution in Britain.

    The rough stem is sparsely hairy and glandular, the prickles are broad-based and usually recurved; the light green leaflets are ovate and taper to a rather fine point. The flowers vary from pale pink to almost white.
Alt. limit: 190 m, Haresceugh (60.42)
Map 406, Plate 73c   47   WFC

Series **Hystrices**

Stems with numerous unequal prickles and pricklets, on the faces as well as the angles, stalked glands sparse to numerous.

35. **R. dasyphyllus** (Rogers) E. Marshall

This is by far the commonest bramble although less frequent in the west. Despite its abundance on the Morecambe Bay limestone, it is absent, like most brambles, from the Asby - Orton limestone country.

    It has a wide British distribution although rare in Scotland apart from the south-west.

    The unshaded stems are reddish with abundant small prickles, glandular pricklets and eglandular and

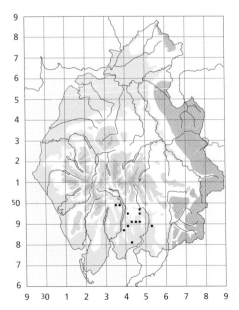

Map 405. *Rubus echinatoides*

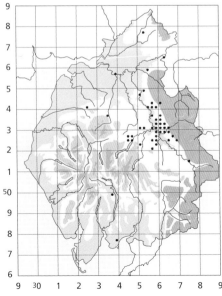

Map 406. *Rubus radula*

a) Rubus echinatoides

d) Rubus dasyphyllus

b) Rubus echinatoides

e) Rubus eboracensis

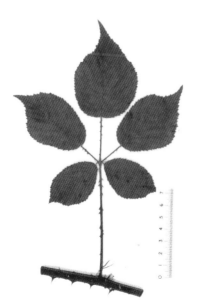

c) Rubus radula

f) Rubus latifolius

Plate 73

glandular hairs, in addition to slender, brown, red-based prickles. These give the stems a distinctly fuzzy appearance. The light green leaves are often ternate with the lower leaflets lobed; the terminal leaflet is elliptic to obovate and rather longly acuminate with the tip somewhat curved to one side. The panicles are small, the sepals reflexed and the petals pale or salmon pink. The berries, like those of *Rubus vestitus*, are particularly sweet.

Alt. limit: 270 m, Alston (70.48)

Map 407, Plate 73d   642   LWFCY

36. **R. hylocharis** W.C.R. Watson

Not uncommon in the area between Windermere and Grasmere, very rare or absent elsewhere, although there are additional Lake District records from the Wast Water (1.0, ESE), mid-Duddon (2.9, ESE) and Hawkshead (3.9) squares and collections by R.S.
Standen in 1908 from Ashness Bridge, Derwent Water (2.1, CGE) and the north side of Keswick (2.2, MANCH).

This species is widely distributed in western England and Wales and is especially frequent in the Isle of Man..

This bramble is characterised by strongly glandular, usually green stems. This latter feature may simply reflect the fact that it is typically a plant of shaded woodland margins. Its most distinctive features are the wide, open panicle, resembling that of *Rubus lindleianus* with its patent branches and fine, pale, patent prickles, but differing in its large pink flowers and long, narrow petals.

Alt. limit: 225 m, Lily Mere, Killington (60.90)

Map 408   11   WFC

Section **Corylifolii**

Leaflets usually 5, the basal pair sessile or almost so. Petals broad and touching. Fruits dull, drupelets few, often imperfect.

Members of this section begin flowering early, before the end of June. They are difficult to identify and consequently are appreciably under-recorded. Perhaps as many as half the specimens are indeterminate, either because they represent unnamed taxa or, more likely, they are of hybrid origin, resulting from crosses with *Rubus caesius* or, less commonly, with members of section Rubus subsection Hiemales. Pink-flowered specimens are particularly troublesome and appear to be not infrequent over much of the north; some may represent an undescribed taxon.

There is also a very distinctive taxon occurring across the Solway Plain. It was collected at Skinburness (12.54, GH, LANC) in 1985 and earlier by C.W. Muirhead in five more tetrads from Grune Point (10.54, PLYP) to Aglionby (44.56, PLYP). The stem has rather few, long, declining prickles, the leaflets are rhomboid-acuminate, markedly and irregularly serrate and the large, open panicle is usually white-flowered.

The map of the combined records of determined and undetermined specimens of this section clearly shows its generally lowland distribution and virtual absence from the Lake District. As mentioned earlier, members of this section are the only ones encountered along the upland limestone lanes of the Asby - Orton area. They are also particularly characteristic of waste ground and railwaysides.

Map 409   217   WFCY

37. **R. eboracensis** W.C.R. Watson

Not uncommon in the upper Eden valley; elsewhere scattered and rare. There are additional records for the Silloth (1.5, AN, 1976) and Brampton (5.6, ESE, herb. E.S.E.) squares. Edees & Newton give further unlocalised ones for the Crosby Ravensworth (6.1) and Cross Fell (6.3) squares; that for the Arnside square (4.7s)

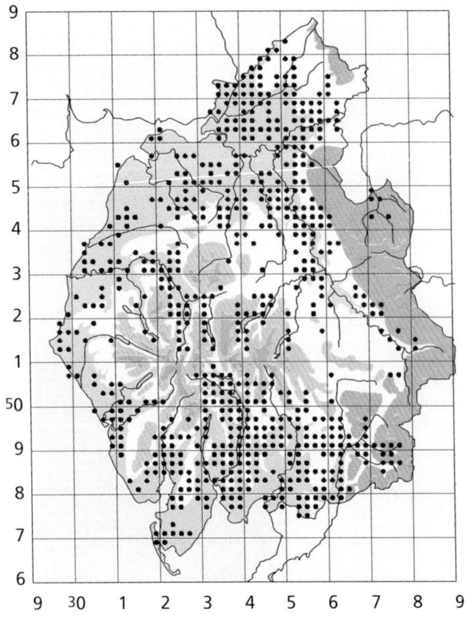

Map 407. *Rubus dasyphyllus*

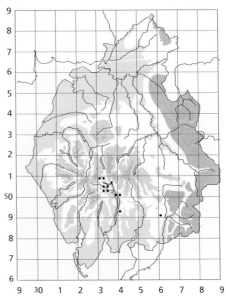

Map 408. *Rubus hylocharis*

refers to Lancashire.

This is a northern species with a British distribution ranging from the Midlands to central Scotland. It is generally eastern although commonest in Cheshire and Staffordshire.

The distinguishing features of this bramble are the smooth to slightly angled stem and the leaflets which are softly hairy beneath, shallowly and rather neatly toothed, with the terminal one broadly rhomboidal and truncate-based. This is the only Cumbrian member of the section which is always pink-flowered.

Alt. limit: 240 m, Waitby (74.06)

Map 410, Plate 73e   16   WC

38. R. latifolius Bab.

Fairly frequent over the northern half of the county, elsewhere scarce. There is also a record for the Distington square (0.2, AN, 1976) and Edees & Newton records for the Solway coast squares (1.6, 2.6).

This is a northern species, frequent in central Scotland and the Borders and reaching its southern limit in Lancashire. It is surprisingly absent from the Isle of Man and Co. Durham.

Its most distinctive feature is the strongly angled stem. The leaflets are large and softly hairy below, often densely so; the terminal one is very broadly acuminate, truncate to slightly cordate and sometimes noticeably biserrate. The panicle has straight or slightly falcate prickles and the flowers are white.

Alt. limit: 270 m, Alston (70.48)

Map 411, Plate 73f   82   WFC

39. R. pruinosus Arrh.

(*R. sublustris* Lees)

Scattered but quite common along the north-west coast. There are also Edees & Newton records from the Kirkbride (2.5, AN, 1976) and Barbon (6.8s) squares; that for the Burton-in-Kendal square (5.7s) refers to Lancashire.

Its distribution in England is wide but patchy. The only Scottish locality is on the coast of Kirkcudbrightshire.

This bramble is distinguished by its smooth, rounded and often shining stem, with rather sparse and usually fine, narrow-based prickles. The terminal leaflets resemble those of *Rubus latifolius* in being broadly ovate, slightly cordate and softly hairy below. Leaves with three leaflets often have the basal pair conspicuously lobed. The flowers are occasionally pale pink.

Alt. limit: at least 180 m, Crosby Ravensworth (60.14)

Map 412   58   WFC

40. R. tuberculatus Bab.

Uncommon but perhaps overlooked. Two of the records, from Meathop (42.78s, GH, 1982, LANC) and Alston (70.46, GH, 1993, LANC), are from railwaysides.

This species occurs in a broad band from Lanc-

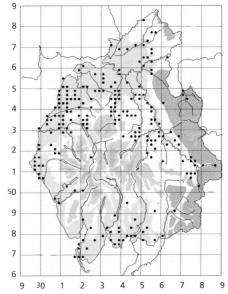

Map 409. *Rubus* section Corylifolii

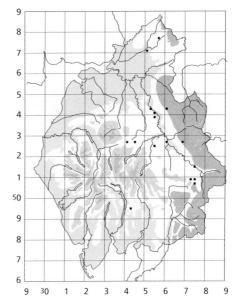

Map 410. *Rubus eboracensis*

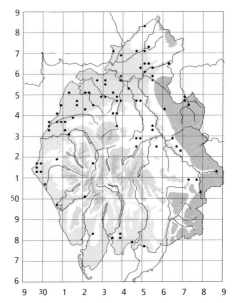

Map 411. *Rubus latifolius*

ashire to the Isle of Wight.

Distinguishing features include the stout stem, with strong, unequal prickles; the leaflets are slightly hairy beneath and the terminal is usually rounded or quadrate. The panicle is narrow, leafy, glandular hairy and furnished with strong, slanting prickles and the petals are large, white and round.

Alt. limit: 260 m, Alston (70.46)

Map 413   8   WFC

41. **R. warrenii** Sudre

Known only from four sites in a limited area to the south of the Solway Plain: Fletchertown, east of Aspatria (20.42, AN, 1976; LANC), by the A591 north of Bewaldeth (20.34, GH, 1988, LANC), south of Bothel (18.36, GH, 1996, LANC), south of Ireby (22.36, AAD, 1993, LANC) and between Bassen-thwaite and the lake (22.30, AAD, 1994, LANC). All the records are from roadside hedges.

This is an uncommon bramble nationally and is largely confined to the north Midlands and Yorkshire.

Particularly diagnostic are the unequal prickles

and the small, neatly toothed and ovate to orbicular terminal leaflets.

Alt. limit: 210 m, Ireby (22.36)

5   C

Section **Caesii**

Stems slender, pruinose, with few, slender prickles. Leaflets 3, glabrous or thinly hairy beneath, ovate, sometimes biserrate and lobed. Panicle few-flowered; the sepals erect in fruit, sometimes with a leafy tip; flowers white; fruits large, well formed and with a whitish bloom. Flowering begins in Cumbria in June.

42. **R. caesius** L.

Probably fairly frequent throughout the lowlands, at least on the limestone. It is especially characteristic of roadside hedges, railwaysides and sand-dunes, but most Survey field records have been omitted from the map because of possible confusion with members of the Corylifolii.

Alt. limit: 320 m, Alston (70.42, 72.42)

Map 414   87   WFC

One evening I sat at table in a company of botanists. They had been out all day after Brambles, not as one might suppose, after Brambles to eat - I doubt if any of them had ever eaten a Bramble - but after what they called the Rubi, the Brambles of which there are more than a hundred species in this country. The talk at table was all about those Rubi, and I sat silent till it occurred to me that I too had a contribution to make to the subject. Turning to my neighbour I said, 'Do you know why Brambles are called Brumliekites in Cumberland?' 'No: why?' he asked. 'Because children eat so many that their kites or bellies rumble,' I replied. He looked at me with surprise, not so much, I imagined, at the information, as at myself.

A Prospect of Flowers, Andrew Young (1945)

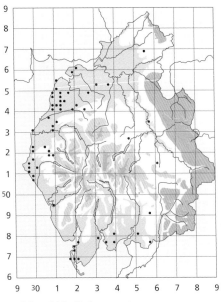

Map 412. *Rubus pruinosus*

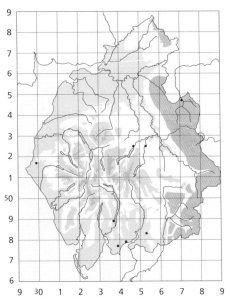

Map 413. *Rubus tuberculatus*

Potentilla fruticosa L.   Shrubby cinquefoil
An extremely rare plant of basic crags and screes in the
Lake District where small populations occur on Pillar
(1.1), Wast Water Screes(1.0) and Fairfield (3.1). It has
long been known on The Screes but the date of its
discovery on Pillar is a mystery - G.A.K. Hervey saw it
there in 1928. It was not until 1975 that it was
discovered on Fairfield by E.G. Hall.

Baker also lists it from above Keppel Cove Tarn,
Helvellyn (3.1) and "near Ulpha". The last locality is
certainly an error. Wilson also mentions Helvellyn,
although commenting that it was probably extinct
there, and Maize Beck, upper Teesdale (8.2), where it
is also extinct.

Outside Cumbria this species occurs in the
British Isles only in the adjoining part of upper
Teesdale and in the Burren, Co. Clare. The nearest
European populations are in the Pyrenees, the Maritime
Alps and the Swedish Baltic islands.
Alt. range: 425 m, Wast Water Screes (1.0) - 700 m,
Pillar (1.1)

                    R 3   WC

P. palustris (L.) Scop.   Marsh cinquefoil
A frequent plant of poor fen communities. It is characteristic of marshy meadows, the richer lowland
mosses and also, especially in the Furness fells and in
the hills north and east of Kendal, of the somewhat
flushed, boggy pools with *Carex rostrata, Hippuris
vulgaris, Menyanthes trifoliata* and *Utricularia* spp.
Alt. limit: 760 m, Little Dun Fell (70.32)
Map 415   492   WFCY

P. anserina L.   Silverweed
Common throughout the lowlands, occurring most
noticeably along roadsides, around farms and on waste
ground. Like a number of other nutrient-demanding
species it is characteristic of strand-line communities. It
occurs also by lake- and tarn-sides and in rather open
marshy areas, usually in fields, often muddy and subject
to drying out.
Alt. limit: 580 m, Hartside (64.40)
Map 416   1364   LWFCY

P. argentea L.
A very rare species of semi-open sandy ground seen
only on cinders at Salthouse Pool, Barrow-in-Furness
(20.68s, J. Chamberlain, 1983, LANC), on a railway
embankment near Aspatria (14.40, ES & MS, 1986,
LANC), at three sites on Lazonby Fell (50.38, 50.40,
52,38, RWMC, 1981, LANC) and two in Westmorland
on the opposite side of the River Eamont to Udford
(56.26, AW, 1981; 56.28, RWMC, 1980, LANC). It
had previously been recorded in this last area in the
1950s independently by N.M. Stalker and G.A.K.
Hervey.

Baker lists it only from the shore of the Leven
estuary (3.7s) and Hodgson from Woodrow, near
Wigton (2.4, E.J. Glaister, 1881, CLE). It was seen in

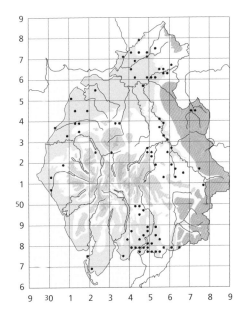

Map 414. *Rubus caesius*

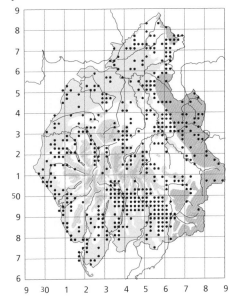

Map 415. *Potentilla palustris*

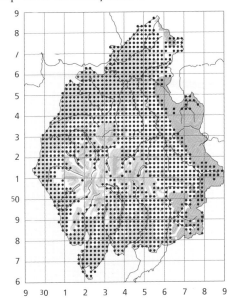

Map 416. *Potentilla anserina*

abundance at Edenhall in 1908 (5.3, H. Britten), recorded by Wilson from Cliburn (5.2) and found at Silloth (1.5) in 1932 by J. Parkin.
Alt. limit: 230 m, Lazonby Fell (50.38)

7   WFC

*P. recta L.

A species of southern and eastern Europe recorded in Cumbria only from Silloth (1.5, T.S. Johnstone, 1910, CLE), where it was subsequently seen by R.H. Williamson in 1923 and C.W. Muirhead in 1937 (BM) and 1939 (CLE).

(C)

*P. intermedia L.

Known only from Silloth where it was seen by G.C. Druce in 1923 (1.5, *BEC Rpt* (1924) 7:182) and R.A. Boyd in 1949 (CLE).

Although widespread in Europe, this species is native only in Russia. It is sometimes confused with *Potentilla norvegica*, but the latter has ternate and not digitate basal leaves.

(C)

*P. norvegica L.

The only Survey records are from the limestone quarry at Meathop (42.78s, KAG & FLW, 1974), probably the same site as G.A.K. Hervey's 1957 Grange-over-Sands record, a slate tip near Ulpha in the Duddon valley (20.94, M.J. Wigginton, 1977, LANC) and by the old railway line east of Canonbie (40.74n, GH, 1979, LANC).

The only record in the Floras for this east European species is in Hodgson's Appendix where he refers to its discovery in 1897 at the Derwent Tinplate Works, Workington (9.2). It was later recorded from a field near Edenhall, Penrith (5.3, J.C. Varty-Smith, undated), Launchy Gill, Thirlmere (3.1, E. & H.

Drabble, 1906, BM), Whitehaven (9.1, M. Cobbe, 1924, *BEC Rpt* (1925) 7:569), and Silloth and a garden at Abbeytown (1.5, C.W. Muirhead, 1932-1935). In Furness it was noted from coppiced woodland at Near Sawrey (3.9, *Lancs. & Cheshire Nat.* (1915-16) 8:57) and near Colwith (3.0, N.E.G. Cruttwell, 1942, *BEC Rpt* (1944) 7:486).

3   W(F)C

P. crantzii (Crantz) G. Beck ex Fritsch

Alpine cinquefoil

A rather rare plant of limestone scars in the Pennines and a few of the more base-rich crags and gullies in the Lake District. As is the case with so many montane species most of the populations are very small and in danger of extinction. That at Pottrigg, Oddendale (58.12), was exterminated during the Survey as a result of unauthorised limestone pavement removal. Records made by DAR during the 1950s from Great End (22.08), Pavey Ark (28.06) and Helvellyn (34.14) require recent confirmation as does one from Longsleddale (46.06, R.J. Birkett, 1962). The *Scarce Plants* record for the Kirkby Thore square (6.2) is an error.

Curiously, the only records in the Floras are Baker's from "Grange Fell", above the Vale of Newlands (2.1), where it has never been refound, and Wilson's from above Stickle Tarn (2.0) and by the Maize Beck, Dufton Fell (8.2).

South of the Scottish Highlands this species is largely restricted in Britain to the northern Pennines, the Lake District and Snowdonia.
Alt. range: 305 m, St John's in the Vale (30.20, DAR, 1956) - 790 m, Helvellyn (34.14, DAR, 1968)

Map 417   S 21   WCY

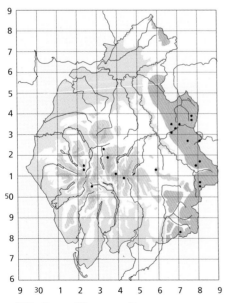

Map 417. *Potentilla crantzii*

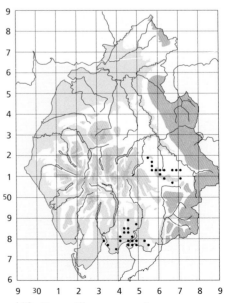

Map 418. *Potentilla neumanniana*

P. neumanniana Reichb.   Spring cinquefoil
(*P. tabernaemontani* Aschers.)
Rather rare but locally frequent to common on the
Morecambe Bay and Asby - Orton limestones where it
grows on very thin soil along the edges of the scars, on
limestone pavement and sometimes on the floors of old
quarries. It prefers south-facing sites and comes into
flower as early as mid-April. Surprisingly there are no
Atlas records for the Asby - Orton limestone.

The only record from v.c.65 is the remarkable
one given by Handley (1898) from the railway bank
near Sedbergh (6.9), close to where the old railway
used to cross the River Rawthey.
Alt. limit: 335 m, Crosby Ravensworth Fell (62.08)
Map 418   34   WF(Y)

P. erecta (L.) Rausch.   Tormentil
Extremely common and an almost ubiquitous species of
acid grassland and moorland, usually associated with
Galium saxatile and *Agrostis capillaris*, but also widespread in limestone grassland.

The only record of subsp. *strictissima* (Zimm.)
A. Richards is from Cumwhitton Moss (50.50, RWMC,
1994, LANC). It is more vigorous and has more
coarsely toothed leaflets and slightly larger petals.
Alt. limit: 915 m, Skiddaw (26.28, Baker).
Map 419   1707   LWFCY

P. erecta x **P. anglica**, **P. anglica** and **P. anglica** x **P.
reptans**
All the cited herbarium specimens of these three taxa
have been checked by B. Harold and the maps of the
first two are based entirely on material submitted to
her.

P. erecta x **P. anglica**
P. x **suberecta** Zimm.
A rare hybrid of roadside banks and tracksides chiefly
in the south. The first definite record for Cumberland is
from Waterside (2.4, E.J. Glaister, 1876, CLE), for
Furness from a roadside grit pile near Wray (36.00,
J. Ellis, 1968, AMDE) and for Westmorland from
Grubbins Wood, Arnside (44.78s, MB, 1976, LANC).

This is a difficult hybrid to identify. Its chief
distinguishing features are the variable number of
petals, the distinctly stalked leaves, although some may
be subsessile, the low pollen fertility (<50%) and
correspondingly few achenes.
Alt. limit: 150 m, Ludderburn (40.92)
Map 420   18   WFC

P. erecta x **P. reptans**
P. x **italica** Lehm. and
P. anglica x **P. reptans**
P. x **mixta** Nolte ex Reichb.
Fairly frequent in the south-east, otherwise scattered
and chiefly along the west coast. It occurs mainly on
roadside verges and hedgerow banks. The only definite
pre-Survey records are from Walney Island (F.A. Lees,

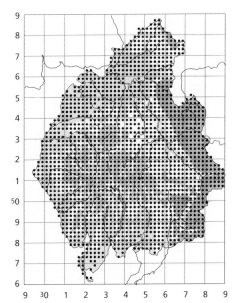

Map 419. *Potentilla erecta*

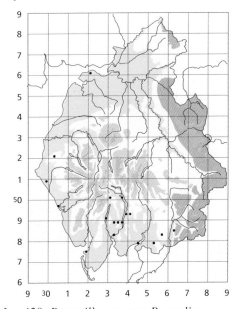

Map 420. *Potentilla erecta* x *P. anglica*

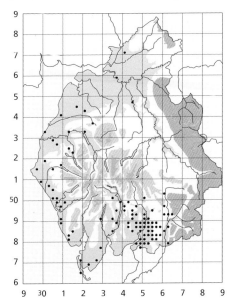

Map 421. *Potentilla erecta/anglica* x *P. reptans*

1895, LANC) and Rigmaden, Barbon (6.8s, A. Wilson, 1937, YRK). It is not possible to determine which of the Survey records was the first for Cumberland.

The large number of Survey records is surprising. The hybrid resembles *Potentilla reptans* in habit but is quite easily identified in the field by being somewhat smaller and more delicate, with at least some leaves with only three or four leaflets and some flowers with only four petals. It is also highly sterile. Like most such hybrids it has a long flowering season.

Alt. limit: at least 180 m, Tebay (60.02) but possibly higher east of Grayrigg (60.96).

Map 421   97   WFCY

P. anglica Laich.

Rare, occurring in similar situations to *Potentilla* x *suberecta*. All the previous Flora records of this species are best disregarded in the absence of herbarium specimens owing to widespread confusion between these two taxa and, to a lesser extent, with *P.* x *mixta*. Specimens collected from a roadside grit pile near Wray (36.00, J. Ellis, 1968, AMDE), growing with *P.* x *suberecta*, are extraordinarily luxuriant, with stems to 70 cm.

This species is best recognised by its high fertility and in having four or five petals. Both this and *P.* x *mixta* tend to have broad leaflets, large stipules and petiolate leaves. Intermediates between the two do occur and probably represent backcrosses.

Alt. limit: 120 m, Firbank (62.92)

Map 422   20   WFC

P. reptans L.   Creeping cinquefoil

Common throughout the lowlands but rather limited in its habitats, occurring in hedgerows and on roadside verges, waste ground and disused railways.

Alt. limit: 415 m, by the A66, Stainmore (88.12)

Map 423   746   LWFCY

P. sterilis L.   Barren strawberry

Generally common but absent from appreciable parts of west Cumberland and the Solway Plain, largely, no doubt, because of the lack of suitable roadside banks. It is in fact one of the most characteristic plants of thinly vegetated roadside and hedgerow banks but it is also frequent in woodland and open rocky ground, both on and off the limestone, and it can also be found in open well-drained *Festuca - Agrostis* grassland in the fells.

This is one of the earliest spring flowers being often in flower at the beginning of March.

Alt. limit: 755 m, Great Dun Fell (70.30) and 790 m on Helvellyn (3.1, Wilson)

Map 424   1258   LWFCY

[Sibbaldia procumbens L.

This species of the Scottish Highlands was reported by C.C. Babington (1874) as having been discovered at High Cup Nick (7.2) by J. Backhouse. There are no subsequent records although G.C. Druce (*BEC Rpt* (1919) 5:501) commented

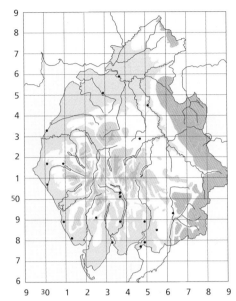

Map 422. *Potentilla anglica*

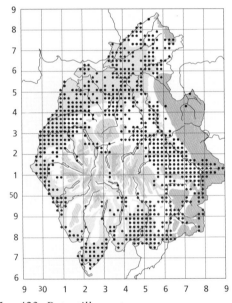

Map 423. *Potentilla reptans*

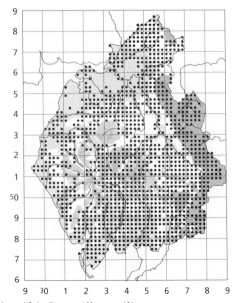

Map 424. *Potentilla sterilis*

enthusiastically on material cultivated by J. Cryer and supposedly originating from the Nick. The various explanations of the record include the misidentification of *Potentilla sterilis*, stunted, ternate-leaved specimens of *P. palustris* or *P. crantzii* "with narrow, obsolescent petals", the last suggestion being an annotation of F.A. Lees in his copy of Babington's *Manual of British Botany* (5th ed. (1862) BM).

(W)]

Fragaria vesca L.　　　　　　　　Wild strawberry
Common on roadside and hedgerow banks, in the richer deciduous woodlands and in the fells in gills and gullies on the more basic rock outcrops.
Alt. limit: 650 m, Crowdundle Beck (68.32)
Map 425　　　　　　1102　　　　　　LWFCY

*F. moschata** (Duchesne) Duchesne Hautbois strawberry
(*F. muricata* sensu D.H. Kent, non Miller)
The only substantiated records determined by A.C. Leslie are a Survey one from the roadside north of Castle Carrock (52.56, FJR, 1982, LANC) and earlier collections from the roadside at Beckfoot, Eskdale (1.0, A. Wilson, [1930s?], YRK) and the roadside above Hawkshead (3.9, G. Wilson, 1949, CLE).

Hodgson gives records from near Latrigg, Keswick (2.2), Grange-in-Borrowdale (2.1), Harrington (9.2), Frizington (0.1), near Cockermouth (1.3) and Harker (3.6). There are later records from Burgh-by-Sands (3.5, W.W. Mason, 1914) and "below the [railway] viaduct", Carlisle (3.5/4.5, C.W. Muirhead, 1948). Baker gives further sites at Newland (3.7s) and between Grayrigg and Oxenholme (5.9) and Wilson from near Bowness and Underbarrow (4.9). There is also an *Atlas* record from the 1950s from the Grizebeck square (2.8) and an unlocalised reference to the Sedbergh area (Handley 1898).

The above records may be substantially correct, but no herbarium specimens have been located and, as the notes by Leslie (1978) indicate, the species is frequently confused with *Fragaria vesca*. Two collections originally named *F. moschata* from the roadside at Wreay (4.4, T.S. Johnstone, 1902, T.W.S. Jones, 1949, both CLE) have in fact been redetermined by him as *F. vesca*.

There are relatively few reliable diagnostic characters but Leslie emphasises the presence of only ascending or appressed hairs on the upper pedicels of *F. vesca*, whereas *F. moschata* has spreading or reflexed hairs. In addition, naturalised plants of the latter almost invariably have functionally dioecious flowers. This species has decreased dramatically this century with the increasing popularity of *F. ananassa*.

1　　　　　　(WF)C(W/Y)

*F. ananassa** (Duchesne) Duchesne　Garden strawberry
An occasional garden escape, usually near houses and well established on railway embankments at Smardale, Ravenstonedale (72.06) and near Harrington (98.24). The only pre-Survey records are those in the *Atlas* for the Milnthorpe (4.8s), Workington (9.2) and Keswick (2.2) squares. The first Furness record is probably that from Cark in 1978 (36.76s, KAG & FLW).
Alt. limit: 240 m, Smardale, Ravenstonedale (72.06)
Map 426　　　　　　27　　　　　　WFC

Geum rivale L.　　　　　　　　Water avens
Very common in the northern part of the county, especially bordering the Lake District. Rather uncommon in the more intensively agricultural areas of the Solway

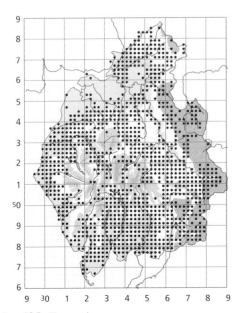

Map 425. *Fragaria vesca*

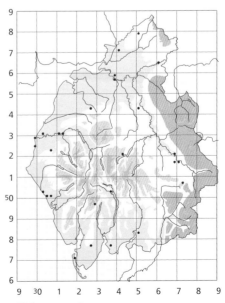

Map 426. *Fragaria ananassa*

Plain and the Eden valley and very scarce on the
Furness peninsula. It occurs, often in profusion, on
damp, roadside verges and in ditches, particularly in
upland areas, and in the fells it is characteristic of wet
rock ledges in sheltered gullies, together with
Alchemilla glabra and *Crepis paludosa*.

A double-flowered variant, perhaps of hybrid
origin and first noted in the county by Lawson and later
by Handley (1898), was occasionally found during the
Survey, and also a form with the receptacular axis
prolonged.

Alt. limit: 865 m, Helvellyn (3.1, Wilson)

Map 427   962   WFCY

G. rivale x **G. urbanum**

G. x intermedium Ehrh.

Frequent in a broad band to the west of the Eden valley
but elsewhere occasional or scarce. It occurs rarely in
the absence of one of the parents. There appear to be no
records for Furness prior to the Survey.

Alt. limit: *c.*450 m, Nenthead (78.42)

Map 428   225   WFCY

G. urbanum L.   Wood avens; Herb Bennet

Very common in the lowlands in deciduous woodland
and scrub, on roadsides and in hedgerows, rarely a
persistent garden weed.

Alt. limit: 450 m, Alston Moor (72.38), although Baker
gives it as 510 m.

Map 429   1205   LWFCY

*G. macrophyllum Willd.

A colony of this scarce alien has persisted since 1982,
despite frequent mowing, on the roadside verge
opposite houses south of Bowness (38.94, CFS & JS;
LANC).

                         1   W

Dryas octopetala L.   Mountain avens

Extremely rare, only a few plants persisting at its well-
known site on crags above Keppel Cove Tarn,
Helvellyn (3.1, J. Backhouse, 1869, E) and at a site in
the Scafell range (1.0) in the west of the Lake District
discovered by DAR in 1957 (Ratcliffe 1960).

It was said by G. Bolam, early this century, to
occur on Mickle Fell (8.2, v.c.65/69) and also on the
Cross Fell ridge above Melmerby (6.3), but neither
record has ever been confirmed, nor also the intriguing
1894 Westmorland record from near Tailbridge Hill
(8.0 (Lees 1938, p.103)).

South of the Scottish Highlands, mountain avens
is known only from Cumbria, North Wales and the
limestone of Teesdale and Wharfedale.

Alt. range: 450 and 760 m

                         S 2   WC

Agrimonia eupatoria L.   Agrimony

Fairly frequent in the lowlands, scarce on the south-
west coast of Cumberland and practically absent from

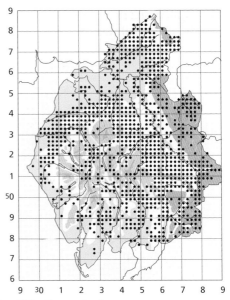

Map 427. *Geum rivale*

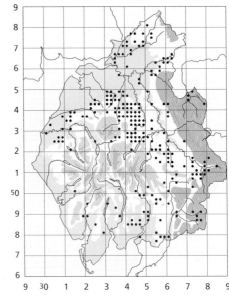

Map 428. *Geum rivale* x *G. urbanum*

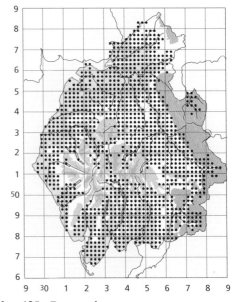

Map 429. *Geum urbanum*

the Lake District valleys and the south-east. It is a plant of roadside verges and railwaysides, of scrub and woodland margins, usually on the better soils. The records have no doubt been confused to some extent with *Agrimonia procera*.

Alt. limit: 300 m, by Argill Beck gill, North Stainmore (80.14)

Map 430   331   LWFCY

A. procera Wallr.

Scattered on the edge of the Solway Plain, in the Eden valley and the south, elsewhere absent or rare. It is no doubt commoner than the records suggest and in the Warcop area (7.0) it is probably more frequent than *Agrimonia eupatoria*. There are no obvious ecological differences between the two species.

Alt. limit: at least 230 m, above Waitby (74.06)

Map 431   59   LWFC

Sanguisorba officinalis L.   Great burnet

A conspicuous, widespread but decreasing plant of damp roadside verges, riversides and of 'unimproved' hay-meadows. There are surprisingly few records from Furness and south-west Cumberland. Although generally common in the county it is virtually absent across the Scottish border.

Alt. limit: 460 m, near Cauldron Snout, Teesdale (80.26)

Map 432   1003   WFCY

S. minor Scop.
subsp. minor   Salad burnet

A common component of short calcareous grassland, chiefly on the limestones of south and north Westmorland but scattered over much of the northern half of the county where it occasionally occurs off the limestone as on the banks of the River Eden below Armathwaite

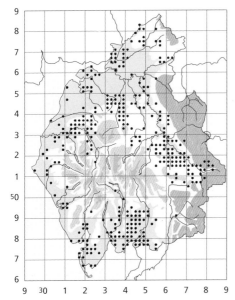

Map 430. *Agrimonia eupatoria*

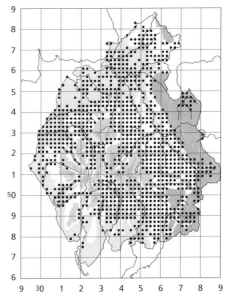

Map 432. *Sanguisorba officinalis*

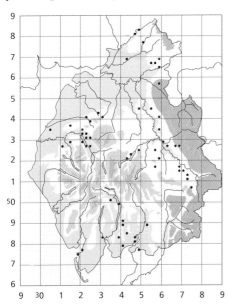

Map 431. *Agrimonia procera*

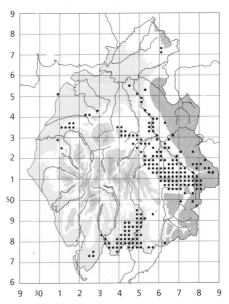

Map 433. *Sanguisorba minor*

(50.48), on a roadside bank at Spadeadam (60.70n, 60.72n) and in dune grassland on the coast at Wolsty, Beckfoot (08.50). There are also early reliable records from Dalston (3.5, T.S. Johnstone, 1901; Miss Oliphant, 1951) and Great Salkeld and Skirwith (respectively 5.3 and 6.3, H. Britten, undated). The basal leaves are sometimes confused with those of *Sanguisorba officinalis* and *Pimpinella saxifraga*. Some of the more isolated records and all early ones from the Lake District probably refer to these species.
Alt. limit: at least 425 m, Stainmore Common (82.18)
Map 433   205   WFC

*subsp. **muricata** (Gremli) Briq.
Recorded from tall herb vegetation, with *Impatiens glandulifera* and *Phalaris arundinacea*, by the River Caldew at Carlisle (38.54, REG, 1985, LANC, *Watsonia* (1986) 16:187) and from the edge of a newly colonised area of salt-marsh at Greenodd (30.82, T.C.G. Rich, 1987, LANC). The only previous record is from Silloth (1.5, CLE), where it was collected by C.W. Muirhead in 1939.
                2   FC

Alchemilla alpina L.   Alpine lady's-mantle
A very common and conspicuous plant of the Lake District fells with isolated occurrences in the Howgill Fells at Cautley Spout (66.96, 68.96) and on the limestone pavement above Sunbiggin (64.08, S.D. Ward & D.F. Evans, 1973), the only authenticated site on limestone in Britain. It was introduced successfully, in about 1956, into two experimental enclosures in the Pennines, at Moorhouse (74.32) and at the head of Knock Ore Gill (70.30). These sites are not included on the map. Handley (1898) gives a record for Dent Crag (6.8s), but it has never subsequently been seen there.

In the Lake District it occurs in open fellside *Festuca - Agrostis* turf, often becoming dominant on flushed slopes below crags; it also grows below and on stone walls, on rock ledges and, together with *Cryptogramma crispa*, it plays an important role in stabilising scree. Although so common over most of the area, it stops abruptly in the south being virtually absent from the fells of the Rothay - Brathay catchment area west of Ambleside, and from the Coniston Fells, the only exceptions being two very small populations on Wetherlam (28.00) and above Coniston (28.98). It is also absent in the north from the Skiddaw fells. There appears to be no geological explanation for this.
Alt. range: 135 m, Ennerdale (14.12) - 870 m, Helvellyn (34.14)
Map 434   144   WFCY

*A. conjuncta** Bab.
An extremely rare garden escape, seen in limestone grassland above Burton-in-Kendal (54.76s, J. Ketchen, 1970s; LANC), on limestone rocks by the road west of Kendal (50.90, GH, 1981; LANC) and on the golf-course (50.92, CEW, 1985).

Baker and Hodgson make dismissive references to a report of this species in Cumberland. This is curious since there is a specimen collected by A. Bennett (1879, BM) from a "root originally from the Buttermere Fells gathered by W. Hodgson, a friend of mine" and supplied by R. Wood. It was earlier collected from Gatesgarth, Buttermere (1.1, W. Borrer, 1844, BM), Ullswater (J. Walton, undated, BM) and Rydal Mount (3.0, Rev. Stillingfleet, all *BEC Rpt* (1930) 9:114). There are also two post-1930 records in the *Critical Atlas* for the Borrowdale (2.1) and Alston (7.4) squares and a collection from the slopes of Arnside Knott (4.7s, P.C. & J. Hall, 1954, BM).

Alchemilla conjuncta is a plant of the Jura and south-western Alps. It is not now generally regarded as native anywhere in Britain although not infrequently cultivated and becoming naturalised.
Alt. limit: 150 m, west of Kendal (50.92)
                3   W(C)

A. glaucescens Wallr.
The only record is from the Crosby Ravensworth area where it was found in 1981 in some quantity in open, damp limestone grassland by the Lyvennet Beck (6.1, FJR, *Watsonia* (1982) 14:191).

With its long silvery hairs, this is an attractive and distinctive species. It was previously known in the British Isles only from its headquarters in the Craven area of Yorkshire, the extreme north-west of Scotland and Ben Bulben, Co. Sligo.
Alt.: 290 m, above Crosby Ravensworth (6.1)
                R 1   W

A. monticola
This species was reported by Bradshaw (1952) from two sites on both sides of the Alston - Middleton-in-Teesdale road (74.40), but several searches by RWMC during the Survey have failed to rediscover it.

This is one of the characteristic hay-meadow and roadside verge species of upper Teesdale and Weardale, its only native locality in Britain.
                R   (C)

A. xanthochlora Rothm.
A common plant of roadside verges, particularly if damp and with some base-enrichment, also in the uplands by limestone becks and in hay-meadows. In the Lake District it is to a large extent replaced by *Alchemilla glabra*.
Alt. limit: 845 m, Great Dun Fell (70.32)
Map 435   935   LWFCY

A. filicaulis Buser
subsp. **filicaulis**
Very rare but easily overlooked. The only Survey records are from short, limestone turf in the Pennines below Melmerby High Scar (64.36, RWMC, 1989, LANC) and to the south (64.34, A.J. Richards, 1974, herb. A.J.R., det. S.M. Walters), on Knock Fell (70.30, A.J. Richards, 1970s) and on Dufton Fell (74.28,

RWMC, 1991), and in the Lake District only from above the Kirkstone Pass (38.08, T. Blackstock, 1978, det. M.E. Bradshaw). There is also a 1968 record from east of High Cup Nick (74.26, A.O. Chater, LANC) and earlier ones from above Red Tarn, Helvellyn (3.1, L.B. Hall, 1921, BM) and the "north-east side of Haweswater" (4.1, C.E. Salmon, 1925, BM) all determined by S.M. Walters.

Alt. limit: 720 m, Knock Fell (70.30)

S 5   WC

subsp. **vestita** (Buser) Bradshaw

A frequent species of short, well-drained limestone turf in south and north Westmorland and the north Pennines, also of cliff ledges in the Lake District. Elsewhere it is rather scattered.

Alt. limit: 740 m, Green Castle, Cross Fell (70.30)

Map 436   433   WFCY

A. minima Walters

Found only on the north-east side of Whernside (7.8, H. Robertson, 1970s, det. S.M. Walters) and between Garsdale Head and upper Dentdale (7.8, GH, 1985, det. M.E. Bradshaw, LANC). At the former site it grows around springs and at the latter intermixed with *Alchemilla filicaulis* in short *Festuca - Agrostis* limestone grassland.

As the name implies this is a minute species. It is very similar to *A. filicaulis* subsp. *filicaulis* differing chiefly in its smaller size, which appears to be retained in cultivation, but also in the fewer leaf-lobes (5) and the entire, untoothed, base of the sinuses between them.

This endemic species is known only from the Ingleborough and Whernside area.

Alt.: 470 m, Garsdale Head (7.8) and 530 m, Whernside (7.8)

R 2   Y

A. glomerulans Buser

This species is represented in Cumbria by a single small population in upper Teesdale (8.2) and just inside the county boundary.

In general habit it resembles *Alchemilla glabra*, but the stems and petioles are densely covered with appressed hairs. It is scattered throughout much of the Scottish Highlands and extends southwards to Teesdale and the Craven area of Yorkshire.

Alt.: 460 m, upper Teesdale (8.2)

S 1   W

A. wichurae (Buser) Stefánsson

A rather rare plant of damp and somewhat base-rich cliff ledges and gullies both in the Lake District and the northern Pennines. The species is no doubt somewhat under-recorded. There are no Survey records for the following sites reported by DAR during the 1950s: Skew Gill, Great End (22.08), Wanthwaite Crags (32.22), Eel Crags, Newlands (22.16), Comb Crags, Helvellyn (32.12) and Gillercombe (20.10).

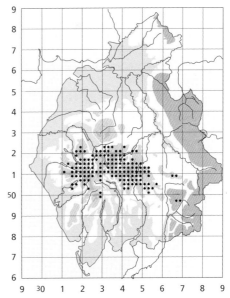

Map 434. *Alchemilla alpina*

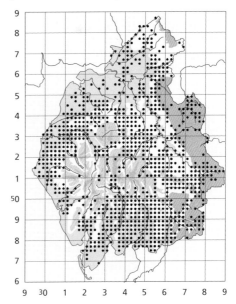

Map 435. *Alchemilla xanthochlora*

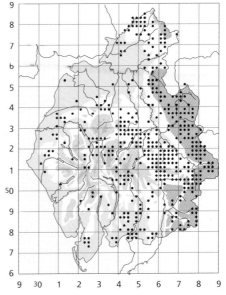

Map 436. *Alchemilla filicaulis* subsp. *vestita*

This species is not always easy to distinguish from *Alchemilla glabra* with which it very commonly occurs. The best characters are the smaller, neater, incurved teeth at the tips of the leaf lobes and the very narrow sinus between the lobes.

This is a plant of the northern and central Scottish Highlands, the Moffat hills, Cumbria, Teesdale and the Craven area of Yorkshire.

Alt. range: 365 m, Honister (20.14) - 790 m, Scafell Pike (20.06)

Map 437   S 30   WC

A. glabra Neyg.

Common except along the south-west coast and on the Solway Plain. It is particularly abundant in the Lake District and it is generally a more upland plant than *Alchemilla xanthochlora*, tolerating appreciably wetter conditions. It is characteristic of wet roadside verges and ditches, mountain gills and wet gullies and is often extremely luxuriant around waterfalls. Small specimens are readily confused with *A. wichurae*.

Alt. limit: at least 740 m, Green Castle, Cross Fell (70.30) but probably higher in the Lake District.

Map 438   1167   LWFCY

*A. mollis (Buser) Rothm.

Rare, with a few scattered localities mainly in the east. It occurs chiefly on disturbed ground, particularly dumps, as in quarries on Lazonby Fell (52.38) and Elliscales, Dalton-in-Furness (22.74, PB, 1988), the first Furness record. The first for Westmorland was from the banks of an artificial pond near Raisbeck (64.06, CEW, 1989, LANC), and for Cumberland one from Wetheral (46.54, C.W. Muirhead, early 1970s).

This species is a native of Romania and Asia Minor. The paucity of records in Cumbria is surprising since it is commonly grown in gardens and readily and persistently reproduces by seed.

Alt.limit: 230 m, Lazonby Fell (52.38)

Map 439   9   WFC

Aphanes arvensis L. sensu lato   Parsley piert
Map 440   301   LWFCY

There appear to be no records of either of the following segregates prior to the *Atlas*.

A. arvensis L. sensu stricto

Probably fairly frequent, occurring in open, neutral or calcareous soil such as quarry floors, rock ledges and field edges. It also occurs on thinly vegetated limestone pavement, flowering and fruiting before the soil dries out in the summer.

Alt. limit: 610 m, Great Dun Fell (70.30)

Map 441   112   LWFCY

A. inexspectata Lippert

(*A. microcarpa* auct.)

Scattered and less common than the last but apprec-

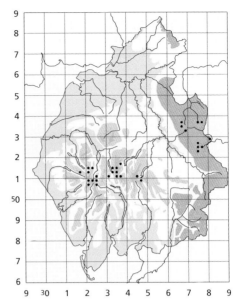

Map 437. *Alchemilla wichurae*

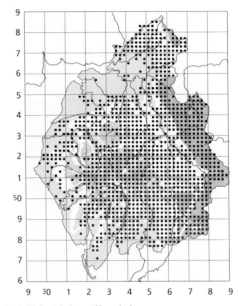

Map 438. *Alchemilla glabra*

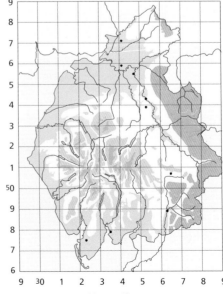

Map 439. *Alchemilla mollis*

iably commoner than earlier records suggest. It is a calcifuge, occurring in open sandy ground, around rabbit burrows, on railway ballast and on quarry debris. The only certain first record is for Cumberland where it was seen at Silloth in 1949 (*BSBI Yr Bk* (1951) p.48).

Alt. limit: 270 m, Midgeholme (62.58)

Map 442   95   WFCY

*Rosa multiflora Thunb. ex Murray
A native of Japan and Korea, this distinctive rose is known only from a hedge near Watermillock (44.22, LANC, *Watsonia* (1980) 13:136) where it was found by J. Taylor-Page in 1979.

Alt.: 155 m, Watermillock (44.22)

          1   C

R. arvensis Hudson   Field rose
Fairly frequent in hedges and scrub around the Kent estuary and on the surrounding limestone, otherwise

sparse, mainly in the west but extending north to Houghton (40.60, MG & JP, 1996, LANC). The only confirmed Survey record from the Eden valley is from Penrith (52.30, RWMC, 1994, LANC), although there are earlier ones from Melmerby (6.3, W.W. Mason, 1918) and Great Salkeld (5.3, H. Britten, 1900-1908; G.A.K. Hervey, 1947, 1952), and also one from Naworth, north-east of Carlisle (5.5, J. Parkin, 1936).

Alt. limit: 170 m, Penrith (52.30)

Map 443   54   WFC(Y)

R. arvensis x **R. canina**
R. x verticillacantha Mérat
Known only from Shoulthwaite Moss (3.2, A.H. Wolley-Dod, BM, 1912, det. A.L. Primavesi).

                (C)

R. pimpinellifolia L.   Burnet rose
Frequent along most of the coast, chiefly on calcareous

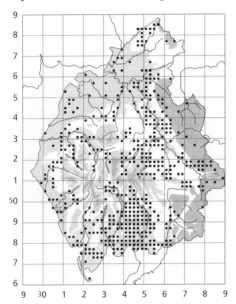

Map 440. *Aphanes arvensis* sensu lato

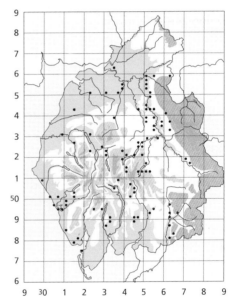

Map 442. *Aphanes inexspectata*

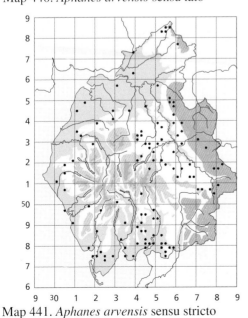

Map 441. *Aphanes arvensis* sensu stricto

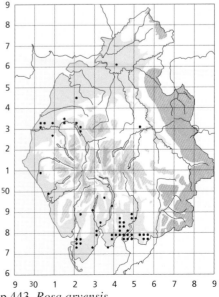

Map 443. *Rosa arvensis*

sand. It grows also on the limestone scars and in lime-stone scrub by Morecambe Bay and in north Westmorland, occasionally in hedgerows and, in the Lake District, on the more basic rock ledges. Spineless plants were seen on a limestone scar at Shap (56.12).

Alt. limit: 610 m, Glaramara (24.10, DAR 1954)

Map 444   114   WFC(Y)

R. pimpinellifolia x **R. canina**
R. x **hibernica** Templeton
Collected at Lorton (1.2, LANC) by E.C. Wallace in 1928 and from a lane at High Lorton, probably the same site, by C.W. Muirhead in 1950 (CLE, PLYP). The specimens were determined as the reciprocal cross by G.G. Graham and R. Melville respectively. Graham considers that a specimen from Kents Bank, Grange-over-Sands (3.7s, W.L. Smith, 1952, LANC) may possibly belong here.

This is a very rare hybrid occurring chiefly in the north of England.

([W]C)

R. pimpinellifolia x **R.mollis**
R. x **sabinii** J. Woods
Known only from near Pooley Bridge (4.2, A.H. Wolley-Dod, 1912, NMW, det. A.L. Primavesi), Kirkby Stephen (7.0, H.J. Riddelsdell, 1903, BM, det. G.G. Graham) and Riggindale, Haweswater (4.1, A. Ley, 1910, LANC, det. G.G. Graham). This last site is now submerged beneath the reservoir.

(W)

R.pimpinellifolia x **R.sherardii**
R. x **involuta** Smith
Found during the Survey on a rocky knoll south of Littlewater, Bampton (50.16, RWMC & TW, 1990, LANC, det. G.G. Graham). There are earlier collections determined by A.L. Primavesi from Keswick (2.2, N.J. Winch, [c.1800], CGE) and south-east of Penruddock (4.2, C. Bailey, 1892, NMW).

Alt. limit: 270 m, Littlewater (50.16)
                        1   W(C)

*R. rugosa Thunb. ex Murray
Frequent along the coast on sand-dunes, cliffs and open ground from Barrow-in-Furness (2.7) north to Grune Point (12.56), rarely inland and then mostly near houses. Its size and exceptionally large flowers and fruits render it very conspicuous. The first record is probably that from Keswick in 1927 (2.2, J.A. Webb, *BEC Rpt* (1928) 8:397). The first Furness and Westmorland records date from the early years of the Survey when it was found at Askam (20.76, KAG, 1974) and near Brough (78.14, LH, *c.*1974), although it had no doubt been present on the coast of Furness for some years before this.

Alt. limit: 200 m, Spadeadam (66.70n)

Map 445   46   WFC

Rosa section **Caninae**
With the exception of the rare and local *Rosa rubiginosa* and *R. micrantha*, plants in this section were

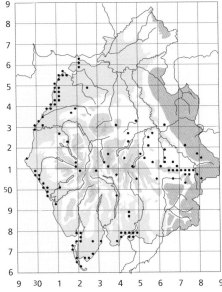

Map 444. *Rosa pimpinellifolia*

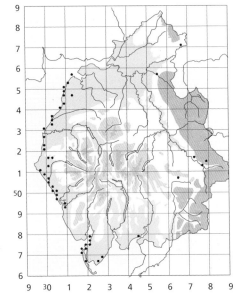

Map 445. *Rosa rugosa*

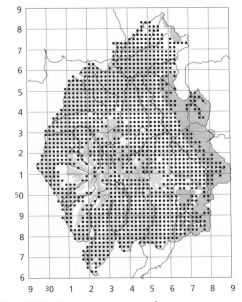

Map 446. *Rosa canina* sensu lato

recorded in the field as either the *R. canina* or *R. mollis* aggregate. Although only four species are involved, extensive hybridisation and introgression can present the recorder with a baffling array of bushes, even in a short length of hedgerow. During two recording meetings in the east led by G.G. Graham, at least half of the bushes were considered to be hybrids. The records of the segregates and their hybrids presented here are patchy, demonstrate the enthusiasm of particular recorders, and do no more than provide a sketchy basis for further study. Unless indicated otherwise, all herbarium specimens have been identified by Graham, and maps, other than for the aggregates, are based almost entirely on material which he has seen.

R. canina L. sensu lato   Dog rose
The following two species and their hybrids were recorded as this during the Survey. As is evident from the map they are extremely widespread and absent only from the higher ground. They occur prominently in hedgerows and along roadside walls, in scrub and woodland margins, and occasionally also in upland gullies and by waterfalls.
Map 446   1440   LWFCY

R. canina L. sensu stricto
No doubt very common. About a quarter of the specimens submitted to Graham were the pubescent form (*Rosa dumetorum* auct.).
Alt. limit: 430 m, above Garrigill (74.38)
Map 447   59   WFCW/Y

R. canina x **R. caesia**
R. x **dumalis** Bechst.
Probably almost as common as *Rosa canina*. Only three collections (54.50, 60.18, 60.20) were thought to involve subsp. *caesia*. There are also records in the *Handbook* for the Hawkshead (3.9), Lorton (1.2), Borrowdale (2.1), Keswick (2.2) and Whitbarrow (4.8s) squares.
Alt. limit: at least 250 m, Little Asby (68.08)
Map 448   27   WFC

R. canina x **R. tomentosa**
R. x **scabriuscula** Smith
Known only from collections by A.H. Wolley-Dod, determined by A.L. Primavesi, from near Dungeon Gill (2.0, 1928, BM), Langdale (2.0/3.0, not 6.0 as shown in the *Handbook*, 1931, BM) and Birkrigg (2.2, 1912, BM).
                                        (WC)

R. canina x **R. sherardii**
R. x **rothschildii** Druce
Seen in 1992 by a track near Long Meg, Little Salkeld (56.36, RWMC, LANC). Material tentatively identified as this was collected previously near Brampton (52.62, MS, 1983). There are earlier collections by A.H. Wolley-Dod from Buttermere (1.1, 1912, BM) and Langdale (2.0/3.0, not 2.1 and 6.0 as shown in the *Handbook*, 1931, BM, CGE).

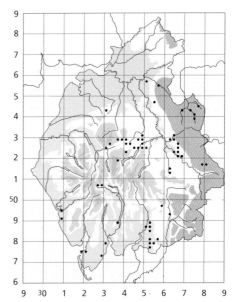

Map 447. *Rosa canina* sensu stricto

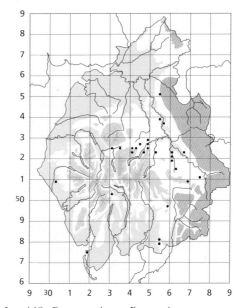

Map 448. *Rosa canina* x *R. caesia*

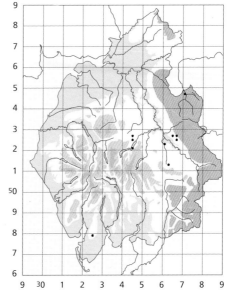

Map 449. *Rosa caesia* subsp. *caesia*

Alt.: 125 m, Little Salkeld (56.36)

2 (W)C

R. canina x **R. mollis**
R. x **molletorum** Heslop-Harrison
The only record, and that tentative, is of a single bush
by the Croglin Water (58.46, RWMC, 1991, LANC).
Alt.: 215 m, Croglin Water (58.46)

1 C

R. caesia Smith
subsp. **caesia**
(*R. coriifolia* Fries)
Apparently rather uncommon and mainly in the middle
and upper Eden valley. There are additional *Handbook*
records for the Scafell (2.0), Matterdale (3.2) and
Martindale (4.1) squares.
Alt. limit: at least 260 m, Alston (70.46) and Little
Asby (68.08)

Map 449 10 WFC

subsp. **vosagiaca** (N. Desp.) D.H. Kent
(subsp. **glauca** (Nyman) G.G. Graham & Primavesi
R. afzeliana Fries)
The commonest member of the *Rosa canina* aggregate.
There is an additional *Handbook* record from the
Ravenstonedale square (7.0). Several of the records are
of *R. afzeliana* var. '*glaucophylla*'.
Alt. limit: 449 m, Tynehead (74.36)

Map 450 92 WFCY

R. caesia
subsp. **caesia** x **R. mollis**
R. x **glaucoides** Wolley-Dod
Known only from two collections from north and east
of Alston (70.48, E. Chicken, 1980, herb. E.C., det.
R. Melville; 74.46, RWMC, 1993, LANC) and a
laneside north of Barbon (62.88s, M. Porter, 1995;
LANC, det. A.L. Primavesi).

Alt.: 350 m, east of Alston (70.48)

3 WC

R. caesia
subsp. **vosagiaca** x **R. sherardii**
Seen by a lane north of Little Salkeld (56.36, RWMC,
1992, LANC), Uldale (24.36, GH & G.G. Graham,
1987) and near Wray (36.00, A.N. Gagg, 1992, det.
A.L. Primavesi).
Alt. limit: 125 m, Little Salkeld (56.36)

2 FC

R. tomentosa Smith
This southern species was said by Wilson to be "rather
common". The only confirmed records are two determinations
by R. Melville of specimens from North Walney (1.7) and the
Duddon valley (2.9), both collected by G. Wilson in 1954,
and specimens collected by A.H. Wolley-Dod from Braithwaite and Wanthwaite (respectively 2.2, 3.2, 1912, BM, det.
A.L. Primavesi).

(FC)

R. tomentosa x **R. sherardii**
R. x **suberectiformis** Wolley-Dod
Collected from a hedge near Whitbarrow in 1965 (4.8s,
E. Hodgson, det. R. Melville, LANC, conf. G.G. Graham).

(W)

R. mollis Smith sensu lato Downy rose
Frequent to very common, especially in the uplands,
where it tends to replace *Rosa canina*, and is characteristically a bush of scars, screes and the sides of waterfalls. However, it is rare in the western Lake District
and along the south-west coast..

Map 451 779 WFCY

R. sherardii Davies
Probably widespread, but less common than *Rosa
mollis*. Several collections were of f. *resinosoides*
(Crépin) Wolley-Dod. There are further *Handbook*

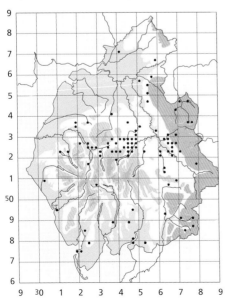

Map 450. *Rosa caesia* subsp. *vosagiaca*

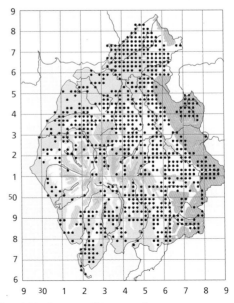

Map 451. *Rosa mollis* sensu lato

records for the Borrowdale (2.1), Martindale (4.1), Matterdale (3.2), Helvellyn (3.1), Melkinthorpe (5.2), Hawkshead (3.9), Crook (4.9) and Borrow Beck (5.0) squares.

Alt. limit: at least 240 m, above Little Asby (68.08)

Map 452   32   WFC

R. sherardii x **R. mollis**
R. x **shoolbredii** Wolley-Dod
This interesting and rare hybrid is represented by two collections of the above combination and two of the reciprocal. These are, respectively: Crosby Garrett (70.10, GH, 1979, LANC, det. R. Melville), Little Langdale (30.02, GH, 1985, LANC), Milburn, Appleby (64.26, GH, 1979, LANC, det. R. Melville), Greystoke (44.28, GH & G.G. Graham, 1987) and below Carlingill in the Lune gorge (60.00, GH, 1981, LANC), the last identification being tentative.

Alt. limit: 260 m, Crosby Garrett (70.10)

4   W

R. sherardii x **R. rubiginosa**
R. x **suberecta** (J. Woods) Ley
Known only from Hartsop (4.1, A. Ley, CGE, det. G.G. Graham) where it was collected in 1910.

(W)

R. mollis Smith sensu stricto
Generally common, except in the south-west, and probably more frequent than *Rosa sherardii* in the east and north. There are additional *Handbook* records for the Keswick (2.2), Helvellyn (3.1), Hawkshead (3.9), Martindale (4.1), Carlisle east (4.5) and Shap (5.1) squares.

Alt. limit: 450 m, Tynehead (76.36)

Map 453   63   WFC

cf. **R. mollis** x **R. rubiginosa**
R. x **molliformis** Wolley-Dod
Material possibly referable to this hybrid was collected in a lane at Flusco, Newbiggin (46.28, RWMC, 1993, LANC). If confirmed, this would be the first British record.

Alt.: 260 m, Flusco (46.28)

1   C

R. rubiginosa L.   Sweet briar
Despite extensive searches the only specimens seen during the Survey were at Hodbarrow (16.78, PB, 1990), on waste ground by the River Ellen at Maryport (02.36, MM, 1986; LANC), newly planted along the Appleby by-pass (68.20, D. Hinchcliffe, 1991, LANC), on a limestone ledge on Burtree Scar, Lowther Park (52.22, RWMC, 1991, LANC), at Newton Reigny (48.30, RWMC, 1992, LANC), newly planted on a lay-by off the A66 road west of Stainton (46.28, RWMC, 1993, LANC), on the roadside between Askam-in-Furness and Dalton (20.74s, 20.76s, PB, 1993, LANC), and on a limestone hillside at Far Arnside (44.76s, N. Webb, 1988, LANC).

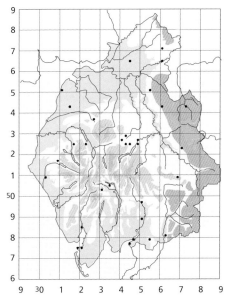

Map 452. *Rosa sherardii*

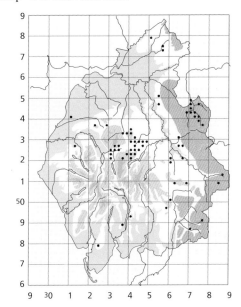

Map 453. *Rosa mollis* sensu stricto

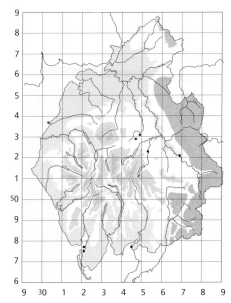

Map 454. *Rosa rubiginosa*

Hodgson gives numerous records for this species, covering at least nine squares and C.W. Muirhead knew of at least a further five records. By contrast, Baker noted that it was "occasionally......in Furness" and Wilson lists it only from Arnside. There are no *Atlas* records for Furness and only two from Westmorland, for the Barbon (6.8s) and Warcop (7.1) squares.

This rose is quite rare in north-west England.
Alt. limit: 230 m, Lowther Park (52.22)
Map 454   9   WFC

R. micrantha Borrer ex Smith
Several bushes of this southern species was found in 1991 in south-facing limestone scrub at Far Arnside (44.76s, MB, 1991, LANC, *Watsonia* (1992) 19:146).
                                   1   W

*Prunus cerasifera Ehrh.   Cherry plum
In hedgerows in a few places in the south, also in the north near Angerton (22.56, REG, 1988) and near Aglionby (44.56, FJR, 1981). The latter record is the first for Cumberland, those for Furness and Westmorland being from Field Broughton (38.80, GH, 1979, LANC) and Witherslack (42.84s, KAG & FLW, 1964), where, unfortunately, the trees have recently been destroyed.

This southern species has probably been in the county for some considerable time and it is surprising that it has been overlooked, especially in view of its early and conspicuous flowering. It flowers earlier than *Prunus spinosa* and in a mild spring at the begining of February. Fruits are very rarely produced.
Map 455   21   WFC

P. spinosa L.   Blackthorn
A common shrub of hedgerows and scrub, more rarely of woodland. It is rather uncommon in the uplands where it grows chiefly on the limestone scars.

The bushes vary appreciably in spininess and may sometimes be difficult to distinguish from *Prunus domestica*.
Alt. limit: 500 m, Ardale, Cross Fell (66.34)
Map 456   1298   LWFCY

*P. domestica L.   Wild plum
Fairly frequent throughout the lowlands and quite common in south Westmorland. It occurs almost exclusively in hedgerows. By contrast, there are relatively few *Atlas* records.

The Cumbrian plums are in need of detailed study. Most bushes have rather glabrous young twigs and would seem to belong to subsp. *domestica*, rather than subsp. *insititia*, which includes bullace and damson. However, many bushes, and lengths of hedgerow, probably represent single, self-incompatible clones and fruit is rare. This is unfortunate as important diagnostic characters of subsp. *domestica* are the rather large fruits with the flesh separating at maturity from

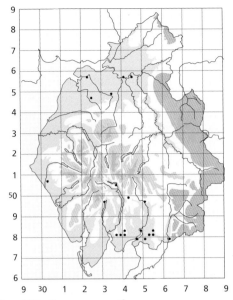
Map 455. *Prunus cerasifera*

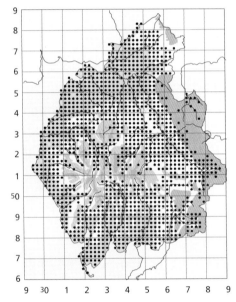
Map 456. *Prunus spinosa*

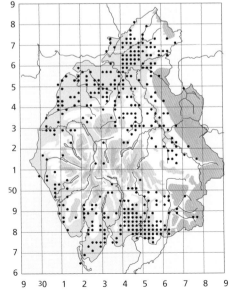
Map 457. *Prunus domestica*

the stone. The fruits of subsp. *insititia* are usually purplish, rounded in bullace but ovoid in damson. Damsons are frequent in south Westmorland, notably in the Lyth valley where they occur chiefly in farm orchards. In the Sedbergh - Dent - Ravenstonedale area plantings are said to date from the late 19th century. A few damson hedges still occur to reward the autumn wayfarer. Some bushes of green-fruited bullace used to grow alongside the A590 at Levens (48.84s) prior to road-widening. Surprisingly, Wilson does not refer to subsp. *domestica* and describes subsp. *insititia* as rare.

Alt. limit: 280 m, north of Alston (70.48)
Map 457, Plate 74   361   LWFCY

P. avium (L.) L.   Wild cherry; Gean
A frequent tree over much of the county and in the valleys of the southern Lake District it is one of the most delightful spring sights when in full bloom. It occurs extensively in deciduous woodland and frequently as a hedgerow tree, being spread very effectively by birds.

    This is a popular amenity tree with local authorities and is also commercially useful for planking and turnery, having the added bonus of being largely immune to attacks by grey squirrels.

Alt. limit: 400 m, north of Garrigill (72.42)
Map 458   829   LWFCY

*P. cerasus L.
A rather rare hedgerow tree almost confined to the north-west of the county. It is easily overlooked as, with regular cutting, it often fails to flower. It is best distinguished from *Prunus avium* by its large, pure-white flowers and dull green leaves, which persist longer and lack the fine autumn colour of that species. It still occurs "in hedges one mile east of Egremont" where it was seen by J. Adair in 1895 (02.10, BM; AAD, 1992, LANC, *BEC Rpt* (1897) 1895:473) and there are also mid-19th century specimens from the Ulverston area (2.7/3.7s, E. Hodgson, BM).

Alt. limit: 170 m, Ousby (60.34)
Map 459   28   WFC

P. padus L.   Bird cherry; Heg; Hags; Heckberry
Common in hedgerows and occasionally in deciduous woodland, particularly on the sides of valleys and in the gills. In hedgerows it is conspicuous during late spring and early summer when the shoots are regularly defoliated and encased in a dense web spun by the caterpillars of the moth *Yponomeuta euonymella*.

The famous Lyth Valley damson blossom will be at least a week later - in an average year April 18 is about the date to see it at its best. For the best effect of the clouds of white on a thousand trees, the blossom should come before the leaf but, after a savage winter, the leaf sometimes comes with the blossom or even before it, making the picture not quite so fine. There is no lovelier route to the Lakes from the south in spring than through the sleepy dale where grows, in almost every garden and orchard, the damson that some people consider the finest in the world - oval, not round, in shape, with a perfect dusty bloom and sharp nutty flavour. And the dale now has another distinction - a certain hostelry with more than 60 malt whiskys on offer and a most erudite barman. You can do them all in a short day - the daffodils, the damson blossom, and a dram or two of your favourite malt.

A Lakeland Mountain Diary, A. Harry Griffin

Plate 74. Damson blossom and Rough Fell sheep, Lyth valley

Interestingly, this cherry is still known, at least in the Ravenstonedale - Dent area, by the old Norse name heg and in the Furness peninsula as hags.

Alt. limit: 650 m, Dove Crag (38.10)

Map 460   866   WFCY

*P. lusitanica L.   Portugal laurel

Scattered around the head of Morecambe Bay but with two outlying sites on the west coast. Some of the records are from estate grounds and few are far from houses. There are no pre-Survey records. The first record for Cumberland is from Muncaster Woods (08.94, MMG, 1990, LANC) but it is not possible to identify that for Westmorland.

Map 461   10   LWC

*P. laurocerasus L.   Cherry laurel

Occasional, nearly always as an undershrub in estate woodland where its deep shade soon eliminates the ground flora. There are no reports of it regenerating. The first records appear to be those shown in the *Atlas* for the (1.0), Hawkshead (3.9) and Barbon (6.8s) squares.

Alt. limit: 140 m, Kentmere (44.00)

Map 462   54   LWFCY

*Pyrus communis L.   Pear

Rather rare, occurring as isolated trees and in hedgerows, usually by roads and often by houses. There is a particularly fine old roadside tree near Howe Farm, Winster (40,92). The fruits are generally small and tart but one of the trees at the abandoned Burton Hall near Warcop (74.18) produces relatively sweet fruit.

No attempt was made during the Survey to distinguish between this tree and *Pyrus pyraster* (L.) Burgsd.

Alt. limit: 220 m, east of Bampton Grange (52.18)

Map 463   35   WFC

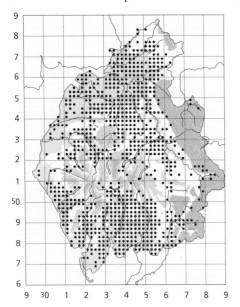

Map 458. *Prunus avium*

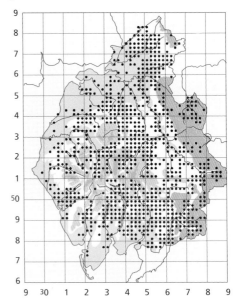

Map 460. *Prunus padus*

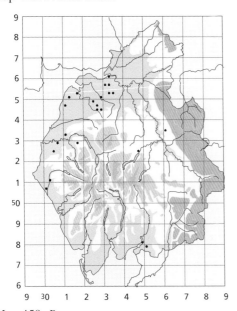

Map 459. *Prunus cerasus*

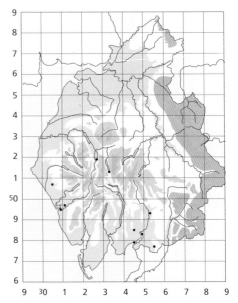

Map 461. *Prunus lusitanica*

Malus sylvestris (L.) Miller   Crab apple
Frequent in hedgerows, some of which date to at least the 17th century, less commonly as a constituent of deciduous woodland and scrub. There was little attempt during the Survey to record this species and *Malus domestica* separately, but there seems no reason to doubt that the map fairly represents the former except perhaps in the south. This agrees with Wilson's comments and the observation of Livermore & Livermore (1987) in the Lancaster District but contrasts with the situation across the Pennines in Co. Durham (Graham 1988).

      In the middle and upper Eden valley, crab was widely planted in the late 18th and early 19th century hedges in a mixture of two thorn to one apple (D. Hinchcliffe *pers. comm.*). There also appears to have been extensive planting in hedges in the north-east of the county.

Alt. limit: 380 m, Swindale, (50.12)
Map 464   975   LWFCY

*M. domestica Borkh.   Apple
(*M. sylvatica* subsp. *mitis* (Wallr.) Mansf.)
Probably not infrequent; largely restricted to hedgerows and generally appreciably less common than crab. Most trees have probably been bird-sown or arisen from discarded cores.
Alt. limit: at least 250 m, north of Lowthwaite (40.24)
Map 465   60   WFC

Sorbus aucuparia L.   Rowan
Very common, occurring in the lowlands in hedgerows and small woods, especially on peaty soils. This is, however, characteristically a tree of the uplands, occurring in gills and on rock ledges wherever grazing permits and seemingly indifferent to exposure. Its

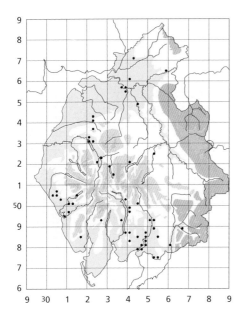

Map 462. *Prunus laurocerasus*

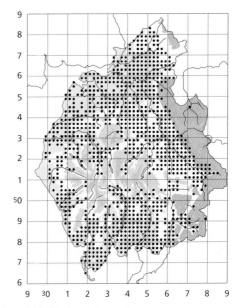

Map 464. *Malus sylvestris*

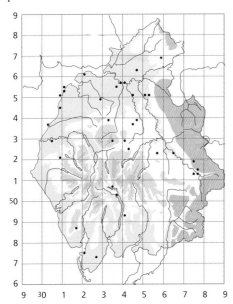

Map 463. *Pyrus communis*

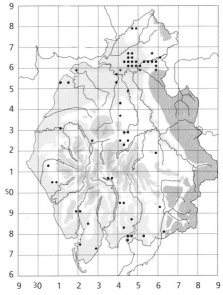

Map 465. *Malus domestica*

scarlet berries epitomise autumn in the fells.
Alt. limit: 870 m (Wilson)
Map 466, Plate 75   1541   LWFCY

*S. aucuparia x **S. aria**
S. x thuringiaca (Ilse) Fritsch
The only record is of a single tree in Riley Plantation,
Aira Force (38.20, MC, 1978, LANC).
Alt.: 215 m, Aira Force (38.20)

                                                           C

*S. intermedia (Ehrh.) Pers.   Swedish whitebeam
Occasionally planted, usually in urban areas and
commonest around Barrow-in-Furness. It is able to
regenerate but not to any significant extent. The earliest
records for Cumberland are probably those given by
Hodgson and Baker repectively from Whitehaven (9.1,
as *Pyrus fennica*) and Wasdale Head (1.0, as *P. semi-
pinnata*). The first record for Furness was from
Roanhead (1.7/2.7, G.A.K. Hervey, 1956) and for
Westmorland probably from Arnside (44.76s, 44.78s,
MB, 1977).
Alt. limit: 380 m, above Leadgate, Alston (70.42)
Map 467   68   WFC

S. aria sensu lato   Common whitebeam
All records of the following three species were included
in the early Floras under *Sorbus aria*. *S. rupicola* was
for long not generally accepted as a British species and
the endemic *S. lancastriensis* was not validly described
until 1957 (Warburg 1957). Their identification in
Cumbria has been made difficult by the inadequacies of
this last description and the misleading maps in the
Critical Atlas which show the last two species as
having mutually exclusive ranges in Cumbria. All three
species may, in fact, occur together on the same
limestone cliff. The situation was greatly clarified in
the paper on *S. lancastriensis* by Rich & Baecker
(1986) from which the following key is taken.
Representative leaves of the three species are shown in
Fig. 16.

1. Lvs ovate, rhombic to elliptic, rarely obovate, the
   base truncate to broadly cuneate (angle with central
   vein 45° or more), rarely cordate, toothed nearly to
   base, with (10.5) 11 or more pairs of veins. Fruit at
   least as long as wide   ....*aria*
1. Lvs obovate, oblanceolate or oblong, cuneate at base
   (angle with central vein less than or equal to 42°
   (47°)), +/- entire in lowest quarter, with 10 (11) or
   fewer pairs of veins. Fruit at least as wide as long
   2. Lvs erect, (1.7-) 1.8-2.5 times as long as wide, with
      (5) 6-9 (10) pairs of veins   ....*rupicola*
   2. Lvs spreading, (1.3-) 1.4-1.75 (-1.9) times as long
      as wide, with (6.5) 8-10 (11) pairs of veins
                                   ....*lancastriensis*

Plate 75. *Sorbus aucuparia*

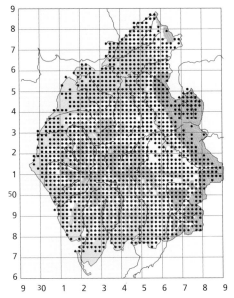

Map 466. *Sorbus aucuparia*

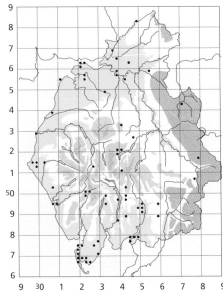

Map 467. *Sorbus intermedia*

S. aria (L.) Crantz

Scattered in the lowlands and not uncommon in south Westmorland, occurring in hedgerows, around the margins of deciduous woodland, as a cosmetic screen around coniferous plantings, in open limestone woodland and on limestone scars where it sometimes occurs with either or both the following species. It has no doubt been extensively planted, but it regenerates easily and is readily bird-sown with the result that at many sites, for example on the Morecambe Bay limestones and on crags in middle Eskdale (16.98, 1600) it looks very native. The *Critical Atlas* suggests, however, that it is only native in southern England.

All Hodgson's localities appear to refer to this species but Petty's could also include records for the following two. Wilson curiously fails to mention it. All the Survey records of *Sorbus aria* from tetrads in which *S. rupicola* or *S. lancastriensis*, or both, occur have been carefully checked.
Alt. limit: at least 335 m, north-west of Nenthead (74.46)
Map 468, Fig. 18a   141   LWFCY

S. lancastriensis E. Warb.

Locally frequent on the Morecambe Bay limestones and extending inland as far north as Cunswick Scar (48.92) and east to Farleton Fell (52.78s). It occurs in rather open woodland on cliffs, screes and, more rarely, in limestone pavement. Rich & Baecker (1986) mention a site on conglomerate rocks at Roughholme, west of Humphrey Head (3.8), from where it was recorded "in great quantity upon the rocks near Rougham" as early as 1700 (Petty 1895 pp.264-265), and it was seen on Silurian slate at Pool Bank (42.86s, GH, 1974, LANC). Rich & Baecker (1992) give the pH of this soil as 6.0-

6.5. They were unable to refind the solitary tree on the limestone pavement of Birkrigg Common (2.7, FLW & KAG, 1982, LANC), the westernmost site for the species. The largest population, estimated by them as in excess of 1,000 plants, is that on Whitbarrow (4.8s).

At some sites it occurs intermixed with *Sorbus rupicola* and some populations on the coast west of Arnside (44.76s, 46.76s) are somewhat intermediate. Indeed, Game (1983) regards all the Morecambe Bay trees as representing a single taxon to which the name *S. lancastriensis* can be applied, an interpretation which accords with Warburg's maps in the *Critical Atlas*.

Originally described from Humphrey Head (38.74s), this endemic species extends only a few kilometres southwards into Lancashire.
Alt. limit: 180 m, Cunswick Scar (48.92)
Map 469, Fig. 18b   R 21   WF

S. rupicola (Syme) Hedlund

Scattered on the Morecambe Bay limestones and often intermixed with the usually commoner *Sorbus lancastriensis*. It also occurs on some of the more base-rich volcanic crags around the northern end of Thirlmere (30.18, 32.18, 32.22) and on the east side of Derwent Water (26.20).

Hodgson appears to have been unaware of these native Cumberland populations. Rich & Baecker (1986) refer Wilson's records from Scout and Cunswick Scars to the previous species.

This whitebeam has a wide if scattered distribution throughout western and upland Britain.
Alt. limit: 365 m, Raven Crag, Thirlmere (30.18)
Map 470, Fig. 18c   S 11   WFC

Fig.18. Leaves of a) *Sorbus aria*, b) *S. lancastriensis*, c) *S. rupicola*, d) *S. croceocarpa* and e) *S. latifolia*.

*S. croceocarpa P.D.Sell

Although trees of the *Sorbus latifolia* complex have long been known to occur in the Arnside area this seems to have escaped the attention of Flora writers. In his recent review of the complex, Sell (1989) describes the new species *S. croceocarpa*, a fairly widespread tree with orange berries. He lists two Cumbrian specimens, one from Hazelslack, Arnside (46.78s, GH, 1986, LANC) and the other from the shore at Sandscale (20.74, MB, 1985, herb. T.C.G. Rich). Rich has also identified material collected from Hodbarrow (16.78, PB, 1991) and two other tetrads around Arnside Knott (44.76s, 44.78s, MB). At least seven individual sites are now known in this latter area.

Sell considers that this species, and the other members of the *S. latifolia* complex, have originated from hybridisation between *S. aria* and *S. torminalis* and nowhere does he believe it to be native although it

readily becomes naturalised.

Fig. 18d   5   WFC

*S. latifolia (Lam.) Pers.

The only record is of a small tree, not apparently planted, on railway property at Cark (36.76s, JA, 1996, LANC, conf. P.J.M. Nethercott).

Fig. 18e   1   F

S. torminalis (L.) Crantz   Wild service-tree

Recorded during the Survey by C.D. Pigott in Copridding (44.76s), Grubbins (44.78s), Meathop (42.78s) and Roudsea (32.82) woods and by the Leven estuary at Plumpton (30.78). It was also found at Far Arnside (44.76s, MB), Middlebarrow (46.76s, J. Leedal) and Humphrey Head (38.74, T.C.G. Rich), where some leaves were seen at the foot of the western cliff. There seems little reason to doubt that the species

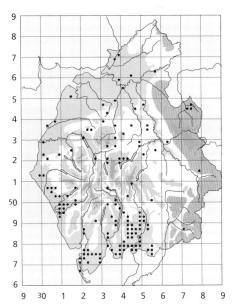

Map 468. *Sorbus aria*

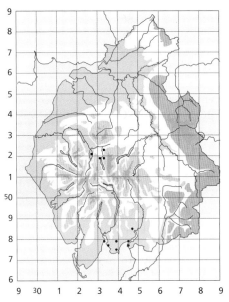

Map 470. *Sorbus rupicola*

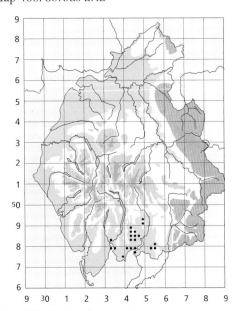

Map 469. *Sorbus lancastriensis*

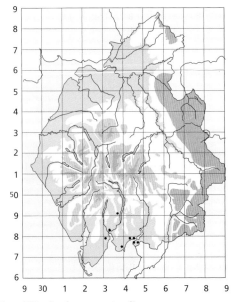

Map 471. *Sorbus torminalis*

is native on the Morecambe Bay limestone despite the indication in the *Atlas* to the contrary. There is also a very interesting find from near Lakeside (36.90, PB, 1987, LANC) where a few stunted trees occur with *Juniperus, Calluna* and *Vaccinium myrtillus* on Silurian rocks by the shore of Windermere. A tree seen near Newbiggin Hall, Wreay (42.50, EHR, not mapped) was presumably planted.

The species has disappeared from Levens Park (4.8s) and Flass, Crosby Ravensworth (6.1), records cited by Baker and Wilson, and also from Blaithwaite quarry (2.4, J. Parkin, 1935), Woodside, Carlisle (4.5, E. Blezard, 1940, CLE), but perhaps the same as the Wreay site, and from near Hesket Newmarket (3.3, Nature Conservancy Council, undated); these records probably all relate to plantings.

Map 471   8   WFC

*Amelanchier lamarckii F.-G.Schroeder

Seen during the Survey in a fen south of Wreaks End, Broughton-in-Furness (22.86, C. Cowan), in some quantity in woodland by the beck at Force Forge, Satterthwaite (32.90, GH), where it had been known for some years, on Lilies of the Valley islands, Windermere (38.96, GH, 1990), near Thurstonfield Lough (30.56, REG, 1978) and, a single tree, on Beacon Hill, Penrith (52.30, RWMC, 1985, LANC, both records *Watsonia* (1986) 16:188).

The first records appear to be those given by Schroeder (1970) from the Windermere islands (T.J. Foggitt, 1891, BM) and mossland at Broughton-in-Furness (K.R. Burgess, 1962, K, *Proc. BSBI* (1966) 6:239). There is also a record from woodland on the west side of Windermere opposite the islands (K. Hollick, 1960).

There has been much confusion concerning the identity of naturalised *Amelanchier* species in Britain but Schroeder concluded that *A. lamarckii* is the only one represented. It is an attractive shrub with its slender, branching sprays of delicate white flowers and purplish young leaves which are silky beneath.

Alt. limit: *c.*240 m, Beacon Hill, Penrith (52.30)

        6   WFC

*A. ovalis Medikus

A Continental species recorded in 1927 by W. Keble Martin from a crag above "Longrigg Brow" (*BEC Rpt* (1928) 8:397).

        (W)

*Photinea davidiana (Decne.) Cardot

(*Stranvaesia davidiana* Decne)
Known only from a single bush, probably planted, in Grubbins Wood, Arnside (44.78s, MB, 1975).

        1   W

Cotoneaster

Although there is only a single record in the early Floras, this genus is now known to be represented in the county by at least 12 species. Probably all but one have appeared during the last 50 years, becoming naturalised from garden throw-outs or by bird-dispersal from gardens, although plantings, especially in the Arnside area, cannot be excluded. With the exception of the three commonest species, *Cotoneaster horizontalis, C. integrifolius* and *C. simonsii*, all records have been checked by J. Fryer. The following key is based on information which she has provided. A further species, *C. rotundifolius* Wallich ex Lindley, is listed in the Addenda.

1. Lvs 0.5-1.5 cm
  2. Lvs sub-orbicular; petals pink; branches herring-bone   ....3. *C. horizontalis*
  2. Lvs obovate-elliptic; petals white
           ....2. *C. integrifolius*
1. Lvs more than 1.5 cm, at least twice as long as wide
  3. Lvs bullate, to 15 cm long
    4. Fruits black
      5. Lvs strongly bullate   ....9. *C. moupinensis*
      5. Lvs not strongly bullate   ....4. *C. villosulus*
    4. Fruits red
      6. Lvs to 7 cm, petioles 3-6 mm; calyx hairy
           ....7. *C. bullatus*
      6. Lvs to 15 cm, petioles 1-3 mm; calyx hairy only on the margins   ....8. *C. rehderi*
  3. Lvs flat
    7. Lvs not tomentose below
      8. Lvs less than 3 cm   ....6. *C. simonsii*
      8. Lvs more than 3 cm
        9. Petals erect; fruits red, becoming black
           ....5. *C. laetevirens*
        9. Petals spreading; fruits remaining red
           ....1. *C. x w* *watereri*
    7. Lvs tomentose below
      10. Corymbs 3-7-fld; calyx lobes long mucronate; lvs 1.5-2.5 cm   ....10. *C. dielsianus*
      10. Corymbs 5-15-fld; calyx lobes acute to acuminate; lvs more than 2 cm
        11. Lvs 2-3 cm; corymb shoots to 4 cm; fruits orange-red, oblong-ovoid ..11. *C. franchetii*
        11. Lvs 2.5-6 cm; corymb shoots to 1.5 cm; fruits subglobose   ....12. *C. sternianus*

*1. C. frigidus Wallich ex Lindley x C. salicifolius Franchet

C. x watereri Exell
A robust and variable hybrid known from Grubbins Wood, Arnside (44.78s, MB, 1988; LANC, *Watsonia* (1989) 17:471), the Kendal golf-course (50.92, CEW, 1987, LANC) and the roadside east of Parton (98.20, AAD, 1992, LANC).

The leaves are elliptic and semi-evergreen and the corymbs are usually many-flowered.
Alt. limit: 135 m, Kendal (50.92)

        3   WC

*2. C. integrifolius (Roxb.) Klotz

(*C. microphyllus* auct.)
Commonly naturalised on the limestone around More-

cambe Bay but also well-established at several sites in the Lake District and elsewhere. In certain limestone quarries it is so rampant as to constitute a threat to the native flora. Although not referred to in earlier Floras, it was present on the limestone at Far Arnside (4.7s, Anon., LANC) as early as 1908. The first Furness record is the post-1930 *Atlas* dot for the Grizebeck square (2.8); there is a possible pre-Survey Cumberland record from the edge of the vice-county at Glencoyne (3.1, G.A.K. Hervey, 1955).
Map 472   41   WFC

*3. C. horizontalis Decne
A frequent garden escape on the limestone around Morecambe Bay and in south Westmorland, otherwise rare. The only first record that can be identified with certainty is that for Cumberland, from a quarry at Friz-ington (02.16, AAD, 1979, *Watsonia* (1980) 13:136). It seems likely, however, that Hodgson's reference to *Cotoneaster vulgaris* (*C. integerima*) on rocks at Hallsteads, Ullswater (Skelly Neb, 4.2) and elsewhere refers to this species since he says it is frequently grown against cottage walls.

This is a very distinctive species with its herringbone branching in one plane and almost prostrate habit.
Alt. limit: 300 m, Matterdale End (38.22)
Map 473   49   WFC

*4. C. villosulus (Rehder & E. Wilson) Flinck & Hylmö
Recorded from Grubbins Wood, Arnside (44.78s, MB, 1983; LANC) but originally identified as *Cotoneaster acutifolius* (*Watsonia* (1989) 17:471).

It is distinguished by its black fruits and some-what bullate leaves.
                            1   W

*5. C. laetevirens (Rehder & E. Wilson) Klotz
Discovered in Red Hills Wood, Arnside (44.78s) by MB in 1983. Although originally identified as *Cotoneaster acutifolius* (*Watsonia* (1989) 17:471), it was redet-ermined by B. Hylmö as *C. laetevirens*.

This is a large-leaved, deciduous species with fruits which turn black on ripening.
                            1   W

*6. C. simonsii Baker
A well established and rapidly spreading garden escape but chiefly on the limestone in the south where it occurs in hedges and quarries and on rock outcrops. There are no pre-Survey records; the only first record that can be identified is the Cumberland one from near Water-millock (42.22, MC, 1975).

Cotoneaster simonsii is generally easily recog-nised by its small leaves (-3 cm) which lack tomentum on the lower surface.
Alt. limit: 335 m, near Watermillock (42.22)
Map 474   61   WFC

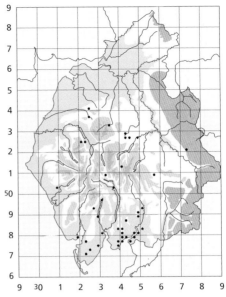

Map 472. *Cotoneaster integrifolius*

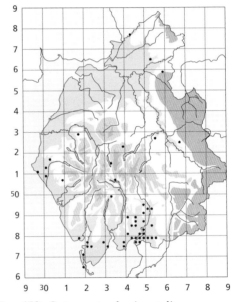

Map 473. *Cotoneaster horizontalis*

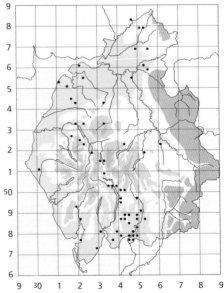

Map 474. *Cotoneaster simonsii*

*7. **C. bullatus** Bois.

Discovered in 1989 in a copse near Arnside station (46.78s, MB; LANC). It is readily confused with *Cotoneaster rehderi*, in fact the record of *C. bullatus* from Grubbins Wood (*Watsonia* (1989) 17:471) was probably based on a misidentification. The best distinguishing characters are the shorter leaves (-7 cm) and longer petioles (3-6 mm).

1   W

*8. **C. rehderi** Pojark.

Known from two woods at Arnside: Beechwood (44.76s, MB, 1988, LANC, *Watsonia* (1990) 18:426) and Grubbins Wood (44.78s, MB, 1989, LANC), near Dalton-in-Furness (22.76, PB, 1989, LANC), by the roadside east of Souterthwaite (24.80, JA, 1994, LANC) and Clints quarry, Egremont (00.12, CCH, 1981; LANC).

This is a tall, robust shrub with bullate leaves up to 15 cm and cymes with up to 25 flowers.

5   WFC

*9. **C. moupinensis** Franchet

Known only from Grubbins Wood, Arnside (44.78s, MB, 1988, LANC).

Although rather similar in appearance to *Cotoneaster rehderi*, this species is smaller, to 3 m high, and has purplish-black fruits.

1   W

*10. **C. dielsianus** E. Pritzel ex Diels

Recorded only from Red Hills Wood and Grubbins Wood, Arnside (44.76s, MB, 1989; 44.78s, MB, 1988, LANC).

This is a small-leaved (-2 cm), deciduous species with red fruits.

Cotoneaster dielsianus and the following two species have rather small leaves which are densely tomentose below.

2   W

*11. **C. franchetii** Bois.

Found on the railway embankment at Arnside (44.78s, MB, 1977; LANC) and on a limestone wall near Dalton-in-Furness (22.74, PB, 1987, LANC, both *Watsonia* (1989) 17:471).

In this species the tomentum is initially brown but becomes silvery with age. The fruits are oblong-ovoid and orange.

2   WF

*12. **C. sternianus** (Turrill) Boom

Found in Red Hills Wood and Grubbins Wood, Arnside (44.76s, MB, 1989; 44.78s, 1977, *Watsonia* (1989) 17:471), on the old railway line at Harrington (98.24, AAD, 1989, LANC, *Watsonia* (1990) 18:426) and in Crag Wood, Cartmel Fell (40.84s, MB, 1989, LANC).

Although rather similar to *Cotoneaster franchetii*, *C. sternianus* has leaves which are more densely tomentose and longer (-6 cm), short-stalked corymbs and orange-red, globose fruits.

4   WC

*****Mespilus germanica** L.   Medlar

The only Survey record is of a small, flowering shrub growing near the sea wall at Maryport (02.36, EEM, 1988, LANC). There is a collection from Black Dyke, Silloth (1.5, E.J. Glaister, CLE) in 1876 and Baker mentions two trees, presumably planted, growing in a hedge between Walney church and North Scale (1.6s).

1   (F)C

Crataegus monogyna Jacq.   Hawthorn

Very common, chiefly as a hedgerow shrub but also in scrub and deciduous woodland, and as an early colonist of formerly grazed land. Mature bushes being relatively resistant to grazing are often the last indicators of former woodland to be seen on the fellsides and dale heads.

Hodgson gives a record of var. *laciniata* from the Duddon (2.8), Baker one from Arnside (4.8s) and there is a later one from Naworth (5.6, *BEC Rpt* (1924) 7:183).

Alt. limit: 590 m, Tynehead Fell (76.36). Rawes (1981) mentions finding seedlings at 750 m west of Moor House (7.2).

Map 475   1603   LWFCY

C. laevigata (Poirret) DC.   Midland hawthorn

Seen during the Survey in lanes between Bleatarn and Little Musgrave (72.12, LH, 1976, LANC), in Helbeck Wood, Brough (76.16, R. Baines, 1980s), Bothel (18.38, REG, 1989) and west of Mealsgate (18.42, EEM, 1994, LANC), in an old hedge at Brampton (50.60, CS, 1980; LANC), at Melmerby (60.38, LR, 1996, LANC) and near the disused railway at Inch in Liddelsdale (40.74n, GH, 1991, LANC). It was first collected in the Inch area by C.W. Muirhead in 1950 (CLE, PLYP, *BSBI Yr Bk* (1951) p.122) and identified by E.F. Warburg as probably the hybrid although much closer to *Crataegus laevigata*. Muirhead subsequently collected similar material from between Inch and Riddings (1962, PLYP).

Petty (1895 p.266) gives a number of unlocalised 19th century Furness records for this southern species, but these were not accepted by Baker. There is, however, a later one from Dalton-in-Furness (2.7, D. Lumb, 1913, *BEC Rpt* (1914) 3:468) and a Westmorland one from Brackenber Wood, Appleby (7.1, F.M. Wilkinson, 1943, *BRC Rpt* (1946) 12:721), the only record in the *Atlas* for the north of England.

The recent records are mainly of single bushes, but with the extensive roadside plantings of hawthorn this century it would be surprising if these did not include some plants of *C. laevigata* or ones showing some evidence of hybridisation.

Alt. limit: 160 m, Bleatarn (72.12)

7   W(F)C

FABACEAE
(*LEGUMINOSAE*)

***Robinia pseudoacacia** L.   False acacia
The only records for this species are of two trees, one dying, by the River Eden below Nunnery, Kirkoswald (52.42, RWMC, LANC, *Watsonia* (1989) 17:468) and a beckside tree at Lazonby (54.38, RWMC, 1989, LANC, *Watsonia* (1991) 18:423). Both had presumably been planted although the former was far from houses.

                              2   C

***Galega officinalis** L.
A Continental legume, widely naturalised in southern England, seen near Penrith station in 1947 (5.3, W. Atkinson & W. Davidson, CLE).

                                                   (C)

Astragalus danicus Retz.   Purple milk-vetch
This pretty vetch is listed by Hodgson from Catlands (2.4) and by Wilson from two sites near Morland (6.2), all presumably on the limestone. The only other record is an undated one from the coast at Grune Point (1.5, H. Stewart).

    In the west of Britain this vetch is now restricted to Tiree, the north side of the Solway and the Isle of Man.
                                                   (WC)

A. glycyphyllos L.
Seen only in a railway cutting at Culgaith where a single plant was seen by RWMC in 1987 (58.28, LANC). This is presumably the site where a large colony was found by A.M. Burkett and G.A.K. Hervey in 1954.

    There are two old Furness records for this species, Baker's from near Cartmel and Wilson's from Humphrey Head (both 3.7s); they probably refer to the same locality. In J. Bevan's copy of *Topographical Botany* there is a note added of a record by W.H. Stansfield from Strickland (5.2, 1903). This could be the source of the undated specimen labelled "Penrith" and collected by J. May (WARMS). Both records may refer to the Culgaith site.

                              1   (WF)C

***Onobrychis viciifolia** Scop.   Sainfoin
Seen on a newly seeded area at Haggs Mine, Nenthead in 1990 (76.44, P. Buchanan, LANC) and the following year on rough ground at Maryport docks (02.36, NB, 1991, LANC, *Watsonia* (1992) 19:144). The specimens at the first site were exceptionally tall, up to 1 m. Although Hodgson gives a record from Nethertown (9.0) he was clearly sceptical about its status as a native plant.

                              2   C

Anthyllis vulneraria L.   Kidney vetch
Locally frequent in dry, base-rich or calcareous grassland, especially around the coast. Inland it occurs on the limestone, on a few Lake District crags, on railwaysides, on the banks of the River Eden between Langwathby and Lazonby (5.3) and the Caldew at Cummersdale (38.52), and on sandy banks near Castle Carrock (54.56). There is also a recent record of its casual occurrence on the recently-seeded verge of the A590 near Levens (46.84s). The railwayside population between Beckermet and Sellafield (00.04) was particularly fine prior to ballast removal.
Alt.limit: 550 m, Honister Crag (2.1)
Map 476   205   WFCY

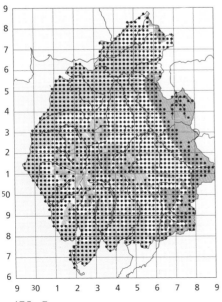

Map 475. *Crataegus monogyna*

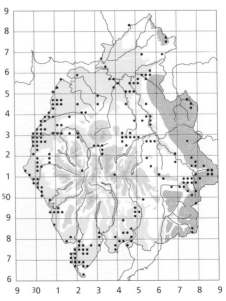

Map 476. *Anthyllis vulneraria*

The following infraspecific records are based on material determined by J. Cullen.

subsp. **vulneraria**
var. **vulneraria**
Probably the commonest of the infraspecific taxa accounting for half of the 22 collections identified by Cullen, who says it also occurs on limestone at Arnside (4.7s). The *Critical Atlas* also shows a record for the Whitehaven square (9.1). Cullen considered two further collections intermediate between this subspecies and subsp. *polyphylla* (DC.) Nyman and another intermediate between subsp. *vulneraria* and subsp. *carpatica*.

Map 477   10   WC

var. **langei** Jalas
Probably restricted to the west coast being recorded only from Workington (98.28, AAD,1992, LANC), north of Seascale (02.00, EEM,1991, LANC) and dunes at Askam-in-Furness (20.76, PB, 1992, LANC).

3   FC

*subsp. **carpatica** (Pant.) Nyman
var. **pseudovulneraria** (Sagorski) Cullen
Fairly widely distributed but less frequent than subsp. *vulneraria* var. *vulneraria*.

Map 478   8   WC

subsp. **lapponica** (N. Hylander) Jalas
Known only from Honister Crag at 550 m (2.1, C.W. Muirhead & D.J. Higgins, 1953, PLYP) and the Wast Water Screes at 425 m (1.0, DAR, 1954, CLE). A record from Eel Crags, Newlands (22.16, DAR, 1955) probably also belongs here. The only three English records in the *Critical Atlas* are from the Ravenstonedale (7.0) and Stainmore (8.1) squares and by the upper River Coquet, Northumberland.

(WC)

*Lotus glaber** Miller   Narrow-leaved bird's-foot trefoil
(*L. tenuis* Waldst. & Kit. ex Willd.)
The only Survey record is from Silloth (10.52) from where it was reported in the early 1970s. Hodgson lists it from St Bees and Whitehaven (9.1), Haile Wood (0.0), Hycemoor (0.8), Clifton (0.2) and between Abbey Town and Silloth (1.5, E.J. Glaister, 1878, CLE). The only Furness record is from a meadow by Humphrey Head in 1871 (3.7s, F.A. Lees, *Botanical Record Club* (1874) 1874:18). There are also post-1930 *Atlas* records for the Distington (0.2) and Maryport (0.3) squares.

It seems unlikely that this species of southern and south-eastern England is native in Cumbria.

1   (F)C

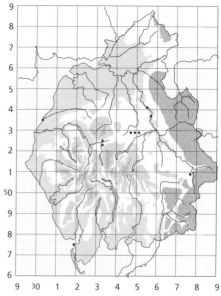

Map 477. *Anthyllis vulneraria* subsp. & var. *vulneraria*

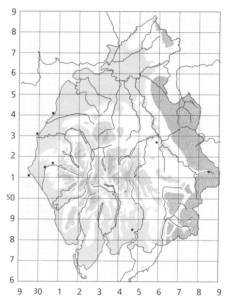

Map 478. *Anthyllis vulneraria* subsp. *carpatica*

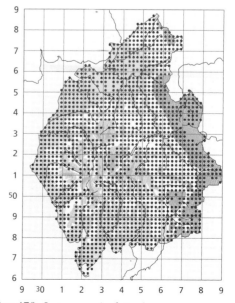

Map 479. *Lotus corniculatus*

L. corniculatus L.   Common bird's-foot trefoil
Very common throughout the lowlands in short, neutral
or calcareous, well-drained grasslands and roadside
verges. Common associates include *Festuca ovina*,
Trifolium dubium and *Briza media*.

   The tall Continental var. *sativus* Chrtkova
occurs on the Ravenstonedale by-pass (70.04, GH,
1990, LANC) and may well appear on other recently
landscaped road improvement schemes.
Alt. limit: 730 m, Green Fell, Cross Fell (66.36)
Map 479   1573   LWFCY

L. pedunculatus Cav.   Greater bird's-foot trefoil
(*L. uliginosus* Schk.)
Very common although slightly less so than *Lotus
corniculatus*. It is characteristic of a wide range of
damp habitats including ditches and poor fen and fen
communities in wet meadows and by tarns and lakes,
riverbanks, roadsides and woodland rides. It commonly
grows with *Achillea ptarmica*, *Galium palustre*,
Filipendula ulmaria and *Lychnis flos-cuculi*. In the
uplands, it occurs in many of the *Juncus acutiflorus -
J. effusus* hillside flushes.

   Most plants are noticeably villous and referable,
as Allen (1984) points out, to the western subsp.
villosus (Lange) Hansen.
Alt. limit: *c.*460 m, north-east of Alston (74.48)
Map 480   1385   WFCY

*L. angustissimus L.
Of the Cumberland records cited by Baker, Hodgson repeats
only his own, from the railway cutting near Bullgill Station
(0.3) where it appeared after the opening of the line but
became extinct *c.*1886. Baker's other records he includes
under *Lotus glaber*.

   This species is now restricted in Britain to the coasts
of south-west England and Kent.
                    R   (C)

Ornithopus perpusillus L.   Bird's-foot
Rather local and chiefly near the coast, occurring in
dry, open sandy ground, in sand and gravel pits, on old
sand-dunes, on heaths, rocky banks and paths.
Alt. limit: 215 m, Penrith golf-course (50.30), although
Baker gives it as 275 m at Penrith.
Map 481   81   WFC

Hippocrepis comosa L.   Horseshoe vetch
Locally frequent and occasionally abundant on the
limestone cliffs around the head of Morecambe Bay and
in north Westmorland on both sides of the upper Eden
valley. It grows usually on rock ledges and crevices
with a southerly aspect and also in thin, rather open
turf.

   A southern species, it reaches its northern limit
in Britain above Hilton (74.20). It occurs as a casual at
Maryport docks (02.36, REG, 1995; LANC) and there
is a similar Hodgson record from Workington (9.2).
Alt. limit: 600 m, Long Fell, Warcop (76.18)
Map 482   57   WFC

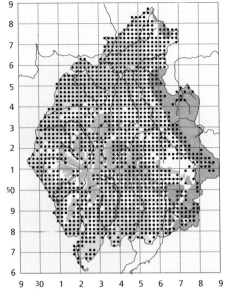
Map 480. *Lotus pedunculatus*

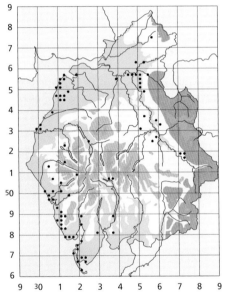
Map 481. *Ornithopus perpusillus*

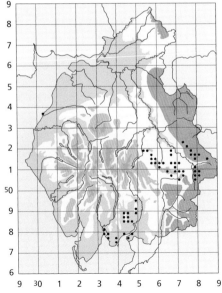
Map 482. *Hippocrepis comosa*

*Securigera varia (L.) Lassen   Crown vetch
(*Coronilla varia* L.)
Seen only at the Barrow-in-Furness docks (18.68s, B. Fisher, 1976), where it was first seen in 1949 by G. Wilson (CLE) who also reported it from nearby at Roose (2.6s).

    The only previous records for this southern species are C.H. Wright's from "Lord W. Gordon's woods, near Keswick", possibly Lord's Island, Derwent Water (2.2), in 1837, Hodgson's from the railway at Longtown (3.6), a later one from Whitehaven (9.1, R.H. Williamson, 1923, CLE) and Wilson's reference to its occurrence in Furness.

1   F(C)

*Scorpiurus muricatus L.
A single specimen of this southern European species was found in a garden at Bassenthwaite village in 1987 (22.30, ES).

1   C

Vicia orobus DC.   Wood bitter-vetch
Recorded during the Survey at only six sites. In the Pennines it was seen near Knock (68.26), where it was growing at the edge of a meadow, in a hayfield at Gale Hall, Melmerby (62.36), where shortly afterwards 'improvement' of the meadow led to its extinction, and in a quarry between Melmerby and Gamblesby (60.38). It was seen also at the well-known meadow site at Howgill in the Lune gorge (62.94), in a laneside hedge by the River Esk (18.00) and, a single plant, on the Wast Water Screes (14.04).

    By contrast, Hodgson reported the vetch as growing plentifully at several places under the Cross Fell range, from Kirkland to Gamblesby (6.3). In his Appendix he gives a further site in the south-west near Boot, in Eskdale (1.0), where it was found by J. Adair who described it as "quite plentiful around the edges of meadows" (Hodgson 1899). There are later records from Unthank (6.4), where it was seen by T.J. Foggitt in 1909 (BM) and 1926 (*BEC Rpt* (1927) 8:111), Renwick (5.4/6.4, J. Parkin (1935), and Hollin How, Matterdale (3.2, B.M. Cunliffe, *c*.1955).

    Wilson gives two further Pennine sites - Dufton Pike and Milburn Grange (6.2), and he refers also to one "north of Ullswater", perhaps mistakenly assuming that the Foggitt locality was the Unthank near Hutton-in-the-Forest (3.4), and another at Low House, Hugill, Staveley (4.9, G.E. & J.A. Martindale, undated, KDL). Finally, there is a puzzling post-1930 *Atlas* record for the Langdale square (2.0).

    At none of the extant sites does the vetch succeed in fruiting regularly, although it flowers and fruits well in a Melmerby garden. The populations are all very small, only one plant in Eskdale, and their future is very uncertain. The situation is apparently very similar in Scotland; only in Wales is the species apparently holding its own.
Alt. limit: 455 m, Wast Water Screes (14.04)

6   WCY

V. cracca L.   Tufted vetch
Very common in the lowlands in the coarse grassland of roadside verges and woodland margins, most characteristically seen scrambling over hedgerows and up stone walls. It also occurs in tall-herb fen and marsh communities, among sand-dunes and on the drift-line.
Alt. limit: 550 m, Moor House, Teesdale (74.32)

Map 483   1313   LWFCY

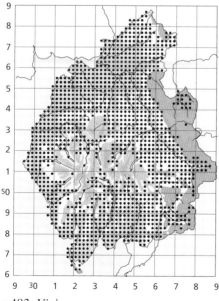

Map 483. *Vicia cracca*

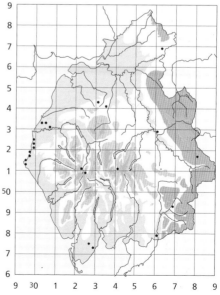

Map 484. *Vicia sylvatica*

V. sylvatica L.   Wood vetch
Rather rare but very fine along the cliffs and in rough
grassland on the west coast from St Bees Head (9.1)
north to Flimby (0.3), and on the railway embankment
at Harrington (98.24). Elsewhere it appears to be
declining and there are now only a few very scattered
sites, chiefly in roadside or riverside woodland, also in
the Lake District in a rocky gully and on a crag,
characteristic habitats of the species in Scotland.
Alt. limit: 610 m, Glaramara (24.08)
Map 484   22   WFCY

*****V. villosa** Roth
Found in 1985 on recently landscaped waste ground at
Silloth (10.52, MMM, LANC).

   This southern vetch was first recorded from
Black Dyke, Silloth (1.5, CLE) by J. Leitch in 1889. It
was growing as a field weed with *Cynosurus echinatus*
and the record is given by Hodgson (p.345) under that
species.

                        1   C

V. hirsuta (L.) Gray   Hairy tare
Frequent in the north but scarce in the south where it is
mainly coastal. It occurs on roadside verges, on waste
ground, railway sites, cultivated land, river and coastal
shingle and also rough grassland. It prefers light, well-
drained soils.
Alt. limit: 310 m, Spadeadam (64.74n)
Map 485   229   WFC(Y)

V. tetrasperma (L.) Schreber   Smooth tare
Seen only by roadsides at Spadeadam (58.70n, GH &
C. Gilham, 1979; 64.72n, MS, 1979, LANC) and
Brough Sowerby (78.12, JMA, 1992, LANC), also near
Roanhead, Dalton-in-Furness (20.74, PB, 1990,
LANC), where it was growing in abundance with *Vicia
sativa* and *V. hirsuta*.

   Hodgson regarded this species as very rare and

gives only two localities: Maryport (0.3) and Westward
(2.4). There is also one from Woodrow (2.4, E.J.
Glaister, 1876, CLE) and from Bowness-on-Solway
(2.6n, C.B. Sharp, 1939, CLE) and an undated *Atlas*
record for the Carlisle square (3.5). Baker lists it from
two places in Furness; Grange-over-Sands (4.7s) and
the Windermere Ferry (3.9, A. Wilson, 1879, YRK;
G.E. Martindale, 1891, KDL).

   Despite the paucity of early records, there are
eight post-1930 *Atlas* dots. It seems probable that at
least some of these refer to *V. hirsuta*.

   This is a southern vetch with few extant sites in
Britain north of the River Humber.
Alt.: 270 m, Spadeadam (58.70n, 64.72n)
                        4   (W)FC(Y)

V. sepium L.   Bush vetch
Very common throughout the lowlands, occurring
chiefly on roadside verges and railwaysides, in hedgerows and in scrub and woodland margins. In the fells it
grows in ungrazed gullies and on rock ledges but
usually below 350 m.
Alt. limit: 650 m, Crowdundle Beck, Cross Fell (68.32)
Map 486   1424   LWFCY

V. sativa L.   Common vetch
Frequent in coarse grassland and meadows, on
railwaysides, roadsides and waste ground and in
hedges.

   The Survey field records for *Vicia sativa* and
V. angustifolia have been used to produce the species'
map but those of the subspecies are based on the
examination of collected specimens.
Map 487   511   LWFCY

subsp. **nigra** (L.) Ehrh.
(*V. angustifolia* L.)
Certainly under-recorded. Probably less common
generally than subsp. *segetalis* but commoner in the

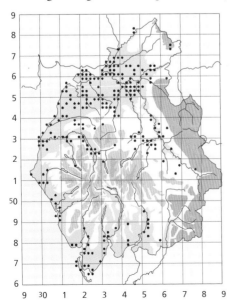

Map 485. *Vicia hirsuta*

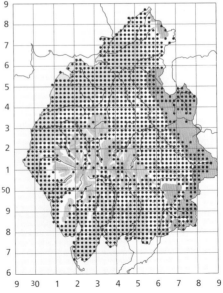

Map 486. *Vicia sepium*

south.
Alt. limit: 330 m, Dent Head (76.82)
Map 488   49   WFCY

subsp. **segetalis** (Thuill.) Gaudin
Probably the commonest subspecies, at least in the
north. It is generally more robust than subsp. *nigra* and
with isomorphous, wider leaflets but plants intermediate between the two subspecies certainly occur.
Alt. limit: 410 m, Garrigill (76.44)
Map 489   33   WFC

subsp. **sativa**
Seen during the Survey north-west of Grayrigg (54.98,
CEW, 1990, LANC). This segregate, now generally
rather rare in Britain, is listed in all the Floras and is
said by Baker to be "frequent in cultivated fields".
                              1   W(FC)

V. lathyroides L.   Spring vetch
A rare and very early flowering vetch of sand-dunes
and dry open ground close to the shore and occurring
from Silloth (10.52) south to Foulney Island (24.64s).
There are a few Survey records from the Furness
peninsula, but in the absence of specimens the
possibility of confusion with small plants of *Vicia
sativa* cannot be ruled out.
Map 490   19   FC

V. lutea L.   Yellow vetch
A rare coastal vetch listed by Hodgson from gravel beds of
the River Derwent above Workington (0.2) and also from
Silloth (1.5, J. Leitch, 1890, CLE). A third record, from
Keswick (2.2), he considered dubious. The only other
Cumberland record is from a field at Anthorn (1.5, C.B.
Sharp, 1939). It was seen in Furness in 1948 by G. Wilson on
the landward side of the Sandscale dunes (1.7, CLE) and by

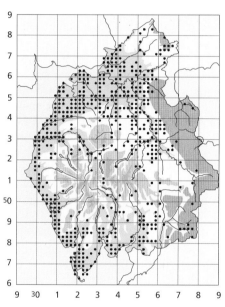

Map 487. *Vicia sativa*

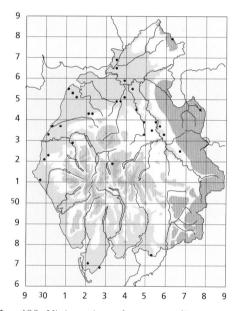

Map 489. *Vicia sativa* subsp. *segetalis*

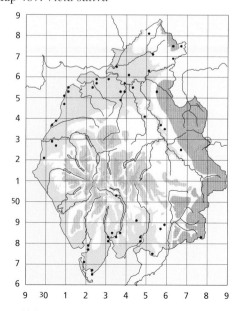

Map 488. *Vicia sativa* subsp. *nigra*

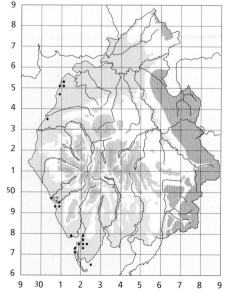

Map 490. *Vicia lathyroides*

others intermittently until 1968.

S   (FC)

*V. bithynica (L.) L.
The only record for this southern species is that given by Wilson from Windermere (4.9, M. Wilkinson).

S   (W)

*V. faba   Broad bean
The only records are of a plant below a hedge and far from houses at St Bees (94.14, AAD, 1993, LANC) and a collection from Silloth in 1889 (1.5, J. Leitch, CLE).

1   C

Lathyrus japonicus Willd.   Sea pea
The only certain record of this attractive coastal pea is a specimen (CLE) collected by J. Curwen from the "West coast of Cumberland" and undated. A single plant was reported by F. Jones in 1975 from the beach at Seascale (02.02, *Watsonia* (1977) 11:394). The area was subsequently disturbed by 'heavy plant' operations and the pea has never been refound. The record was not checked at the time and it still seems best to regard it with some doubt.

Hodgson lists it from St Bees (9.1), Harrington and Parton (9.2) and he quotes the opinion of the Rev. H. Friend that they "are an error for the last, I think", referring to *Lathyrus sylvestris*, although there are no definite records of that species as far south as St Bees.

The only extant record of this pea on the west coast of Britain is from Caernarvonshire.

S   (C)

L. linifolius (Reichard) Bässler   Bitter vetch
(*L. montanus* Bernh.)
Frequent, but characteristically a plant of the dales where it occurs in similar habitats to *Lathyrus pratensis*, also in heathland and on the fells where it

flourishes on ungrazed gillsides and rock ledges.

The distinctive narrow-leaved variant, 'var. *tenuisectus*', is not infrequent.
Alt. limit: 610 m, Helvellyn (34.14) and Black Door, Cross Fell (66.34)
Map 491   657   WFCY

L. pratensis L.   Meadow vetchling
Very common in the lowlands by roadsides, in rough grassland, and in hedgerows, scrub and woodland margins.
Alt. limit: 845 m, Great Dun Fell (70.32)
Map 492   1394   LWFCY

[**L. palustris** L.   Marsh pea
The St Bees (9.1) record given by Baker was regarded by him with suspicion and omitted by Hodgson. It almost certainly must refer to *Lathyrus sylvestris*.

(C)]

*L. tuberosus L.
A widely distributed Continental species known in Cumbria only from the ironworks at Askam-in-Furness (2.7, D. Lumb, 1912, YRK, *BEC Rpt* (1913) 3:210) and a pre-1930 *Atlas* record from the Great Strickland square (5.2).

(WF)

*L. grandiflorus Smith
Found in 1978 growing abundantly in a hedge south of Whitehaven (98.12, CCH, LANC, *Watsonia* (1979) 12:351), where it is still flourishing. It has also been seen near Embleton (16.30, CCH, 1985), in a roadside hedge north of Kendal (52.94, GH, LANC) and near Castle How, Wythop (20.30, REG, 1990).

It is a native of the eastern Mediterranean occurring occasionally in Britain as a garden escape.

4   WC

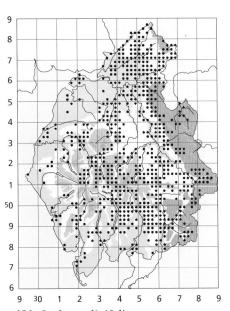

Map 491. *Lathyrus linifolius*

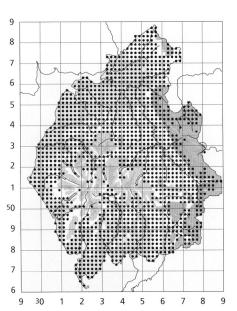

Map 492. *Lathyrus pratensis*

L. sylvestris L.

Still locally frequent in Hodgson's localities along the coastal cliffs between Whitehaven and Harrington (9.2), from where it was first recorded by Lawson in 1690, as well as by the railway north of Siddick (00.30). It also occurs further north at Maryport (02.36) and Silloth (10.52), where it was discovered by J. Parkin in 1935.

Map 493   8   C

*L. latifolius L.

Very rare, recorded from rough grassland, roadside hedgerows, waste places and the shore at a few mainly coastal sites from Bardsea (30.74s) north to Silloth (10.52, EEM, 1992), and with an isolated railwayside site at Upper Denton (60.64). It has since disappeared from the Silloth site, the first Cumberland record, and also from nearby Overby (12.46) following human disturbance.

There are no records in the Floras of this garden escape although it is listed for Cumberland in the *Comital Flora*. The first definite record is from 1957 when it was found in the Dalton-in-Furness square (2.7, K. Jones).

Alt. lmit: 150 m, Upper Denton (60.64)

Map 494   10   FC

*L. sativus L.

A species of southern and south-eastern Europe found by Hodgson on the Maryport ballast tips (0.3) in 1894.

(C)

*L. sphaericus Retz.

This Continental species was recorded by Hodgson (p.93) by the River Derwent just above Workington (0.2) in 1896.

(C)

*L. hirsutus L.   Hairy vetchling

A Mediterranean species recorded in 1935 by M. Cross at Thwaites, near Millom (1.8), and shortly afterwards by C.B. Sharp in a field at Anthorn (1.5/2.5, 1939, CLE).

(C)

*L. nissolia L.   Grass vetchling

A plant of southern and south-east England found in rough grassland by the River Kent on the south side of Kendal (50.90, R. Dalton, 1982; LANC, *Watsonia* (1983) 4:422), on roadside verges south of Brough (78.12, P.M. Atkins, 1990, *Watsonia* (1991) 18:423) and near Countess Pillar, Penrith (54.28, MG & JP, 1993). Baker and Hodgson give a record from Irton (1.9/1.0) but regard it as dubious.

3   W[(C)]

*L. aphaca L.   Yellow vetchling

Seen once during the Survey in a cornfield near Brough (78.14, LH, 1971, *Watsonia* (1977) 11:393). It was recorded by Hodgson from Workington (9.2), and by the River Derwent a short distance upstream (0.2), and also from Silloth and Grune Point (1.5, E.J. Glaister, 1890, 1900, CLE). There is a 1902 record from Grinsdale by the River Eden (3.5, T.S. Johnstone, CLE), one from Bowness-on-Solway (2.6n, C.B. Sharp, 1939, CLE), and later B. Cunliffe noted single specimens in cornfields at Lowca (9.2, 1950s) and The Luham, Langwathby (5.3, 1960s). The only Furness record is from Kents Bank, Grange-over-Sands (3.7s, W.L. Smith, 1911, LANC, *Watsonia* (1977) 11:393).

Like *Lathyrus nissolia*, this attractive pea is native only in south-east Britain.

Alt.: 165 m, Brough (78.14)

S 1   W(FC)

*Pisum sativum L.   Garden pea

A relict of cultivation, recorded from a beckside at

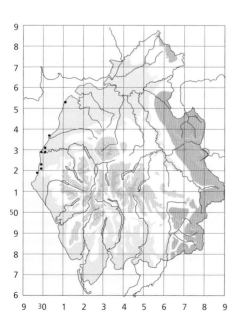

Map 493. *Lathyrus sylvestris*

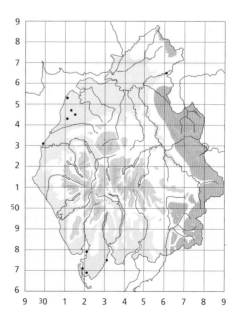

Map 494. *Lathyrus latifolius*

Seascale (02.00, CCH, 1970s) and the roadside north-west of Grayrigg (54.98, CEW, 1990, LANC). The only previous record is from by the pond at Naworth (5.6, *BEC Rpt* (1923) 6:725) where it was seen by the Countess of Carlisle and Mrs Bodley in 1922.

2   W(C)

Ononis reclinata L.   Small restharrow
There is a single record from Seascale (0.0, J. Curwen, CLE) where it was collected *c*.1890. This species is now restricted in Britain to the Channel Islands, Devon, Wales and the Mull of Galloway.

R   (C)

O. spinosa L.   Spiny restharrow
Rather local and largely restricted to the Solway coast where it is characteristic of the stiff, saline clays of the reclaimed, grazed marshes around the head of the Firth where the species reaches its northern limit on the west coast. There are a few inland occurrences on roadside verges in the upper Eden valley. Records from the coast of Morecambe Bay, including those in the *Atlas*, probably all refer to *Ononis repens* var. *horrida*, although there is a genuine record from Silverdale (4.7s, Anon., 1910, LANC), just over the border in Lancashire.
Alt. limit: 190 m, near Maulds Meaburn (62.16)
Map 495   27   W[(F)]C

O. repens L.   Common restharrow
Fairly common around the entire coast and frequent in the Eden valley, otherwise rather uncommon. It is a plant of dry, light, sandy or stony soils on base-rich roadside, field and river banks. It is particularly abundant on the coastal dunes and dune-slacks and extremely beautiful when in full flower.

There are a few records of spinose plants, the predominantly coastal var. *horrida* Lange, from the Duddon estuary eastwards. These are best distinguished from *Ononis spinosa* by having the young stems uniformly hairy all round, instead of with two distinct rows of hairs, and by having pods which do not generally exceed the mature calyx.
Alt. limit: 280 m, near Sunbiggin (64.08)
Map 496   290   WFCY

Restharrow crawls on the sand, throwing up its little pink sweet-peas, but it is a sticky plant and gets clart up with sand, so there is no pleasure in gathering it.
Cumberland and Westmorland, Norman Nicholson

*Melilotus altissima Thuill.   Tall melilot
Rather rare and virtually restricted to coastal sites at the head of the Solway Firth, the Barrow-in-Furness area and around the Kent estuary. It occurs on waste ground, by field edges, on sandstone cliffs at Rockcliffe (34.60) by the River Eden, and on the railway embankment at Grange-over-Sands (42.78s).
Map 497   19   WFC

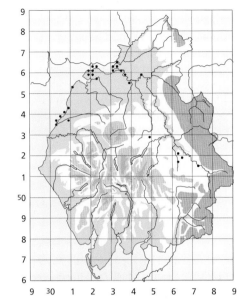

Map 495. *Ononis spinosa*

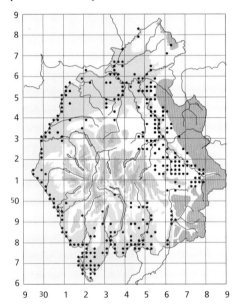

Map 496. *Ononis repens*

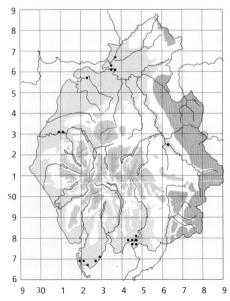

Map 497. *Melilotus altissima*

*M. albus Medikus   White melilot
A rare casual of waste ground at a few places along the
coast, persisting only in the Barrow-in-Furness, Workington - Maryport (0.3) and Silloth (1.5) areas. It has
also been reported inland near Cocklakes (44.50),
Albyfield (54.52), Kirkby Thore (62.24) and on a disturbed roadside verge at Gawthrop, Dentdale (68.88s). It
was first seen in Westmorland in 1888 at Sandside
(4.7s, J.A. Martindale, KDL) and in Furness in 1912 at
Askam ironworks (2.7, D. Lumb, YRK, *BEC Rpt*
(1913) 3:244).
Alt. limit: 160 m, Albyfield (54.52)
Map 498   14   WFCY

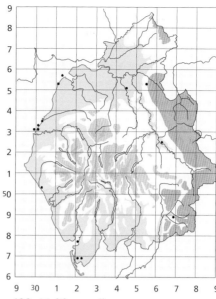

Map 498. *Melilotus albus*

*M. officinalis (L.) Lam.   Common melilot
Rather rare and coastal like the previous species and
occuring in similar habitats. Inland records probably
indicate only casual occurrences, such as that on the
new A590 dual-carriageway at Haverthwaite (34.84).
Map 499   35   WFC

*M. indicus (L.) All.   Small melilot
The only Survey record is from disturbed ground in a
field west of Levens (48.84s, CEW, LANC) where it
appeared briefly in 1989.

   This Mediterranean species is reported by
Hodgson from the Derwent Tinplate Works, Workington (9.2), the River Derwent upstream of Workington (0.2, p.380), Maryport (0.3) and Floshgate,
Ullswater (4.2). In 1889 it was seen at Silloth (1.5,
J. Leitch, BM, CLE), in 1899 at Braithwaite station
(2.2, H. Friend, *Naturalist* (1899) 1899:300) and in
1932 at Silloth by C.W. Muirhead (PLYP). There are
also records from Melmerby (6.3, W.W. Mason, 1924)
and Baker's from Ulverston (2.7).

   The post-1930 *Atlas* records are for the mid-
Walney (1.6s), Distington (0.2) and Barbon (6.8s)
squares. The Walney record dates from 1954 and was
contributed by G. Wilson and the Barrow-in-Furness
N.H.S., who also noted the plant near Barrow (2.6s)
and Ulverston (2.7).
                    1   W(FC)

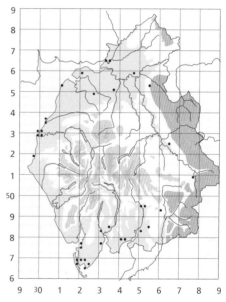

Map 499. *Melilotus officinalis*

*Trigonella corniculata (L.) L.
A Mediterranean species known only from Hodgson's 1889
record from the Derwent Tinplate works, Workington (9.2).
                                        (C)

*T. monspeliaca L.
A species of southern and central Europe found at Silloth in
1889 (1.5, J. Leitch, CLE).
                                        (C)

*T. polyceratia L.
An Iberian species found by Hodgson in abundance at the
Derwent Tinplate Works, Workington (9.2) in 1889 and
1890.
                                        (C)

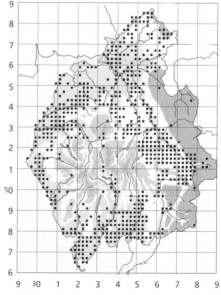

Map 500. *Medicago lupulina*

***T. caerulea** (L.) Ser.
A Continental fodder plant found on the gravel beds of the River Eden below Carlisle in 1902 (3.5, T.S. Johnstone, CLE).

(C)

***T. foenum-graecum** L.   Fenugreek
A fodder plant found at Silloth docks in 1891 (l.5, J. Leitch, CLE).

(C)

Medicago lupulina L.   Black medick
Fairly common over most of the lowlands but scarcely extending into the Lake District valleys. It is a plant of roadsides, disturbed and cultivated ground in towns and villages, of lawns and short turf and of limestone out-crops.
Alt. limit: 440 m, Nenthead (76.42)
Map 500   632   LWFCY

***M. sativa** L.
subsp. **falcata** (L.) Arcang.   Sickle medick
Seen during the Survey only on sand-dunes and waste ground at Silloth (10.52, R. Young, 1970, AMDE; GH, 1986, LANC) and at Barrow-in-Furness docks (18.68s, B. Fisher), where it persisted until 1993. It was recorded from Silloth by Hodgson and also from the Derwent Tinplate Works, Workington (9.2). There is a record from Whitehaven (9.1, R.H. Williamson, 1922) and *Atlas* records contributed by G. Wilson from the North Walney and Dalton-in-Furness squares (1.7, 2.7) and Rampside (2.6s, 1949, CLE).

This subspecies is native in Britain only in East Anglia.

S 2   (F)C

subsp. **falcata** x subsp. **sativa**
subsp. **varia** (T. Martyn) Arcang.
Known only from Silloth where it was first reported from the docks by G.C. Druce and R.H. Williamson (10.52, CLE, *BEC Rpt* (1924) 7:117) in 1923. It was seen on the dunes there by R. Young in 1970 (AMDE) and by A. Cannell in 1988.

1   C

subsp. **sativa**   Lucerne
A rare species of rather open and often disturbed habitats, chiefly on the west coast and especially in the Maryport area (0.3) and around Barrow-in-Furness. It is now rarely grown for fodder but is occasionally present in grass seed. The first Furness record is probably G. Wilson's *Atlas* record from mid-Walney Island (1.6s).
Alt. limit: 375 m, below Nenthead (76.44)
Map 501   15   WFC(Y)

***M. minima** (L.) Bartal.
The only records are from Silloth (1.5), where it was seen by Hodgson and in 1935 by J. Parkin, and Seascale (0.0,

R. Hellon, 1920, *BRC Rpt* (1921) 6:118). It was thought to have drifted ashore at the latter site on the carcasses of sheep or rabbits from a torpedoed vessel!

This species is native in Britain only in south-east England.

S   (C)

***M. polymorpha** L.
The only Survey record is from the banks of the River Mint above Kendal (52.94, CEW, LANC, *Watsonia* (1984) 15:129) where it appeared in 1983.

Hodgson gives records from Workington (9.2), Maryport (0.3) and Silloth (1.5, J. Leitch, 1890, CLE) and there are later Cumberland ones from Great Salkeld (5.3, H. Britten, 1900-1908), Stainton, Carlisle (3.5, T.S. Johnstone, 1910), allotments at Carlisle (3.5/4.5, H. Stewart, 1926, CLE), Seascale beach with *Medicago minima* (0.0, R. Hellon, 1920, *BEC Rpt* (1921) 6:118), Thwaites, near Millom (1.8, M. Cross, 1929) and Silloth (J. Curwen, *c.*1890s, CLE).

This species is native only along the coasts of south and south-east England.

S 1   W(C)

***M. soleirolii** Duby
A Crimean species collected at Silloth (1.5) by J. Leitch in 1889 (CLE).

(C)

***M. arabica** (L.) Hudson   Spotted medick
Recorded by Hodgson from the usual alien localities at Workington (9.2), Maryport (0.3) and Silloth (1.5). It was seen again at Silloth docks by W.A. Sprott and J. Parkin between 1931 and 1936. There is an undated *Atlas* record for the Brampton square (5.6). The only Westmorland record dates from 1844 when it was collected between Greenside and Viver, Milnthorpe (5.8s, "S.H.H.", KDL).

This is another predominantly southern species which in the north occurs only as a casual.

(WC)

[Trifolium ornithopodioides L.
Listed with reservations by Baker from Workington Warren (9.2), a record subsequently disregarded by Hodgson. Wilson gives only one locality, Windermere (4.9). These two sites are well to the north of the species' northern limit which lies on the North Welsh coast and the records are best regarded as dubious.

(WC)]

T. repens L.   White clover
One of the commonest plants in Cumbria occurring in nearly every type of short grassland - natural, improved and leys. In the fells it occurs in the better and rather moister *Festuca - Agrostis* grassland, often associated with *Bellis perennis*, and is absent only from the highest ground and areas of continuous blanket peat.
Alt. limit: 845 m, Great Dun Fell (70.32)
Map 502   1741   LWFCY

*T. hybridum** L.   Alsike clover
Fairly frequent though chiefly in the east. It is often
included in ley seed mixtures and is commonly seen on
re-seeded roadside verges.
Alt. limit: at least 350 m, Stainmore (86.12)
Map 503   301   WFC

*T. retusum** L.
A clover of central and southern Europe seen at Silloth docks
in 1891 and 1892 (1.5, J. Leitch, CLE).
                                        (C)

T. fragiferum L.   Strawberry clover
Known only from a few coastal sites on Walney Island
and the adjacent mainland and from a single,
interesting inland site in damp turf by a ditch at

Caldbeck (32.38, REG, 1977, LANC), the species'
northern limit in western Britain. The Furness records
are mostly close to the shore, just above or at the upper
limit of the salt-marsh, where it grows in *Festuca rubra*
- *Agrostis stolonifera* turf with *Oenanthe lachenalii*
and *Trifolium repens*. The only previous record was
from Grune Point (1.5, E.J. Glaister, 1899, CLE).
Alt. limit: 165 m, Caldbeck (32.38)
Map 504   8   FC

*T. resupinatum** L.
Hodgson records this southern clover from the Derwent
Tinplate Works Workington (9.2), Silloth (1.5, J. Leitch,
1891, CLE) and Floshgate, Ullswater (4.2), and H. Friend
(*Naturalist* (1899) 1899:300) from Braithwaite station (2.2).
                                        (C)

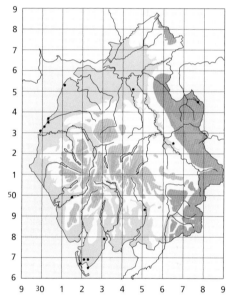

Map 501. *Medicago sativa* subsp. *sativa*

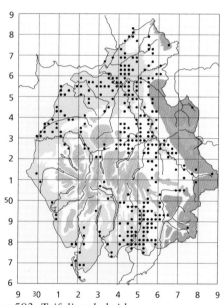

Map 503. *Trifolium hybridum*

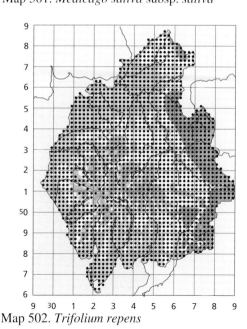

Map 502. *Trifolium repens*

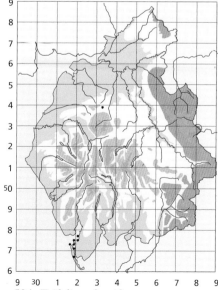

Map 504. *Trifolium fragiferum*

*T. aureum Pollich

Recorded by Baker as a field weed at Witherslack (4.8s) and Plumpton, Ulverston (3.7s), by E.J. Glaister at Black Dyke, Silloth (1.5, 1878, CLE) and by W.W. Mason at Melmerby (6.3, 1913-1916).

　　　This clover is not native in Britain although widespread on the Continent.

(WFC)

T. campestre Schreber　　　　　　　　Hop trefoil

Fairly frequent over most of the lowlands occurring on dry grassy banks, by roads and railways and preferring the more neutral or basic soils.

Alt. limit: 275 m, Spadeadam (64.72n)

Map 505　　　　　　374　　　　　　WFCY

T. dubium Sibth.　　　　　　　　Lesser trefoil

A common lowland species of short, dry and usually somewhat open grassland, and of disturbed ground, waste places and railway tracks. It is appreciably less demanding in its requirements than *Trifolium campestre* and commoner than *Medicago lupulina* with which it is sometimes confused.

Alt. limit: 530 m, south-west of Garrigill (76.40)

Map 506　　　　　　1105　　　　　　LWFCY

T. micranthum Viv.　　　　　　　Slender trefoil

The only substantiated Survey record is from the gardens by the station at Grange-over-Sands (40.78s, S.M. Coles, 1977, LANC, *Watsonia* (1978) 12:170). Interestingly, according to CFS all six recent records from the Fylde, Lancashire, are from garden and park lawns. There are also records from Humphrey Head (38.74s, FLW) and Meathop (42.78s, FLW) but in view of the the frequent confusion of this species with depauperate forms of *Trifolium dubium*, it seems best to regard these last records as doubtful although quite plausible, similarly the unconfirmed records of G. Wilson and the Barrow-in-Furness N.H.S. in the 1950s from the Rampside square (2.6s) and from Bardsea (3.7s) and Broughton-in-Furness (2.8).

　　　Neither Hodgson nor Wilson mention this species and Baker states that despite its occurrence in several lists he had failed to find it in the Lake District.

　　　There are also early unsubstantiated records from Melmerby (6.3, W.W. Mason, 1913) and Blaithwaite, Wigton (2.4, J. Parkin, 1935), and there is a post-1930 *Atlas* record for the Kirkby Lonsdale square (6.7s) but this is probably in Lancashire.

　　　　　　　3　　　　　　[(W)]F[(C)]

T. pratense L.　　　　　　　Red clover

Very common in the lowlands and extending well up the hillsides. It occurs in a wide range of grasslands but usually on the better soils and it is particularly frequent on roadside verges and in 'unimproved' grasslands. A variety of selected strains are common components of ley mixtures.

Alt. limit: 550 m, Moor House, Teesdale (74.32),

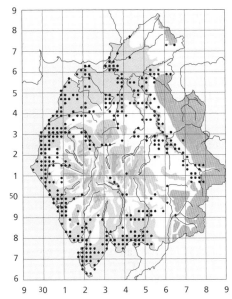

Map 505. *Trifolium campestre*

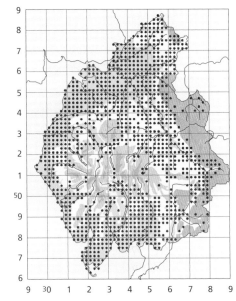

Map 506. *Trifolium dubium*

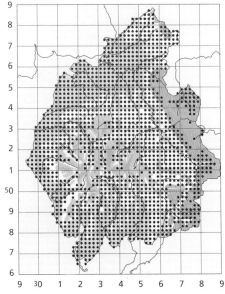

Map 507. *Trifolium pratense*

although certainly introduced.

Map 507   1490   LWFCY

T. medium L.   Zigzag clover

Frequent except in the Lake District and the south-west. It occurs in similar habitats to *Trifolium pratense* but is usually indicative of the richer soils. There are also records from flushed riverside grassland, from rocky gills and rock outcrops in the fells.

Alt. limit: 610 m, Comb Crags, Helvellyn (32.12)

Map 508   672   WFCY

*T. incarnatum L.

Noted by Hodgson as occurring at the Derwent Tinplate Works, Workington (9.2) and at Flimby (0.3). The only other records are from Black Dyke and East Cote, Skinburness (1.5, E.J. Glaister, 1881, 1900, CLE) and a cornfield at Great Salkeld (5.3, H. Britten, 1904).

        Although a southern species, it is widely cultivated and naturalised over most of Europe.

(C)

*T. lappaceum L.

A southern European species recorded by Hodgson from the Derwent Tinplate Works, Workington (9.2) in 1890 and collected in 1910 at Stainton, Carlisle (3.5, T.S. Johnstone, CLE).

(C)

T. striatum L.

Very rare. This southern species was seen during the Survey at Silloth (10.52, A. Newton, 1976), by the River Caldew near Dalston (36.48, EEM, 1985) and Cummersdale (38.52, REG, 1988, LANC), in a lawn at Aglionby (44.56, C.W. Muirhead, 1971, PLYP), on a limestone bank on the north side of the River Eamont near Stainton (48.26, PT, 1979, LANC), growing with *Aphanes arvensis, Vulpia bromoides, Koeleria macrantha* and *Helianthemum nummularium*, and in Mayburgh Henge, Eamont Bridge (50.28, RWMC, 1977). In the south it occurs by Old Park Wood, Holker (32.78s, P. Gibson, 1996) and there are early Survey records from limestone grassland at Wartbarrow (38.76s, KAG & FLW, 1976), possibly Baker's Grange-over-Sands site, and Far Arnside (44.76s, MB, 1983).

        Hodgson also lists it from St Bees (9.1) and Grune Point (1.5, E.J. Glaister, 1877, CLE). It was recorded near Great Salkeld (5.3, H. Britten) in 1903 and later by G.A.K. Hervey from 1947 to 1952. It was seen by G. Higgin in 1951 on Walney Island (1.6s/1.7, herb. P. Reiss, KDL), where it probably still persists.

Alt. limit: 135 m, Stainton, Penrith (48.26)

Map 509   9   WFC

T. arvense L.   Hare's-foot clover

Not uncommon on sandy and disturbed ground along the coast but rare elsewhere. It was reported from railway tracks at Carlisle (38.56) and Kendal (50.92).

Alt. limit: *c.*200 m, south of Kirkby Stephen (76.04) and Cockley Beck (24.00)

Map 510   61   WFC

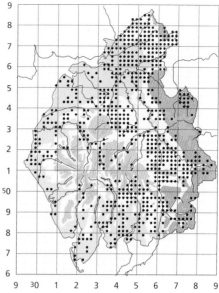

Map 508. *Trifolium medium*

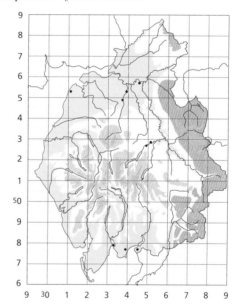

Map 509. *Trifolium striatum*

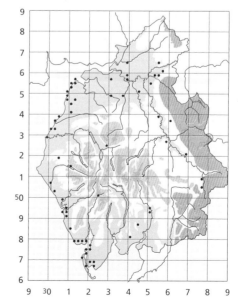

Map 510. *Trifolium arvense*

*T. squamosum L.
A maritime clover of southern Britain listed with reservation by Hodgson from the shore at Braystones (0.0). However, there is a later substantiated record from Stainton Cross, Carlisle (3.5, T.S. Johnstone, 1910, CLE).

S   (C)

*T. echinatum M. Bieb.
Hodgson comments, under *Trifolium maritimum*, that he collected this Balkan species at Floshgate, Ullswater (4.2) in 1882.

(C)

*T. subterraneum L.   Subterranean clover
A species of southern and south-eastern England recorded from Grune Point (1.5, E.J. Glaister, CLE) in 1900.

(C)

*Lupinus angustifolius L.
A single plant of this southern alien was found by the roadside at Brunstock, Carlisle in 1992 (42.58, PB, LANC).

1   C

*L. arboreus Sims   Tree lupin
Well distributed and increasing along the west coast from the sand-dunes of Silloth (10.52) south to Walney Island, where it was found, new to Furness, in old gravel workings in 1979 (16.70, A.O. Chater & GH, *Watsonia* (1980) 13:134). It was first recorded in Cumberland by C.W. Muirhead at Silloth about 1940.

This Californian species is now widely distributed around the coasts of Britain but chiefly in the south.
Map 511   25   FC

*L. arboreus x L. polyphyllus Lindley   Garden lupin
L. x regalis Bergmans
A garden escape well established by disused railways at Newtown, Carlisle (38.56, REG, 1977) and Kirkbride (22.56, EEM, 1994) and by the main railway line at Upperby, Carlisle (40.52, REG, 1984).

3   C

[L. nootkatensis Donn ex Sims
There is a single undated record of this North Pacific species from the "upper Duddon" where it was seen by J.W. Haines (*c*.1920s). Since this lupin is virtually restricted in Britain to the valleys of the eastern Scottish Highlands the record seems very doubtful.

(C)]

*Laburnum anagyroides Medikus   Laburnum
Scattered and chiefly in the Eden valley, occurring by roadsides and in woodland. Surprisingly the only record prior to the Survey is from near Cockermouth (1.3, T. Brown, 1896, LIV). It is not possible to identify the first Westmorland and Furness records.

The records represent spontaneous occurrences from seed of garden origin as well as plantings. Some of the Survey field records may refer to *Laburnum alpinum* or the hybrid, with which it is easily confused.
Alt. limit: *c*.180 m, Sleagill (60.18)
Map 512   64   WFCY

*L. anagyroides x L. alpinum
L. x watereri (Wettst.) Dippel
Known from five tetrads in a limited area of west Cumberland: Boonwood, Gosforth (06.04, 1992), near Calder Bridge (02.06, 1992; 04.06, 1991), Haile (04.08, 1992) and west of Lamplugh (06.20, 1991; all AAD, LANC, det. A.O. Chater).

This hybrid is intermediate in corolla length (18-21 mm); the leaves have sparse long hairs on the lower surface (*Laburnum alpinum* is more or less glabrous except on the veins) and the fruit is more shortly keeled (-1 mm) than in that parent (A.O. Chater *in litt.*).

5   C

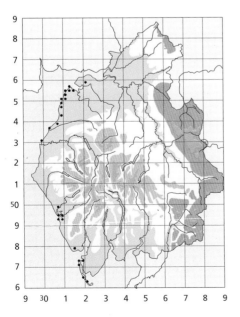

Map 511. *Lupinus arboreus*

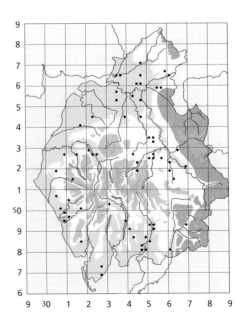

Map 512. *Laburnum anagyroides*

*L. alpinum (Miller) J.S. Presl

The only records for this southern European species are Survey ones from Crosby Ravensworth (62.14, CW, 1984), the roadside west of Pooley Bridge (46.24, RWMC, 1983, *Watsonia* (1984) 15:129), and east of Kershopefoot (48.82n, REG, 1986, *Watsonia* (1992) 19:144, all LANC, det. A.O. Chater).

   Laburnum alpinum differs from *L. anagyroides* in having very few hairs on the underside of the leaves, except on the veins, and also shorter racemes, paler flowers and pods with a distinct dorsal keel.

Alt. limit: 180 m, Crosby Ravensworth (62.14)

                       3   WC

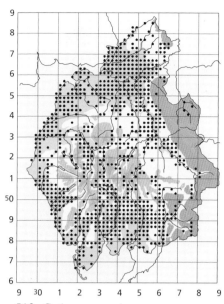

Map 513. *Cytisus scoparius*

Cytisus scoparius (L.) Link   Broom

A frequent lowland shrub of roadside verges, where it is sometimes planted, railway banks, rock outcrops, stony riversides, heath and scrub but absent from the limestone.

Alt. limit: *c*.600 m, Sergeant Man (28.08)

Map 513   847   WFCY

Genista tinctoria L.   Dyer's greenweed

Predominantly a plant of cliffs and rough, neutral or basic grassland along the coast from Walney Island to Maryport (0.3). Inland it is very scattered although locally common and most frequent in the middle Eden valley.

Alt. limit: 250 m, above Milburn (66.28)

Map 514   146   WFC

[G. pilosa L.

Although Baker gives a record for this species as "frequent on rocks in High Furness" he rightly considers it to be an error for *Genista tinctoria*.

                                        (F)]

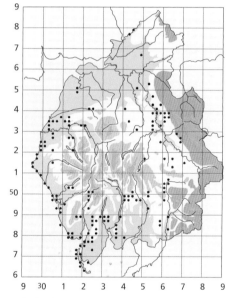

Map 514. *Genista tinctoria*

G. anglica L.   Petty whin

Scattered in the Eden valley, otherwise very rare. It is a plant of acid heath and mossland and very vulnerable to agricultural improvement. It is appreciably less common than is indicated in the *Atlas* and has all but disappeared from Furness and the Solway Plain. During the Survey a colony near Calder Bridge (04.06) became extinct and many of the roadside verge populations must have an uncertain future.

   Particularly interesting are the two pre-Survey Lake District records of the unarmed var. *subinermis* (Le Grand) Rouy: Scar Crags, Coledale (2.2, G.A.K. Hervey, 1928, CLE) and Yew Crags, Honister (2.1, F. Jackson, 1950s).

Alt. limit: 270 m, above Littlewater, Bampton (50.16) but perhaps 640 m at the early Scar Crags site (2.2).

Map 515   32   WFC

*G. hispanica L.   Spanish gorse

Seen since 1987 on a roadside wall-top at Newby Bridge (36.86, PB, LANC).

                       1   F

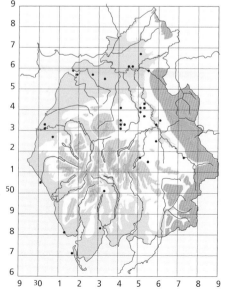

Map 515. *Genista anglica*

Ulex europaeus L.   Gorse; Whin
Plentiful almost everywhere but mostly below 300 m. It
occurs chiefly in areas of rough and neglected grassland, moorland, woodland margins, in hedges,
especially in the north and west, and by railwaysides. It
is also a feature of the landward edge of sand-dunes
and salt-marshes, especially by the Solway Firth.
Although often thought of as calcifuge, it is frequent on
the Morecambe Bay limestones. In severe winters
bushes may suffer considerable die-back but they are
seldom killed.

    Although it commonly grows with *Ulex gallii*,
there are no reports of hybrids.
Alt. limit: 450 m, Nenthead (78.42)
Map 516   1300   WFCY

U. gallii Planchon   Western gorse
Widely distributed but appreciably less common than
Ulex europaeus, particularly in the east although
curiously common in Cumberland along the western
edge of the Pennines. It too avoids the uplands but it
tends to occupy an intermediate altitudinal range
between the higher fells and the agricultural lowlands.
This is well shown in the western fells of the Lake
District, for example in the Lorton area (1.2), where the
steep hillsides are golden in late August and September. It generally occurs on noticeably poorer soils
than *U. europaeus* and, unlike that species, it avoids
the limestone, occurring on it only on the sea-cliffs at
Far Arnside (44.76s). Both species suffer much from
burning and increased grazing.

Alt. limit: 535 m, Yarlside, Shap Fells (52.06)
Map 517   502   WFCY

U. minor Roth   Dwarf gorse
Known only from a limited area west of Carlisle where
it occurs in rough pasture, in hedgerows and on mossland. During the Survey it was found by REG at Corn-
hill (36.56, BM, LANC), south-west of Grinsdale
(34.56, LANC), Sowerby Wood (36.50) and Kingmoor
(38.58), and by GH at Finglandrigg (26.56, LANC). It
was first seen at Kingmoor and Sowerby Wood by C.W.
Muirhead (CLE) in 1947 and 1948 respectively and at
Finglandrigg by DAR in 1961. The first record was in
1944 when it was found at Hosket Hill (3.5), south of
Kirkandrews-on-Eden by M. Oliphant; it was last seen
there by M.C.F. Proctor in 1961 (Proctor 1965). It has
also disappeared from a hedgerow at Aglionby (4.5,
C.W. Muirhead, 1960, PLYP).

    The Finglandrigg population consists of several
hundred bushes scattered over several acres of mossland dominated by *Calluna* and *Molinia*. All the other
extant populations are small and fragmentary; that at
Kingmoor, for example, consists now of only two
isolated bushes. The peak flowering time is the latter
half of September, more than a month later than in the
south of England.

    The occurrence of this species in Cumbria is
curious in view of its south-eastern distribution in
England. Although there are a few records of presumably introduced plants in mid-Scotland, there seems no
reason to doubt its native status in Cumbria.
Plate 17   5   C

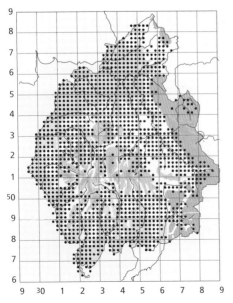

Map 516. *Ulex europaeus*

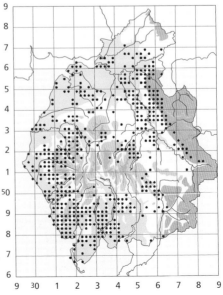

Map 517. *Ulex gallii*

ELEAGNACEAE

*Hippophae rhamnoides L.   Sea buckthorn
Scattered along the coast but only really frequent around the Ravenglass estuary (0.9). Most of the sites are sand-dunes, but it also occurs on waste ground and by sea-walls. The populations have presumably all originated from plantings or were bird-sown. The earliest record for Cumberland is from Silloth (10.52, C. Bailey, 1896, OXF), for Furness from Furness Abbey (2.7, W.H. Pearsall, 1913, *BEC Rpt* (1914) 3:393), and for Westmorland from north of Kirkby Lonsdale (6.8s, C. Snow, 1959). The *Scarce Plants* record for the Arnside square (4.7s) is from v.c.60.

Sea-buckthorn has been widely planted for stabilising sand-dunes and it provides valuable cover and food for migrating birds. Unfortunately it is very invasive and for this reason was eliminated at North Walney (16.72) and Sandscale (18.74) during the Survey. At Eskmeals (0.9), the bushes are now quite old and tall, up to 7 m, and with a stem diameter of up to 16 cm; the original dense cover has considerably diminished.
Map 518, Plate 76   S 21   (W)FC

HALORAGACEAE

[Myriophyllum verticillatum L.   Whorled water-milfoil
Listed by Baker and Wilson from Windermere, where it was said to be abundant, Coniston Water (2.9/3.9), Brigsteer Moss (4.8s), a ditch at Howtown, Ullswater (4.1), in Naddle Beck, Threlkeld (2.2), Lodore, Derwent Water (2.1), a pond near Edenhall (5.3) and the River Eden.

Although this is a southern species of base-rich waters, neither author expressed any reservations about these records. It has proved impossible to locate any confirmatory specimens and it seems very likely that the records refer to one of the following two species, most probably *Myriophyllum alterniflorum*.

(WC)]

M. spicatum L.   Spiked water-milfoil
Locally frequent, chiefly near the coast, and occurring in eutrophic and sometimes brackish pools, tarns and rivers. It is this species which is commonly stocked by garden centres and it seems likely that a number of newly created ponds in and around the upper Eden valley have been colonised, intentionally or by bird dispersal, from this original source. There are no recent records from any of the lakes and it seems likely that Baker's from Windermere and Watendlath Tarn (2.1) and several of the *Atlas* records refer to the following species, the two being frequently confused.

A study by Lawson (1992) of Cumbrian material of the two species indicates that the mean number of leaf segments in *Myriophyllum spicatum* varies from 19-33 and in *M. alterniflorum* from 11-15.
Alt. limit: 295 m, Ravenstonedale (72.96). Wilson's figure of 520 m from Conny Pot Beck, on Stainmore (8.2), must refer to *M. alterniflorum*.
Map 519   71   WFC

M. alterniflorum DC.   Alternate water-milfoil
Very common in nutrient-poor waters but scarce towards the coast. It occurs in all the lakes, with the exception of Thirlmere and Haweswater, and all the main river systems apart from the Irthing, the middle and lower Eden and most of the Petteril, although there is an old record from the River Eden at Carlisle (4.5, T.S. Johnstone, 1901, CLE).

The morphological differences between this species and *Myriophyllum spicatum* are given under the latter. The species occur together at a few sites, for

Plate 76. *Hippophae rhamnoides*, Eskmeals

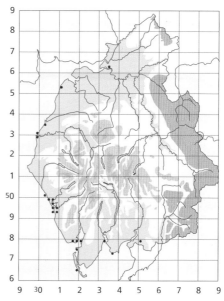

Map 518. *Hippophae rhamnoides*

example in Helton Tarn in the Winster valley (40.84s, 42.84s) and in the River Eden at Great Ormside (68.16, 70.16). *M. alterniflorum* is not strictly calcifuge as is suggested in a number of Floras: in Potts Beck (68.08), east of Sunbiggin Tarn, it thrives in water of pH 7.3.

Alt. limit: 615 m, Brown Cove, Helvellyn (34.16)

Map 520   289   WFC

LYTHRACEAE

Lythrum salicaria L.   Purple loosestrife

A prominent and frequent member of the tall-herb community of ungrazed riversides and lakesides, marshes, fens and ditches occurring over most of the lowlands and commonly associated with *Filipendula ulmaria* and *Lysimachia vulgaris*.

Alt. limit: 200 m, Greystoke (46.30)

Map 521   289   WFC

*L. hyssopifolium L.

Baker cites records of this species from Wasdale (1.0) and Derwent Water (2.1) but regards them as errors. There is, however, a 1962 collection in CLE from Glinger Bank, Longtown (3.7n, Mrs J. Westall). This is a species of central and southern England and now all but extinct.

     R   (C)

L. portula (L.) D. Webb   Water purslane
(*Peplis portula* L.)

Generally uncommon but well scattered over most of the county. It is typically a lowland plant of open, mesotrophic communities, tarn margins and seasonal field ponds. It is particularly characteristic of the draw-down zones of reservoirs, often carpeting the ground pink and frequently associated with *Rorippa palustris*.

   The *Critical Atlas* shows Cumbrian records of both subspp. *portula* and *longidentatum* (Gay) Sell. These two taxa are said to differ in the length of the

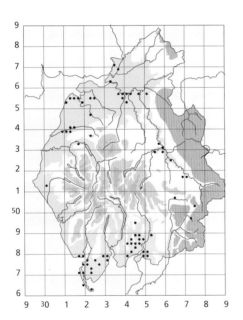

Map 519. *Myriophyllum spicatum*

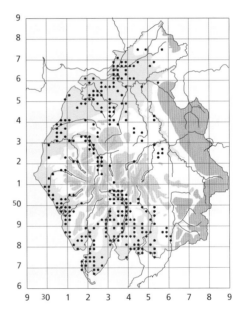

Map 521. *Lythrum salicaria*

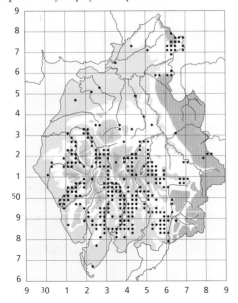

Map 520. *Myriophyllum alterniflorum*

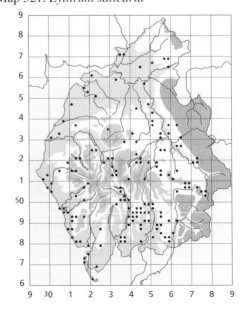

Map 522. *Lythrum portula*

epicalyx segments, those of the latter exceeding 1.5 mm. However, the few herbarium specimens examined from the Survey span the whole range.

This species has a pronounced southern and western distribution in Britain.

Alt. limit: 340 m, Knipe Moor (52.18, Wilson)

Map 522   154   WFCW/Y

THYMELAEACEAE

Daphne mezereum L.   Mezereon

Very rare, known only from a few species-rich limestone woods around the Kent estuary, where it is probably native, and also from outlying sites in woods at Gosforth (06.02), near Sowerby Row (38.38), where it was earlier recorded by C.W. Muirhead in 1953 (PLYP), and above Staveley (46.00, TW, 1990). In these last sites it was presumably bird-sown from gardens. Although mezereon has been much in demand for gardens, even the early Floras comment on its rarity. Other outlying sites include Hodgson's record for the Wigton square (2.4) and Wilson's from near Windermere (4.9) and Kings Meaburn (6.2).

Alt. limit: 300 m, above Staveley (46.00)

Map 523   S 12   WFC

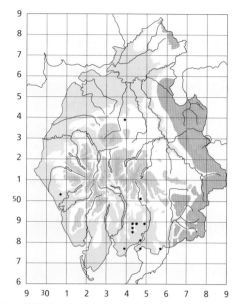

Map 523. *Daphne mezereum*

D. laureola L.   Spurge laurel

Like the last, this predominantly woodland species is largely restricted to limestone in the valley and estuary of the River Kent, where it is probably native and from where it was curiously not recorded by Wilson. There are also a number of records from around Carlisle, the Solway Plain and south-west Cumberland. It may be declining having been unconfirmed during the Survey in seven of the post-1930 *Atlas* squares.

Alt. limit: 120 m, Staveley (44.98)

Map 524   27   WFC

ONAGRACEAE

Epilobium hirsutum L.   Great hairy willow-herb

Common in the lowlands and a characteristic member of tall-herb marsh and fen communities in meadows and ditches, and by rivers, tarns and lakes. It may, however, occur in surprisingly dry situations on roadside verges and urban waste sites.

Alt. limit: *c.*300 m, Helbeck Wood (78.16) and Swindale, Brough (80.16)

Map 525   856   WFCY

[E. hirsutum x E. montanum
E. x erroneum Hausskn.
The record from Boltonfellend (*Watsonia* (1988) 17:189) was based on a misidentification of the following hybrid.

C]

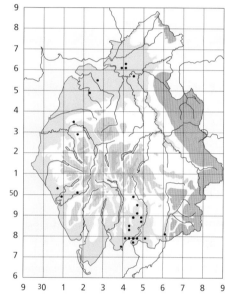

Map 524. *Daphne laureola*

*E. hirsutum x E. ciliatum

E. x nova-civitatis Smejkal

A few plants were found in 1987 growing in a roadside

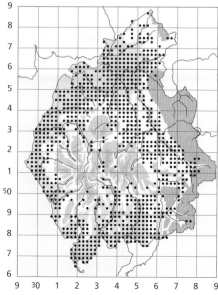

Map 525. *Epilobium hirsutum*

ditch at Boltonfellend (46.68, GH, LANC, det. T.D. Pennington). According to Stace (1975), this hybrid is known only from six other vice-counties.
Alt.: 105 m, Boltonfellend (46.68)
1   C

E. parviflorum Schreber
Small-flowered hairy willow-herb
Fairly frequent in the lowlands but curiously patchy, being most frequent in the upper Eden valley, the north-east, west Cumberland and the Furness peninsula. It grows in more open habitats than *Epilobium hirsutum*, often in places subject to poaching by cattle.
Alt. limit: at least 350 m, Garrigill (74.42)
Map 526   400   WFCY

E. parviflorum x **E. montanum**
E. x **limosum** Schur
Listed by Wilson from Dufton (6.2).
(W)

E. parviflorum x **E. obscurum**
E. x **dacicum** Borbás
Listed by Wilson from east of Dufton (7.2).
(W)

E. parviflorum x **E. palustre**
E. x **rivulare** Wahlenb.
Listed by Stace (1975) for Cumberland, but no other information is available.
(C)

E. montanum L.   Broad-leaved willow-herb
Very common, occurring in the lowlands on waste sites, by pavements and roadsides, on walls, in hedgerows, gardens and in woodland margins. In the fells it is common in wooded gills, wet gullies and on sheltered rock ledges.
Alt. limit: 845 m, Great Dun Fell (70.32)
Map 527   1482   LWFCY

E. montanum x **E. obscurum**
E. x **aggregatum** Čelak.
Despite the abundance of both parents, this hybrid has been recorded only from the Midgey allotments, Whitehaven (98.16, CCH, 1981) and the roadside at Dunmail Raise (32.10, PB, 1983, both LANC, det. T.D. Pennington).
2   WC

*****E. montanum** x **E. ciliatum**
Collected in 1981 by A.J. Silverside at Castlerigg, Keswick (28.22, LANC) in a damp, disturbed pasture growing with both parents.
1   C

E. montanum x **E. alsinifolium**
E. x **facchinii** Hausm.
This rare hybrid was collected in Sandbed Gill, Helvellyn (32.20, MC, 1975) and subsequently identified by C.A. Stace who had earlier (1975) listed it from only three vice-counties. It was growing in some quantity by small waterfalls in the bed of the gill. Attempts to refind it have failed. (See Addenda)
Alt.: *c.*240 m, Sandbed Gill, Helvellyn (32.20)
1   C[(Y)]

[*E. tetragonum* L.   Square-stemmed willow-herb
Hodgson gives records for this species from Workington (9.2), Westward (2.4), near Silloth (1.5) and Edenhall (5.3), and there are other Cumberland ones from Blaithwaite (2.4, E.J. Glaister, 1878), Ravenglass (0.9, J. Parkin, 1935), Gilsland (6.6, G.A.K. Hervey, 1945, v.c.67/70) and a later one from Edenhall (5.3, M. Cross, 1949). Baker makes no

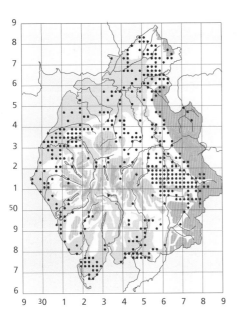

Map 526. *Epilobium parviflorum*

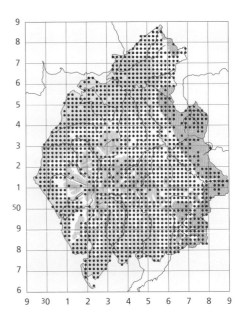

Map 527. *Epilobium montanum*

reference to the species and Wilson lists only an old record from Staveley (4.9). There are no Survey records.

In view of the frequent confusion between this species and *Epilobium obscurum*, it seems very likely that all the above records, and the post-1930 *Atlas* record for the Hawkshead square (3.9), refer to the latter species. Surprisingly, Livermore & Livermore (1987) describe the species as frequent in the Lancaster district.

(WFC)]

[E. tetragonum x E. obscurum
E. x semiobscurum Borbás
Stace's (1975) reference to this hybrid in Cumberland is probably based on a misidentification since there are no reliable records of *Epilobium tetragonum* from the county.

(C)]

E. obscurum Schreber

A frequent species of waste places, gardens, riverbanks and ditches, generally preferring rather open, damp situations on heavy soils.
Alt. limit: 500 m, above Cow Green Reservoir, Teesdale (76.30)
Map 528   953   LWFCY

*E. obscurum x E. ciliatum

Material possibly of this hybrid was collected by CCH from the Midgey allotments, Whitehaven (98.16, 1981, det. T.D. Pennington, LANC).

C

E. obscurum x E. palustre
E. x schmidtianum Rostkov

Material tentatively identified as this hybrid was collected from Brathay (36.02, R. Young, 1969, det. T.D. Pennington, AMDE) and Raisbeck (64.06, GH, 1992, LANC).

WF

E. roseum Schreber

The only substantiated Survey records for this southern species are from Brathay Hall (36.02, GH, 1986, LANC), where it had been reported some years previously by A.J. Richards, and near Westward, Wigton (28.44, REG, 1986, LANC, both records *Watsonia* (1988) 17:189), also from the Midgey allotments, Whitehaven (98.16, CCH, 1989, LANC), a pavement at St Bees (96.10, CCH, 1979; LANC) and Roanhead (20.74, PB, 1988, LANC).

The only previous definite record is from Carlisle (3.5/4.5, R.H. Williamson, 1926, CLE) and the only reference in the Floras to this southern species is Hodgson's doubt regarding a record from near Workington (9.2).

5   FC

*E. ciliatum Raf.   American willow-herb

This North American species was discovered in Britain in 1891. It spread rapidly early this century in south-eastern England and was first recorded in Furness and Westmorland c.1969 when it was noted by R. Young as being common in the Brathay - Ambleside area (3.0) and by J. Barkham in Brigsteer Park (4.8s). In Cumberland it was seen a few years later by C.W. Muirhead in Carlisle (4.5, 1973, PLYP, *Watsonia* (1979) 12:354). It is now well dispersed throughout the lowlands but is still rather scarce in Furness. It thrives as a garden weed and on waste ground around towns and villages, but it is still uncommon around Barrow-in-Furness and Workington (9.2).
Alt. limit: *c*.450 m, above Garrigill (76.38)
Map 529   317   WFCY

E. palustre L.   Marsh willow-herb

A common upland plant being particularly characteristic of *Juncus effusus* flushes and of the bryophyte-

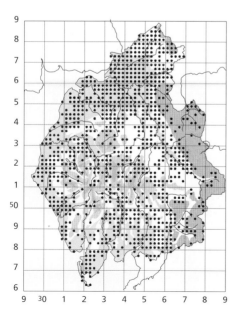

Map 528. *Epilobium obscurum*

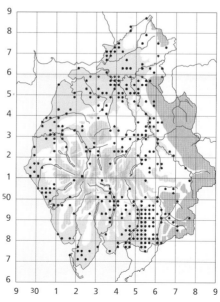

Map 529. *Epilobium ciliatum*

rich *Chrysosplenium oppositifolium - Montia fontana* community around upland springs. It also occurs in the more base-rich flushes and at low altitudes in the tall-herb communities of wet meadows and the margins of lakes and tarns.

Alt. limit: 700 m, above Knock Ore Gill (70.30)

Map 530   1337   WFCY

E. palustre x E. alsinifolium
A rare hybrid known only from High Cup Nick (7.2, E.G. Baker, 1894, BM) and Fisher Gill, Helvellyn (3.1, herb. A.H. Wolley-Dod, 1912, BM). There is also an unconfirmed report from Cautley (Lees 1938, p.213).

(WCY?)

E. anagallidifolium Lam.   Alpine willow-herb
An inconspicuous and often overlooked willowherb of high-level bryophyte-rich flushes and sykes throughout the Cross Fell range. It flowers about a month earlier than *Epilobium alsinifolium* with which it is sometimes confused.

These Cumbrian sites are the only ones in Britain south of the Moffat Hills, the *Atlas* records from the Ravenstonedale square (7.0) and from Nidderdale in Yorkshire being almost certainly errors.

Alt. range: 540 m, Ousby Fell (68.38) - 850 m, Cross Fell (68.34)

Map 531   24   WC

E. alsinifolium Villars   Chickweed willow-herb
Frequent in the northern Pennines, sparse but well-dispersed over the Howgill Fells and frequent also in the eastern fells of the Lake District. It is a colourful and conspicuous plant of upland bryophyte-rich spring vegetation, of rather open hillside flushes, sheltered wet gullies and dripping rock ledges. There are no Survey records for two sites in the central Lake District: on the western side of Styhead Pass (2.0, C.W. Muirhead, 1940, PLYP) and on Glaramara (22.10, DAR, 1956), or confirmation of an anonymous, undated *Scarce Plants* record for the Lorton square (1.2), Baker's record for Shap (5.1) or of Hodgson's statement that it is "common in upper Ennerdale" (1.1/2.1). The *Scarce Plants* record for the Borrow Beck square (5.0) is an error for Wilson's record for Rundale and Swindale (7.2).

This species extends south down the Pennines to the Craven area of Yorkshire and has outlying sites in Snowdonia.

Alt. range: 150 m, Wanthwaite Crags (32.22) - 825 m, Cross Fell (68.34)

Map 532, Plate 81   S 127   WCY

*E. brunnescens (Cockayne) Raven & Engelhorn
(*E. pedunculare* Cunn. pro parte; *E. nerteroides* auct., non Cunn.)   New Zealand willow-herb
Introduced at Edinburgh in 1904, this small willowherb is a thoroughly established feature of the Cumbrian hills, creeping over bare ground, shingle and rocks by

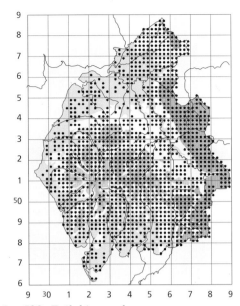

Map 530. *Epilobium palustre*

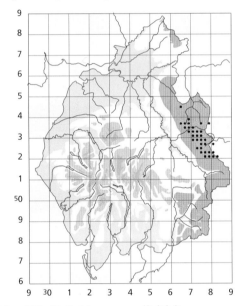

Map 531. *Epilobium anagallidifolium*

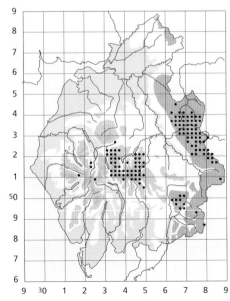

Map 532. *Epilobium alsinifolium*

gills, becks and drainage ditches and preferring mineral ground. At low altitudes it is not infrequently a persistent weed of rock gardens.

The first record dates from *c.*1923 when it was recorded by G. Bolam from the roadside at Priorsdale, above Garrigill (7.4, *Trans. nat. Hist. Soc. Northumb.* (1923-1926) 6:83). It was seen in Westmorland a few years later at Grasmere (3.0, L. Baker, 1929, *BEC Rpt* (1930) 9:116) and in Furness in Tilberthwaite Gill (3.0, J.P.M. Brenan, 1945; PLYP, *BEC Rpt* (1947) 13:58). Surprisingly, although it was recorded in 1949 near Furness Abbey (2.7, G. Wilson, CLE), there are still extremely few records from the Furness peninsula.
Alt. limit: 845 m, Great Dun Fell (70.32)

Map 533   580   WFCY

Chamerion angustifolium (L.) Holub

Rosebay willow-herb; Fireweed
Very common, occurring often in large stands on roadside verges, in waste places and in coarse grassland, particularly on railwaysides. In the fells it grows on sheltered rock ledges and in gullies and these populations are often considered to represent the original, native, non-weedy biotypes, although there do not appear to be any correlated morphological differences. Many of these montane populations are very remote, such as that by the Force Burn in Teesdale (76.30) at 530 m. It is, however, evident from the comments of the early botanists that the species was at least locally common last century at low altitudes; Hodgson, for example, describes it as abundant on the cliffs at St Bees (9.1).

Some of the early records are of the sterile garden escape earlier known as *Epilobium angustifolium* var. *brachycarpum* (Leight.) Druce.
Alt. limit: 855 m, Scafell (20.06, DAR)

Map 534   1566   LWFCY

*****Oenothera**   Evening primroses
The Survey records of this genus are all based on material determined by J.C. Bowra or K. Rostański.

*****Oenothera glazioviana** Micheli ex C. Martius
(*O. erythrosepala* Borbás)
Like the following hybrid this is a plant of dock sites, sand-dunes and waste ground near the coast. The only inland record is from waste ground at Kendal.

The first confirmed record for Westmorland is from Kendal (50.90, CEW, 1986), for Furness from Barrow (20.68, GH, 1983, LANC) and for Cumberland from Workington docks (98.28, EEM, 1985).

Oenothera glazioviana is distinguished chiefly by its large petals, up to 5 cm long.

Map 535   11   WFC

*****O. glazioviana** x **O. biennis**
O. fallax Renner
Rare and largely restricted to coastal urban sites and sand-dunes but with a few scattered inland localities,

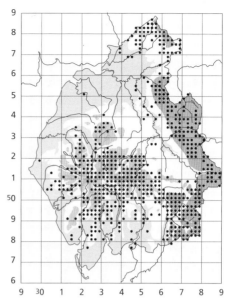

Map 533. *Epilobium brunnescens*

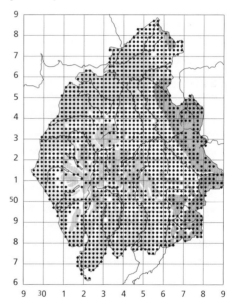

Map 534. *Chamerion angustifolium*

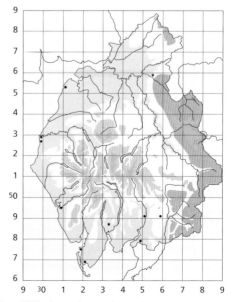

Map 535. *Oenothera glazioviana*

especially on roadside grit-piles, having been introduced with the grit from the west coast.

The first substantiated record for Westmorland
was from a quarry between Sizergh and Heaves
(48.86s, GH, 1975, LANC, *Watsonia* (1982) 14:192),
for Furness from South Walney (22.62s, GH, 1979,
LANC) and for Cumberland from Beckfoot (08.48,
MMM, 1981, LANC).

This hybrid can easily be separated from the
following species by its smaller petals which rarely
exceed 3 cm in length.
Alt. limit: 260 m, east of Grayrigg (60.98)
Map 536   13   WFC

[O. biennis L.   Common evening primrose
The records given by Hodgson and Wilson probably refer to
Oenothera glazioviana or the hybrid.

(WC)]

*O. cambrica Rostański
Material collected by A. Cannell in 1986 at Silloth
docks (10.52; LANC, *Watsonia* (1989) 17:472) was
identified by him as this species and this was later
confirmed by J.C. Bowra. Bowra has since identified
material collected at Kirkbride in 1877 (1.5, E.J.
Glaister, CLE) as *cf. Oenothera cambrica* x *O. biennis*.
Some recent collections made by Bowra at Silloth are
not entirely typical and may be the same hybrid.

This is a small-flowered plant which differs
from the previous two in having green sepals. This
Cumbrian site represents a major extension of range as
the species was not previously known in the west north
of Wales, but it has since been reported from the
Glasgow area.

1   C

*Oenothera sp.
A specimen collected at Silloth (1.5, J. Leitch, CLE) in 1890
and labelled "*Oenothera tenella*" has been referred by J.C.
Bowra to Series Devriesia.

(C)

*Fuchsia x **hybrida** hort. ex Vilm.
A single low shrub of an unidentified garden fuchsia
was discovered growing amongst rocks on the east side
of the A591 road south of Dunmail Raise (32.10, PB,
1993, LANC).

1   W

Circaea lutetiana L.   Enchanter's nightshade
Frequent but rather unevenly distributed in the lowlands occurring in moist deciduous woodland, more
rarely in hedgerows, and preferring the more base-rich
soils. Like many common woodland plants it is scarce
on the Solway Plain.
Alt. limit: *c.*370 m, Garrigill (72.42)
Map 537   805   LWFCY

C. lutetiana x **C. alpina**
C. x **intermedia** Ehrh.
                    Intermediate enchanter's nightshade
Scattered along the middle Eden valley and from the
northern part of the Lake District south to beyond
Sedbergh. Many of the records are from wooded
riversides, as in Garsdale and by the lower River Eden,
and by lakesides, particularly Windermere and
Ullswater. Other habitats include an abandoned railway
line and wall-tops, and in a garden at its isolated,
westernmost site at Cleator Moor. It has not been
refound at Scale Wood, Crummock Water (1.1, C.W.
Muirhead, 1949, PLYP).

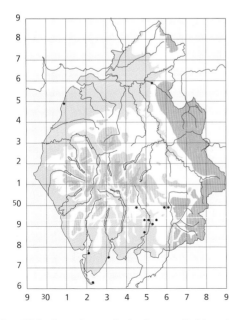

Map 536. *Oenothera glazioviana* x *O. biennis*

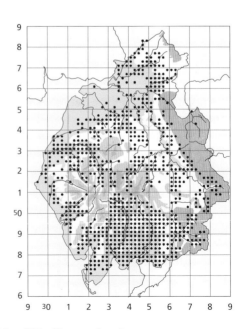

Map 537. *Circaea lutetiana*

Alt. limit: at least 250 m, upper Dentdale (76.84)
Map 538   93   WFCY

C. alpina L.   Alpine enchanter's nightshade
Rare and largely confined to the head of Borrowdale
(2.1), the south-west end of Ullswater (3.1) and the
head of Haweswater (4.1). There are interesting
outlying sites above Tilberthwaite Gill (28.00, Anon.,
1973, LANC), near Wray (34.00, R. Atkins, 1989,
LANC), in upper Kentmere (46.06, T.C.G. Rich, 1982)
and on Great Mell Fell (40.24, MC, 1978; LANC)
where it occurs in some quantity. The Survey records
are all supported by herbarium specimens and confirmed by D.E. Boufford. There are no recent records
from the west shore of Windermere or from Swindale,
both localities cited by Raven (1963) in his revision of
the genus in Britain.

Raven pointed out that many of the records
previously attributed to this species actually refer to
Circaea x *intermedia*. *Circaea alpina* is far less
common, being restricted to a few localities in western
Scotland and Wales and Cumbria. The earliest Furness
record is from between Wray Castle and the ferry (3.9,
R.S. Adamson, 1920, BM), erroneously listed by Raven
under v.c.60; that for Westmorland is from Swindale
(4.1/5.1, H.C. Watson, 1856, K), and for Cumberland:
from Derwent Water (2.2, D. Turner, 1806, K). Only
three of the Cumbrian records which he lists are post-
1930 and the pre-1930 dot on his map for the Carlisle
square (4.5) is an obvious error and is not based on any
cited locality. The record from Knock Fell (7.3, H.G.
Proctor, *Proc. BSBI* (1963) 5:133; Rawes (1981);
LANC) is, as listed by Raven, *C.* x *intermedia*. The
species has now disappeared from the site near
Rawthey Bridge above Sedbergh (7.9).

Although many of the records are from periodically wet and rather bare seepage areas in deciduous
woodland, there are other fellside ones from shady gills

and gullies as well as block scree. At several sites it
occurs under bracken and in Greenup Gill (28.10)
amongst *Sphagnum* at the foot of boulders.

Circaea alpina is largely restricted in Britain to
Cumbria, Arran and Kintyre but with outlying sites in
the Scottish Highlands and Wales.
Alt. limit: 755 m, Knock Fell (70.30)
Map 539   19   WFC(Y)

CORNACEAE

Cornus sanguinea L.   Dogwood
Not uncommon in south Westmorland but otherwise
very local. Around the Kent estuary it occurs in
limestone scrub, along woodland margins and in old
hedgerows. As Wilson says, this probably represents
the species' northern limit on the west coast. Records
from the north, mostly from hedges, are probably of
plantings.

A small tree with predominantly pentamerous
flowers and lacking the usual red bark was discovered
in a lane between Stanwix and Tarraby, Carlisle (40.56,
REG, 1989; CGE, LANC).
Alt. limit: 185 m, east of Kendal (54.90)
Map 540   46   WFC

*C. sericea L.
A popular shrub in gardens and parks, this North
American species is naturalised in several places,
mostly in the south of the county where it occurs in
hedgerows, in large estates and luxuriantly along the
River Brathay (3.0). The first Cumberland records date
from the 1970s, that for Furness was from by the River
Brathay (34.02, R. Young, 1969, AMDE) and for West-
morland: from Underley, Kirkby Lonsdale (60.80s, GH,
1975).
Alt. limit: at least 165 m, Askham (50.22)
Map 541   27   WFC

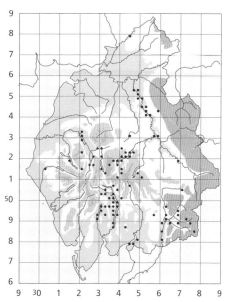

Map 538. *Circaea lutetiana* x *C. alpina*

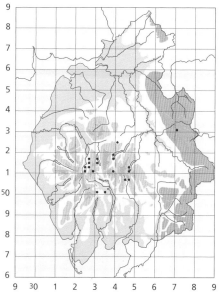

Map 539. *Circaea alpina*

VISCACEAE
(*LORANTHACEAE*)

Viscum album L.   Mistletoe
Rare, seen only in gardens around the Kent estuary, at
two sites near Carlisle and in the garden of a long-
abandoned farm at Keldhead, Bampton (48.18). Apple
was usually the host although MB also noted it on
hawthorn and lime at Heversham (48.82s). G.A.K.
Hervey recorded it on elm at Levens (4.8s, 1954) and
Wilson refers to it on lime at Levens Hall. However, the
map includes several records from the north of the
county arising from the 1969/70 B.S.B.I. Mistletoe
Survey (Perring 1973). The Silloth square (1.5) is the
only one from this last survey for which there is no
localised record.
Alt. limit: 295 m, Keldhead, Bampton (48.18)
Map 542   16   WC

CELASTRACEAE

Euonymus europaeus L.   Spindle
Common in hedgerows and deciduous woodland on the
limestone around Morecambe Bay, otherwise very
scattered and usually only in hedges. Some of the
northern records probably represent native sites, for
example in the limestone woodland of the Sebergham
gorge (34.40), and lie close to the species' northern
native limit. Others may be plantings or bird-sown.
Alt. limit: 305 m, Yew Crag, Ullswater (40.20) and
Helbeck Wood, Brough (78.16)
Map 543   98   WFC

***E. japonicus** L. fil.
Recorded from Drigg (04.98, A.J. Richards, 1973), by
the west coast road at Saltpans, south of Allonby
(06.40, AAD, 1992, LANC) and below a cliff on
the east side of the River Eden near Armathwaite

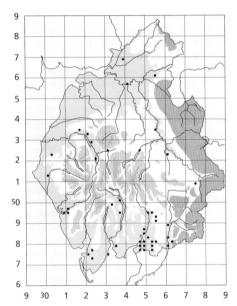

Map 540. *Cornus sanguinea*

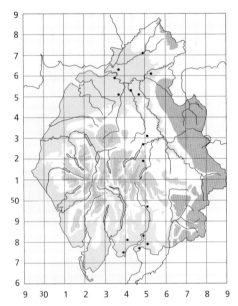

Map 542. *Viscum album*

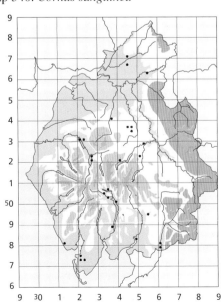

Map 541. *Cornus sericea*

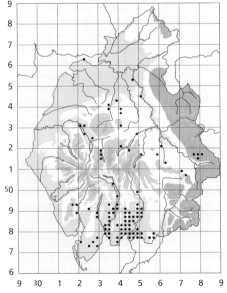

Map 543. *Euonymus europaeus*

(50.44, RWMC, 1994).

3   C

AQUIFOLIACEAE

Ilex aquifolium L.   Holly

Common but like most Atlantic species thinning out noticeably towards the Pennines. It occurs in hedgerows and in deciduous woodland, also on rocky outcrops and gullies in the fells.

Alt. limit: 600 m, Eel Crags, Newlands (16.22)

Map 544   1295   LWFCY

BUXACEAE

*Buxus sempervirens L.   Box

Rather rare and mostly in the south and north-east. The first record for Westmorland was from Levens deer-park (50.84s, CEW, 1987) and for Cumberland from Corby Woods (46.52, MS, 1970s). It occurs by lane-sides far from houses and in estate woodlands, usually as isolated bushes, but it is prolific in Levens deer-park (48.84s) where there are hundreds of seedlings. Probably all the occurrences originated as plantings and this would account for the absence of literature records and for likely under-recording during the Survey.

Alt. limit: 250 m, Keskadale (20.18)

Map 545   R 27   LWC

EUPHORBIACEAE

Mercurialis perennis L.   Dog's mercury

A common herb of lowland deciduous woodland, especially of oak and ash woods, also in old hedgerows, perhaps often persisting from woodland, and in the grikes of limestone pavement. It prefers the better, damper soils and is often dominant over extensive areas. Although a shade plant, it often wilts severely during dry periods. It is surprisingly resistant to trampling by cattle in heavily grazed woodland.

Alt. limit: 460 m, above Hayes Water (42.12)

Map 546   1127   LWFCY

*M. annua L.   Annual mercury

A southern species recorded by Baker and Hodgson only from Dalegarth, Eskdale (1.0) and Maryport (0.3, W. Hodgson, 1890, CLE), and by Handley (1898) from the Sedbergh area. There are no recent records despite the cluster of *Atlas* dots in south Furness. Those, from the North Walney (1.7), Dalton-in-Furness (2.7) and Grizebeck (2.8) squares were contributed by the Barrow-in-Furness N.H.S. during the 1950s. The sources of the northern records in the *Atlas*, from the Hesket Newmarket (3.3), Sebergham (3.4), Skelton (4.3), Southwaite (4.4), Kirkoswald (5.4) and Alston (7.4) squares are not known. The occurrence of so many records well to the north of the main range of the species is suspicious.

(FCW/Y)

While the villains......of High Furness were charged with the care of flocks and herds, to protect them from the wolves which lurked in the thickets, and in winter to browze them with the tender sprouts of hollies and ash. This custom was not till lately discontinued in High Furness; and holly-trees were carefully preserved for that purpose when all other wood was cleared off; large tracts of common being so covered with these trees, as to have the appearance of a forest of hollies. At the Shepherd's call, the flocks surrounded the holly-bush, and received the croppings at his hand, which they greedily nibbled up, bleating for more.

The antiquities of Furness, Thomas West

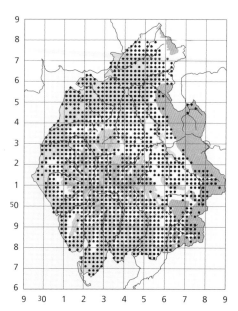

Map 544. *Ilex aquifolium*

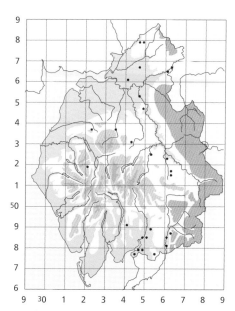

Map 545. *Buxus sempervirens*

*Euphorbia oblongata Griseb.

Two collections from roadsides in and to the north of
Barrow-in-Furness (20.72, JW, 1980; 22.74, PB, 1991,
both LANC) have been identified as this species by E.J.
Clement. Unfortunately the plant can no longer be
found at either site. It is a native of the southern
Balkans.

2 F

E. helioscopia L.   Sun spurge

A fairly frequent lowland weed of cultivated and waste
land: gardens, allotments, arable fields and rubbish
tips.

Alt. limit: 355 m, below Nenthead (74.44)

Map 547   421   WFCY

*E. lathyris L.   Caper spurge

Rather rare, occuring in gardens, farmyards and by
roadsides, probably now always originating from bird-

seed and, like *Bupleurum subovatum*, much commoner
than in the past. The first Westmorland record was
from a garden at Underbarrow (4.9, G.A.K. Hervey,
1948), and for Furness from Grange-over-Sands
(40.70s, KAG, early 1970s).

Alt. limit: 360 m, Nenthall (74.44)

Map 548   31   WFC

E. exigua L.   Dwarf spurge

Recorded only from two sites: on a garden path at
Arnside (46.76s, R. Dalton), where a single plant was
seen in 1977, and in some quantity on the slag tip at
Askam-in-Furness (20.76, PB, 1987, LANC), where
part of the colony has survived a major landscape
'improvement' scheme.

It is listed by Hodgson from Nethertown (9.0),
Lowca (9.2), Clifton (0.2), the railway line near Bull
Gill (0.3, W. Hodgson, 1880, CLE) and Aspatria (1.4),
and by Wilson from Arnside and Hazelslack (4.7s),

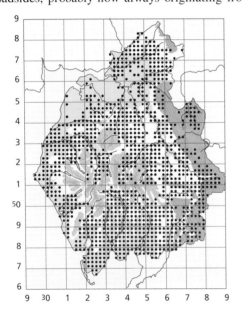

Map 546. *Mercurialis perennis*

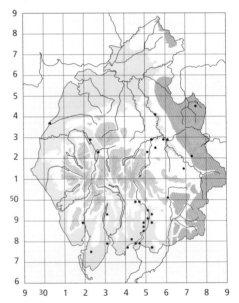

Map 548. *Euphorbia lathyris*

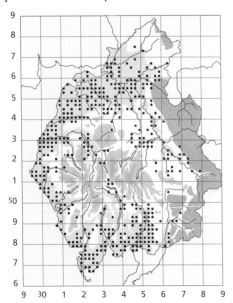

Map 547. *Euphorbia helioscopia*

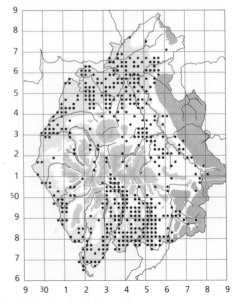

Map 549. *Euphorbia peplus*

Ulpha and Sizergh (4.8s) and near Bolton (6.2). There are also records from Silloth docks (1.5, J. Leitch, 1883, CLE) and Skinburness (1.5, J. Parkin, 1935) and Petty (1897 p.330) gives one from Rampside (2.6s). The *Atlas* has post-1930 dots for the Distington (0.2), Rampside (2.6s), Dalton-in-Furness (2.7) and Cark (3.7s) squares, all of which date from the 1950s.

2 WF(C)

E. peplus L. Petty spurge
Very similar in distribution and ecology to *Euphorbia helioscopia* but less frequent in the west and more frequent in gardens, on waste ground and by walls, roadsides and pavements.
Alt. limit: 215 m, Helbeck, Brough (78.14)
Map 549 462 LWFCY

E. portlandica L. Portland spurge
Scattered along the coast north to Seascale, occurring on sandy shores and dunes, also on cliffs as at Humphrey Head (38.74s) and in Elliscales quarry, Dalton-in-Furness (22.74, PB, 1987). There are no recent records from Silloth (1.5) where it was seen in 1935 by J. Parkin. The first Furness record was from Walney Island (Henry 1897).
Map 550 S 19 FC

E. paralias L. Sea spurge
Fairly frequent and conspicuous on sandy shores and dunes from Walney Island to Seascale and with outlying sites at Beckfoot (08.48) and Silloth (10.52). There is also a pre-Survey record from Harrington (9.2, C.W. Muirhead, 1949, 1951) and a *Scarce Plants* record from the Maryport square (0.3) dating from the 1950s.
Map 551 S 24 FC

*E. waldsteinii (Sojak) Czerep.
(*E. virgata* Waldst.& Kit.)
Material collected from Silloth in 1988 (1.5, REG, MMM, LANC) has been identified by A. R.-Smith as this eastern European species. The material agrees with earlier collections from behind the docks (C.B. Sharp, 1937, C.W. Muirhead, 1947, R.A. Boyd, 1949, M. Horsfall, 1955, all CLE). These appear to be the earliest British records and currently the only English ones.

Wilson's record from by the River Lune at Rigmaden, Barbon (6.8s) may refer to *Euphorbia esula*.

1 C

*E. waldsteinii x **E. cyparissias**
E. x gayeri Boros & Soó
The only record is from Egremont (0.1, J. Adair, BM, LANC, YRK, det. A.R.-Smith), where it was collected during the 1890s. This is presumably the plant referred to by Hodgson under *Euphorbia cyparissias*. It may be the first British and only English record.

(C)

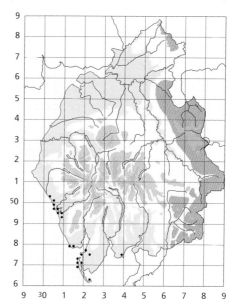

Map 550. *Euphorbia portlandica*

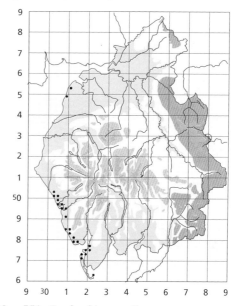

Map 551. *Euphorbia paralias*

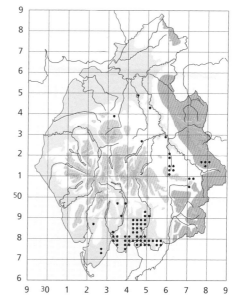

Map 552. *Rhamnus cathartica*

*E. esula L.

Seen only on the roadside near Inglewood Bank, Penrith (52.34, RWMC, 1978, LANC, det. A.R.-Smith), from where it was recorded by Hodgson, and from shingle by the Birk Beck, a tributary of the River Lune north of Tebay (58.06, GH & CW, 1975, LANC, det. A.R.-Smith).

Hodgson also lists it from Workington (9.2) and Wilson from Kirkby Stephen (7.0) but in the absence of specimens the identity of these plants must be considered uncertain.

Alt. limit: 190 m, Inglewood Bank (52.34) and Birk Beck (58.06)

          2   WC

E. cyparissias L.   Cypress spurge

Seen during the Survey only at Ramsden Dock, Barrow-in-Furness (20.68s, B. Fisher, 1982), by a car-park at Askam-in-Furness (20.76, PB, 1990, LANC), the verge of the A6 at Hale, Milnthorpe (50.78s, GH, 1980, LANC), where it grows on the opposite side from and some distance north of the parent colony which is outside a garden, by a track near Dundraw (18.48, MP, 1987, LANC, *Watsonia* (1988) 17:190) and at Grune Point (12.56, A. Cannell, 1991).

It was recorded by Hodgson from High Lorton (1.2), Seatoller (2.1), Ulpha (1.9/2.9), Beckermet (0.0) and Aspatria (1.4). As mentioned above, his record from Egremont probably refers to the hybrid with *Euphorbia waldsteinii*. Baker has additional records from Whitbarrow (4.8s) and "Jackland Tarn", Ulverston (*cf.* 2.7) and Wilson two, from Lyth (4.8s, Anon., KDL) and the railway bank at Appleby (6.2, A. Sprott, 1935, YRK). There are also records made by the Barrow-in-Furness N.H.S. during the 1950s from the mid-Walney Island (1.6s), Rampside (2.6s) and Dalton-in-Furness (2.7) squares.

          5   WFC

*E. amygdaloides L.   Wood spurge

A species of southern England seen only amongst scrub by a lane at Levens (48.86s, MB, 1995; LANC) but not far from houses. Baker lists it from woods near Milnthorpe (4.7s/4.8s) and there is a recent record from the Lancashire part of the Arnside square (4.7s, Livermore & Livermore (1988).

          1   W

RHAMNACEAE

Rhamnus cathartica L.   Buckthorn

Frequent on the wooded limestone hills of south Westmorland, where it fruits abundantly, and sparingly on the limestone in the north. Elsewhere it is very rare. Like *Euonymus europaeus* and *Acer campestre,* it is a southern, warmth-demanding shrub which in south Cumbria reaches the limit of its native distribution in western Britain.

Alt. limit: 300 m in Helbeck Wood (78.16) and Swindale, Brough (80.16)

Map 552   71   LWFC

Frangula alnus Miller   Alder buckthorn

Locally frequent in the south, rare elsewhere. Most of the localities are on lowland mosses such as Hale Moss (50.76s) where it is the dominant undershrub beneath the pines. It has been reported from hedgerows in the north and there are recent plantings in the south along the A590 west of Levens (44.84s). It appears to tolerate a wide range of soil pH and moisture, from the very calcareous conditions on Hale Moss to the acid peat of Nichol's Moss (42.82s), across the Kent estuary, and the Solway mosses in the north, and from the water's edge on the Windermere islands (38.96) to the well-drained limestone slopes of Arnside Knott (44.76s).

Map 553   62   WFC

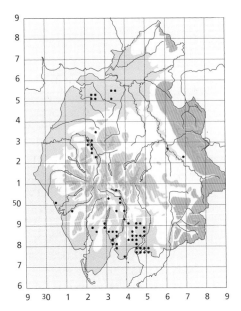

Map 553. *Frangula alnus*

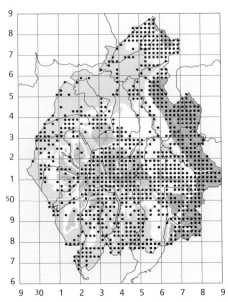

Map 554. *Linum catharticum*

VITACEAE

*__Parthenocissus quinquefolia__ (L.) Planchon

Virginia creeper
A familiar urban creeper and garden throw-out found
in 1978 well established on the railway embankment at
Caldewgate, Carlisle (38.54, REG, LANC, *Watsonia*
(1980) 13:134). It has since disappeared following
disturbance of the site, but it was subsequently recorded
in Furness at a woodland edge, since destroyed, north-
west of Dalton-in-Furness (20.74, PB, 1987, LANC),
also in a nearby quarry (22.74, PB, 1986), on the beach
between Newbiggin and Aldingham (26.68s, PB, 1989)
and at Salthouse Pool, Barrow in Furness (20.68s,
CEW, 1987).

5 F(C)

LINACEAE

Linum bienne Miller
Known only from the record given by Hodgson from Shawk
quarries, Curthwaite (3.4) and one from Greenroad station,
Millom (1.8, M. Cross, 1907).

This predominantly coastal species of southern
England reaches its northwestern limit on the Isle of Man.

(C)

*__L. usitatissimum__ L. Flax
Seen only on waste ground in Carlisle, (38.54, 38.56,
40.54, REG, 1978, 1989), where it was growing in
some quantity, on disturbed roadside verges to the
north near Smithfield (44.64, GH, 1979, LANC) and
Brunstock (42.58, PB, 1992), by Ullswater (46.22, MG
& JP, 1996), and, in the south, at The Green, Millom
(16.84, JA, 1996) and by Killington reservoir (58.90,
CEW, 1990, LANC).

Flax was fairly widely cultivated during the
nineteenth century, particularly near the Cumberland
coast, and several mills were devoted to its processing.
Although it persisted for some time as a casual, it is
now all but extinct but there are signs of its renewed
cultivation. Hodgson gives no individual localities but
there is a record from Whinseyrigg, Abbeytown (1.5,
E.J. Glaister, 1877, CLE) and later ones for Grinsdale,
Carlisle (3.5, T.S. Johnstone, 1901, CLE), Workington
(9.2, J. Curwen, undated, CLE), Silloth (1.5, C.W.
Muirhead, 1932) and Melmerby (6.3, W.W. Mason,
1916). Baker gives records for Newland, Ulverston
(2.7/3.7s), Arnside (4.7s) and Cliburn (5.2) and Wilson
lists Gilpin Bridge, Levens (4.8s) and Crosby Ravensworth (6.1). It was seen by W.H. Pearsall in 1913 at
Barrow-in-Furness docks (1.6s/2.6s, *BEC Rpt* (1914)
3:402) and during the 1950s the Barrow N.H.S.
recorded it from several places in the Dalton-in-Furness
(2.7), Grizebeck (2.8) and Cartmel (3.7s) squares.

8 W(F)C(Y)

L. perenne L. Perennial flax
(*L. anglicum* Miller)
Extremely rare, known only from five tetrads in the

Shap (5.1), Crosby Ravensworth (6.1) and Kirkby
Stephen (7.0) squares. Four are in unimproved limestone grassland and on rock outcrops, the fifth is in the
cutting of a disused railway, probably the site shown by
M. Mason to G.C. Druce in 1930 (*BEC Rpt* (1931)
9:339). Unfortunately, during the period of the Survey
it has disappeared from this last site following scrub
invasion and also from one in the Shap square. The
remaining three populations are small, overgrazed and
vulnerable, fluctuating precariously from year to year.

This is a scarce species of eastern England, from
Cambridgeshire to Co. Durham. The only sites west of
the Pennines are the Cumbrian ones and one in Kirk-
cudbrightshire.
Alt. range: 240 m, near Shap (5.1) - 340 m, near
Crosby Ravensworth (6.1)

S 5 W

L. catharticum L. Fairy flax; Purging flax
Frequent, particularly in the south and east, in short,
base-rich grassland, also in the hills on cliff ledges and
in flushes. In the west it is characteristic of abandoned
railway lines. As in the Isle of Man (Allen 1984), it is
certainly not confined, as asserted in so many Floras, to
dry habitats.
Alt. limit: *c.*750 m, Great Dun Fell (70.30)

Map 554 948 LWFCY

Radiola linoides Roth Allseed
A rare, inconspicuous but locally plentiful plant of
seasonally damp, sandy and open habitats, usually by
the coast, and now almost entirely restricted to the
south-west.

It was once appreciably more widespread being
recorded by Hodgson from several localities from all of
which it has now disappeared: Harrington (9.2), Sellafield and Egremont (0.1), Cleator Moor (0.1), Keswick
(2.2), Birker Moor (1.9), Salta Moss (0.4) and Little
Salkeld (5.3), and W.W. Mason lists it from between
Armathwaite and Cotehill (4.4). There are later records
from Silloth and the pond at Newby West (respectively
1.5, 3.5, T.S. Johnstone, 1904, 1899, CLE), Carlisle
Castle walls (3.5, D. Stewart, 1928), Kirksanton (1.8,
M. Cross, 1908), Haverigg (1.7, J. Parkin, 1936) and
Drigg dunes (0.9, E.M. Satow, 1954).

For Westmorland, Baker gives records from
Clifton and Cliburn (5.2) and Foulshaw Moss (4.8s)
and Wilson one from Brougham (5.2). It was seen on
Lazonby Fell (5.3) by G.A.K. Hervey from 1948-1956
but it has probably disappeared from there following
afforestation. There are also post-1930 *Atlas* records
from the Hawkshead (3.9) and Alston (7.4) squares.
This last seems highly unlikely in view of the altitude,
at least 260 m.

This dramatic decline is part of the national
picture. The reasons for it, at least in Cumbria, are far
from obvious, damp, open, sandy habitats still being
frequent around the entire coast.

Map 555 10 (W)FC

POLYGALACEAE

Polygala vulgaris L.   Common milkwort
Fairly frequent in short, neutral or basic grassland but it is likely that the species is somewhat over-recorded, recorders having missed the opposite basal leaves which characterise *Polygala serpyllifolia.*

No reliable information is available on the occurrence of subsp. *collina* (Reichb.) Borbás.
Alt. limit: at least 550 m, east of Warcop Fell (80.20)
Map 556   445   WFCY

P. serpyllifolia Hose   Heath milkwort
Very common in all the upland areas but apparently less frequent than the previous species in the lowlands. It is a plant of rather short, acid grassland and of the drier areas of lowland heaths and upland moors where it is commonly associated with *Calluna vulgaris, Agrostis capillaris, Deschampsia flexuosa, Potentilla*

erecta and *Galium saxatile.*
Alt. limit: 850 m, Cross Fell (6.3, Baker)
Map 557   792   WFCY

P. amarella Crantz
Extremely rare, known only from a site near Orton (6.0) where it was first discovered by R. Clarke in 1965 (BM), and nearby in the Crosby Ravensworth square (6.1). At both sites it grows in short, damp limestone grassland. The populations vary in size from year to year and the species may be decreasing. With its intensely blue flowers it is a most attractive plant.

Prior to its discovery in Cumbria it was thought to be restricted in Britain to the chalk of the north Downs, the limestone of the Craven area of Yorkshire and the sugar limestone of upper Teesdale.
Alt.: 290 m, near Orton (6.0), 305 m near Crosby Ravensworth (6.1)
                    R 2   W

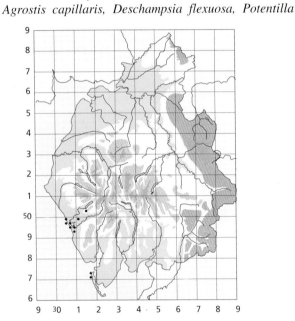

Map 555. *Radiola linoides*

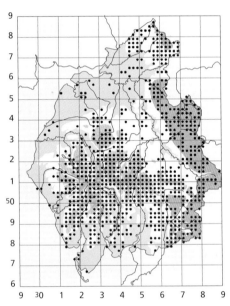

Map 557. *Polygala serpyllifolia*

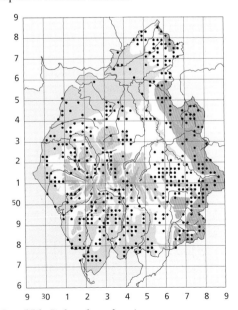

Map 556. *Polygala vulgaris*

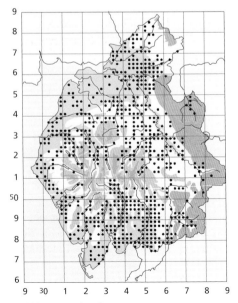

Map 558. *Aesculus hippocastanum*

STAPHYLEACEAE

*Staphylea pinnata L.   Bladdernut

Recorded during the Survey only from a hedge near Broughton-in-Furness (22.90, MMG, 1988, LANC), where it was seen by E. Hodgson in 1965, and by a roadside plantation in Storth, Arnside (46.78s, B. Fisher, 1991).

It was earlier listed by Baker from a roadside at Rydal (3.0), in woods at Finsthwaite (3.8) and hotel grounds at Furness Abbey (2.7), and by Hodgson from Whicham (1.8), Gilgarran (0.2) and Armathwaite (5.4). Presumably all the records are of plantings.

                2   WF(C)

HIPPOCASTANACEAE

*Aesculus hippocastanum L.   Horse chestnut

A very commonly planted amenity tree, characteristic of open spaces, parks and roadsides in towns and villages, also of country estates. Natural regeneration appears to be rare. Despite the considerable age of many of the trees there appear to be no published records prior to the *Atlas*.

Alt. limit: 505 m, Ashgill, Garrigill (76.38)

Map 558   649   LWFCY

ACERACEAE

*Acer platanoides L.   Norway maple

Scattered but not uncommon in south Westmorland and in the Ullswater area, occuring almost exclusively as a hedgerow, roadside or parkland tree, strikingly crimson coloured in the autumn and with very conspic-uous yellow-green flowers in spring. There are a few records of natural regeneration.

Although there are trees which must be over 60 years old, Wilson does not mention this species. The first Furness record appears to be from near Wray (38.98, K.G. Messenger, *c*.1960) and for Cumberland from Naworth (5.6, G.A.K. Hervey, 1945). It is not possible to identify the first Westmorland record.

Alt. limit: 340 m, south-east of Alston (72.44)

Map 559   128   WFCY

A. campestre L.   Field maple

Well scattered throughout the lowlands but probably native only in the south of the county, perhaps as far north as the north Westmorland limestone. Whether native or not most of the records are from hedgerows and probably represent plantings. Interesting records of possible native occurrences in the north are from by the River Lyne near Kirklinton (4.6, T.S. Johnstone, 1900), on south-facing sandstone cliffs by the River Eden at Rockcliffe (34.60, EHR, 1989) and on a limestone crag at Blindcrake (14.34, RS, 1979).

Alt. limit: at least 180 m, Tebay (60.04)

Map 560   84   WFC

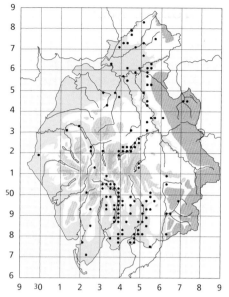

Map 559. *Acer platanoides*

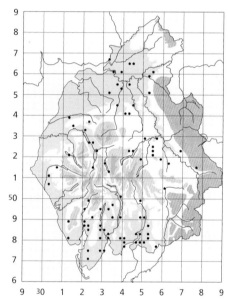

Map 560. *Acer campestre*

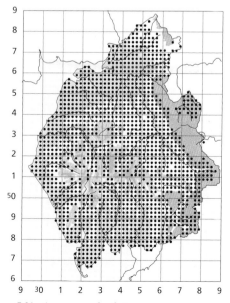

Map 561. *Acer pseudoplatanus*

*A. pseudoplatanus L.   Sycamore
A very common, fast-growing tree, widely planted in towns and villages, roadsides and woodland, freely regenerating and thoroughly naturalised. Few areas of deciduous woodland now lack sycamores and in many semi-natural woods on both base-rich and siliceous soils it is competing all too successfully with the native oak and ash. Sycamore, oak and beech are the only broadleaved trees that are grown commercially in Cumbria to any significant extent. The wood is used for veneers, planking and turnery.

In the valleys of the Lake District, and elsewhere in the uplands, the sheltering clumps of sycamores around farmsteads are now an indispensable and conspicuous feature of the landscape. It is, however, not very successful in colonising sheltered rock-ledges and gullies on the fellsides. On the coast there is no other broad-leaved deciduous tree so successful at resisting exposure and surviving the salt-laden gales.
Alt. limit: 580 m, Dowgang Hush, Nenthead (76.42)
Map 561   1486   LWFCY

*A. rubrum L.
There is a record of a single tree, presumably planted, growing by the River Brathay near Skelwith Fold (34.02, R. Young, det. K, 1969). It is a native of eastern North America.

1   F

ANACARDIACEAE

*Rhus typhina L.   Stag's-horn sumach
A single bush was seen by a hedge north of Burton-in-Kendal (52.76s, CEW, 1991, LANC) and there is a record the same year from the north on derelict land near Mealsgate (20.42, MP, *Watsonia* (1992) 19:144).

2   WC

OXALIDACEAE

*Oxalis corniculata L.
A rather rare garden weed but apparently spreading. Hodgson gives only a single record, from Carlisle (3.5/4.5), for this southern European species, but there are later ones from Wigton (2.4, J. Parkin, 1935), Great Salkeld (5.3, H. Britten, 1900-1908), Thwaites, near Millom (1.8, M. Cross, 1907) and Bellevue, Carlisle (3.5, E. Oliphant, 1946). The *Atlas* shows a pre-1930 dot for the Dalton-in-Furness square (2.7) and it was reported from the Furness peninsula (2.6s, 2.7, 2.8) by the Barrow-in-Furness N.H.S. in the 1950s. The first Westmorland record dates from 1974 when it was seen at Levens Hall (44.84s, J. Hewson).
Alt. limit: 190 m, Unthank End, Skelton (44.34)
Map 562   32   WFC

*O. exilis Cunn.
This Australasian species has been recorded only four times: from a farmyard near Whicham (14.84, CFS & JS, 1984), a gravel drive at Maryport (04.36, MMM, 1988, LANC), north-east of Oxen Park (32.88, CEW, 1990, herb. C.E.W.) and on a Kendal pavement (50.92, CEW, 1993, LANC).

Although superficially similar to *Oxalis corniculata,* it can easily be recognised by its smaller leaflets (3-5 mm), solitary flowers and shorter capsules (5-7 mm).

4   WFC

*O.stricta L.
(*O. europaea* Jordan)
An occasional garden weed, chiefly in the Grasmere, Windermere and Kendal areas.

Young (1958) gives three Cumbrian records for

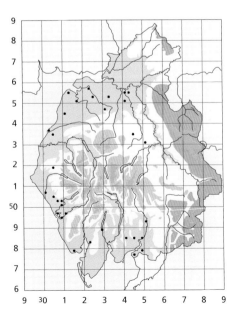

Map 562. *Oxalis corniculata*

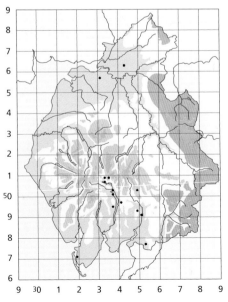

Map 563. *Oxalis stricta*

this species: one dated 1867 from Ulverston (2.7, E. Hawks, CGE), a 1915 record from The Craig, Windermere (*cf*.4.9, F. Long, BM) and a later one from Near Sawrey (3.9, M.R. Gilson, 1956, herb. D.P. Young), the last site being mistakenly included under Cumberland. These records are probably all from gardens as is that of J. Lund's from Wray Castle (3.0, 1946, FBA). Prior to the Survey, the only Cumberland records are ones from Thwaites, Millom (1.8, M. Cross, undated) and Great Salkeld (5.3, G.A.K. Hervey, 1956).

The specimens from Near Sawrey, Brathay (36.02, R. Young, 1969, AMDE) and from near Kendal (48.92, R. Young, 1971) are of the anthocyanic var. *rufa* (Small) D.P. Young.

A native of North America and eastern Asia, this species differs from *Oxalis corniculata* in having non-rooting, erect stems, erect fruiting pedicels and glabrous capsules.

Alt. limit: *c*.350 m, west of Kendal (48.92)

Map 563   14   WFC

O. acetosella L.   Wood-sorrel
Very common practically everywhere, though most typical of damp, peaty banks and hedgerows, acid woodland, both deciduous and coniferous, shaded cliff ledges and gills. It sometimes occurs in quite open situations and indeed it is surprisingly a very characteristic plant of dry limestone scree, associated with *Teucrium scorodonia* and *Geranium robertianum*.

There are several Survey records of the lilac-flowered variant noted by both Baker and Hodgson.

Alt.limit: *c*.900 m, Scafell Pike (2.0, Hodgson)

Map 564   1514   LWFCY

*O. incarnata L.
Known only from a single collection from Boothby, near Brampton (5.6, B.H.S. Russell, 1957, det. D.P. Young, *Proc. BSBI* (1959) 3:294).

This is a native of South Africa and has an erect, leafy stem and pale lilac flowers.

(C)

GERANIACEAE

*Geranium endressii Gay
Although less common than *Geranium phaeum*, this attractive species is the second most commonly naturalised geranium in the county. Surprisingly there are no records of it in the Floras. The first for Cumberland are the post-1930 *Atlas* ones for the Threlkeld square (3.2) and a 1952 collection from Watermillock (4.2, N.M. Stalker, CLE). The first for Westmorland is probably that from Skelwith Bridge (3.0, undated but pre-1950, herb. L.B. Hall, BM), and for Furness probably that from a marshy field at Pull Wyke, Brathay (36.02, H. Robertson, 1979).

Alt. limit: *c*.300 m, Leadgate, Alston (70.42)

Map 565   30   WFC

*G. endressii x G. versicolor
G. x oxonianum Yeo
The only definite records are Wilson's from between Troutbeck and Low Wood (3.0/4.0) and Survey ones from the roadside between Wood Broughton and Haverthwaite (36.80, KR, 1979, LANC), from where it has since disappeared, by the roadside between Wreay and Dacre (44.24, CEW, 1991, LANC), Spital, Kendal (52.94, CEW, 1991, LANC) and Storth, Arnside (46.78s, REG, 1992, LANC); all have been determined by P.F. Yeo. The Storth plant is the distinctive narrow-petalled var. *thurstonianum*.

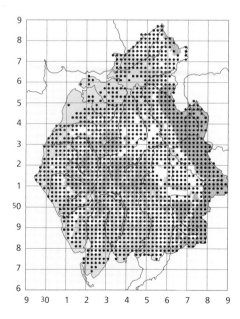

Map 564. *Oxalis acetosella*

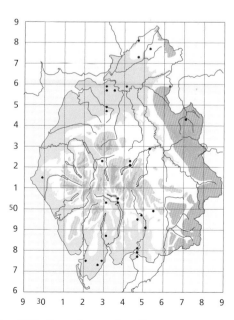

Map 565. *Geranium endressii*

This hybrid is said to be often commoner than the parents and frequently confused with them. It is best distinguished by the petals, which are of a paler pink than *Geranium endressii*, more emarginate and with the dark veins of *G. versicolor*.

4   WFC

*G. versicolor L.
A widespread garden escape in southern and south-western England but in Cumbria known only from Hodgson's localities at Flimby (0.3, C.H. Wright, 1837, CLE) and Floshgate, near Pooley Bridge (4.2, W. Hodgson, 1883, CLE), and W.H. Pearsall's record from Haverthwaite (3.8s, 1913, BM, YRK). All three collections have been determined by P.F. Yeo. Hodgson comments that it occurs in "several cottage gardens" in the Ullswater area. There are no substantiated Survey records.

(FC)

*G. nodosum L.
The only Survey record is from Hutton-in-the-Forest (46.34, 1979, RS, herb. R.S.) where it is naturalised in estate grounds.

This garden escape is recorded by Baker from the north side of Ullswater (4.2) and from Thirlmere (3.1). Hodgson makes no mention of the latter and equates the former with the Floshgate record of *Geranium versicolor*. Wilson's only records are from Casterton, Kirkby Lonsdale (6.8s) and Skelwith (3.0). The only definite record from Furness is from Kents Bank, Grange-over-Sands (3.7s, W.L. Smith, 1913, LANC).

The petals of this species are deeply emarginate as in *G. versicolor* but they are light pink or violet. In addition the leaves are frequently only 3-lobed, the lobes being less deeply toothed and incised.

1   (WF)C

[**G. rotundifolium** L.
Reported only from Quarry Flat, Holker (34.78s) by the Barrow-in-Furness N.H.S. in 1949, and the Grange-over-Sands N.H.S. in 1962, also from Humphrey Head (38.74s) by K.G. Messenger *c*.1960. These are both limestone sites.

This species is native only in the southern half of England and Wales and in the absence of specimens its occurrence in Cumbria is best considered doubtful.

(F)]

G. sylvaticum L.   Wood cranesbill
Frequent, especially in the uplands, but with a distinct eastern tendency. It occurs often in profusion along road and lanesides and no other species so epitomises early summer in the dales. Although a typical hay-meadow species, it has now largely disappeared from such habitats as a result of the excessive use of fertilisers and the change to silage. It occurs occasionally in relict fragments of ungrazed damp woodland and quite commonly at higher altitudes in moist, sheep-free gullies, in a rich-herb flora which includes *Alchemilla glabra*, *Valeriana officinalis* and *Angelica sylvestris*.

A variant (Plate 77) with petals pale pink with dark veins, var. *wanneri* Briq., was seen by a lane west of Sedbergh (60.90) and in hay-meadows near Ravenstonedale (72.02, 74.02).

This is a northern species which in Britain is almost confined to the area between Craven and the central Scottish Highlands.
Alt. limit: 870 m, Helvellyn (34.14)
Map 566, Plates 35, 42, 77   653   WFCY

G. pratense L.   Meadow cranesbill
Common throughout the lowlands occurring usually on roadside verges but also on riversides and in rough grassland, particularly on railwaysides. It comes into

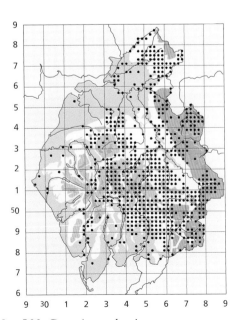

Map 566. *Geranium sylvaticum*

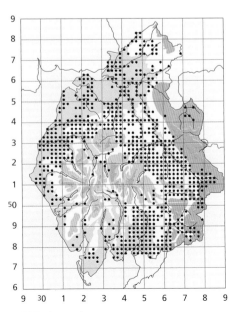

Map 567. *Geranium pratense*

flower about three weeks earlier than *Geranium sylvaticum*. Like that species it was also a frequent hay-meadow species prior to the excessive use of nitrogenous fertilisers.

Alt. limit: 375 m, above Leadgate, Alston (70.42)

Map 567   732   LWFCY

G. sanguineum L.   Bloody cranesbill
Fairly frequent around the coast, usually on sand-dunes, beaches and cliffs, being particularly abundant on the limestone at Humphrey Head (38.14) where in early summer it is spectacular. It also extends on to the limestones inland of the Kent estuary and is very common on those of north Westmorland.

The well-known variant with flowers pale pink with darker veins, var. *striatum* Weston (var. *lancastrense* (Miller) Gray), occurs in a few places on Walney Island (16.72, 18.66s) though now very much rarer than in the past. Lawson (Whittaker 1986, p.149),

for example, commented "thousands hereof I found in the Isle of Waney ...". There is a record from Seascale (0.0, A. Templeman, 1921, *BEC Rpt* (1922) 6:721) and it still occurs in that area on the old railway line near Beckermet (00.04, CCH, 1979). Where the variant occurs it is present at a very low frequency, about 1%, although up to 10% in one population. *Geranium sanguineum* shows quite a range of colour forms on Walney Island even in areas where the extreme form, var. *striatum*, is absent (Plate 78). There are no records of white-flowered specimens.

This is quite a popular garden plant and it is possible that some of the Survey records may be of naturalised populations.

Alt. limit: 335 m, Oddendale (58.12)

Map 568, Plates 11 & 78   97   WFC

G. columbinum L.   Long-stalked cranesbill
Fairly frequent in dry, open limestone soils around the

Plate 77. *Geranium sylvaticum* var. *wanneri*

Plate 78. *Geranium sanguineum*: normal flowers and var. *striatum* (arranged)

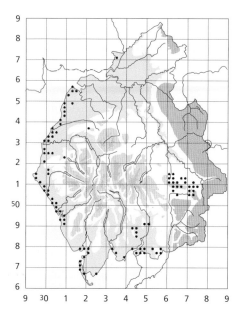

Map 568. *Geranium sanguineum*

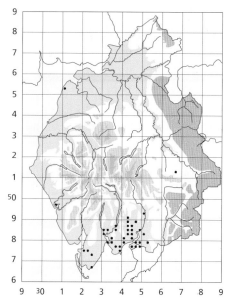

Map 569. *Geranium columbinum*

head of Morecambe Bay but otherwise rare and now far less widely distributed than the *Atlas* records suggest. Some of these records may be errors for *Geranium dissectum*; one, for example, for the Alston square (7.4, G. Bolam, 1924) seems very unlikely for this southern species.

Alt. limit: 210 m, Great Asby (66.12)

Map 569   39   WFC(W/Y)

G. dissectum L.   Cut-leaved cranesbill

Frequent in the lowlands on roadside verges and hedgebanks, in quarries and open woodland, and in more disturbed and cultivated sites such as tips and allotments.

Alt. limit: *c*.300 m, north of Alston (70.48)

Map 570   563   WFC

*G. ibericum Cav. x **G. platypetalum** Fischer & C. Meyer

G. x magnificum N. Hylander

An arresting garden escape, first noted, though apparently long established, in 1987 in a quarry near Kirkby Stephen (74.04, JF, herb. J.F.). It was found later that year in Cumberland on a roadside near Great Salkeld (52.34, RWMC, LANC) and in Furness at Thwaite Flat, Dalton-in-Furness (20.74, PB, herb. P.B.).

Alt. limit: 325 m, near Kirkby Stephen (74.04)

Map 571   12   WFC

*G. collinum Stephan ex Willd.

A very rare alien and native of southern Russia and Romania collected in a meadow at Caldbeck in 1949 (3.3, Mrs Ashridge, det. A.J. Wilmott, P.F. Yeo, CLE).

      (C)

*G. pyrenaicum Burm. f.   Mountain cranesbill

A rare but widely scattered plant of waste places and roadside verges. Most of the records, including the pre-Survey ones, are from the west coast. The Silloth (10.52) population is particularly large and persistent though vulnerable to redevelopment schemes. The first Furness records are those of G. Wilson's for the mid-Walney Island (1.6s) and Dalton-in-Furness (2.7) squares shown in the *Atlas* and dating from 1954.

Alt. limit: 140 m, Brampton, Appleby (68.22)

Map 572   22   WFC

G. pusillum L.   Small-flowered cranesbill

Rather rare and elusive and, on account of its similarity with *Geranium molle*, possibly under-recorded. The records are mostly from disturbed and cultivated ground and they represent the north-western limit in Britain for this south-eastern species.

Alt. limit: 105 m, west of Langwathby (54.32)

Map 573   38   WFC

G. molle L.   Dove's-foot cranesbill

A rather frequent lowland species, much commoner than *Geranium pusillum*. It occurs in a very similar range of habitats although frequently also in open,

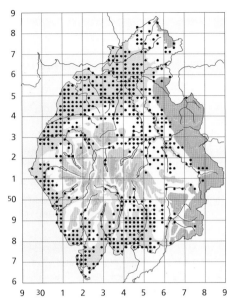

Map 570. *Geranium dissectum*

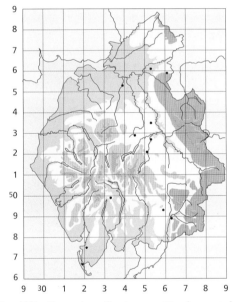

Map 571. *Geranium ibericum* x *G. platypetalum*

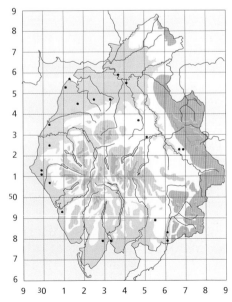

Map 572. *Geranium pyrenaicum*

shallow limestone soils and disturbed grassland.
Alt. limit: 550 m, Moor House, Teesdale (74.32)

Map 574   730   LWFCY

G. lucidum L.   Shining cranesbill

A locally common plant of roadside banks and limestone outcrops and particularly of limestone and
mortared walls, also on river shingle at Cummersdale
(38.52) and railway tracks at Tindale (60.58). Although
generally widespread in the lowlands, it is curiously
absent from apparently suitable limestone sites near the
west coast. The shiny leaves, turning scarlet in late
summer, elicited Sir J.E. Smith's description of it as "a
brilliant ornament to the romantic dales of Westmore-
land".

Alt. limit: 610 m, Melmerby High Scar (64.36) and
Great Dun Fell (70.30)

Map 575   482   LWFCY

G. robertianum L.   Herb Robert

Very common almost everywhere, absent only from the
highest ground, upland peat mosses and the most
intensively farmed areas. It occurs chiefly by hedgerows
and woodland margins, in damp woodland, on wall-
tops, on limestone pavement and scree, where it is often
abundant, and on coastal shingle.

Baker (1956) gives a description of subsp. *maritimum* (Bab.) H.G. Baker. This has a wide coastal distribution in Britain, chiefly on shingle beaches. It is
characterised by procumbent or ascending flowering
stems and fruits which are glabrous, or almost so.
However, the only Cumbrian specimen which he cites
is one collected by R.S. Adamson in 1919 at Slack
Head, Arnside (4.7s, BM), an inland site. The specimen

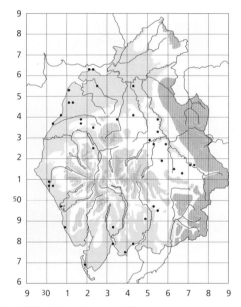

Map 573. *Geranium pusillum*

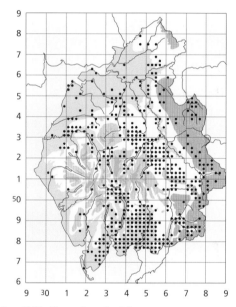

Map 575. *Geranium lucidum*

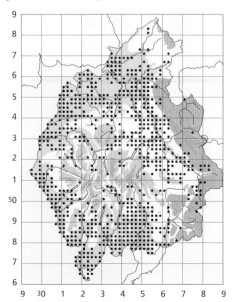

Map 574. *Geranium molle*

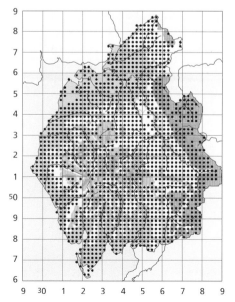

Map 576. *Geranium robertianum*

has not been traced.
Alt. limit: *c*.700 m, Great Dun Fell (70.30)
Map 576   1556   LWFCY

*G. phaeum L.   Dusky cranesbill
A rather rare but persistent garden escape, occurring
usually on roadsides and seldom far from houses.
Alt. limit: 200 m, near Crosby Ravensworth (62.14)
Map 577   33   WFCY

[Erodium maritimum (L.) L' Hér.   Sea storksbill
Both Baker and Hodgson give a record from St Bees (9.1) but
regard it as an error for *Erodium cicutarium*. Petty (1895
p.56) puts the same interpretation on records from Bardsea
and Cark (3.7s). In the absence of any authenticated record
this species is best omitted from the Cumbrian list although it
occurs at several sites on the Isle of Man and on the Mull of
Galloway, its northern limit on the west coast.
                                              (FC)]

Erodium moschatum (L.) L' Hér.   Musk storksbill
Of the two records cited by Baker, that from "upland pastures
in High Furness" is no doubt erroneous; the other is from
near Ambleside (3.0). There is also a collection from
Bowness-on-Windermere (4.9, C.A. Stevens, 1837, CLE) and
Hodgson mentions it as "reported as growing wild in
Borrowdale" (2.1/2.2). Wilson's only locality is Biggar Bank,
Walney Island (1.6s). There is a 1935 record from Siddick,
Workington (0.3), where it was seen by J. Parkin, and a 1958
one from the Ravenglass square (0.9, Mrs G. Satow).
                              S   (WFC)

E. cicutarium (L.) L' Hér.   Common storksbill
Scattered in open sandy ground along most of the
coasts. It is frequent east and south-east of Carlisle on
bare, sandy ground and it extends far up the Eden
valley. There are also scattered records from the lime-

stone north of the Kent estuary, and from the Lake
District, including ones from roadside grit on the Kirkstone Pass (38.06, 40.06).

Buchanan (1996) has examined populations on
sand-dunes in the south-west of the county and found
the plants to be very variable in respect of size,
glandulosity and flower colour. Small plants have in
the past been refered to subsp. *dunense* Andreas.
Alt. limit: 420 m, Kirkstone Pass (40.06)
Map 578   146   WFC

E. lebelii Jordan
Locally frequent to rare in open ground on sand-dunes
at Drigg (04.96, 04.98, 06.94, 06.96), Haverigg (12.78,
14.78), Sandscale (18.74, 20.74), North Walney (16.70,
16.72) and Rampside (24.66s).

All the populations have been studied by Buchanan (1996). The plants are densely glandular, with
pale pink or, very rarely, white flowers, purplish
anthers (yellow in *Erodium cicutarium*) and the mericarps have the apical pit not separated by a prominent
groove and ridge as in that species.
Map 579   11   FC

LIMNANTHACEAE

*Limnanthes douglasii R. Br.
Several plants of this North American garden plant
were found established on the upper salt-marsh at
Sandscale in 1992 (20.74, PB, LANC) and subsequently on dredged silt at Salthouse, Barrow-in-
Furness (20.68s, PB, 1993). A single plant was also
seen on shingle by the River Irthing below Birdoswald
(60.64, GH, 1994, LANC).
                              3   FC

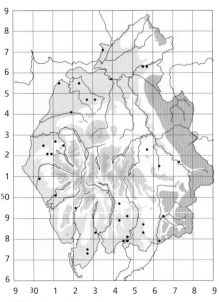

Map 577. *Geranium phaeum*

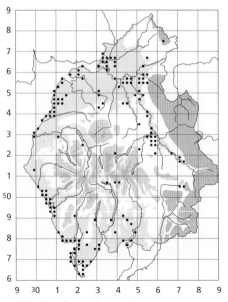

Map 578. *Erodium cicutarium*

BALSAMINACEAE

Impatiens noli-tangere L.   Touch-me-not

Locally quite frequent in the Lake District from Derwent Water and Matterdale south to Newby Bridge occurring in damp, lowland woods and usually by streams as in Stock Gill, Ambleside (38.04). It is probably commoner here than anywhere else in Britain and is probably native in Britain only in the Lake District, the Dolgellau area of North Wales and the Montgomeryshire - Shropshire border.

There are no recent records to confirm Hodgson's for Gilgarran (0.2) and Crummock Water (1.2), G.A.K. Hervey's for Buttermere (1.1, 1929) or others from Skinburness (1.5, J. Parkin, 1935), Great Salkeld (5.3, 1917), where H. Britten reported it as a garden weed, or Wilson's for Kent Force (*cf*.4.0). The *Scarce Plants* records for the Arnside square (4.7s) and the Burton-in-Kendal square (5.7s) are from v.c.60.

The largest population, 1500 plants, seen by P.E. Hatcher during his recent survey of the netted carpet moth (*Eustroma reticulata*) (Hatcher & Alexander 1994), is between Windermere and Troutbeck Bridge (40.98). This moth is now restricted to the Lake District and the larvae feed exclusively on the balsam.

Alt. limit: 125 m, Windermere (40.98)

Map 580   S 43   WFC

*I. parviflora DC.   Small balsam

Frequent in the Great Langdale - Windermere area with outlying stations north-east of Keswick (2.2), at Longtown (36.68) and by Morecambe Bay. It occurs in damp woodland with *Impatiens noli-tangere* in Stock Gill, Ambleside (38.04), by the shore of Windermere, and commonly along roadsides. It was also seen in an Ulverston timberyard (30.76s). This introduced species is probably still spreading. The many Survey records compare with the single records given by Hodgson and Wilson and three others from Booth (*cf*.3.8, J. Haines, 1925, *BEC Rpt* (1926) 7:870), a Penrith garden (5.2/5.3, J. Varty-Smith, 1911) and Skinburness (1.5, N.M. Stalker, 1948). The records for this species and the last from Skinburness could well refer to the same species.

The first Furness record is from the track between Ferry House and Balla Wray (3.9) where it was seen in 1959 by K.G. Messenger who ten years later noted that the population had spread considerably.

Alt. limit: at least 175 m, west of Killington (58.88s)

Map 581   49   WFC

*I. glandulifera Royle   Policeman's helmet

An abundant, colourful and sometimes dominant plant of riversides in many parts of the county but especially along the Eden and Eamont, the lower Caldew, between Carlisle and Cummersdale, and also the Derwent and Kent, and by lakes and rivers around Keswick and the Ullswater basin. It is very successful in colonising dredged material and silt deposits on river

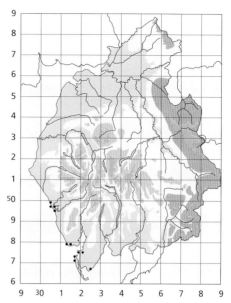

Map 579. *Erodium lebelii*

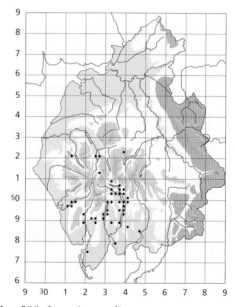

Map 580. *Impatiens noli-tangere*

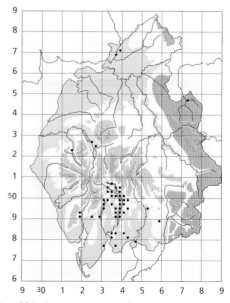

Map 581. *Impatiens parviflora*

banks. It also occurs in ditches and damp woodland and on wet roadside verges.

It is easy to forget that this familiar plant, a native of the Himalayas, was not recorded in Cumbria until 1921 when it was seen at Buttermere (1.1) by G.A.K. Hervey. The first Westmorland record is Wilson's from the River Kent between Levens and Sedgwick (4.8s, 1934, YRK), but it was not seen in Furness until 1968 when FLW noted it by the River Ea between Cark and Cartmel (34.76s). It is still evidently spreading.

The flowers vary in colour from deep magenta to almost white with pale pink the most frequent.

Alt. limit: 215 m, Tindale Tarn (60.58)

Map 582   293   LWFCY

ARALIACEAE

*Hedera colchica (K. Koch) K. Koch   Persian ivy
Found on a sandstone wall of a plantation by the Fellfoot road, Barbon (62.80s, GH, 1981, LANC, *Watsonia* (1983) 14:425) and presumably of garden origin.

                               1   W

H. helix L.
subsp. **helix**   Ivy
Common, although generally absent from the higher fells and moorlands, most noticeably in the east where it is restricted to rock outcrops and screes. It occurs most commonly on the ground below hedgerows, scrambling through hedges and as a climber on walls, rocks and trees. There is considerable variation in the extent to which the vegetative leaves are lobed.

Alt. limit: 560 m, Pavey Ark (28.08)

Map 583   1346   LWFCY

*subsp. hibernica (Kirschner) D. McClint.   Irish ivy
Fairly frequent on the Furness lowlands, elsewhere rather rare. Most of the records are from estate woodland, shrubberies and hedgerows and most probably represent plantings. The very large, broad and shallowly-lobed leaves are characteristic of the popular horticultural arboreal clone of this species. There are as yet no records of the wild plant. This is generally more similar to subsp. *helix* but is more robust, has paler, somewhat curled leaves and the hairs on the under-surface have adpressed rather than spreading rays. It is also tetraploid. It can grow in rank vegetation and is strongly oceanic, replacing subsp. *helix* in many parts of western Britain including the Isle of Man (Allen 1984). McAllister & Rutherford (1990) report it from the Arnside square (4.7s, v.c.60/69) and it should be looked for along the coasts of Morecambe Bay and south-west Cumberland.

The first record for Westmorland was from Kendal (52.92, Anon., 1970s), for Furness from Field Broughton (38.82, JA, 1982) and for Cumberland from Whins Pond, Edenhall (54.30, RWMC, 1978).

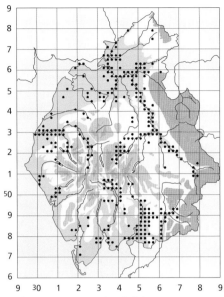

Map 582. *Impatiens glandulifera*

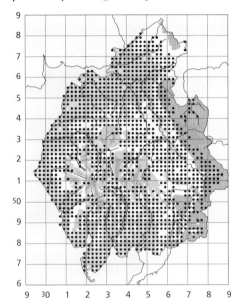

Map 583. *Hedera helix* subsp. *helix*

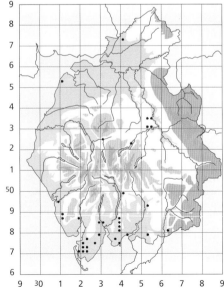

Map 584. *Hedera helix* subsp. *hibernica*

Alt. limit: 110 m, Whins Pond, Edenhall (54.30)
Map 584   34   WFC

APIACEAE
(*UMBELLIFERAE*)

Hydrocotyle vulgaris L.   Marsh pennywort
Frequent, particularly in the south and south-west, occurring, often abundantly, by pools, tarns and ditches, in swamps and marshy fields and in seasonally wet depressions as in coastal dune-slacks. By contrast it is also frequent in base-poor flushes at low altitudes in the fells, although scarce in the Pennines.
Alt. limit: 460 m, Thornthwaite (20.26)
Map 585   674   LWFCY

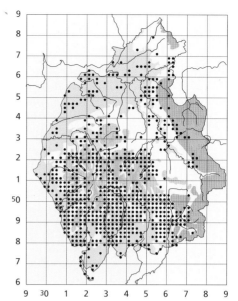

Map 585. *Hydrocotyle vulgaris*

Sanicula europaea L.   Sanicle
Locally common in usually damp base-rich deciduous woodland, especially in south Westmorland, the Eden valley, south of Wigton and, in the north-east, in the many wooded cloughs along the rivers Liddel, Lyne and Irthing. It is also characteristic of grikes in upland treeless limestone pavement as in north Westmorland.
Alt. limit: 430 m, Fell End Clouds, Ravenstonedale (72.96) and south-east of Garrigill (76.40)
Map 586   561   LWFCY

***Astrantia major** L.
An occasional and ephemeral garden escape. There are Survey records from riverside rocks at Ambleside (36.02, R. Young, 1969, AMDE), from Ackenthwaite (50.82s, J.A. Wood, 1975), Shap Abbey (54.14, CW, 1980) and Bassenthwaite (22.32, REG, 1996, LANC), and earlier ones from Brownrigg, Abbeytown (1.5, E.J. Glaister, 1877, CLE) and Great Easby, Brampton (5.6, G.A.K. Hervey, 1943).
Alt. limit: 230 m, Shap Abbey (54.14)
                     4   WW/F(C)

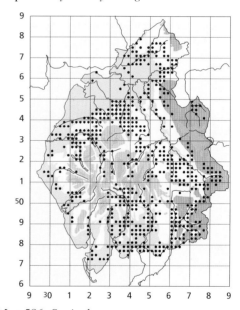

Map 586. *Sanicula europaea*

***Eryngium giganteum** M. Bieb.
An impressive garden escape which appeared from 1986 to 1993 in a field at Far Arnside (44.76s, MB).
                     1   W

E. maritimum L.   Sea holly
Frequent along the strand-line of sand and shingle beaches and characteristic of the younger dunes. It becomes quite rare north of the Solway Firth.
Map 587   41   FC

*E. x **oliverianum** D. Delaroche
A garden plant of uncertain origin found in the docks at Barrow-in-Furness in the early 1980s (20.66s, B. Fisher, det. A.O. Chater, LANC).
                     1   F

[Echinophora spinosa L.
A Mediterranean species listed by Baker from Roosebeck (2.6s), from where it was originally reported by Lawson, and

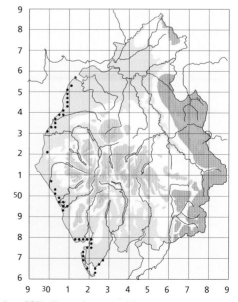

Map 587. *Eryngium maritimum*

Plate 79. *Chaerophyllum hirsutum*

from near Ulverston and Flookburgh (3.7s). These records probably refer to either *Salsola kali* or *Eryngium maritimum.*

(F)]

*Chaerophyllum hirsutum L.

An attractive pink-flowered umbellifer, a native of southern and central Europe, found well-established by a laneside wall east of Windermere (40.96, KAG, *c.*1970; LANC, det. C.A. Stace). It was apparently cultivated in the nearby garden in 1962 but was later eradicated and dumped in the lane. The only other British record is from Lanarkshire.

Plate 79   1   W

C. temulum L.   Rough chervil
C. temulentum L.
Common in the lowlands but, like other southern species, rarely getting into the dales. It is very characteristic of hedgerows and wooded margins, coming into flower as *Anthriscus sylvestris* finishes.
Alt. limit: 250 m, Swindale, Brough (80.16)
Map 588   650   LWFC(W/Y)

Anthriscus sylvestris (L.) Hoffm.   Cow parsley
A common lowland umbellifer, being particularly conspicuous along roadside verges and hedgerows in late spring but also common on waste ground, in 'unimproved' meadows, coarse grassland and woodland margins. In the fells, but chiefly in the Pennines, it occurs with other tall herbs in damp sheltered gullies and on ledges.
Alt. limit: 845 m, Great Dun Fell (70.32)
Map 589   1352   LWFCY

*A. cerefolium (L.) Hoffm.
The only records are that given by Baker from a lane near Patterdale church (3.1, 1864) and a slightly later one from Silloth (1.5, R. Wood, 1880, CLE).

(WC)

A. caucalis M. Bieb.   Bur chervil
An extremely rare and decreasing species found at only four sites during the Survey: at Rockcliffe (34.60, REG, 1979), where it is locally frequent on sandstone ledges above the River Eden, below a roadside hedge near Dundraw (20.48, MS, 1983, LANC) and, after a long search, in a laneside near Holme St Cuthbert, Mawbray (10.46, EEM, 1992, LANC), an area where Hodgson says it was the prevailing umbellifer, and in the car-park at Sandscale (20.74, PB, 1993, LANC).

Hodgson's sites are mostly along the coastal fringe: St Bees (9.1), Workington (0.2), Cockermouth (1.3), Holme St Cuthbert (1.4), upper High Pow, Wigton (2.4) and Bowness-on-Solway and Port Carlisle (2.6n). It seems to have been well dispersed throughout this area and even locally common. There are later records from Dalston (3.5, T.S. Johnstone, 1901), Cumwhitton (5.5, C. Platt, 1900-1908), Skinburness (1.5, W.W. Mason, 1921) and Dundraw (2.4/2.5, C.W. Muirhead, 1932). It was last seen at Silloth by C.W. Muirhead in 1940. There are additional post-1930 *Atlas* records for the Distington (0.2) and Dalston squares. The only record from the south of the county is an unlocalised one from the Dalton-in-Furness square (2.7, K. Jones, 1957).

4   FC

Scandix pecten-veneris L.   Shepherd's needle
Baker considered this species to be not uncommon as an arable weed and Hodgson lists it from Harrington (9.2), near Workington (9.2), Maryport (0.3), Cockermouth (1.3), Silloth (1.5, J. Leitch, 1891, CLE) and Carlisle (4.5). There are later records from by the River Eden at Grinsdale, Carlisle (3.5, T.S. Johnstone, 1900, CLE) and Silloth (A. Boyd, 1949, CLE) and two post-1930 *Atlas* dots for the west Carlisle (3.5, DAR, 1950s) and Penrith (5.3) squares.

Handley (1898) records it from the Sedbergh area (6.9) and Wilson from near the Dungeon Ghyll Hotel, Langdale (2.0), Burneside and Windermere (4.9, E. Walker, KDL), and near Tebay (6.0). The only localised Furness record, and probably the last sighting in the county, is from Sandscale, Dalton-in-Furness (1.7, K. Jones, 1960s).

This is a rapidly decreasing umbellifer of southern and eastern England.

S   (WFCW/Y)

*Myrrhis odorata (L.) Scop.   Sweet Cicely
Frequent in the lowlands and extending far into the dales. This lush, robust and eye-catching umbellifer is easily recognised by the smell of aniseed, the leaves which are usually blotched with white, and the large dark fruits. It occurs along roadsides, by walls, in damp gills and along riversides, but usually close to houses and farms, reflecting its earlier culinary and medicinal use. It is still used occasionally to sweeten sour fruit.
Alt. limit: 430 m, south of Garrigill (74.38)
Map 590   703   LWFCY

*Coriandrum sativum L.   Coriander
A single specimen was seen by Hodgson on the sandhills at Silloth (1.5) in 1890.

(C)

*Smyrnium olusatrum L.   Alexanders
An extremely rare, attractive and very rare early-flowering coastal umbellifer. During the Survey a small colony was found growing characteristically by a streamside on the coast at Anthorn (18.58, REG, 1984, LANC, *Watsonia* (1986) 16:189) and it was seen shortly afterwards on the roadside at Eskmeals (08.92, O.M. Stewart, 1986). This latter site was subsequently destroyed by road-widening. It was also noticed by a garden wall in Blennerhasset (16.40, MP, 1990).

The only previous records are one from Aspatria (1.4), regarded by Hodgson as dubious, and Wilson's from Grasmere (3.0), which seems similarly unlikely.

It is curious that these should be the only sites between Lancashire and Ayrshire when the species is so widespread on the Isle of Man.

3   (W)C

Conopodium majus (Gouan) Loret   Pignut
A very common umbellifer of roadside verges, coarse grassland, pastures and haymeadows, also in open woodland and on the fellsides in *Agrostis - Festuca* grassland and in more open areas of bracken. Although reported from the limestone, it prefers fairly deep, brown-earth soils.
Alt. limit: 600 m, south-west of Nenthead (76.42)
Map 591   1455   LWFCY

Pimpinella major (L.) Hudson

Greater burnet saxifrage
Extremely rare, seen at only five sites: by a car park on the south side of Arnside Knott (44.76s, MB, 1975), where it appears to be increasing, in the disused station yard at Kendal (50.92, CEW, 1985, LANC), by the roadside east of Lindale (42.80s, CFS & JS, 1985), on an islet in the River Lune near Lowgill (62.96, CFS & JS, 1987) and at two places on the roadside above

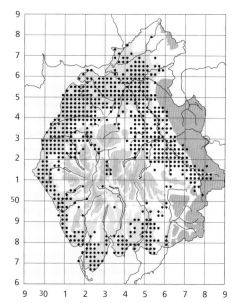
Map 588. *Chaerophyllum temulum*

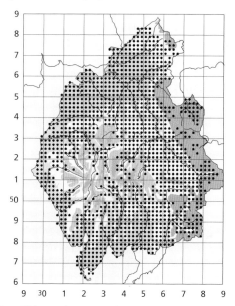
Map 589. *Anthriscus sylvestris*

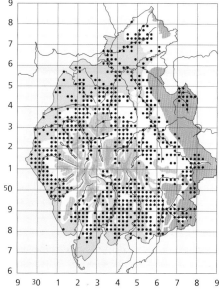
Map 590. *Myrrhis odorata*

Brough (80.14, S. Murphy, 1994, LANC). There is also a thriving population in a ditch by the A6 just outside the county boundary south of Hale (50.76s).

The only previously published records of this predominantly eastern species are the 19th century one from near Arlecdon (0.1), cited somewhat doubtfully by Hodgson, and Wilson's own from a roadside near Brough (7.1/8.1, 1935, YRK), probably the same as the Survey site. There are later Cumberland records from Great and Little Salkeld and Briggle Beck, Lang-wathby (5.3, H. Britten, 1900-1908), Drumburgh (2.5, J. Parkin, 1935) and a garden at Carlisle (4.5, D. Stewart, 1939).

Alt. limit: 235 m, east of Brough (80.14)

                         5   W(C)Y

P. saxifraga L.   Burnet saxifrage

Common on the limestones and towards the west coast but otherwise rather scattered. It is characteristic of fairly short, dry grassland on the better soils occurring on rock outcrops, hillsides, roadsides and railwaysides. It is also frequent in the fells on sheltered, somewhat base-rich ledges and in gullies.

Alt. limit: 810 m, Dollywaggon Pike (34.12, Wilson)

Map 592   543   WFCY

*Aegopodium podagraria L.

             Ground elder; Bishop's weed; Goutweed

Abundant in the lowlands, usually forming dense colonies on damp roadside verges and on waste land. It is also a pernicious weed of gardens, shrubberies and allotments. Its close association with houses and ruins is a reflection no doubt of its earlier medicinal and culinary importance, although it appears very native in several places in woodland by the lower River Eden where it is associated with *Gagea lutea* and *Chryso-splenium alternifolium*.

Alt. limit: 450 m, above Nenthead (76.44)

Map 593   1198   LWFCY

[Sium latifolium L.

A species predominantly of eastern and southern England and not surprisingly listed with misgivings by Baker from Stock-gill, Ambleside (3.0).

                                         (W)]

Berula erecta (Hudson) Cov.

             Narrow-leaved water parsnip

Present in a few localities, mostly near the coast, in the south of the county, elsewhere distinctly rare and probably decreasing, at least in the west to judge by the number of localities mentioned by Hodgson. It has probably been substantially over-recorded in the *Atlas*. It occurs in ditches and by streams, small pools and rivers. It appears to have a higher nutrient requirement than *Apium nodiflorum* with which it is often confused. A reliable distinguishing character for vegetative material is the ring mark on the petiole below the lowest pair of leaflets.

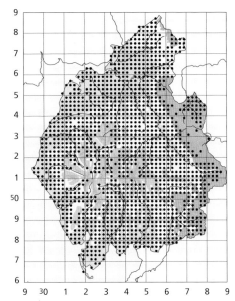

Map 591. *Conopodium majus*

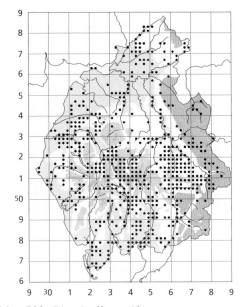

Map 592. *Pimpinella saxifraga*

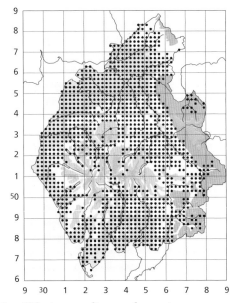

Map 593. *Aegopodium podagraria*

Alt. limit: 260 m, south of Croglin (58.44)
Map 594   25   WFC

Crithmum maritimum L.   Rock samphire
A rare plant of coastal cliffs and scars by the Kent
estuary and along the west coast from Silecroft to
Ravenglass (08.86, 08.96, 10.82, MMG) and St Bees to
Whitehaven. It is most frequent on the sandstone cliffs
of St Bees Head (9.1) and the limestone at Arnside
(44.76s) and Humphrey Head (38.74). Only three
plants were recorded between Silecroft and Ravenglass,
two on a shingle beach, the other on the sea wall.

There are also old west coast records from
"Silloth in Allonby direction" (*cf*.0.4, W.H. Youdale
(1892)) and near Coulderton (9.0, C.D. Pigott, 1944).
Baker cites Lawson's record from Dunnerholme (2.7), a
limestone promontory in the Duddon estuary, and the
Atlas has a pre-1930 dot for the South Walney square
(1.6s).

Cumbria is close to the northern limit of this
species which, like *Tilia cordata*, appears to set viable
fruit only in exceptionally warm summers.
Map 595   12   WFC

Oenanthe fistulosa L.   Water dropwort
Known only from a marshy field at the north end of
Derwent Water (24.22, P. Taylor, 1982; LANC), where
it grows with *Filipendula ulmaria, Juncus effusus,
Lythrum salicaria* and *Oenanthe crocata*.

Hodgson lists it from near Allonby (0.4) and
Blackdyke (1.5). There is a further record from Monk-
hill Lough (3.5, P. Shepherd, 1884, CLE), where it was
seen by J. Parkin in 1935. It was discovered at Temple
Sowerby Moss (6.2) in the 19th century and collected
there by Wilson in 1934 (YRK). There is a single
Furness record, cited by Baker, from Grange-over-
Sands (4.7s).

                         1   (WF)C

[O. pimpinelloides L.
Hodgson gives an 1854 Workington (9.2) record for this
southern species. This is not a plant likely to appear as a
casual and the record is best treated as an error.

                                               (C)]

O. lachenalii C. Gmelin   Parsley water dropwort
A common and characteristic species of the uppermost
salt-marshes and of ditches and brackish marshes
behind the sea-walls, growing in tall herb communities
often associated with *Lotus pedunculatus, Elytrigia
atherica* (in the south), *Juncus maritimus* and *Bolboschoenus maritimus*.
Map 596   71   WFC

O. crocata L.   Hemlock water dropwort
A generally western and south-western species in
Britain, this species is frequent to common in the lowlands, growing often very luxuriantly along riversides,
streams and ditches, in carr and marsh communities
and extending far up the dales. Like *Apium nodiflorum*,

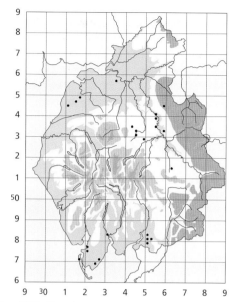

Map 594. *Berula erecta*

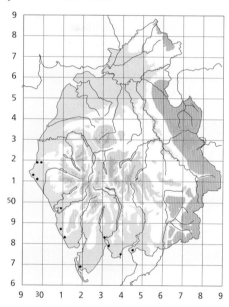

Map 595. *Crithmum maritimum*

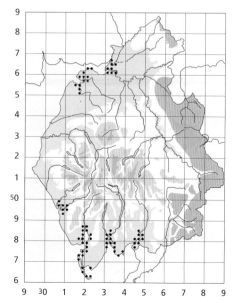

Map 596. *Oenanthe lachenalii*

it is largely absent from the northern rivers although frequent along the Eden.

Well known to be extremely poisonous, there are still instances of cattle dying through eating the succulent tubers which have been dredged out of dykes and left on the banks. Hodgson refers to them being "masticated by cattle with avidity" and he believed, probably erroneously, that Keswick derives its name from the withered stalks of umbellifers, including this species, which were known locally as keshes, a name still occasionally used.

Alt. limit: 230 m, above Crosby Ravensworth (60.12). There is an improbable record in the *Atlas* for the Alston square (7.4) which would be at least 260 m.

Map 597   648   WFCY

O. aquatica (L.) Poiret   Fine-leaved water dropwort
The only Survey find, and the only substantiated one this century, is from Irthington (48.60, CS, 1979, LANC, *Watsonia* (1980) 13:137), where it grows abundantly in the vallum of the Roman Wall. This represents the northernmost limit of the species in western Britain.

The only previous records are the four cited by Baker: from a bog near Portinscale (2.2), a stream at Grange-in-Borrowdale(2.1), Allonby (0.4) and, in the south, Brigsteer Moss (4.8s). There is also a problematical record in C.W. Muirhead's card-index "(?)Skinburness Marsh, Miss Parkin, 29/7/1935". This, and that from Allonby, may be due to confusion with *Oenanthe lachenalii* and the Portinscale record may refer to *O. fistulosa*.

1   (W)C

Aethusa cynapium L.   Fool's parsley
Surprisingly sparse, largely restricted to the west coast, the Eden valley and the south where it occurs as a weed of waste ground, gardens and allotments, but also occasionally in hedgebanks. This is a rather rare

species in Scotland and it may be that in parts of Cumbria it is only an infrequent casual.

Alt. limit: 260 m, Shap (56.14)

Map 598   232   WFCY

*Foeniculum vulgare Miller   Fennel
A very rare casual, recorded from waste ground at Hawkshead (34.96, R. Young, 1970, AMDE), Bowness-on-Solway (20.62n, 22.62n, EHR, 1976, *Watsonia* (1977) 11:395), near Wabberthwaite (10.94, REG, 1996), a roadside verge at Cummersdale (38.52, REG, 1977) and near the Concle Inn, Rampside (22.62s, B. Fisher, 1985). Most of the sites are relatively close to houses.

Hodgson lists it from St Bees (9.1), Maryport (0.3), Allonby (0.4) and Silloth (1.5) and Friend from Braithwaite station (2.2, *Naturalist* (1899) 1899:300) in 1899. It was seen at Stainton, Carlisle (3.5, T.S. Johnstone, CLE) in 1908 and half a century later at Barrow-in-Furness (1.7, K. Jones, *c.*1965), on a site which was subsequently bulldozed.

5   FC

Silaum silaus (L.) Schinz & Thell.   Pepper saxifrage
Quite common in a belt extending from the upper Eden valley northwards almost to Dalston (3.5), but virtually absent elsewhere except for a cluster of records behind Maryport (0.3) and two sites in the south where it occurs in damp woodland near Cunswick Tarn (48.92) and near Beetham (44.78s). Its usual habitat is rather coarse, base-rich or calcareous grassland and scrub.

This is a south-eastern species which reaches its northern limit in Roxburghshire.

Alt. limit: 305 m, Hardendale (58.14)

Map 599   105   WC

Meum athamanticum Jacq. Spignel; Meu; Bald money
A local plant of the Lune gorge between Sedbergh and Tebay and of the tributary valleys to the west and

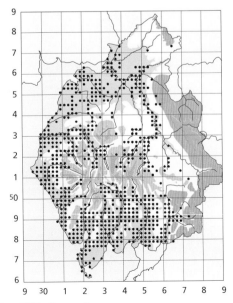

Map 597. *Oenanthe crocata*

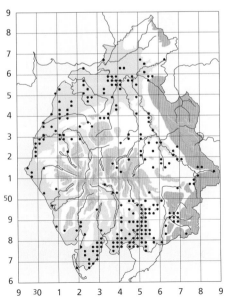

Map 598. *Aethusa cynapium*

north-west. There are a few isolated occurrences north of Kendal and in the Kentmere valley and an interesting outlier near Torver (28.94, E.M. Lind, 1967; R. Young, 1969, AMDE). Like *Conopodium majus,* it prefers deep brown-earth soils and occurs most commonly in coarse grassland by roadsides and lanes and, especially, along unimproved banks and knolls in hayfields.

Meum is of particular interest having been recorded in 1597 by Gerard from Roundthwaite, Tebay (6.0), still one of the best sites. Its British distribution is concentrated in a band from Ayrshire to Aberdeenshire. In England and Wales it is known outside Cumbria only from a few sites in Wales and one in Northumberland. Its present status in the county is fairly healthy and the Survey revealed a remarkable number of new localities although it seems to have disappeared

from the Keswick (2.2), Dunmail Raise (3.1), Ennerdale (0.1) and Barbon (6.8s) sites mentioned by Baker and Hodgson and from Wilson's at Hugill, Staveley (4.9, G.E. Martindale, 1886, KDL) and Docker (5.9). The marvellous field at Dillicar illustrated by Wilson became an M6 construction site in the early 1970s.

According to Gerard the plant was used as a diuretic. The leaves have a powerful and interesting astringent smell and it seems likely that at some time it has been used as a culinary herb. C.D. Pigott has suggested that Dillicar could be a corruption of dill acre. The fruits also are extremely pungent, smelling of curry powder. The equally strong-smelling roots of 'The Westmorland Herb' were once sold in London as a sexual stimulant.

Alt. limit: 300 m, Borrow Beck (52.04)

Map 600, Plate 80   S 23   WF(C)Y

Plate 80. *Meum athamanticum*

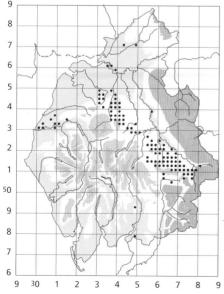

Map 599. *Silaum silaus*

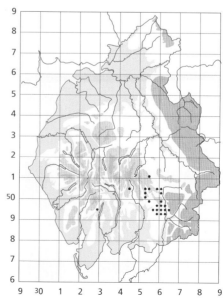

Map 600. *Meum athamanticum*

Conium maculatum L.   Hemlock
Common and very conspicuous in the north but distinctly local in the south. It is a plant of waste ground, railway sidings, roadside verges, hedges and riverbanks, often growing in profusion.
Alt. limit: 250 m, near High Ireby (22.36), near Caldbeck (28.36) and above Brough (78.14)
Map 601   304   WFC(W/Y)

*****Bupleurum rotundifolium** L.   Thorow-wax
An umbellifer of central and southern Europe listed by Hodgson from near a corn-mill at Lorton (1.2), Maryport (0.3, p.380) and Silloth docks (1.5, E.J. Glaister; J. Leitch, 1889, det. T.G. Tutin, CLE). The only other records are one from Stanwix, Carlisle (3.5/4.5, T.S. Johnstone, 1906), that cited by Wilson from Arnside (4.7s, 1894), and more recent ones from near Coniston (3.9, E.V. Watson, 1955, *Proc. BSBI* (1954) 2:32) and Askham (5.2, G.A.K. Hervey, 1956).

This species has been much confused with the following and all records unsupported by specimens are best considered doubtful.
                         R   ([WF]C)

*****B. subovatum** Link ex Sprengel   False thorow-wax
An occasional bird-sown alien and native of southern Europe recorded during the Survey mainly from gardens. The first record for Furness is from Brathay, Ambleside (36.02, R. Young, 1969, AMDE) and for Westmorland from Windermere (40.96, M. Milligan, 1971, AMDE).

There are no previously published records and the species was thought to be a relative newcomer in the county, but several specimens in CLE, collected between 1876 and 1906, have been identified as this species by T.G. Tutin, the first being from Silloth (1.5, J. Leitch, 1876).

This species differs from the last in having longer leaves, about three times as long as wide, and umbels with only two or three rays.
Alt. limit: 215 m, above Brough (78.14)
Map 602   9   WFC

[*Trinia glauca* (L.) Dumort.
Listed by Hodgson from Tallentire Hill, Cockermouth (1.3) but rightly regarded by him as an obvious error.
                                             (C)]

Apium graveolens L.   Wild celery
Very rare, seen only at a few sites in the Anthorn area (1.5, 2.5; 18.58, REG, 1983, LANC), growing usually in ditches and runnels at the top of the salt-marsh, in a similar habitat on both sides of the Leven estuary (30.78s, GH, 1986, LANC; 32.80, P. Adam, 1970s) and inland on the Solway Plain near Black Dub Beck, Westnewton (12.44, MP, 1988).

This species was once widely distributed in coastal marshes and salt-marshes, being recorded by Hodgson from Braystones and Seascale (0.0), Workington (9.2), Anthorn (1.5) and Kirkbride (2.5) and, inland, at Edenhall, Langwathby (5.3). There are later records from Monkhill Lough (3.5, D. Stewart, 1928), Anthorn (Mrs Eastwood, 1949, CLE) and Kirkbride (C.W. Muirhead, 1943). In the south, Baker lists it from Arnside (4.7s), Foulshaw and Brigsteer mosses and Milnthorpe (4.8s) and in Furness from Cark, Holker, Cartmel and Plumpton (3.7s).

The Barrow-in-Furness N.H.S. recorded it in the 1950s from several places in the Leven estuary and at Humphrey Head (3.7s), also on South Walney (2.6s) and in the Duddon estuary (2.7, 2.8). There are other post-1930 *Atlas* records for the Millom (1.8), Ravenglass (0.9), Seascale (0.0) and west Carlisle (3.5) squares.

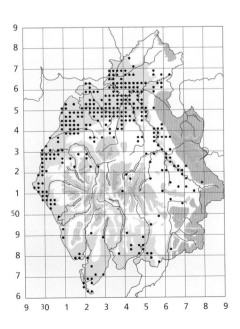

Map 601. *Conium maculatum*

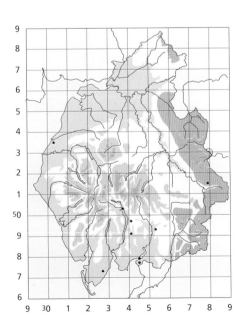

Map 602. *Bupleurum subovatum*

Coastal defence works, improved drainage and intensive grazing are no doubt responsible for the decline of this species in Cumbria. There are very few west coast records north of the Solway Firth.
Map 603   9   (W)FC

A. nodiflorum (L.) Lag.   Fool's water-cress
A frequent lowland plant of ditches and streamsides, pools and the margins of tarns. By comparison with the Eden, it is surprisingly scarce along the northern rivers Irthing, Lyne and Esk. This species is rather easily confused when vegetative with *Berula erecta*.
Alt.limit: 290 m, Mallerstang (78.04)
Map 604   283   LWFC(W/Y)

A. inundatum (L.) H.G. Reichb.
Rare in the north, fairly frequent in the south, particularly in the low hill country to the south of the Lake District. This is a plant of base-poor tarns and pools and of sheltered lake-margins, occasionally occurring in abundance.
Alt. limit: 335 m, east of Shap (58.12/58.14)
Map 605   123   WFC(W/Y)

*Petroselinum crispum (Miller) Nyman ex A.W. Hill
                                        Parsley
Extremely rare, recorded only from Rockcliffe (34.60, EHR, 1979, LANC), where it grows on sandstone ledges above the River Eden, and in the south on limestone rocks by the Kent estuary at Castlehead and Low Meathop (42.78s), where it was discovered by the Grange-over-Sands N.H.S.in 1964, and on a recently disturbed area of sand and mud by the River Crake at Greenodd (30.82, GH, 1985, LANC).

The only previous records are those given by Petty (1896 p.107) from waste ground at Barrow-in-Furness and the shore at Grange-over-Sands (4.8s) and a later one from Rockcliffe (4.6, J. Parkin, 1935).

All of the extant populations are of plants with non-crisped leaves.
                          3   WFC

*Sison amomum L.   Stone parsley
The only record for this southern species is that cited by Hodgson from near Penrith (5.3).
                                        (C)

[Cicuta virosa L.   Cowbane
Hodgson gives a doubtful record from Thursby (3.5) and a further reference to its introduction with clover around Cockermouth (1.3). For a marsh and semi-aquatic species this seems most unlikely and the records are best treated as errors although the species does occur on the north side of the Solway Firth.
                                        (C)]

*Ammi majus L.
A southern European species found only twice, at Hayton sewage works, Aspatria (10.40, AM, 1982, LANC, *Watsonia* (1983) 14:425) and in a garden at

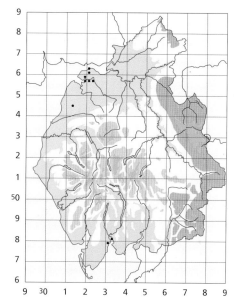

Map 603. *Apium graveolens*

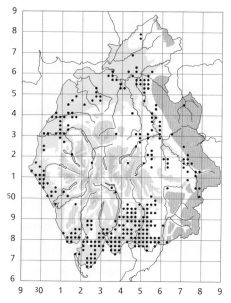

Map 604. *Apium nodiflorum*

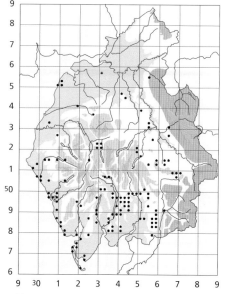

Map 605. *Apium inundatum*

Bassenthwaite village (22.32, ES, 1986, LANC). The only previous record is Hodgson's from the Derwent Tinplate works, Workington (0.2).

2   C

*A. visnaga (L.) Lam.
A native of the Mediterranean found on the Hawkshead rubbish dump in 1970 (34.96, R. Young, AMDE).

1   F

*Carum carvi L.   Caraway
Extremely rare, seen during the Survey only in a meadow at Clinthead, Great Corby (48.52, MS, 1980, herb. R.E.G., *Watsonia* (1981) 13:335), where it was originally shown to MS, on the roadside between Flimby and Maryport (02.34, A. Mitchell, *c.*1981, LANC) and by a farm road near Gilts, Crosby Ravensworth (62.10, CW, 1970s). It has never been refound at the first two sites.

Baker gives records from Shap (5.1) and St Bees (9.1) and Hodgson additional ones from Workington (9.2) and gardens at Haile (0.0) and Dean (0.2). It was also seen at Silloth in 1890 (1.5, J. Leitch, CLE) and in 1935 by J. Parkin. Wilson's only record is his own 1935 collection (YRK) from Shap.
Alt. limit: 270 m, above Crosby Ravensworth (62.10)

3   WC

C. verticillatum (L.) Koch   Whorled caraway
Listed by Hodgson from the Vale of St John (3.2) and Kingmoor, Carlisle (3.5). It was known at the latter site from 1837 until 1946 when it was seen by C.W Muirhead (CLE). There are also records from pastures on Wan Fell (5.3, H. Britten, 1900-1908) and an early 19th century one from Newby Cross (3.5). Wilson was surely right in regarding a record from Low Borrowbridge, Tebay (6.9, M.A. Rogers, 1906) as an error for *Meum athamanticum*.

This species of marshes and wet pastures is strongly

western in Britain and practically restricted in England to Cornwall. It is, however, quite frequent on the Scottish side of the Solway Firth.

(C)

Angelica sylvestris L.   Wild angelica
Very common in wet places: in ditches, tall herb communities in ungrazed fields, the margins of pools, tarns, lakes and carr. In the fells it is a characteristic member of herb-rich sheltered wet gullies.
Alt. limit: 855 m, Helvellyn (34.14)
Map 606   1397   LWFCY

*Levisticum officinale Koch   Lovage
A single plant of this rare casual was found in 1990 on the beach at Beckfoot, Mawbray (08.48, P. Tuffen, LANC).

1   C

[Peucedanum palustre (L.) Moench
Listed with reservations by Baker from west of Flookburgh (3.7s) and north of Cartmel (3.8). This is a local species of eastern England and it seems likely, as suggested by Petty (1896 p.198), that there has been confusion with *Oenanthe lachenalii*.

(F)]

*P. ostruthium (L.) Koch   Masterwort
Restricted now to Cumberland where it is rare and widely scattered. It still occurs at Rockcliffe (34.62, U.K. Duncan, 1949, CLE) but it may have disappeared during the Survey from Hodgson's locality at Dowthwaite Head, Matterdale (36.20).

Hodgson also lists it from Aira Beck near Ullswater and Watermillock (4.2), the north end of Thirlmere (3.1), Mungrisdale (3.3) and by the Irthing at Gilsland (6.6). Handley (1898) records it from Dent (7.8) and Wilson from Overend, Kentmere (4.0, J.A.

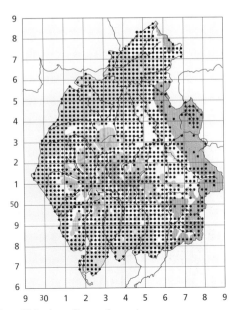

Map 606. *Angelica sylvestris*

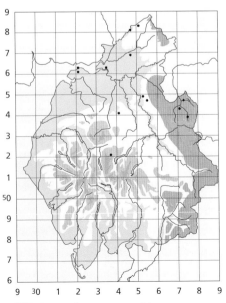

Map 607. *Peucedanum ostruthium*

Martindale, 1912, KDL) and Windermere. The only Furness record is from Ormsgill, Barrow-in-Furness (1.7, W.H. Pearsall, 1913, *BEC Rpt* (1914) 3:387).

Most sites are near houses or farms, reflecting its previous culinary and veterinary importance. At Dowthwaite Head it used to be associated with *Rumex alpinus*, which was similarly highly valued, and the two also grow together at Mungrisdale (40.40).
Alt. limit: 390 m, Leadgate, Alston (72.46)
Map 607   13   (WF)C(Y)

Pastinaca sativa L.   Wild parsnip
A species with a pronounced south-eastern distribution in Britain and probably only native in Cumbria around the Furness peninsula. It occurs in rather open sites, particularly on waste ground in urban areas. Judging from the *Atlas* it appears to be extending its range. Wilson gives no records, the first definite one for Westmorland was from Meathop in 1955 (4.8s, Grange-over-Sands N.H.S.). The first for Cumberland was from Silloth (1.5, J. Leitch, 1887) and there is an interesting pre-Survey record from Gilsland (6.6, v.c.67/70, G.A.K. Hervey, 1943).

During the Survey it was noted on a number of trunk-road verges, for example the A590 near Newby Bridge (38.84), the A591 near Kendal (48.94) and, a particularly vigorous colony, along the M6 north of Crooklands (54.82s). It was also seen on the disused railway near Penruddock (40.26). It still thrives in the Workington - Maryport area, where it was seen by Hodgson and was probably originally associated with basic slag.
Alt. limit: 380 m, Stainmore (88.12)
Map 608   72   WFC

Heracleum sphondylium L.   Hogweed
subsp. **sphondylium**
Very common, occurring by roadsides and hedges, in coarse grassland and 'unimproved' meadows, in tall-herb marsh communities and, like *Angelica sylvestris*, ascending high in the fells in sheltered gullies.

Var. *angustifolium* Hudson, with narrow leaf segments, was seen during the Survey at Lamplugh (08.20), near Gilcrux (12.38), Maryport (02.36), Little Crosthwaite (22.26), Field Broughton (38.80s) and Gill Head, Winster (38.92).

The flowers of hogweed attract a surprisingly large number of insects: "118... but no butterflies" according to Handley (1898)!
Alt. limit: 845 m, Great Dun Fell (70.32)
Map 609   1469   LWFCY

*H. sphondylium x **H.mantegazzianum**
Recorded only at Wetheral (4.5, C.W. Muirhead) in the early 1970s but very probably overlooked.

C

*H. mantegazzianum Sommier & Levier
                    Giant hogweed
Common along the wooded banks of the middle and lower Eden and the lower Kent, otherwise rather local. It also occurs by roadsides, sometimes in profusion as along the Lindale by-pass (42.80s), and in waste places. The only published record, and the only one in the *Atlas*, is Wilson's from the River Kent (5.8s). The first Cumberland record dates from 1967 when it was found at Wetheral (46.54, C.W. Muirhead) and for Furness from 1977 (38.78s, Cartmel, KAG).

Despite its size and recent spread, this Caucasian species is still regarded in the county more as a curiosity than as a threat to the native flora, unlike the situation along, for example, the River Tweed where systematic efforts are being made to control it.

There have been very few reports in the county of the well-known and very painful photo-allergic response to the plant but the spraying of roadside plants

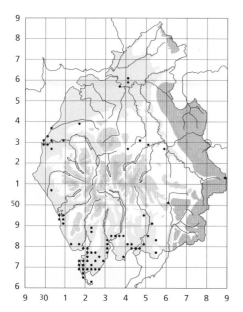

Map 608. *Pastinaca sativa*

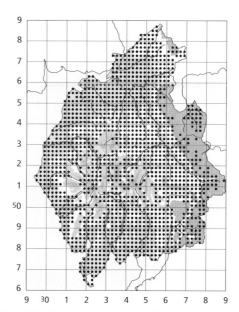

Map 609. *Heracleum sphondylium*

is becoming increasingly common.
Alt. limit: 180 m, Blencow (44.32)
Map 610   63   WFC

*H. canescens Lindl.
A native of the Himalayas found by Hodgson at Silloth docks
(1.5) in 1869.

(C)

Torilis japonica (Houtt.) DC. ʼ Upright hedge-parsley
A common, if rather inconspicuous, late-flowering
umbellifer throughout the lowlands, occurring in
usually rather dry habitats such as hedgebanks, scrub
and woodland margins, and in rough grassland, particularly on roadside verges and railwaysides.
Alt. limit: 420 m, Ash Gill, Garrigill (74.40)
Map 611   1206   LWFCY

[T. arvensis (Hudson) Link
A decreasing plant of southern and south-eastern England. It
is listed for v.c.69 in the *Comital Flora* but the record has
never been substantiated and is best treated as dubious.

(W/F)]

T. nodosa (L.) Gaertner   Knotted hedge-parsley
Recorded at only three sites during the Survey, all on
limestone around the Kent estuary: at Far Arnside
(44.76s, AF, PT & P. Jepson, 1973), Milnthorpe
(50.82s, J.A. Wood, 1974; LANC, *Watsonia* (1977)
11:395) and Allithwaite (38.76s, FLW, 1976), where it
was last seen in 1976.
     This rare southern umbellifer is listed by
Hodgson from St Bees (9.1), Silloth (1.5) and
Bewaldeth (2.3). There are later records from by the
River Eden at Stainton, Carlisle (3.5, T.S. Johnstone,
1908, CLE), Great Salkeld (5.3, H. Britten, 1905),
Silloth (C.W. Muirhead, 1932) and a hedgerow at St
Bees (C.D. Pigott, 1944). The only previous records
from the south are from The Haggs, Dalton-in-Furness
(2.7, D. Lumb, 1912, YRK) and the pre-1930 *Atlas* dot
for the Rampside square (2.6s).

3   WF(C)

*Caucalis platycarpos L.   Small bur-parsley
Recorded by Hodgson as a garden weed at Whitehaven (9.1),
at Silloth docks (1.5) and on gravel by the River Derwent
above Workington (0.2). The only subsequent records are
from gravel by the River Eden at Grinsdale, Carlisle (3.5,
T.S. Johnstone, 1901) and from Silloth (A. Sprott, 1933,
1934).
     This species has been decreasing nationally for many
years and is now all but extinct.

R   (C)

*Turgenia latifolia (L.) Hoffm.   Great bur-parsley
A southern European umbellifer recorded in 1905 at Willow
Holme, Carlisle (3.5, T.S. Johnstone, 1905, CLE).

(C)

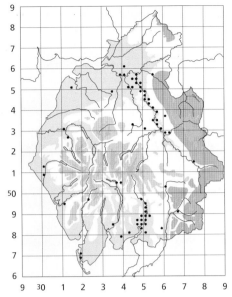

Map 610. *Heracleum mantegazzianum*

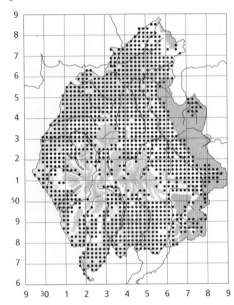

Map 611. *Torilis japonica*

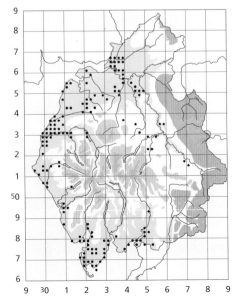

Map 612. *Daucus carota*

*Orlaya grandiflora (L.) Hoffm.

The only record for this uncommon alien is from Silloth (1.5, W.W. Mason, 1930).

This species has a wide distribution on the Continent from the Mediterranean northwards to the English Channel.

(C)

Daucus carota L.   Wild carrot

Fairly common around the coast with a scattering of inland sites, chiefly along the Eden valley. This is similar to its British distribution which, away from the south-east, is strongly coastal. It grows on sea cliffs and dunes, waste ground and dry, usually calcareous banks, as by the railway in the Wigton area (2.4).

There are no records of the maritime subsp. *gummifer* (Syme) J.D. Hook.

Alt. limit: 160 m, Great Ormside (68.16)

Map 612   152   WFC(W/Y)

GENTIANACEAE

Centaurium erythraea Rafn   Common centaury

A fairly frequent species of usually dry coastal habitats but penetrating far up the Eden valley where it appears to favour the dry, sandy soils of the Permian sandstone. It is, however, equally at home on thin limestone soils, on sand-dunes and in damp dune-slacks.

Var. *capitatum* (Willd. ex Cham.) Meld., with stamens attached at the base of the corolla tube, has been reported from the Lancashire part of the Arnside square (4.7s) and the Scottish side of the Solway Firth.

Alt. limit: 220 m, Crosby Ravensworth (60.12)

Map 613   165   WFC

C. erythraea x **C. littorale**

C. x intermedium (Wheldon) Druce

Extremely rare. It occurs with both parents in dune-slacks at Sandscale (18.74), where it was first tentatively identified by R. Young in 1967. R. Ubsdell

has subsequently confirmed this identification on material collected during the Survey (GH, 1984, LANC) and she has also identified as hybrids specimens collected from the upper edge of the salt-marsh near Anthorn on the Solway coast (18.56, REG, 1983, LANC). *Centaurium littorale* was absent at this last site.

2   FC

C. littorale (Turner ex Smith) Gilmour

Very local, occurring in similar situations to *Centaurium pulchellum*, in open, sandy or muddy ground by the estuaries but in the north known only from north of Anthorn (18.60n). It has also been reported from a slag heap at Askam-in-Furness (20. 76).

The species appears to be declining there being no Survey records from Hodgson's localities in the Gosforth square (0.0), and Wampool (2.5) and Cardurnock Point (1.5), where it was last seen in 1949 (N.D. Simpson, BM). The *Scarce Plants* records for the Bowness-on-Solway (2.6n) and Burton-in-Kendal (5.7s) squares refer to Dumfriesshire and Lancashire respectively. Hodgson's record for Ulpha is that cited by Baker (C. Bailey, 1874, BM) but this was an error since Baker clearly indicates that it is the Westmorland Ulpha (4.8s), and not that in the Duddon valley.

Map 614   S 9   WFC

C. pulchellum (Sw.) Druce

Rather local, on open, muddy or sandy ground by the estuaries, in the north only recorded from near Anthorn (18.56, 18.60n), the most northerly site in western Britain and the first Cumberland record (U.K. Duncan, 1949, *BSBI Yr Bk* (1951) p.49). There are no Survey records from Walney Island (1.6s) or Sandscale (1.7), where it was recorded by A.W. Westrup and G. Wilson respectively in the mid-1950s.

Baker's record for Ulpha, and also possibly Far Arnside, is based on a specimen collected by C. Bailey

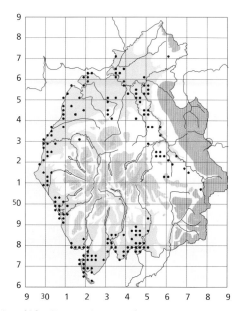

Map 613. *Centaurium erythraea*

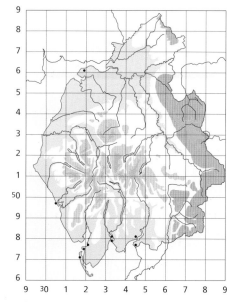

Map 614. *Centaurium littorale*

from "Ulpha moss edge, opposite Arnside" (4.8s, 1874, BM). As with *Centaurium littorale*, this record was erroneously included by Hodgson.

Map 615   16   WFC

Blackstonia perfoliata (L.) Hudson   Yellow-wort
Extremely rare, being seen only on the slag tip at Askam-in-Furness (20.76, B. Fisher, 1980). The only previous record is of a site at Cark (3.7s) mentioned by Petty (1897 p.92). There is now a flourishing colony at Askam where it was earlier threatened with extinction by a 'reclamation' scheme.

                            1   F

Gentianella campestris (L.) Börner   Field gentian
Rather rare and rapidly declining, as over much of England. It is a plant of short, open turf, usually on the limestone,as around the Kent estuary, but it also occurs on sand-dunes, as at Wolsty (08.50) on the Solway Firth, and the Permian sandstone north of Penrith (5.3) where it grows with *Gentianella amarella, Leontodon saxatilis, Viola canina* and *Echium vulgare.*

Cumbria now has more sites for this gentian than any other area south of the Border but even here it has disappeared from half of the 30 post-1930 *Atlas* squares. The major losses are from the Lake District, the limestone on its northern rim and, surprisingly, Walney Island.
Alt. limit: 410 m, Knock Ore Gill (68.28)

Map 616   51   WFCY

G. amarella (L.) Börner   Felwort
subsp. **amarella**
Frequent in short, rather open limestone grassland and in quarries, chiefly in the Pennines, mid-Westmorland

and around Morecambe Bay but, like *Gentianella campestris,* also in sand-dunes and dune-slacks, as by the Duddon estuary and near Silloth (1.5). In the north-east, at Spadeadam (6.7n) it is locally frequent along forestry tracks. It has disappeared from a number of *Atlas* squares, for example Ireby (2.3) and Matterdale (3.2) in the north and Barbon (6.8s) in the south-east.

The plants are usually small with pale pink, pale lilac or whitish flowers, but there are populations on the Furness peninsula with deep purple flowers. One of these, near Dalton-in-Furness (24.72, GH, 1986, LANC), is of exceptionally tall plants, up to 40 cm.
Alt. limit: 650 m, west of Cross Fell (66.34)

Map 617   156   WFC(Y)

Gentiana verna L.   Spring gentian
Very rare, although locally frequent, in seven tetrads on the limestone of the high Pennines and Teesdale from Little Fell and Mickle Fell (7.2) north to Alston Moor (7.3).

This gentian, the best-known member of the 'Teesdale flora', was recorded from the Westmorland side of the Tees in 1805. It was found on Alston Moor, new to Cumberland, by G. Bolam in 1912, but it seems likely that Nicolson's record of *Gentiana pneumonanthe* from there refers to *G. verna.* There are additional tantalising reports from Gilderdale (6.4) and Geltsdale (5.5/6.5) further north. Hodgson's Egremont (0.1) record and Baker's from the Furness fells must, as the latter suggests, be errors.

It grows in short *Festuca - Agrostis* grassland, usually associated with *Thymus polytrichus, Galium sterneri, Minuartia verna* and, in the south, *Myosotis alpestris.* It has probably decreased in recent years as a result of increased stocking and the flowering heads are

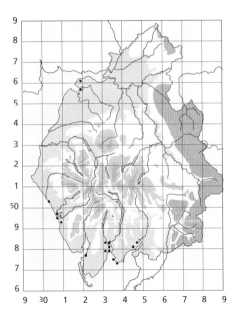

Map 615. *Centaurium pulchellum*

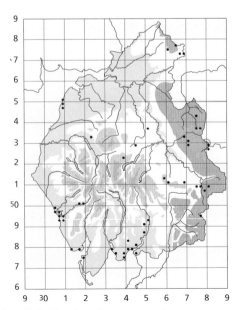

Map 616. *Gentianella campestris*

frequently bitten off by sheep.

Outside the general Teesdale area, this gentian is known in the British Isles only from the limestone of mid-west Ireland. Unlike most Cumbrian montane species it is absent from Scandinavia and Iceland.

Alt. range: 425 m, Alston Moor (76.36) - 730 m, Little Fell (78.20)

Map 618   8   W[(F)]C

G. pneumonanthe L.   Marsh gentian

Known now only from its old site on Walney Island (1.6s), where it is flourishing, and one in a marshy field in east Furness (3.8) discovered in 1975.

With the exception of Walney Island, this beautiful gentian has long disappeared from all the localities cited in the Floras. Hodgson lists Maryport (0.3), Baker Roosecote (2.6s) and Foulshaw Moss (4.8s), where it was discovered in 1744, and Wilson gives additional sites at Gamelands, Tebay (6.0) and Milburn (6.2). Petty (1897 p.229) mentions a site at Coniston (2.9/3.9) and there is an old collection from Rosley (3.4, R. Wood, 1884, CLE), but the reference in Nicolson's Flora to "Aulston Moor" probably refers to *Gentiana verna*.

The two Cumbrian populations, three in south Lancashire and the Wirral, and one in west Yorkshire are now the only remaining ones in north-west England.

S 2   (W)F(C)

APOCYNACEAE

*Vinca minor L.   Lesser periwinkle

A not infrequent garden escape, naturalised chiefly on roadsides and lanesides but avoiding the more acid soils, the east and the extreme north. It is mentioned in all the Floras but it is clearly now much commoner, especially in the south.

Variants with white and lilac flowers occur intermixed with normal plants by the River Eden near Holmwrangle (50.48, GH, 1985, LANC).

Alt. limit: 240 m, east of Raisbeck (66.06)

Map 619   87   WFCY

*V. major L.   Greater periwinkle

Not uncommon in the south of the county, particularly near the Kent estuary where it occurs by roadsides and paths but never far from houses. It is considerably less common in gardens and shrubberies than *Vinca minor* and only rarely can it be considered effectively naturalised.

Wilson's record from Scarfoot, Underbarrow (4.9) may be an error for *Vinca minor* which now occurs there.

Alt. limit: at least 180 m, Dacre (44.26)

Map 620   27   WFC

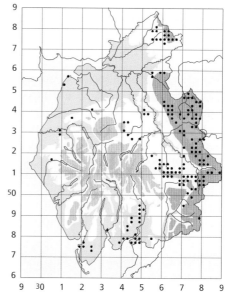

Map 617. *Gentianella amarella*

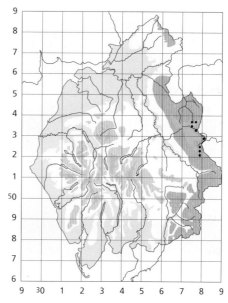

Map 618. *Gentiana verna*

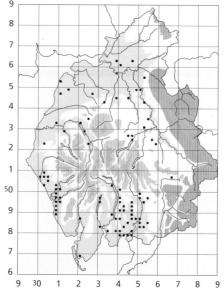

Map 619. *Vinca minor*

SOLANACEAE

***Lycium barbarum** L. and **L. chinense** Miller

Duke of Argyll's tea-plant

Not infrequent around the south coast, otherwise rare and probably always planted, most of the sites being close to houses. At Silloth (10.52) it is well established on the dunes and at Rockcliffe (34.60) on sandstone cliffs above the River Eden. The *Atlas* shows only one locality, Bute, further north on the west coast.

No attempt has been made to make the very tenuous distinction between the two species.

Alt. limit: 280 m, Melmerby (62.36)

Map 621   51   WFCY

Atropa belladonna L.   Deadly nightshade

Occasional on the limestones around the head of Morecambe Bay. Hodgson gives five sites, at some of which it would appear to have been introduced, and including Nicolson's record from the Carlisle city walls (3.5). The only subsequent records from Cumberland are from Blaithwaite, Wigton (2.4, J. Parkin, 1935), a railwaybank in Carlisle (4.5, D. Stewart, 1939, E. Blezard, 1940, CLE), where it was last seen in 1946, and, more recently, from the Eskmeals dunes (08.92, A. Warburton, c.1965). There is no recent record from its old site at Furness Abbey (2.7), but it is encouraging to report that despite extensive quarrying it still survives at Holme Park quarry (52.78s), where it was first seen in 1744.

Map 622   14   WFC

Hyoscyamus niger L.   Henbane

Known now only from a few coastal sites, where it occurs on bare ground, stabilised shingle and sand. It is quite common on Walney Island, probably on account

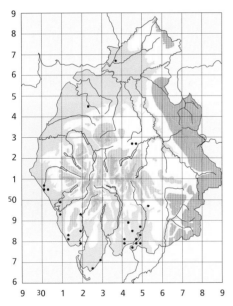

Map 620. *Vinca major*

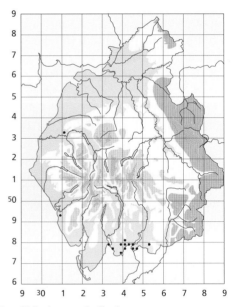

Map 622. *Atropa belladonna*

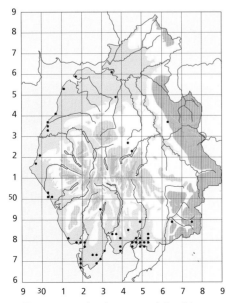

Map 621. *Lycium barbarum* and *L. chinense*

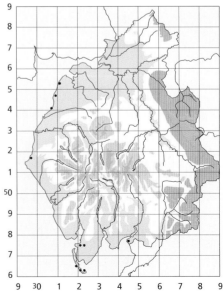

Map 623. *Hyoscyamus niger*

of its unpalatability to rabbits. It appeared at Arnside (44.76s, MB) in 1977 but, as on the shore south of Allonby (06.40) and at Mawbray (08.46), it has failed to persist. There are old inland records from the Eden valley (5.3, 6.3), the Carlisle area (3.5) and the Wigton (2.4) and Kirkby Lonsdale (5.7s) squares, but these probably represent only casual occurrences.

      Like many southern species, henbane becomes increasingly coastal in the north of its range.

Map 623   10   WFC

Solanum nigrum L.   Black nightshade
A rather local plant of the coast and the Eden valley. Inland, it occurs usually as a weed of cultivated and waste ground and on manure heaps, while on the coast it grows, like *Hyoscyamus niger*, in nitrogen-enriched soils around gulleries and along the strand-line.
Alt. limit: *c*.130 m, Morland (60.22)
Map 624   30   WFC

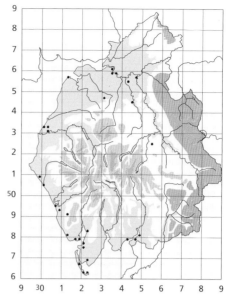

Map 624. *Solanum nigrum*

S. dulcamara L.   Bittersweet; Woody nightshade
Common in the lowlands where it occurs in the fringing vegetation of riverbanks and the Lancaster - Kendal canal, often with its stems trailing in the water, also in hedgerow ditches and willow carr and occasionally in drier situations as on the sand-dunes at Sandscale (18.74) and on cindery railway land at Carlisle (38.54).

      The schoolboy habit, described by Hodgson, of keeping "a stock of twigs in their pockets, which they chew as their elders do tobacco" has presumably long since disappeared!
Alt. limit: 150 m, Flitholme, Warcop (76.14)
Map 625   494   LWFCY

*S. rostratum Dunal
(*S. cornutum* auct.)
A yellow-flowered North American species found during the Survey by roadworks on the A66 north-east of Keswick (26.24, G. Wilson, 1975, LANC) and in a garden at Ulverston (28.78, M.A. Bushby, 1977, LANC).

                  2   FC

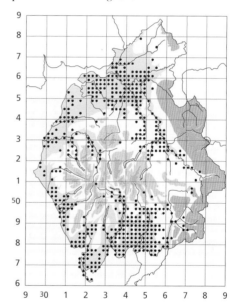

Map 625. *Solanum dulcamara*

*Datura stramonium L.   Thorn-apple
A rare and striking plant appearing usually following disturbance of long-buried seed. At Langwathby (58.32) it appeared after ploughing in a field which had been undisturbed for at least 40 years and, coincidentally or not, there were several sightings during the hot summer of 1995.

      The only pre-Survey records are from Great Salkeld (5.3, H. Britten, 1900-1908), Newtown, Carlisle (3.5, *cf.* C.W. Muirhead, 1945) and Carlisle waterworks (4.5, J. Parkin, 1945), Troutbeck (3.0, G.A.K. Hervey, 1955), Kirkby Stephen (76.08, Anon., 1940) and the *Atlas* dot for the Allonby square (0.4). The first Furness record was from a potato field between Newton and Roose (20.70, K. Jones, *c*.1970).

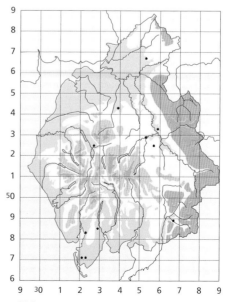

Map 626. *Datura stramonium*

Alt. limit: 120 m, lower Dentdale (66.88s))
Map 626   11   WFCY

CONVOLVULACEAE

Convolvulus arvensis L.   Field bindweed
Fairly frequent in the lowlands, except in the north and east, and occurring as a weed of cultivated ground, chiefly gardens and allotments, and in waste places and open habitats, especially along the coast.

Like *Echium vulgare*, this is another south-eastern species which becomes progressively more coastal northwards.
Alt. limit: 300 m, Stennerskeugh, Ravenstonedale (74.00), appreciably higher than that given by Wilson.
Map 627   226   WFC(W/Y)

Calystegia soldanella (L.) R. Br.   Sea bindweed
Common along the coast, although almost absent from Morecambe Bay, occurring on sand-dunes and just above the strand-line on sand and shingle beaches.
Map 628   47   FC

C. sepium (L.) R. Br.   Hedge bindweed
subsp. **sepium**
Very common throughout the lowlands and extending into the lower fell country. It occurs chiefly in hedge-rows, especially around villages, and on fencing around overgrown gardens, waste sites and railwaysides. It is much commoner than *Calystegia silvatica*, a situation which is the reverse of that on the east side of the Pennines in Co. Durham (Graham 1988).

The variant with flowers with five white stripes, f. *colorata* (Lange) Dörfler, has been reported by CEW from Heversham (48.84, 50.82s, 1991, LANC)) and near Staveley (44.98, 1994), and by PB from Sandscale (18.72, 1995, LANC).
Alt. limit: *c*.280 m, Alston (70.46)
Map 629   782   LWFCY

subsp. **roseata** Brummitt
Seen during the Survey climbing over a garden fence at Wetheral (46.54, C.W. Muirhead, 1975) and over streamside alder bushes at Studholme, Kirkbride (24.56, GH, 1985, LANC). The only other record for this pink-flowered subspecies is that in the *Critical Atlas* for the Cleator Moor square (0.1).
                                   2   C

*C. sepium x **C. silvatica**
C. x **lucana** (Ten.) Don
Despite the frequency of the two parent species, the only record of this hybrid is from Etterby, Carlisle (3.5, CLE), where it was found by C.W. Muirhead in 1951.
                                              (C)

*C. pulchra Brummitt & Heyw.
A fairly common garden escape but only rarely found at any distance from houses. This is the pink-flowered

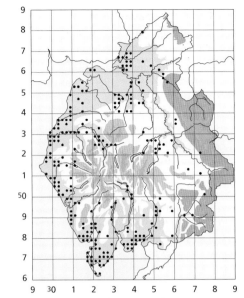

Map 627. *Convolvulus arvensis*

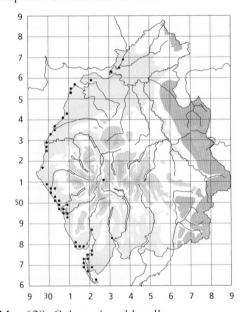

Map 628. *Calystegia soldanella*

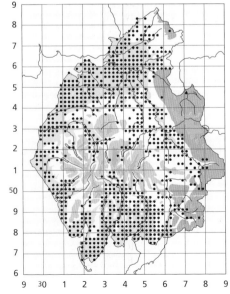

Map 629. *Calystegia sepium*

variant referred to as a garden plant by Baker and Hodgson. Wilson makes no reference to it. There are records from 1949 from: Scotby (4.5, R.A. Boyd, CLE), Stanwix, Carlisle (3.5/4.5, C.W. Muirhead, CLE) and Dalston (3.5, D. Graham, CLE) and several in the *Critical Atlas* from the south of the county.

Alt. limit: 170 m, Brough (78.14)

Map 630   130   WFCY

*C. silvatica (Kit. ex Schrader) Griseb.

Frequent in the lowlands, particularly in the south. It is noticeably less common and more lowland than *Calystegia sepium* and is less successful in establishing itself away from habitation.

Alt. limit: 350 m, Ravenstonedale (72.04)

Map 631   274   LWFCY

*Ipomoea hederacea Jacq.   Ivy-leaved morning glory

Found by G. Wilson in 1975 by the partly constructed Keswick by-pass (26.24, 1975, LANC).

                              1   C

CUSCUTACEAE

[Cuscuta europaea L.

This is listed for Westmorland in the *Comital Flora* but the source of the record has never been traced. Hodgson appears to accept that the Cumberland record from Greysouthen (0.2) probably refers to *Cuscuta epithymum*.

      Cuscuta europaea is a decreasing species now largely confined in England to the south.

                                  (W/FC)]

*Cuscuta epilinum Weihe

A widespread Continental species known only from the record cited by Hodgson from Westward, Wigton (2.4) and a later one from Melmerby (6.3, W.W. Mason, 1928, *BEC Rpt* (1929) 8:748).

                                  (C)

C. epithymum (L.) L.   Dodder

Listed by Hodgson only from Blackdyke, Silloth (1.5, E.J. Glaister, 1874, CLE) and Westward, Wigton (2.4), but the record of *Cuscuta europaea* from Greysouthen (0.2) probably belongs here. There is also an undated *Atlas* record for the Milnthorpe square (5.8s).

      Although the commonest British species, this too is declining and is now largely confined to south of a line from the Wash to the Bristol Channel.

                                  (WC)

MENYANTHACEAE

Menyanthes trifoliata L.   Bogbean

Frequent, particularly in the Lake District, in nutrient-poor or mesotrophic tarns, pools and flushes, also amongst very wet *Sphagnum* in valley mires and blanket peat. It often completely fills the shallow pools and it is a marvellous sight when in full flower.

Alt. limit: 760 m, east side of Little Dun Fell (70.32)

Map 632   418   WFCY

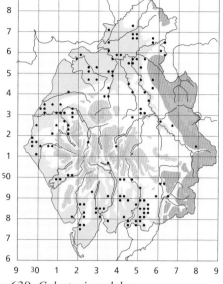

Map 630. *Calystegia pulchra*

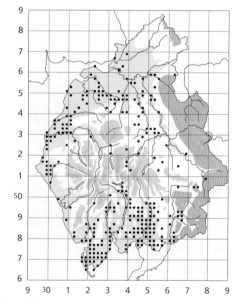

Map 631. *Calystegia silvatica*

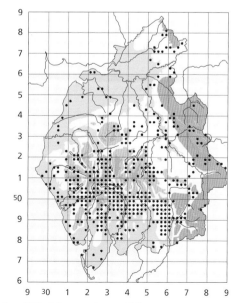

Map 632. *Menyanthes trifoliata*

***Nymphoides peltata** Kuntze   Fringed waterlily
Discovered in 1987 in a pond at Silecroft (12.80,
R. Jerram) and subsequently in a roadside ditch at
Broughton Moor (04.32, CCH & AD, 1988, LANC,
both records *Watsonia* (1989) 17:474), in an old clay-
pit at Barrow-in-Furness (18.70, PB, 1991, LANC,
Watsonia (1992) 19:148) and in a remote fish pond in
woodland north of Longtown (34.70n, KR, 1991,
LANC). Three of these sites are some distance from
houses, but it seems likely that the plants were intentionally introduced. Lawson's record from the river
near Hawkshead (3.9) and cited by Baker under
Villarsia nymphoides, is, as suggested by Whittaker
(1986 p.200), almost certainly *Nuphar lutea*.

    This species is native in Britain only in the Fens
and the Thames valley, but the above finds, and its
discovery in 1988 in the Lancaster - Preston canal
(Livermore & Livermore 1989), suggest that it may be
spreading and not entirely by human agency.
Alt. limit: 115 m, Broughton Moor (04.32)

S 4   FC

POLEMONIACEAE

Polemonium caeruleum L.   Jacob's ladder
Known only from a few sites, mainly in the east
although there are several records in the Floras from
other parts of the county. Most of the Survey finds have
been from roadsides and all presumably represent
garden escapes. Yet it is considered in the *Atlas*,
though not by Graham (1988), to be native in its former
Teesdale site in Co. Durham. Baker says "Truly wild in
the Westmorland portion of Teesdale" and Wilson
gives an 1842 record from "Between Greenburn Lead
Mine and Dun Fell Hush". If it really was native on the
limestone there, as the finder, J. Bell, stated, then
perhaps the present concentration of records in the

Alston area may be due to the transfer by miners of
wild plants down to their gardens.

    The only extant British localities of presumed
native plants are in Northumberland, the Craven area
of Yorkshire and the Peak District.
Alt. limit: 440 m, Nenthead (78.42)
Map 633   R 22   W(F)C

HYDROPHYLLACEAE

***Phacelia** *cf.* **ciliata** Benth.
A North American species recorded from Silloth (1.5,
J. Leitch, 1889, 1890, CLE; W. Muirhead, 1930, CLE, all
det. E.J. Clement; H. Stewart, 1930, BM).

(C)

***P. tanacetifolia** Benth.
Seen at Winster (40.90, H. Smith, 1996, LANC) in a field
apparently sown with bird seed. Otherwise only known from
a collection in 1901 from the gravel beds of the River Eden
below Grinsdale, Carlisle (3.5, T.S. Johnstone, CLE, det. E.J.
Clement (Thomson 1902)).

(C)

BORAGINACEAE

***Lithospermum purpurocaeruleum** L.   Purple gromwell
A handsome species, native in Britain only in the south and
known in Cumbria only from the record given by Hodgson
from a Workington garden (9.2) and a later one from Metal
Bridge, north of Carlisle (3.6, J. Parkin, 1935).

R   (C)

L. officinale L.   Gromwell
A very local plant of the limestone around the head of
Morecambe Bay, occurring in hedgerows, woodland
margins, rides and scrub. There are also post-1930
Atlas records from the Dalton-in-Furness (2.7), Dist-
ington (0.2) and Maryport (0.3) squares where it was

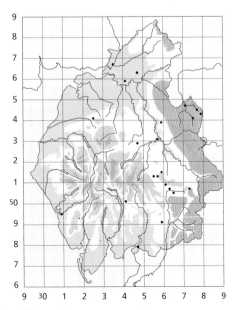

Map 633. *Polemonum caeruleum*

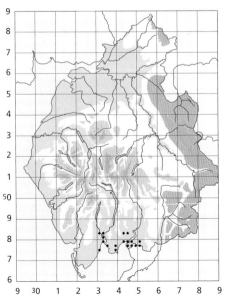

Map 634. *Lithospermum officinale*

probably only a casual.

These are the most north-westerly records in Britain for this predominantly south-eastern species.

Map 634　　　　　　18　　　　　　WF(C)

*L. arvense L.　　　　　　　　　　Corn gromwell
Recorded by Hodgson and Wilson from eleven sites dating from last century and there are later records from the River Eden near Grinsdale, Carlisle (3.5, T.S. Johnstone, 1901), Carlisle waterworks (4.5, J. Parkin, 1935) and Thwaites, Millom (1.8, M. Cross, 1937).

This is another south-eastern species which has probably never been other than a rare casual in Cumbria.

(WC)

Echium vulgare L.　　　　　　Viper's bugloss
Rather rare, in dry sandy places and on cliffs and in quarries, usually near the coast but also at a few inland sites at some of which it appears to persist. Nowhere is it as common as at the South End of Walney Island (22.62s) where in midsummer it transforms the old gravel workings.

Like so many south-eastern species, viper's bugloss becomes increasingly coastal towards the north and west of Britain.

Alt. limit: 200 m, Wan Fell, Penrith (50.36), although there is an *Atlas* dot for the Alston square (7.4)

Map 635　　　　　　22　　　　　　WFC

*E. italicum L.
A southern European species found on the gravel beds of the River Eden at Grinsdale, Carlisle in 1902 (3.5, T.S. Johnstone, CLE).

(C)

*E. plantagineum L.　　　　Purple viper's bugloss
A rare species of south-west England recorded in 1876 at Silloth (1.5, E.J. Glaister, CLE).

R　　　　　　　　　　　　(C)

*Pulmonaria officinalis L.　　　　　　Lungwort
This popular garden plant is widely if sparsely distributed throughout the county, occurring by lanesides and paths but rarely far from houses. Like *Pulmonaria longifolia*, it was listed without localities as an occasional garden escape by both Baker and Hodgson. It was collected at Ireby (2.3, 1876, CLE) by E.J. Glaister and in 1935 it was seen in the same square by J. Parkin at Lake Wood, Bassenthwaite. There is a 1908 record from Thwaites, Millom (1.8, M. Cross) and 1949 ones from Buckman Brow, Broughton-in-Furness (1.8, C.W. Muirhead) and Penton Linns (4.7n, J.H. Creed & C.W. Muirhead, CLE). Wilson lists it from Underbarrow and Crook (4.9) and Appleby (6.2). There are also pre-1930 *Atlas* records from the Cockermouth (1.3), Arnside (4.7s) and Burton-in-Kendal (5.7s) squares and a later one from Spark Bridge, Greenodd (3.8, G. Wilson, 1948).

Alt. limit: 200 m, Orton (62.06)

Map 636　　　　　　42　　　　　　WFC

*P. longifolia (Bast.) Boreau　　Narrow-leaved lungwort
A rare garden escape listed without localities by Baker and Hodgson, but recorded by E.J. Glaister from High Ireby (2.3, 1876, CLE) and noted by Wilson from Nook (5.8s).

S　　　　　　　　　　　　(WC)

Symphytum officinale L.　　　　　　Comfrey
Apparently rather rare and chiefly in the south where it occurs usually on roadsides and near houses. It probably has a more southerly distribution in the country as a whole than *Symphytum* x *uplandicum*, but rather surprisingly Livermore & Livermore (1987) report it as frequent in the Lancaster area. The records of the early Flora writers include those of *S.* x *uplandicum* but despite this the species was regarded as infrequent.

Diagnostic features include the long-decurrent

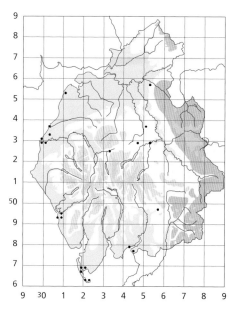

Map 635. *Echium vulgare*

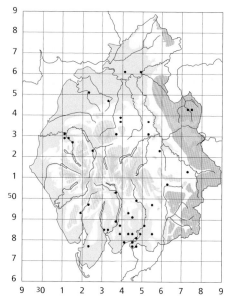

Map 636. *Pulmonaria officinalis*

leaves and usually cream or pale pink flowers.
Alt. limit: 305 m, Leadgate, Alston (70.44)
Map 637   7   WFCY

*S. officinale x **S. asperum** Lepechin   Russian comfrey
S. x **uplandicum** Nyman
Frequent in the lowlands where it is a characteristic
plant of riversides, ditches and moist roadside verges
but, like so many one-time herbs and medicinal plants,
generally not far from houses.

  The only references to purple-flowered plants in
the early Floras are Hodgson's mention, under *Symphytum officinale,* of var. *patens* (Sibth.) Lindley near
Workington (9.2), where it was mixed with typical
plants, Petty's (1897 p.231) from near Woodland,
Broughton-in-Furness (2.8/2.9), and one from Belle
Vue, Carlisle (3.5, T.S. Johnstone, 1899). There are
also records (under *S. peregrinum*) from Calder Bridge
(0.0, R.H. Williamson, 1923), Melmerby and Kirkoswald (respectively 6.3, 1920; 5.4, 1921, W.W.
Mason), Naworth (5.6, G.C. Druce, 1922, *BEC Rpt*
(1924) 7:200), between Knock and Milburn (6.2, A.J.
Wilmott, 1921, *BEC Rpt* (1948) 13:303) and Dalton-in-
Furness (2.7, W.H. Pearsall, 1913, *BEC Rpt* (1914)
3:390).

  It is curious that there should be so few early
records for such a widespread and familiar plant. This
contrasts markedly with the hybrid's present relatively
wide distribution, the Survey map showing it as
occurring in more than twice as many squares as are
indicated in the *Atlas.*
Alt. limit: 365 m, north-east of Alston (72.48)
Map 638   384   LWFCY

S. tuberosum L.   Tuberous comfrey
Rather local, chiefly in the west but with an isolated
site in the upper Eden valley in Flakebridge Wood,
Appleby (70.20, D. Hinchcliffe, 1980). It grows in
similar situations to *Symphytum* x *uplandicum* forming
extensive colonies by rhizomatous spread. Some
colonies have definitely been planted and there is some
doubt as to its native status in the county. There is a
conspicuous colony between the two carriageways of
the A590 on the west side of Ulverston (28.76).
Alt. limit: 180 m, Flakebridge Wood,Appleby (70.20)
and Roadhead (50.74n)
Map 639   42   WFCY

*S. grandiflorum DC.
(*S. ibiricum* Stevens)
Seen during the Survey on roadsides at Arnside
(44.76s, MB, 1984, LANC; 46.78s, MB, 1983), in the
grounds of an hotel near Levens (48.86s, MB, 1984;
LANC), near The Green, Millom (18.84, A. Cannell,
*c.*1990) and effectively established on roadsides at
Storth (46.80s, CEW, 1990, LANC), near Old Hutton
(56.88s, CEW, 1990; LANC, *Watsonia* (1991) 18:428)
and Blindcrake (14.34, MP, 1993, LANC).

  This is an early, yellow-flowered, rhizomatous

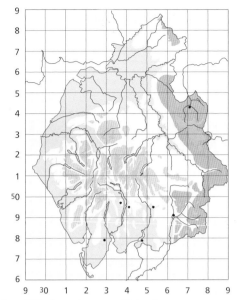

Map 637. *Symphytum officinale*

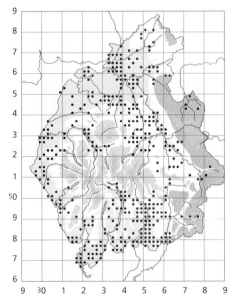

Map 638. *Symphytum officinale* x *S. asperum*

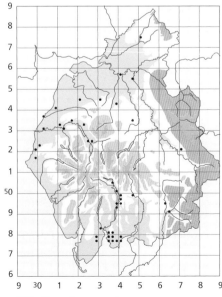

Map 639. *Symphytum tuberosum*

species which is often confused with *Symphytum tuber-osum*. It is a native of the Caucasus.

Alt. limit: 175 m, Old Hutton (56.88s)

Map 640   7   WC

[*S. caucasicum* M. Bieb.
Listed by Hodgson from Egremont but this record could well refer to *Symphytum* x *uplandicum*.

(C)]

*Brunnera macrophylla (Adams) I.M. Johnston
Discovered in 1984 well naturalised in woodland by the River Lyvennet in the grounds of Crossrigg Hall, Cliburn (60.24, RWMC, LANC, *Watsonia* 1988) 17:191) and in a lane at Cockermouth (12.30, A. Cannell, 1990).

Alt. limit: 110 m, Cliburn (60.24)

                    2   WC

*Anchusa undulata L.
A Mediterranean plant collected with the previous species at Grinsdale, Carlisle (3.5, T.S. Johnstone, CLE) in 1902.

(C)

*A. officinalis L.
A few plants were found during the early 1980s by A. Warburton on a garden site at Eskmeals which had been abandoned for at least 70 years (08.92; LANC), and in 1988 a single plant, which presumably originated from the garden colony, was seen by MMG on the dunes. The only other records are from Seascale (0.0, R.H. Williamson, CLE) where it was collected in 1923 and 1925.

                    1   C

*A. azurea Miller
A species of southern Europe collected from gravel beds by the River Eden at Grinsdale, Carlisle in 1902 (3.5, T.S. Johnstone, CLE).

(C)

A. arvensis (L.) M. Bieb.   Bugloss
(*Lycopsis arvensis* L.)
A not uncommon arable weed of the light soils of the Eden valley and the Solway coast, also on the sand-dunes, sandy car parks and gravel workings on Walney Island and by the Duddon estuary. Elsewhere it is very rare and probably only a casual.

        Its coastal and Eden valley distribution is typical of many species with a predominantly south-eastern British distribution.

Alt. limit: 140 m, Brampton golf-course (52.58) and Warcop (74.14)

Map 641   72   WFC(W/Y)

*Pentaglottis sempervirens (L.) Tausch ex L. Bailey
                                        Green alkanet
Widespread in the lowlands, particularly in the south where it is a prominent feature of roadside verges and hedges. Like comfrey, it was widely cultivated for its medicinal properties, hence its close association with

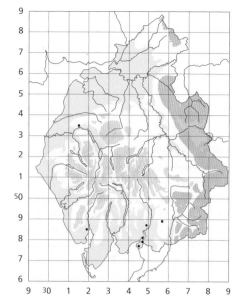

Map 640. *Symphytum grandiflorum*

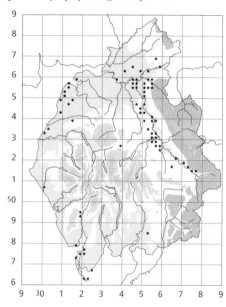

Map 641. *Anchusa arvensis*

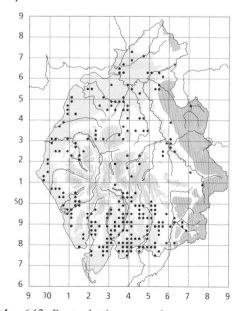

Map 642. *Pentaglottis sempervirens*

houses and villages.
Alt. limit: 300 m, north of Alston (70.46)
Map 642   234   LWFCY

*Borago officinalis L.   Borage
An occasional garden escape probably never occurring other than as a casual and near houses. Hodgson gives several sites from St Bees to Carlisle and Ullswater, and Wilson three in south Westmorland.
Alt. limit: 425 m, south of Alston (72.42)
Map 643   14   WFC

*Trachystemon orientalis (L.) Don
Found in 1985 by MMG well naturalised in the Muncaster woods (08.94) and subsequently at three other nearby sites (08.96, 10.94, 10.00). The habitats were damp woodland and a long-abandoned garden. The only other record is that given by Hodgson from the roadside near Portinscale (2.2).

This species of south-east Europe and Asia Minor is now only rarely cultivated.
                                                  4   C

Mertensia maritima (L.) Gray   Oyster plant
An extremely rare plant of west coast shingle beaches. The largest populations were those on Walney Island, and latterly at the South End (22.62s), and on Foulney Island opposite (24.64s), but these disappeared following exceptionally severe storms in the late 1970s. On the Cumberland coast it was seen independently during the 1970s by C.W. Muirhead and CCH in the Workington square (9.2), and this was followed during the 1980s by further finds of a single plant in the Allonby square (0.4, J. Ratcliffe, 1985) and a small population in the Bootle square (0.8, MMG, 1984).

The populations are usually small, mobile, fluct-

uating and capricious in their appearance from year to year, sometimes persisting for only a few years before succumbing to the ravages of winter storms.

This species has a long recorded history in Cumbria. Baker and Hodgson's localities include Walney Island and virtually the whole coastline from Bootle north to Maryport. The earliest records are the 16th century ones in Nicolson's Flora from Walney and Parton (9.2). There is also a pre-1768 record from 'Abbey Holm' which probably refers to the coast at Silloth (1.5, Rev. W. Richardson). The *Atlas* record for the Broughton-in-Furness square (2.8) is based on a report from the River Duddon (*Phytologist* (1861) n.s. 5:222) and probably refers to the North Walney square (1.7). By the end of the last century Hodgson found it necessary to comment on its decline, which still continues both locally and nationally.

The Bootle population is probably the most southerly extant one in Europe since the species has now disappeared from its last site on the north coast of Wales.
                                                  S 5   F†C

*Amsinckia lycopsoides (Lehm.) Lehm.
A yellow-flowered North American species known only from Silloth (1.5, CLE, conf. J.M. Mullin), where it was collected by J. Leitch in 1889.
                                                                    (C)

*A. micrantha Suksd.
Seen only twice during the Survey, in 1988 on dumped soil at Tindale, Hallbankgate (60.58, REG, LANC) and in 1992 in an old sandpit at Great Salkeld (52.36, JMA & MP, LANC), both collections being identified by J.M. Mullin.
Alt. limit: 230 m, Hallbankgate (60.58)
                                                  2   C

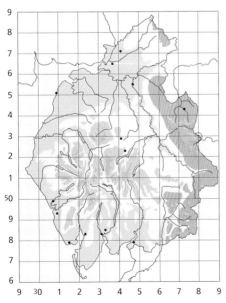

Map 643. *Borago officinalis*

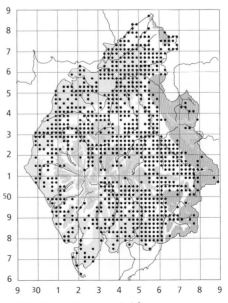

Map 644. *Myosotis scorpioides*

*Asperugo procumbens L.   Madwort
A widespread Continental species recorded by Hodgson from
Maryport (0.3), Silloth (1.5, J. Leitch, 1891, CLE) and Kirk-
bride (2.5) at the end of the 19th century.

(C)

Myosotis scorpioides L.   Water forget-me-not
Common in the lowlands and extending far up the
dales. It occurs by ponds, tarns and lakes, and in
swamps, marshy fields and ditches. It is probably more
demanding in its nutrient requirements than *Myosotis
laxa*.
Alt. limit: at least 600 m, Moor House, Teesdale
(72.32)
Map 644   891   LWFCY

M. scorpioides x **M. laxa**
M. x suzae Domin
This rather rare or overlooked hybrid was discovered in
1986 growing with both parents in a marshy field near
Frizington (02.14, CCH, LANC, *Watsonia* (1988)
17:191). Its identity was confirmed the following year
by P.M. Benoit. The hybrids were not entirely sterile
and the population included plants which were probably backcrosses to the parents.

1   C

M. secunda Al. Murray   Creeping forget-me-not
A common plant of becksides and the poorer flushes in
the dales and on the fellsides but rather uncommon or
scarce in the lowlands proper. It is often mat-forming
and this, combined with the medium-sized blue flowers
and spreading, almost hispid hairs on the lower part of
the stem, usually makes it easily identifiable.
Alt. limit: 540 m, Moor House, Teesdale (74.32). Baker
gives it as 730 m above Red Tarn, Helvellyn (3.1), but

this could refer to *Myosotis stolonifera*.
Map 645   638   WFCY

M. stolonifera (DC.) Gay ex Leresche & Levier
(*M. brevifolia* Salmon)
Not uncommon in the fells, occurring in base-poor
springs and seepage areas, soakaways and flushes on the
fellsides and in the valleys by pools, in ditches and in
the backwaters of streams. Common associates include
Cardamine pratensis, Montia fontana and *Stellaria
uliginosa*. It occurs in much the same habitat as
Myosotis secunda but ascends higher. The first Cumberland record was from Fisher Gill, Thirlmere (3.1,
herb. A.H. Wolley-Dod, 1912, BM) and for Furness
that from Coniston Old Man (26.96, Anon., 1970s).
Scarce Plants shows a record for the Langdale Pikes
(2.0) but there is no Survey record for that square.

    Myosotis stolonifera has been much overlooked
in the past and the map shows it as occurring in substantially more squares than does the *Atlas*. It was first
collected in Britain by A. Wilson near Cautley Crag
and Skeggleswater, Kentmere (respectively 6.9, 1892,
4.0, 1894, both YRK) although it was not recognised
until 1926 (Salmon 1926). It is fairly easily recognised
by its small, almost white flowers, short leaves and,
towards the end of the summer, the dense mat of shoots
arising from the lower leaf axils.

    This forget-me-not is restricted in Britain to
northern England, from Cumbria across to the North
Yorkshire Moors, and a number of sites in the Southern
Uplands. On the Continent it is known only from the
mountains of northern Portugal and Spain.
Alt. range: 130 m, above Barbon (62.82s) - 820 m,
Cross Fell (68.34)
Map 646, Plate 81   S 204   WFCY

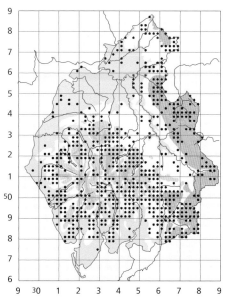

Map 645. *Myosotis secunda*

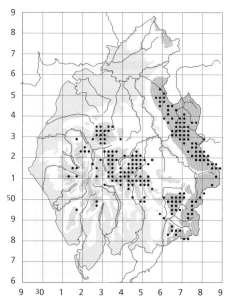

Map 646. *Myosotis stolonifera*

M. laxa Lehm.   Tufted forget-me-not
(*M. caespitosa* Schultz)
Common in the lowlands occurring, like *Myosotis scorpioides*, by ponds and tarns and in marshy fields and ditches. It is sometimes confused with the more upland *M. secunda* but the appressed hairs on the lower stem are diagnostic.
Alt. limit: 580 m, above High Cup Nick (74.26), although Wilson gives it as 610 m in Swindale, above Knock (7.2).
Map 647   911   WFCY

M. alpestris F.W. Schmidt   Alpine forget-me-not
This attractive and much-sought-for member of the 'Teesdale Flora' still occurs in its well-known sites on Mickle Fell in Yorkshire and in Westmorland on Little Fell (7.2). There is also a third and much smaller pop-ulation further north near Great Dun Fell (7.3) and a tantalising report of a single plant from Rigg Beck gill above Winton (8.0, G.A.K. Hervey, 1950).

Like *Gentiana verna*, it grows in short limestone turf, usually around outcrops and associated with *Saxifraga hypnoides, Draba incana, Thymus poly-trichus* and *Minuartia verna*. Flowering is often rather sparse in dry summers and the flowering stems are usually heavily grazed.

The only other area in Britain where this central European species occurs is in the central Scottish Highlands around Ben Lawers. The Cumbrian plants are appreciably smaller than the Scottish, being mostly only 3-6 cm high. The Cumbrian sites are also drier.
Alt. range: 700-750 m, Little Fell (7.2) and Great Dun Fell (7.3)
Plate 82   R 5   W

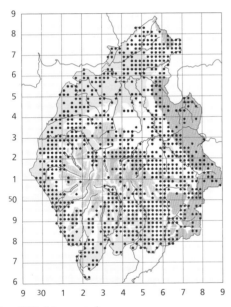

Map 647. *Myosotis laxa*

Plate 82. *Myosotis alpestris*

Plate 81. *Myosotis stolonifera* and *Epilobium alsinifolium*

M. sylvatica Hoffm.   Wood forget-me-not
Frequent in the Eden valley and south Westmorland,
otherwise rather uncommon. It usually occurs in light
shade in moist hedges and woodland margins and
prefers the richer soils. As Livermore & Livermore
(1987) observe, many of the records from roadsides and
waste ground are probably of plants of garden origin.
Alt. limit: 420 m, Garrigill (74.40), although Wilson
gives it as 520 m in Swindale, above Knock (7.2).
Map 648   329   LWFCY

M. arvensis (L.) Hill   Field forget-me-not
Very common in the lowlands as a weed of cultivated
ground and waste places, roadsides, woodland
clearings, hedgerow banks, sand-pits and quarries and
sandy ground along the coast.
          There are no reports of the larger-flowered
woodland var. *sylvestris* Schldl.

Alt. limit: 610 m, above Nenthead (78.44)
Map 649   1306   LWFCY

M. ramosissima Rochel   Early forget-me-not
Rather scarce, and virtually restricted to dry, well-
drained, usually sandy soils on the south-west coast, the
Solway Plain and the Eden valley. On the coast it is a
characteristic member of the spring sand-dune flora.
The first Furness record is unlocalised (1.7, D. Lumb,
BEC Rpt (1914) 3:390).
Alt. limit: at least 430 m, above Swindale, Brough
(82.18)
Map 650   64   WFC(W/Y)

M. discolor Pers.   Yellow and blue forget-me-not
Frequent in the north, less so in the south and very
scarce in the Lake District. Although generally considered to be a plant of dry, open habitats, for example

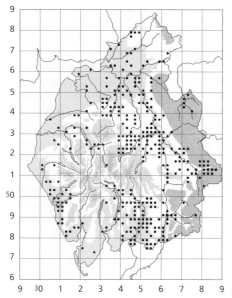

Map 648. *Myosotis sylvatica*

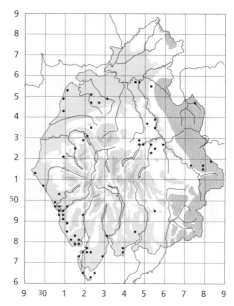

Map 650. *Myosotis ramosissima*

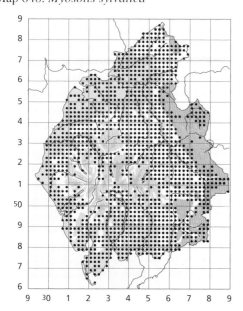

Map 649. *Myosotis arvensis*

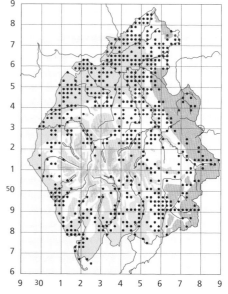

Map 651. *Myosotis discolor*

sandy fields and open hedgerow banks, many of the Cumbrian sites are damp roadside verges, 'unimproved' hay-meadows, damp woodland tracks and even river-banks and marshes.
Alt. limit: at least 610 m, Little Fell (78.20), more than twice that given by Wilson.
Map 651   595   WFCY

[M. persoonii Rouy
A species of the Iberian peninsula recorded in 1921 from Melmerby (6.3, W.W. Mason, *BEC Rpt* (1922) 6:390, under *Myosotis lutea* (Cav.) Pers.). This seems very unlikely and the record is best treated as an error.
                                                                                (C)]

*Lappula squarrosa** (Retz.) Dumort.
A widespread Continental species known only from the late-19th century records cited by Hodgson from the ballast-tips of Workington (9.2), Maryport (0.3) and Silloth (1.5, J. Leitch, 1883, 1889, 1890, CLE) and a collection from the gravel beds of the River Eden at Grinsdale, Carlisle (3.5, T.S. Johnstone, 1901, CLE (Thomson 1902)).
                                                                                (C)

*Omphalodes verna** Moench   Blue-eyed Mary
An early-flowering species frequently cultivated for ground cover but the only Survey records of naturalised plants are from Eggerslack Woods, Grange-over-Sands (40.78s, I.R. Bonner, 1979, *Watsonia* (1982) 14:194) and laneside scrub at Levens (48.86s, MB, 1995). Baker mentions that it occasionally escapes and Hodgson records it from "pleasure grounds" at Clifton and Gilgarran, Workington (0.2) and Ullswater (*cf*.4.2). In 1944 it was seen in limestone woodland at the south end of Whitbarrow (4.8s, M. Cross, *BEC Rpt* (1947) 13:62), in 1949 on nearby Holme Island (4.7s, G. Wilson, CLE) and in 1950 at Blaithwaite, Wigton (2.4, J. Parkin, 1950).
                                2   (W)F(C)

Cynoglossum officinale L.   Hound's tongue
Rare, although locally fairly frequent, occurring on the coast from Walney Island to the Duddon estuary, by the Ravenglass estuary and at Maryport (02.36), and inland on the disturbed limestone below the southern end of Whitbarrow (44.84s). The coastal records are all from sand-dunes or shingle and the plant thrives partic-ularly around rabbit warrens and gulleries. There is a 1954 record from Foulney Island (2.6s) and post-1930 *Atlas* dots for the Cartmel (3.7s), Arnside (4.7s, v.c.60/69) and Silloth (1.5) squares.
        This species reaches its northern limit on the west coast in Cumbria.
Map 652   16   WFC

*Cerinthe minor** L.
A Continental species recorded in 1902 from the gravel beds of the River Eden at Grinsdale (3.5, T.S. Johnstone, CLE).
                                                                                (C)

VERBENACEAE

Verbena officinalis L.   Vervain
As elsewhere in northern England this southern species is in decline. Although there are Survey records from Silloth (10.52, MP, 1980), the old railway line at Rampside (22.66s, GH, 1981) and roadsides at Witherslack (42.84s, KAG & FLW, 1966) and Newland, Ulverston (30.78s, K. Jones, 1976), it can no longer be found at any of these sites. Its disappearance from the last two is the result of verge-cutting. It also appeared briefly on an area of tipped garden rubbish near Lanercost (56.62, EHR, 1986). The only extant sites, and the most northerly on the west coast of Britain, appear to be on coastal shingle at Far Arnside (44.76s, MB, 1975) and on a railway embankment at Low Meathop (42.78s, J. Ketchen, 1975).
        There are eleven sites given in the Floras, all prior to 1910. The only subsequent pre-Survey records are from Silloth (10.52) where it was seen by W.W. Mason in 1921 and by C.W. Muirhead at intervals up to the 1970s, and a pre-1930 *Atlas* dot for the Kirkby Lonsdale square (6.7s, v.c.60/69).
                                8   WF(C)

LAMIACEAE
(*LABIATAE*)

Stachys officinalis (L.) Trev. St Leon   Betony
Very common throughout most of the lowlands but appearing to prefer rather leached and mildly acidic loamy soils. It is a plant of roadside verges, 'unimproved' meadows, heathy grassland and wood-land margins. It is surprisingly rare north of the Solway Firth.
Alt. limit: 460 m, Birkdale, Teesdale (80.26)
Map 653   995   LWFCY

*S. grandiflora** (Willd.) Benth.
A garden escape and native of the Caucasus recorded from a beckside at Hartley, Kirkby Stephen (7.0, I.F. Gravestock & B. Sydenham, *BSBI News* (1989) 52:30) in 1965 and for a few years subsequently.
                                                                                (W)

*S. germanica** L.
A southern species recorded in 1885 by Hodgson from ballast tips at Maryport (0.3).
                                R   (C)

* **S. byzantina** K. Koch
A garden escape found by A. Cannell at Grune Point (12.56) in 1994.
                                                                                C

S. sylvatica L.   Hedge woundwort
Very common in the lowlands occurring on roadside verges and in hedgerows, around villages, sometimes as

a weed of gardens and waste ground, and in deciduous woodland.

   With its deep claret-coloured flowers, long-petiolate, broadly ovate leaves and unpleasant smell, it is unlikely to be confused wth *Stachys* x *ambigua*.
Alt. limit: 845 m, Great Dun Fell (70.32)
Map 654   1378   LWFCY

S. sylvatica x S. palustris
S. x ambigua Smith
Scattered, in similar habitats to *Stachys palustris*, the parent it most closely resembles. Like that species it often forms extensive colonies. It frequently occurs in the absence of both parents. It is best distinguished by its paler flowers, low fertility and shortly but distinctly petiolate leaves.

   C.D. Pigott (*in litt.*) has noticed its frequent abundance close to old cottages and farmland and suggests it was formerly cultivated.
Alt. limit: 215 m, Tindale Tarn, Hallbankgate (60.58)
Map 655   151   WFCY

S. palustris L.   Marsh woundwort
Generally common in the lowlands where it is characteristic of the sides of ditches, tarns and rivers, also of ungrazed marshy fields. There are also records from relatively dry roadside banks.
Alt. limit: 540 m, Moor House, Teesdale (74.32)
Map 656   814   WFCY

*S. annua (L.) L.
A southern-central European plant listed by Hodgson as occurring in considerable quantity on ballast tips at Silloth (1.5, J. Leitch, 1889, CLE) in 1890. It was also collected at Arnside in 1911 (4.7s, Anon., LANC).

(WC)

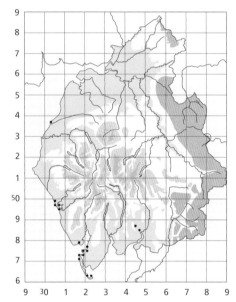

Map 652. *Cynoglossum officinale*

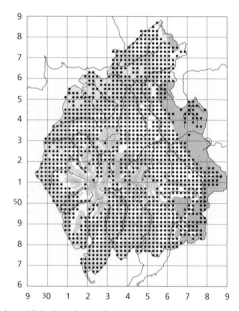

Map 654. *Stachys sylvatica*

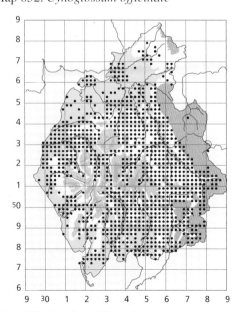

Map 653. *Stachys officinalis*

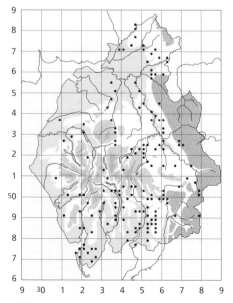

Map 655. *Stachys sylvatica* x *S. palustris*

S. arvensis (L.) L.   Field woundwort
A fairly frequent weed of arable land, allotments and
gardens on the west coast and the Solway Plain, otherwise very local. Although generally considered to be
decreasing, the Survey records are from nearly double
the number of squares in the *Atlas*.
Alt. limit: 250 m, south of Kirkby Stephen (76.04)
Map 657   152   WFC

Ballota nigra L.   Black horehound
Very rare and apparently decreasing. It was refound during the Survey at Hodgson's original site
at Harrington (98.24, CCH, 1970s), but it has subsequently been lost by landscape 'improvements'. It was
also seen on the old railway line near Great Broughton
(04.30, A. Chandler, 1980), with *Salvia verticillata* at
Silloth (10.52, NB, 1991), by a hedgerow at Haverigg
(14.78, D. Moss, 1979) and, in the east, on church
walls at Great Ormside (70.16, W. Robson, *c.*1976).
The most extraordinary find was by a newly restored
path up the west side of Helvellyn (32.16, R. Wall,
1996; LANC).

It is listed in the Floras from waste ground and
railways at several localities and there are further
Cumberland records from Great Salkeld (5.3,
H. Britten, 1903, G.A.K. Hervey, 1958) and Silloth
(1.5, J. Parkin, C.W. Muirhead, 1930s). There are no
records from the following post-1930 *Atlas* squares:
Millom (1.8), Distington (0.2), Aspatria (1.4), Langwathby (5.3) and Arnside (4.7s).

Hodgson makes the interesting observation,
attributed to the Rev. Hilderic Friend, that "People go
many miles to Beaumont on the Eden to obtain the
herb, which is very scarce here".
Alt. limit: 135 m, Great Ormside (70.16), but, no doubt
as a casual, at *c.*480 m. on Helvellyn (32.16)
                          7   WC

*Leonurus cardiaca L.   Motherwort
Listed by Hodgson from Workington Row (9.2), Peel Wyke,
Bassenthwaite (1.3/2.3), Dundraw (2.4), Curthwaite (3.4),
Carlisle Abbey and Grinsdale (3.5, J. Leitch, 1891, CLE, T.S.
Johnstone, 1900, CLE), Lessonhall (2.5) and the Nunnery,
Kirkoswald (5.4), and by Wilson from Whitbarrow (4.8s) and
Kendal (5.9). It was also collected at Rosley (3.4, J. Leitch,
1889, CLE), but the only records this century appear to
be from Grinsdale, where it was last seen by C.W. Muirhead
in 1943, and Peel Wyke (J. Robinson, 1927, J. Curwen,
*c.*1890s, CLE).

                                    (WC)

Lamiastrum galeobdolon (L.) Ehrend. & Polatschek
                          Yellow archangel
The Survey produced numerous records of the naturalised variegated garden plant, subsp. *argentatum*
(Smejkal) Stace, most of the specimens being checked
by A. McG. Stirling or A. Rutherford. The only non-
variegated plants were from the Muncaster woods
(08.94, MMG, 1986) and a collection from a streamside at Torpenhow originally found by R. Bennet

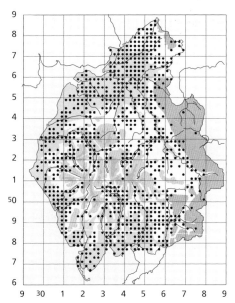

Map 656. *Stachys palustris*

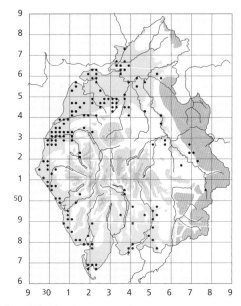

Map 657. *Stachys arvensis*

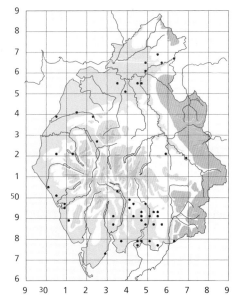

Map 658. *Lamiastrum galeobdolon* subsp. *argentatum*

(22.38, 1980, herb. R.B., *Watsonia* (1981) 13:338; LANC). The former was identified by Stirling and Rutherford as subsp. *montanum* (Pers.) Ehrend. & Polatschek.

Subsp. *argentatum* is not known to occur in the wild. The first records for Cumberland are from Millbeck, near Keswick and near Durdar (respectively 24.26, EEM, 1986; 38.50, REG, 1987, *Watsonia* (1989) 17:477), for Westmorland from Scroggs Wood, Kendal and east of Kendal (respectively 50.90, 54.90, 1986, CEW, *Watsonia* (1989) 17:476), and for Furness near Oxen Park (32.86, JA, 1984).

There are rather few records of this species in the Floras. Most are from Cumberland. Baker mentions one from Coniston (3.9), but with some reservation, and Wilson records it from Elleray Woods and Orrest Head, Windermere (4.9) and Melkinthorpe (5.2). His herbarium sheet of two specimens of subsp. *montanum* from Elleray (1907, KDL) is annotated "all that was left of it"! It was seen in the Wetheral - Warwick Bridge area (4.5) up to 1936 by J. Parkin. There is a 1949 record from Side Wood, Ennerdale (1.1, N.D. Simpson) which seems so unlikely that one suspects an error in labelling. The only other relatively recent record is from the Ravenglass woods (0.9, Mrs H. Ford, 1955) which probably refers to the Muncaster plant, subsp. *montanum*. Most of the early records no doubt also refer to this subspecies.

Alt. limit: 170 m, Old Hutton (56.86s)

Map 658   48   LWFC

Lamium album L.   White dead-nettle
Common on the west coast and in a broad band across to the Eden valley, frequent in the south but elsewhere rather uncommon and absent from many of the Lake District valleys. Although present in Westmorland since at least 1790, this species has increased considerably since Wilson's Flora in which it is described as "rather rare". It is a plant of roadside verges, waste ground and dumps, and railwaysides as, for example, along the old line at Hallbankgate (58.58n). It is seldom seen far from houses.

Alt. limit: 345 m, Garrigill (74.40)

Map 659   468   LWFCY

*L. maculatum (L.) L.   Spotted dead-nettle
An occasional garden throw-out, usually growing on roadside verges and in hedgerows near houses. It appears to be slowly increasing, at least in the east.

Alt. limit: 490 m, above Garrigill (76.40)

Map 660   67   WFC(Y)

L. purpureum L.   Red dead-nettle
Very common throughout the lowlands in waste places and on disturbed and cultivated ground, being frequent both as a garden and arable weed.

Alt. limit: *c*.450 m, Birkdale, Teesdale (80.28)

Map 661   817   LWFCY

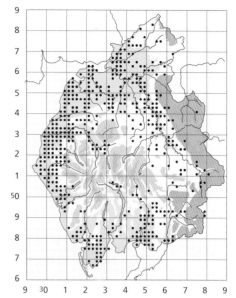

Map 659. *Lamium album*

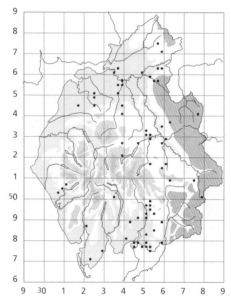

Map 660. *Lamium maculatum*

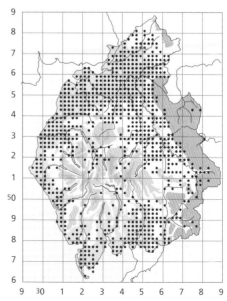

Map 661. *Lamium purpureum*

L. hybridum Villars   Cut-leaved dead-nettle
A local plant of waste and cultivated ground chiefly in
the east and south-east. Near Milnthorpe it occurs on
the limestone of Capon Crag (50.82s).
Alt. limit: 280 m, High Borrow Bridge (54.04)
Map 662   89   WFC(W/Y)

L. confertum Fries   Northern dead-nettle
(*L. molucellifolium* auct.)
A northern species reported by Handley (1898) from the
Sedbergh area apparently having been introduced with
corn. There is an *Atlas* record from the Arnside square (4.7s,
v.c.60/69).

                                              (W?W/Y)

L. amplexicaule L.   Henbit
A rather uncommon plant of waste and cultivated
ground chiefly along the coast where, as at Flimby
(0.3), it may even be locally common. There are a
number of records from the east but none more recent
than 1921 (5.3, Langwathby, W.W. Mason).

      All the plants have the small corolla charac-
teristic of self-pollinated individuals.
Map 663   42   WFC(W/Y)

*****L. garganicum** L.
A southern European species collected in 1937 outside an
abandoned farmhouse near Haweswater (4.1/5.1, Mrs Pem-
berton-Pigott, det. J.F. Chapple & A.J. Wilmott, *BEC Rpt*
(1938) 11:498).

                                              (W)

Galeopsis angustifolia Ehrh. ex Hoffm.
                         Narrow-leaved hemp-nettle
Extremely rare. It was seen by the path below Whitbarrow Scar (44.84s, KAG & FLW) from 1966 until
*c.*1970 when it appears to have been eliminated by
limestone removal. It was first seen there by G. Wilson
in 1951 (CLE). It was also reported in limestone pavement near Hale, Milnthorpe (50.78s, S.D. Ward &
D.F. Evans) in 1973. These records represent the
species' north-western limit in Britain.

      Hodgson lists it from Coulderton (9.0), St Bees
(9.1), Maryport (0.3) and Silloth (1.5, J. Leitch, 1889,
CLE), Baker from Hawkshead (3.9) and Wilson from
Arnside Knott (4.7s), Brigsteer (4.8s), Scout Scar (4.9)
and Great Strickland (5.2). There are unlocalised
Scarce Plants records for the Rampside (2.6s) and
Newby Bridge (3.8) squares and a 1959 record from the
Ravenglass square (0.9, Mrs G. Satow).
                         S 2   W(FC)

G. speciosa Miller   Large-flowered hemp-nettle
Fairly frequent north of Carlisle, occasional in the
south around the Kent estuary but elsewhere rare or
absent. Almost all the records of this very attractive and
decreasing weed are from arable fields or recently
disturbed sites. Improvements to the A590 road west of
Levens (4.8s) and the temporary cultivation for potatoes
of adjacent old peaty meadows produced a substantial

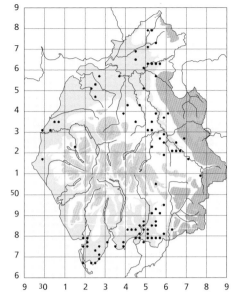

Map 662. *Lamium hybridum*

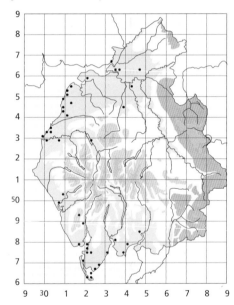

Map 663. *Lamium amplexicaule*

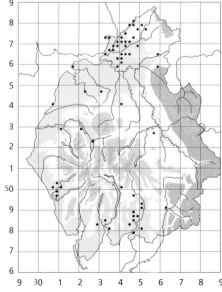

Map 664. *Galeopsis speciosa*

but short-lived population in the late 1970s.
Alt. limit: 130m, Bewcastle (52.76n)
Map 664   60   WFCY

G. tetrahit L. and **G. bifida** Boenn.
                        Common hemp-nettle
Common throughout the lowlands, chiefly as a weed of
cultivated and waste ground and rubbish dumps, but
also of roadside verges, scrub and woodland margins.

As far as it is possible to use the tenuous distinctions between the two species, *Galeopsis bifida*
would appear to be appreciably commoner.
Alt. limit: 390 m, north of Alston (72.48)
Map 665   1062   LWFCY

*Phlomis russeliana (Sims) Benth.   Turkish sage
Found on the railway embankment at Lang-
wathby (56.32, RWMC, 1990, LANC, *Watsonia* (1992)
19:150), where it was earlier seen by G.A.K. Hervey in
1956 and F. Lawson in 1957 (CLE). It also occurs by
an artificial fishpond at Raisbeck (64.06, CEW, 1996)
where it was no doubt planted.

This is a handsome south-west Asiatic sage with
large, cordate basal leaves and dense yellow-flowered
spikes.

                        2   C

*Marrubium vulgare L.   White horehound
Seen only at Haverigg (14.78, GH, 1981, LANC,
Watsonia (1982) 14:195), where there is a well-established colony by a farm to the west of the village. This
is probably now the north-western limit of the species
in Britain.

It was noted by Hodgson from Workington (9.2),
Maryport (0.3), Aspatria (1.4), Silloth (1.5) and West-
ward, Wigton (2.4), and by Baker from "Jacklands
Tarn", Ulverston (*cf.*2.7). There are also records for
Beckfoot and Allonby (0.4, E.J. Glaister, CLE;
J. Leitch, 1877, CLE) and pre-1930 *Atlas* ones for the
Arnside (4.7s, v.c.60/69) and Penrith (5.3) squares. It
was seen by C.W. Muirhead at Silloth from 1931 until
1943.

                        S 1   (W?F)C

*Sideritis montana L.
A southern European species found at Silloth in 1889 (1.5,
J. Leitch, CLE).

                                        (C)

Scutellaria galericulata L.   Skull-cap
Scattered, though locally frequent in the Lake District
and in the south. Like *Lycopus europaeus*, it is a
typical species of tall-herb fen vegetation by lakes, tarns
and rivers and in ungrazed marshy fields. There are no
Survey records for a number of west coast *Atlas* squares
and this may represent a real decline.
Alt. limit: 240 m, English Kershope (52.84n), but
Baker (1894) gives it as 365 m in High Cup Gill.
Map 666   132   LWFC

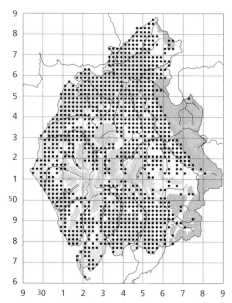

Map 665. *Galeopsis tetrahit* and *G. bifida*

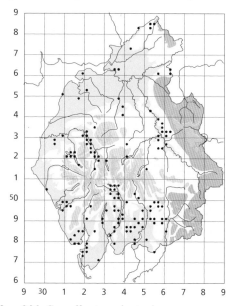

Map 666. *Scutellaria galericulata*

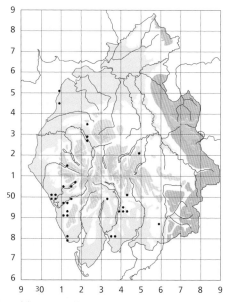

Map 667. *Scutellaria minor*

S. minor Hudson   Lesser skull-cap
A rather rare south-western species which, like *Hypericum elodes*, is decreasing both locally and nationally. It has gone from most of the Westmorland sites given by Wilson and also from the Furness peninsula, yet surprisingly it was found at a new site on the eastern fringe of the Lake District on Helton Fell (48.20, RWMC, 1992, LANC).

This is a lowland plant of slightly mesotrophic communities in marshy fields and rough grazing, by sluggish becks, small tarns and oligotrophic lakes such as Ennerdale and Wast Water.
Alt. limit: 310 m, Helton Fell (48.20)
Map 667   31   WFC

Teucrium scorodonia L.   Wood sage
Frequent to common or even, in the south-west, very common. This is a plant of rather dry hedgebanks and of siliceous rocks, crags and gills where it commonly grows with *Erica cinerea* and *Hypericum pulchrum*. Yet it also thrives on shattered limestone pavement and scree, usually growing with *Geranium robertianum*.
Alt. limit: at least 450 m, Tyne Head farm (76.36)
Map 668   987   LWFCY

Ajuga reptans L.   Bugle
Widespread and common in damp deciduous woodland and scrub, on roadside verges and wet meadows and in flushed fellside grassland.
Alt. limit: 610 m, Yad Moss (78.36)
Map 669   1169   LWFCY

A. pyramidalis L.   Pyramidal bugle
Known on Ill Bell, Kentmere (4.0) since 1869 when it was discovered by J. Backhouse. According to Baker it was then "very fine on precipitous rock". The population is now reduced to only two small plants and these, unfortunately, are accessible to sheep.

Elsewhere this species is restricted in Britain to the Scottish Highlands, particularly the extreme north, the Hebrides and one locality in the Moffat hills.
                           S 1   W

*****Nepeta cataria** L.   Cat-mint
Seen during the Survey growing among roadside gravel at Low Crosby (44.58, MP, 1991, LANC). The only previous records are Baker's from the beach at Rampside (2.6s), waste ground at Dalemain (4.2) and the roadside between Bell Bridge and Sebergham Hall (3.4), also Nicolson's from Carlisle Abbey and walls (3.5), and Wilson's from Furness Abbey (2.7).
                           1   (F)C

Glechoma hederacea L.   Ground ivy
Common throughout the lowlands, occurring in hedgerows, scrub and deciduous woodland, usually on the better drained, richer and often calcareous soils.
Alt. limit: 365 m, above Brough (78.16, Wilson)
Map 670   952   LWFCY

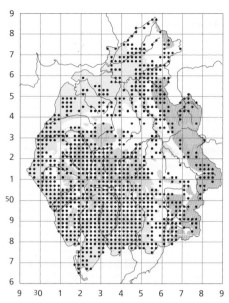

Map 668. *Teucrium scorodonia*

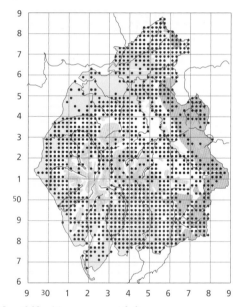

Map 669. *Ajuga pyramidalis*

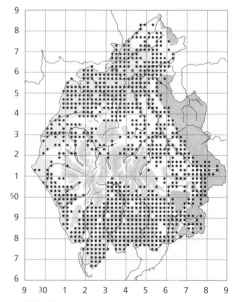

Map 670. *Glechoma hederacea*

Prunella vulgaris L.   Self-heal

Extremely common virtually everywhere except for the most intensively farmed parts of the Solway Plain and the highest parts of the Lake District. It occurs usually in rather short, damp grassland, on roadside verges, meadows and pastures, as a weed of lawns, in damp deciduous woodland and, with *Bellis perennis*, around seepage areas on the fells.

Alt. limit: 845 m, Great Dun Fell (70.32)

Map 671   1679   LWFCY

*Melissa officinalis L.   Balm

The only Survey records of this southern species are of brief occurrences on disturbed ground by woodland near Mirehouse, Whitehaven (98.14, CCH, 1979, *Watsonia* (1980) 13:140) and by a footpath, but near a garden, east of Arnside (46.78s, MB, 1981).

     The earliest record was from Arnside (4.7s, J.F. Pickard, YRK) in 1937. In 1945 it was seen in Cumberland by H. Clipshaw (site unknown) and five years later near houses at Dalston (3.5, R. Wilson & T. Gray, 1950, CLE). It was also recorded in the 1950s by A.W. Westrup at Grange-over-Sands and nearby Hampsfield Fell (4.7s, 4.8s).

     All the above records probably represent garden escapes.

     1   W(F)C

Clinopodium ascendens (Jordan) Samp.

(*Calamintha sylvatica* Bromf.)   Common calamint

There are a few records of this species in the Floras and a further ten this century from Cumberland but none later than 1945 when it was seen at Gilsland (6.6, v.c.67/70) by G.A.K. Hervey. This is the northernmost British record.

     (WFC)

C. vulgare L.   Wild basil

Frequent on the limestones around Morecambe Bay and in mid-Westmorland, elsewhere rare. It is a plant of base-rich or calcareous coarse grassland, old quarries, railwaysides, hedgerows and woodland margins. It is absent from the middle section of the River Eden but otherwise its distribution is generally similar to that of *Origanum vulgare*. Like that species it appears also to be declining being unrecorded during the Survey from a third of the *Atlas* squares and it is now scarce in west Cumberland where Hodgson described it as frequent and "in some parts very abundant".

Alt. limit: 200 m, Swindale, Brough (80.14)

Map 672   132   WFCY

C. acinos (L.) Kuntze

(*Acinos arvensis* (Lam.) Dandy)

A rare southern and decreasing species surviving in Cumbria in only seven tetrads and in three of these it is associated with railways. It was seen in the south by the railway at Grange-over-Sands (42.78s, KAG & FLW, 1966), where it survives precariously, and on the railway embankment at Arnside (44.78s), although it has disappeared during the Survey from the adjacent tetrad (46.78s). In north Westmorland it occurs on a railway bank near Warcop (74.14, Anon., 1978) and in short turf by the railway near Soulby (72.10, JMA, 1990); in Cumberland it was found in an old sand-pit at Lazonby (54.38, RWMC, 1992, LANC) and on a dry bank by the River Caldew at Cummersdale (38.52, EHR, 1990). These are the northernmost sites in western Britain.

     By contrast, there are numerous records in the Floras and additional Cumberland ones from Thwaites, Millom (1.8, M. Cross, 1932), Bromfield (1.4, J. Parkin, 1935), New Cowper (1.4, J. Curwen, c.1890,

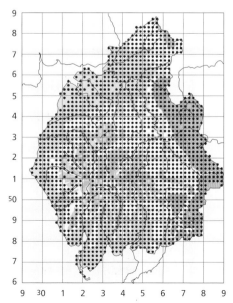

Map 671. *Prunella vulgaris*

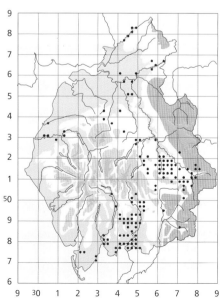

Map 672. *Clinopodium vulgare*

CLE), Armathwaite (5.4, W.W. Mason, 1914) and Hayton (5.5, T.S. Johnstone, 1906, CLE). Interestingly, three of the records given by Baker and Hodgson were from railwaysides.

Alt. limit: 210 m, Soulby (72.10)

	7		WF(CW/Y)

Origanum vulgare L.   Marjoram

Fairly common in freely draining base-rich and calc-areous habitats, as on the southern and eastern lime-stones and along the flushed, silty banks of the middle section of the River Eden. It is absent from shortly grazed grassland, but it is very much a feature of old limestone quarries, scrub and woodland margins.

      It appears to have decreased in its outlying stations during the last 50 years, there being no Survey records from nearly a third of the *Atlas* squares.

Alt. limit: 370 m, Swindale, Brough (80.16)

Map 673   91   WFC(W/Y)

***Thymus pulegioides** L.

Known only from a sand-pit near Cumwhitton (50.54, FJR, LANC, *Watsonia* (1980) 13:140) where it was discovered in 1979. This is a remarkable occurrence of a species which is largely restricted to the limestone and chalk of southern and eastern England. There is, however, an authenticated record from the coast of Wigtownshire.

      According to C.D. Pigott (*in litt.*), this species used to be accidently introduced with grass seed from the Continent.

	1		C

T. polytrichus A. Kerner ex Borbás   Wild thyme
(*T. praecox* Opiz)

Common except in the more intensively farmed low-lands, the high-level marginal hill land between the rivers Kent and Lune and the forested uplands of the north. It occurs, often abundantly, in short *Festuca ovina* or *Festuca - Agrostis* grassland wherever there is any base-enrichment and on outcrops and shallow soils over limestone, slates and volcanic rocks. One of the pleasures of walking the limestone grasslands of the Pennines on a warm summer's day is the all-pervading scent of thyme. Its presence in fellside grassland below gullies or tumbling over rock ledges is usually a clear indication of basic outcrops. Along the coast it is a common feature of sandy cliff-top turf.

Alt. limit: 915 m, Scafell Pike (20.06, Hodgson).

Map 674   1098   LWFCY

Lycopus europaeus L.   Gipsy-wort

Scattered but not infrequent in the central and southern Lake District occurring in tall-herb fen communities by tarns and rivers and in marshy fields and neglected ditches.

Alt. limit: 150 m, Bewcastle (54.76n)

Map 675   90   WFC

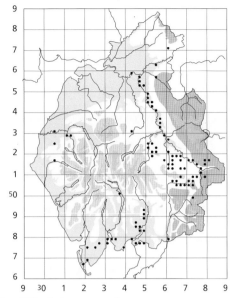

Map 673. *Origanum vulgare*

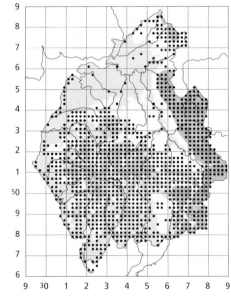

Map 674. *Thymus polytrichus*

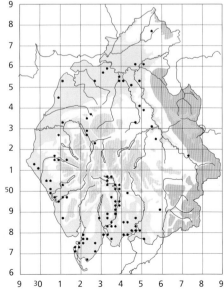

Map 675. *Lycopus europaeus*

Mentha arvensis L.   Corn mint

Scattered throughout the west, becoming commoner towards the east and very common in north Westmorland. Although a weed of fields and open ground its habitat range overlaps that of *Mentha aquatica* and it is probable that vegetative plants of *M. arvensis* from ditches and pond margins have been misidentified. *M. aquatica* does have a more pungent scent but there are no reliable vegetative characters for distinguishing between the two species.

Alt. limit: 290 m, above Ravenstonedale (74.00)

Map 676   418   LWFCY

M. arvensis x **M. aquatica**

M. x **verticillata** L.

Well distributed in the south and east but rather rare elsewhere. The concentration of records in the Ullswater area reflects the enthusiasm of the local recorder. The hybrid grows in similar situations to *Mentha aquatica*. It is robust and attracts attention by its usually vigorous and ascending lower branches; most of the records have been checked by R.M. Harley. The *Critical Atlas* records for the Newby Bridge (3.8) and Hawkshead (3.9) squares are probably the first for Furness.

Alt. limit: 300m, south of Alston (70.42)

Map 677   80   WFCY

*M. arvensis x M. aquatica x M. spicata

M. x **smithiana** R.A. Graham

The only Survey record for this rare hybrid was by a pool in a timberyard at Hutton Roof (56.76s, GH, 1983, LANC, det. R.M. Harley), but this has since been filled in. The material shows well the characteristic exserted anthers which distinguish this hybrid from *Mentha* x *gracilis*.

     This hybrid is listed by Wilson (under *M. rubra*) from the Barbon Beck (6.8s) and the River Lune at Kirkby Lonsdale (6.7s), and by Handley (1898) from the Lune near Sedbergh (6.9). The first two may be the bases of the dots in the *Critical Atlas*, which also refers to an unlocalised Cumberland record. Baker gives a single record from by the River Crake (2.8/3.8). In the absence of specimens these records should be treated as doubtful, especially those from Kirkby Lonsdale and Barbon since these are two of the Survey sites for *M.* x *gracilis*.

Alt.: 130 m, Hutton Roof (56.76s)

                             1   W[(FCW/Y)]

*M. arvensis x M. spicata

M. x **gracilis** Sole

(*M.* x *gentilis* auct.)

A very rare but no doubt overlooked mint of river and becksides. Baker lists it from St John's in the Vale (3.2) and Wilson gives 19th century records from Skelwith Bridge (3.0) and by the River Eamont (4.2/5.2). The only other pre-Survey ones are from a ditch by the River Eden at Cargo, Corby and the beck at Skirwith

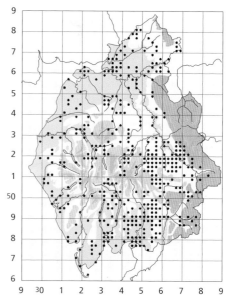

Map 676. *Mentha arvensis*

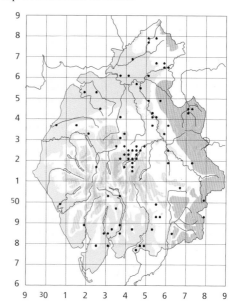

Map 677. *Mentha arvensis* x *M. aquatica*

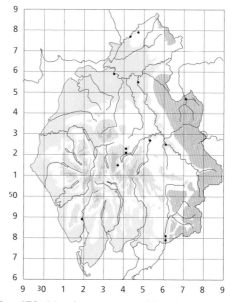

Map 678. *Mentha arvensis* x *M. spicata*

(respectively 3.5, 4.5, 6.3, C.W. Muirhead, 1949, BM,
CLE), an additional *Critical Atlas* dot for the Skelton
square (4.3) and specimens collected by C.M. Rob in
1949 from by the River Lune at Kirkby Lonsdale (6.7s,
YRK) and on an island in the River Lowther near Shap
Abbey (5.1, YRK). As mentioned under *Mentha x
smithiana*, Wilson's records of *M. rubra* from Kirkby
Lonsdale (6.7s) and Barbon (6.8s) and Handley's
(1898) from by the Lune near Sedbergh (6.9) probably
all refer to *M. x gracilis*. All the Survey records have
been determined by R.M. Harley.
Alt. limit: 260 m, north of Alston (70.46)
Map 678   13   WC(W/Y)

M. aquatica L.   Water mint
Common in wet places throughout the lowlands,
especially in ditches, by the margins of ponds and tarns
and in marshy fields.
Alt. limit: 430 m, south-east of Garrigill (76.40)
Map 679   1103   LWFCY

*****M. aquatica** x **M. spicata**   Peppermint
M. x piperita L.
Scattered and chiefly in the east. Apparently absent
from Furness. It occurs as a garden escape by road-
sides and in ditches, often at some distance from
houses. The only record of the hairy plant, f. *hirsuta*
Fraser, is from Drigg (0.9, R.H. Williamson, 1924,
PLYP).
Alt. limit: 300 m, Leadgate, Alston (70.42))
Map 680   74   WCY

*****M. spicata** L.   Spearmint
Fairly frequent in the east, rather uncommon or even
rare in the west. Like *Mentha x piperita*, it occurs as a
garden escape in ditches and on damp roadside verges
but also on tips and waste ground.
Alt. limit: at least 350 m, south of Alston (72.44)
Map 681   194   WFCY

*****M. spicata** x **M. longifolia** (L.) Hudson
M. x villosonervata Opiz
A very rare garden escape recorded during the Survey
from only seven tetrads, mostly from roadsides. Baker's
record of *Mentha sylvestris* from Furness Abbey (2.7)
may refer to this hybrid, but the first authenticated
record is from a roadside near Ireleth (22.76, JW, 1980,
LANC). The first Westmorland record is from a roadside near Kendal (5.9, T.J. Foggitt, 1934, BM, YRK,
det. R.A. Graham) and for Cumberland one from Orton
Rigg (32.52, REG, 1982, LANC, det. R.M. Harley,
Watsonia (1989) 17:476).

This hybrid is distinctly hairier than *M. spicata*,
in fact often quite grey, and the teeth of the leaves are
coarser and patent-recurved.
Alt. limit: 280 m, Waitby, Kirkby Stephen (74.06)
Map 682   7   WFCY

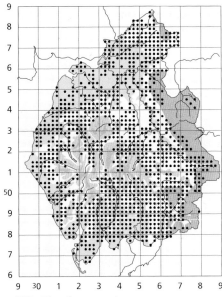

Map 679. *Mentha aquatica*

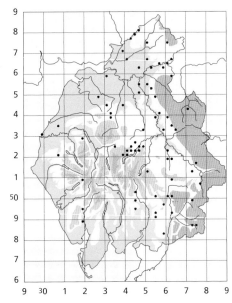

Map 680. *Mentha aquatica* x *M. spicata*

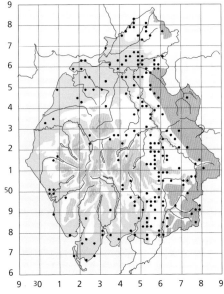

Map 681. *Mentha spicata*

*M. spicata L. x **M. suaveolens**   Apple mint
M. x **villosa** Hudson
An occasional garden throw-out, particularly in the lower Eden valley, occurring on roadsides, in ditches and waste places and often attracting attention by its vigorous growth and broadly ovate, coarsely toothed leaves. All the material submitted to R.M. Harley has been referred to var. *alopecuroides* (Hull) Briq. Most of the early records of *Mentha rotundifolia* refer to this hybrid, which is characterised by more coarsely-toothed leaves and flowering spikes 10 mm or more wide.
Alt. limit: at least 150 m, Patterdale (38.14), near Dacre (44.24) and north of Bewcastle (54.78n), although Wilson gives it as 245 m for *M. rotundifolia*.
Map 683   59   WFC

*M. suaveolens Ehrh.   Round-leaved mint
Seen only by the side of Boardale Beck on the south side of Ullswater (42.18, GH, 1983, LANC), on road-side verges north and north-east of Aspatria (12.46, A. Cannell, *c*.1990; 16.42, 18.42, NB & EEM, 1992, LANC) and south-west of Torver (26.92, PB, 1993, LANC). All the specimens were determined by R.M. Harley. The records of *Mentha rotundifolia* in the Floras probably mostly refer to *M.* x *villosa*, although early collections from "Dent Wood" (P. Shepherd, 1884, CLE) and Waterside (2.4, E.J. Glaister, 1876, CLE) are of *M. suaveolens*.

This garden escape is a smaller plant than *M.* x *villosa*. The leaves are more finely-toothed and the flowering spikes less than 10 mm wide.
Alt.: 165 m, Boardale, Ullswater (42.18)
                    5   WFC

*M. pulegium L.   Penny Royal
Listed by Hodgson from Salt Cotes and between Silloth and Skinburness (1.5) and from Low Holme Mire (5.4), and by Baker from Dalton-in-Furness (2.7). There are no records this century for this southern species which has now disappeared

from most of its British sites.
                    S   (FC)

*M. requienii Benth.   Corsican mint
A diminutive and very aromatic mint of Corsica and Sardinia collected in 1890 by Hodgson at the Derwent Tinplate Works cinder tip at Workington (9.2).
                    (C)

[Salvia pratensis L.   Meadow clary
Hodgson's only record is of a collection by J. Leitch from Silloth (1.5). It seems likely that it refers to his specimen (1889, CLE) labelled *Salvia sylvestris* which has been redetermined by I.C. Hedge as *S. nemorosa*. This is true also of C.W. Muirhead's records from between 1939 and 1943.
                    (C)]

*Salvia nemorosa L.
A very rare British casual reported from Silloth by G.C. Druce and R.H. Williamson in 1924 (1.5, 1923, CLE, *BEC Rpt* (1925) 7:590) but first collected there by E.J. Glaister in 1876 (CLE) and last seen by C.W. Muirhead from 1931 until 1943 (1939, CLE). All the specimens have been determined by I.C. Hedge.
                    (C)

[S. verbenaca L.   Wild clary
A species of southern and eastern England listed by Hodgson from the "ballast hills" at Maryport (0.3) and Silloth (1.5, E.J. Glaister, 1876, CLE). It was reported from Silloth by J. Parkin in 1935 and by C.W. Muirhead in 1939. The only subsequent record is from Little Salkeld (5.3) in 1951 where it was noted by G.A.K. Hervey. The pre-1930 *Atlas* dot for the Arnside square (4.7s) refers to a site in Lancashire. Since the only named herbarium specimen, one collected at Silloth by E.J. Glaister in 1876 (CLE), was redetermined by I.C. Hedge as *Salvia nemorosa*, it seems best to regard all the records as doubtful.
                    (C)]

*S. viridis L.
A southern European annual collected on a tip at Silloth in 1889 and 1890 (1.5, J. Leitch, CLE).
                    (C)

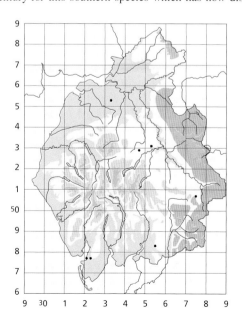

Map 682. *Mentha spicata* x *M. longifolia*

Map 683. *Mentha spicata* x *M. suaveolens*

*S. verticillata L.   Whorled clary
Known only from Silloth, where it was rediscovered in
1989, after a lapse of nearly 50 years, growing in some
quantity behind the docks (10.52, M. Armstrong, 1989,
LANC). It was first collected there in 1889 (J. Leitch,
CLE) and in 1896 and 1907 from Skinburness and
Silloth respectively by W.H. Youdale (LIV, Hofman *et
al.* (1990)) as *Salvia verbenaca*. It was seen again at
Silloth in 1926 (N. Stewart, CLE) and by C.W. Muirhead from 1932 until 1943. There are also records from
the Barrow-in-Furness docks in 1913 (1.6s/2.6s, W.H.
Pearsall, *BEC Rpt* (1914) 3:402) and from a mill yard
at Whitehaven in 1924 (9.1, R.H. Williamson, CLE)
where it persisted into the 1930s.

    These are probably the most northerly British
records.

        1   C

HIPPURIDACEAE

Hippuris vulgaris L.   Mare's-tail
Rather local and with a distinct south-eastern tendency. Most of the records are from lowland tarns and
small pools. These are often quite eutrophic but in the
fells it also occurs in slow-flowing becks, as in Connypot Beck on Stainmore (80.20), and in *Sphagnum*
'lawns', these sites being in marked contrast to its
frequently stated preference for base-rich waters.
Alt. limit: 540 m, Connypot Beck, Stainmore (80.20)
Map 684   52   WFC

CALLITRICHACEAE

Callitriche hermaphroditica L.
            Autumnal water-starwort
Rather rare and largely confined to the Solway Plain,
Bassenthwaite Lake and south Westmorland, occuring
in small mesotrophic tarns and in the Lancaster -
Kendal canal. Most of the records are from shallow
water but in Bassenthwaite J.M. Lock found it in 2 m of
water with stems up to 1 m long.

    The first definite record for Cumberland is from
Moorthwaite Lough, Wigton (28.48, GH, 1976, LANC,
Watsonia (1978) 12:172) and for Furness from
Esthwaite Water (3.9, T.J. Foggitt, 1917, BM).
Alt. limit: 125 m, Talkin Tarn (54.58)
Map 685   19   WFC

C. stagnalis Scop.   Common water-starwort
Common, occurring in practically all fairly shallow and
calm freshwater communities and over a wide range of
nutrients and pH. It grows high on the fells in peaty
pools and flushes and it is frequent in the lowlands on
the dried-up mud of rutted farm tracks, the margins of
ponds, damp forestry rides and seepage areas at the
upper limit of salt-marshes.
Alt. limit: at least 610 m, near Great Dun Fell (70.32),
although Wilson gives it as 715 m on Helvellyn (3.1).
Map 686   1136   LWFCY

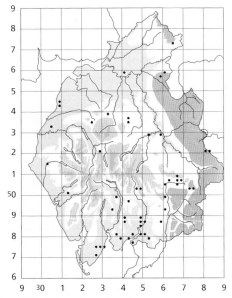

Map 684. *Hippuris vulgaris*

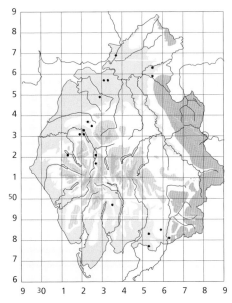

Map 685. *Callitriche hermaphroditica*

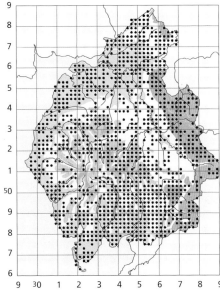

Map 686. *Callitriche stagnalis*

C. platycarpa Kütz.

Apparently rare but no doubt appreciably overlooked. It was seen during the Survey at four sites each in Cumberland and Furness, the earliest being from Southerfield Moss, Abbeytown (1.4, C.W. Muirhead, 1948, CLE, det. N.F. Stewart) and near Wray Castle (36.00, A.J. Richards, 1971) respectively.

This species can be distinguished only with difficulty from *Callitriche stagnalis*, the most reliable characters being the chromosome number ($2n=20$, instead of 10) and the variable shape of the pollen. As a rule the mericarps are less strongly winged. These distinctions have only recently been appreciated and the records given by Baker and Hodgson are best disregarded.

This species could well be quite widespread, particularly in the south of the county since it is considered by Graham (1988) to be relatively widespread in Co. Durham and by Livermore & Livermore (1987) to be frequent in the Lancaster district.

Map 687   10   FC

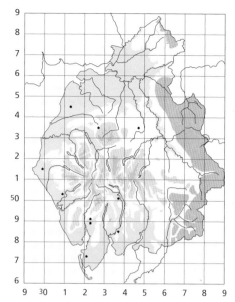

Map 687. *Callitriche platycarpa*

C. obtusangula Le Gall

Very rare, mainly restricted to the Solway Plain, Low Furness and south Westmorland but, like *Callitriche platycarpa*, certainly under-recorded. It is characterised by rhomboid leaves and unwinged or narrowly winged mericarps with patent styles. The first record for Westmorland is Wilson's from Meathop Marsh (4.7s/4.8s, 1935, YRK, conf. N.F. Stewart), for Cumberland from Monkhill Lough, west of Carlisle (32.58) and Longtown gravel pits (36.68, both RS, 1977, herb. R.S., det. P.M. Benoit, *Watsonia* (1978) 12:171-172), and for Furness from Helton Tarn in the Winster valley (40.84s, RS, 1977, herb. R.S., det. P.M. Benoit).

Map 688   9   WFC

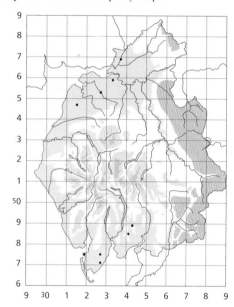

Map 688. *Callitriche obtusangula*

C. hamulata Kütz. ex Koch

                    Intermediate water-starwort

Appreciably more widespread than is suggested by the *Atlas* but common only in the Lake District, south Westmorland and Furness. It grows in generally deeper water than *Callitriche stagnalis*, forming beautiful undulating sheets of vegetation in the rivers and long delicate ascending filaments in the deep dubs and tarns. Although apparently flourishing best in oligotrophic waters, it is quite common on the limestone around the upper Eden valley and between the upper Lune (6.0) and Crosby Ravensworth (6.1).

There are no records of the related *C. brutia* Petagna with only weakly emarginate leaves and smaller, more obviously winged fruits.

Alt. limit: 820 m, Cross Fell (68.34)

Map 689   399   WFCY

PLANTAGINACEAE

Plantago coronopus L.   Buck's-horn plantain

Very common around almost the entire coast but pre-

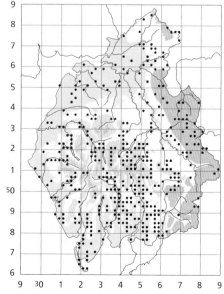

Map 689. *Callitriche hamulata*

ferring somewhat open, dry, stony or sandy habitats including paths and the stonework of sea-walls and embankments. Wilson gives a single inland site at Asby (6.1, H. Britten) and there are Survey records from roadside verges east of Crook (48.94, CEW, 1987, LANC) and the Brampton by-pass (54.60, MG & JP, 1995).

Alt. limit: 130 m, east of Crook (48.94)

Map 690   119   WFC

P. maritima L.   Sea plantain

Like *Plantago coronopus,* this is very common around the coast occurring in the middle and upper zones of the salt-marshes and in coastal turf, often with *Armeria maritima.*

It is also frequent on the limestone around the upper Eden valley where it also grows as an adventive along roadsides and is apparently increasing. Other inland sites include a species-rich pasture with *Genista anglica* and *Serratula* near Dean (06.26), the limestone near Blindcrake (14.36), flushes with *Primula farinosa* at Johnby (40.32) and *Kobresia simpliciuscula* in Teesdale (80.26, 80.28), along the forestry road by Ennerdale Lake (1.1), on Honister Crag (20.14) and Hartside (64.40), and by Cartmel racecourse (36.78s).

Alt. limit: 575 m, Hartside (64.40)

Map 691   176   WFC

P. major L.   Greater plantain

Extremely common, absent only from the highest mountains and areas of blanket peat. This species is extremely resistant to mechanical damage and hence is characteristic of roadsides, pavements, footpaths and waste sites, also of grazed lowland patures and intake fields, football pitches and lawns.

The only records for subsp. *intermedia* (Gilib.) Lange are from Askam-in-Furness where it was seen in a salt-marsh (20.76, PB, 1993, LANC) and on a coastal woodland path.

Alt. limit: 845 m, Great Dun Fell (70.32)

Map 692   1621   LWFCY

P. media L.   Hoary plantain

A common species of the limestone, picking out remarkably well the limestones to the north and west of the Lake District and around Morecambe Bay yet quite rare on the Pennine scarps. The rather few occurrences on base-rich soil off the limestone proper include the Sebergham area (3.4), the Duddon estuary (2.7), a pasture by the River Eden near Rockcliffe (36.60) and rocks of Silurian slate by the River Lune near Sedbergh (6.9). Most of the records are from grazed *Festuca ovina* grassland and roadside verges.

Alt. limit: 480 m, east of Hartside (68.42) and South Stainmore (88.08)

Map 693   273   LWFCY

P. lanceolata L.   Ribwort plantain; Ribgrass

Abundant, one of the commonest species, occurring in

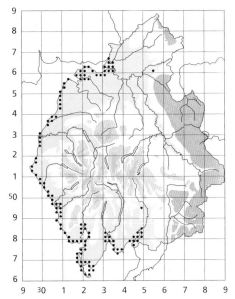

Map 690. *Plantago coronopus*

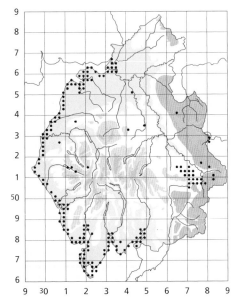

Map 691. *Plantago maritima*

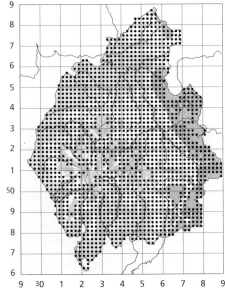

Map 692. *Plantago major*

all but the most acid grassland and particularly characteristic of roadside verges, meadows and pastures, short,
base-rich upland grassland, flushed seepage areas in
the fells and rock ledges and crevices on most of the
less acid rock outcrops. It was once valued as a
constituent of hay-meadows, partly because, being
deep-rooted, it was relatively rich in minerals.

Alt. limit: 770 m, south of Knock Fell (72.30)

Map 694   1722   LWFCY

*P. arenaria Waldst. & Kit.
A Continental species known only from Hodgson's records
from Workington (9.2), Maryport (0.3) and Silloth saltworks
(1.5, J. Leitch, 1889, CLE), and one from the docks at
Barrow-in-Furness in 1913 (1.6/2.6s, W.H. Pearsall, *BEC Rpt*
(1914) 3:402).

                                                    (FC)

Littorella uniflora Aschers.   Shore-weed
A common species of the margins of pools, tarns and
lakes, also of quiet backwaters of rivers, often forming
dense carpets and superficially resembling *Isoetes*. It
grows in shallower water than the latter, seldom more
than a metre deep, and it is capable of tolerating the
fairly long periods of exposure which usually induce
flowering.

Alt. limit: 490 m, Angle Tarn, Patterdale (40.14)

Map 695   223   WFC

BUDDLEJACEAE

*Buddleja davidii Franchet
Well established, vigorous and spreading in a few,
mainly urban sites where it grows on walls, in car parks
and on waste land, railway stations and sidings.

        There are no pre-Survey records and it is
impossible to distinguish first records.

Alt. limit: 210 m, Spadeadam (60.70n)

Map 696   33   WFC

*B. globosa J. Hope
An uncommon garden shrub seen by a footpath near
Hawkshead in 1989 (34.98, CEW, LANC).

                        1   F

OLEACEAE

*Forsythia suspensa (Thunb.) M. Vahl x **F. viridissima**
Lindley
F. x **intermedia** Zabel
A bush, probably of this hybrid, was seen in a roadside hedge near Cartmel (38.78s, CEW, 1992, LANC),
numerous bushes were found around an old mine pool
near Askam-in-Furness (20.76, PB, 1993, LANC) and
a large bush was seen by the A590 near Sedgwick
(50.86s, GH, 1995, LANC). This popular garden shrub
may well be established elsewhere with the increasing
dumping of surplus garden material.

                        3   WF

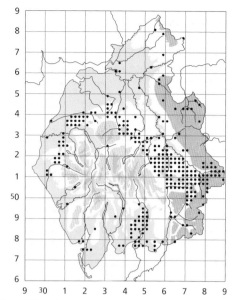

Map 693. *Plantago media*

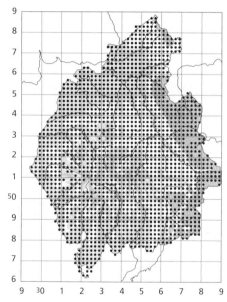

Map 694. *Plantago lanceolata*

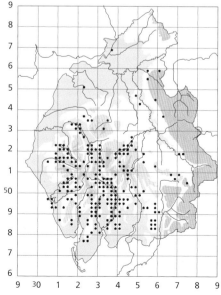

Map 695. *Littorella uniflora*

Fraxinus excelsior L.   Ash

Probably the commonest tree in the county, absent only from the highest fells, peatlands and areas of blanket afforestation. It is the characteristic hedgerow tree, being readily laid, but if left alone easily recognised from afar, whether in leaf or not, by its pale bark and the ascending tips to its drooping branches. It is usually the dominant tree in the species-rich limestone woods, such as those around the Kent estuary and above Brough (7.1, 8.1), and with oak it picks out those areas of richer soils, usually gillsides, in the Lake District fells. The extent to which the forest cover has disappeared around the dale heads can be gauged by the old, isolated ash-trees, now often far removed from the lower valleyside woods, as, for example, at Haweswater and Kentmere. Ash is frequently used in mixed amenity plantings but rarely commercially except on the richer soils of the southern limestone.

Pollarded ash trees, locally called 'cropping ashes', are a conspicuous feature of field boundaries and around old farms in most Lake District valleys. C.D. Pigott points out (*in litt.*) that the straight branches were cut at 4-5-year intervals and used within living memory to make wattle fences known, at least in Borrowdale, as 'stake-and-rise'. The stakes were driven into the ground and the long twigs known as rise were woven between them. At an earlier period rise was almost certainly used as winter fodder, probably cut leafy and dried, as was still the practice in the mountain valleys of Scandinavia in this century. Not only the leaves but the green bark of ash are both palatable and nutritious for stock.

Wilson gives a single record for the entire-leaved var. *diversifolia* Aiton from Staveley (4.9, G.E. Martindale, 1886, BM; KDL) and during the Survey it was discovered in a hedgerow at Strands, Millom (18.84, MMG, 1989, LANC).

Alt. limit: 490 m, above Garrigill (76.38)

Map 697, Plate 83   1544   LWFCY

Observe that the tops of the ash trees were lopped; and was informed that it was done to feed the cattle in *Autumn*, when the grass was on the decline; the cattle peeling off the bark as food. In Queen Elizabeth's time the inhabitants of *Colton* and *Hawksheadfells* remonstrated against the number of forges in the county, because they consumed all the loppings and croppings, the sole winter food for their cattle.

A Tour of Scotland and Voyage to the Hebrides in 1772
Thomas Pennant

Plate 83. Pollarded ash tree in field wall, Underbarrow

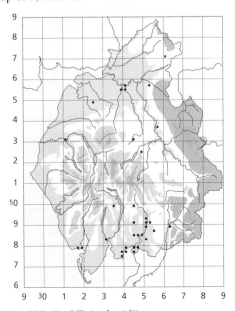

Map 696. *Buddleja davidii*

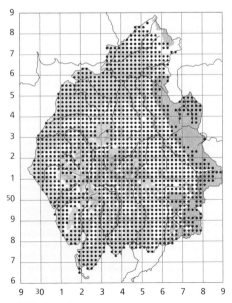

Map 697. *Fraxinus excelsior*

*F. ornus L.   Manna ash
A southern European tree reported by J.A Martindale as planted on a field-edge at Benson Hall, Kendal (5.9, 1905, KDL).

(W)

*Syringa vulgaris L.   Lilac
Occasional but apparently extensively planted in hedges across north Cumberland. It is also a feature of urban waste sites and may sometimes be found well established far away from present or former gardens. Recorders have probably been reluctant to record it from the gardens of many upland abandoned farms in which it is a frequent feature and where it will doubtless long persist.

      The only pre-Survey record is from near Egremont where it was noted as long ago as 1895 by J. Adair (00.10, *BEC Rpt* (1897) 1897:490) and where it still occurs. The first Furness record was from Colton (30.86, JA, 1978) and for Westmorland from Mealbank, Kendal (54.94, CEW, 1984).
Alt. limit: 205 m, railway embankment near Grayrigg (58.96)
Map 698   91   WFCY

Ligustrum vulgare L.   Wild privet
A fairly frequent species of hedges in the lowlands, where it is probably mostly planted, but also of woodland, especially on the limestone. It is particularly fine on the exposed western cliffs of Humphrey Head (38.74s).
Alt. limit: *c*.260 m, Shap (56.14)
Map 699   440   LWFCY

*L. ovalifolium Hassk.   Garden privet
Scattered, but perhaps rather frequently unrecorded in view of its usual occurrence in hedges near houses or persisting as a garden throw-out on waste ground. The only pre-Survey record is from Lodore (2.1, R.H. Williamson, 1926, CLE); the first for Westmorland is from Casterton (62.80s, GH, 1983), but it is not possible to identify that for Furness.
Alt. limit: at least 150 m, Soulby (74.10)
Map 700   56   WFC

SCROPHULARIACEAE

*Verbascum blattaria L.   Moth mullein
A widespread Continental species known in Cumbria only from the single record given by Hodgson from Aspatria churchyard (1.4).

(C)

*V. virgatum Stokes   Twiggy mullein
The only Survey records are of casual occurrences at Arnside (44.76s, MB, 1975, 1985), from the roadside between Grange-over-Sands and Cartmel (38.78s, E.J. Harling), where a single plant was seen in 1978, and on recently landscaped ground at Silloth (10.52, NB & MMM, LANC, *Watsonia* (1986) 16:190), where it

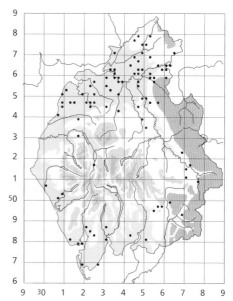

Map 698. *Syringa vulgaris*

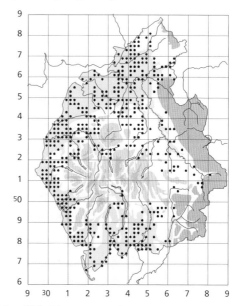

Map 699. *Ligustrum vulgare*

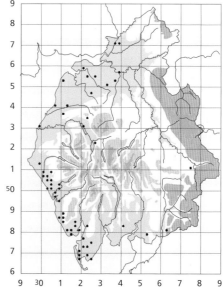

Map 700. *Ligustrum ovalifolium*

appeared briefly in 1985.

It is listed by Baker from Roosebeck (2.6s) and by Hodgson from Workington (9.2). There is a post-1930 *Atlas* record for the Distington square (0.2) and in 1953 it was found by G.A.K. Hervey at Kendal (5.9).

S 3   WFC

V. thapsus L.   Great mullein

A conspicuous and fairly frequent plant throughout the lowlands, particularly along the coast and in the lime-stone country around the Kent estuary. It occurs on waste ground, dumps, gravel workings. quarries and occasionally in gardens, as well as in undisturbed habitats such as limestone scars and pavements. It may well be commoner than in the past.

Alt. limit: 305 m, Alston (70.44)

Map 701   283   LWFCY

The mullein would be one of the most striking of our wild flowers if all its flower heads came out together. Instead they come out first at the bottom, as if a lamplighter were climbing its fat stalk, and as each flower head lights up, the one below goes out.

Cumberland and Westmorland, Norman Nicholson

*V. nigrum L.   Dark mullein

There are Survey records of casual occurrences on the roadside at Clappersgate (36.02, R. Young, 1970, AMDE) and nearby at Ambleside (38.00, Anon.), Arnside (44.76s, MB, 1977), by the roadside west of Sedbergh (60.92, CEW, 1988, LANC), on the old rail-way embankment between Middleton and Sedbergh (62.88s, PB, 1992, LANC) and on waste land east of Heversham (50.82s, CEW, 1986), the two plants at the last site being white-flowered.

Hodgson records it from Workington (9.2), Westward, Wigton (2.4) and Overwater (2.3), and Wilson from near Temple Sowerby (6.2). It was also seen at Workington in 1928 by J.C. Varty-Smith (*BEC Rpt* (1929) 8:749) and there is an *Atlas* dot for the Silloth square (1.5) and a record from South Walney (2.6s, J. Buch, 1965).

Alt. limit: 230 m, west of Sedbergh (60.92)

6   W(FC)

Scrophularia nodosa L.   Figwort

Common throughout the lowlands, on roadside verges, in hedgerows, scrub and woodland and by becks and riversides. It tolerates considerable shade and seems to prefer the heavier, wetter soils. It seems likely that many of the *Atlas* records of *Scrophularia auriculata* belong here.

Alt. limit: 480 m, Angle Tarn, Patterdale (40.14)

Map 702   1042   LWFCY

S. auriculata L.   Water figwort

Rather rare; most frequent in the south, especially around Morecambe Bay and characteristic of ditches

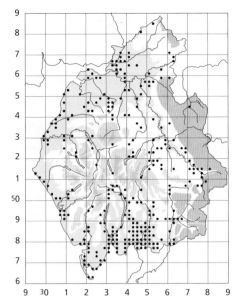

Map 701. *Verbascum thapsus*

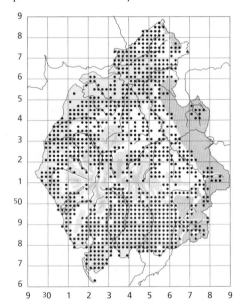

Map 702. *Scrophularia nodosa*

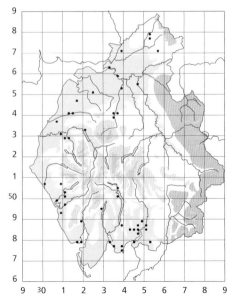

Map 703. *Scrophularia auriculata*

rather than the sides of tarns and rivers.

By contrast, Hodgson gives numerous localities, all by, or close to, the River Eden. Baker records it from the Furness coast and Kirkby Lonsdale (6.7s) and Wilson describes it as rather rare. He also cites it from along the Eden as well as from Windermere (4.9), Far Arnside (4.7s), Levens and Brigsteer (4.8s). It has also been reported in Cumberland from Bromfield (1.4, J. Parkin, 1925) and Isel Woods, Cockermouth (1.3, C.W. Muirhead, 1948).

There seems little doubt that most if not all the Eden records belong to *Scrophularia umbrosa*, although C.W. Muirhead does mention the two growing together at Grinsdale, Carlisle (3.5, 1939). If the Eden records are excluded, the general picture from these early sources agrees with the Survey records. It seems very likely that the many *Atlas* records indicate confusion both with *S. nodosa* and *S. umbrosa*.
Alt. limit: 135 m, north-west of Bewcastle (52.78n)
Map 703   48   WFC(W/Y)

*S. umbrosa Dumort.

Well distributed and locally frequent along the River Eden, from the Scandal Beck down to Carlisle.

The first records date from 1924 when it was found by H.H. Harvey on the banks of the Eden at Temple Sowerby (6.2, *BEC Rpt* (1925) 7:586) and 1939 when C.W. Muirhead noted it at Grinsdale, Carlisle (3.5, CLE). However, as mentioned under *Scrophularia auriculata*, it seems likely that most if not all the Eden records given for that species by Hodgson and Wilson belong to *S. umbrosa*.

The important diagnostic features are the leaves, which are never cordate at the base, and the staminode, which is broader than long and somewhat two-lobed.
Alt. limit: 225 m, Ravenstonedale (70.04)
Map 704   33   WC

*S. vernalis L.

The only Survey record is from a public garden at Arnside (46.78s, MB) where it has persisted since 1985. It was also reported by Wilson as a garden escape on a railway bank at Arnside and there is an undated specimen from Levens (4.8s, J.A. Martindale, KDL).

1   W

*Mimulus moschatus Douglas ex Lindley   Musk

Rare, and mainly in the south and west. Most sites are fairly close to houses and farms, one, in upper Bannisdale (50.04), is the yard of an abandoned farmstead. It was seen in some quantity by the Haweswater Beck immediately below the dam (50.14) and between paving stones near the centre of Carlisle (40.56). It is interesting that it still persists at Mosedale (34.30) where, a century earlier, Hodgson had noted that it was "not likely to become permanent"!

Hodgson also mentions a well-established colony of this North American species by the mouth of the River Calder (0.0) and one at Pooley Bridge (4.2), the last site possibly being in Westmorland. There is an early Westmorland record from Rydal (3.0, W.M. Rogers, 1903, LANC), probably the same site as a later one from between Grasmere and Ambleside (3.0, T.B. Blow, 1923, *BEC Rpt* (1924) 7:203). The earliest Furness records are those of the Barrow N.H.S. for the Broughton-in-Furness (2.8) and Duddon (2.9) squares in the early 1950s.
Alt. limit: 210 m, Haweswater (50.14)
Map 705   19   WFC

*M. guttatus DC.   Monkey flower

It is clear that most of the plants previously included under this western North American species are of the hybrid with *Mimulus luteus*. The true species, which is fertile, is rather uncommon but is certainly under-

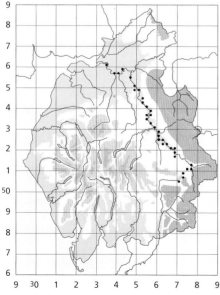

Map 704. *Scrophularia umbrosa*

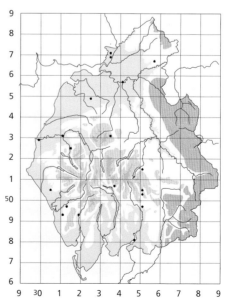

Map 705. *Mimulus moschatus*

recorded and probably occurs throughout the range of the hybrid. Like the latter, it is a plant of becksides and riversides.

It is most readily distinguished from the hybrid towards the end of the flowering period when the swelling capsules are evident within the calyces.

Alt. limit: 270 m, Hen Hill, Spadeadam (60.72n)

Map 706   10   WC

M. guttatus x **M. luteus**

M. x **robertsii** Silverside

Very common throughout the upland areas of the eastern half of the county where it is a colourful sight by nearly every beck and extending well down the rivers into the lowlands. The map is substantially correct although it no doubt includes some records of *Mimulus guttatus*.

Like most sterile hybrids it has a long flowering season. Vegetative spread by detached pieces of stem clearly seems to be extremely efficient. The flowers vary greatly in the intensity and extent of blotching but this appears to have no taxonomic significance.

Alt. limit: *c.*440 m, above Garrigill (76.40)

Map 707   445   WFCY

M. guttatus x **M. cupreus** Dombrain

M. x **burnetii** S. Arn.

Very rare, known only from St John's Beck at Legburthwaite and along one of its tributaries near Wanthwaite (30.20, 30.22, MC, 1984, herb. M.C.; LANC), from Lowthwaite (40.22, MC, 1980, LANC), and by the River South Tyne below and above Garrigill (72.40, GH, 1975, LANC; 70.42, 72.42, 74.40), the River Irthing above Gilsland (62.68, MG, 1990, LANC; 68.72n) and the River Mint above Patton Bridge, north-east of Kendal (54.00, MB & R. Dalton, 1984; LANC).

This hybrid appears to have been first collected from Dale Head, presumably that by Thirlmere (3.1, A.H. Wolley-Dod, BM) in 1931. There are no subsequent records from there, but it was collected in the Vale of St John during the 1960s by Mrs E.M. Satow, the material being identified as the hybrid, and not, as originally thought, *Mimulus cupreus*.

Alt. limit: 415 m, above Garrigill (74.40)

Map 708   10   WC

M. guttatus x **M. cupreus** x **M. luteus**

x **M. polymaculus** Silverside ined.

Recorded in 1982 by T.C.G. Rich from the shore of Grasmere (32.06, LANC, det. A.J. Silverside), where it had earlier been noted by L. Holmes in 1974.

This is a very rare British plant, most of the records being from South Wales.

1   W

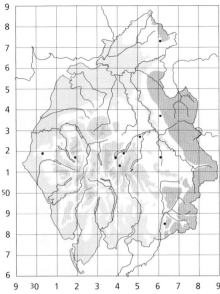

Map 706. *Mimulus guttatus*

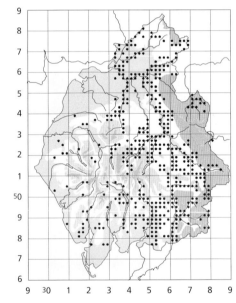

Map 707. *Mimulus guttatus* x *M. luteus*

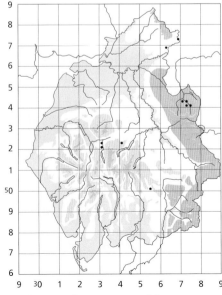

Map 708. *Mimulus guttatus* x *M. cupreus*

*M. guttatus x **M. luteus** x **M. variegatus** Jaume St-Hil.
M. x **caledonicus** Silverside ined.
Found by a streamside near Grizebeck (24.84, GH, 1978, LANC) and by the River Leven at Newby Bridge (36.86, GH, 1991, LANC). The only other record is from near Boltongate (24.40, E.J. Glaister, 1879, CLE), probably the first English record. All three collections have been identified by A.J. Silverside.

This handsome triple hybrid, previously confused with the preceding, is widely scattered in Scotland but in England is known only from north Yorkshire, Co. Durham and Northumberland.

2   WF(C)

[**M. luteus** L.
There are no authenticated records of this species from Cumbria and it seems likely that those shown in the *Atlas* refer to either *Mimulus guttatus* or the hybrid. The species is in fact extremely rare in Britain.

(FC)]

Limosella aquatica L.   Mudwort
Extremely rare, known now only from the margin of a pond in the Silloth square (1.5) where it was particularly abundant during the dry summer of 1995. It was first discovered there by C.W. Muirhead in 1946. The only previous records are two given by Hodgson from Newby (3.5, P. Shepherd, 1884, CLE) and Thurston-field Lough (3.5), where it was seen by DAR in the 1950s, and ones from Seascale (0.0, Anon., undated, BM) and Cardurnock (1.5, A. Wallis, 1914, BM).

This species has declined considerably in Britain during the present century.

S 1   C

*Antirrhinum majus L.   Snapdragon
A native of south-west Europe very commonly grown in gardens and reported during the Survey on walls at Carlisle (40.54, REG, 1985) and Great Corby (46.54, REG, 1985), and on waste ground at a number of other sites.

Baker gives it as naturalised at Town End, Witherslack (4.8s) and Hodgson as a casual at Flimby (0.3) and Aspatria (1.4). It was also collected in 1901 from Dalston station (3.5, T.S. Johnstone, CLE).
Alt. limit: 190 m, Hesket Newmarket (34.38)
Map 709   10   WC

Chaenorhinum minus (L.) Lange   Small toadflax
Common on the ballast of used and disused railway lines and persisting long after their demise. It also occurs along the south-west shore of Bassenthwaite Lake having presumably spread from the adjacent disused railway. It seems to have been particularly effective in spreading along the old Penrith - West Coast line and up the Liddel valley into Scotland. Elsewhere it is rather rare, but there are records from industrial waste ground, road and tracksides, piles of roadside grit, gardens and the gravel of a sewage bed at Tindale (60.58). It has also appeared along remote forestry tracks in the Kershope and Spadeadam forests.
Alt. limit: 425 m, Kirkstone Pass (40.06)
Map 710   192   WFC

*Misopates orontium (L.) Raf.   Weasel's snout
A southern weed of gardens and arable land known only from Hodgson's localities, Braystones and Beckermet (0.0), Nethertown (9.0), Frizington (0.1) and Flimby (0.3), a record from Silloth (1.5, R. Wood, 1880, CLE) and later ones from Kirkoswald (5.4, M. Cross, 1907), Skinburness (1.5, "D.S.", 1928) and again at Braystones in 1944 (C.D. Pigott).

(C)

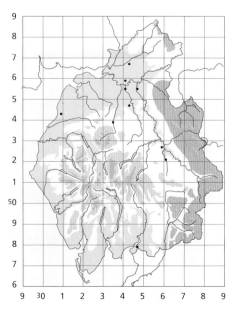

Map 709. *Antirrhinum majus*

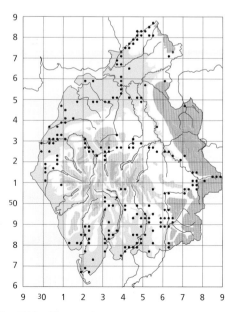

Map 710. *Chaenorhinum minus*

***Cymbalaria muralis** P. Gaertner, Meyer & Scherb.

Ivy-leaved toadflax

Very common on mortared walls and buildings, both of stone and brick. Plants appear sensitive to severe and prolonged frost and are not infrequently killed, particularly in the east.

Alt. limit: 450 m, west of Garrigill (72.40)

Map 711   649   LWFCY

C. pallida (Ten.) Wettst.

A very rare plant of walls and garden paths, similar habitats to those of *Cymbalaria muralis*, but it appears to be less successful than that species in escaping from habitation. It was first recorded by Clapham *et al.* (1952) from the shingle beach at Bardsea (30.74s). The original source of this record is unknown but the plant continued to be seen there up to 1977. It was seen on waste ground at Hawkshead Hill (32.98, CEW, 1989) and at Hallbankgate (58.58, FJR, 1989, LANC), and on walls at Castle Carrock (54.54, MG, 1990, LANC), around Sandwith and at Loweswater (96.14, 98.14, 10.22, CCH, 1979, herb. C.C.H., *Watsonia* (1981) 13:337), the first Cumberland records. The only Westmorland records are unlocalised Biological Records Centre ones for the Shap (5.1) and Warcop (7.1) squares.

It is most easily distinguished from *C. muralis* by its much larger corolla (15-25 mm, incl. spur).

Alt. limit: 200 m, Hallbankgate (58.58)

6   (W)FC

Kickxia elatine (L.) Dumort.

A rapidly declining weed of southern and central England known in Cumbria only from the two sites listed by Hodgson, Powhill and Kirkbride (2.5, E.J. Glaister, 1882. CLE), and a late 17th century Nicolson record from Great Salkeld (5.3).

(C)

Linaria vulgaris Miller   Toadflax

Frequent in the lowlands, especially in the north and south-west but curiously scarce in the south-east where, as by the A590 east of Newby Bridge (38.84), the M6 in the Lune gorge (60.96) and along the A685 converted railway line between Gaisgill and Ravenstonedale (6.0), it appears to be a relatively recent colonist. It is a plant of roadside verges and hedgerows, of urban waste sites and railwaysides.

Alt. limit: at least 360 m, north-east of Alston (72.48)

Map 712   250   WFCY

L. purpurea (L.) Miller   Purple toadflax

An occasional garden escape, establishing itself in villages and built-up areas by pavements and on walls, waste-tips and industrial sites.

The only previous records are the two given by Hodgson and the *Atlas* dots for the Dalton-in-Furness (2.7) and Barbon (6.8s) squares.

Alt. limit: 200 m, Knock (68.26)

Map 713   49   WFC

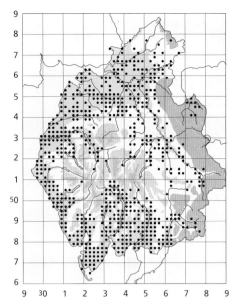

Map 711. *Cymbalaria muralis*

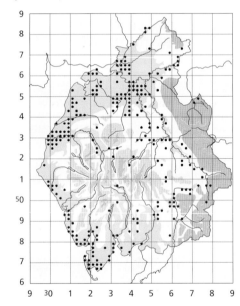

Map 712. *Linaria vulgaris*

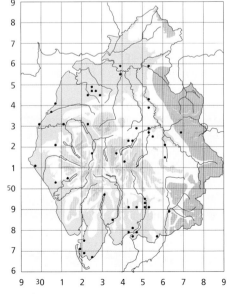

Map 713. *Linaria purpurea*

*L. purpurea x **L. repens**
L. x dominii Druce
This hybrid, wrongly referred to in the *Check-list* as *Linaria repens* x *L. vulgaris*, occurs in some quantity around Glenridding (38.16, MC, 1977, LANC, *Watsonia* (1984) 15:135) growing on walls and by the Grisedale track. It is very variable, but no pure *L. repens* was seen. There are further records from the grounds of North West Water's premises, Greenside, Kendal (50.92, GH, 1982, LANC) and Dalemain House (46.26, MC, 1978).
Alt. limit: 150 m, Glenridding (38.16)

3	WC

L. repens (L.) Miller   Pale toadflax
Occasional, occurring in rather dry open ground, by pavements, tracks and roadsides and on walls, railway ballast and sandy ground by the coast. Only rarely is it found away from houses, as on roadside grit-piles in the Lune gorge (60.98, 60.00) where it has presumably been introduced with the grit from the west coast.
Alt. limit: 300 m, Dillicar, Grayrigg (60.98)

Map 714	72	WFC

*L. arenaria DC.
A native of western France collected by Wilson from a railway track near Sedbergh (6.9, YRK) in 1894.

(Y)

Digitalis purpurea L.   Foxglove
Very common everywhere except for areas of predominantly calcareous soil and continuous blanket peat. It prefers acid, siliceous soils and is particularly characteristic of oakwoods, rocky becksides, roadsides and railwaysides. As Hodgson observes it "springs up in myriads" following the clear-felling of woodland, and sometimes also after bracken clearance. A major factor in its success is undoubtedly its unpalatability to grazing animals.
Alt. limit: 625 m, Pavey Ark (28.06)

Map 715	1561	LWFCY

*Erinus alpinus L.   Fairy foxglove
An occasional garden escape, chiefly in the south, occurring on mortared or limestone walls and rock outcrops but seldom far from houses. It is particularly fine on the limestone walls by Ulverston station (28.78) and on castle walls, as at Muncaster (10.96), Greystoke (42.30) and Naworth (56.62). Surprisingly, a few plants occur on limestone ledges near the summit of Scout Scar (48.90, GH, 1988), but these may well have been planted. It was first recorded for Cumberland from the Roman Wall near Brampton (5.6, G.A.K. Hervey, 1951), for Westmorland from Shap Wells (56.08, G.A.K. Hervey, 1961) and for Furness from a quarry in Roudsea Wood (32.82, G. Wilson, 1951, CLE).
Alt. limit: 350 m, north of Alston (72.48)

Map 716	37	WFC

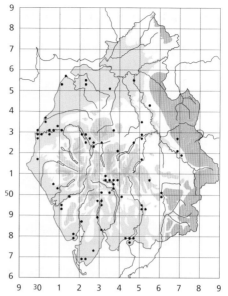
Map 714. *Linaria repens*

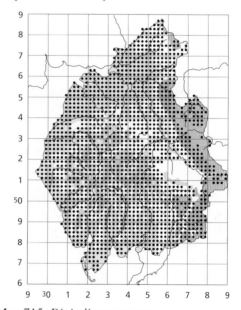
Map 715. *Digitalis purpurea*

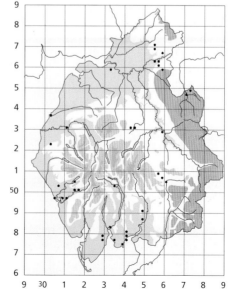
Map 716. *Erinus alpinus*

Veronica serpyllifolia L.   Thyme-leaved speedwell
Common in short, damp grassland, occurring in 'unimproved' pastures, on roadside verges and hedge-banks, in flushed grassland by streams and becks and, in the fells, by waterfalls and on damp rock ledges. It occasionally occurs as a weed of lawns and cultivated ground.

Subsp. *humifusa* (Dickson) Syme was reported last century from the Helvellyn cliffs (3.1), but the record has never been substantiated nor has any convincing herbarium material been seen to support the other *Critical Atlas* records for the Lorton (1.2) and Duddon (2.9) squares. This montane race is characterised by fewer and larger, deep blue flowers and pedicels which appreciably exceed the calyx.
Alt. limit: 825 m, Cross Fell (68.34)
Map 717   1421   LWFCY

V. officinalis L.   Common speedwell
Common to very common, especially in the Pennines, in short, well-drained *Festuca - Agrostis* grassland. It tends to avoid both the more acid and calcareous soils.
Alt. limit: 720 m, Dufton Fell (74.28)
Map 718   1065   LWFCY

V. chamaedrys L.   Germander speedwell
Extremely common, perhaps the commonest of all the hedgerow flowers, occurring also in 'unimproved' grassland and on roadside verges. It is mainly lowland in the Lake District, but it is noticeably commoner on the generally richer soils of the Pennine hills, occurring far up the becksides, around waterfalls and on rock outcrops.
Alt. limit: 720 m, Dufton Fell (74.28)
Map 719   1591   LWFCY

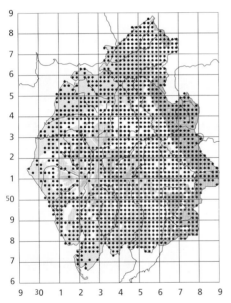

Map 717. *Veronica serpyllifolia*

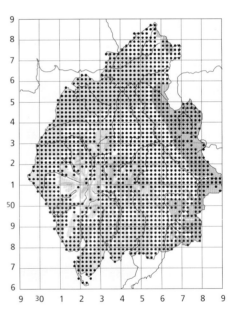

Map 719. *Veronica chamaedrys*

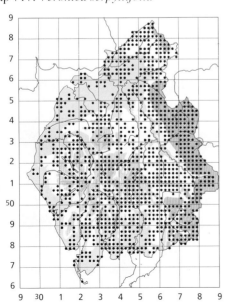

Map 718. *Veronica officinalis*

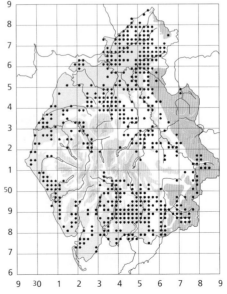

Map 720. *Veronica montana*

V. montana L.   Wood speedwell
Frequent in the lowlands on moist, usually flushed and
rather heavy soils in deciduous woodland. It is
commonest on the band of calcareous soils to the north
of the Lake District, in the extreme north along the
Liddel and Lyne rivers and their associated cloughs,
and in the woods of the south-eastern Lake District and
Morecambe Bay.
Alt. limit: 305 m, Alston (70.46)
Map 720   552   LWFCY

V. scutellata L.   Marsh speedwell
Occasional on slightly flushed but generally base-poor
peat, often occurring as scattered plants in taller
vegetation in ditches, around mires and in *Juncus
effusus* soakways on the fellsides.
　　　Allen (1984) mentions the wide variation in
petal colour in the Isle of Man. Cumbrian plants are
usually pale pink, but very occasionally white.
Alt. limit: 780 m, Cross Fell (68.34)
Map 721   386   WFCY

V. beccabunga L.   Brooklime
Very common in the lowlands in ditches and dykes, by
ponds and tarns, by slow-flowing becks and rivers and
also in hillside runnels, but surprisingly rare in the
Lake District.
Alt. limit: 845 m, Great Dun Fell (70.32)
Map 722   1344   LWFCY

V. anagallis-aquatica L.   Blue water-speedwell
A lowland plant occurring in similar habitats to
Veronica beccabunga but only really frequent in the
north, the Eden valley, Low Furness and around the
Kent estuary.
Alt. limit: at least 250 m, Hardendale Fell (58.10)
Map 723   158   LWFCY

V. anagallis-aquatica x **V. catenata**
V. x lackschewitzii J.Keller
Known only from Gleaston Beck, Urswick (2.7, D. Lumb,
1912, BM, CGE, NMW, det. J.H. Burnett) and a 1950 BRC
record from the Whitbarrow square (4.8s). It should be
looked for around Mere Tarn, Aldingham (2.7) where both
parents occur.

(W)

V. catenata Pennell   Pink water-speedwell
A rare but perhaps overlooked species of muddy pond
and stream margins on the Solway Plain and by Morecambe Bay.
　　　The only previous records are, for Cumberland,
from Briggle Bank, Langwathby (5.3, C.W. Muirhead,
1947, CLE) and a pre-1930 *Atlas* record for the Silloth
square (1.5), for Westmorland the post-1930 *Atlas* one
from the Tebay square (6.0), and for Furness A.W.
Westrop's from the Dalton (2.7, 1953) and North
Walney (1.7, 1953) squares.
Map 724   18   WFC

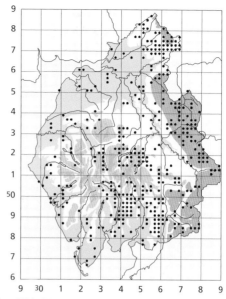
Map 721. *Veronica scutellata*

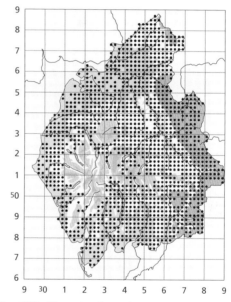
Map 722. *Veronica beccabunga*

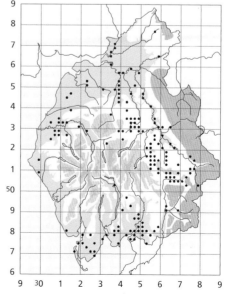
Map 723. *Veronica anagallis-aquatica*

V. arvensis L.   Wall speedwell
A frequent lowland weed of gardens and cultivated
ground, also occurring on wall-tops and in other dry,
open habitats such as sand-dunes and quarries, limestone scars and pavements.
Alt. limit: 780 m, Knock Fell (70.30)
Map 725   976   LWFCY

*V. peregrina L.
Rare, occurring almost exclusively as a garden weed,
sometimes in considerable quantity as at a Grasmere
garden centre (32.06, K.M. Hollick, 1955, DBY;
LANC), the first Westmorland record. The first for
Cumberland was from a garden at Boothby, Brampton
(5.5, Mrs B.H.S. Russell,1954-1957, CGE).

     This American species is probably spreading as
the above are the only Cumbrian records featured on
Bangerter's (1966) map.

Alt. limit: 190 m, Tebay (60.04)
Map 726   13   WCY

V. agrestis L.   Field speedwell
Fairly frequent in the lowlands, occurring in much the
same habitats as *Veronica arvensis*, in open communities on walls and tracksides and in cultivated
gardens and fields.

     It differs from *V. polita*, with which it is often
confused, in having capsules with only sparse glandular
hairs and a style which is shorter than the sinus.
Alt. limit: at least 255 m, near Shap (54.14)
Map 727   198   WFCY

V. polita Fries   Grey speedwell
A rather rare plant of rather dry, open and usually
cultivated ground. It is appreciably less common than

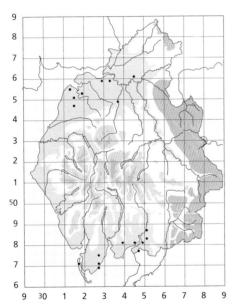

Map 724. *Veronica catenata*

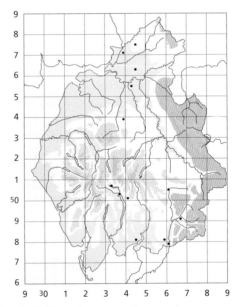

Map 726. *Veronica peregrina*

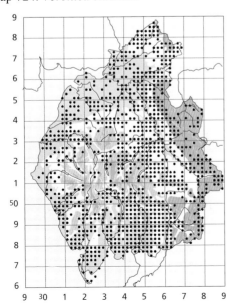

Map 725. *Veronica arvensis*

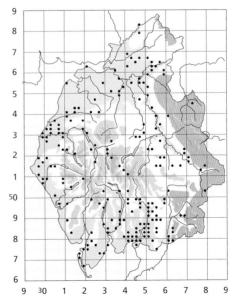

Map 727. *Veronica agrestis*

Veronica agrestis and it is certainly not as frequent now as was suggested by Wilson.

The most reliable diagnostic characters separating it from *V. agrestis* are the capsules, which have a style longer than the sinus and a mixture of long glandular and short, crisped, eglandular hairs.
Alt. limit: *c*.350 m, near Heggerscales, Kaber (82.08)
Map 728   48   WFCY

*V. persica Poiret   Buxbaum's speedwell
Frequent in the lowlands as a weed of cultivated and waste ground. It is particularly common as an arable weed.
Alt. limit: at least 350 m, south-west of Leadgate, Alston (68.42)
Map 729   496   LWFCY

*V. filiformis Smith   Slender speedwell
A common and colourful early-flowering species of south Westmorland, the Eden valley and much of the north but still scarce along the west coast and over much of Furness.

The spread of this sterile Caucasian speedwell has been as dramatic in Cumbria as in Britain as a whole. The first record for Furness dates from 1941 when it was seen on a roadside at High Wray (36.98, T.G. Tutin, *Watsonia* (1950) 1:255). In Cumberland it was reported from Wreay Woods (4.4/4.5, E. Oliphant) in 1945 and in Westmorland from near Appleby (6.2, Miss P. Bewlay, BM) in 1953.

It occurs usually as a weed of lawns, parks, village greens and roadside verges, transforming them into a blue haze for a brief period in April and May.
Alt. limit: *c*.450 m, Nenthead (78.42)
Map 730   492   LWFCY

V. hederifolia L.   Ivy-leaved speedwell
Frequent along the Eden valley and in the south, elsewhere sparse or absent. Wilson actually refers to it as "rather rare". It is a plant of cultivated ground, especially gardens, but also of hedgerows and deciduous woodland on the richer soils.

Subsp. *lucorum* (Klett & H. Richter) Hartl is by far the commoner. It has larger, pale lilac flowers and a shorter style than the blue-flowered subsp. *hederifolia*, but the differences are not always convincing. The latter has been reported only from coastal scrub at Anthorn (18.58, REG,1985), on sandy banks at Cumwhitton and Moorthwaite (50.50, FJR, 1989, LANC), a garden at Little Salkeld (56.34, RWMC, 1986) and at two sites at Penrith (50.30, 52.30, RWMC, 1987).
Alt. limit: 200 m, Brough (78.14)
Map 731   338   LWFCY

*V. longifolia L.   Garden speedwell
A large flowering clump was found in 1989 established on the verge of the Bull Pot fell road above Casterton (64.78s, K. Butler; LANC, *Watsonia* (1991) 18:429). It presumably originated as a garden throw-out. The

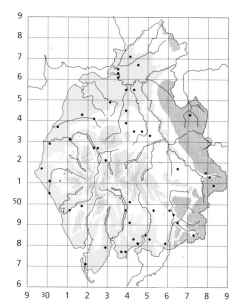

Map 728. *Veronica polita*

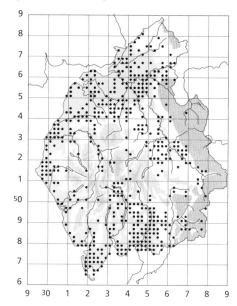

Map 729. *Veronica persica*

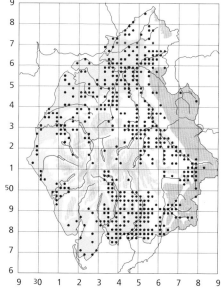

Map 730. *Veronica filiformis*

material may possibly represent the hybrid with *Veronica spicata*.

l   W

V. spicata L.   Spiked speedwell
Known only from the limestone cliffs on the west side of Humphrey Head (38.74s) and two sites in limestone grassland near Arnside (4.7s). The present sites are renowned for their rarities, the Humphrey Head cliffs also harbouring *Aster linosyris* and *Hypochaeris maculata*. All three species here reach the northern limit of their very discontinuous distributions on the limestones of western Britian. The only other site in northern England for this *Veronica*, a Nationally Protected Species, is in the Ingleborough area.

The Cumbrian plants are usually included in the western subsp. *hybrida* (L.) Gaudin, but the distinctions between this and the eastern subsp. *spicata* are unconvincing. The Humphrey Head population shows genetic variation in size, plants from the lower cliffs being appreciably taller in cultivation and having longer spikes than those from the cliff-top. The latter, in turn, are taller than those at Arnside.

The records given by Baker and Hodgson from Penny Bridge, Ulverston (3.8) and Eskdale Green (1.0) presumably refer to garden escapes of this or the preceding species.

S 2   WF

[Sibthorpia europaea L.   Cornish moneywort
Both Baker and Wilson regarded the 19th century record from Longsleddale (4.0/5.0) as an error, the species being virtually confined in Britain to south-west England.

(W)]

*****Melampyrum cristatum** L.   Crested cow-wheat
Reported by Hodgson as having been collected at Silloth (1.5)

in 1875. This distinctive and beautiful south-eastern species can hardly have been misidentified; it may, perhaps, have been introduced with corn.

S   (C)

M. pratense L.   Common cow-wheat
Fairly frequent and a very characteristic herb of valley-side oakwoods on siliceous soils, especially in the Lake district, also in the open at higher altitudes among *Calluna* and *Vaccinium myrtillus*.

As with *Rhinanthus minor*, no attempt was made to record the segregates, but it seems likely that all the material is of subsp. *pratense* and that the deep golden-yellow var. *hians* Druce is less common than the pale yellow var. *pratense*. The presence of the two variants was noted as long ago as 1822 when Dawson Turner (Hodgson 1898) commented on them both in Lanthwaite Wood, Crummock Water (1.2). A population with yellow, pink and white-flowered individuals was seen on the fellside above Kirkland (64.34) growing amongst bilberry.
Alt. limit: 715 m, Dove Crag (36.10)
Map 732   231   WFCY

[M. sylvaticum L.   Wood cow-wheat
Although still known to occur on the Yorkshire side of the river in upper Teesdale, there are no authenticated records from Cumbria and it seems likely that those cited in the early Floras refer to *Melampyrum pratense* var. *hians*.

(WFC)]

Euphrasia   Eyebrights
This account is based largely on a substantial collection of Survey and early herbarium material examined by A.J. Silverside in 1991 and 1993. Field records have also been accepted for the four commonest species:

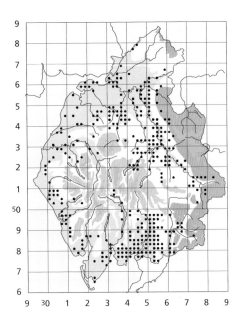

Map 731. *Veronica hederifolia*

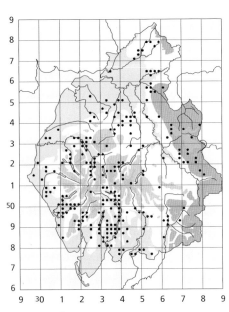

Map 732. *Melampyrum pratense*

Euphrasia nemorosa, E. confusa, E. scottica and *E. officinalis* and for these no attempt has been made to identify first records. The general picture is now fairly clear, but certain areas, notably the Pennines, from Dentdale to Alston, and the central Lake District need more attention.

Silverside's (1991) treatment of the *E. officinalis* group is adopted in preference to that of Stace (1991).

Map 733, all records   1249   WFCY

Key

1. Upper stem with long glandular hairs ....*officinalis*
  2. Stem and branches usually flexuous, lower internodes less than 1.5 times as long as leaves; corolla not more than 8 mm long   ....subsp. *anglica*
  2. Stem and branches erect or divergent; lower internodes more than 1.5 times as long as the leaves; corolla 8-12 mm long
    3. Lowest fls at nodes 2-6; stem with 0-3 pairs of branches   ....subsp. *monticola*
    3. Lowest fls at nodes 6-10; stem with up to 5 pairs of branches   ....subsp. *rostkoviana*
1. Stem eglandular or with short glandular hairs
  4. Upper stem with short glandular hairs (Lake District)   ....*rivularis*
  4. Upper stem with crisped eglandular hairs only
    5. Stem less than 3 cm; lvs with hairs on the surface (Lake District)   ....*ostenfeldii*
    5. Stem usually more than 3 cm; lvs glabrous on the surface
      6. Stem unbranched or with less than 3 pairs of branches
        7. Floral lvs not sharply toothed; capsule emarginate (Lake District)   ....*frigida*
        7. Floral lvs sharply toothed; capsule truncate

        8. Corolla predominantly white; branches, if any, from the base   ....*scottica*
        8. Corolla predominantly lilac; branches, if any, in the upper half   ....*micrantha*
      6. Stem with 3 or more pairs of branches
        9. Lvs longer than the internodes; fl spike very congested (coastal)   ....*tetraquetra*
        9. Lvs shorter than the internodes
          10. Stem with slender, ascending flexuous branches from the base   ....*confusa*
          10. Stem with straight, erect or divergent branches
            11. Floral lvs *c*.6 mm, ovate, eglandular   ....*nemorosa*
            11. Floral lvs 6-9 mm, ovate to orbicular, occasionally glandular   ....*arctica*

E. officinalis L.
(*E. rostkoviana* Hayne)
An occasional and probably decreasing species of unimproved grassland and hay-meadows and now largely confined to the western and southern Lake District, the Howgill Fells and the extreme north-east. There are additional pre-Survey records from the Cleator Moor (0.1), mid-Duddon (2.9), Helvellyn (3.1), Matterdale (3.2), Shap (5.1), Barbon (6.8s), Dentdale (7.8), Alston (7.4), and Stainmore (8.1) squares. The following subspecies are not always readily distinguishable.
Alt. limit: 370 m, Honister Pass (2.1, A.J. Silverside (1994))

WFC(Y)

subsp.**rostkoviana** (Hayne) F. Towns.
(*E. rostkoviana* subsp. *rostkoviana*)
This is the commonest of the three subspecies. It flowers later than subsp. *monticola* and has more (6-10) non-flowering internodes which are also shorter

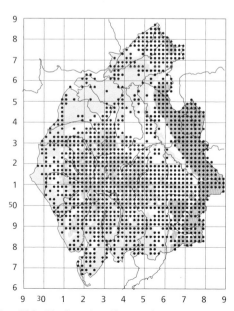

Map 733. *Euphrasia*, all records

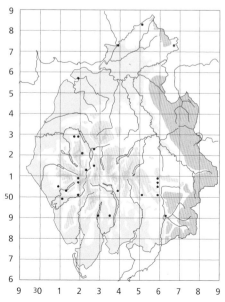

Map 734. *Euphrasia officinalis* subsp. *rostkoviana*

(to twice as long as the leaves). Silverside (1994) states that hybrids between this and subsp. *monticola* occur in the Lake District

This subspecies is largely restricted in Britain to Wales, north-west England and southern Scotland.
Map 734   S 25   WFC

subsp. rostkoviana x E. rivularis
An extremely rare hybrid listed by Yeo (1975) for Cumberland and recorded during the Survey from Honister Pass (22.12, A.J. Silverside, 1981, herb. A.J.S.) and Pillar (18.10, T.C.G. Rich, 1982, LANC).
Alt. limit: 670 m, Pillar (18.10)
                                2   C

***cf.* subsp. rostkoviana x E. confusa**
This rather distinctive hybrid was found on Murton Fell, near Ennerdale Bridge (08.18, Anon., *c.*1980, herb. A.A.D.), at two sites in damp turf above Wythop Mill (16.28, REG, 1980, LANC), in a meadow near Irton Road Station (14.98, MMG, 1991, LANC), at Buttermere (18.14, AAD, 1992, LANC), by the River Gilpin west of Crook (42.94, GH, 1992, LANC) and on Blawith Common (28.90, PB, 1992, LANC). The only other record is from Hobcarton Crag (1.2 (Yeo 1975)).
Alt. limit: at least 500 m, Hobcarton Crag (1.2)
                                6   WFC

subsp.monticola Silverside
(*E. rostkoviana* subsp. *montana* (Jordan) Wettst.)
Rather less common than subsp. *rostkoviana* but with no obvious difference either in distribution or ecology.

This is nationally much less common than subsp. *rostkoviana* being virtually restricted to the Lake District, the northern Pennines and the Borders.
Map 735   S 12   WFC(Y)

subsp. anglica (Pugsley) Silverside
(*E. anglica* Pugsley)
Extremely rare. This southern eyebright is known with certainty only from a 1947 collection from below Whinlatter Pass (1.2/2.2, K.J.F. Park, BM). A specimen from a beckside west of Crook (42.94, CFS & JS, 1982) was determined by P.F. Yeo as subsp. *anglica* but no specimen was retained. As later material from the same site has been determined by Silverside as *Euphrasia officinalis* x *E. confusa* the record is best regarded with some doubt.

The *Critical Atlas* gives records from the Duddon (2.9) and Hawkshead (3.9) squares and Wilson mentions his own record, verified by H.W. Pugsley, from by the River Lune near Sedbergh (6.9). No specimens have been traced but subsp. *rostkoviana* was collected during the Survey from the same area (62.92, GH, 1991, LANC).

This subspecies differs from the previous two in having dull, rather grey leaves, lower leaves which are as large or larger than the upper, and an elongate spike of smaller flowers.

                                ([WF]C)

subsp. monticola x E. confusa
A single population of this hybrid was found in rough grassland west of Strands (10.04, GH, 1994, LANC).
                                C

subsp. monticola x E. arctica
Extremely rare, known only from Buttermere (1.1, PLYP) and Watendlath Tarn (2.1, PLYP), where fine material of the hybrid and parent species was collected by C.W. Muirhead in 1950.
                                (C)

E. rivularis Pugsley
A small but conspicuous species of rocky flushes, seepage areas and wet rock ledges in the central Lake District and with an outlying new site in the east by Blea Tarn, High Street (44.10, GH, l987, LANC). The

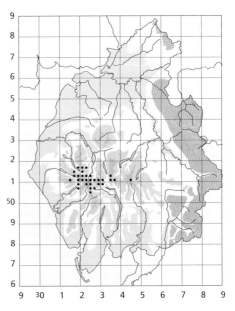

Map 735. *Euphrasia officinalis* subsp. *monticola*   Map 736. *Euphrasia rivularis*

first Cumberland and Westmorland records were from Sticks Pass (3.1, 1912, ex herb. A.H. Wolley-Dod, BM, det. P.F. Yeo) and Langdale Pikes (2.0, H.F. Pugsley, 1943).

Material from Scafell Crag (2.0, DAR, 1955, CLE) is considered by AJS to be either this species or a hybrid with *Euphrasia officinalis* subsp. *rostkoviana*.

The delicate habit, short glands and seemingly disproportionately large, pale violet flowers make this an easily recognisable species. It is endemic to North Wales and the Lake District.

Alt. range: 380 m, Langstrath (26.10) - 700 m, Pillar (l6.12)

Map 736, Plate 44   R 31   WC

E. arctica Lange ex Rostrup

Occasional but perhaps often wrongly recorded as *Euphrasia nemorosa*. Most of the Survey records are from dry banks, unimproved pasture, riverbanks, and also from a disused railway line. The first record for Cumberland is from north Bassenthwaite Lake (2.3, J. Cosmo-Melville, 1901, BM), for Westmorland from Rydal (3.0, C. Bailey, 1886, BM) and for Furness from Hawkshead Moor (2.7, W.H. Pearsall, 1917, BM). Exceptionally fine specimens, to 30 cm, were collected by C.W. Muirhead at Buttermere and Watendlath (respectively 1.1 and 2.1, both 1950, PLYP).

Most of the specimens appear intermediate between subsp. *arctica* of the Shetlands and Orkney and the widespread northern and western subsp. *borealis* (F. Towns.) Yeo, the leaves usually lacking glandular hairs, as in the former, but having the straight, non-flexuous branches of the latter. The plants are typically robust with long internodes, short lower cauline leaves and a rather dense, wide, 'top-heavy' spike with very broad, strongly toothed floral leaves.

Alt. limit: 330 m, Eycott Hill (38.30) but recorded in the *Critical Atlas* (as *E. brevipila*) from the Alston square (7.4) and by Rawes (1981) from Middle Tongue, Cross Fell (6.3).

Map 737   44   WFCY

E. arctica x E. nemorosa

The following collections are referred by AJS to this hybrid, which is probably not infrequent within the area of *Euphrasia arctica*: Barrow Docks (18.66s, JW, *c*.1980, LANC), Threlkeld (32.24, 1987), Newbiggin, Penrith (46.28, 1985) and a roadside bank west of Penrith (48.28, 1982, all MC, herb. M.C.). The plants in the last collection are exceptionally tall, up to 50 cm. There are also earlier collections from Hawkshead Moor (3.9, W.H. Pearsall, 1917, YRK) and Talkin Tarn (5.5, C.W. Muirhead, 1950, PLYP). The specimens from the last site are very tall and sparingly branched.

Alt. limit: 185 m, Newbiggin, Penrith (46.28)

                         4   FC

E. arctica x E. confusa

Known only from Walney Island (16.66s, JW, *c*.1980,

LANC), Rather Heath, Crook (48.94, GH, 1975, LANC), Seatoller (24.12, CCH, 1982, herb. A.A.D.), English Kershope (52.84n, MG & JP, 1994, LANC) and an early collection from the railway embankment at Wreay (4.4, D. Stewart, 1925, CLE).

Alt. limit: at least 110 m, Seatoller (24.12)

                         4   WFC

E. arctica x E. micrantha
E. x difformis F. Towns.

This rare hybrid is known only from a single Survey collection from a flushed area by Cuddyshall Bridge, Kershope (52.80n, GH, 1981, LANC).

Alt.: 155 m, Kershope (52.80n)

                         1   C

E. arctica x E. scottica
E. x venusta F. Towns.

Known only from a roadside flush between Hawkshead and Coniston (32.98, PB, 1994, LANC) and material, tentatively assigned by AJS to this hybrid, collected in short, flushed turf below Cold Fell (60.54, FJR, 1980, LANC).

Alt. limit: 500 m, Cold Fell (60.54)

                         2   FC

E. tetraquetra (Bréb.) Arrond.

Extremely rare, recorded only from Askam slag bank (20.76, PB, 1992, LANC) and dunes at Haverigg (16.78, JW, 1980, LANC) and Drigg (04.98, AAD, 1992, LANC; 06.96, CCH, 1979, LANC). Most records from coastal sites probably belong to one of the following two hybrids. There are also *Critical Atlas* dots for the Walney Island (1.6s), Millom (1.8) and Cartmel (3.7s) squares.

This is a predominantly western maritime eyebright, occurring on sea-cliffs and sand-dunes.

                         4   FC

E. tetraquetra x E. nemorosa

Known only from Survey records from sand-dunes, dune-slacks and coastal grassland at Sandscale (18.74, GH, 1984, LANC), North Walney (16.72, GH, 1981, LANC) and Haverigg (14.78, GH, 1981, LANC).

                         3   FC

E. tetraquetra x E. confusa

The only records are an undated one from Drigg (0.9, A. Wilson, YRK), Survey ones from coastal grassland at Askam-in-Furness (20.76, PB, 1992, LANC) and a tentative identification from coastal grassland near Allonby (06.40, CCH, 1982, herb. A.A.D.).

                         1   F(C)

E. nemorosa (Pers.) Wallr.

Frequent, occurring on heaths and dunes, in rough grassland and moorland but possibly less frequent in the uplands than *Euphrasia confusa*.

This is the commonest species and usually

readily recognisable by its robust branching habit, the absence of glands and the narrow spikes. It varies in height from 2-50 cm. Small plants from open, exposed, or heavily grazed sites may be impossible to distinguish from *E. confusa*. The map should be interpreted with caution as some of the field records may refer to the hybrid or, particularly in the uplands, *E. confusa* .

Alt. limit: 825 m, Cross Fell (6.3)

Map 738   525   LWFCY

E. nemorosa x E. confusa

Plants forming a complete continuum between these two species are not uncommon and perhaps as common in Cumbria as anywhere. These have either been recorded in the field as the aggregate or assigned to one of the parents. Mapping the hybrid would be highly subjective.

                                   WFC

E. nemorosa x E. scottica

The only records are from flushes at Wythburn (30.10, FJR, 1985, LANC), Side Fell, Bewcastle (58.72n, GH, 1981, LANC), a roadside bank west of Penrith (48.28, MC, 1982, herb. M.C.), a collection from Ambleside (38.04, AJS, l981, herb. A.J.S.) and one from Hawkshead Moor, 34.96, PB, 1994, LANC). Material from Dufton Fell (74.28, RWMC, 1991, LANC) probably belongs here.

Alt. limit: 720 m, Dufton Fell (74.28)

                         5   WFC

E. confusa Pugsley

Frequent or locally common in the uplands but absent from the lower Eden valley and Solway Plain and much of the west coast. It appears to be generally more upland than *Euphrasia nemorosa* where, as stated above, the latter may well be over-recorded. Wilson's specimen of *E. pseudokerneri* Pugsley from Arnside Tower (4.7s, 1934, YRK) is *E. confusa*.

     This is usually a smaller plant than *E. nemorosa* and distinguished by the flexuous, slender branches from the base of the stem and leaves with forwardly-directed teeth.

     Euphrasia confusa is essentially a plant of short grassland, both on the well-drained Asby - Orton and Pennine limestones and in the Lake District.

Alt. limit: at least 660 m, east of Cross Fell (70.32)

Map 739   458   WFCY

E. confusa x E. scottica

Probably not infrequent although unrecorded prior to the Survey. The first records are: for Westmorland from Hutton Roof crags (54.78s, GH, l970, LANC), for Furness from Brantwood, Coniston (30.94, CEW, 1989, herb. C.E.W.) and for Cumberland from Priest's Crag, Ullswater (42.22, MC, l975, LANC).

Alt. limit: 380 m, above Melmerby (62.38)

Map 740   22   WFCY

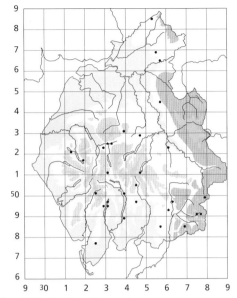

Map 737. *Euphrasia arctica*

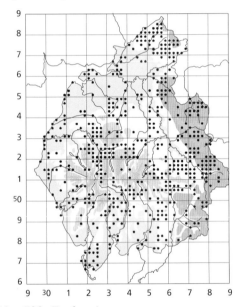

Map 738. *Euphrasia nemorosa*

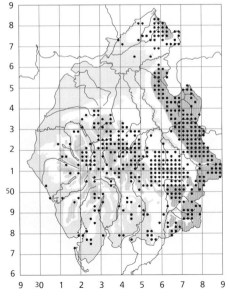

Map 739. *Euphrasia confusa*

E. frigida Pugsley

An extremely rare eyebright of rock ledges in the Lake District where it was recorded from Piers Gill, Scafell Pike (20.06, GH, l988, LANC), Red Screes, Kirkstone (38.08, RWMC, 1984, LANC), Link Cove, Fairfield (36.10, GH, 1978, LANC), Dollywaggon Pike (34.12, T.C.G. Rich, 1982, LANC) and Brown Cove, Helvellyn (32.14, RWMC, l989, LANC), where it was also collected by DAR in 1954 (CLE). There are other early collections made by C.W. Muirhead from Great End (2.0, 1952, BM, PLYP) and Coniston Old Man (2.9, 1951, BM, PLYP). No information is available concerning the *Critical Atlas* record for the Ennerdale square (1.1). The Red Screes specimens are particularly well developed being up to 12 cm high and with the characteristic rounded floral leaves and emarginate capsules. Material (GH, LANC) possibly referable to this species, or hybrids with it, has been collected from Red Gill, Glaramara (24.10, 1995), Keppel Cove, Helvellyn (34.16, 1993) and Catstycam (34.16, 1995).

This species also occurs in the Scottish Highlands and Southern Uplands.
Alt. range: 535 m, Red Screes (38.08) - 700 m, Piers Gill (20.06)

5   W(F)C

E. frigida x E. scottica

A rare hybrid found during the Survey in Brown Cove, Helvellyn (32.14, T.C.G. Rich, 1982, LANC) and above Low Water, Coniston Old Man (26.98, GH, 1992, LANC), and collected by C.W. Muirhead in 1951 on Honister Crag (2.1, PLYP).
Alt. limit: 685 m, Low Water (26.98) but no doubt higher in Brown Cove (32.14).

2   WF(C)

[E. cambrica Pugsley
This dwarf montane species is recorded by Pugsley (1930) from the Kirkstone Pass (3.1/4.1, Ridley, 1881). This is discounted in the *Critical Atlas* where the species is treated as endemic to North Wales.

(W)]

E. ostenfeldii (Pugsley) P.F. Yeo

An extremely rare species of rock ledges in the Lake District, at two of the sites growing with *Alchemilla alpina*. The only Survey record is from Eel Crags, Newlands (22.16, GH, 1991, LANC), but there are earlier collections from the head of Kentmere (4.0, A. Wilson, 1932, YRK), Brandreth (2.1, C.W. Muirhead, 1941, PLYP), above Sprinkling Tarn (2.0, K.J.F. Park, c.1948, BM) and from Comb Crags and Nethermost Cove, Helvellyn (3.1, DAR, 1959, LANC).

Euphrasia ostenfeldii is a small and usually unbranched plant with hairy leaves, very small, pale flowers and without the enlarged floral leaves of *E. frigida*. The Kentmere specimens are less than 1 cm high and were referred by Wilson to *E. curta* var. *piccola*. Most of the numerous literature records for *E. curta* (Fries) Wettst. probably refer to *E. nemorosa* or *E. confusa*.

A collection of taller plants, also from Comb Crags (DAR, 1955, CLE), was determined by E.F. Warburg as *E. curta* var. *rupestris*. This is a variant of *E. ostenfeldii*, possibly of hybrid origin, and AJS considers that the specimens agree well with material from Snowdonia.

This is a western, coastal or montane eyebright of open, sandy or stony sites in North Wales, the Lake District and northern Scotland.
Alt.: 620 m, Eel Crags (22.16)
Plate 40   S 1   (W)C

E. scottica Wettst.

A frequent species of hillside flushes and wet moorland and particularly common in the eastern fells of the

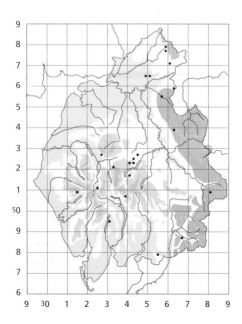

Map 740. *Euphrasia confusa* x *E. scottica*

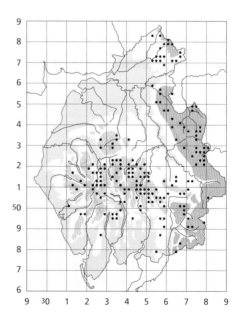

Map 741. *Euphrasia scottica*

Lake District, the high Pennines and the Bewcastle Fells.

This is an easily recognised eyebright with its small, predominantly white flowers and erect, elongate stems which may be unbranched or with a few short branches at the base. It seems likely that field records of *Euphrasia micrantha* relate to this species and they have therefore been included in the *E. scottica* map.
Alt. limit: at least 620 m, Great Dun Fell (70.34, 72.30, 72.32)
Map 741   176   WFCY

E. micrantha Reichb.
Apparently extremely rare. The only Survey records are from *Calluna* heath near the coast at Beckfoot, Wolsty (08.50, GH, 1986, LANC), a similar habitat further along the coast at Mawbray Bank (08.46, GH, 1992, LANC), although the last collection shows a slight influence of *Euphrasia confusa*, and one from sandy ground by the Black Burn, Midgeholme (62.58, FJR, 1982, herb. C.C.H). Of several collections in BM previously identified, or tentatively identified, as this species, only the following are considered by AJS as convincing: Bigland Tarn, Newby Bridge (3.8, W.H. Pearsall, 1919), the type of Pugsley's var. *simplex*, and rough pasture, Hoff Lunn, Appleby (6.1, N.D. Simpson, 1938). Wilson's records from the central Lake District, south Westmorland and Cauldron Snout, Teesdale (8.2) probably refer to *E. scottica*, as do those in the *Critical Atlas*.

The distinguishing characters of this species are the small, purple flowers, the general purplish colour of the stem and leaves and the tendency to branch, if at all, in the upper half of the stem.
Alt. limit: 260 m, Midgeholme (62.58)
                       3   (WF)C

Odontites vernus (Bellardi) Dumort.   Red bartsia
subsp. **vernus**
Fairly frequent in the lowlands, flowering late in the summer and characteristic of short, often trampled grass, for example tracksides, field entrances and roadside verges. It is also frequent on disused railway lines.

All the specimens examined by CCH from the Whitehaven area and by CEW from around Kendal were of subsp. *vernus* and the vast majority of Survey records no doubt belong here.
Alt. limit: at least 540 m, Nenthead (76.40)
Map 742   702   LWFCY

subsp. **serotinus** (Syme) Corbière
Material of this subspecies, characterised chiefly by several pairs of divergent branches and intercalary leaves, was collected in a salt-marsh at Askam-in-Furness (20.76, PB, 1992, LANC) and on old dunes at Sandscale (18.74, PB, 1994, LANC).
                       2   F

subsp. **litoralis** (Fies) Nyman
The only collection, and perhaps the first English record of this northern segregate, was from a salt-marsh at Askam-in-Furness (20.76, PB, 1994, LANC, conf. C.A. Stace). The specimens were mostly unbranched with subacute calyx teeth shorter than the tube.
                       1   F

Bartsia alpina L.   Alpine bartsia
Extremely rare, known now only from two sites in the Orton (6.0, 6.1) area where one of the populations is very small and declining. As mentioned by Wilson, this attractive plant was reported, new to Britain, by John Ray in 1670 from near Orton (6.0).

The plants grow in open, calcareous, flushed turf and, although cattle are instrumental in keeping the turf open, increased grazing by both cattle and sheep virtually prevents flowering. The small, slender shoots produced by the close-cropped plants seldom manage to flower and those that do are usually eaten. Associated species include *Pinguicula vulgaris, Blysmus compressus* and *Schoenus nigricans*.

The only other localities south of the Scottish Highlands for this striking arctic-montane species are near Malham Tarn, Yorkshire and upper Teesdale.
Alt. range: 245 m, (6.0) - 275 m, (6.1)
                       2   W

*Parentucellia viscosa (L.) Caruel   Yellow bartsia
A southern species seen during the Survey only on North Walney (16.70, GH, 1978, LANC; 16.72, P. Kirkland, 1990s), where there are long-established colonies by an old gravel pit and on disturbed ground, and in the landscaped grounds of the new Furness Hospital at Barrow (20.70, GH, 1983, *Watsonia* (1984) 15:135). It has probably disappeared from this last site.

The earliest records are the post-1930 *Atlas* dot for the North Walney square and N.M. Stalker's from the sand-dunes at Silloth (1.5, *Proc. BSBI* (1954) 1:59).
                       3   F(C)

Rhinanthus angustifolius C. Gmelin
Hodgson gives eight localities for this rapidly disappearing arable weed. There are no subsequent records and no herbarium specimens are known.
                       R   (C)

R. minor L.   Yellow-rattle
subsp. **minor**
A common plant of 'unimproved', rather poor pastures and hay-meadows, rough grassland and species-rich roadside verges.

Unfortunately little attention was paid during the Survey to the segregates of this difficult complex but it is fair to assume that most Survey records belong here.
Alt. limit: 570 m, Moor House, Teesdale (74.30)
Map 743   971   LWFCY

subsp. **stenophyllus** (Schur) O. Schwarz
Known only from two Survey collections, from Dubbs Moss, Eaglesfield (10.28, GH, 1984, LANC) and coastal grassland at Askam-in-Furness (20.76, PB, 1992, LANC) but it is certainly under-recorded. The *Critical Atlas* shows it as occurring in several Westmorland squares, one in Furness and two in Cumberland. It is characterised by flowering rather later than subsp. *minor* and having long, ascending flowering branches from the lower part of the stem.

2 (W)FC

subsp. **monticola** (Stern.) O. Schwarz
A plant of somewhat basic rock ledges in the Lake District and recorded during the Survey from Honister Crag (20.14, GH, 1993), Glaramara (24.10, GH, 1995, LANC), Helvellyn (32.14, RWMC, 1970s; 34.12, T.C.G. Rich, 1982), Red Pike, Wasdale (16.10, JS, 1995, LANC) and High Street (44.10, RWMC, 1970s).

There are additional records made by DAR in the 1950s from Eel Crags (22.16), Dale Head and Yew Crag (22.14), Ashness Gill (26.18) and the head of Glencoynedale (34.18).

The distinguishing characters of this subspecies are the short basal internodes and few or no flowering shoots from the lower nodes.
Alt. range: 500 m, Yew Crag (22.14) - 855 m, Helvellyn (32.14)

5 WC

Pedicularis palustris L. Marsh lousewort
Fairly frequent, in flushed but not particularly calcareous areas around mires and in tall herb vegetation by tarns and in marshy fields. Specimens up to 1 m high have been reported from Cliburn Moss (56.24).
Alt. limit: at least 500 m, Tynehead (76.36)
Map 744 278 WFCY

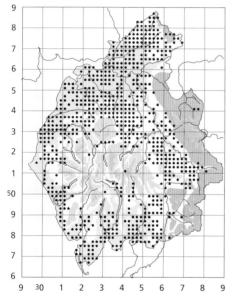

Map 742. *Odontites vernus*

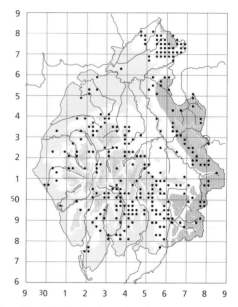

Map 744. *Pedicularis palustris*

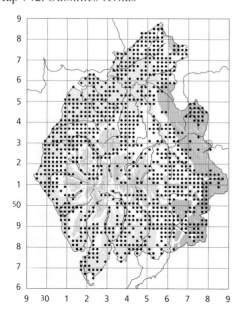

Map 743. *Rhinanthus minor*

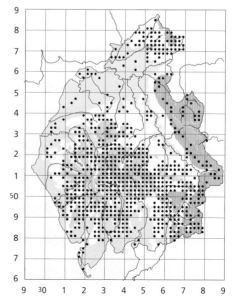

Map 745. *Pedicularis sylvatica*

P. sylvatica L.   Lousewort
Frequent in non-calcareous flushes and in wet
moorland but absent from blanket peat. Common
associates include *Erica tetralix, Agrostis capillaris,
Nardus stricta and Polygala serpyllifolia.*

The western subsp. *hibernica* D. Webb, characterised by a hairy calyx, was found at Butterburn
Flow in 1994 and is apparently new to England
(66.74n, M.I. Trotman (1995). It was growing intermixed with subsp. *sylvatica.*
Alt. limit: 475 m, Tynehead (74.36)
Map 745   725   WFCY

OROBANCHACEAE

Lathraea squamaria L.   Toothwort
Scattered but locally not infrequent as, for example, in
the woodlands around Sebergham (3.4), Melkinthorpe
(5.2) and the Kent estuary. It prefers the richer soils,
particularly those on the limestone, and it is absent
from the west. It usually parasitises hazel but it has also
been seen on aspen, elm, hawthorn and *Populus* x
canadensis.
Alt. limit: 350 m, Alston (72.44)
Map 746   94   WFCY

*Orobanche purpurea Jacq.   Purple broomrape
This striking broomrape is known only from Maryport
(02.34, 02.36) where it was probably first found by
A. Old in 1983 but it was not until 1985 that it was
independently seen and identified by MMM and others
(*Watsonia* (1986) 16:191). It occurs on waste ground
with *Orobanche minor* and that year there were three
neighbouring populations with about 100 flowering

spikes. A similar number was seen in 1986 but not all
at the original sites. The host appears to be *Achillea
millefolium.* In 1992 a second colony of *c.*80 plants was
discovered by A. Pickering in an adjacent tetrad.

This is an extremely rare plant of southern
Britain where it is known now only from a few places
in Dorset, the Isle of Wight and the Norfolk coast. The
origin of the Maryport colony is a mystery.

Unfortunately a redevelopment scheme resulted
in the destruction of most of the area and many of the
plants. This occurred despite an agreement to protect
the plants. The site, an S.S.S.I., was recently the subject
of an application for speculative building, but this was
fortunately rejected following a Public Inquiry in 1992.
Plate 84   R 2   C

O. rapum-genistae Thuill.   Greater broomrape
Extremely rare; seen during the Survey in 1984 at two
sites near Kendal, one to the west near Burneside
(50.96, R. Stringer, 1970s) and the other to the north-
east near Mealbank (52.94, CEW, 1984). About 20
plants were present at the first site, only two at the
second. The former population persisted until 1989
when the host plant, *Cytisus scoparius*, was grubbed
out and the same fate befell the plants at the second
site. Fortunately a few plants were subsequently seen at
both localities a short distance from the original sites.

This southern species has suffered a drastic
decline in Britain this century and this is true in
Cumbria also. Hodgson lists it from near Whitehaven
(9.1), Mawbray (0.4), Abbeytown (1.5, E.J. Glaister,
1876, CLE), near Wigton (2.4) and Carlisle (3.5/4.5),
the only later Cumberland records being from Thornby,
near Thursby (2.5, G.A. Field, 1934, CLE) and between

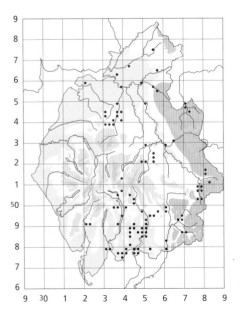

Map 746. *Lathraea squamaria*

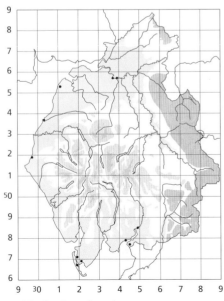

Map 747. *Orobanche minor*

Plate 86. *Orobanche minor*, Maryport

Plate 85. *Orobanche rapum-genistae*, Burneside

Plate 84. *Orobanche purpurea*, Maryport

Eaglesfield and Distington (0.2, DAR, 1950s). Wilson's records are from Windermere (4.9), Skelsmergh and Mint Bridge, Kendal (5.9, A. Wilson, 1878, YRK), near Sedbergh (6.9), and between Milburn Grange and Kirkby Thore (6.2). The Skelsmergh site may be the same as one reported in 1952 south of Garnett Bridge (5.9, E. Hyde). There is also in Wilson's herbarium a specimen collected by him from Howgill, Sedbergh (6.9, 1903, YRK).

Plate 85   S 2   W(CY)

[O. elatior Sutton
The only record for this south-eastern species is that cited by Hodgson from near Silloth (1.5, E.J. Glaister, 1877, CLE). The specimen has recently been redetermined by F. Rumsey as *Orobanche rapum-genistae*.

                                                    (C)]

*O. minor Sm.   Common broomrape
Very rare, but regularly seen during the Survey on waste land and railway sites at Barrow-in-Furness, west coast towns and Carlisle (3.5). The first Westmorland record was from Low Meathop (40.78s, FLW, 1976), where it grows on the railway embankment parasitising *Pastinaca sativa*. Its infortunate predilection for such vulnerable sites is curious, but it possibly owes its spread to the movement of railway ballast. The species has also been found in gardens at Levens (48.86s), parasitising a cultivated *Senecio,* probably *S. grayi* Hook., and Far Arnside (44.76s). These sites represent the species' northern limit on the west coast of Britain.

Surprisingly, it was not discovered in the county until 1954 when it was seen by G. Wilson at Barrow-in-Furness and nearby at Thwaite Flat, Dalton (2.7), and the following year it was found by A.W. Westrup on Walney Island (1.6s).

Map 747, Plate 86   11   WFC

GESNERIACEAE

*Ramonda myconi (L.) Reichb.
A single plant of this Pyrenean alpine plant was found on a limestone scar by the upper River Rawthey in 1990 (72.96, FJR, LANC).

                    1   Y

LENTIBULARIACEAE

Pinguicula vulgaris L.   Common butterwort
Common in the uplands in base-poor as well as base-rich or calcareous open flushes, crevices of irrigated rocks and rock ledges, often associated with *Selaginella selaginoides*. It has probably disappeared from most of the lowlands, particularly the Solway mosses, through drainage although it still persists locally, as at Hale Moss, Milnthorpe (50.76s).
Alt. limit: at least 600 m, Little Fell (78.20), although Wilson gives it as at least 825 m on Helvellyn (3.1).
Map 748   714   WFCY

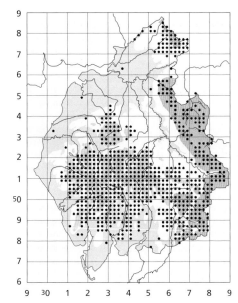

Map 748. *Pinguicula vulgaris*

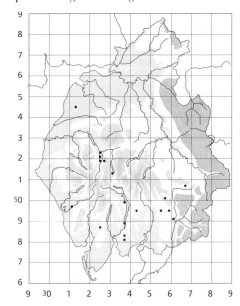

Map 749. *Utricularia australis*

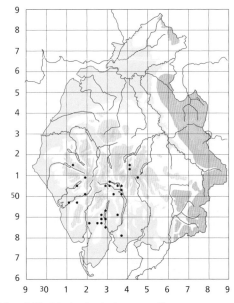

Map 750. *Utricularia intermedia*

Utricularia vulgaris L. sensu stricto and
U. australis R. Br.   Greater bladderwort
A rare and decreasing plant of slightly mesotrophic, poor-fen pools and shallow tarns, rarely, as by Derwent Water, of lake margins. Baker noted that it was "not infrequent" and Hodgson refers to its abundance on Newton Reigny Moss (4.3) where it was "one of the chief ornaments of the bog during the month of August". Unfortunately it has long been extinct there. It seldom flowers, but the few Survey records of flowering material are all of *Utricularia australis* and it is doubtful if the more southern *U. vulgaris* occurs in Cumbria.

Although this and the following species are absent from the north of the county, both were recorded in 1944 by G.A.K. Hervey across the border at Greenlee in Northumberland.
Alt. limit: 335 m, Lambrigg Fell (58.94)
Map 749   18   WFC(Y)

U. intermedia Hayne sensu lato
                    Intermediate bladderwort
Rather rare and confined to the Lake District, where it is commonest in the south and west, growing, like *Utricularia australis*, in shallow, somewhat mesotrophic pools and shallow tarns and base-poor flushes, often with *Sphagnum* and *Menyanthes trifoliata*. It occurs in some quantity in Blelham fen (36.00). There are Survey records from 11 squares, compared with four in the *Atlas*. However, like all bladderworts it is easily overlooked and has no doubt been under-recorded in the past. There are no records of flowering material.

Fresh Cumbrian material was examined during the Survey by Brown (1990) who demonstrated that most can be assigned to *U. stygia* Thor. This is characterised by having 2-9 teeth on the ultimate leaf segments, each often with several bristles. It also has the widest angle between the shorter pair of arms of the

quadrifid hairs found inside the bladders. He found one possible specimen of *U. ochroleuca* R. Hartman at Subberthwaite, Blawith (26.86). This segregate has longer and narrower leaf segments, with 0-5 teeth and fewer bristles, and the angle of the shorter arms of the quadrifid hairs is intermediate between *U. stygia* and *U. intermedia* s.s.. The latter has not yet been found in Cumbria; it is the only member of the complex which apparently never has bladders on the green stems. The only definite record of *U. ochroleucha* is from ponds near Walna Scar (2.9, H.T. Mennell, 1891, BM, det. K. Norcott).

There are no Survey records from any of the lakes although it was seen in Derwent Water (2.2, W.H. Pearsall, BM) in 1916 and earlier in Coniston (2.9/3.9, M.A. Temperley, 1893, HAMU; J. Comber, BM, *BEC Rpt* (1914) 3:111, under *U. ochroleuca*). Comber's material was collected in 2 m of water.

Utricularia intermedia s.l. is very rare in Britain south of the Scottish Border, being known only from the Lleyn peninsula in North Wales, Dorset and the New Forest, Norfolk and the Lake District.
Alt. limit: 480 m, Angle Tarn, Patterdale (40.14)
Map 750   27   WFC

U. minor L.   Lesser bladderwort
Uncommon and largely confined to the Lake District where it occurs in similar habitats to *Utricularia vulgaris* and *U. intermedia*. It frequently grows intermixed with the former, but it appears to be more tolerant of oligotrophic conditions. It is also the only species to flower at all frequently.

This bladderwort is largely restricted in England and Wales to the Lake District and the Purbeck and New Forest heaths.
Alt. limit: 505 m, Dalehead Tarn (22.14)
Map 751   54   WFC

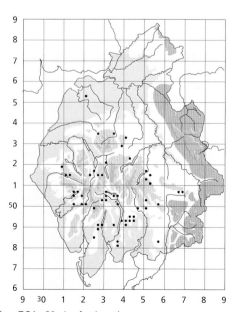

Map 751. *Utricularia minor*

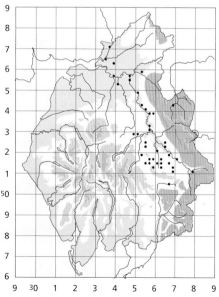

Map 752. *Campanula glomerata*

CAMPANULACEAE

*Campanula patula L.
A declining species of southern and central England known in Cumbria only from Hodgson's record from Ullock Moss, Keswick (2.2) and the inclusion of v.c.69 by Druce (1932) in his *Comital Flora*.

(W/FC)

*C. lactiflora M. Bieb.
An attractive tall, pale blue-flowered bellflower from the Caucasus found in 1987 well established along lanesides on the Goody Bridge side of Grasmere but always near houses (32.08, GH, LANC, det. D.H. Kent & J.M. Mullin, *Watsonia* (1989) 17:477).

W

*C. persicifolia L.
Seen during the Survey at Low Cunsey (36.92, R. Young, 1969, AMDE), near Greenroad Station, Millom (18.82, MMG, 1991), briefly in shady woodland by the River Caldew, Buckabank, Dalston (36.48, REG, 1985, LANC, *Watsonia* (1988) 17:192), a trackside east of Langwathby (58.34, MG, 1990, LANC), Ramsden dock, Barrow-in-Furness (20.68s, B. Fisher, 1993) and at Crosby Ravensworth (62.14, CW), where it has persisted for at least ten years.

The only records in the Floras are Hodgson's single, dubious and unlocalised one and Wilson's from Far Arnside (4.7s). The first localised Cumberland find was from Pardshaw (0.2, G. Templeman, 1922, *BEC Rpt* (1923) 6:736).
Alt. limit: 190 m, Crosby Ravensworth (62.14)

R 7   WFC

C. glomerata L.   Clustered bellflower
A frequent sight on the mid-Westmorland limestone especially on ungrazed roadside verges, and extending north to Stainton (48.28). North of here it occurs chiefly on the flushed sandstone banks of the River Eden, sometimes associated with *Galium boreale*, but also, as a probable garden escape, near Hayton (52.58), on a roadside verge at Glassonby (58.38), on the railwayside near Metal Bridge, north of Carlisle (34.64) and on the old line at Stainton, Carlisle (38.56). Wilson's two records from the south of the county, A.W. Westrup's from Roanhead (1.7, 1955) and J. Parkin's from Skinburness (1.5, 1933) no doubt also represent escapes.
Alt. limit: 355 m, Oddendale (58.12)
Map 752   40   WC

*C. poscharskyana Degen
A garden escape discovered in 1986 at Hackthorpe (54.22, RWMC, LANC), on a roadside rock at Grasmere (32.10, W. Hay & CEW, 1985), by the River Leith near Great Strickland (54.24, CEW, 1990, LANC), by a lane at Natland (50.88s, CEW, 1994, LANC) and on a traffic island of the A590 road at Arrad Foot (30.80, PB, 1993, LANC). The first two collections were determined by A.L. Grenfell.

5   WF

C. latifolia L.   Giant bellflower
Common in the Eden, Irthing, Liddel and Lyne valleys, along the southern edge of the Solway Plain and in south Westmorland, otherwise rather local or scarce. This is a plant of woodland and wooded riversides and damp hedgerows. The deep blue-flowered plants are particularly handsome and conspicuous but pale blue or whitish flowers preponderate.
Alt.limit: 300 m, Argill, Brough (82.14)
Map 753   437   LWFCY

*C. trachelium L.   Nettle-leaved bellflower
The only Survey records are from Ambleside (36.04, Miss Gregory, 1970s), a builder's yard at Grange-over-Sands (40.76s, L. & P. Livermore, 1972), and a farm

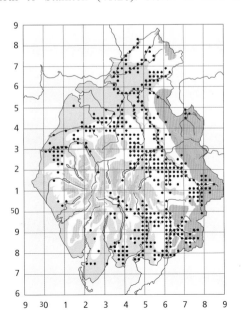

Map 753. *Campanula latifolia*

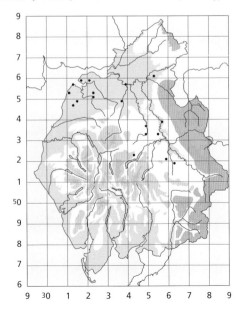

Map 754. *Campanula rapunculoides*

track north of Gosforth (06.04, MMG, 1992, LANC).

The only Flora records are those given by Wilson from Levens (4.8s, G.E. Martindale, 1882, KDL), Scarfoot, Underbarrow (4.9, T. Gough, 1855, KDL) and Kirkby Stephen (7.0). The only Cumberland ones are from Gilsland (6.6, G.A.K. Hervey, 1943, v.c.67/70) and St Helens, Cockermouth (1.3, C.D. Pigott, 1944, 1945). These records are far to the north of the species' presumed native limit from the Mersey to the Humber.

4   WFC

*C. rapunculoides L.   Creeping bellflower
A rare garden escape largely confined to the north where it is not infrequent in the Silloth area. Most records are from roadside verges and banks but one is from a disused railwayline. There are a few records in the Floras from the south and one from Dent (7.8, Anon., *c.*1910, LANC) but only one from Furness, where it was noted with other aliens at Dalton in 1913 (2.7, D. Lumb, *BEC Rpt* (1914) 3:403).
Alt. limit: 160 m, south of Kings Meaburn (62.18)
Map 754   18   W(F)C(Y)

*C. rhomboidalis L.
A rarely cultivated mountain plant of southern Europe found in 1983 in a sunken laneside near Knock (68.26, FJR, det. A.L. Grenfell, LANC, *BSBI News* (1986) 43:17) and still present in 1995. This appears to be only the second British record.

1   W

C. rotundifolia L.   Harebell
Very common, absent only from the more extensive areas of blanket peat and conifer plantations. It tolerates a wide range of soil pH, appearing equally at home on limestone and siliceous soils provided they are freely draining. It flowers best in ungrazed, rocky gill-sides,

on limestone scars, roadside banks and sand-dunes.
Alt. limit: 915 m, Helvellyn (34.14, Wilson)
Map 755, Plate 87 1604   LWFCY

*Legousia hybrida (L.) Delarbre Venus's looking-glass
A plant of eastern and south-eastern England seen since 1990 by the edge of a motorway bridge at Burton-in-Kendal (52.74s, GH, LANC).

1   L

Wahlenbergia hederacea (L.) Reichb.

Ivy-leaved bellflower
This attractive and delicate bellflower was reported by W.H. Pearsall from "both sides of a rill at Winster over the boundary line of North Lancashire, so that it is in both v.c.69 and v.c.69b" (4.9, 1925, *BEC Rpt* (1926) 7:883). No herbarium specimens are known and the site has never been refound. There is another record, which is best treated as doubtful, of a find in the Coniston Fells (2.9) made during the B.S.B.I. field meeting in 1951.

The only extant sites in north-west England are in west Durham and the Forest of Bowland.

(WF)

Jasione montana L.   Sheep's-bit
Locally frequent on dry, siliceous soils, especially in coastal turf, on stabilised sand-dunes and roadside banks, on the rocky Silurian outcrops to the south of the Lake District and on the sandstone of the Eden valley. It is very much a feature of dry, acid hedgebanks on the Solway Plain. Its frequency is in marked contrast to its rarity east of the Pennines in Co. Durham.
Alt. limit: at least 395 m, Knock Ore Gill (70.30)
Map 756   328   WFCY

Lobelia dortmanna L.   Water lobelia
A frequent and characteristic plant of the stony shores of tarns and lakes, requiring at least 30 cm of water and

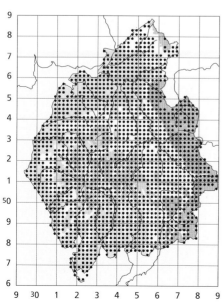

Map 755. *Campanula rotundifolia*

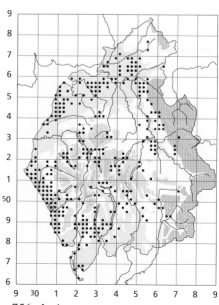

Map 756. *Jasione montana*

almost invariably growing with *Littorella uniflora*. It is understandably absent from the Haweswater and Thirlmere reservoirs but its present absence from Ullswater is very odd. John Ray (1677, under *Gladiolus lacustris*) contributed the first British record from there: "*In a Pool or lake called Hulls water that divides Westmorland from Cumberland, three miles from Pereth plentifully*" and Wilson refers to it as "still in Ullswater". The Atlas dot for the Ravenglass square (0.9) seems very improbable in the absence of any substantial tarns.

South of the Scottish Highlands, this species is restricted in Britain to the Lake District and Wales.

Alt. limit: 505 m, Dale Head Tarn (22.14)

Map 757   100   WFC

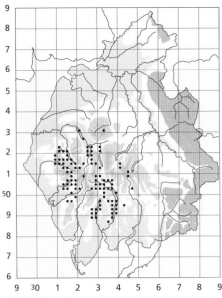

Map 757. *Lobelia dortmanna*

RUBIACEAE

Sherardia arvensis L.   Field madder

Fairly frequent around Morecambe bay, otherwise scattered in the Eden valley and in the west. It requires dry, open habitats and it grows particularly well in the shattered limestone debris left by pavement removal and also on quarry floors. It appears to be decreasing, not having been seen during the Survey in nearly half the *Atlas* squares.

Alt. limit: at least 300 m, Hardendale, Shap (58.14)

Map 758   123   WFC

Asperula cynanchica L.   Squinancywort

Occasional in short limestone turf around the head of Morecambe Bay where it reaches its northern limit in Britain. There is, however, an interesting outlying record given by Wilson from north Stainmore (8.1).

Alt. limit: 230 m, Hutton Roof Crags (54.78s)

Map 759   22   LWF

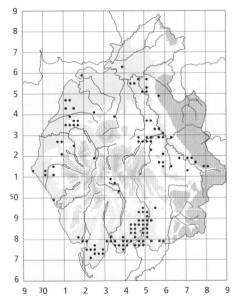

Map 758. *Sherardia arvensis*

Plate 85. *Campanula rotundifolia*

[A. taurina L.   Pink woodruff
A southern European species listed by Wilson as having been
seen by C.C. Babington at Casterton, Kirkby Lonsdale (6.7s).
It seems quite possible that this was a hasty misidentification
of *Galium cruciata*.

(W)]

*A. arvensis L.
A Continental species listed by Hodgson from Camerton (0.3)
and Silloth (1.5, J. Leitch, 1889, CLE) and seen by H. Britten
in 1926 at Beck Mill, Skirwith (6.3, *BEC Rpt* (1927) 8:116).

(C)

Galium boreale L.   Northern bedstraw
Not uncommon in the Lake District on base-rich rock,
occurring in the gullies and on the crags. It is also
fairly frequent on the limestone scars of the Pennines. It
suffers much from grazing and tends to be restricted to
the less accessible ledges and crevices, where the plants
may be quite luxuriant, but in open, grazed limestone
grassland the shoots may be only 2-3 cm high and
difficult to detect.

The montane populations have, presumably,
given rise to the larger and more luxuriant ones among
riverside rocks, as, for example, along the middle
sections of the Lune and Eden where the rocks,
although non-calcareous, are subject to flushing and silt
deposition. There are also small populations on the
rocky shores of the lakes, notably Windermere and
Ullswater. Other sites include the sandstone walls of
old salmon coops in the grounds of Corby Castle
(46.52, CS, 1978, LANC), where it is particularly
abundant.
Alt. limit: 885 m, Helvellyn (34.14)
Map 760   155   WFCY

G. odoratum (L.) Scop.   Sweet woodruff
Frequent in the south, elsewhere rather scattered and
absent from much of the Solway Plain. It occurs in
shady hedgerows, but it is essentially a plant of the
richer woodlands and copses, particularly on the limestone, and of river and gillside woods. It is, for
example, very much a feature of the Liddle and Lyne
valleys in the north and the Lune in the south.
Alt. limit: 550 m, above Rampsgill, Martindale (44.16)
Map 761   449   LWFCY

G. uliginosum L.   Fen bedstraw
Fairly frequent in the east, otherwise rather scarce. It
occurs in generally similar situations to *Galium palustre* but it is more characteristic of calcareous or base-
rich sites. In the uplands it grows in somewhat open
communities, often with *Eleocharis quinqueflora*,
Carex hostiana and *C. panicea*.

With its slightly narrower, darker leaves and
stiffer, shorter, rougher stems, it is usually easily distinguished from *Galium palustre*.
Alt. limit: 750 m, Cross Fell (70.34)
Map 762   461   WFCY

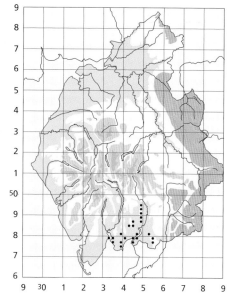

Map 759. *Asperula cynanchica*

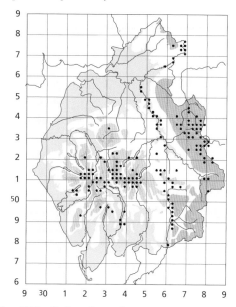

Map 760. *Galium boreale*

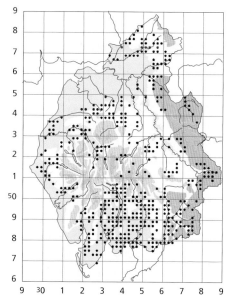

Map 761. *Galium odoratum*

G. palustre L.   Marsh bedstraw

Very common in ditches, marshy fields, fens, the margins of tarns and lakes, also in the fells in *Juncus effusus* soakways.

No attempt was made during the Survey to distinguish between subspp. *palustre* and *elongatum* (C. Presl) Arcang.

Alt. limit: 825 m Cross Fell (68.34)

Map 763   1487   LWFCY

G. verum L.   Lady's bedstraw

Very common on the limestones and in the usually lime-enriched sandy grassland and dunes along the coast, otherwise rather scattered. It is a species of dry, well-drained base-rich or calcareous grassland, often much suppressed by grazing and when in flower most prominent on roadside verges.

Alt. limit: 610 m, Melmerby High Scar (64.36)

Map 764   825   LWFCY

G. verum x **G. mollugo**

G. x **pomeranicum** Retz

Scattered and surprisingly rare considering the frequency with which the two parent species occur together, particularly in north Westmorland. The first Cumberland record was from an old limestone quarry near Greystoke (44.28, Anon.) and the only previous one for Westmorland was Wilson's from Sandford Bog (7.1).

Alt. limit: 320 m, near Ravenstonedale (74.02)

Map 765   10   WC

G. mollugo L.   Hedge bedstraw

Very common in the lowlands in the north and east, rather uncommon elsewhere and also on the Scottish side of the Solway Firth. It is curious that this essentially southern species should be so scarce over much of the south and south-west of the county. It occurs in hedgerows and scrub, against roadside walls and in

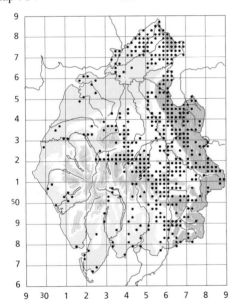

Map 762. *Galium uliginosum*

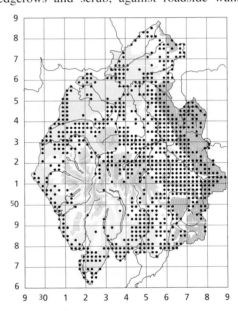

Map 764. *Galium verum*

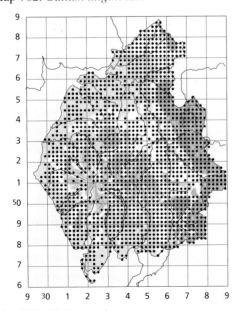

Map 763. *Galium palustre*

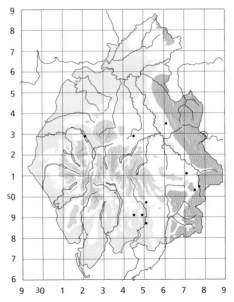

Map 765. *Galium verum* x *G. mollugo*

coarse grassland on waste sites and along railwaysides.

This species appears noticeably less calcicole in Cumbria than in the south of England and probably only the more widespread subsp. *mollugo* is present.
Alt. limit: 570 m, High Cup Nick (74.26)

Map 766	842	WFCY

G. sterneri Ehrend.
Common in the three major limestone areas of Morecambe Bay, Asby - Orton and the Pennines, occurring everywhere on the scars and in short *Festuca ovina - Thymus* grassland, also sometimes in abundance in limestone quarries and areas of pavement removal. All records from the Lake District should be regarded as suspect.

This bedstraw is probably commoner in Cumbria than anywhere else in Britain.
Alt. limit: 760 m, Knock Fell (70.30)

Map 767	210	LWFCY

G. saxatile L.   Heath bedstraw
Very common nearly everywhere though rather sparse on the more intensively farmed areas of the Solway Plain, the Eden and Kent valleys and parts of the Furness peninsula. It is most characteristic of acid grassland, where it is associated with *Potentilla erecta, Agrostis capillaris, Nardus stricta, Polygala serypyll-ifolia* and *Calluna*. It is frequent even in the limestone areas in pockets of non-calcareous drift and leached soil.
Alt. limit: 950 m, Helvellyn (34.14).

Map 768	1485	LWFCY

G. aparine L.   Cleavers; Goosegrass
Extremely common in the lowlands and valleys, in fact wherever there are hedges. It occurs in woodland, copses and carr, as an arable weed and in nitrogen-enriched habitats along the coast, especially on shingle beaches.

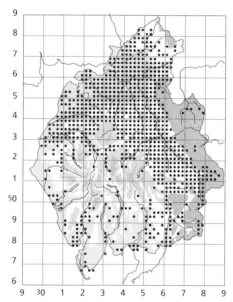

Map 766. *Galium mollugo*

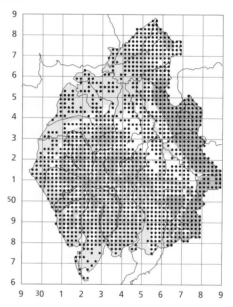

Map 768. *Galium saxatile*

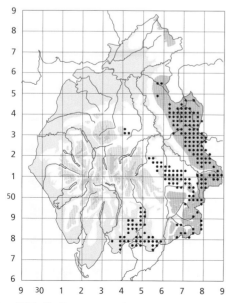

Map 767. *Galium sterneri*

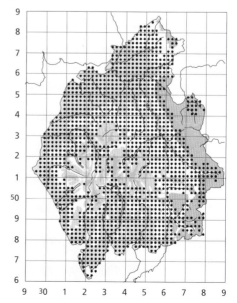

Map 769. *Galium aparine*

Alt. limit: *c.*430 m, Nenthead (78.42)

Map 769   1389   LWFCY

*G. tricornutum Dandy   Corn cleavers
Recorded during the Survey only from a laneside west
of Dalton-in-Furness (22.74, PB, 1986). Unfortunately
the site was later damaged and the plant failed to
reappear.

This very distinctive southern bedstraw is listed,
with some reservation, by Baker and Hodgson from
Brookfield, Wigton (2.4) and also by the latter from
Silloth (1.5, J. Leitch, 1890, CLE). In 1905 it was seen
at Grinsdale gravel beds, Carlisle (3.5, T.S. Johnstone,
CLE) and Wilson states that W.H. Pearsall found it on
railway ballast at Cavendish Dock, Barrow-in-Furness
(2.6s). Pearsall also collected it at Dalton-in-Furness
(2.7, BM ex hort.) in 1915.

R †1   F(C)

Cruciata laevipes Opiz   Crosswort
(*Galium cruciata* (L.) Scop.)
Very common and even abundant in the south-east and
most of the north yet oddly very scarce along the south-
west coast and in the Lake District, except around
Keswick (2.2). Particularly puzzling is its absence from
the Ambleside - Grasmere - Langdale area. It is very
characteristic of damp roadside verges, banks, hedgerows and 'unimproved' meadows.

Alt. limit: 550 m, south-west of Garrigill (76.40)

Map 770   1101   LWFCY

CAPRIFOLIACEAE

*Sambucus racemosa L.
Rare, known only from a few sites, mostly by roads and
tracks near estates, but at least some bushes are likely
to have been bird-sown, as, for example, at Jemmy
Crag, Windermere (38.94, CFS & JS, 1970s). The first
Furness record is from Pull Garth wood, Brathay
(36.02, R. Young, 1969, AMDE), for Westmorland
from Barbon Park (62.82s, 64.82s, CFS & JS, 1960s)
and for Cumberland from Hutton John (4.2, E. Blezard,
1947, CLE).

Alt. limit: 210 m, Kershope Forest (52.84n)

Map 771   14   WFC

S. nigra L.   Elder
A common hedgerow shrub, especially around villages
and old farmsteads where it has possibly been planted
for medicinal and culinary purposes. It prefers enriched
soils and forms areas of scrub on old dumps and valley-
side tips; it is also frequent in the more open limestone
woodlands.

There are a few records of plants with variegated
leaves and one of a plant with laciniate leaves. There
are no Cumbrian records of the related *Sambucus canadensis* L. which has been planted on railway property

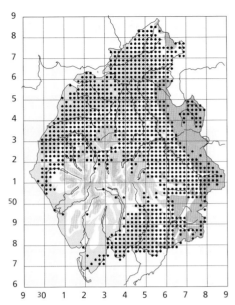
Map 770. *Cruciata laevipes*

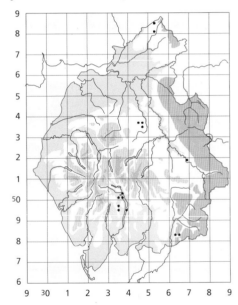
Map 771. *Sambucus racemosa*

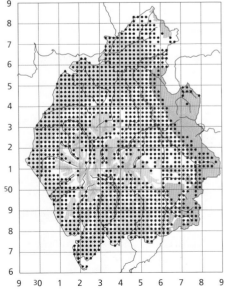
Map 772. *Sambucus nigra*

in Co. Durham (Graham 1988). This species has more attenuate and finely and more sharply toothed leaflets.
Alt. limit: 470 m, above Nenthead (78.44)

Map 772   1349   LWFCY

*S. ebulus L.   Danewort
Seen during the Survey only at Wilson's earlier site at Gaisgill (64.04) and from roadside verges near Oulton (24.50, REG, 1990, LANC, *Watsonia* (1991) 18:430), Brougham (54.28, AW, 1980) and Newby (58.20, D. Hinchcliffe, 1988). It has since disappeared from the last two sites as a result of injudicious management of the verge and the Oulton site is similarly under threat. The Floras give several scattered localities.
Alt. limit: 200 m, Gaisgill (64.04)

       4   W(F)C(W/Y)

Viburnum opulus L.   Guelder rose
A frequent undershrub in the richer, and usually damper, lowland woodlands, also in hedgerows where perhaps it is nowhere more colourful in the autumn than by the A590 south-east of Newby Bridge (36.84).
Alt. limit: at least 400 m, south of Garrigill (74.38)

Map 773   718   LWFCY

*V. lantana L.   Wayfaring tree
Recorded from a number of sites around Arnside (44.78s), in an old roadside hedge close to the county boundary south of Burton-in-Kendal (52.74s), on a limestone scar near Lowther (50.22, 52.22), where it is probably bird-sown, and, obviously planted, on the embankment of a road crossing the M6 near Thrimby (56.18).

       It is listed by Baker and Hodgson as obvious introductions at Gilgarran (0.2), Cockermouth (1.3) and Ullswater (4.2). It was seen in 1956 at Great Salkeld (5.3, G.A.K. Hervey) and during the 1950s by the Barrow-in-Furness N.H.S. in the Dalton (2.7), Newby Bridge (3.8) and Hawkshead (3.9) squares. The

Atlas records of presumed introductions also include the Whitehaven (9.1), Blindcrake (1.3) and Cartmel (3.7s) squares.
Alt. limit: 240 m, Thrimby (56.18)

       6   LWFC

*Symphoricarpos albus (L.) S.F. Blake   Snowberry
(*S. rivularis* Suksd.)
A frequent and conspicuous feature of lowland hedges and shrubberies but seldom far from houses. Surprisingly, this North American shrub is not mentioned in any of the earlier Floras. The first records appear to be those shown in the *Atlas* and the species seems to have spread appreciably since the 1950s.
Alt. limit: at least 360 m, Nenthead (74.44)

Map 774   438   WFCY

*S. orbiculatus Moench x **S. microphyllus** Kunth
S. x chenaultii Rehder
An eastern North American shrub recorded from a plantation ride near Catlowdy (48.78n, REG, 1990), laneside scrub at Levens (48.86s, GH, 1995, LANC, det. E.J. Clement) and by a quarry at Sandside (48.80s, CEW, 1996, LANC).

       3   WC

*Leycesteria formosa Wallich   Himalayan honeysuckle
This popular garden shrub was found in 1982 naturalised on the disused railway line at Coniston (28.96, JA, LANC, *Watsonia* (1983) 14:.428) and in 1994 in woodland near Windermere (40.98, CEW, LANC).

       2   WF

*Lonicera nitida E. Wilson
A rare garden escape. The first record for Furness appears to be that from Force Forge, Satterthwaite (32.90, I. Bonner, 1987), for Westmorland from by the River Brathay near Elterwater (32.04, GH, 1988) and for Cumberland from by the coast road south of

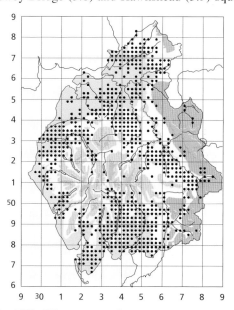

Map 773. *Viburnum opulus*

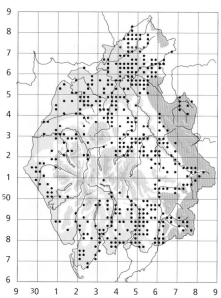

Map 774. *Symphoricarpos albus*

Allonby (06.40, AAD, 1992).
Alt. limit: 110 m, Satterthwaite (32.92)
Map 775   13   WFC

***L. involucrata** (Richardson) Banks ex Sprengel
Clumps of this North American shrub were seen during
the Survey on sand-dunes at Silloth (10.52, REG, 1993)
and nearby in rough grassland near East Cote lighthouse (10.54, REG, 1990; LANC, *Watsonia* (1991)
18:430).

                           2   C

***L. xylosteum** L.   Fly honeysuckle
Seen during the Survey at Grange-over-Sands (38.76s,
FLW, 1970s), a wood at Lindale (40.78s, FLW, 1984),
probably the same site as that given by Petty (1896
p.200), the west side of Windermere (38.96, FLW,
1970s), Plumptonfoot, Plumpton (48.38, I. Mortemore,
1978), where it grows on the railway embankment, and
in woodland on Soulby Fell, Dacre (44.24, CEW, 1992,
LANC). It also still occurs nearby in the churchyard at
Watermillock (42.22, LANC) where it was seen by
G.A.K. Hervey in 1950.

     Hodgson lists it from St Bees (9.1), Workington
Hall Park (9.2), Yeorton Hall (0.0) and St Herbert's
Isle, Derwent Water (2.2). Baker mentions additional
sites at the head of Coniston Water (3.9), Plumpton
woods and Conishead Priory (3.7s) and Wilson the
Windermere islands and a wood near the ferry (3.9),
also Singleton Park, Kendal (5.9), Middlebarrow
Wood, Arnside (4.7s) and woods near Patterdale (3.1).
There are additional Cumberland records from Distington (0.2, J. Adair, 1894, YRK) and Naworth Castle
(5.6, G.C. Druce, 1923, *BEC Rpt* (1924) 7:187).

     Although widespread on the Continent, this
honeysuckle is not native in Britain except perhaps in
the extreme south.
Alt. limit: 230 m, Soulby Fell (44.24)
                    R 5   (W)FC

L. periclymenum L.   Honeysuckle
Very common in the lowlands but thinning out noticeably in the east. This is essentially a woodland plant,
present in almost all except the wettest types of deciduous woodland but probably most familiar scrambling
over old hedges. It occurs also in the fells on sheltered,
moist rock ledges, associated with *Geranium sylvaticum*, *Heracleum* and *Anthriscus sylvestris*.
Alt. limit: 610 m, Fairfield (36.10) and Piers Gill,
Scafell Pike (20.08)
Map 776   1292   LWFCY

***L. caprifolium** L.   Perfoliate honeysuckle
A species of southern and eastern Europe listed by Hodgson
from near Lorton Hall (1.2) and Wetheral churchyard (4.5). It
was presumably planted at both sites as also in the garden at
Great Salkeld (5.3) where it was seen by H. Britten between
1900 and 1908.

                                              (C)

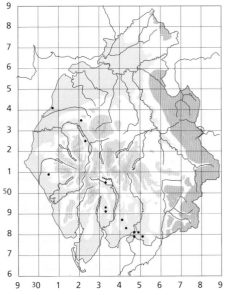

Map 775. *Lonicera nitida*

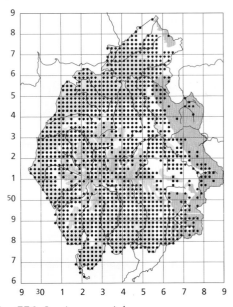

Map 776. *Lonicera periclymenum*

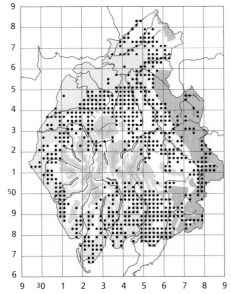

Map 777. *Adoxa moschatellina*

ADOXACEAE

Adoxa moschatellina L.   Moschatel; Townhall clock
Fairly frequent in the richer lowland woods but particularly on hedgerow banks where it is often conspicuous in early spring. However, like most woodland species it is absent from the Solway Plain. No information is available regarding the frequency of fruiting. It dies down very suddenly and is rarely recorded after May. It occurs also in the open on the fells in sheltered gills, chiefly on the limestone.
Alt. limit: 760 m, in a limestone pot-hole, Knock Fell (70.30)
Map 777   701   LWFCY

VALERIANACEAE

Valerianella locusta (L.) Laterr.
                                    Corn salad; Lamb's lettuce
Scattered, in dry, often disturbed habitats chiefly along the coast but extending inland, noticeably up the Eden valley. It occurs on dry hedgebanks, in disused quarries and as a weed of sandy arable land, urban waste sites and streets.
     The dwarf variant, var. *dunensis* D.E. Allen, with a condensed inflorescence, is characteristic of the sand-dunes on the south-west coast, but its distribution is not precisely known. It was first recorded in 1948 from Sandscale (18.74, G. Wilson, CLE).
Alt. limit: 340 m, Shap Fell quarries (54.08)
Map 778   132   WFCY

V. carinata Lois.
A rare southern species seen on open, sandy ground between Langwathby and Culgaith (58.30, LR, 1996, LANC). The only previous records are from Silloth saltworks (1.5, J. Leitch, 1890, CLE) and Gilsland (6.6, G.A.K. Hervey, 1946, v.c.67/70).

C

V. rimosa Bast.
Listed by Hodgson from Upperby, Carlisle (4.5) and by Wilson from Arnside (4.7s). There are also records from Bassenthwaite (2.2/2.3, E.J. Glaister, 1876, CLE), a cornfield at Warwick Road, Carlisle (4.5, T.S. Johnstone, 1905, CLE), Buttermere (1.1, G.A.K. Hervey, 1930) and a railway embankment at Gilsland (6.6, G.A.K. Hervey, 1945, v.c.67/70). There are no recent records and the above represent the northern limit in Britain of this rapidly disappearing species.

(WC)

V. dentata (L.) Pollich
Recorded by Hodgson from Nethertown (9.0), Frizington (0.1), Great Broughton (0.3), Aspatria and Thrushgill (1.4), Cummersdale station (3.5, P. Shepherd, 1884, CLE) and Ullswater (4.2). Wilson lists it for Hazelslack and Arnside (4.7s, Miss Jowitt, 1923, YRK) and Baker from "Furness, in cultivated fields". There is an additional Cumberland record from a gravel pit at Dalston (3.5, T.S. Johnstone, 1901, CLE) and a Westmorland one from Staveley (4.9, G.E. Martindale, 1885, KDL). There are no Survey records for this decreasing species, but rather surprisingly there are five post-1930 records: the *Atlas* dots for the Maryport (0.3), Dalton-in-Furness (2.7), and Arnside (v.c.60/69) squares and records from Humphrey Head (3.7s, A.W. Westrup, 1951, HIWNT) and the Carlisle and Orton area (3.5, DAR, 1950s).
     This is a decreasing cornfield weed now largely restricted in Britain to the south and east.

(WFC)

Valeriana officinalis L.   Valerian
Common throughout the lowlands occurring in wet woodland, in ditches, on damp roadside verges and in tall-herb marsh and fen communities, also in upland hay-meadows, in gills and on relatively sheltered and base-rich rock ledges.
     Most if not all of the Survey records are presumably of the tetraploid subsp. *sambucifolia* (Mikan. f.) Čelak. The only reference to the diploid subsp. *collina* (Wallr.) Nyman is Wilson's record from Dungeon

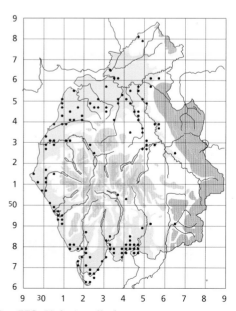

Map 778. *Valerianella locusta*

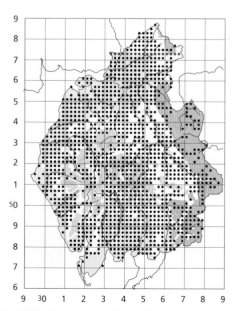

Map 779. *Valeriana officinalis*

Ghyll (2.0). The latter is a plant of drier habitats; it lacks epigeal stolons and has the terminal leaflet no wider than the middle lateral ones.

Alt. limit: 730 m, above Sprinkling Tarn, Borrowdale (22.08, Wilson (1956))

Map 779   1261   LWFCY

*V. pyrenaica L.   Pyrenean valerian

Known only from the north where it occurs at several sites around the lower Esk and its tributary the Liddel Water and also from by the River Eden at Rockcliffe (34.60). It is locally abundant, growing usually in wet woodland although one site is around an abandoned farm building.

The records in the early Floras indicate a wider distribution. Hodgson lists it from Eskdale (1.9/1.0) and Silverhill Wood, Longtown (3.7n), on the 'Debatable Ground', and Wilson from Ambleside (3.0). In 1935 it was found at Netherby (3.7n, J. Parkin), on the River Esk, and in 1946 by R. Martindale and T.S. Johnstone on the Liddel Water at Penton Linns (4.7n, BM).

It is easily distinguished from *Valeriana officinalis* by its large, cordate basal leaves.

As the name indicates, this is a native of northeast Spain and the Pyrenees. In Britain it is more or less confined to the area between Cumbria and the Moray Firth.

Map 780   8   (W)C

V. dioica L.   Marsh valerian

Frequent to common in the east and south-east, elsewhere rather scarce. It occurs typically in upland flushes and seepage areas where it is often vegetative and easily overlooked. It is a frequent associate of *Primula farinosa*. In the lowlands it is a plant of marshy fields, fens and rich, damp roadside verges as, for example, in the Sebergham - Greystoke area.

Alt. limit: 780 m, Cross Fell (68.34)

Map 781   539   LWFCY

*Centranthus ruber (L.) DC.   Red valerian

Rather rare, but locally frequent at a number of coastal sites, usually on roadside walls near houses and, by the Kent estuary, on limestone sea walls, embankments and roadside cliffs. There are only two recent inland records, from Ambleside (36.04) and Torver (28.94), although several are shown in the *Atlas*. All the populations are no doubt of garden origin. White-flowered plants are not uncommon.

Map 782   34   WFC(W/Y)

DIPSACACEAE

Dipsacus fullonum L.   Teasel

Scattered, usually only as a casual, appearing on tracksides and waste ground, in gardens and, at Grange-over-Sands (42.78s) on the golf-course.

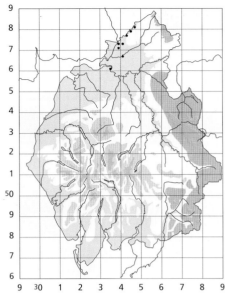

Map 780. *Valeriana pyrenaica*

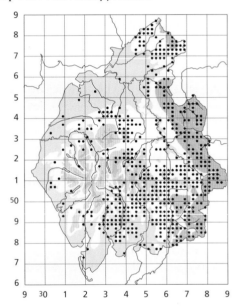

Map 781. *Valeriana dioica*

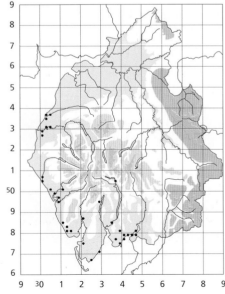

Map 782. *Centranthus ruber*

Alt. limit: 365 m, Garrigill (76.40)
Map 783   116   WFCY

Knautia arvensis (L.) Coulter   Field scabious
Frequent in the south, throughout the Eden valley and
down the coast to St Bees Head, elsewhere rather
scarce. It may be decreasing as there are no Survey
records from about a quarter of the *Atlas* squares. This
is typically a plant of rather dry, fertile roadside verges,
railwaysides and ungrazed limestone grassland.

     The entire-leaved variant, 'var. *indivisa*', was
recorded in 1924 from Melmerby (6.3, W.W. Mason,
BEC Rpt (1925) 7:574) and found during the Survey
nearby at Renwick (60.42, LR, 1994, LANC).
Alt. limit: at least 350 m, Garrigill (74.42)
Map 784   410   LWFCY

***Cephalaria gigantea** (Ledeb.) Bobrov.
A garden throw-out found in 1991 and still persisting
on the roadside verge between Broughton-in-Furness
and Grizebeck (22.86, PB, LANC).
                                              1   F

Succisa pratensis Moench   Devil's-bit scabious
Common but noticeably thinning out in the more
intensively farmed lowland areas. It is a species of wet
grassland, flourishing in ungrazed rough grassland and
marshy fields, wet roadside verges and haymeadows,
and apparently tolerating a very wide range of soil pH,
being equally at home in calcareous fens and in base-
poor seepage areas and rock ledges on the fells. It is
less common than formerly as a meadow plant as a
consequence of agricultural 'improvement' and this
decline is mirrored by that of the marsh fritillary
(*Euphydryas aurinia aurinia*).
Alt. limit: 755 m, Knock Fell (70.30)
Map 785   1214   WFCY

Scabiosa columbaria L.   Small scabious
Frequent in the upper Eden valley and along the band
of calcareous soils around the north of the Lake
District, also on the Morecambe Bay limestone.
Elsewhere it is either absent or very rare. It grows
chiefly in limestone grassland, particularly on ungrazed
roadside verges, on limestone scars and in quarries.
Near Sedbergh it is on the flushed soils among Silurian
slate rocks by the River Lune (62.90) and at Dalston
(36.48) in flushed grassland by the River Caldew. It
also occurs on a few of the Lake District crags, as on
Raven Crag, Hartsop (40.10) and at the head of
Derwent Water (24.18), on coastal sand-dunes at
Eskmeals (08.92), and in turf on the sandstone cliff at
Rockcliffe (34.60).

     Like *Knautia arvensis*, this southern species
becomes appreciably rarer across the Solway Firth.
Alt. limit: 640 m, Cross Fell (64.36)
Map 786   205   WFCY

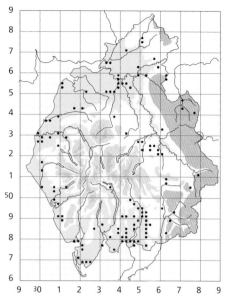

Map 783. *Dipsacus fullonum*

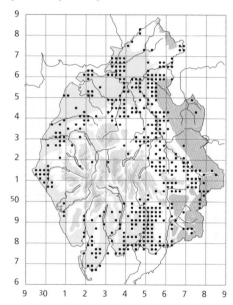

Map 784. *Knautia arvensis*

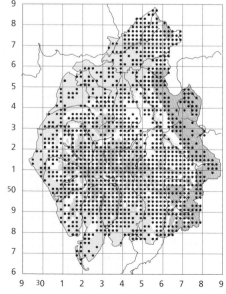

Map 785. *Succisa pratensis*

ASTERACEAE
(*COMPOSITAE*)

*Echinops bannaticus Rochel ex Schrader   Globe thistle
A garden escape found during the Survey at Far
Arnside and near Moorhouse, Carlisle (respectively
44.76s, CEW, and 32.56, REG, 1986, LANC, *Watsonia*
(1988) 17:193), in Furness near Leece (24.68s, PB,
1988, LANC, *Watsonia* (1989) 17:478, both under
Echinops sphaerocephalus), and in Kendal (52.90,
A. Boucher, 1996. All the sites were by roads or tracks.
4   WFC

Carlina vulgaris L.   Carline thistle
Frequent on the limestones, especially around the head
of Morecambe Bay but present around the coast from St
Bees to the Kent estuary. It is also well distributed, if
sparsely, throughout the Lake District. It occurs in
well-drained, short, calcareous or base-rich grassland,
particularly in the more open vegetation verging on
scree below limestone scars, and on coastal sand-dunes.
The Lake District habitats include quarries, rather dry
rock ledges, dry stony slopes and the richer areas of
mineral spoil heaps.
Alt. limit: 455 m, above Hayeswater (4.1, Wilson)
Map 787   203   LWFCY

Arctium minus (Hill) Bernh.   Lesser burdock
Frequent to very common in the lowlands, a familiar
sight on waste land, roadsides and railwaysides, and by
woodland margins, clearings and paths.

Almost all the specimens examined critically
were subsp. *nemorosum* (Lej.) Syme with large, shortly-
peduncled capitula. Subsp. *minus*, with mature heads
not more than 25 mm wide, was seen at Roanhead
(20.74), Aldingham (28.70) and Brigsteer (46.88)
but it has no doubt been overlooked. The southern
subsp. *pubens* (Bab.) Arenes, with terminal peduncles
exceeding 10 mm, should also be looked for.
Alt. limit: *c*.390 m, near Nenthead (76.44)
Map 788   1115   LWFCY

Saussurea alpina (L.) DC.   Alpine saw-wort
A rare species of damp, base-rich rock ledges across the
central part of the Lake District. Many of the sites
found or confirmed during the Survey do not feature in
the Floras, but this attractive plant, which is restricted
in England to the Lake District, Ingleborough and the
Cheviots, appears to be decreasing. There are old
records for Floutern Tarn (1.1), the head of Easedale
(3.0) and Stake Pass (2.0), and the Survey failed to
refind it on Crinkle Crags (24.04), Hanging Knotts
(24.06) and Tarn Crags, Grisedale (34.12), where it
was seen by DAR during the 1950s.

Hodgson's only Cumberland record is that from
Floutern Tarn, but he said it had not been refound
there. If the record is erroneous, the first for the vice-
county is that from Esk Hause in 1914 (22.08, J.W.
Haines, *BEC Rpt* (1946) 12:733).

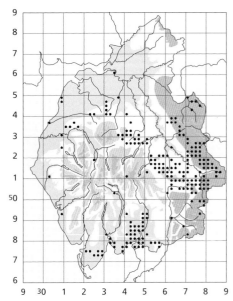

Map 786. *Scabiosa columbaria*

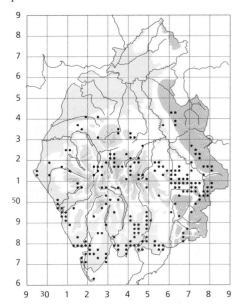

Map 787. *Carlina vulgaris*

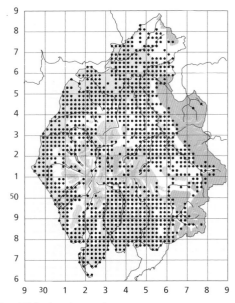

Map 788. *Arctium minus*

Alt. range: 550 m, Pillar (16.12) - 870 m, Helvellyn (34.14)

Map 789   15   WC

Carduus tenuiflorus Curtis   Slender thistle

A rare and apparently decreasing plant of sandy, waste ground close to the shore. It disappeared during the Survey from its Harrington foreshore site (98.24) following an 'improvement' scheme.

It is listed by Hodgson (as *Carduus pycno-cephalus*) from Beckermet (0.0), Egremont (0.1), Workington (9.2), Little Broughton and Maryport (0.3) and Silloth (1.5), also by Baker from Barrow-in-Furness and Humphrey Head (3.7s). It was seen at Allonby (0.4, W.W. Mason) in 1916 and in 1943 at Silloth and Grune Point by C.W. Muirhead, who also saw it at Carlisle "post-1945". There are post-1930 *Atlas* records for the Ravenglass (0.9) and Grizebeck (2.8) squares, but the dots for the Langdale (2.0) and Ambleside (3.0) squares and Handley's record (1898) for the Sedbergh area are probably errors.

Map 790   12   [W]FC[W/Y]

C. crispus L.   Welted thistle

(*C. acanthoides* auct.)

Common throughout the Eden valley and in the south but rare or absent elsewhere. It is a plant of tall grassland, occurring on roadside verges and railwaysides and in hedgerows, scrub and woodland margins.

Alt. limit: 245 m, near Crosby Ravensworth (60.12)

Map 791   208   LWFCY

C. crispus x **C. nutans**

C. x dubius Balbis

The only record of this hybrid is a specimen collected by J. Leitch at Silloth docks (1.5, CLE) in 1882. Both the parent species used to occur in the area but there are no Survey records from there. It is surprising that the hybrid has not been reported from the upper Eden valley where the parents are frequent.

                                     (C)

C. nutans L.   Musk thistle

Almost entirely restricted to the limestones around the upper Eden valley where it is common and very conspicuous in overgrazed open grassland and ascending to over 500 m. It also occurs on roadside verges, disturbed ground and in quarries. The records from outlying areas mostly represent casual roadside occurrences.

These are the most northerly records in western Britain.

Alt. limit: *c*.530 m, High Cup Nick (74.24)

Map 792   44   W(F)C

Cirsium vulgare (Savi) Ten.   Spear thistle

Extremely common, occurring in coarse grassland, as a weed of arable land and meadows, pastures and fellsides, on waste sites and tips and on burnt areas in woodland clearings. It avoids only the wetter and more

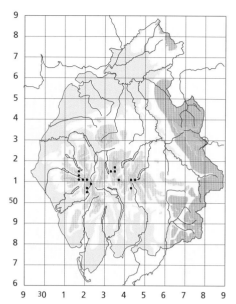

Map 789. *Saussurea alpina*

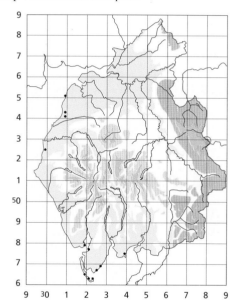

Map 790. *Carduus tenuiflorus*

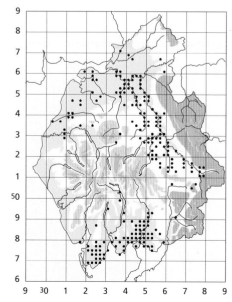

Map 791. *Carduus crispus*

acid soils.

Alt. limit: 845 m, Great Dun Fell (70.32)

Map 793   1668   LWFCY

C. heterophyllum (L.) Hill   Melancholy thistle
(*C. helenioides* auct.)

A magnificent thistle of upland hay-meadows and road-side verges, streamsides and woodland margins. It is becoming less common, particularly in the Lake District, as a result of grassland 'improvement' and increased silage-making. In upper Kentmere, for example, meadows in which it was once frequent now have only a few riverside colonies struggling for survival half-buried under dredged gravel. Fortunately it is still quite common in the east; particularly impressive are the flourishing colonies along the realigned Tebay - Kirkby Stephen road (6.0). The only sites on the Furness peninsula are the cemetery at Penny Bridge (30.82), where it was probably planted, and the roadside between Newby Bridge and High Newton (38.84), although there is an *Atlas* dot for the Dalton square (2.7). Outlying Cumberland populations such as those near Micklethwaite (26.48, 28.50) and Hayton (48.58) may also have been planted. Rather surprisingly there is only one record from crags in the Lake District, from Haystacks (18.12, DAR, 1952).

This species appears to set almost no viable seed.

Alt. limit: 490 m, between Garrigill and Nenthead (74.40)

Map 794, Plate 43   240   WFCY

*C. oleraceum (L.) Scop.

A Continental, yellow-flowered thistle found by C.J. Graham in 1975 growing in a small embayment off the A591 road near Wansfell Holme, Ambleside (38.02, LANC). The colony flourished for a number of years but had disappeared by 1980.

1   W

[C. acaule (L.) Scop.   Stemless thistle
The two 19th century records for Hard Knott (2.0) and Ennerdale (0.1/1.1) are rightly dismissed by Hodgson as errors, and are probably misidentifications of *Carlina vulgaris*.

C]

C. palustre (L.) Scop.   Marsh thistle

Virtually ubiquitous, growing in a wide range of wet grassland habitats from the tall herb vegetation of wet meadows, fens and lake and riversides to the upland *Juncus effusus* flushes and relatively dry, thin fellside soils, both on and off the limestone. It also occurs by woodland margins and along tracks.

White-flowered plants are common, at least in the fells, and, according to Mogford (1974), they are preferentially pollinated.

Alt. limit: 760 m, Cross Fell (70.34)

Map 795   1706   LWFCY

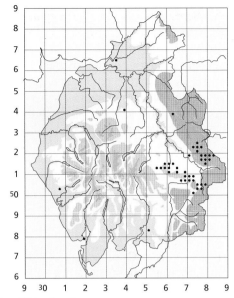

Map 792. *Carduus nutans*

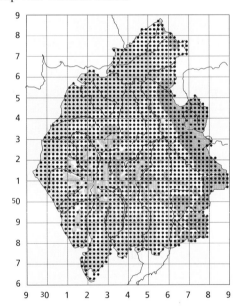

Map 793. *Cirsium vulgare*

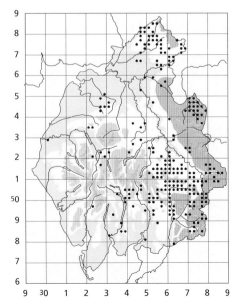

Map 794. *Cirsium heterophyllum*

C. arvense (L.) Scop.   Creeping thistle
Extremely common, absent only from the higher areas
of blanket peat. It occurs in most grassland communities, except on the more acid soils, often forming
dense stands on roadside verges, railwaysides, waste
sites and tips. Despite the long-continued use of
herbicides, it is still a serious persistent weed of arable
fields and pastures. In the uplands it occurs frequently
with *Cirsium palustre* in the fellside *Agrostis - Festuca*
grasslands.
Alt. limit: 845 m, Great Dun Fell (70.32)
Map 796   1596   LWFCY

*C. rivulare (Jacq.) All.
A flourishing colony of this central European species
was found in 1988 near Catterlen, Penrith (48.32,
RWMC, LANC). It was about 50 m from the nearest
house although it appears not to have been cultivated
there.
                                1   C

*Onopordum acanthium L.   Scotch thistle
A very rare casual occurring usually by roadsides, but
as a garden weed at Rampside (22.66s, L. Cowcill,
1979-1985). Most plants are probably of garden origin.
Alt. limit: *c.*330 m, north of Alston (72.48)
Map 797   9   WFC

*Silybum marianum (L.) Gaertner   Milk thistle
Listed by Hodgson as planted at Workington (9.2) and
growing in several places in gravel by the River Esk near
Longtown (3.6). There is a post-1930 *Atlas* dot for the
Heversham square (5.8s) and a 1934 record from Dalston
(3.5, G.A. Field, CLE).
        Although native in Europe only in the south, this
thistle is now very widely naturalised.
                                (WC)

Serratula tinctoria L.   Saw-wort
A species with a curiously patchy distribution, locally
frequent from Sebergham north to Bewcastle, on the
Asby - Orton limestone and on the shores of Ullswater
and Windermere, but scarce or absent elsewhere. It
grows in a wide variety of habitats: damp, often calcareous or base-rich roadside verges and open woodland,
herb-rich hay-meadows and coarse grassland, often
with *Sanguisorba officinalis*, *Cirsium palustre* and
Carex spp., dry limestone grassland and on wooded,
stony or rocky lake shores.
        This is a characteristic member of a lakeside
community which includes *Thalictrum minus, Trollius
europaeus* and *Galium boreale* but its restriction to
Windermere, Bassenthwaite and Ullswater is curious.
        Although frequent in the north of the county this
species is very rare across the Border in southern
Scotland.
Alt. limit: *c.*300 m, Swindale, Brough (80.16)
Map 798   182   WFCY

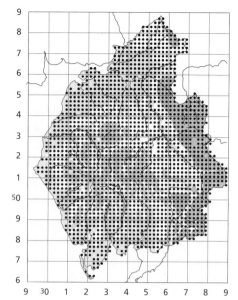

Map 795. *Cirsium palustre*

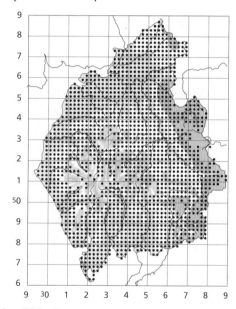

Map 796. *Cirsium arvense*

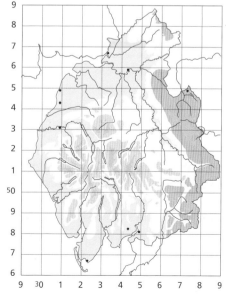

Map 797. *Onopordum acanthium*

*Centaurea atropurpurea Waldst.& Kit.
A native of the Balkans, this, or possibly *C. kotskyana*
Heuffel ex Koch, is probably the plant listed by Hodgson as
Centaurea calocephala and found on the Maryport (0.3)
ballast tips in 1887.

(C)

C. scabiosa L.   Greater knapweed
Frequent in the Eden valley, particularly in the upper
part, also along the Morecambe Bay coast, but otherwise rare, and in some areas only as a casual. It occurs
in well-drained grassland, often on limestone, for
example on roadside verges and, noticeably, on railwaysides, also on limestone scars and in quarries and
waste sites.

Many of the plants at Ramsden Dock, Barrow-
in-Furness (20.68s) are white-flowered.

These Cumbrian sites are the northernmost ones
in western Britain.
Alt. limit: 260 m, above Crosby Ravensworth (62.12)
but introduced below Nenthead (76.44) at 375 m.
Map 799   75   WFC(W/Y)

*C. montana L.
An occasional but persistent and conspicuous garden
escape occurring almost always by roadsides and
usually near houses. Surprisingly, there appear to be no
pre-Survey records.
Alt. limit: *c.*380 m, near Garrigill (74.42)
Map 800   70   LWFCY

C. cyanus L.   Cornflower
Seen during the Survey at only five sites and then only
briefly: on the newly-seeded verge of the A590 near
Town End, Witherslack (44.82s, GH, 1984, LANC), on
roadside verges near Barrow-in-Furness (18.72, PB,
1993, LANC) and near Broughton Moor (06.34, NB,
1972, *Watsonia* (1992) 19:157), on newly-seeded
ground on the Brampton by-pass (54.60, FJR, 1991,
LANC), and on the ethylene pipeline on Wan Fell,
Plumpton (50.36, T.R. Harwood, 1992).

Hodgson describes it as "rare and probably
always introduced" and he lists it from Egremont (0.1),
Flimby (0.3), Workington (9.2), Aspatria and Blennerhasset (1.4), Silloth (1.5) and Grinsdale and Kingmoor
(3.5). Baker gives an additional record from Kirkby
Lonsdale (6.7s) and Wilson one from Sandside (4.8s).
There are later Cumberland records from Skelton (4.3,
1895, A. Wilson, YRK), Culgaith (6.2, H. Britten,
1903), Thwaites, Millom (1.8, M. Cross, 1908), Grune
Point (1.5, J. Parkin, 1935) and Great Salkeld (5.3,
G.A.K. Hervey, 1957), and post-1930 *Atlas* dots for the
Penrith (5.3) and Melmerby (6.3) squares. There is also
a 1950 record from Walney Island (G. Higgin, herb.
P. Reiss, KDL).

S 5   WFC(Y)

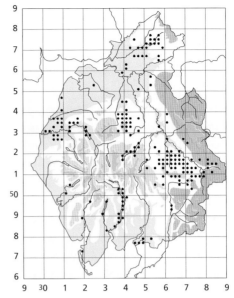

Map 798. *Serratula tinctoria*

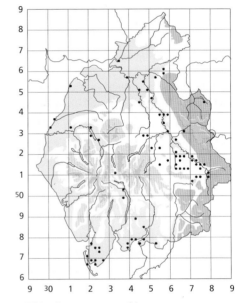

Map 799. *Centaurea scabiosa*

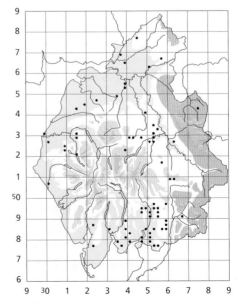

Map 800. *Centaurea montana*

*C. calcitrapa L.   Star thistle
A southern species known in Cumbria only from the 19th
century records cited by Hodgson from the shore at Workington (9.2) and at Silloth (1.5), and a slightly later one from
the gravel of the River Eden at Grinsdale, Carlisle (3.5, T.S.
Johnstone, 1902, CLE).

                        R   (C)

*C. solstitialis L.   St Barnaby's thistle
A southern species, like *Centaurea calcitrapa*, collected at
Silloth (1.5, CLE) by J. Leitch in 1890 and by D. Stewart in
1927.

                                                 (C)

*C. melitensis L.
A southern European species listed by Hodgson from Work-
ington (9.2), Egremont (0.1), Maryport (0.3), Silloth (1.5,
J. Leitch, 1889, CLE) and Floshgate, Ullswater (4.2).

                                                 (C)

C. nigra L.   Lesser knapweed
Extremely common, absent only from the higher hills
and areas of peatland. It is a plant of coarse grassland,
particularly roadside verges, railwaysides and woodland
margins, also of sea cliffs on the west coast and occasionally of fellside ledges. It tolerates a wide range of
soil moisture and acidity, occurring in marshy fields
and on the ledges of limestone scars.
Alt. limit: 845 m, Great Dun Fell (70.32)
Map 801   1504   LWFCY

*Carthamus lanatus L.
A Continental species collected at Silloth in 1891 (1.5,
J. Leitch, CLE).

                                                 (C)

*Cichorium intybus L.   Chicory
Now a rare casual of waste ground and roadsides,
usually near houses. Hodgson refers to its probable
introduction in the past with clover and grass seed and
its cultivation near Aspatria (1.4) and Baker mentions
it as being frequent on field borders in Furness. It was
recorded during the Survey in only a few of the *Atlas*
squares.
Alt. limit: 148 m, Brampton, Appleby (68.22)
Map 802   16   WFCY

Lapsana communis L.   Nipplewort
Very common in the lowlands occurring in woodland,
scrub and hedgerows, less commonly on waste ground
and roadside verges. It apparently thrives over a wide
range of soil acidity and moisture.
Alt. limit: *c*.390 m, near Nenthead (76.44)
Map 803   1254   LWFCY

Hypochaeris radicata L.   Cat's ear
Very common in the lowlands but with numerous gaps,
too many to be dismissed as due to under-recording. It
is typically a plant of light, sandy soils, occurring on
heaths and in hedgerows, acid grassland and as a weed
of lawns and waste ground.

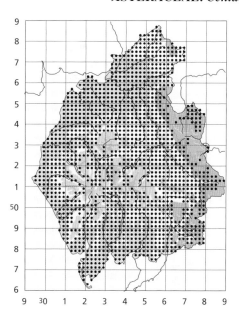

Map 801. *Centaurea nigra*

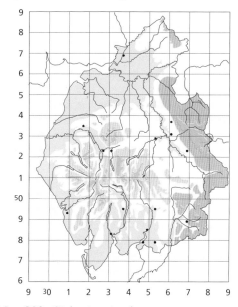

Map 802. *Cichorium intybus*

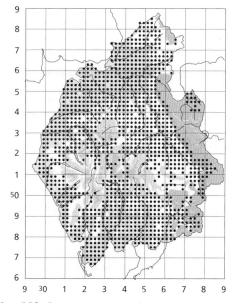

Map 803. *Lapsana communis*

Alt. limit: at least 530 m, south of Garrigill (76.36)
Map 804   1460   LWFCY

H. glabra L.   Smooth cat's ear
A rare and inconspicuous plant of open and somewhat
acid sandy ground and sand-dunes along the coast from
South Walney (22.68s) to the Ravenglass estuary. It
was first discovered by M. Bailey and C.W. Muirhead
in 1969 on the Drigg dunes (06.96). It was first found
on South Walney, the only Furness record, in the early
1970s (PT; LANC).
Map 805   S 13   FC

H. maculata L.   Spotted cat's ear
Known only from its classic site on the limestone cliffs
of Humphrey Head (38.74s) where it was first found by
I. Hall two centuries ago. Baker describes the population as "nearly or quite extinct" but,despite its
conspicuousness, it still survives although there are
only about 30 plants. Baker also gives a record from
between Kendal and Ambleside. This could conceivably
be correct, referring to a site on Cunswick or Underbarrow Scars, but the likelihood is that it is an error.

    Like the other specialities of the headland,
Veronica spicata and *Aster linosyris*, this is the most
northerly population in Britain, the nearest being on
the limestone of Great Orme's Head on the North
Welsh coast.
Plate 88   R 1   [(W)]F

Leontodon autumnalis L.   Autumnal hawkbit
Extremely common but rather surprisingly patchier
than *Hypochaeris radicata* in the lowlands and
ascending noticeably higher. It grows in a wide range
of mainly short grassland communities, on roadside
verges and lawns, in limestone grassland, where it is
appreciably commoner than *Hypochaeris*, also in hay-
meadows and waste places. It is also more tolerant of
wet conditions growing by the margins of lakes and
tarns, and in hillside flushes.

    An interesting population in flower in early June
occurs with the early-flowering, dwarf *Solidago
virgaurea* on the slate rocks of the River Lune near
Sedbergh (62.92). There are no records of the Scottish
montane subsp. *pratensis* (Hornem.) Gremli.
Alt. limit: 750 m, east side of Cross Fell (70.34)
Map 806   1668   LWFCY

L. hispidus L.   Rough hawkbit
Common over much of the lowlands but sparse in the
Lake District and on the more intensively agricultural
areas of the lower Eden valley and the Solway Plain.
This is a good indicator species of neutral, base-rich or
calcareous grassland, being most abundant around the
upper Eden valley. Characteristic habitats include limestone grassland and rock ledges, roadside verges, railwaysides and hay-meadows.
Alt. limit: 550 m, Hartside (64.40) and Moor House,

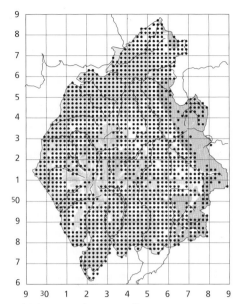

Map 804. *Hypochaeris radicata*

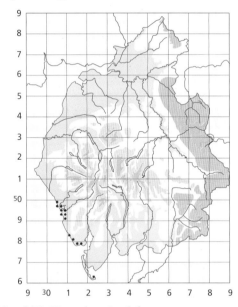

Map 805. *Hypochaeris glabra*

Plate 88. *Hypochaeris maculata*

Teesdale (74.22)
Map 807   1073   LWFCY

L. saxatilis Lam.   Lesser hawkbit
(*L. taraxacoides* (Villars) Mérat)
Not infrequent in short, rather open limestone turf and sandy banks in north Westmorland, also on dry sandy soils along the south and south-west coasts where it is particularly a feature of the early summer flora of the dunes. It has also been reported from a lawn at Penrith (50.30). Elsewhere it is scarce or absent although this small plant, quaintly called "Deficient hawkbit" by Hodgson, is easily overlooked, especially on dull days.
Alt. limit: 310 m, Spadeadam (62.74n)
Map 808   105   LWFCY

Picris echioides L.   Bristly ox-tongue
Extremely rare, seen during the Survey only on the driftline at Eskmeals (06.92/06.94) during a Liverpool University survey in the early 1970s, in a Kendal street and garden (50.92, CFS & JS, 1978; 52.92, CEW, 1989), by the verge of the new Lindale A590 by-pass (42.80s, CEW, 1988, LANC, *Watsonia* (1989) 17:479) and nearby on the roadside at Sampool Bridge (46.84s, CEW, 1990, LANC). It was also noticed in a Whitehaven garden (98.18, EEM, 1990) having probably originated from bird seed.

It is listed by Hodgson from Workington (9.2) and Egremont (0.1) and by Baker from Oxenfell, north of Coniston (3.0). There is also a record from Curthwaite (3.4, R. Wood, 1882, CLE), one from the station yard at Sedbergh (Handley 1898), a pre-1930 *Atlas* one from the Ulpha square (1.9) and a post-1930 one from the Heversham square (5.8s). The only Furness record is G. Wilson's from mid-Walney (1.6s) in 1954.

6   WFC(Y)

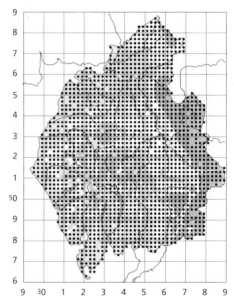

Map 806. *Leontodon autumnalis*

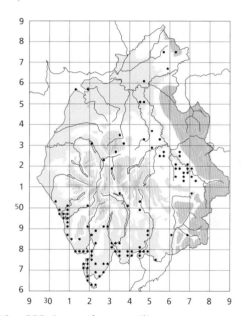

Map 808. *Leontodon saxatilis*

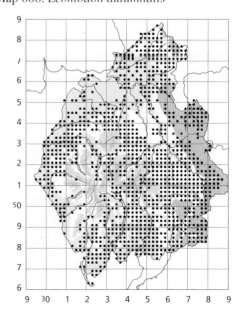

Map 807. *Leontodon hispidus*

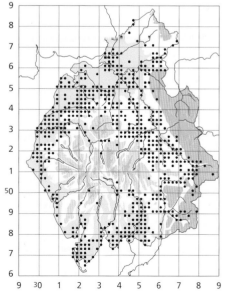

Map 809. *Tragopogon pratensis*

P. hieracioides L.   Hawkweed ox-tongue
The only record is of a plant in a Penrith garden which
presumably originated from bird seed (50.30, K.J. Corner,
1990, LANC). Hodgson records it from Egremont (0.1),
Gilgarran (0.2), Keswick (2.2) and Silloth (1.5, E.J. Glaister,
1881, CLE) and Wilson from Appleby (6.2) and Crosby
Garrett (7.0). In addition it was seen at Bullgill, Maryport
(0.3, J. Parkin) in 1935. There are six post-1930 *Atlas* dots,
of which two are on the Furness peninsula and one on
Walney Island (1.7, G. Wilson, 1954).

                                                            (WFC)

Tragopogon pratensis L.
                        Goat's beard; Jack-go-to-bed-at-noon
Frequent in the lowlands, less so in the more intensively agricultural areas of the Solway Plain and very
scarce in the Lake District. It is a plant of tall grassland
and with the disappearance of the traditional hay-
meadows it is becoming increasingly restricted to
roadside verges and railwaysides. It also occurs on
coastal dunes at Beckfoot (08.50).
            All the records refer to subsp. *minor* (Miller)
Wahlenb.
Alt. limit: *c.*320 m, Hardendale (58.14)
Map 809   567   LWFCY

*****T. porrifolius** L.   Salsify
A Mediterranean species recorded by Hodgson from churchyards at Workington (9.2) and Aspatria (1.4), Rose Castle
(3.4) and near Burgh-by-Sands (3.5, J. Leitch, 1876, CLE).
Nowhere does he regard it as other than a relic of cultivation.
                                                            (C)

Sonchus arvensis L.   Perennial sow-thistle
Common in the coastal lowlands, generally sparse in
the Eden valley and the north-east, although frequent in
the area of the upper Lyne, and very rare in the Lake
District. This is a prominent plant of nutrient-rich
coastal sites, occurring along the strandline on sand
and shingle beaches and in the uppermost drift zone of
salt-marshes. Away from the shore, it occurs most
frequently by roadside hedgerows and ditches, as a
weed of arable land and also around farms and waste
tips.
Alt. limit: *c.*300 m, Helbeck Wood, Brough (78.16)
Map 810   449   LWFCY

S. oleraceus L.   Smooth sow-thistle
A common lowland weed of gardens, waste places, tips,
quarries and arable fields, also of disturbed or trampled
roadside verges.
Alt. limit: *c.*350 m, east of Alston (74.46)
Map 811   787   LWFCY

S. asper (L.) Hill   Prickly sow-thistle
A very common lowland weed, occurring in similar
habitats to *Sonchus oleraceus* but tolerating wetter
conditions, hence its greater frequency in the uplands.
Alt. limit: *c.*380 m, near Nenthead (76.44)
Map 812   1159   LWFCY

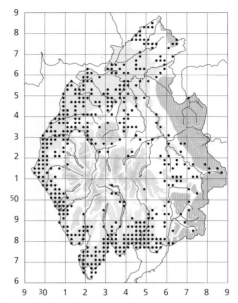

Map 810. *Sonchus arvensis*

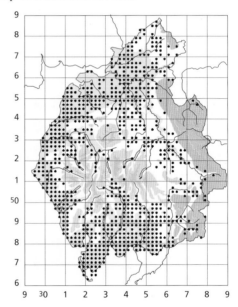

Map 811. *Sonchus oleraceus*

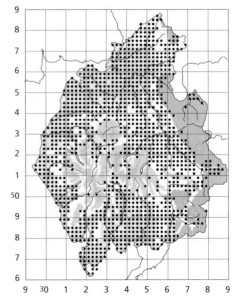

Map 812. *Sonchus asper*

Lactuca virosa L.

Extremely rare. It occurs on the Sandscale dunes (18.74), where it has been known since at least 1955, and amongst ponds and woodland immediately inland (20.72, 20.74, 22.72, PB, 1986). Most of these populations appear to be large and thriving. There is an unconfirmed report from across the Duddon estuary at Hodbarrow (18.78, Anon., 1988) and a single robust plant was found on the bridge at Santon Bridge (10.00, TW & KW, 1992, LANC).

Hodgson lists it from St Bees (9.1) and ballast tips at Maryport (0.3). Baker records it as "very fine" on the walls of Furness Abbey (2.7) and adjacent quarries. It was seen in 1950 by R. Martindale on a railway bank at Carlisle (4.5, CLE) and there are post-1930 *Atlas* dots for the Gosforth (0.0) and North Walney (1.7) squares.

5   FC

*L. tatarica (L.) C. Meyer

An eastern European species recorded by A.W. Westrup on waste land by the docks on Walney Island (1.6s) in 1952.

(F)

*Cicerbita macrophylla (Willd.) Wallr.

Blue sow-thistle

This conspicuous species is now known from several sites, but it appears to be particularly concentrated around Kendal (5.9). It is a plant of roadside and waste ground and most of the colonies presumably originate from garden material. The plants appear to be quite sterile.

According to P.D. Sell (1986), British plants belong to subsp. *uralensis* (Rouy) Sell, which, as the name implies, is a native of the Urals. He gives the first Cumbrian record as from the roadside at Glenridding (4.1, C. Bucknall) in 1915 and it persisted there at least until 1939 (A.E. Ellis, LANC). W.W. Mason found it nearby at Patterdale in 1922 and at Melmerby (6.3) in 1924. In 1927 it was seen at Silloth (1.5, Mrs H. Stewart, CLE) and in 1950 in Carlisle (3.5, M. Watson, CLE). The first Furness record was from Brathay, Ambleside (36.02, A.J. Richards, 1970s).

The reference to the occurrence of the very rare *Cicerbita alpina* by Ullswater (A.K. Swaine, *BEC Rpt* (1934) 10:532) presumably refers to *C. macrophylla*. Following the scheduling of *C. alpina* as a protected species in 1975, the Cumbria County Surveyor was alerted by several worried botanists anxious to protect their roadside populations from verge-cutting and warning him of the dire penalties he might incur.

Alt. limit: 250 m, Ravenstonedale (72.02)

Map 813   36   WFC

Mycelis muralis (L.) Dumort.   Wall lettuce

Common in the lowlands over most of the south and east, becoming scarcer northwards and absent from the extreme north and west. It is perhaps most commonly seen on and by mortared roadside walls and bridges and on urban sites, but it is also a frequent plant of rocky, wooded gills and, by contrast, of open limestone pavement and screes.

Alt. limit: at least 500 m, Alston Moor (72.36)

Map 814   515   LWFCY

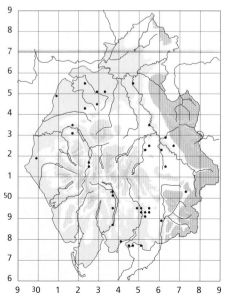

Map 813. *Cicerbita macrophylla*

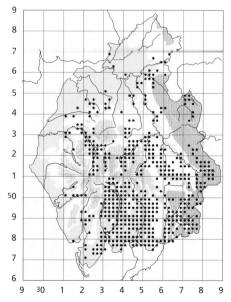

Map 814. *Mycelis muralis*

Taraxacum   Dandelions

[This account is substantially that prepared by Chris Haworth in 1989 shortly before his untimely death. It has been revised by Andrew Dudman to incorporate subsequent records and recent taxonomic work (Dudman & Richards 1997).]

The genus *Taraxacum* poses considerable problems for the Flora writer. Like *Hieracium* and *Rubus*, this largely apomictic group abounds in species (microspecies), about 250 being currently recognised in the British Isles. This fact alone creates obvious problems of coverage as well as of identification. Dandelions are, however, even commoner and more widespread than hawkweeds or brambles, and the weedy nature of many species in some of the sections compounds the recording problem. Yet this very abundance demands some serious treatment in a Flora.

However, there are further difficulties, which have led to the genus still being by far the least understood of the 'difficult' genera that are represented in these islands. It seems worth outlining why. Firstly, in contrast to the intensive attention that most of our flora has received for a hundred years and more, *Taraxacum* has been strangely neglected. British botanists were reluctant (some still are!), despite the obvious genetic variation of these ubiquitous plants, to accept the special taxonomic problems present. The treatment in the *Atlas of the British Flora (Critical Supplement)* (Perring & Sell 1968), for example, actually impeded progress. Secondly, the material is not 'easy': the collecting season is short and only practicable during the spring flush from mid-March (in sheltered lowland places) to the end of June (in the uplands), collecting and herbarium preparation must be done with great care, and there is a daunting phenotypic plasticity. Thirdly, there appear to be many rare species (including some yet unpublished). Many are doubtless local endemics, but, with over 2000 species worldwide, there is a large pool of potential casuals, especially in section Ruderalia. These rarities cloud the identification process. Reflecting these difficulties is the fact that Richards's (1972) *Taraxacum Flora* dealt with only about half the species now known in Britain. Nevertheless, A.J. Richards and I now agree that there are about 100 'important' species in Great Britain, i.e. those that are common, at least regionally, and are well understood taxonomically. Many of these have been recorded in Cumbria.

All these factors, then, have influenced the treatment here. Fig. 19 indicates the number of species recorded in each 10 km square. Clearly the cover is inadequate and uneven. No records from v.c.65 (N.W. Yorkshire) are included and only some squares in west Cumbria can claim to approach a representative coverage. What the map does suggest, however, is that one might expect the average number of species per square to be about 50. The total number of species in this account is 123, and this accords with the picture for other well-studied areas.

It is inevitable that there will be some errors of omission and commission with such a fluid and complex genus. Nevertheless, the writer is convinced that this report represents a fairly accurate picture of Cumbrian dandelions. It is unlikely that any significant species has been overlooked. Further work could doubtless change the picture for some of the apparently uncommon species and there will always be problems of misidentification and misinterpretation. The author has made the majority of records (many in the field) and determinations; most of the other records are those of A.J. Richards and A.A. Dudman. Collections of modest size have been made by a number of Survey workers. These contributions have been included here and the author is grateful for these efforts. Species with more than ten 10 km square records are mapped. Representative material is in OXF and LANC.

Regarding the Sections, it should be understood that, of their nature, the concept is 'multidimensional'. In other words, anyone using the keys should use as many characteristics as possible. Ecological ones are given more weight than is perhaps usual. It should be noted that some species occur at the boundaries of sections. This underlies the somewhat arbitrary nature of sections, which are meant to reduce the genus to more manageable groups, as well as to engender taxonomic meaning.

Maybe these finer points will not concern the average Flora user, but are made both to underline the relative youth of British Taraxacology, and perhaps more importantly, to act as a stimulus to encourage botanists to meet the many challenges yet to be encountered in this, the last major frontier of British floristic botany.

The task of constructing keys to the species has, so far, proved beyond the author's skill and energies! Attempts to use the familiar and simple, one-character dichotomous keys provided by workers in other large and critical genera have convinced him of the limitations to such keys. One might suggest that they are at best often misleading, and at worst mischievous, where they can give the average botanist what may be false concepts, which may then prove difficult to eradicate. To be more positive, however, most of the following accounts include a description of some distinguishing features of each species. Most attention has been given to those species which the author feels warrant this. At the end of the day, if the student admits that there is something that can be said beyond '*hic sunt dentes leonis*', the writer will be gratified!

Finally, a few acknowledgments: Without the constantly stimulating encouragement over the years from my co-worker John Richards, on whose shoulders all British taraxacologists stand, most of the advances evinced below would never have occurred. My friend Andrew Dudman has been ever supportive in ways too numerous to mention, both in the field and in particular with the production of this account. My wife Bertha has proved to be a patient word-processor in the face of outrageous demands and changes of mind. In addition, without her domestic succour through some difficult times, the account would never have appeared. Words, being inadequate for thanks, the following account will, I trust, suffice instead. I lay a firm claim to the errors of all types which certainly remain.

C.C.H. October, 1989

(Chris Haworth died December 2, 1989)

Key to Sections

For details of the terminology used, see Dudman & Richards's *Handbook* (1997). The species are arranged alphabetically within each section, the sections following the sequence in the *Handbook* and in Kent (1992). Besides leaf morphology and achene colour, the outer involucral bracts (here referred to just as 'bracts') are very important. In the newly opened inflorescence their orientation is particularly characteristic for each species. Three species are keyed out, these being the only representative of their sections in Cumbria.

Only one of the nine British sections, Section *Taraxacum*, is not represented in Cumbria. It contains very few British species which are known only from high Scottish mountains, but it is not totally inconceivable, though unlikely, that small relict populations may lurk in the Welsh or Lake District hills.

1. Small plants with small capitulae <3 cm; leaves usually highly dissected; inner and outer bracts often with corniculate appendages
   2. Achenes grey-brown, pyramidal cone ≤0.5 mm; lvs often with ≥ 6 pairs of lateral lobes
      ....*T. obliquum*, Section 2. **Obliqua**
   2. Achenes usually reddish to yellowish-brown, cylindrical cone *c*.1 mm; lvs rarely with > 6 pairs of lateral lobes ....Section 1. **Erythrosperma**
1. Plants not as above, generally more robust and with achenes with a shorter conical cone
   3. Leaves grass-like; outer bracts broad, tightly adpressed and with a scarious border, plants of basic, wet habitats *T. palustre*, Section 3. **Palustria**
   3. Plants not as above
      4. Plants small but capitulae relatively large; leaves ovate to lanceolate, almost entire or with many recurved lobes, often spotted; bracts ovate and erect to adpressed; achenes (with cone) at least 5 mm; oblong; very variable plants of moist or boggy habitats
         ....*T. faeroense*, Section 4. **Spectabilia**
      4. Plants generally more robust and with achenes rarely more than 4 mm and then more conical
         5. Lvs usually conspicuously spotted above and hairy; bracts rarely reflexed; pollen sometimes absent; robust plants of damp, grassy habitats, often upland ....Section 5. **Naevosa**

5. Lvs very rarely spotted, if so, then glabrous; bracts various; pollen usually present
   6. Lvs with red or colourless petiole and midrib; upper surface of midrib lacking green and coloured streaks, lf shape complex; bracts rarely erect; pollen nearly always present ....Section 8. **Ruderalia**
   6. Lvs with coloured petiole and midrib, upper surface of midrib with interwoven streaks (use lens), lvs generally rather flat and simple in shape; bracts seldom long or reflexed; pollen sometimes absent
      7. Lf lobes usually hamate (hooked) and rather obtuse, petiole seldom either green or brilliantly purple/red; bracts arched, short, dark, often suffused purple and pruinose; pollen always present; rather weedy in habit ....Section 7. **Hamata**
      7. Lf lobes more triangular, patent and acute; petiole often brightly coloured; pollen sometimes absent; plants of more natural, damp habitats ....Section 6. **Celtica**

Note: In the following account the number of squares refers to 10 km squares, and not to tetrads.

Section 1. **Erythrosperma** (Plate 89a)
This largely native section is well represented in Cumbria. The main characteristics and habitats are indicated in the above key.

T. arenastrum A. Richards
A widely scattered but local and rare native species. Found on sand-dunes, chalk and limestone grassland, usually near the sea. The leaf-shape combines the expanded lobe bases of *Taraxacum lacistophyllum* and the recurved lobe apices of *T. haworthianum*.

3 squares (0.4, 1.7, 5.4)   C

T. argutum Dahlst.
The inrolled ligules, with reddish backs, and the never-opening capitulae render this a readily identified species. Little or no pollen. It is largely confined to western Britain, especially in limestone areas, and was thought to be endemic until its recent discovery in Bohemia!
10 sqs (1.7, 3.7s, 5.7s, 6.7s, 0.1, 0.4, 1.3, 2.1, 6.3, 7.1)WFC

T. brachyglossum (Dahlst.) Raunk.
This is a widespread British species. It has characteristically spreading to erect purple external bracts. and rather short ligules. Unlike *Taraxacum argutum*, to which it seems to be related, it produces pollen. It is rather catholic in its habitat.
Map 815   28 squares   WFC

T. falcatum Brenner
This brown-achened species is scattered through Britain but is readily overlooked as it tends to grow in

grass. The diminutive form and laciniate leaves render it even more inconspicuous. Hence it is likely to be under-recorded. It is not a well-understood species.

2 squares (0.0, 0.1)   C

T. fulviforme Dahlst.
Although generally common, there appear to be few Cumbrian records for this species with cinnamon achenes and without pollen. The green external bracts are spreading and rather broad.

5 squares (0.9, 4.8s, 0.1, 3.0, 6.2)   WFC

T. fulvum Raunk.
Like *Taraxacum fulviforme,* with its cinnamon achenes and the absence of pollen, but readily distinguished by its slender, reflexed bracts.

4 squares (9.2, 2.7, 4.8s, 0.1)   FC

T. haworthianum Dudman & A. Richards
A recently published endemic species. It is widespread in Britain in sand-dune grassland and commonest in the north and west. It is probably more widespread than the only two Cumbrian records suggest.

2 squares (1.7, 3.7)   W

T. inopinatum C. Haworth
This is a recently published endemic species described from a plant growing in a disused limestone quarry at Dean (06.24, OXF). It is present in most similar quarries in west Cumberland. The species appears to be common in parts of Wales. It is a relatively robust, red-achened plant with no pollen. The external bracts are very dark, erect and pruinose. But its most distinctive character is the presence of spots on the upper leaf surface: a feature unique in the British species of this section.

7 squares (9.1, 0.0, 0.1, 0.5, 1.4, 1.7, 5.0)   C

T. lacistophyllum (Dahlst.) Raunk.
This species is probably the most widespread Erythrosperma species in both Britain and Cumbria. The expanded bases to the multiple leaf-lobes make it a neat and easily identified species. Like *Taraxacum brachyglossum,* it is polliniferous and has brick-red achenes.
Map 816, Plate 89a   31 squares   WFC

T. oxoniense Dahlst.
This is a generally widespread and abundant species on dunes and limestone and has been well-recorded in Cumbria. The leaf-shape can be variable but the bright green colour contrasting with the narrow vivid red/purple petiole, together with the erect, ovate and white-bordered bracts render it readily identifiable. The achenes are cinnamon, and it seems to be variable in pollen production.
Map 817   27 squares   WFC

T. proximiforme Soest in Lambinon & Soest
This is a generally uncommon brown-achened species. There may be some doubts regarding its conspecificity with Continental material.

3 squares (0.2, 1.3, 5.2)   WC

T. proximum (Dahlst.) Raunk.
This is a native species, distinguished by the unique colour of the achenes, which are dark brown to purple-brown or puce.

4 squares (2.2, 7.2, 7.3, 7.5)   WC

T. rubicundum (Dahlst.) Dahlst.
This tiny and pretty species is abundant on the limestone around Morecambe Bay. It also grows in dunes. It has erect, dark green bracts and violet achenes and is easy to identify.

7 squares (0.9, 4.8s, 4.9, 1.3, 1.7, 4.2, 6.0) WFC

T. scoticum A. Richards
An endemic species scattered in a number of sandy localities in northern England and Scotland but recorded in Cumbria only from the middle of Kirkby Lonsdale (60.78s).

1 square (6.7s)   W

Section 2. **Obliqua**
This section has few species and is seemingly very discrete and natural. The species have a superficial resemblance to section Erythrosperma but the leaves have a different texture and the achenes are very different.

T. obliquum (Fries) Dahlst.
A small plant with highly dissected leaves, orange-yellow capitula and grey-brown achenes, found on dune-slacks and grey dunes on the coasts of Scotland and northern England. Although recorded only from South Walney (22.62s), it seems possible that it is more frequent in Cumbria than this would suggest.

1 square (2.6s)   W

Section 3. **Palustria** (Plate 89b)
There are only a handful of British species in this section which morphologically is very distinct. All are uncommon or rare and grow in wet, nutrient-rich places and there is little doubt that many British populations have been extinguished over the years with the draining of suitable habitats.

T. palustre (Lyons) Symons
A small colony was found by REG in 1965 growing in a bare stony flush near Little Asby (68.10). Despite searching, it has not been seen since. The species should be looked for in similar habitats nearby.
Plate 89b   1 square (6.1)   W

a) Section Erythrosperma: *Taraxacum lacistophyllum*   b) Section Palustria: *Taraxacum palustre*

c) Section Spectabilia: *Taraxacum faeroense*

Plate 89. d) Section Naevosa: *Taraxacum pseudolarssonii*

a) Section Celtica: *Taraxacum duplidentifrons*

b) Section Hamata: *Taraxacum hamatum*

Plate 90. c) Section Ruderalia: *Taraxacum ancistrolobum*

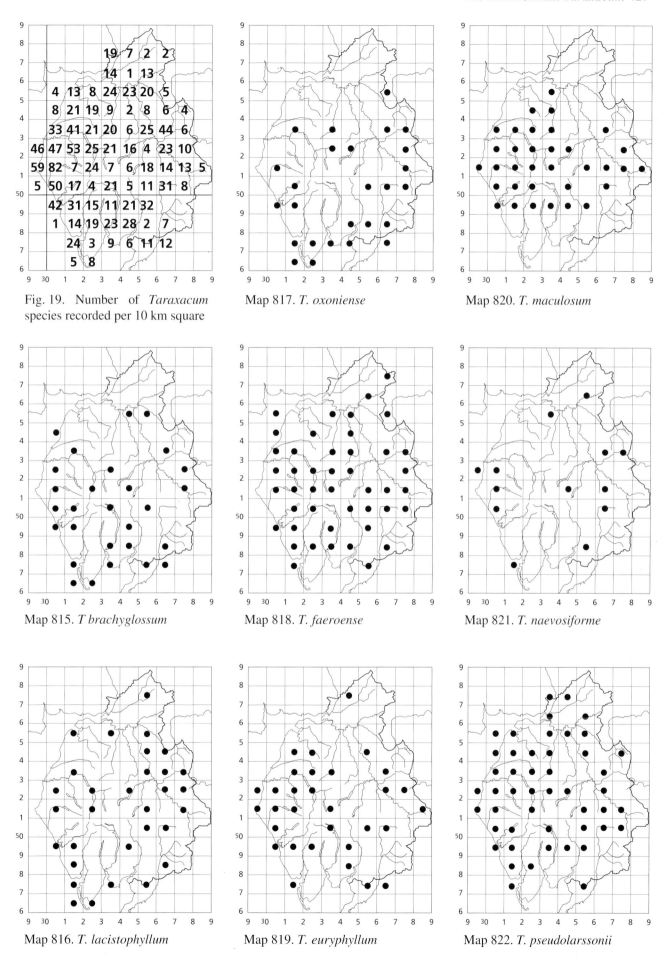

Fig. 19. Number of *Taraxacum* species recorded per 10 km square

Map 817. *T. oxoniense*

Map 820. *T. maculosum*

Map 815. *T brachyglossum*

Map 818. *T. faeroense*

Map 821. *T. naevosiforme*

Map 816. *T. lacistophyllum*

Map 819. *T. euryphyllum*

Map 822. *T. pseudolarssonii*

Section 4. **Spectabilia** (Plate 89c)
Recent work has suggested a narrower concept than hitherto for members of this section. The achene shape and size alone clearly separate the species from those of other sections. A study of holotypes and the plasticity of British material through experimental work has suggested that there is essentially only one very variable but very common species in Britain with, at most, a couple of very restricted rarities. Plants previously called *Taraxacum eximium* Dahlst. and *T. spectabile* Dahlst. are now all included within *T. faeroense*.

T. faeroense (Dahlst.) Dahlst.
This is one of Cumbria's most widespread species being found in grassy wetlands of all types. The key includes the species characteristics.
Map 818, Plate 89c   45 squares   WFC

Section 5. **Naevosa** (Plate 89d)
This is a largely native section with many endemic species. These are mainly confined to western and northern Britain in grassy places in the uplands. Not surprisingly then, there are some common Cumbrian species. With a few exceptions (e.g. *Taraxacum subnaevosum*), the plants are robust with large capitulae. Pollen is often absent. The chief morphological characteristic of this section is the presence of pigment spots on the upper leaf surface. The unwary can confuse these with disease or the midrib blotch of many species belonging to other sections.

T. euryphyllum (Dahlst.) Hjelt
This is a robust species, virtually without pollen, with hairy leaves with many patent lobes and a broad purple petiole wing. It is by far the most weedy in habit of all members of this section though it is probably native. It is most frequent on roadsides where it can be abundant.
Map 819   31 squares   WFC

T. maculosum A. Richards
An endemic species which is abundant in grassy upland places. It never has pollen. The narrow, very dark external bracts in two rows, the upper erect and the outer more spreading, are very characteristic. Also, the generally hairy leaves with large end lobes and sinuose margins to the distal side of the leaf lobes render this an easy species to identify.
Map 820   33 squares   WC

T. naevosiforme Dahlst.
The plants are robust, with generally yellow- or grey-green, dentate foliage with broad, recurved leaf-lobes, and abundant pollen. It is particularly common on the limestone cliffs of Melmerby Scar and Cross Fell (6.3).
Map 821   13 squares   WC

T. naevosum Dahlst.
This is a very robust, dark green, hairy, polliniferous species. It is a plant of old hay-meadows and it also grows on scars by Crowdundle Beck, Cross Fell (6.3).
        6 squares (5.9, 0.1, 0.2, 5.6, 6.2, 6.3)   WC

T. pseudolarssonii A. Richards
Until the discovery in 1986 of a single isolated site in Denmark, this was thought to be a British endemic species. In fact it has a relatively restricted British distribution which is clearly centred on northern England, where it is an abundant species and in some meadows the dominant dandelion. It occurs equally commonly in country lanes and by moorland tracks. Sand-dunes are another habitat. With *Taraxacum duplidentifrons*, it is perhaps the most characteristic and abundant Cumbrian dandelion. It is rather plastic in leaf morphology but distinctive in the heavy spotting which can coalesce into blotches covering more than half the leaf surface. Late season plants seem to have the pigment 'washed out', again giving a distinctive appearance. The spreading arcuate bracts are pale above and usually with some reddish pigmentation. Pollen is abundant.
Map 822, Plate 89d   48 squares   WFC

T. richardsianum C. Haworth
This newly published endemic species was described from material collected at Haile, near Egremont (02.08), where it was growing on a grassy bank near woodland. The species seems to be closely related to *Taraxacum euryphyllum* and shares its lack of pollen (or sometimes with a very small amount in the outer florets) with that species. The bracts are, however, shorter, darker, and more crowded and erect. The leaves are dark and shiny, and the leaf-lobes are more numerous and more recurved and acute. There are now many records in Britain, but mainly in western and northern areas.
    9 sqs (0.9, 0.0, 0.1, 2.4, 4.0, 4.4, 5.5, 6.0, 6.2)   WFC

T. stictophyllum Dahlst.
This species can be robust and its prostrate, dentate, grey-green leaves and capitulae with crowded, large, patent to erect bracts mark it out. It is sometimes rather faintly spotted. It produces pollen. This is a common species in Scotland and has scattered records in northern England and in Wales. There are few Cumbrian records but it is suspected that it may be commoner than this suggests, especially in the taraxacologically relatively neglected Pennine area.
        6 squares (1.9, 5.9, 0.1, 1.2, 2.1, 2.2)   WFC

T. subnaevosum A. Richards
This is generally a small plant growing in open and often rocky areas. It never has pollen. Although, depending on habitat, there is some variation in the orientation of the bracts, these are generally spreading to recurved and are the most tenuous of any member of this Section. The leaf is always a pale green and often has a yellow hue. The leaf spotting is punctate and

often obscure.

Map 823   16 squares   WC

Section 6. Celtica (Plate 90a)

As the name suggests, the species of this section are concentrated in oceanic areas and thus Cumbria is well represented by members, some of which are abundant. All the species are thought to be native and some are endemic. It is a somewhat heterogeneous section and merges into several others, in particular section Hamata. However, it seems to be a very convenient grouping which serves to reduce the genus to more manageable chunks.

T. berthae C. Haworth

First described from Cumbria, there are now about a dozen British records for this endemic species. It seems to belong to a group of species around *Taraxacum nordstedtii* (see above). It is a small species with no pollen, erect bracts and with dark green, spotted leaves with large end lobes.

          7 squares (9.1, 0.1, 0.3, 5.6, 6.0, 6.4, 7.1)   WC

T. bracteatum Dahlst.

Although with rather few records in Cumbria, this is a widespread British species and may be under-recorded. It always grows in damp habitats. The dark green leaf has a contrasting brightly coloured petiole. The bracts are also very dark, crowded, and erect.

Map 824   13 squares   WFC

T. britannicum Dahlst.

This is an elegant species and a typical member of this section. The foliage has a bluish-green hue with brightly coloured petioles and midribs. The bracts are erect, black and with a distinct white margin. The achenes are often barely spined - a useful diagnostic feature. So far, it seems, in our area, to be confined to the limestones of west Cumberland.

          5 squares (9.1, 0.9, 0.0, 0.1, 2.3)   C

T. cambricum A. Richards

Fairly recently described from Wales, this was unknown elsewhere until the discovery of a large colony in 1987 in a sunken lane near Embleton church (16.28). Such sites are often rich in native species, and this one is no exception. *Taraxacum cambricum* is close to the much commoner *T. gelertii*, but is more robust generally; the end lobe is more rounded, and the lobation more sinuose.

          1 square (1.2)   C

T. celticum A. Richards

An endemic species scattered in well-drained soils, on walls etc., in Wales and the south-west. The only Cumbrian record is from near Newbiggin (64.24).

          1 square (6.2)   W

T. duplidentifrons Dahlst.

This species can perhaps lay claim to being the commonest British dandelion, though, in common with other members of the Section, numbers are concentrated to the north and west of these islands. The often rather crispate foliage which is dull, hairy, and with triangular lobes ending in a very wispy point make it easily known with practice.

Map 825 (Plate 90a)   55 squares   WFC

T. excellens Dahlst.

This is a very variable taxon of somewhat puzzling sectional affinity. It is scattered mainly through the western side of Britain. It is doubtless more common than the records suggest.

          4 squares (9.1, 1.8, 3.8, 0.1)   WFC

T. gelertii Raunk.

This is an abundant species in western Britain and is a typical member of the section with its neat, simple leaf form of patent triangular lobes. The external bracts are dark green and erect. It is common throughout Cumbria in grassy natural or semi-natural habitats.

Map 826   31 squares   WFC

T. haematicum Hagl. ex Oellg. & Witzell

This is a species of rich, damp, basic grassland. It is possibly commoner than the records suggest.

          2 squares (1.7, 0.1)   C

T. hesperium C. Haworth

An endemic described by Chris Haworth and published posthumously. The species lies morphologically between sections Celtica and Hamata, occurring mainly in the west of Britain, but only in 'good' habitats.

          3 squares (9.2, 0.1, 1.2)   C

T. inane A. Richards

This is an endemic species common and almost confined to Scotland and northern England where it is common. The lack of pollen is a useful character as is the broad shining purple petiole. For a Celtica species, the external bracts are unusually long and spreading.

Map 827   15 squares   WC

T. lancastriense A. Richards

An endemic species related to *Taraxacum nordstedtii*, occurring mainly in the west of Britain and in Cumbria known only from below Hartside (62.40. It has conspicuous borders to the external bracts, a long hastate terminal leaf lobe, and no pollen.

          1 square (6.4)   C

T. landmarkii Dahlst.

There are scattered records for this delicate and elegant species though some of the older determinations may be suspect. It has long ligules and dark erect bracts. It is a

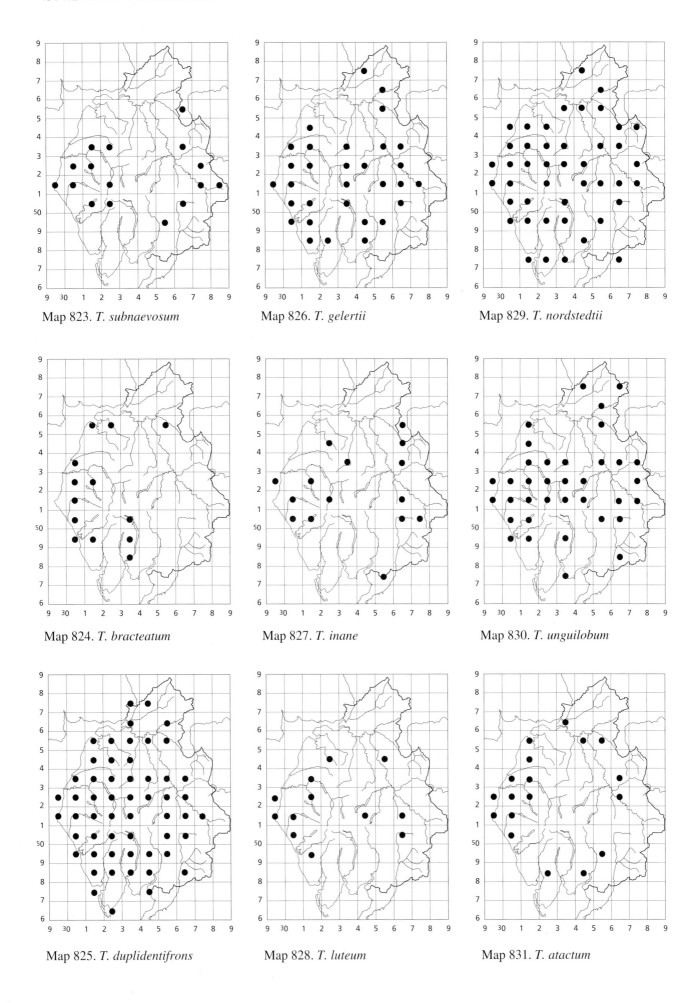

Map 823. *T. subnaevosum*

Map 826. *T. gelertii*

Map 829. *T. nordstedtii*

Map 824. *T. bracteatum*

Map 827. *T. inane*

Map 830. *T. unguilobum*

Map 825. *T. duplidentifrons*

Map 828. *T. luteum*

Map 831. *T. atactum*

very 'western' species, being abundant on the coasts of Wales and Scotland.

10 sqs(9.1, 1.9, 6.7s, 0.1, 0.2, 0.3, 1.0, 2.1, 6.0, 6.3)WC

T. luteum C. Haworth & A. Richards

This species seems to be centred on Cumbria although there are scattered records from Scotland, Wales and Ireland. One feature which renders it unique in our flora and which gives the plant its specific epithet is the complete absence of any coloured stripe to the ligules. This makes the plant immediately recognisable. It was described from material growing in a dandelion-rich sunken lane at Egremont (02.10). This has, sadly, been destroyed in recent road-widening improvements, though the species is still to be found in the near neighbourhood.

Map 828, Plate 91   12 squares   WC

T. nordstedtii Dahlst.

This is one of the most interesting and puzzling of all dandelion species. It is a plant of moist places, abundant in Britain, especially in western and northern areas and is very variable in leaf morphology; perhaps more than any other species. It is nevertheless usually readily recognised by its crowded and imbricate, erect, ovate and ciliate bracts and the very short ligules with rich-brown stripes. Anthocyanin development varies a lot and in western Britain the leaves can be spotted. Another unusual feature is that it possesses, unlike other British dandelions, six sets of chromosomes (the others ranging from three to five sets), and this may contribute to its problematical taxonomic nature. There may be arguments for putting it into a section of its own, for it has a number of related species, including *Taraxacum berthae* and *T. pseudonordstedtii*.

Map 829   44 squares   WFC

T. olgae A. Richards

A relative of *Taraxacum nordstedtii*, but with conspicuously white-bordered bracts and yellow-green leaves. It is an endemic known only from Cleator (00.12) and Kirkcudbrightshire.

1 square (0.1)   C

T. ostenfeldii Raunk.

Scattered records throughout Britain and probably not native. It has yellow styles, small capitula with short, crowded florets, and no pollen.

3 squares (9.1, 5.7s, 0.1)   WC

T. pseudonordstedtii A. Richards

A close relative of *Taraxacum nordstedtii*, it is distinguished from it by its leaf-shape with 4-6 short, recurved lobes, its dark purplish leaves and its dwarf, prostrate stature. It is an endemic centred on upper Teesdale, with outlying records from Cumberland and north Lancashire.

2 squares (0.1, 4.2)   WC

T. subbracteatum A. Richards

For a species which is common in both Scotland and Wales, this seems oddly uncommon in Cumbria.

5 squares (9.1, 1.5, 3.0, 5.6, 6.3)   C

T. tamesense A. Richards

There is just one record for this endemic 'mead' species, but the locality, Lowca Beck (98.20), suggests that it is a casual.

1 square (9.2)   C

T. unguilobum Dahlst.

This is another northern and western species, being particularly abundant in Scotland where, with the

Plate 91. *Taraxacum luteum*

ubiquitous *Taraxacum duplidentifrons*, they must form at least half of the native dandelion population. It is one of the most easily identified of all dandelions with its crowded, dentate, highly recurved lobes to the leaves. The erect, glaucous, white-bordered bracts are also distinctive. It never has pollen. The achenes are rather small and of a reddish hue.

Map 830   36 squares   WFC

Section 7. **Hamata** (Plate 90b)
This section forms a fairly coherent grouping morphologically. The species also share a chromosomal peculiarity. Their status in Britain is somewhat difficult to ascertain as most tend to behave in a weedy fashion. Some, if not most, are nevertheless undoubtedly native. They include some of Britain's most abundant species and the Cumbrian picture mirrors this. Collectors (and referees!) should avoid small and immature material with even greater determination than usual, for this is the most 'critical' of the sections.

T. atactum Sahlin & Soest
This species may well have been over-recorded due to some confusion in the past with *Taraxacum sahlinianum*. Both have a large end lobe with one or two large teeth, but *T. atactum* is a neater, less dentate species.

Map 831   18 squares   WFC

T. boekmanii Borgv.
This species has a somewhat hairy leaf of spathulate shape with a rounded end lobe and, for this section, the brightest of all midrib colours: on the underside of the leaf the bright red coloration extends to the very tip. It is difficult to say whether it is a native or an introduction, being found commonly in waste places but also in open areas in natural woodland.

Map 832   14 squares   WFC

T. hamatiforme Dahlst.
Characteristically, this species has brightly coloured petioles and rather narrow and asymmetrical leaf-lobing. It is common throughout Great Britain and equally so in Cumbria. It may be native.

Map 833   24 squares   WFC

T. hamatum Raunk.
The inky involucre and rather more erect bracts than those of other species in this section, together with the rather blunt and hooked leaf lobes, render this a generally easy plant to identify. The plants are rarely more than medium sized. It is certainly native, preferring damp grassland and even tolerating some shade in woodland. Like most dandelion species, however, it can also be found on roadsides.

Map 834, Plate 90b   41 squares   WFC

T. hamiferum Dahlst.
The most distinctive character of this species is the irregular splitting of the inner bracts as the capitulum opens. It is doubtfully native.

Map 835   15 squares   FC

T. kernianum Soest, Hagend & Zevenb.
This is perhaps the least characteristic member of the section. The lobing is irregular, the lateral lobes of the inner leaves being characterised by a long linear apical part.

4 squares (9.1, 0.1, 1.2, 2.2)   C

T. lamprophyllum M. Christiansen
This is a heavily midrib-blotched species with large external bracts and big-toothed leaves. Again, it seems to be an introduced, though well-established, species.

Map 836   18 squares   WFC

T. marklundii Palmgren
A native species, clearly related to *Taraxacum subhamatum* but characterised by the very dentate leaf-lobes.

3 squares (2.9, 0.1, 6.3)   WC

T. pseudohamatum Dahlst.
This is a very robust weed and is the commonest member of this section and indeed one of Britain's commonest dandelions. It seems almost confined to these islands but is nevertheless more characteristic of roadsides than of native haunts. It has the largest capitulae and bracts in the section. It is our earliest flowering dandelion and can, in favourable locations, be in flower by mid-March.

Map 837   40 squares   WFC

T. quadrans Oellg.
This is the neatest Hamata species. Its rather blue-green foliage grows very erect and the leaves bear rather regular 'squared-off;' lobes. It is probably not native.

Map 838   17 squares   WFC

T. sahlinianum Dudman & A. Richards
A recently described endemic, closely related to *Taraxacum atactum*, but distinguished from that species by its untidy, dirty green leaves, which are broader, more dentate and more broadly winged.

6 squares (2.9, 0.1, 0.3, 3.2, 6.2, 6.3)   WC

T. subhamatum M. Christiansen
The very pruinose involucres are a useful character and the leaf shape is characteristic though difficult to describe easily. The end lobe tends to be apiculate. It is a very common species and certainly seems to be native in Scotland and northern England where it is often abundant.

Map 839   22 squares   WFC

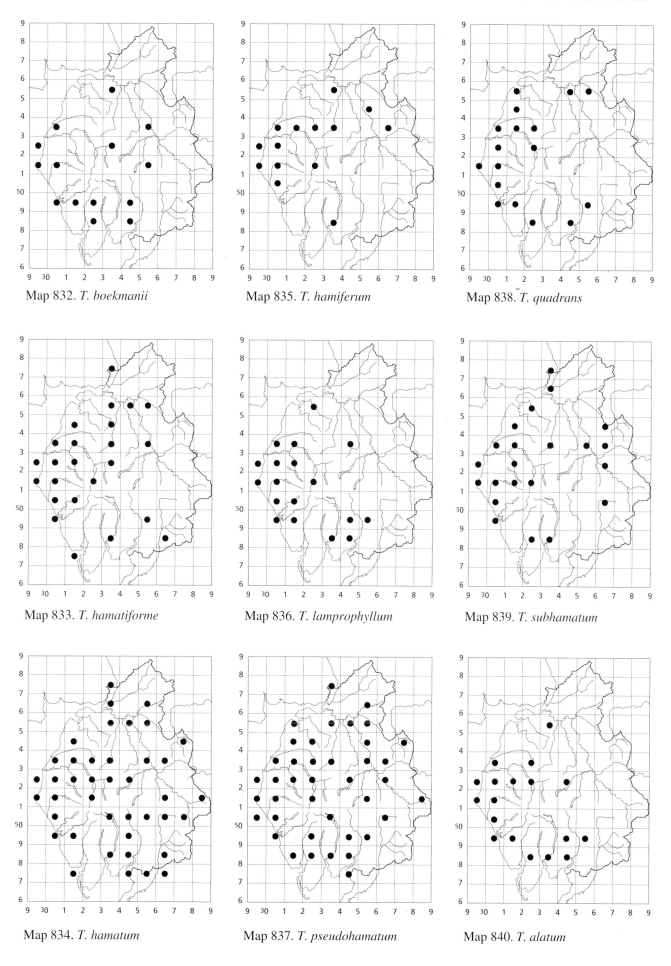

Map 832. *T. boekmanii*

Map 835. *T. hamiferum*

Map 838. *T. quadrans*

Map 833. *T. hamatiforme*

Map 836. *T. lamprophyllum*

Map 839. *T. subhamatum*

Map 834. *T. hamatum*

Map 837. *T. pseudohamatum*

Map 840. *T. alatum*

Section 8. **Ruderalia** (Plate 90c)

On the open road on a sunny mid-April day, the carpet of yellow along road-verges and on roundabouts speaks to the botanist and layman alike, that spring is truly here again. For the taraxacologist, in addition, there is challenge - even potential panic. For almost all these dandelions will belong to the section Ruderalia. Since this section contains more than half the known (and unknown!) British dandelions, and there being a considerable pool from the Continent to draw upon as introductions, it is perhaps not surprising that difficulties of identification loom large. Moreover, their generally robust size can understandably deter even the most ardent collector. Even when not so deterred, the authority to whom the student sends his collection all too often finds himself faced with poorly presented and ill-dried material. Small wonder then that this section poses most problems for the taraxacologist and flora-writer.

Yet much can be said to redress this gloomy introduction. The really common British (and Cumbrian) dandelions are relatively few and quite well understood. There is little doubt that many of these are native too, for whilst straying into ruderal habitats, they present certain clues which suggest this status.

In order then to simplify the matter, the author has decided to treat this section in a different way from the others. The first part will deal with the more significant species (as far, that is, as the author feels them to be so). These will generally be common. Most too are probably native and can be assumed to be so unless stated otherwise. The second part will deal much more briefly with a motley collection comprising species with few records, or those which are not well-understood, or even in some cases as yet unpublished. In this part it can be assumed that the species is probably introduced unless otherwise indicated.

For most dandelions (and indeed for many British plants) there are problems of ascertaining the status of each species and the whole spectrum of problems can be illustrated in this section. However, some are known to be pan-European (e.g. *Taraxacum alatum*); some are certainly common natives, though not confined to these islands even though we may be the main centre of distribution (e.g. *T. cordatum*); some appear to be widespread endemics (e.g. *T. stenacrum*); some, one can be fairly certain, are introductions even if widespread in Britain, for they never seem to be found away from the edge of main roads (e.g. *T. undulatiflorum*). In other cases determining the status is more difficult: for example, one does not know whether one is looking at the remnants of a relict native species or a casual introduction. In other instances (e.g. *T. fasciatum* and *T. sellandii*), the species can be abundant on roadsides but is occasionally found in semi-natural grassland. Further studies should help elucidate at least some of these problems.

Part One

T. aequilobum Dahlst.
This species is scattered throughout Britain and Cumbria. It is best recognised by its many-lobed leaves and the long, reflexed and twisted bracts.
9 sqs (9.1, 0.9, 2.8, 0.1, 0.3, 1.2, 3.7n, 5.1, 5.3)   FC

T. alatum Lindb. f.
This is a widespread species in Cumbria and in Europe generally. Its long, winged, green petiole and large, reflexed, glaucous bracts with a pink tip, make it comparatively easy to identify.
Map 840   18 squares   WCF

T. ancistrolobum Dahlst.
This is a very robust and abundant dandelion generally. It has green-petioles and a 'spinach-like' leaf with very few, large, very blunt lobes. It is abundant on roadsides and is readily identified.
Map 841, Plate 90c   39 squares   WFC

T. cordatum Palmgren
This is another common species which often grows on cliffs, in sand-dunes and other natural places, as well as on waysides. The cordate end lobe and regular, patent, and rather blunt lobes to the leaf, together with broad, spreading, dark bracts, are all useful diagnostic features. It is an important member of the British flora.
Map 842   35 squares   WFC

T. croceiflorum Dahlst.
This is a green-petioled species with a small, triangular end lobe to the leaves and a broad, green, winged petiole, but the leaf shape is somewhat undistinguished. The heads are orange-yellow due to the ligule-teeth and inner florets being tipped orange-red. It is a widespread plant of uncertain status.
Map 843   20 squares   WC

T. cyanolepis Dahlst.
One of a number of common Scottish species which seem oddly uncommon in Cumbria. It is a relatively easy species to identify, as it is the only common species with green petioles and lurid violet bracts. It is probably native.
2 squares (9.1, 1.9)   C

T. dahlstedtii Lindb. f.
This species is particularly common near the coast. It has undistinguished foliage, but the narrow, reflexed bracts and very red petiole are useful markers.
Map 844   15 squares   WC

T. ekmanii Dahlst.
A widespread species, often abundant where found. Its chief features are the pink petioles and often rather chlorotic foliage with very varied leaf-shape. It is one of

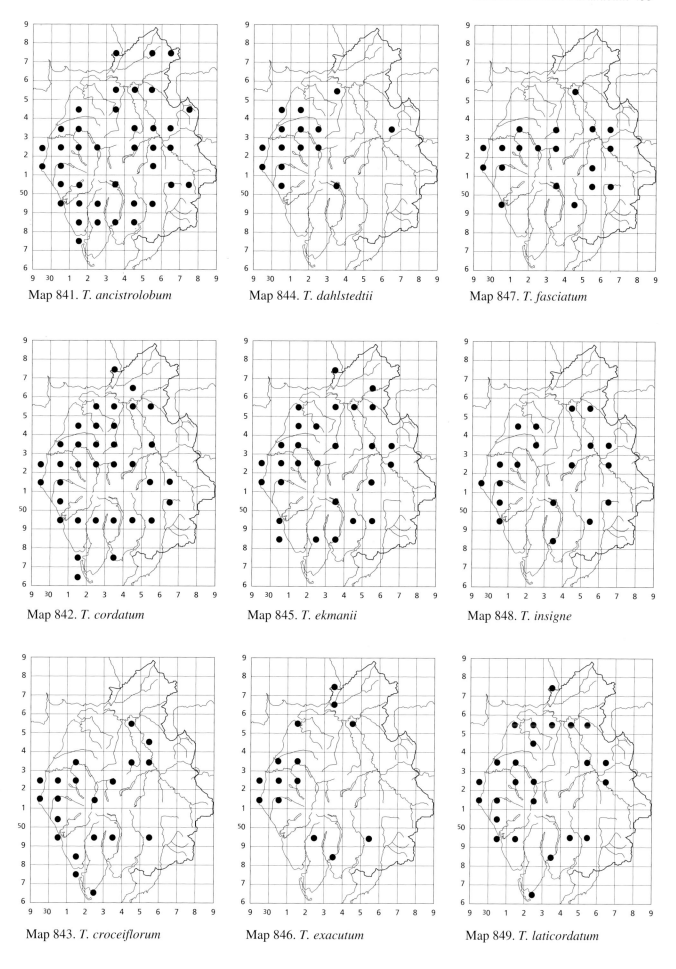

Map 841. *T. ancistrolobum*

Map 844. *T. dahlstedtii*

Map 847. *T. fasciatum*

Map 842. *T. cordatum*

Map 845. *T. ekmanii*

Map 848. *T. insigne*

Map 843. *T. croceiflorum*

Map 846. *T. exacutum*

Map 849. *T. laticordatum*

the commonest and most widespread British species.
Map 845   28 squares   WFC

T. exacutum Markl.
This tends to grow in urban areas and on rubbish dumps. Many of the dandelions in Carlisle and other Cumbrian towns are this species. It has very squat capitulae with spreading, imbricated bracts which taper to an elegant point, giving a stellate appearance to the capitulum. The end lobe is drawn out into a long apiculate point. An introduced but well-established British species.
Map 846   14 squares   WFC

T. expallidiforme Dahlst.
(See *Taraxacum subcyanolepis*)

T. fasciatum Dahlst.
Often a very abundant species. Almost invariably found on roadsides. This species has a heavily blotched midrib to the leaf, and capitulae with characteristically dense bracts. Its status is uncertain.
Map 847   19 squares   WC

T. insigne Ekman ex M. Christiansen & Wiinst.
A species of grassland and old verges. The external bracts which radiate like the separate spokes of a wheel and are lilac-coloured make this species instantly recognisable. It is common in Cumbria.
Map 848   19 squares   WFC

T. laticordatum Markl.
Although very common, this species is never found away from ruderal habitats and the presumption must be that it is introduced. It has green-petioled leaves which are but little incised and with lobes that overlap.
Map 849   25 squares   WFC

T. lingulatum Markl.
One of the commonest of British species. It is a robust green-petioled plant, somewhat variable, with wing-like lateral lobes and long, pale green, highly reflexed bracts.
Map 850   27 squares   WFC

T. oblongatum Dahlst.
This is a medium-sized species with pink petioles and uncomplicated few-lobed leaves, with a rounded terminal lobe.
Map 851   14 squares   FC

T. pannulatum Dahlst.
This species poses something of a problem since there are two allied species which have not always been clearly distinguished by us in the past. So, although this species is a common and important one, some of the records here may refer to *Taraxacum nigridentatum* or to *T. pannulatiforme* (see also under Part Two).
Map 852   14 squares   WC

T. pectinatiforme Lindb. f.
This can be abundant where found, but is scattered. It is an elegant species with leaves bearing long, narrow side-lobes, giving the plant its name. It has pink petioles and midribs.
6 squares (9.1, 4.8s, 0.1, 1.2, 2.2, 3.2)   FC

T. polyodon Dahlst.
This is one of our commonest species. The petiole is purple and the leaf oblong in shape and well supplied with black-tipped teeth, especially at the interlobes which are often also blotched. The narrow, recurved, untidy purple bracts are distinctive too. It is relatively easy to identify. It may well be a very successful introduced species.
Map 853   28 squares   WCF

T. rhamphodes Hagl.
This species has rather tongue-shaped leaf lobes. The lobes (especially the terminal ones) have rounded incisions. The bracts are distinctly grey. It is probably commoner in Cumbria than the records suggest, but it is certainly an introduction.
5 squares (9.1, 0.9, 1.8, 0.0, 0.1)   C

T. sellandii Dahlst.
This is one of the commonest of dandelions. It is a robust, squat plant, whose leaf form is not easy to describe, but the arching lobes and subdivided uppermost leaf lobe give the plant, when once known, a very distinctive appearance. The broad, spreading and leaden bracts are a notable feature. The dirty green colour to the middle of the leaf is characteristic. Although an aggressive weed, it is very possibly a native species.
Map 854   25 squares   WCF

T. stenacrum Dahlst.
This green-petioled species is best recognised by the strictly patent upper leaf lobes being deeply bilobed and the end lobe being inclined and narrowly linguate. It is one of the very few endemic species in this section and is well-scattered throughout most of mainland Britain.
5 squares (9.0, 0.0, 1.2, 3.6, 4.5)   C

T. subcyanolepis M. Christiansen
It is only very recently that the distinctions between what may be two common species, namely this and *Taraxacum expallidiforme* have become apparent. Both are difficult to describe as well as difficult to separate, but they are both green-petioled. Since most records are field ones, only further studies will clarify the position of these two species in Cumbria and indeed Britain as a whole. The initial impression is that *T. subcyanolepis* is commoner generally and the voucher material from Whitehaven is certainly this species. In the meantime the records for both species are dealt with here.
Map 855   24 squares   WCF

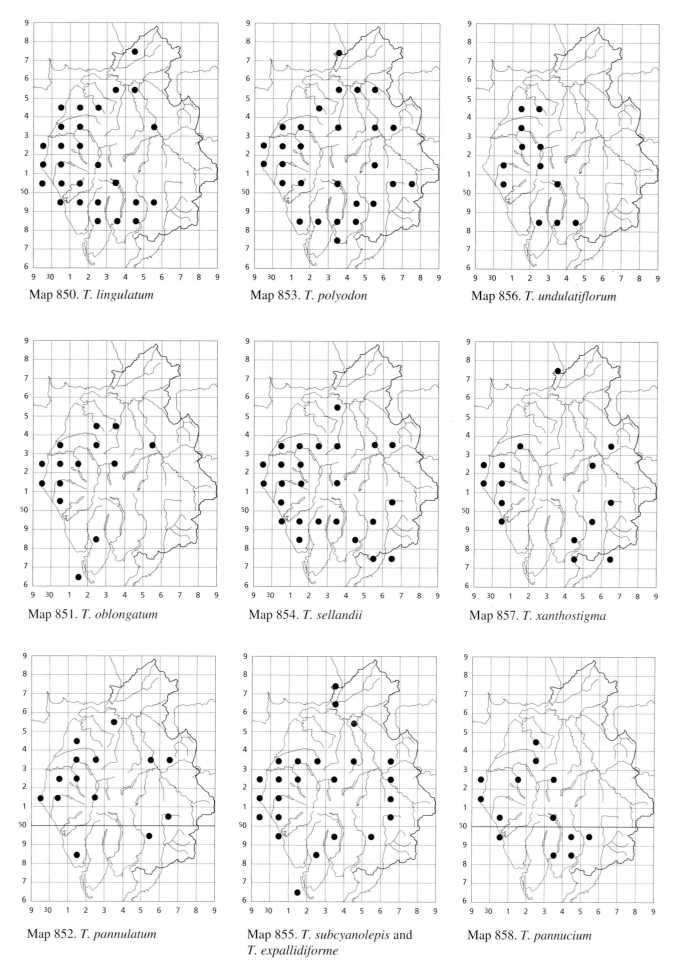

Map 850. *T. lingulatum*

Map 853. *T. polyodon*

Map 856. *T. undulatiflorum*

Map 851. *T. oblongatum*

Map 854. *T. sellandii*

Map 857. *T. xanthostigma*

Map 852. *T. pannulatum*

Map 855. *T. subcyanolepis* and *T. expallidiforme*

Map 858. *T. pannucium*

T. sublaeticolor Dahlst.
This is almost certainly under-recorded. It is a rather small species belonging to a group of natives all with green narrow petioles, small capitulae, and rather dark erect to suberect bracts.

2 squares (0.1, 6.3) C

T. subundulatum Dahlst.
Known only from Ravenstonedale (74.02) where it grows in herb-rich meadows, its typical habitat. This is a native species which tends to keep good botanical company. The bracts are dark and rather erect. The leaf lobation is complex, often with some lobes pointing upwards. It has broad red petioles.

1 square (7.0) W

T. undulatiflorum M. Christiansen
This is not a particularly characterful species but is common on roadsides throughout Britain. The most striking feature is the large hastate end lobe. It is an introduced species.

Map 856 12 squares WCF

T. xanthostigma Lindb. f.
This is a generally common species with very marked leaf interlobe blotching. The yellow-orange stigmas are unusual in this section and help to identify it.

Map 857 15 squares WCF

Part Two

T. aberrans Soest, Hagend. & Zevenb.
Found only at Workington (98.28). This uncommon British casual is characterised by its long, broad, apiculate end lobe .

1 square (9.2) C

T. adiantifrons Ekman ex Dahlst.
A distinctive species with a very 'square' end lobe.

5 squares (0.0, 0.1, 1.2, 3.7, 6.3) C

T. anceps Oellg. ined.
A somewhat obscure and uncommon unpublished species, a relative of *Taraxacum xanthostigma*.

3 squares (9.2, 3.8, 0.0) FC

T. aurosulum Lindb. f.
This is one of the most robust of all dandelions with huge capitulae. There are few Cumbrian records of this scarce but widespread species.

4 squares (9.1, 9.2, 3.8, 0.0) FC

T. cophocentrum Dahlst.
An endemic species, closely related to *Taraxacum oblongatum* but with a large, rounded, entire end lobe.

8 squares (9.1, 9.2, 1.9, 0.1, 1.2, 3.5, 3.6, 6.3) C

T. densilobum Dahlst.
An uncommon British casual seen only north of Netherby (38.72n). It is related to *Taraxacum alatum*.

1 square (3.7n) C

T. dilaceratum M. Christiansen
A widespread British casual, characterised by the lateral leaf-lobes with linear processes often forward-pointing.

3 squares (1.7, 2.9, 6.3) WC

T. dilatatum Lindb. f.
This is a generally uncommon but seemingly native species of rich grasslands. It is rare in Cumbria but grows in some quantity around Eaglesfield. It is easily recognised by the parallel-sided green petiole contrasting sharply with the red mid-rib.

3 squares (0.2, 3.6, 6.3) C

T. exsertum Soest, Hagen. & Zevenb.
A widespread but uncommon British species, probably introduced.

10 squares

T. fagerstroemii Såltin
Presumably a casual, there is just one record on a lay-by near Thursby for this uncommon but striking species.

1 square (3.4) C

T. horridifrons Rail.
This species is a relative of *Taraxacum alatum*, from which it is distinguished by its hairy leaves.

5 squares (9.1, 0.1, 1.2, 5.1, 5.3) WC

T. huelphersianum Hagl.
A probable introduction. Widespread in Britain but not common.

3 squares (9.2, 5.9, 6.3) WC

T. incisum Oellg.
Scattered through Britain, common only in London.

4 squares (9.2, 4.8s, 0.1, 6.2) WC

T. interveniens Hagl.
Found only at Cleator Moor (02.14). This is a widespread but uncommon British dandelion, except in cities such as London, Edinburgh, Glasgow.

1 square (0.1) C

T. laciniosum Dahlst.
A scarce British species found in west Cumberland near Embleton (16.28) and Lamplugh Green (08.20). Its status is unclear.

2 squares (0.2, 1.2) C

T. laeticolor Dahlst.
A widespread introduction in Britain but known only from two sites in the north of the county.

2 squares (1.5, 3.6) C

T. latisectum Lindb. f.
A widespread but very local British species.
1 square (6.3)   C

T. lepidum M. Christiansen.
This is probably a native species, belonging to the same group as *Taraxacum sublaeticolor*. It is probably more widespread than the apparent confinement to west Cumberland suggests. Most records were made under its old name of *T. subpraticola*.
4 squares (0.9, 9.1, 0.0, 0.1) C

T. longisquameum Lindb. f.
A probable native which is nowhere common but seems to be widespread over much of Britain, especially on dunes or near the sea.
3 squares (9.2, 1.9, 0.5) C

T. lucidum Dahlst.
Found only near Embleton (16.28). This is recorded from a few widely scattered sites in England and is no doubt introduced. It has large erect bracts, bright purple petioles and large helmet-shaped terminal leaf lobes.
1 square (1.2)   C

T. macrolobum Dahlst.
An uncommon casual.
3 squares (9.1, 9.2, 6.2) WC

T. necessarium Oellg.
Found only near Shap summit.
1 square (5.1)   W

T. nigridentatum T. Edmondson
There are three records for this important and recently described British species. Unfortunately it has been confused in the past with *Taraxacum pannulatum* and *T. pannulatiforme*, so it is difficult to know the local picture. It is a local, western species in Britain which could be an endemic.
2 squares (0.1, 0.2) C

T. nitidum Soest, Hagend. & Zevenb.
This was recorded near Thursby (30.48) with *Taraxacum fagerstroemii* (see above).
1 square (3.4)   C

T. pachylobum Dahlst.
This is yet another species growing in the rich sunken lane at Embleton (16.28) with *Taraxacum cambricum*. There are very few British records. It is a most striking species and could well be a relict native.
1 square (2.0)   C

T. pachymerum Hagl.
This species has very crowded overlapping leaf lobes and broad, coloured petioles, a combination which makes it relatively easy to identify. It is particularly

well established on roadsides around Cockermouth.
6 squares (0.1, 0.2, 1.2, 1.3, 3.2, 5.3) C

T. pallidipes Markl.
This is a rather undistinguished species and may have been confused with others in the past. It therefore may be commoner than the records suggest.
7 squares (1.7, 1.2, 2.2, 3.6, 5.3, 6.0, 6.2) WC

T. pannucium Dahlst.
A widespread and locally common casual. It has a long green petiole and a long extenuate terminal lobe.
Map 858 13 squares WFC

T. pannulatiforme Dahlst.
Owing to problems of taxonomic distinction, this species, which is certainly present in Cumbria, needs reassessment in order to clarify its distribution. See *Taraxacum pannulatum* (Part One) and *T. nigridentatum* (Part Two).

T. piceatum Dahlst.
This species is present on roadsides throughout Britain. It is very variable in leaf morphology but the leaden-violet spreading bracts are distinctive, as is the combination of white-green petioles with coloured midribs. It is doubtless commoner than the records suggest.
3 squares (0.1, 0.2, 5.9) CW

T. procerisquameum Oellg.
Found only at Cleator Moor (02.14). This species is related to *Taraxacum lingulatum* but is greyer-green in foliage and has even longer bracts. It can be a huge plant, and is never found away from roadsides.
1 square (0.1)   C

T. sagittipotens Dahlst. & R. Ohlsen ex Hagl.
This is a distinctive species with a very sagittate end lobe; there are scattered British records, though it tends to be commoner in the east.
3 squares (2.6s, 2.8, 5.1) WF

T. scotiniforme Dahlst. ex Hagl.
An uncommon introduction, scattered mainly in the south of England.
2 squares (9.1, 4.0) WC

T. sinuatum Dahlst.
The only record for this species is from the Lyth valley. The highly sinuate lobes to the leaves and the red petioles are striking. It is probably native and is commonest in southern and western Britain.
1 square (4.8s)   W

T. subexpallidum Dahlst.
A widespread and locally common British species; many early records were made under the old name *Taraxacum linguatum*.
6 squares (9.1, 9.2, 0.9, 0.0, 0.1, 2.5) C

T. tanyphyllum Dahlst.
A rare British plant of road verges and disturbed ground.

1 square (0.1)　　　　　　　　　　C

T. trilobatum Palmgr.
A striking, if untidily ugly, plant, scattered throughout Britain

2 squares (0.0, 2.5)　　　　　　　　C

T. tumentilobum Markl. ex Puol.
A widespread but uncommon casual.

2 squares (0.0, 0.1)　　　　　　　　C

T. vastisectum Markl. ex Puol.
An uncommon but widespread introduced British dandelion.

3 squares (3.6, 3.7n, 6.2)　　　　　WC

Crepis paludosa (L.) Moench　　　Marsh hawk's-beard
Common in the uplands, becoming scarce towards the coast and rare around Morecambe Bay. It is a characteristic plant of rocky, wooded gills where it often occurs in profusion on wet ground close to the water and where it is commonly mistaken for a hawkweed by the uninitiated. It is also frequent in sheltered mountain gullies and at lower altitudes in fens, damp hay-meadows, roadside verges and ditches, being particularly abundant in the Alston area (7.4) and the extreme north of the county.
Alt. limit: 640 m, Dove Crag (36.10, Wilson)
Map 859　　　　　864　　　　　LWFCY

C. mollis (Jacq.) Asch.　　　Northern hawk's-beard
Extremely rare and apparently decreasing. The only Survey records are from open gillside woodland, limestone pavement and hay-meadows above Crosby Ravensworth (6.1, CW; LANC), by a woodland path by the River Nent east of Alston (7.4, GH, 1988, LANC)

and at three sites below Nenthead (7.4, P. Buchanan, 1991, LANC). It seems likely that it now survives at only one of the Crosby Ravensworth sites.

The only records in the Floras are Wilson's from between Keld and Shap Abbey (5.1), Asby (6.1) and near Appleby (*cf*.6.2). There are, however, specimens in BM collected from Barras (8.1, H.J. Riddelsdell, 1903), from the lane leading from Bardsea to Birkrigg Common (2.7, Dr Collins, 1909), a most surprising lowland site, and from meadows near Nenthead (7.4, C. Waterfall, 1921). In 1965 it was seen in hay-meadows on the Cumberland side of the Ayle Burn north of Alston (7.4, G.A. & M. Swan) but subsequent search during the Survey failed to refind it. The pre-1930 *Atlas* dot for the Lorton square (1.2) seems very doubtful, as also J. Parkin's records from Barugh (2.4), Wigton(2.4, 1933) and Workington (9.2, 1934). The *Scarce Plants* record for the Borrow Beck square (5.0, 1964) is anonymous and unlocalised.

This is essentially a species of the northern Pennines, Northumberland and south-east Scotland. Although it appears to be generally declining, several new sites have recently been discovered in the Scottish Borders, mostly on rather dry grassy slopes associated with *Primula veris* and *Leontodon hispidus*.
Alt. limit: 400 m, below Nenthead (7.4)

S 7　　　　　　　　　W(F)C

C. biennis L.　　　　　　　　Rough hawk's-beard
Scattered on the Solway Plain, where it seems to be increasing, and with four sites along the Eden valley and one in the south at Arnside (46.78s, MB; LANC). It was seen at this last site in the mid-1970s, although known for some time from just across the county boundary. Baker lists it from the east side of Kendal (5.9) but considered the record doubtful.

Most of the records are from the sides of roads or lanes where perhaps it benefits from road improvement schemes. It has also been seen on a slag heap near

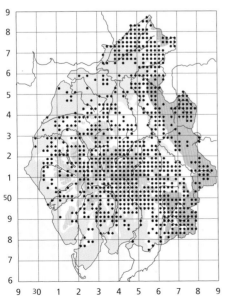

Map 859. *Crepis paludosa*

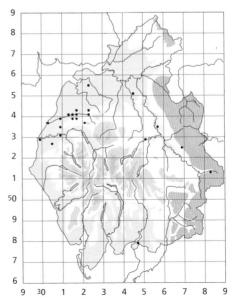

Map 860. *Crepis biennis*

Aspatria (12.40), another locality considered as dubious by Baker, and at the British Gypsum plant at Cocklakes (44.50). The first definite Cumberland record is one from Little Salkeld railway station (5.3, W.W. Mason, *BEC Rpt* (1925) 7:579) in 1924, close to where it still occurs. The only Furness record is a statement in a letter by S.L. Petty attached to a sheet of *Crepis mollis* from Birkrigg (2.7, Dr Collins, 1909, BM) saying that Collins found both species in the same lane.

Alt. limit: 200 m, near Dufton (68.24)

Map 860   22   W(F)C

*C. tectorum L.
A rare southern casual which was seen in a newly-sown lawn at High Hesket (46.42, MS, 1990, LANC), by the new Brampton by-pass (52.58, MG, LANC) and by the A590 near Barrow-in-Furness (18.72, PB, 1993, LANC).

                           3   C

*C. nicaeensis Balbis
A Mediterranean plant once found in a field between Seatoller and Seathwaite, Borrowdale (2.1, I. Adams, *BEC Rpt* (1920) 5:664).

                                                (C)

C. capillaris (L.) Wallr.   Smooth hawk's-beard
A frequent to common grassland species occurring throughout the lowlands on roadside verges and railwaysides, in 'unimproved' pastures and meadows and also on waste sites, always on well-drained soils.

Var. *glandulosa* Druce, with glandular involucral bracts, has not been systematically recorded but is probably frequent.

Alt. limit: *c*.430 m, south of Garrigill (74.38)

Map 861   700   WFCY

*C. setosa Haller f.   Bristly hawk's-beard
A very rare southern casual seen on a roadside verge at Cark (36.76s, KAG, 1976, LANC, *Watsonia* (1979) 12:358), on business sites at Greenodd (30.82, PB, 1988, LANC, *Watsonia* (1989) 17:480), the old steelworks site at Barrow-in-Furness (18.70, PB, LANC, 1992), in a garden at High Hesket (46.42, MS, 1992, LANC), at Kingstown, Carlisle (38.58, FJR, LANC, 1989) and on the newly-seeded verge of the Brampton by-pass (54.60, FJR, 1991, LANC).

The only previous record was from a field opposite East Cote lighthouse, Skinburness (1.5, J. Leitch, CLE) in 1890.

Alt. limit: 125 m, High Hesket (46.42)

                           6   FC

C. praemorsa (L.) F. Walther
A wide-ranging Eurasiatic species found in 1988 new to Britain and western Europe at the edge of a hay-meadow in north Westmorland (6.0, GH, LANC (Halliday 1990)). The plants were on a low bank of limestone drift and associated species included *Helicto-*

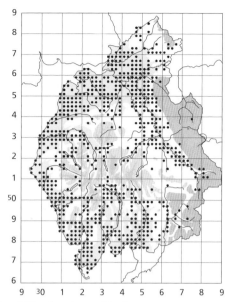

Map 861. *Crepis capillaris*

Plate 92. *Crepis praemorsa*, north Westmorland

trichon pubescens, Carex flacca, Geranium sanguineum and *Scabiosa columbaria,* a habitat apparently similar to that described from Scandinavia. The species was not previously known further west than easternmost France. The undisturbed nature of the site and its remoteness, coupled with the presence not far away of rare and indisputably native species, such as *Bartsia alpina* and *Carex capillaris,* strongly suggest that the *Crepis* too is native, despite the scepticism of Stace (1991) and Kent (1992).

Numerous sterile rosettes were present but only a few flowering scapes. These are tall, to 65 cm, with several, rather small, almost glabrous capitula, which scarcely open in dull weather but in bright sunlight are up to 3 cm in diameter. No viable fruits are produced. The leaves are also diagnostic being long, obovate, entire and minutely apiculate.

Alt.: 240 m, north Westmorland

Plate 92   R 1   W

Pilosella officinarum F. Schultz & Schultz-Bip
(*Hieracium pilosella* L.)
Common in short, dry grassland, occurring both on siliceous and calcareous soils though commoner on the latter.
Alt. limit: at least 640 m, Dufton Fell (74.28) and Little Fell (78.22)

Map 862   1336   LWFCY

An attempt has been made to assign the 142 collections to the subspecies recognised by Sell (1988). His key is not always easy to use, the distinction between dark hairs and pale hairs with dark bases being somewhat arbitrary. Despite Stace's (1991) comment that the subspecies are best relegated to varietal status, it seems worthwhile presenting the results of this survey for others to dismiss or build upon. Nearly half of the samples have been determined by A.O. Chater. The presence of an enthusiast in the Matterdale (3.2) and Pooley Bridge (4.2) squares is very evident!

subsp. **micradenia** (Naeg. & Peter) Sell & C. West
Very common. As in west Wales (A.O. Chater *in litt.*) and Co. Durham (Graham 1988), this and subsp. *trichosoma* are the most frequent segregates. The involucral bracts are densely covered with more or less equal glandular hairs up to 0.7 mm; simple hairs are inconspicuous or absent.

Map 863   55   WFC

subsp. **euronota** (Naeg. & Peter) Sell & West
Occasional. It is similar to subsp. *micradenia* but has unequal, somewhat flexuous glandular hairs, many exceeding 0.7 mm.

Map 864   13   WFC

subsp. **officinarum**
Occasional. The involucral bracts have a mixture of glandular hairs and pale, but dark-based, simple hairs.

Map 865   15   WFC

subsp. **trichosoma** (Peter) Sell & C. West
Frequent. It differs from subsp. *officinarum* in having uniformly dark simple hairs.

Map 866   27   WFC

subsp. **tricholepia** (Naeg. & Peter) Sell & C. West
Occasional. The involucral bracts have only pale, dark-based simple hairs.

Map 867   15   WFC

subsp. **melanops** (Peter) Sell & C. West
Occasional. It differs from subsp. *tricholepia* in having uniformly dark simple hairs.

Map 868   14   WFC

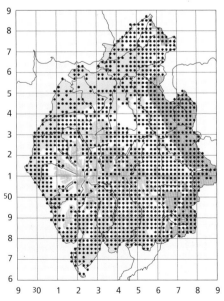

Map 862. *Pilosella officinarum*

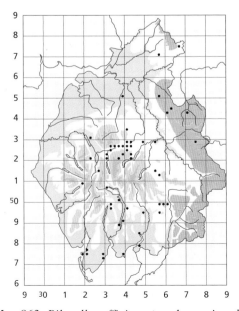

Map 863. *Pilosella officinarum* subsp. *micradenia*

subsp. **trichoscapa** (Naeg. & Peter) Sell & West
Very rare, known only from Dockray (38.20), Pooley Bridge (46.24) and north-west of Grayrigg (54.98).

It differs from subsp. *melanops* in having appreciably longer (-5 mm) simple hairs.

3   WC

[***P. lactucella** (Wallr.) Sell & C. West
Material from 'Westmorland' (Anon., 1814, LIV, det. J.N. Mills, *Watsonia* (1971) 8:309) and from near Ingleborough (1818, *ibid.*) is currently being examined by P.D. Sell. Superficially neither collection appears to belong to this species otherwise known in Britain only from one site in Wiltshire early this century.

(W)]

P. flagellaris (Willd.) Sell & C. West
*subsp. flagellaris
A very rare British plant recorded in Cumbria only from the banks of Newlands beck, Braithwaite (24.22,

CCH, 1983, LANC, det. P.D. Sell).

1   C

*P. aurantiaca (L.) F. Schultz & Schultz-Bip
Fox and cubs
subsp. **aurantiaca**
Known only from the two *Critical Atlas* records for the Keswick (2.2) and the Carrock Fell (3.3) squares.

(C)

subsp. **carpathicola** (Naeg. & Peter) Soják
(*Hieracium brunneocroceum* Pugsley)
Frequent in the south, elsewhere scattered. It is a successful colonist of roadsides, railwaysides and churchyards. Rather surpringly Wilson gives only one record, from Grasmere. The first Furness record is probably that in the *Critical Atlas* for the Hawkshead (3.9) square.
Alt. limit: 445 m, Nenthead (76.42)
Map 869   185   WFCY

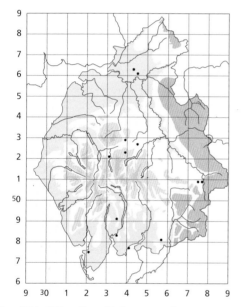

Map 864. *Pilosella officinarum* subsp. *euronota*

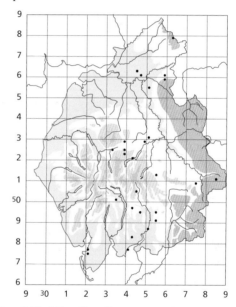

Map 866. *Pilosella officinarum* subsp. *trichosoma*

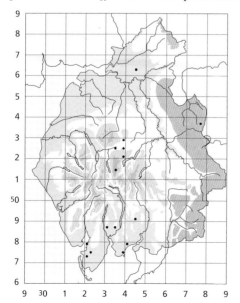

Map 865. *Pilosella officinarum* subsp. *officinarum*

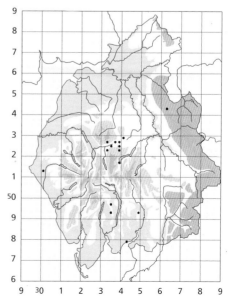

Map 867. *Pilosella officinarum* subsp. *tricholepia*

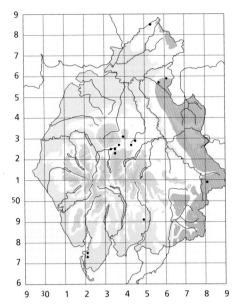

Map 868. *Pilosella officinarum* subsp. *melanops*

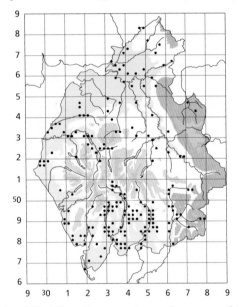

Map 869. *Pilosella aurantiaca* subsp. *carpathicola*

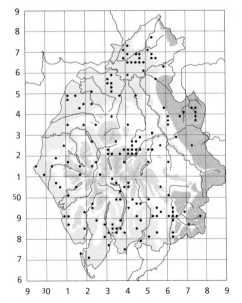

Map 870. *Hieracium sabaudum*

Hieracium   Hawkweeds

As novices, provokingly ignorant of hawkweed problems, and without a description of *decolor* (it is not indexed in Linton's work and so we missed it) we ran this plant down to *H. Leyi* and feel encouraged. But what a liquid genus must *Hieracium* be if a plant can be put back into a 'better hole' behind fifty intervening species (see *Lond. Cat.*) without the ruffling of a ligule!
                      Bucknall & White, *BEC Rpt* (1920) 5:827

No serious attempt was made during the Survey to investigate the hawkweed flora until 1985 although numerous specimens had been collected, in a rather desultory way, during the previous ten years. Since then much additional material has been accumulated and, together with specimens seen at BM and CGE, considerably more information is now available than ever before. Preparation of this account has revealed a number of errors in the *Critical Atlas*, errors arising from misplaced dots, omissions and misidentifications. A particularly acute problem has been the unrealistically large number of species previously reported from some sites. Familiarity with the plants in the wild suggests that environmental variation is of more significance than is generally allowed for; it also engenders healthy caution in relying too much on leaf-spotting. For all its undoubted shortcomings this account at least sets out the current position and indicates the nature of the remaining unsolved problems. Considerable help was received during the early years of this study from P.D. Sell and J. Bevan, and latterly from D. McCosh, who has seen the bulk of the herbarium material. Ultimate responsibility for the taxonomic conclusions must, however, rest with the author.

As with *Rubus*, earlier published accounts, for example Ley & Linton (1905) as well as the previous Floras, pose serious problems of interpretation. This account is therefore based entirely on Survey material and herbarium specimens. With the exception of the rarer species, records prior to the Survey ones are only given where they are from additional 10 km squares.

The account lists 62 species which are considered to occur or to have occurred in the county. Although seven fewer than are shown in the *Critical Atlas*, this number still leaves Cumbria with by far the largest hawkweed flora of any county south of the Scottish Highlands. Much remains to be done. There is at least one possible new species which requires further study and many unvisited crags and gullies still beckon. The upland hawkweeds have undoubtedly declined considerably this century, chiefly as a consequence of ever increasing sheep stocking, but it is encouraging that the Survey rediscovered the elusive *Hieracium itunense* and, briefly, *H. mirandum,* failing only with *H. subintegrifolium*. These last two species should now be considered extinct.

The following key is an attempt to facilitate identification of most of the species seen during the Survey.

1. Stem with more than 5 stem lvs; basal rosette usually absent at flowering
  2. Involucral bracts densely glandular, lvs strongly amplexicaul ....16. *prenanthoides*
  2. Inv. bracts eglandular or with sparse glandular hairs, or microglands
    3. At least the lower lvs with a tapering base, upper lvs with a rounded but not amplexicaul base
      4. Lvs narrow, with recurved margins, toothed; inner inv. bracts glabrous ....5. *umbellatum*
      4. Lvs ovate-lanceolate, inv. bracts rarely glabrous
        5. Lvs entire or weakly toothed
          6. Inv. bracts with very few glandular hairs; peduncles lacking long hairs ....4. *vagum*
          6. Inv. bracts with many microglands and long simple hairs ....1. *sabaudum*
        5. Lvs distinctly toothed
          7. Infl. of many heads, dense, subumbellate ....3. *salticola*
          7. Heads usually <5, if more, infl. elongate
            8. Lvs spotted ....15. *sparsifolium*
            8. Lvs not spotted
              9. Basal rosette of lvs present at time of flowering; inv. bracts ± glabrous ....14. *lissolepium*
              9. Basal rosette absent at flowering; inv. bracts hairy
                10. Inv. bracts with long, simple hairs only ....11. *placerophylloides*
                10. Inv. bracts with glandular and simple hairs ....9. *eboracense*
    3. Base of upper lvs semi-amplexicaul
    11. Styles yellow ....8. *latobrigorum*
    11. Styles dark
      12. Inv. bracts with mainly simple hairs ....6. *strictiforme*
      12. Inv. bracts with rather few simple and glandular hairs ....7. *subcrocatum*
1. Stem with less than 6 lvs; basal rosette present when flowering
  13. Inv. bracts with conspicuous long simple hairs
    14. Heads large, rarely solitary
      15. Lvs with stellate hairs on upper surface ....58. *flocculosum*
      15. Lvs without stellate hairs on upper surface
        16. Stem lf petiolate; inv. bracts with conspicuous stellate hairs ....57. *ampliatum*
        16. Stem lf sessile; inv. bracts without conspicuous stellate hairs ....56. *anglicum*
    14. Heads solitary; lvs and inv. bracts with long, simple hairs
      17. Lvs entire; heads shaggy with long white hairs ....62. *holosericeum*
      17. Lvs slightly toothed; heads not shaggy, hairs dark ....61. *subgracilentipes*
  13. Inv. bracts lackingconspicuous long simple hairs

18. Inv. bracts with long, flexuous glandular hairs only; lvs spotted ....59. *clovense*
18. Not as above
  19. Lvs often glaucous; heads medium to large, usually few; styles usually yellow
    20. Stem lvs 2-4; lvs rough above with scattered short hairs ....43. *orimeles*
    20. Stem lvs 1 or 2; lvs glabrous above
      21. Lvs spotted
        22. Robust plant; lvs often broadly ovate; heads rather large ....52. *hypochaeroides*
        22. Less robust; lvs lanceolate to oblong-lanceolate; heads medium
          23. Lf base subtruncate; lvs narrowly ovate, strongly spotted ....53. *saxorum*
          23. Lf base narrowed; lvs oblong-lanceolate, somewhat obtuse, lightly spotted ....47. *sommerfeltii*
      21. Lvs not spotted
        24. Lvs strongly toothed ....46. *vagense*
        24. Lvs entire or weakly toothed
          25. Lvs narrowly lanceolate or elliptic, glaucous; stem lf narrow, patent ...45. *argenteum*
          25. Not as above
            26. Inner inv. bracts obtuse, tips often flushed pink 44. *caledonicum*
            26. Inner inv. bracts acute, tips not pink
              27. Lvs with long setose hairs on margin and both surfaces........50. *lasiophyllum*
              27. Upper leaf surface glabrous
                28. Lvs with prominent, neat, short setaceous teeth; inv. bracts long and pointed ....54. *dicella*
                28. Not as above
                  29. Lvs glaucous; inv. bracts with conspicuous white margin of stellate hairs ....48. *decolor*
                  29. Lvs rarely glaucous; inv. bracts without conspicuous white margin
                  30. Lvs with stellate hairs beneath ....55. *stenopholidium*
                  30. Lvs without stellate hairs beneath ....49. *leyi*
  19. Lvs rarely glaucous; heads small to medium; styles usually dark
    31. Stem lvs 3 or more; peduncles and inv. bracts with glandular hairs
    32. Inv. bracts with dense stellate hairs ....21. *acuminatum*
    32. Inv. bracts with few if any stellate hairs
      33. Lvs ovate; inv. bracts usually with glandular hairs only ....23. *diaphanum*
      33. Lvs ovate-lanceolate; inv. bracts and peduncles with a few simple hairs ....22. *diaphanoides*
    31. Stem lvs usually 1 or 2
      34. Inv. bracts with short, black glandular hairs only; stem lf 1

35. Lvs glabrous, entire, often purplish below
..........37. *pellucidum*
35. Lvs not as above
  36. Lvs softly hairy, strongly toothed,the lower often deflexed ....41. *grandidens*
  36. Lvs subentire to shortly toothed
    37. Styles yellowish; lvs subentire40.*sublepistoides*
    37. Styles fuscous or dark; lvs often large and somewhat toothed at the base ...39. *subcrassum*
34. Inv. bracts with simple and usually some glandular hairs
  38. Lvs spotted
    39. Inv. bracts with long white simple hairs, few glandular hairs, and many stellate hairs
..........25. *maculoides*/29. *maculosum*
    39. Inv. bracts and peduncles conspicuously glandular hairy
      40. Peduncles with dense stellate hairs and conspicuous simple ones ....34. *scotostictum*
      40. Peduncles with few stellate or simple hairs
..........24. *glanduliceps*
  38. Lvs not spotted
    41. Inv. bracts with many glandular hairs
      42. Glandular hairs rather long and flexuous
        43. Lvs narrow, elliptic; heads small35. *duriceps*
        43. Lvs oblong-lanceolate, often strongly toothed at the truncate base heads medium ....32. *pseudostenstroemii*
      42. Glandular hairs short
        44. Glandular hairs pale; stem lf broadly triangular ....27. *oistophyllum*
        44. Glandular hairs dark; stem lf narrow
          45. Lvs oblong, with spreading or deflexed teeth at the base; panicle lax; heads with wide base ....30. *auratiflorum*
          45. Lvs more pointed, weakly toothed at base; panicle rather narrow; heads narrowed to base ....31. *crebridentiforme*
    41. Inv. bracts not conspicuously glandular hairy
      46. Lvs small, boat-shaped with stellate hairs on upper surface ....33. *cymbifolium*
      46. Lvs not boat-shaped, lacking stellate hairs above
        47. Older lvs broad, obtuse, with a subtruncate base; panicle often umbellate28. *silvaticoides*
        47. Lvs not as above
          48. Inv. bracts with short, crisp simple hairs
..........19. *vulgatum*
          48. Inv. bracts with long white, simple hairs
            49. Inv. bracts without glandular hairs; heads medium ....18. *cravoniense*
            49. Inv. bracts with glandular hairs; heads large
              50. Inv. bracts with glandular hairs in upper half; stem leaf rarely strongly toothed
..........*rhomboides*
              50. Inv. bracts with glandular hairs confined to lower half; stem leaf often strongly toothed ....20. *rubiginosum*

Section **Sabauda**

Stems very leafy and often hairy, lower leaves tapering at base, the upper rounded; heads many; involucral bracts obtuse; styles dark.

1. **H. sabaudum** L.
(*H. perpropinquum*)
Very common, especially in the lowlands but not infrequent in the fells. It favours shady woodland margins, hedgerows and roadsides, and beck- and riversides. It is probably commoner than *Hieracium vagum* but, because of the similarity of the two, no attempt was made to record them separately in the field. The maps therefore show both as substantially under-recorded.

  Hieracium sabaudum, like the following four species, flowers relatively late, from mid-August and through September. It is characterised by its hairy lower stem and leaves, its elongate panicle with flexuous lower branches and hairy peduncles, and dark involucral bracts which are more or less uniformly covered with stellate hairs, rather numerous long simple hairs and medium to short (microglands) glandular hairs.

  It is frequent throughout most of England and Wales but rather local in Scotland.
Alt. limit: 565 m, High Cup Nick (72.24)
Map 870   152 (813 for aggregate)   WFCY

2. **H. rigens** Jordan
Known only from a specimen collected by the roadside south-west of Penruddock (40.26, MC, 1987, LANC) and an earlier record from river shingle at Sebergham (3.4, Anon., 1966, not traced).

  This species differs from the last in having rather densely glandular involucral bracts with at most a few simple hairs. The mid-stem leaves are irregularly serrate.

  This is predominantly a plant of south-east England.
Alt.: 280 m, Penruddock (40.26)
           1   C

3. **H. salticola** (Sudre) Sell & C. West
Seen only in old sand workings at Roose, Barrow-in-Furness (22.68, GH, 1978) and a roadside at Spadeadam (62.68, MG, 1991, LANC). The only other record is from Low Borrowbridge, Tebay (6.0, F. Houseman, 1965, MANCH).

  The most obvious features of this hawkweed are the corymbose inflorescence and the strongly toothed leaves. The involucral bracts are generally stellate hairy and lack both glandular and eglandular hairs.

  This is an introduced plant of waste ground and roadsides with a scattered distribution mainly from south-central England to central Scotland.
Alt. limit: 200 m, Spadeadam (62.68)
           2   (W)FC

4. **H. vagum** Jordan

Common, occurring in very similar habitats to *Hieracium sabaudum* but perhaps more characteristic of disturbed habitats and waste places. The first certain records are for Furness from Haverthwaite (3.8, Anon., 1912, herb. *cf.* S.H. Davey), for Westmorland from Patterdale (3.1, J.C. Melville, 1900, BM) and for Cumberland from Lodore (2.1, Anon., 1870, BM).

Although very similar to that species in general appearance, typical plants have glabrescent involucral bracts, with glandular hairs very few or absent. They also lack the long simple hairs on the peduncles and upper stem.

Although this species is widespread in central and northern England and North Wales it is quite rare in Scotland.

Alt. limit: 610 m, Fleetwith Pike (20.12)

Map 871, p.443   73   WFCY

Section **Hieracioides**

Stems very leafy; leaves rather narrow with recurved margins; inflorescence umbellate; involucral bracts more or less glabrous; styles yellow.

5. **H. umbellatum** L.

Not infrequent on low ground in the north, especially around Carlisle and along the Eden valley. It occurs usually on roadside banks and hedgerows, by railways and in heathy areas. The isolated west coast records are from railwaysides (04.98, 04.00, 98.30 (where it was seen by Hodgson)) and among sand-dunes on Walney Island (16.70, GH, 1978, LANC), this last being the only Furness record.

The most important distinguishing characters are the narrow, parallel-sided leaves and recurved outer involucral bracts. All the material belongs to the widespread subsp. *umbellatum*.

Alt. limit: 260 m, Orton (62.08)

Map 872   40   WFC

Section **Foliosa**

Stems very leafy, the upper leaves semi-amplexicaule; heads many; involucral bracts obtuse.

6. **H. strictiforme** (Zahn) Roffey

Rare, seen only in Red Gill, Glaramara (24.08, GH, 1987), Dovedale (38.10, M. Richards, 1988), by Bleawater Beck, Haweswater (46.10, GH, 1993) and Tarn Beck, Seathwaite (22.96, J. Bevan & GH, 1988), and on a roadside verge in upper Garsdale (76.90, GH, 1987, all LANC). There are also collections from Stock Gill, Ambleside (3.0, S.H. Haslam, 1843, KDL, *Watsonia* (1988) 17:194), the first Westmorland record, and Chapel Stile (3.0, A. Ley, 1905, LIV). The *Critical Atlas* record for the Sedbergh (6.9) square is based on a specimen of *Hieracium subcrocatum* (see below).

This hawkweed is not easily distinguished from *H. subcrocatum*. The stems are strictly erect, the mid-leaves oblong with a truncate base, and the involucral bracts have more simple hairs. It is essentially a plant of the Scottish Highlands.

Alt. limit: 670 m, Glaramara (24.08)

       5   WFCY

7. **H. subcrocatum** (E.F. Linton) Roffey

Rather rare and restricted to the central and eastern Lake District and the south-east of the county but with an isolated site by the upper River Irthing (64.70n). It occurs mostly by the sides of lanes, lakes, becks and rivers. There is also a record from Taithes Gill, Cautley (6.9/7.9, A. Wilson, 1894, BM, YRK) and a *Critical Atlas* dot for the Kirkby Thore square (6.2). The many early records (BM, CGE, LIV, YRK) from the Haweswater, Ullswater, Coniston and Sedbergh areas suggest that the species has become much rarer. Presumably the ever-increasing pressures, from tourists and grazing, on river- and lakeside habitats have taken their toll. The earliest records for Cumberland and Westmorland were from Lorton (1.2) and Patterdale (3.1, both D. Oliver,

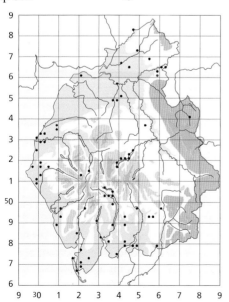

Map 871. *Hieracium vagum*

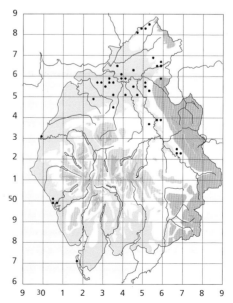

Map 872. *Hieracium umbellatum*

1854, BM) respectively, and for Furness from the lakeside near Coniston (3.9, H.T. Mennell, 1891, LIV).

The leaves are characteristically amplexicaul, subequal, somewhat glaucous and not noticeably hairy. The inflorescence is corymbose and the involucral bracts have a few long simple hairs and shorter glandular ones. *Hieracium sabaudum* grows in similar habitats but has noticeably hairy leaves, the lower being distinctly petiolate and larger, and involucral bracts with microglands. It also flowers a few weeks later, towards the end of August. The similarity between the two species may be heightened by the occurrence of plants of *H. subcrocatum* with long, flexuous flowering branches from the axils of the upper leaves (*H. neocorymbosum* Pugsley).

This hawkweed is widely distributed throughout Scotland and the uplands of England and Wales.
Alt. limit: 305 m, Lowthwaite (40.22)
Map 873   13   WFCY

8. **H. latobrigorum** (Zahn) Roffey

Very rare: it was seen by the upper Borrow Beck north of Kendal (54.04, GH, 1978) growing with *Hieracium subcrocatum*, also above Aira Force (40.20, GH, 1991) and by the River Nent east of Alston (72.46, TW & ES, 1988, all LANC). The only previous records are those cited by Pugsley (1948) from Alston (7.4, J.G. Baker, undated), Ullswater (in Ray's *Historia*, 5:1 (1686)) and Pooley Bridge (4.2, W. Borrer, 1845, BM, CGE), and others from Lodgegill Sike, Teesdale (7.3, J. Tatham, undated, BM) and the north-east end of Coniston Water (3.9, D.E. Allen, 1951, CGE).

This species differs from the closely related *H. subcrocatum* in having pure yellow styles.

This is the commonest member of the Section in Scotland. It has a few stations in northern England and an isolated one in North Wales.
Alt. limit: 305 m, east of Alston (72.46)
                    3   WC

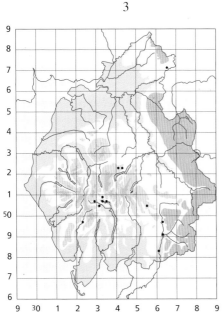

Map 873. *Hieracium subcrocatum*

Section **Tridentata**
Stems very leafy; leaves toothed and narrowed at the base; heads many.

[H. ornatilorum Sell & C. West
Specimens from a roadside verge near Dalton-in-Kendal (54.74s, C. Harris & GH, *Watsonia* (1971) 8:410) in 1970 were identified as this species by P.D. Sell and C. West. However, subsequent collections (LANC) from this site and a search by West revealed only *Hieracium eboracense*. Unfortunately the original material is lost and it seems best to regard the record as unconfirmed.

                    L]

9. **H. eboracense** Pugsley
An uncommon hawkweed of hedgerows and roadsides, chiefly in the south-east but with a few sites north of Ullswater and on the Solway Plain. It is commonest by the river in the Lune gorge. The first record for Furness was from north-east Coniston Water (3.9, E.J. Gibbons, 1901, CGE), for Westmorland from Miller Ground, Windermere (4.9, J.A. Martindale, 1909, KDL) and for Cumberland from Priest's Crag, Ullswater (42.22, MC, herb. M.C.) in 1975. There is also a record from Whitbarrow (4.8s, V. Gordon, 1964, LIV).

This is usually a tall, vigorous plant, readily recognised by the many, sharply-toothed, ovate-lanceolate leaves, the large corymbose inflorescence and the green involucral bracts. The main flowering period is in August, some weeks earlier than that of *Hieracium sabaudum*. Some of the plants by the Lune are relatively slender, with fewer heads. Although similar plants from the Yorkshire Dales have been referred to *H. calcaricola* (F. Hanb.) Roffey, they cannot realistically be separated from *H. eboracense*.

This hawkweed is rather thinly scattered from southern England to south-central Scotland.
Alt. limit: 285 m, south of Penruddock (40.26)
Map 874   19   LW(F)CY

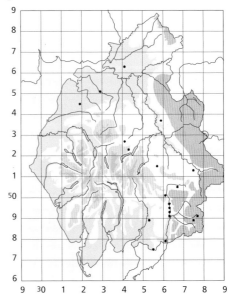

Map 874. *Hieracium eboracense*

10. H. subintegrifolium Pugsley
A British endemic hawkweed known only from "grassy banks" at Glenridding where it was discovered by H.W. Pugsley in 1927 (3.1, BM). The only person to refind it was J. Raven in 1955 (CGE ex hort.). This is a distinctive plant which differs from the following in having long, narrow, subentire, roughly hairy and distinctly reticulate-veined leaves, and a widely spreading and arching panicle.

(W)

11. H. placerophylloides Pugsley
Seen only by Borrow Beck, Shap Fells (54.04, GH, 1978), where a few plants were growing with *Hieracium subcrocatum*, in hayfields at Crosby Ravensworth (60.14, CW, 1987) and Orton (62.08, GH, 1990), by the South Tyne below Garrigill (74.40, RWMC & LR, 1989) and north of Alston (70.46, LR, 1993, all LANC). There are earlier records from Yewdale Beck, Coniston (3.9, W.R. Linton, 1905, LIV), the River Brathay from Little Langdale Tarn to Skelwith Bridge (3.0, A. Ley & W.R. Linton, 1905, BM, LIV) and Dentdale (7.8, D. Oliver, 1902, CGE).

Unlike most species of this Section, *Hieracium placerophylloides* is characterised by leaves which are subentire and narrowed below and by involucral bracts clothed only with conspicuous white simple hairs.

This is a riverside species of Wales and northern England with a few isolated records from southern Scotland.
Alt. limit: 335 m, Garrigill (74.40)

5   W(FY)

12. H. gothicoides Pugsley
Recorded only from Pooley Bridge (4.1, C. West, 1950, MNE). The specimen is labelled 'Cumberland' but a later grid-reference is in Westmorland!

This species has yellow styles and broadly ovate, sharply toothed lower leaves which give way abruptly to narrow, lanceolate upper leaves. Particularly diagnostic are the eglandular involucral bracts with more or less numerous simple hairs

It is a hawkweed of central and northern Scotland with a few records from the south and from northern England

(W/C)

13. H. uiginskyense Pugsley
Known only from rocks in Tarn Beck above Seathwaite in the Duddon valley (22.96, J.N. Mills & C. West, 1969, MNE, *Watsonia* (1971) 8:309; CGE, LANC). This is the only locality between the Scottish Highlands and Wales.

This species is similar to *Hieracium lissolepium* but it differs in having numerous pale, simple hairs and micro-glands on the involucral bracts and yellow styles.

1   F

14. H. lissolepium (Zahn) Roffey
An occasional plant of rocky becksides chiefly in the Lake District but with outlying sites in the Howgill Fells and Dentdale. Tall, robust plants, such as those collected in some quantity by A. Ley early this century from Grasmere and along the River Brathay (3.0), are now rarely seen, probably a consequence of increased grazing of the riverside pastures. The plants are now often dwarfed, growing in narrow rock crevices, sometimes in the middle of waterfalls. There are also records from Buttermere (1.1, Anon., 1928, BM) and for Crowdundle Beck, Cross Fell (6.3, W.E. Richardson, 1952, PLYP?). The earliest Westmorland record is that from Patterdale (3.1/4.1, S. Boult, 1864, LIV) and for Cumberland from Kirk Fell (*cf.* 1.1, H.E. Fox, 1888, BM).

This is a distinctive species with its dark green, rather stiff leaves, the lower somewhat crowded, and the dark, almost glabrous involucral bracts.

The headquarters of this upland hawkweed is in the northern Pennines and the Lake District but it is not infrequent on the better rocks in North Wales and

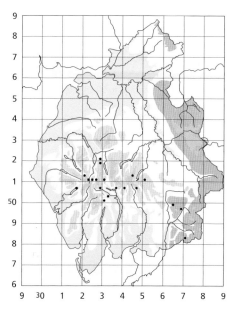

Map 875. *Hieracium lissolepium*

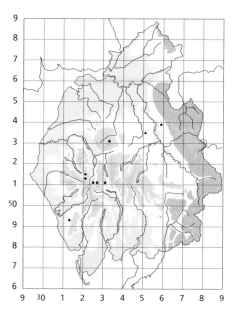

Map 876. *Hieracium sparsifolium*

southern Scotland north to Perthshire.
Alt. limit: 390 m, Raven Crag, Glaramara (24.10)
Map 875   19   WFCY

15. **H. sparsifolium** Lindeb.

Rare, in damp gullies and by streams in the central
Lake District with outlying sites on Corney Fell (12.92)
in the south-west and, in the north, at Bowscale
(32.30), Bowscar (50.34) and east of Glassonby (58.38,
RWMC, 1986, LANC), where the population has
unspotted leaves. There are further records from Skel-
with Bridge (3.0, A. Ley, 1905, BM, CGE) and above
Seatoller (22.14, C. West, 1958, MNE). It has not been
refound in a gill "on the back of Kirk Fell" (*cf.*1.1,
H.E. Fox, 1888, BM) nor in High Cup Nick (7.2,
W. Leach, 1894, BM), the first Cumberland and
Westmorland records.

The leaves are narrowly elliptic, weakly toothed
and purple blotched and the styles are yellow.

Although widespread in Scotland and Wales it is
confined in England to Cumbria and Northumberland
Alt. limit: 450 m, Rough Crag, Green Burn (30.10)
Map 876   9   W(W/F)C

Unnamed taxon
Eight collections from the Summerfield (4.3) and Asby
- Orton (6.0) limestone and by becksides in the Shap
Fells and Haweswater area (5.0, 5.1), and Deepdale,
Dentdale (7.8, all GH, LANC) show a general resemblance to *Hieracium eboracense* but differ in having
involucral bracts with appreciably more glandular
hairs, including numerous microglands, and somewhat
less strongly toothed leaves.
                    8   WCY

Section **Prenanthoidea**

Stems with numerous, more or less amplexicaul leaves;
heads many, small, densely glandular.

16. **H. prenanthoides** Villars

Seen only at Cautley Spout, in the Howgill Fells (68.96,
GH, 1984, LANC). The only pre-Survey records are
from "Cumberland" ([*cf.* W. Borrer], 1845, CGE),
Wythburn (3.1, E. Drabble, 1906, LIV) and an island
in Ullswater (3.1, J.N. Mills, 1960, not traced).

This is a distinctive plant with its numerous,
rather short, amplexicaul, dull grey-green leaves, and
peduncles and involucral bracts densely covered with
black, glandular hairs. The leaf-base is noticeably
widened.

It is characteristic of upland streamsides from
South Wales and the Peak District northwards. Its
rarity in Cumbria is surprising.
Alt.: 240 m, Cautley (68.96)
                    1   (WC)Y

Section **Alpestria**

Similar to Foliosa but leaves fewer, more amplexicaule
and more remote.

17. **H. mirandum** Sell & C. West

Recorded during the Survey from a long abandoned
limestone quarry area between Gaisgill and Newbiggin-
on-Lune (68.04, M. Atkinson, 1985, LANC) destroyed
shortly afterwards by the dumping of farm refuse.
Examination of A. Wilson's herbarium (YRK) revealed
a 1935 collection, confirmed by P.D. Sell, from "a
grassy lane above Orton" (6.0). There is also a
problematical 1969 record from "by a small brook" on
the Furness side of the River Duddon above Seathwaite
(2.9, J.N. Mills, MANCH, *Watsonia* (1971) 8:309).
Several searches of this last area have proved abortive
and it is possible that Mills' specimen was mislabelled.

This, the only English member of the Section,
was previously known from two sites in Yorkshire and
one in Derbyshire, all post-dating Wilson's find.

This hawkweed is very distinctive. The 3-6 stem
leaves are elliptic, almost entire and abruptly narrowed
at the base. The lower are shortly petiolate and the
upper semi-amplexicaule.
Alt.: 245 m, Newbiggin-on-Lune (68.04)
                    1†   W[(F)]

Section **Vulgata**

Stem scape-like or with up to 6 stem leaves, basal
rosette of leaves present at flowering, leaves rarely
glaucous; inflorescence of several medium to small
heads; styles discoloured.

18. **H. cravoniense** (F. Hanb.) Roffey

Several collections, originally identified as this species
by P.D. Sell , were redetermined by D. McCosh and GH
as *Hieracium vulgatum*. The only convincing specimen,
having involucral bracts with long simple hairs and no
glandular hairs, was a later one from by the Cash Burn
on Alston Moor (70.38, LR, 1995, LANC). Pugsley
(1948) cites specimens from Dentdale (6.8/7.8),
Sedbergh (6.9) and Colwith Force (3.0).
Alt.: 470 m, Cash Burn (70.38)
                    1   (F/W)C(Y)

19. **H. vulgatum** Fries

Probably the most frequent hawkweed and conspicuous
from early summer in waste places, on walls, roadsides
and railwaysides, and in the fells by becks and on rock
ledges, especially on the limestone. The first record for
Westmorland was from Fairfield (3.1, J.C. Melville,
1865, herb. H.W. Borrer), for Furness from Furness
Abbey (2.7, C. Bailey, 1869, BM), and for Cumberland
from Piers Gill, Scafell Pike (2.0, A. Ley, 1876, CGE).

It is readily recognised by its dark, toothed,
elliptic leaves, which are often purple beneath; the
involucral bracts are densely stellate hairy below and
have short, crisp, simple hairs and rather few, short
glandular ones.

This is a very common species from Scotland
south to the north Midlands and North Wales.
Alt. limit: 660 m, Crowdundle Beck, Cross Fell (68.32)
Map 877   481   LWFCY

20. H. rubiginosum F. Han**b.**

Not infrequent on the limestone in the south and east.
Authenticated records from off the limestone are rare:
the only Survey ones from the Lake District are from
Dungeon Gill (28.06, M. Richards, 1989, LANC) and
above Troutbeck (40.06, GH, 1992). It was also found
by the River Irthing west of Gilsland (60.66, GH, 1995,
LANC). There are earlier records from Levers Water,
Coniston Old Man (2.9, P.M. Newby, 1951, RNG), the
first Furness record, Helvellyn (3.0, A. Ley, 1905,
CGE), Firbank, Sedbergh (6.9, E. Milne-Redhead,
1960, K), Garsdale (7.9, C. West, 1958, CGE) and
Humphrey Head (3.7, C. West 1962, CGE, MNE). The
first Westmorland record, of cultivated material from
Scout Scar (4.9, E.F. Linton, BM, LIV), dates from
1894, that for Cumberland is from Ardale Beck, Cross
Fell (6.3, C. West, 1952, MNE). A problematical plant
from St Bees Head (94.14, CCH, 1987, LANC) may
belong here.

   Although having a general resemblance to
robust specimens of *Hieracium vulgatum*, this species
has typically very strongly and coarsely toothed stem
leaves, larger heads and involucral bracts with longer
white simple hairs and more glandular hairs.

   This is a generally montane species occurring
throughout Scotland and southwards to the limestone of
the Peak District and South Wales.

Alt. limit: 660 m, Crowdundle Beck, Cross Fell (68.32)

Map 878   27   W(F)C(Y)

21. H. acuminatum Jordan

(*H. strumosum* (Ley ex W.R. Linton) Ley)

Occasional to frequent and chiefly in the limestone
country of the south-east.. It occurs along road- and
railwaysides, in quarries and, more rarely, in more
natural habitats such as limestone pavement and scars
and on riverside rocks. There are earlier records for
Barrow Docks (1.6/2.6s, D.E. Allen, 1951, CGE), Little
Langdale, Colwith Bridge and the River Rothay (3.0,
W.R. Linton, 1905, LIV) and Long Marton, Appleby
(6.2, C. West, 1966, MNE). The Linton records
probably include material from the Westmorland side
of the river and so constitute the first records. The first
definite ones for Furness are from Coniston, Yewdale
Beck and Tilberthwaite, (3.9, 3.0, A. Ley, 1905, BM,
CGE) and for Cumberland from Clints quarry,
Egremont (00.12, B.N.K. Davies, 1975, ABRN).

   Hieracium acuminatum resembles the following
two species but it differs from both in having involucral
bracts with many dense stellate hairs intermixed with
the abundant glandular hairs.

   This hawkweed replaces *H. vulgatum* as the
commonest species in southern England and Wales.

Alt. limit: 610 m, Middle Tongue, Little Dun Fell
(68.32)

Map 879   43   WFCY

22. H. diaphanoides Lindeb.

Rare, recorded only from seven sites: by the Scandale

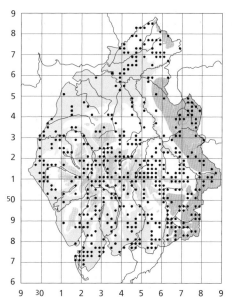

Map 877. *Hieracium vulgatum*

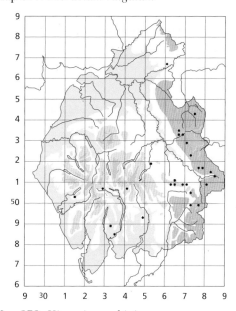

Map 878. *Hieracium rubiginosum*

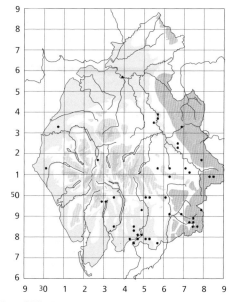

Map. 879. *Hieracium acuminatum*

Beck, Ambleside (36.06, M. Richards, 1988), by the River South Tyne above and below Garrigill (74.36, GH, 1981, the first Cumberland record; 70.40, 70.42, LR, 1996), on a wall west of Talkin (52.56, GH, 1988), on a railway embankment at Appleby (68.20, GH, 1993) and on islands in the River Lune near Sedbergh (62.92, GH, 1988, all LANC). There are additional early records from Helbeck scars above Brough (A. Wilson, 1934, YRK), the River Rawthey above Sedbergh (6.9/7.9) and riverside rocks in Dentdale (7.8, both A. Ley, 1902, CGE; BM), Yewdale, Coniston (3.0, A. Ley, 1905, LIV) and Whelter Crags, Haweswater (4.1, A. Ley, 1910, CGE, wrongly mapped in the *Critical Atlas* as 5.1). These last two records are respectively the first for Furness and Westmorland.

This species can be very difficult to distinguish from *Hieracium diaphanum*. The chief differences are the usually longer, narrower leaves and the occasional simple hairs on the involucral bracts and peduncles.

It is essentially an upland species of northern England, south-west Scotland and the Highlands.
Alt. limit: 455 m, Tyne Head (74.36)

7   W(F)C(Y)

23. **H. diaphanum** Fries

Apparently rather scarce, several of the Survey collections originally thought to be this species proving to be *Hieracium acuminatum*. The records are mainly from the Ambleside (3.0) and Sedbergh (6.9) areas, where it grows on rocks by the Rivers Brathay, Lune and Rawthey. It also occurs as a garden weed at Ambleside, on roadside rocks west of Scales (32.26, MC, 1981, herb. M.C.), in a quarry at Kirkcambeck (56.68, MG & JP, 1989, LANC) and by the railway in Dentdale (76.84, 76.86, FCS & JS, 1987, LANC). There are additional records from Hobcarton Crag, Grizedale Pike (1.2, W.B. Waterfall, 1884, BM), the first Cumberland record, Measand Beck, Haweswater (4.1, A. Webster, 1910, CGE), and Appleby (6.2, E. Milne-Redhead & N.D. Simpson, 1947, BM, CGE). Material collected by A. Ley from Little Langdale and the River Rothay (2.0, 3.0, 1905, BM, CGE) could be the first Furness record. The first Westmorland record was from Far Easedale (3.0, H.E. Fox, 1888, BM).

Hieracium diaphanum has leaves which are generally shorter than those of either of the two preceding species. The involucral bracts are more or less densely covered with dark glandular hairs and lack both stellate and simple hairs. It occurs chiefly in Wales and central England.
Alt. limit: 305 m, Carlin Gill, Howgill Fells (64.98)
Map 880   19   WFCY

24. **H. glanduliceps** Sell & C. West

Rather rare; known in the south on limestone at Underbarrow Scar (48.88s, 48.90, 1988), Farleton Fell (54.80, 1988) and on a roadside wall at Bardsea (30.74s, 1984), and in the east on limestone around the Eden valley: by Crowdundle Beck, Cross Fell (68.32,

1987), the only Cumberland record, on Murton Pike (72.22, 1987), in High Cup Gill (72.24, 1987; 74.24 1988; 74.26, 1987), Great Rundale (70.26, 1995), and above Brough (78.16, 1992, all GH, LANC), as well as above a conglomerate riverside cliff.at Kaber (78.12, GH, 1987, LANC).

This spotted hawkweed, previously only known from west Yorkshire, resembles *Hieracium maculatum* but it is distinguished by the truncate to wedge-shaped leaf base and the dense, unequal and rather fine glandular hairs of the peduncles and involucral bracts.
Alt. limit: 660 m, Crowdundle Beck (68.32)
Map 881   12   WFC

[*H. maculatum* Smith
The specimen collected by T.J. Foggitt at Appleby (6.2, BM) in 1929 and mapped as this species in the *Critical Atlas* is actually *Hieracium scotostictum*.

(W)]

25. **H. maculoides** Sell & C. West

This early-flowering limestone species was first described from material from Ingleborough and the Arnside - Silverdale (v.c.60) area. The material from the latter area in CGE has rather small (4-6 cm long) leaves which are, at the most, only lightly spotted. However, Survey material (all GH, LANC) from Arnside to the east side of the Leven estuary suggests that the CGE specimens are not really typical. There are Arnside plants (44.76s, 1988; 44.78, 1988) with appreciably larger (-9 cm) spotted leaves and plants with very large, broadly ovate (11 x 7 cm) and strongly spotted leaves from the Leven estuary (32.78s, 1987). Spotted, large-leaved plants also occur a little way inland on the screes at the south end of Whitbarrow Scar (44.84s, 1988). Smaller-leaved plants,similar to *Hieracium maculosum*, also occur at this last site. There are additional records from Humphrey Head (38.74s, O.M. Stewart, 1987) and the railway embankment at Meathop (42.78s). The first record was from Far Arnside in 1909 (J.A. Martindale, KDL).

Both species have involucral bracts with prominent white simple hairs, marginal stellate hairs and rather few, short glandular hairs.
Alt. limit: 445 m, Whitbarrow Scar (44.84s)

6   WF

26. **H. rhomboides** (Stenström) Johansson

Rare, known only from four tetrads in the Pennines, mostly on limestone: Crook Burn, Yad Moss (78.34, RWMC, 1992), Black Door, Ardale (66.34, GH, 1987), Middle Tongue Beck and Crowdundle Beck (68.32, GH, 1987), and by the Swindale Beck above Knock (70.28, GH, 1987, all LANC). The first record for Cumberland was from Crowdundle Beck (C.E. Salmon, 1919, BM) and for Westmorland from High Cup Nick (7.2, J.A. Whellan, 1946, BM). The *Critical Atlas* record for the Shap square (5.1) is an error for Swindale, Knock. A collection from the whinstone above Knock Ore Gill (70.30, GH, 1989, LANC) may belong here but the involucral glands are sparser and the simple hairs longer. This species resembles *Hieracium rubiginosum* in having involucral bracts with both long, simple hairs and weak, glandular ones but there are more of the latter which also continue down the peduncles. The leaves are broadly elliptic, often blotched with purple and less strongly toothed.

This hawkweed is known elsewhere in England only from Teesdale, Ingleborough and Northumberland. There are a few scattered localities in Scotland. Alt. limit: 660 m, Crowdundle Beck, Cross Fell (68.32)

4   WC

27. **H. oistophyllum** Pugsley
Occasional, chiefly throughout the Pennines and in the extreme north, occuring on damp rocks, especially limestone. It was even found growing epiphytically on branches over the River Lyne (40.74n). The first record

for Cumberland is probably one from Garrigill (7.4, C. Waterfall, 1921, CGE) and for Westmorland from the Maize Beck, Teesdale (7.2, GH, 1959, LTR).

This is an attractive plant with pale ligules and small heads, the involucral bracts having long, flexuous simple hairs and fewer, shorter, dark glandular hairs. There is usually a single, broadly triangular stem leaf.

It is predominantly a hawkweed of northern England, southern and central Scotland.
Alt. limit: 660 m, Crowdundle Beck, Cross Fell (68.32)
Map 882   24   WCY

28. **H. silvaticoides** Pugsley
Frequent on the limestone hills behind Morecambe Bay but with a single outlying site on the limestone of Ash Gill, Garrigill (74.40), where it was first discovered by J.E. Raven in 1953 (CGE). There is also an old record from Middle Tongue beck, Great Dun Fell (6.3/7.3, D. Welch & C. West, 1966, MNE). The first record for

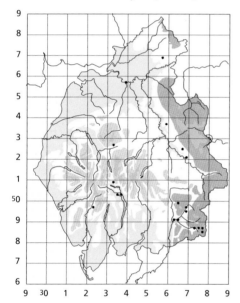

Map 880. *Hieracium diaphanum*

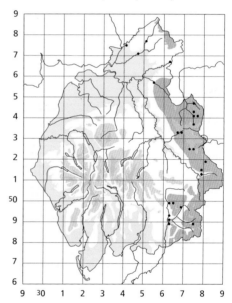

Map 882. *Hieracium oistophyllum*

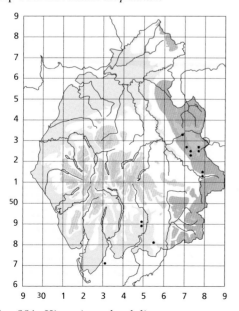

Map 881. *Hieracium glanduliceps*

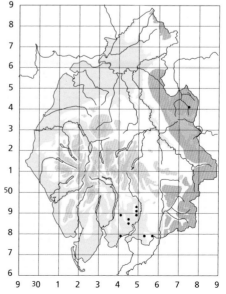

Map 883. *Hieracium silvaticoides*

Westmorland was from Meathop, Grange-over-Sands (42.78s, C. Bailey, 1881, BM).

This species, with its broad, rounded leaves, often purplish beneath, is easily confused with *Hieracium auratiflorum*. However, the panicle is frequently umbellate with a short-stalked primary head; the involucral bracts often exceed 10 mm and have numerous long, white-tipped hairs and rather fewer, shorter black glandular hairs, the tips are distinctly hairy.

Outside the Yorkshire Dales and Cumbria, this species is known only from a few sites in Scotland.

Alt limit: 435 m, Garrigill (74.40)

Map 883   11   WC

29. H. maculosum Dahlst. ex Stenström

Seen only on limestone scars around Morecambe Bay: Farleton Fell (52.78s, 1988), Hutton Roof Crags (56.78s, 1988), Whitbarrow (42.88s, 44.84s, 1988) and Underbarrow (48.90, 1988), and on the Pennines: Wild Boar Scar, Cross Fell (68.32, 1989), Great Rundale (70.26, 1991) and Hilton Fell (76.22, 1988, all GH, LANC). The only previous record is one from Knock Ore Gill (7.3, D. Welch, 1962, herb. Eng. Nat., "Whinsill Fell" of the *Critical Atlas*), and the only other British records in the *Atlas* are two from the Ingleborough area.

This is a rather distinctive hawkweed with its neatly toothed, lightly spotted leaves, which are purple beneath, and the markedly stellate hairy involucral bracts with conspicuous white, simple hairs. However, as mentioned under *Hieracium maculoides*, it does not seem possible to distinguish between some of the Whitbarrow plants and those referred to *H. maculoides* on the coastal limestone around Morecambe Bay.

Alt. limit: 610 m, Wild Boar Scar (68.32)

Map 884   8   WC

30. H. auratiflorum Pugsley

A frequent hawkweed on the Pennine limestone north of Stainmore and on the whinsill of High Cup Nick (7.2), with an outlying site on the conglomerate banks of the River Lune above Casterton (60.80s, GH, 1995, LANC). The *Critical Atlas* dots for the Shap (5.1) and Kirkby Thore (6.2) squares are misplaced. The record for Piers Gill, Scafell Pike (2.0, H.E. Fox, 1888, BM) is so incongruous that one suspects mislabelling. That for Whitbarrow (4.8s, A. Wilson, 1927, YRK) is based on a misidentification of *Hieracium silvaticoides*. The first Westmorland record was from High Cup Nick (T.J. Foggitt, 1905, BM) and the first definite one for Cumberland was from Melmerby High Scar (64.36, A.J. Richards, 1973, NCE).

The leaves of this species are broad, oblong, obtuse or apiculate, somewhat toothed at the base and resemble those of *H. silvaticoides*. The peduncles and involucral bracts are, however, conspicuously glandular hairy in *H. auratiflorum* and the heads are smaller, the involucral bracts being usually appreciably less than 10 mm. *H. crebridentiforme* has ovate-lanceolate leaves and fewer, less spreading heads, which also taper to the base.

Outside Cumbria, this hawkweed is known only from the adjacent parts of Lancashire, Yorkshire, Co. Durham and two sites by the River Tyne in Northumberland.

Alt. limit: 615 m, Melmerby High Scar (64.36)

Map 885   14   WC

31. H. crebridentiforme Pugsley

Rather scarce and virtually restricted, like *Hieracium auratiflorum*, to limestone in the Pennines. It was seen by the River South Tyne above Garrigill and at Tynehead (74.36, 1981; 76.36, 1993), by Ardale and Crowdundle Becks, Cross Fell (66.34, 1987; 68.32, 1987), by the Maize Beck (74.26, 1988), in Knock Ore Gill (70.30, 1989), on Middle Fell, Warcop (76.16, 1992) and in Swindale, Brough (80.16, 1995, all GH, LANC). Specimens from Cautley Spout, in the Howgill Fells

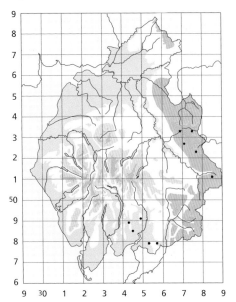

Map 884. *Hieracium maculosum*

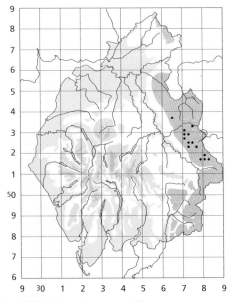

Map 885. *Hieracium auratiflorum*

(6.9, A. Wilson, 1902, YRK) and Dentdale (7.8, A. Ley, 1909; D. Oliver 1910, both CGE), the basis of the *Critical Atlas* records, do not belong here and are perhaps best placed under *Hieracium stenopholidium*.

This is typically a robust species with large, broadly obovate, subentire to sparsely toothed leaves which resemble the more robust Pennine plants of *H. decolor*. The latter, however, has glandular hairs only on the upper parts of the peduncles and the involucral bracts have conspicuous marginal stellate hairs and fewer glands. The heads and peduncles of *H. crebridentiforme* more closely resemble those of *H. auratiflorum* being conspicuously glandular. The differences between the two species are given under the latter.

This species has a restricted range from the limestone of west Yorkshire to Northumberland.
Alt. limit: 660 m, Crowdundle Beck, Cross Fell (68.32)

8　　　　　　　　　　WC

32. **H. pseudostenstroemii** Pugsley

Known only from the top of a conglomerate cliff near Kaber (78.12, GH, 1987, LANC) and a limestone cliff in Swindale, Brough (80.16, GH, 1995, LANC). Earlier records from Crowdundle Beck and Knock Ore Gill (6.3, A. Eddy, 7.3, Anon., 1960s, herb. Eng. Nat.) are based on inadequate material and a collection from riverside rocks in Dentdale (7.8, A. Ley, 1902, 1909, 1910, CGE) is puzzling but does not belong here.

With its involucral bracts clothed with a mixture of prominent, unequal, slender, black glandular hairs and simple hairs, this species is close to *Hieracium duriceps*. It differs chiefly in its longer and somewhat wider leaves and larger heads (involucral bracts >10 mm).

Most of the records for this hawkweed are from the Craven area of Yorkshire.
Alt. limit: 275 m, Swindale, Brough (80.16)

2　　　　　W[(C/WY)]

33. **H. cymbifolium** Purchas

Rare; seen at five sites, all on the Pennine limestone: Murton Crag (72.22, RWMC, 1988), Swindale, Hilton (76.20, GH, 1987), where it was first collected by N.D. Simpson in 1938 (BM), Scordale, Hilton (76.22, GH, 1987), Barras (84.10, GH, 1987) and above the Augill Beck, North Stainmore (82.16, GH, 1988, all LANC).

This distinctive hawkweed usually has rather small, stiff and somewhat cucculate leaves which have stellate hairs on the upper surface. The involucral bracts have a mixture of white-tipped simple hairs and abundant shorter glandular ones. It is essentially a plant of the Pennine limestone.
Alt. limit: 430 m, Murton Crag (72.22)

5　　　　　　　　　　W

[**H. piligerum** (Pugsley) Sell
There is an old but undated specimen in HAMU collected from rocks at Gilsland (6.6, J. Thompson). There is nothing

to indicate whether it was from v.c.67 or v.c.70.

This is a spotted-leaved hawkweed of the Scottish Highlands.

(C?)]

34. **H. scotostictum** N. Hylander

Known only from the middle Eden valley where it occurs chiely on railway embankments and bridges as at Appleby (68.20, GH, 1972, LANC), where it was first noticed in 1929 by T.J. Foggitt (BM), and Long Marton (66.22, 66.24, 1995, GH, LANC), also near Great Ormside (68.16, RWMC, 1978, herb. C.C.H.), on a roadside verge near Flitholme (76.14, CW, 1982, LANC) and in Cumberland on the wooded bank of the Eden near Armathwaite (50.48, GH, 1985, LANC, *Watsonia* (1988) 17:194).

The strong purple blotching of the ovate, toothed, long-stalked leaves is very distinctive. The stellate-hairy peduncles and involucral bracts are densely covered with white-tipped eglandular hairs and unequal black, glandular hairs.

This introduced species is still largely confined to the London area where it was first seen in 1920.
Alt.limit: 155 m, Flitholme (76.14)

6　　　　　　　　　　WC

35. **H. duriceps** F. Hanb.

Rather rare; known mostly from rocky river- and beck-sides. In the north it occurs on sandstone rocks by the Liddel (42.76), in the Pennines from Alston Moor to Dentdale, mostly on limestone but also on the whinstone of High Cup Nick (74.24), and in the Lake District at a few sites in the centre and east. There are additional records from limestone by the River Gelt (5.5, C.W. Muirhead, 1962, PLYP), Keswick (2.2, J.C.D. Melvill, 1901, HWB), Ennerdale (1.1, H.E. Fox, 1888, *ex hort*. BM, placed by Pugsley (1948) under *Hieracium praetenerum*), the first Cumberland record, and Loweswater (1.2, E. Hayward, 1921, CGE), wrongly mapped as 1.3 in the *Critical Atlas*. The *Critical Atlas* record for the Kirkby Thore square (6.2) is based on a misidentification of *H. scotostictum*. The first record for Furness was from Yewdale, Coniston (3.9, A. Ley, 1910; BM) and for Westmorland from Maize Beck, Teesdale (7.2, E.G. Baker, 1894, BM).

This hawkweed typically has numerous small heads on short peduncles; the involucral bracts and peduncles have dense black glandular hairs mixed with shorter, black simple ones; the leaves are ovate-lance-olate, subentire and darkish green.

Hieracium duriceps is not uncommon in the upland regions of Scotland and in England it extends south to the Yorkshire Dales.
Alt. limit: 580 m, Maize Beck (74.36)
Map 886　　　　16　　　　WFCY

36. **H. itunense** Pugsley

A Cumbrian endemic, first collected from limestone rocks near South Stainmore by H.J. Riddelsdell and

G.C. Druce in 1903 (8.1, BM, OXF) and named by W.R. Linton *Hieracium cumbriense*. The site was visited by T.J. Foggitt in 1929 (BM) and by Druce in 1930 and 1931 but then 'lost' for a further half century until 1982 (GH, LANC; BM). The population is small and very vulnerable.

This hawkweed has leaves which are ovate to ovate-oblong, subentire or somewhat toothed towards the base, and softly hairy beneath. The stem is robust, usually with a single leaf, and the peduncles and bracts are densely covered with unequal dark glandular hairs interspersed with a few longer dark, simple ones.
Plate 95   1   W

37. **H. pellucidum** Laest.
Occasional on limestone scarps and pavement in the Pennines and around Morecambe Bay, also on shaded riverside rocks, usually of slate, by the River Lune and its tributaries and in the east of the Lake District. There are earlier records from Kirkby Lonsdale (6.7s, E.B. Terras, 1888, E), the first Westmorland record, Fleetwith Pike, Honister (2.1, C. West & J.W. Cardew, 1954, MNE), Yewdale, Coniston and Tilberthwaite (3.9/3.0, A. Ley, 1905, CGE), the first Furness records, and Moor House, Teesdale (7.3, R.J.F. Park, 1954, BM), wrongly shown in the *Critical Atlas* as 7.4, and a *Critical Atlas* record for the Arnside square (4.7s). The first Cumberland record is from Ardale Beck, Cross Fell (6.3, C. West, 1952, MNE).

This species is sometimes mistaken for another montane, small-headed, glandular plant, *Hieracium duriceps*. However, the latter has more pointed, narrower leaves and the dense glandular hairs of the involucral bracts are noticeably intermixed with longer, simple hairs. *Hieracium pellucidum* is, in fact, closer to the *H. exotericum* group from which it differs in its almost glabrous leaves which are often purplish below.

This is predominantly a hawkweed of northern England, the Peak District and South Wales.

Alt. limit: 600 m, head of Glencoynedale (34.18)
Map 887   42   WFCY

[*H. hjeltii* Norlin ex T. Saelan, W. Nylander & T.S. Nylander The only *Critical Atlas* record for this plant is from the railwayside at Silverdale (4.7s, v.c.60, J. Cryer, 1912, CGE). Material was subsequently collected from limestone rocks by the coast road at Storth (4.8s, J. Bevan & P.D. Sell, 1980, CGE, *Watsonia* (1982) 14:196). Later investigations showed this taxon to be common at Silverdale towards the county border and on the abandoned railway at Storth. The styles are discoloured and the leaves glabrescent. The plants vary appreciably in size, the smaller having subentire and finely denticulate leaves, the larger are more coarsely toothed in the lower half. The leaf tips vary from rounded-obtuse or acute. There seems no obvious reason why they should not be placed under *Hieracium subcrassum*; indeed, one of the later Silverdale specimens (MB, 1982, LANC) was so named by P.D. Sell. *Hieracium hjeltii* is otherwise known only from a single site in Finland.
Plate 93d   W]

H. exotericum group, nos. 38 - 42
The following six species are characterised by small heads, involucral bracts with black glandular hairs only and leaves varying in hairiness and toothing (Plate 93).

38. **H. integratum** (Dahlst. ex Stenström) Dahlst.
Known only from Arnside Knott (44.76s, M. McC. Webster, 1965, CGE; LANC) where there is a small and apparently declining population on wooded scree. The plants are tall and rather weak. This, and the long, slender petiole, may, however, be a response to shading and the plant is possibly only a shade form of the following species. The leaves are entire, broadly elliptic, with a cuneate base, the upper surface is glabrous; there is a single ovate stem leaf and the styles are dark.

The only other British site is on Ingleborough.
1   W

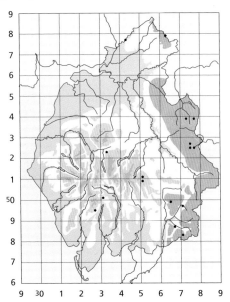

Map 886. *Hieracium duriceps*

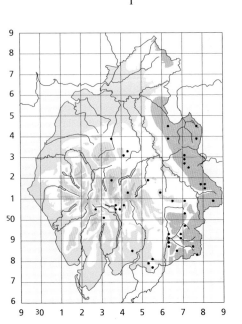

Map 887. *Hieracium pellucidum*

39. H. subcrassum (S. Almq. ex Dahlst.) Johansson
Rare. It was first recorded from a steep wooded slope in Grisedale, Patterdale (38.14, CGE) by J. Bevan and P.D. Sell in 1980, the population being estimated at over 1000 plants. It was later found in the nearby churchyard and by the Goldrill Beck (38.16, CFS & JS, 1987; GH 1984, LANC). In the south of the county it occurs in a tributary valley off Barbondale (66.84s, CFS & JS, 1987, LANC), on coastal limestone on Holme Island, Grange-over-Sands (42.78s, GH, 1991, LANC) and at Storth (46.80s, J. Bevan & P.D. Sell, 1980, CGE, det. Sell as *Hieracium hjeltii*); LANC), where it also grows on the limestone cutting and ballast of the old railway. The only other British sites are two in the Welsh Borders and two near Glasgow.

The leaves are glabrescent, obtuse or acute, often rather large (-13 x 6.5 cm), varying from subentire or finely denticulate in small plants to rather coarsely but not deeply toothed, especially towards the subtruncate base. There is usually a large, ovate-lanceolate, serrate stem leaf. The styles are fuscous or dark.
Alt. limit: 300 m, Barbondale (66.84s)
Plate 93c   4   WF

40. H. sublepistoides (Zahn) Druce
Rare, but not uncommon on railway bridges and railwaysides in the upper Eden valley as at Waitby (74.06, 76.08, GH, 1993) and Smardale (72.06, GH, 1989). In the north it occurs on a roadside wall at Armathwaite Hall, Bassenthwaite (20.32, 1977, GH), where it was first collected by H.L.G. Stroyan in 1952 (CGE), the first Cumberland record, by the roadside between Bassenthwaite village and Keswick (22.28, 1982, GH) and Applethwaite, Keswick (28.28, A. Franks, 1989). In the south it occurs in a forestry area near Satterthwaite (34.90, J. Williamson, 1980), by the roadside on Cartmel Fell (40.88s, GH, 1995) and Underbarrow (48.92s, GH, 1995), also by the river in mid-Dentdale (76.86, CFS, 1997, all LANC). The only

previous Westmorland records were from Arnside (4.7s, D. Oliver, 1912, BM) and the railway at Appleby (6.2, J. Raven, 1953, LANC).

This species is characterised by leaves which are ovate and obtuse, subentire or finely and shortly denticulate.

Elsewhere in Britain this hawkweed ranges in a broad belt from south Lancashire to the south coast with a few isolated sites in western Scotland.
Alt. limit: 250 m, Smardale (72.06)
Map 888, Plate 93b   13   WFC

[H. severiceps Wiinst.
Known only from two collections: Clints limestone quarry, Egremont and old mine workings at Bowthorn, Cleator Moor (00.12, 1987, 00.16, 1980, both CCH, LANC, *Watsonia* (1979) 12:358), material from the last site being determined by P.D. Sell.

This material is characterised by having leaves which are moderately hairy, rather small and narrow, subentire or finely denticulate, and styles which are yellow. The original description gives the leaf shape as being broadly obovate, which hardly agrees with the Cumbrian material, and the styles as being dark. It is in fact difficult to see how the specimens can be distinguished from some of those included here under *Hieracium sublepistoides*. It seems best to treat the identifications as doubtful.

The only other British records are from Kent, Cheshire, Derbyshire and the Silverdale area of Lancashire, with a few isolated ones from Scotland.
Plate 93e   C]

41. H. grandidens Dahlst.
The commonest member of the group occurring throughout the lowlands, usually in ruderal situations but sometimes far from houses as by the Croglin Water (60.48). There is an additional record from Risehow colliery, Maryport (0.3, C.W. Muirhead, 1951, CGE). It was apparently not recorded until 1948 when it was collected at Armathwaite Hall, Bassenthwaite (2.3, C.W. Muirhead, K). The first record for Westmorland

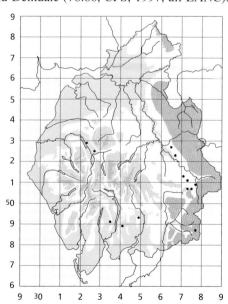

Map 887. *Hieracium sublepistoides*

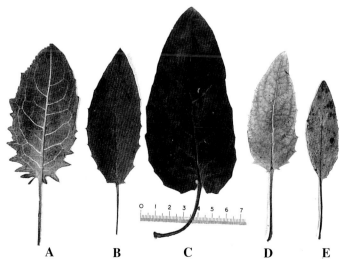

Plate 93. Leaves of a) *Hieracium grandidens*, b) *H. sublepistoides*, c) *H. subcrassum*, d) *H. hjeltii* and e) *H. severiceps*

was from Winton (7.1, C. West, 1960, MNE) and for Furness from Graythwaite, north of Newby Bridge (3.9, GH, 1975, LANC).

This is a distinctive plant with its softly hairy, somewhat glaucous, prominently toothed leaves, the lower teeth often deflexed behind the truncate base; it has a large, triangular stem leaf and dark styles.

Hieracium grandidens is widely scattered throughout Britain, usually in waste places and on roadsides. It is usually regarded as an introduction and it is probably still spreading in Cumbria, having been recorded from appreciably more squares (31) than the five shown in the *Critical Atlas*.
Alt. limit: *c*.500 m, above Swindale, Brough (80.16)
Map 889, Plate 93a   53   WFCY

42. H. exotericum Jordan ex Boreau

Apparently very rare, having been collected only from a roadside bank at Grange-over-Sands (40.78s, GH, 1977), the roadside south of Beetham (50.78s, GH, 1993), limestone scree at Dalton Crags (54.76s, GH, 1970) and on a railway bridge and riverside in upper Dentdale (76.84, CFS & JS, 1987; 74.86, GH, 1995; all LANC). The styles are yellow, the leaves are less hairy than in *H. grandidens* and less strongly toothed, though more so than in *H. sublepistoides*. It also differs from the latter in having subacute leaves.

This is a rare hawkweed and largely restricted to south Wales and southern England.
Alt. limit: 310 m, upper Dentdale (76,84)
Map 889, Plate 93   5   LFC

Section **Oreadea**

Stems scape-like or with two to several stem leaves; leaves often glaucous; heads one to several, medium to large; styles usually yellow.

43. H. orimeles F. Hanb. ex W.R. Linton

Rather rare, seen only below Black Force in the

Howgill Fells (64.98, GH, 1984, LANC) and in six tetrads in the Lake District: Cawdale, Bampton Common (46.16, RWMC, 1990), above Blea Tarn, Haweswater (44.10, GH, 1987), Red Screes, Kirkstone (38.08, M. Richards, 1988), Scrubby Crag, Fairfield (36.10, GH, 1978), Brown Cove, Helvellyn (34.14, AAD, 1988) and by the Hayeswater Beck, Hartsop (40.12, 1988, GH, all LANC). There are also records from below Pavey Ark (2.0, W.R. Linton, 1905, LIV) and Rigg Head, Seatoller (2.1, C. West, 1958, CGE), the only Cumberland record. The first Westmorland record was from Fairfield (3.1, Anon., 19th C, BM). The *Critical Atlas* dot for the Shap square (5.1) is an error for the Helvellyn square (3.1).

This species is characterised by having usually two or three heads, elliptic, weakly-toothed basal leaves and usually two smaller stem leaves of similar shape with one or two very small setaceous leaves above these. The leaves are typically rough with short, sparse hairs noticeably over the upper surface. The peduncles and involucral bracts have a variable amount of stellate hairs with white, dark-based simple hairs and fewer short glandular hairs, the inner bracts have pale margins and a narrow median line of hairs; the tips are somewhat obtuse and tipped with hairs.

Hieracium orimeles is frequent in the Outer Hebrides and the Shetlands but is otherwise very scattered in Scotland, northern England and Wales.
Alt. limit: *c*.750 m, Brown Cove, Helvellyn (34.14)
                              7   WC

[*H. subrude* (Arv.-Touv.) Arv.-Touv.
The only collection is one from by the Hayeswater Beck, Hartsop (4.1, *cf.* A. Ley, 1910, BM, CGE). This is the basis of the *Critical Atlas* record and the only one from England off the Northumberland whinstone. The specimen is not typical of *Hieracium subrude* and material from this site collected during the Survey proved to be *H. orimeles*.

(W)]

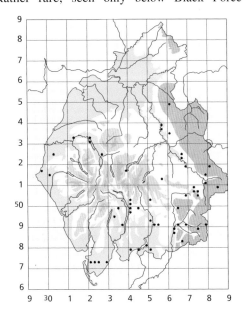

Map 889. *Hieracium grandidens*

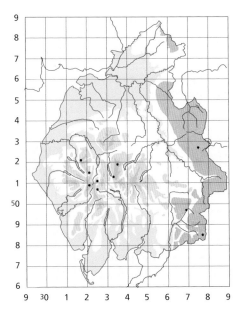

Map 890. *Hieracium caledonicum*

44. **H. caledonicum** F. Hanb.

Scattered, mainly on crags and gullies in the central Lake District. There are Pennine records from the whinstone of High Cup Nick (74.26, GH, 1987, LANC), from limestone by the Artengill Beck, Dentdale (76.84, CFS & JS, 1987, LANC), and from the Howgill Fells at Cautley Spout (68.96, GH, 1984, LANC) and where it was earlier seen at Black Force (6.9, A. Wilson, 1892, YRK). The first record for Cumberland was from Pillar (1.1, C.W. Muirhead, 1949, PLYP), for Furness from Coniston Old Man (24.96, GH, 1989, LANC), and for Westmorland from the head of Glencoynedale (34.18, GH, 1986, LANC). Material from coastal rocks at Arnside (4.7s, C. West, 1962, CGE), the *Critical Atlas* record, does not appear to belong to this species.

Cumbrian plants of this variable and widespread upland species have rather long, elliptic, entire or toothed leaves, often purplish below and one or two well-developed stem leaves. The involucral bracts have long, simple hairs, fewer and shorter glandular hairs, and stellate hairs; the inner bracts are obtuse, flushed pink above, hairy-tipped but otherwise only sparsely hairy.

Alt. limit: 700 m, Piers Gill, Scafell Pike (20.08)

Map 890   10   WFCY

[*H. angustatum* (Lindeb.) Lindeb.

The only *Critical Atlas* records of this species are from Cumbria. Pugsley (1948) cites three collections: Dollywaggon Pike (3.1, H.E. Fox, 1890, BM), Helvellyn (3.1, R.P. Murray, 1881, BM), and Ennerdale (1.1, F. Addison, 1867, BM). In his description he mentions that the leaves are glabrescent above and slightly hairy beneath and that the involucral bracts have only scattered simple hairs and usually no glandular ones. These three collections are heterogeneous and only Fox's can be said to match the description. Several collections from the central Lake District have since been referred to this species. P.D. Sell and C. West, for example have determined material from Dove Crags (3.1, H.E. Fox, 1890, BM), Catstycam (3.1, F. Rose, 1956, CGE) and an anon-ymous collection from Helvellyn (KDL). None are convincing. There is a 1980 collection by J. Bevan and Sell from Dollywaggon Pike (CGE), which matches a specimen coll-ected from the same gully by GH (1990, LANC). This is distinctive in its densely stellate hairy peduncles and dark involucral bracts with many dark simple hairs. Its identity is uncertain. It seems best to conclude that there is insufficient evidence for the occurrence of *Hieracium angustatum* sensu Pugsley in Britain.

The *Critical Atlas* record for the Haweswater square (4.1) is an error for Angle Tarn, Cumberland (2.0, A. Ley, 1905, CGE) and that for 4.9 is based on a specimen of *H. argenteum* from Scout Scar (H.E. Fox, 1888, BM; CGE).

W(C)]

45. **H. argenteum** Fries

A rare plant of mountain gullies and of lake- and beckside rocks, with a preference for dry siliceous rocks. There is a particularly fine luxuriant lowland population by a forestry track in Dodd Wood, Keswick

(22.26). It rarely occurs in any quantity and may be decreasing since there are no Survey records from six of the ten *Critical Atlas* squares: Goat Crag Gill, Buttermere (1.1, J.E. Raven, 1954, CGE), Hobcarton Crag, Grisedale Pike (1.2, J. Groves, 1884, BM), Coniston (2.9/3.9, H.T. Mennell, not traced), the only Furness record, Watendlath (2.1, A. Wilson, 1902, YRK), Fleetwith Pike, Honister (2.1, C. West, 1954, CGE), Snaka Rocks, Ambleside (3.0, A. Ley & W.R. Linton, CGE, LIV) and Scout Scar (4.9, H.E. Fox, 1888, BM, W.R. Linton, 1905, LIV), the first Westmorland record and the only one from limestone. The first Cumberland record was from Grange-in-Borrowdale (2.1, J.G. Baker, 1870, BM). The *Critical Atlas* record for the Pooley Bridge square (4.2) is probably an error for records from the head of Ullswater (3.1).

This is usually an easily recognised hawkweed with its narrowly lanceolate or elliptic, somewhat glaucous, subentire leaves and usually a narrow, patent stem leaf. The involucral bracts are sparsely hairy, with medium glandular hairs and longer simple hairs.

It is widely distributed throughout western and northern Britain.

Alt. limit: 610 m, Ruddy Gill, below Great End (22.08)

Map 891   11   W(F)C

46. **H. vagense** (F. Hanb.) Ley

Known only from the Lake District where it has been seen at three sites: Rampsgill Head, High Street (44.12, T.C.G. Rich, 1982, LANC), on granite in Great Hall Gill, Wasdale (14.02, CCH, 1982, LANC) and in a gully below Great End (22.08, GH, 1988, LANC).

This is a distinctive hawkweed with its deeply toothed leaves, the narrow teeth tapering to setaceous points. The peduncles have a dense covering of stellate hairs and long, dark glandular hairs.

The discovery of this plant in Cumbria helps to fill the gap between its Welsh localities and isolated stations in Lancashire, Dumfriesshire and Kirkcud-brightshire.

Alt. limit: *c.*700 m, Rampsgill head (44.12)

3   WC

47. **H. sommerfeltii** Lindeb.

Very rare, only seen during the Survey at Black Sail, Ennerdale (18.10, CCH, 1982, herb. C.C.H.), Brown Cove, Helvellyn (32.14, GH, 1989, LANC) and Ashstead Fell, Borrow Beck (54.02, GH, 1979, LANC). There are earlier records from Dove Crag (3.1, H.W. Pugsley, 1926, BM), the first Westmorland record, above the Glenridding lead mine (3.1, J.N. Mills & C. West, 1968, MNE), the first Cumberland record, and Coniston Old Man (2.9, R.S. Adamson, 1920, BM). A collection from Crowdundle Beck, Cross Fell (6.3, M. Rawes & C. West, 1967, MNE) could well be immature *Hieracium rhomboides*, and one from Fleetwith Pike, Honister (2.1, C. West, 1954, MNE) is

H. saxorum; that from Coniston Old Man is probably the population referred to under this last species.

Although the leaves are almost entire and blotched, like those of *H. saxorum*, they are noticeably longer, more oblong and obtuse, and have a tapering base; the peduncles and involucral bracts are distinctly less glandular-hairy.

These are the only records of this hawkweed south of the Scottish Highlands apart from isolated localitites in Dumfriesshire and Wigtownshire.

Alt. limit: *c*.730 m, Brown Cove, Helvellyn (32.14)

3   W[(F)]C

[H. carneddorum Pugsley
The only record outside Snowdonia is from Fleetwith Pike, Honister (2.1, C. West, 1954, CGE, MNE). It is erroneously shown in the *Critical Atlas* for the Ennerdale square (1.1).

The material was originally identified as *Hieracium sommerfelti*. However, the involucral bracts match those of *H. saxorum* and the leaves lack the long dense hairs on the underside and margins which are a feature of the Welsh plant. The specimen is probably only a taller, shaded form of *H. saxorum*, which is very common there.

(C)]

48. **H. decolor** W.R. Linton) Ley

Scattered on the limestone scars and pavements around the upper Eden valley with outlying records from the limestone at Greystoke (40.30, GH, 1991), Dentdale (70.82, GH, 1993), limestone pavement at Hutton Roof Crags (54.76s, GH, 1979) and coastal limestone at Arnside Knott (44.76s, MB, 1980, all LANC) and Meathop (42.78, GH & J.M. Mullin, 1987, BM). The first record for Westmorland was from Scout Scar (4.9, G.E. Martindale, 1891, KDL) and for Cumberland from Melmerby Low Scar (62.38, GH, 1989, LANC). The *Critical Atlas* record for the Ennerdale square (1.1) is an error for Crummack Dale, Yorkshire; those for the Ambleside square (3.0, Snaka Rocks, A. Ley, 1908,

CGE *ex hort*.) and Helvellyn square (3.1, Dollywaggon Pike, A. Ley, 1906, CGE) refer to specimens of *Hieracium leyi*.

Identification of this variable hawkweed is not helped by the absence of an adequate description based on the species' current interpretation. There is clearly a close general similarity to *H. leyi* and *H. stenopholidium* and it may be a convenient oversimplification that, as interpreted here, *H. decolor* is only on the limestone and the others mainly in the Lake District. The involucral bracts are very similar although those of *H. decolor* have conspicuously more stellate hairs, particularly along the margin. The styles vary from yellow to discoloured. The leaves of *H. decolor* are usually glaucous and vary from broadly ovate-oblong with a somewhat truncate base, resembling *H. auratiflorum*, to elliptic, and from entire to finely or coarsely denticulate.

This is a plant of the Pennine limestone, from the Peak District to Cumbria.

Alt.limit: 660 m, Crowdundle Beck, Cross Fell (68.32)

Map 892, Plate 94   27   WC

Plate 94. Leaves of typical *Hieracium decolor* (left) and the broad-based variant (right).

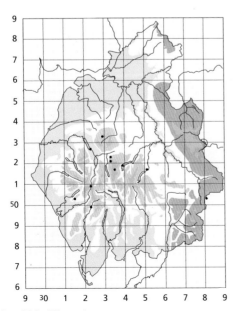

Map 891. *Hieracium argenteum*

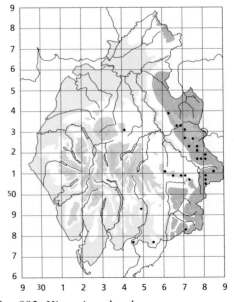

Map 892. *Hieracium decolor*

49. H. leyi F. Hanb.

Fairly frequent in the central and eastern Lake District and extending east into the Howgill Fells. The only additional squares are from above Wast Water (1.0, H.T. Mennell, undated, BM), probably the first Cumberland record, Scout Scar (4.9, G.E. Martindale, 1891, KDL), which is based on a specimen of *Hieracium decolor*, and Cross Fell (6.3, L.E. Hall, 1925, BM; A.J. Richards, 1973, NCE). This last material is poor and it seems unlikely that the species occurs in the Cumbrian Pennines. There are no reliable records from limestone. The first record for Furness was from Coniston Old Man (2.9, A. Ley, 1891, BM, CGE) and for Westmorland from Snaka Rocks, Ambleside and Dollywaggon Pike (3.0, 3.1, A. Ley, 1905, BM, CGE).

For much of the Survey this species was 'lumped' with *Hieracium stenopholidium* for, in the absence of a description of the latter based on British material, no convincing differences between the two were evident. However, P.D. Sell (*verb. comm.*) has latterly drawn attention to the presence of stellate hairs between the veins on the underside of the leaves of *H. stenophollidium* and the absence of these in *H. leyi*. These hairs may be somewhat sparse but they are used here to separate the two taxa. The leaves of *H. leyi* are usually somewhat oblong and entire or with a few teeth at the base but plants with more strongly toothed, ovate-lanceolate leaves typical of *H. stenopholidium* also occur. Both have usually yellow styles and involucral bracts with simple hairs and numerous short glandular ones but the bracts of *H. leyi* are longer and more pointed; those of *H. decolor* have a more distinct white margin of stellate hairs.

The species appears to be largely restricted to the Lake District and North Wales.
Alt. limit: 600 m, Coniston Old Man (26.96) but no doubt higher on Helvellyn.
Map 893   22   WFCY

50. H. lasiophyllum Koch

The only Survey collections which agree with the descrip-

tion of this species in having long setose hairs not only on the leaf margin but also extending onto the upper surface are from the Wast Water Screes (14.04, CCH, 1977, LANC) and Barf, Keswick (20.26, CCH, 1977, LANC) and there are earlier collections from Glaramara (2.1, H.E. Fox, 1888, BM) and Whiteside, Lorton (1.2, D. Oliver, 1854, BM). With the above exceptions, and possibly of a collection from Piers Gill, Scafell Pike (2.0, J.N. Mills, 1965, MANCH), which has not been re-examined, all the *Critical Atlas* records are best placed under *Hieracium stenopholidium* (P.D. Sell *in litt.* to DAR 1958). An early but undated collection from Gilsland (6.6, Anon., HAMU) could be from v.c.67 or 70.

The distribution of this rather handsome and local montane species extends from South Wales to the central Scottish Highlands.
Alt. limit: 395 m, Wast Water Screes (14.04
            2   C

51. H. brigantum (F. Hanb.) Roffey

A large-headed hawkweed known only from the limestone above Brough (7.1, C.E. Salmon, 1892, BM) and from Scout Scar (4.9, C. West, 1947). Both localities have been very well botanised and it is surprising that these are the only collections. The *Critical Atlas* record for the Whitehaven square (9.1) is based on a misidentification.

This species has a general resemblance to *Hieracium anglicum* but the styles are pure yellow.

The two Cumbrian records are the only ones outside its headquarters on the limestone of the Ingleborough area.
                            (W)

[**H. schmidtii** Tausch
The only *Critical Atlas* record is based on a specimen from rocks above Low Borrow Bridge, Tebay (60.98, J.N. Frankland, 1935, LIV). This hawkweed is otherwise restricted to the limestone of the Cheddar Gorge and its occurrence in Cumbria seems highly unlikely. No alternative determination of the population has yet been made.
                           W)]

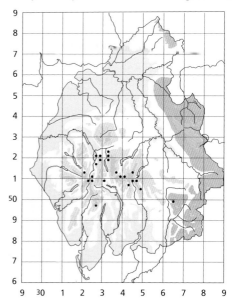

Map 893. *Hieracium leyi*

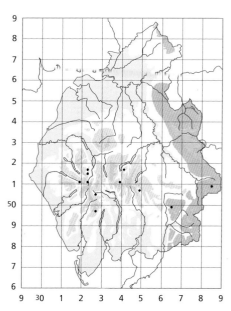

Map 894. *Hieracium stenopholidium*

52. **H. hypochaeroides** Gibson

Very rare, seen only on limestone grassland along Whitbarrow and Scout Scars (42.88s, 48.90, GH, 1988, LANC) and above Kirkby Stephen (80.04, 1988, 80.06, GH, 1985, LANC). It has also been recorded from Helbeck Wood, Brough (7.1, H.K. Airy Shaw, 1938, K). The first record was from Scout Scar (W.R. Linton, 1871, BM). The *Critical Atlas* record from Wild Boar Fell, Cross Fell (6.3) is of *Hieracium maculosum* and that for Haystacks (1.1, C.W. Muirhead, 1948, CGE, PLYP) in the Lake District is an error for *H. leyi*.

Distinctive features are the very strongly spotted leaves, often strongly toothed towards the base, and large heads, the involucral bracts having many white simple hairs and a few short glandular ones.

This is a local hawkweed of the Yorkshire and Cumbrian limestone with a few isolated records from Wales and Ireland.
Alt. limit: 440 m, Rigg Beck, Kirkby Stephen (80.04)

4 W

53. **H. saxorum** (F. Hanb.) Sell & C. West

Largely restricted to crags in the central Lake District, where it is known from the Honister area (20.14, 22.14) Glaramara (24.10), Middlesteads Gill and Dunmail Raise (28.16, 32.10), Wanthwaite Crags (32.22), the head of Glencoynedale (34.18) and Fairfield (36.10), also on either side of the Lune gorge (60.98, J.N. Frankland, 1935, LIV; 64.98, 1984, all GH, LANC). There are also early records from Scandale, Ambleside (3.0, A. Ley, 1905, CGE), the first Westmorland record, Dove Crag (36.10, C. West, 1954, MNE) and Whelter Crags and Riggindale, Haweswater (4.1, A. Ley, 1910, CGE). C.W. Muirhead's 1954 collection from Yew Crag, Honister (2.1, E) is the first Cumberland record. The *Critical Atlas* record for the Arnside square (4.7s) is based on a misidentified specimen of *Hieracium decolor*.

It is less robust than *H. sommerfeltii*. The leaves are small, oblong-lanceolate and with a subtruncate base; the peduncles and involucral bracts have usually many short glandular and stellate hairs as well as simple ones. A population on the west side of Coniston Old Man (26.96, M. Richards, 1988, LANC), although similar to *H. saxorum*, has involucral bracts with few simple hairs and distinctly hairy-tipped ligules.

This is a montane species and most frequent in the eastern Scottish Highlands.
Alt. limit: 600 m, Honister Crag (20.14, 22.14) and the head of Glencoynedale (34.18)

10 WC

[H. subplanifolium Pugsley
Specimens collected from the limestone quarry at Meathop (42.78, J.M. Mullin, 1987, BM; LANC, *Watsonia* (1988) 17:194) and determined by P.D. Sell as *Hieracium subplanifolium* are the same as earlier and recent collections from this site identified as *H. decolor*.

W]

54. **H. dicella** Sell & C. West

Seen only on the limestone cliffs of Whitbarrow Scar (44.84, GH, LANC), Cunswick Scar (48.92, E.F. Greenwood, 1970, LIV) and Black Door, Ardale (66.34, GH, 1987, LANC) and on whinstone in High Cup Gill (74.24, 74.26, RWMC & GH, 1988, LANC), where it was first collected in 1894 by J.G. Baker (BM). There are also doubtful records from Dale Head crags, Newlands (2.1, H.G.P.[roctor?], 1957, untraced) and Humphrey Head (3.7s, P.M. Newby, 1951, untraced).

Hieracium dicella is often confused with *H. decolor* but the leaves are more broadly oblong-ovate and have regular, prominent, short, setaceous teeth and the involucral bracts are very narrowly lanceolate, with fewer stellate hairs.

It is locally frequent in the the Peak District, the Yorkshire Dales and Teesdale, and is widely scattered in Scotland, and on the limestone of North and South Wales.
Alt. limit: 550 m, High Cup Nick (74.24, 74.26)

3 W[FC]

55. **H. stenopholidium** (Dahlst.) Omang

Occasional in the Lake District and, like *Hieracium leyi*, with outlying sites in the Howgill Fells. The two species can look very similar and the main points of difference are given under *H. leyi*. All the *Critical Atlas* records are post-1950. It seems likely that the first records will be found amongst early material of *H. leyi* or *H. lasiophyllum*.

Practically all the British records are from the Lake District.
Alt. limit: 600 m, Coniston Old Man (26.96)
Map 894 11 WFCY

Section **Cerinthoidea**

Stems scape-like; leaves cuneate-based, hairy above; heads large, densely hairy; styles usually dark.

[H. iricum Fries
Recorded only from Dollywaggon Pike (3.1, A. Ley, 1905, CGE), the Maize Beck, Teesdale (7.2, E.S. Marshall, 1883, CGE), Cross Fell (7.3, A.J. Wilmott (Pugsley 1948) and Crowdundle Beck, Cross Fell (6.3, D. Welch (Rawes 1981)). P.D. Sell now considers that neither of the herbarium specimens is typical and that all the records are best regarded as suspect, although there are authentic records from the Co. Durham part of upper Teesdale.

(WC)]

56. **H. anglicum** Fries

Fairly frequent on the Pennine limestone scarps and on the limestone between Asby and Orton but with rather few records from the Lake District and a single site in the Howgill Fells (64.98). However it has probably been under-recorded. The *Critical Atlas* records for the Kirkby Thore (6.2) and Dentdale (7.8) squares have not been traced; that for the Lorton square (1.2) may refer to a record from Haystacks (1.1, C.W. Muirhead, 1954, CLE). The first records for Westmorland are from Dove

Crag, Fairfield and Helvellyn (3.1, H.E. Fox, 1890, BM) and for Cumberland from Ennerdale and Piers Gill, Scafell Pike (1.1, 2.0, H.E. Fox, 1888, BM).

This species has often been confused with *Hieracium ampliatum*, especially in the Lake District. Both species may occur together, for example on Harter Crag (44.10) in the Lake District and in Swindale, above Hilton (76.20) in the Pennines.

This is a handsome and conspicuous large-flowered plant. It differs from *H. ampliatum* in having fewer heads, more hairy and often glaucous leaves, sessile stem leaves and involucral bracts with more and longer simple hairs and few if any stellate hairs.

This species is widespread in the Scottish High-lands and extends south to the Craven area of York-shire, with an outlying site in Montgomeryshire.
Alt. limit: 660 m, Crowdundle Beck, Cross Fell (68.32)
Map 895   34   WC(Y)

57. **H. ampliatum** (W.R. Linton) Ley

Frequent in the central and eastern Lake District where it is the commonest and most conspicuous hawkweed. It also extends across to the limestone of Scout Scar (48.92) and Whitbarrow (42.88), the Howgill Fells and the limestone scars of the Pennines where it is probably commoner than *Hieracium anglicum*. There are also records from Kirk Fell (*cf.*1.1, Anon., 1894, BM), the earliest Cumberland record, Walla Crag, Keswick (2.2, H.W. Pugsley, 1903, BM), Crowdundle Beck, Cross Fell (6.3, A. Eddy, 1960, herb. Eng. Nat.) and Moor House (7.3, D. Welch, 1966, herb. Eng. Nat.), and untraced *Critical Atlas* ones for the Kendal (5.9) and Melkinthorpe (5.2) squares. The first record for Furness was from Coniston Old Man (2.9, H.T. Mennell, 1891, BM) and for Westmorland from Long-sleddale (4.0, J.A. Martindale, 1888, KDL).

This is a very variable species. It is best distinguished from *H. anglicum* by the usually thinner,

less glaucous and less hairy leaves, the often distinctly petiolate stem leaf and the conspicuous stellate hairs on the involucral bracts, especially along their margins. Stylose plants, with poorly developed ligules, are not infrequent and the leaves are occasionally spotted.

Its British distribution extends from the Scottish Highlands to the Craven area of Yorkshire.
Alt. limit: 810 m, Dollywaggon Pike (34.12)
Map 896   61   WFCY

58. **H. flocculosum** Backh.

Known only from the limestone scars above Brough (78.16, LH, 1971; 82.18, LH, 1972) and Barras (80.10, GH, 1997) in the Pennines, and from the cliffs above Blea Water, High Street, (44.10, F.J. Roberts, 1980, all LANC) in the Lake District.

These plants all have a moderate amount of stellate hairs on the upper surface of the leaves and must presumably be referred to this species. In other respects it is very similar to *Hieracium anglicum* and has therefore probably been under-recorded.

There are the only records in Britain south of Fife.
Alt. limit: *c.*650 m, above Blea Water (34.12)
                         4   W

Section **Subalpina**

Stem scape-like; leaves with a few microglands chiefly on the margins; heads usually more than one.

[H. cumbriense F. Hanb.
This supposed Cumbrian endemic hawkweed is shown in the *Critical Atlas* as occurring in five squares. The sources for these, all determined by P.D. Sell and C. West, are: above Buttermere (1.1, P.H. Oswald, 1954, herb. J. Raven), Piers Gill, Scafell Pike (2.0, H.E. Fox, 1888, CGE), Ruddy Gill, Great End (2.0, A. Wilson, 1896, LIV), Dollywaggon Pike (3.1, H.E. Fox, 1890, BM; A. Ley, 1905, 1910, CGE; J. Bevan & P.D. Sell, 1980, CGE), Dove Crag (3.1, H.E. Fox, 1890, CGE) and, from the Pennines, Crowdundle Beck,

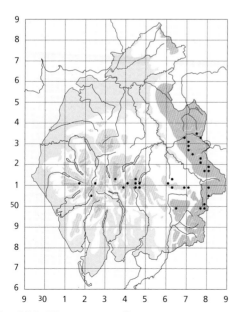

Map 895. *Hieracium anglicum*

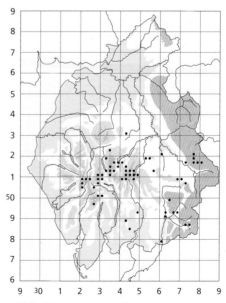

Map 896. *Hieracium ampliatum*

Plate 95. *Hieracium itunense*

Plate 96. *Hieracium subgracilentipes*

Cross Fell (6.3, herb. Eng. Nat.) and Rough Sike, Moorhouse (7.3, Rawes (1981)). Pugsley's description refers to the the glabrous-tipped ligules and rather sparse stellate hairs on the peduncles. Examination of the above specimens, except the first and last suggests that they can be referred to *Hieracium ampliatum,* since all have pilose ligules. Recent material in LANC from the same gully on Dollywaggon Pike as Bevan and Sell's specimen is typical *H. ampliatum.* Ley's 1910 collection is a mixture of *H. ampliatum* and *cf. H. leyi.* It seems safe to conclude that *H. cumbriense* should be deleted from the British list.

W(C)]

59. H. clovense E.F. Linton

Occasional and restricted to gullies and rock ledges in the central and eastern Lake District with an outlying station in the Shap Fells (54.02, GH, 1979, LANC). The only additional records are from Pillar (16.12, DAR, 1954, CLE, wrongly shown in the *Critical Atlas* as 0.1), the first Cumberland record, Keppel Cove, Helvellyn (3.1, J.E. Raven, 1954, CGE), Catstycam (3.1, J.N. Mills, 1968, MANCH) and Dove Crag (3.1, H.E. Fox, 1890, BM; C. West, 1954, MNE). The first Westmorland record was from Striding Edge (3.1, J. Backhouse, 1847, BM). These are the only records south of the central Scottish Highlands.

This hawkweed is characterised by large, ovate, sinuously toothed and spotted leaves and dark involucral bracts, the only hairs being dense, long, black and glandular.

Alt. limit: 790 m, Nethermost Cove, Helvellyn (34.14)

Map 897   12   WC

60. H. vennicontium Pugsley

Seen only by a beck in the Uldale Fells north of Skiddaw (28.32, GH, CGE, LANC), but previously collected by C. West from by the Ardale Beck (6.3, 1952, CGE, MNE) and Crowdundle Beck, Cross Fell (6.3, 1967, MNE).

The Cumbrian plants are tall, with long-petiolate and strongly toothed leaves and several medium-sized heads. The involucral bracts have a mixture of predominantly black simple hairs and shorter glandular hairs which extend down the peduncles; the styles are livid.

These records and two in Yorkshire are the only ones south of the central Scottish Highlands.

Alt.: 470 m, Uldale Fells (28.32)

1   C

Section Alpina

Stems scape-like; leaves hairy, microglands present; 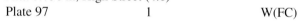 heads usually single, very hairy.

61. H. subgracilentipes (Zahn) Roffey

A Lake District endemic known from several mildly basic crags and gullies in the Helvellyn (3.1) and High Street (4.1) ranges and with an outlying station to the west on Pillar (16.12, GH, 1993, LANC). Four of the Survey records are from new sites. It was first collected by J. Backhouse in 1854 on Helvellyn (3.1, BM) and from a gully on Glaramara (2.0, D. Oliver, 1854, BM), but it has not been seen at the latter site since.

This distinctive hawkweed has an erect, single-headed stem, long, narrow and slightly toothed leaves and dark involucral bracts with dense, long, black simple hairs and short black glandular ones.

Alt. limit: 810 m, Brown Cove, Helvellyn (32.14)

Map 898, Plate 96   7   WC

62. H. holosericeum J. Backh.

Seen only on the High Street range (4.1, GH, 1987, LANC) where it was originally discovered by J. Backhouse in 1869 (BM). He first found it in 1847 on the summit of Glaramara (2.1, 1847, BM, CGE), but it was never refound, and the same year on Striding Edge, Helvellyn (3.1), where it was apparently last seen by G.A.K. Hervey in 1959. Attempts to refind it at the following sites have also failed: Boat How, Kirk Fell (1.1, J.E. Raven, 1953, CGE), Coniston Old Man (2.9, DAR, 1959, CLE) and the Langdale Pikes (2.0, H.W. Pugsley, 1926, BM).

Unlike most rare montane hawkweeds, this species is characteristic of rather acid and otherwise unproductive crags and rock ledges.

It is a most beautiful plant, with its dense rosette of leaves, usually short stems and single heads covered in long, white silky hairs. It is known south of Scotland only from Snowdonia and the Lake District.

Alt.: *c*.750 m, High Street (4.1)

Plate 97   1   W(FC)

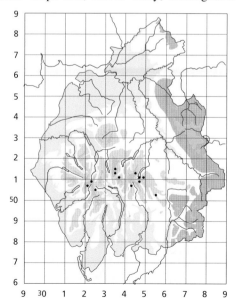

Map 897. *Hieracium clovense*

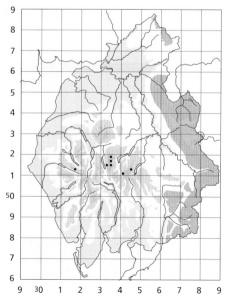

Map 898. *Hieracium subgracilentipes*

Filago vulgaris Lam.   Cudweed
Rare and decreasing, now known only from a few sites
along the Eden valley and near Anthorn (16.58, 20.58).
It occurs in open, sandy ground and by field edges, on
sandstone outcrops and, at Coombs Wood, Lazonby
(50.44), on a gravel forestry track with *Gnaphalium
sylvaticum* and *Filago minima*, with which it is
frequently confused. In the absence of any herbarium
material from the south of the county, it seems best to
regard the *Atlas* records from there and the few Furness
ones given by Petty (1896 p.204) with some scepticism.
Alt. limit: 230 m, Lazonby Fell (50.38)
Map 899   13   W[(F)]C

F. minima (Smith) Pers.   Small cudweed
Occasional in the Eden valley, on the Solway coast and
down the west coast from Drigg (04.98) to the Duddon
estuary. It occurs in open, sandy ground, in sand and
gravel pits, on sand-dunes, forestry tracks and on sandstone outcrops.
Alt. limit: 230 m, Lazonby Fell (50.38)
Map 900   43   WFC

Antennaria dioica (L.) Gaertner Mountain everlasting
Rather uncommon in the uplands but locally frequent
on the Asby - Orton limestone; very rare in the lowlands. It is a plant of short base-rich or calcareous
grassland and in the Lake District of rock faces and
ledges. One of the most interesting sites is at 825 m on
the north-west side of Raise (34.16) where it grows in
Festuca ovina turf with *Salix herbacea*, *Persicaria
vivipara*, *Carex bigelowii* and *Selaginella selaginoides*.

    Its decline in recent decades, particularly in the
lowlands, is a consequence of grassland 'improvement'.
Even during the period of the Survey it has disappeared
from an S.S.S.I. at Far Arnside (44.76s). It is very
susceptible to grazing and in the uplands flowering is
rare and torn-off shoots are all too evident. It has gone
from most of the sites listed by Wilson and despite
frequent searches around Cross Fell and in the Lake
District it could not be refound at 13 of the sites where
it was seen by DAR during the 1950s. This decline has
been repeated nationally with the result that present
maps give the false impression that this is essentially a
montane species.
Alt. limit: 825 m, Raise, Helvellyn (34.16)
Map 901   101   WFCY

*****Anaphalis margaritacea** (L.) Benth.
                    Pearly everlasting
The only Survey records of this attractive garden escape
are from waste ground by the old paper mill by
Cavendish Dock, Barrow-in-Furness (20.68s, 22.68s,
B. Fisher, 1977, LANC), on a bridge at Watermillock
(44.22, MC, 1978-1985), by a forestry track in Brund-
holme Woods, Keswick (28.24, GH, 1986, LANC), in
dune-grassland at Mawbray Bank on the west coast
(08.48, REG, 1986, LANC) and by the new pipeline

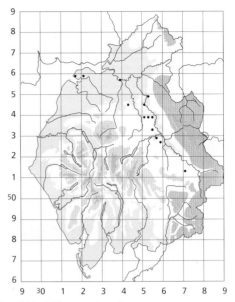

Map 899. *Filago vulgaris*

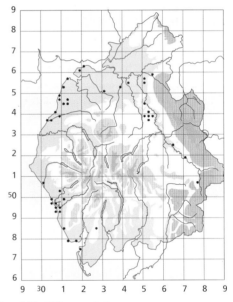

Map 900. *Filago minima*

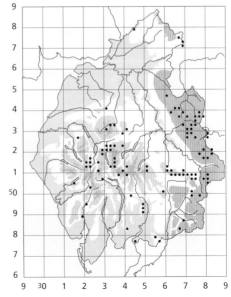

Map 901. *Antennaria dioica*

above Killington (58.90, T.R. Harwood, 1996, LANC).

It is recorded by Hodgson from Egremont (0.1) and between Dearham and Bullgill stations (0.3) and by Wilson from between Bowness and Ambleside (3.9/4.9). Wilson also collected it in Westmorland from near Sedbergh (6.9, undated, YRK).

Alt. limit: 155 m, Watermillock (44.22)

6		WFC(W/Y)

Gnaphalium sylvaticum L.   Heath cudweed

Not infrequent in the middle and lower Eden valley, very rare elsewhere. Most of the recent records are from forestry tracks where it grows in peaty sand and gravel, also on heavy wet clay. There are also records from disturbed ground over a pipeline and from the sandstone outcrop at Common Holme Bridge, Cliburn (56.24). The only Survey records from the south are from a woodland track near Beetham (48.78s, CFS & JS) in 1971 and 1974, and a farm track near Ulpha in the Duddon valley (18.92, 1996, AAD, LANC).

This species has declined very considerably, the *Atlas* showing 28 post-1930 squares and the Survey map only twelve. This decline has occurred chiefly down the west coast and in Furness, although it must be said that some at least of the *Atlas* records for Furness may, like those of *Filago vulgaris*, be the result of mis-identification.

Alt. limit: 400 m, Kershope Forest (56.84n)

Map 902   27   W(F)C(W/Y)

G. uliginosum L.   Marsh cudweed

Fairly common in the lowlands, occurring in damp. open, often trodden habitats such as rutted tracks, field entrances, and around ephemeral pools and the drawdown zone of reservoirs. It is also frequent as a weed of damp arable fields.

Alt. limit: 425 m, Stainmore Summit (88.12)

Map 903   772   WFCY

*Inula helenium L.   Elecampane

The only Survey records for this striking south-east European species are from a roadside verge at Cark (36.76s, REG, 1973), by a field wall at Saltcoats (06.96, MMG) where it is long established, on a railwaybank at Workington (98.26, AD, 1990) and in a lane at Brigflatts, Sedbergh (64.90, TW, 1989; LANC).

It was recorded by Hodgson from Coulderton (9.0), Egremont and Arlecdon (0.1) and Loweswater and Lorton (1.2); Baker gives additional records from the Dalton-in-Furness area (2.7) and near Cunswick (4.9), and Wilson one from Lyth Fell (4.8s, G.E. Martindale, 1889, KDL).

There are four post-1930 *Atlas* records, including one from Culgaith (5.2/6.2, G.A.K. Hervey, 1954) and there is a further one from Glenridding (3.1, D. Stewart, 1947, CLE).

4		(W)FCY

I. conyzae (Griess.) Meikle   Ploughman's spikenard

A southern calcicole which reaches its northern limit in Britain in Cumbria where it occurs quite commonly on the limestone screes and scars around Morecambe Bay but with an outlying site on the Pennines above Brough (78.16). It occurs also on waste ground by the old mine workings around Millom (18.78), on the dunes at Sandscale (18.74), on the opposite side of the Duddon estuary, and around the gravel workings at the South End of Walney Island (22.62s).

The first Cumberland record was from Haverigg and Hodbarrow (16.78, N. Nicholson, 1955, *Watsonia* (1979) 12:357).

Alt. limit: *c.*300 m, above Helbeck Wood, Brough (78.16), probably the highest site in Britain.

Map 904   66   LWFC

*I. hookeri C.B. Clarke

A large colony of this garden plant was found in limestone woodland on the north-east side of Whitbarrow in

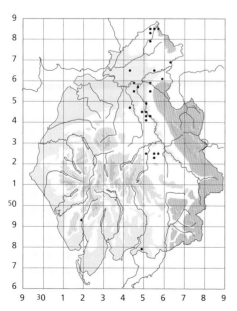

Map 902. *Gnaphalium sylvaticum*

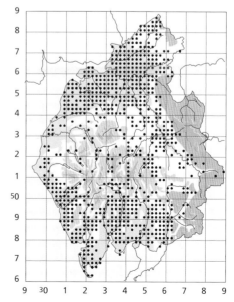

Map 903. *Gnaphalium uliginosum*

1995 (44.88s, CEW, LANC, det. E.J. Clement).

1   W

Pulicaria dysenterica (L.) Bernh.   Fleabane
Occasional on the Furness peninsula, rare elsewhere
although locally frequent, as amongst the gorse and
bramble scrub on Brough Sowerby common (80.12,
GH, 1983, LANC). Most of the sites are amongst tall
grass in wet meadows, on roadside verges and on waste
sites, for example Salthouse Pool, Barrow-in-Furness
(18.68, 20.68s), and between the A590 carriageways at
Arrad Foot, Ulverston (30.80).

This is a markedly southern species in Britain
which reaches its northern limit on the west coast in
south-west Scotland.
Alt. limit: 215 m, Brough Sowerby (80.12)
Map 905   26   WFC

*Grindelia squarrosa (Pursh) Dunel
A rare North American casual collected at Silloth in 1889
(1.5, J. Leitch, CLE).

(C)

Solidago virgaurea L.   Goldenrod
Common along the west coast, in the Lake District and
in the south-east, scattered in the Pennines and Eden
valley and sparse on the Solway Plain and the Furness
peninsula. In the lowlands it occurs in rough grassland, usually by hedgerows, on railwaybanks, woodland margins and by riversides. In the uplands it is very
common on rock ledges and ungrazed gills. It tends to
avoid the limestone.

Wilson mentions early-flowering dwarf plants
on the slate rocks by the River Lune near Sedbergh and
on rocks at high altitudes refering them to var.
cambrica Hudson. Cultivated plants from the Lune and
Rawthey (60.02, 62.92, 62.94) have flowering stems up
to 30 cm and branching about the middle. They are in
flower by early June, three months before the tall plants
of the adjacent riverside woodland. Perhaps, like the
Luneside *Galium boreale*, they have been derived from
montane populations but the latter seem generally to be
actually taller. In his note on *Viola canina*, Corner
(1989) considers that the isolated lowland inland
populations are a relict of a formerly widespread race of
open habitats. This could also be true of the Luneside
Solidago which grows with the violet. The only known
population of dwarf plants is on Crag Hill, east of
Grasmoor (18.20, G.A.K. Hervey, 1950). These remain
dwarf in cultivation with flowering stems to 5 cm and
they are similar to the dwarf plants found on many
Scottish mountains.
Alt. limit: 825 m, Helvellyn (34.14)
Map 906, Plate 98   615   WFCY

*S. canadensis L. and ***S. gigantea** Aiton
For most of the Survey the records of these two North
American garden plants were pooled. The more recent
records, several of which have been checked by D.H.

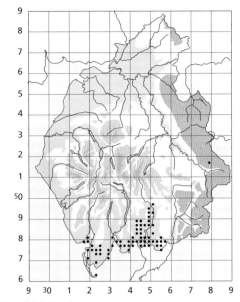

Map 904. *Inula conyzae*

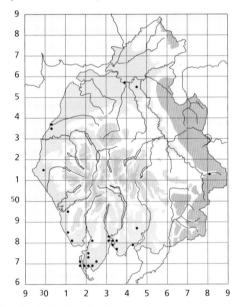

Map 905. *Pulicaria dysenterica*

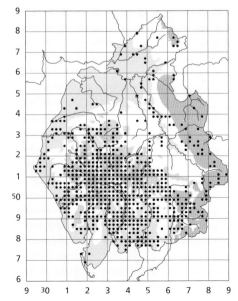

Map 906. *Solidago virgaurea*

a) by the River Lune, Sedbergh, early June

c) montane plant, below Great End

b) cultivated plant from the River Lune

d) with *Salix herbacea*, summit of Crag Hill

Plate 98. *Solidago virgaurea*

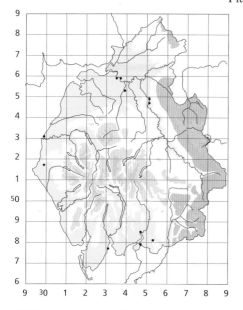

Map 907. *Solidago canadensis*

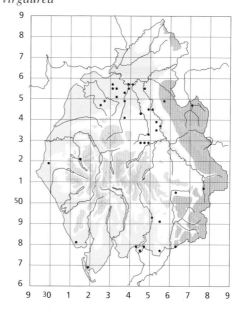

Map 908. *Solidago gigantea*

Kent, suggest that *Solidago gigantea* is by far the commoner. These garden escapes are a frequent sight by roadsides and on railwaysides and waste ground, usually near villages and urban areas. Although these species appear to be long established, there seem to be no pre-Survey records.

Solidago canadensis is best distinguished by its broader panicle, with the outer branches often deflexed, and by the pubescent stems.

The single species maps are based only on authenticated specimens and represent about a third of the total records.

S. canadensis

Map 907   11   WFC

S. gigantea

Alt. limit: *c*.300 m, Alston (70.46)

Map 908   38   WFC

*S. graminifolia (L.) Salisb.

A North American garden plant found in 1983 growing on the rocky south side of the River Brathay at Clappersgate (36.02, GH, LANC, *Watsonia* (1984) 15:137).

1   F

*Aster - naturalised taxa   Michaelmas daisies

These are frequent and conspicuous garden escapes of North American or garden origin. They occur in a wide variety of habitats: in the tall herb vegetation fringing lakes, tarns and rivers, by roadsides and on damp waste ground in towns and villages.

They are often difficult to identify as a result of hybridisation and subsequent backcrossing. Consequently they are likely to have been substantially under-recorded. According to Baker, Michaelmas daisies were well established by Ullswater, Derwent Water and Windermere by the 1880s, but in the absence of herbarium specimens it is difficult to

establish any first records prior to the Survey. All the Survey records cited below or mapped have been determined by P.F. Yeo.

*A. novi-angliae L.

A striking plant with hairy stems and purple ligules recorded in 1970 on waste ground in Ambleside (36.04, R. Young, AMDE).

'   1   W

*A. laevis L.

Recorded in 1988 by Ormsgill reservoir, Barrow-in-Furness (18.70, PB, LANC, *Watsonia* (1995) 20:432).

Its leaves are glaucous above and distinctly clasping, and the outer involucral bracts are less than half as long as the appressed inner ones.

1   F

*A. laevis x **A. novi-belgii**

A. x versicolor Willd.

Very rare, known only from nine widely scattered tetrads. The first definite records are: for Cumberland from reedswamp at the north end of Bassenthwaite Lake (20.32, REG, 1984, LANC) and for Furness from waste ground at the Barrow-in-Furness docks (20.68s, PB, 1982, LANC).

This hybrid has clasping leaves <5 times as long as wide, large capitula with the outer involucral bracts at least half as long as the inner and bluish ray florets.

Alt. limit: 345 m, Garrigill (74.40)

Map 909   9   FC

*A. novi-belgii L.

Occasional. The first definite record for Cumberland was from Newtown, Carlisle (38.56, REG, 1978, herb. R.E.G.), and for Westmorland and Furness from by the river at Brathay (36.02, R. Young, 1959, 1970, AMDE, det. F. Davis).

This species has been much confused with both

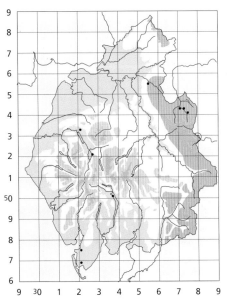

Map 909. *Aster laevis* x *A. novi-belgii*

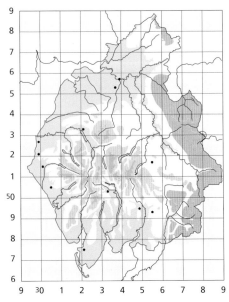

Map 910. *Aster novi-belgii*

the preceding and the following hybrids. It has leaves 4-10 times as long as wide, usually blue ligules, and the outer involucral bracts are widest at or above the middle and only loosely appressed in their upper half.
Alt. limit: 275 m, Shap (54.16)
Map 910   12   WFC

*A. novi-belgii x **A. lanceolatus**
A. x salignus Willd.
Apparently the commonest taxon being particularly frequent in the Ambleside area (3.0) and along the shores of Windermere.

The first certain record for Cumberland was from the shore at Ullswater (40.20, MC, 1975, LANC), for Westmorland from Grasmere (32.06, RS, 1980, herb. R.S.) and for Furness from Low Wray (36.00, K. Dalby, 1949, FBA).

The ligules are usually white, as in *Aster lanceolatus*, but the leaves are wider. It differs from *A. novi-belgii* chiefly in the involucral bracts which are widest below the middle, appressed in the upper half, somewhat lax and tapering to a rather long point.
Alt. limit: 305 m, Hardendale Fell (56.12)
Map 911   30   WFC

*A. lanceolatus Willd.
A North American garden escape seen on the Cumberland side of the river at Duddon Bridge (18.88, GH, 1975, LANC, *Watsonia* (1988) 17:193), by the River Bela at Milnthorpe (48.80s, CEW, 1989, LANC, both det. P.F. Yeo, *Watsonia* (1995) 20:433), roadsides at Endmoor and Lupton (52.84s, 1988; 54.80, 1994, both CEW, LANC) and a lakeside swamp at Portinscale (24.22, A. Cannell, *c*.1991, LANC).

It has a more open panicle than *Aster* x *salignus*, with smaller capitula (involucral bracts <5.5 mm), entirely white ray florets, and narrowly lanceolate (usually <1 cm wide), non-clasping upper leaves.
                                5   WC

A. tripolium L.   Sea aster
Fairly frequent around the coast growing in the upper parts of salt-marshes and on the banks of tidal creeks and estuaries, also in brackish ditches and by pools behind the sea-walls. Although sensitive to grazing it appears to be more resistant than, for example, *Halimione* and species of *Limonium*. North of Mawbray (08.48) it occurs on the upper part of the sandy shore and extends on to the fore-dunes.

There are no records of the rayless var. *discoideus* Reichenb. f.
Map 912   102   WFC

A. linosyris (L.) Bernh.   Goldilocks aster
In the Postscript to his Flora, Baker comments that C.C. Babington showed him a specimen collected by W. Nixon "in Furness near Hampsfield". No date is given but this is presumably the first record and no doubt refers to the well-known site on the limestone cliffs of Humphrey Head (38.74s) where one or two plants still persist. This is the northernmost British locality. The other sites, all on limestone, are on the coasts of Devon, Somerset and North and South Wales.
                    R 1   F

Erigeron acer L.   Blue fleabane
A rather uncommon coastal plant of limestone rocks and quarries, open sandy places and urban waste sites in the south, although it occurred briefly on a roadside verge near Hartley, Kirkby Stephen (78.08, JCF, 1985). These are the most northerly extant sites in western Britain.

Hodgson's record from The Green, Dalston (3.4) prior to 1887 is probably the basis of the *Atlas* record for the Carlisle square (3.5). The first Furness record dates from 1913 when it was seen at Sandscale (18.74, D. Lumb, *BEC Rpt* (1914) 3:387).
Alt. limit: 170 m, Hartley, Kirkby Stephen (78.08)
Map 913   27   WFC

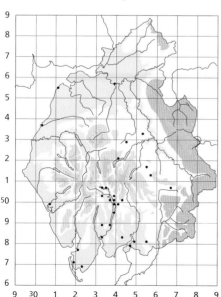

Map 911. *Aster novi-belgii* x *A. lanceolatus*

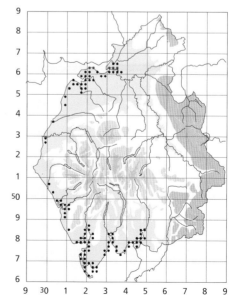

Map 912. *Aster tripolium*

*Conyza canadensis (L.) Cronq.   Canadian fleabane
(*Erigeron canadensis* L.)
Seen during the Survey only on railway tracks at
Carlisle (38.54, REG, 1987, LANC, *Watsonia* (1988)
17:193; 40.54, REG, 1992).

   The only previous records of this casual species,
here at its northern limit in western Britain, are Baker's
from Grange-over-Sands (4.8s), W.H. Pearsall's from
Barrow-in-Furness docks (2.6s, 1913, *BEC Rpt* (1914)
3:402) and one from a railway bank near Scotby,
Carlisle (4.5, R. Martindale, 1950, CLE).

                                    2   (F)C

Bellis perennis L.   Daisy
Extremely common. In the lowlands it occurs as a weed
of open waste ground, of lawns and playing fields. It is
also common both in the lowlands and the uplands on
open roadside verges and embankments, also along
lanes and tracks. It prefers rather heavy damp soils and
it is a characteristic plant of short grassland around
fellside seepage areas and springs, associated with *Poa
annua, Agrostis stolonifera* and *Stellaria uliginosa.*

   Rayless plants were noted by W.W. Mason at
Melmerby (6.3) in 1921 and were seen during the
Survey at Penrith (50.30) and Ravenstonedale (74.02).
Alt. limit: 760 m, Knock Fell (72.30)
Map 914   1677   LWFCY

*Tanacetum parthenium (L.) Schultz-Bip.   Feverfew
Widespread and fairly frequent in the lowlands,
especially in the Eden valley and on the Solway Plain.
It occurs by roadside walls, hedgerows and banks and
on waste ground, nearly always close to houses, a
reflection of its earlier medicinal use.
Alt. limit: 380 m, above Alston (72.48)
Map 915   518   LWFCY

T. vulgare L.   Tansy
A lowland species, common along the Rivers Eden and
Duddon and on the Solway Plain, not infrequent on the
west coast and in the south but rare elsewhere. Many of
the records are from rather dry riverbanks but it also
occurs by roadsides, on waste sites and on old railway
lines.
Alt. limit: 375 m, below Nenthead (76.44))
Map 916   249   WFCY

Seriphidium maritimum (L.) Polj.   Sea wormwood
(*Artemisia maritima* L.)
'Very rare and apparently declining having been seen
during the Survey only in muddy tidal creeks on the
landward side of Walney Island (18.64s, 18.66s) and
Eskmeals (08.92, 08.94). The populations at Eskmeals
have been much reduced in recent years.

   Baker lists it from Coulderton Point (9.0),
Muncaster and Ravenglass (0.9), the Duddon estuary,
Walney Island and Rampside (2.6s). There is a
specimen collected by Wilson on Walney Island in his
herbarium (1913, YRK) and in his Flora he refers to an

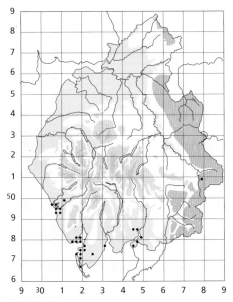

Map 913. *Erigeron acer*

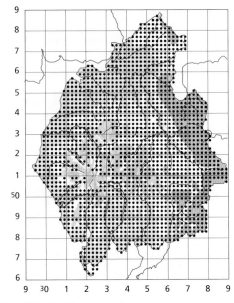

Map 914. *Bellis perennis*

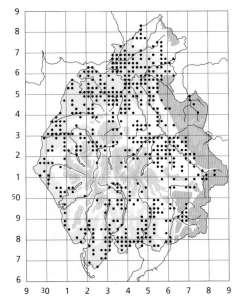

Map 915. *Tanacetum parthenium*

undated record of W.H. Pearsall's from the Leven viaduct (3.7s), where it was later seen by the Barrow-in-Furness N.H.S. during the 1950s.

4   FC

Artemisia vulgaris L.   Mugwort
A frequent plant over most of the lowlands although rather sparse in the east and along the south-west coast of Cumberland and absent from much of the Lake District. This is essentially a plant of waste sites and tips, quarries, railwaysides and disturbed ground along roadsides. It is very much a feature of the central reservation of the M6 and of roadside grit piles.
Alt. limit: 350 m, north of Shap summit (54.06)
Map 917   447   LWFCY

A. absinthium L.   Wormwood
Rare and decreasing, known only from a handful of sites on the south and west coasts, from Coniston (30.96, R. Young, 1970, AMDE) and a few in the Carlisle area and Dalston (36.48). Only two plants survive at the last site where Hodgson described it as abundant. Most of the sites are urban waste land and river shingle.

Since the 1950s this southern species has declined markedly on the Furness peninsula and has disappeared from Walney Island. It has also gone from all the post-1930 *Atlas* squares along the west coast and in the east of the county.
Map 918   8   (W)FC

*****A. ludoviciana** Nuttall
A North American species found by W.H. Pearsall in 1912 at Dalton-in-Furness (2.7, *BEC Rpt* (1913) 4:16).

(F)

*****A. stellerana** Besser
A garden escape known for some years on the north side of the Solway Firth and found in 1985 on the Cumbrian shore north of Mawbray (08.46, MMM; LANC), where a few plants were growing in sand and shingle just above the strand-line and among marram grass. In 1986 it was seen at Allonby (08.42, GH) and there are later records from near Beckfoot (08.48, REG, 1991), the dunes at Wolsty (08.50, REG, 1994) and the shore at Blitterlees (10.52, A. Cannell, 1993).

5   C

*****A. biennis** Willd.
A North American - Asiatic species seen briefly on a landscaped area at Eden Bridge, Lazonby (54.40, REG, LANC, 1984, det. A.L. Grenfell). The only previous records are from Silloth in 1917 (1.5, A. Wallis, *BEC Rpt* (1918) 5:112), Barrow-in-Furness docks in 1913 (2.6s, W.H. Pearsall, *BEC Rpt* (1914) 3:402) and Whitehaven in 1924 (9.1, M. Cobbe, 1924, *BEC Rpt* (1925) 7:577).

1   (F)C

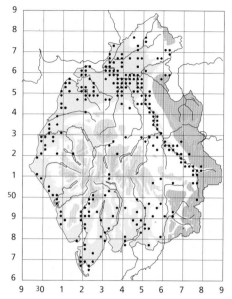

Map 916. *Tanacetum vulgare*

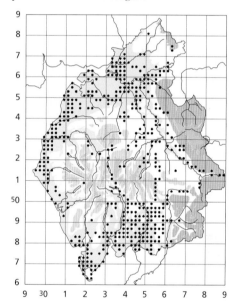

Map 917. *Artemisia vulgaris*

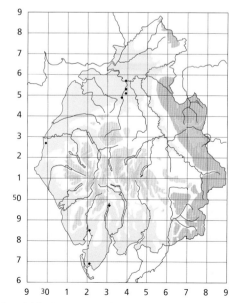

Map 918. *Artemisia absinthium*

***A. scoparia** Waldst. & Kit.
Noted by Hodgson in the early 1890s at the Derwent Tin-plate works, Workington (9.2) and at Maryport (0.3). It is a native of central and eastern Europe.

(C)

[Otanthus maritimus (L.) Hoffmansgg. & Link
The record in Baker's Flora from Grange-over-Sands (4.7s) is corrected in his Postscript to *Filago vulgaris*. An 1842 Cumberland record from Flimby (0.3) is similarly dismissed by Hodgson as an error.

(FC)]

Achillea ptarmica L.   Sneezewort
A common species of tall herb vegetation in marshy fields, fens and in seepage areas in the fells, also on wet roadside verges. Common associates include *Succisa pratensis, Filipendula ulmaria* and *Juncus* species.
Alt. limit: 770 m, Cross Fell (68.34)
Map 919   1299   LWFCY

***A. ligustica** All.
A Mediterranean species recorded by Hodgson from the Derwent Tinplate Works, Workington (9.2) during the 1890s.

(C)

A. millefolium L.   Yarrow
Extremely common, absent only from the highest fells and areas of blanket peat. It occurs in a wide range of grasslands but shuns the most acid and prefers the coarse, ungrazed grasslands, particularly roadsides, railwaysides, 'unimproved' meadows and scrub.

Plants with flowers of varying shades of pink are frequent.
Alt. limit: 825 m, Cross Fell (68.34)
Map 920   1699   LWFCY

***Chamaemelum nobile** (L.) All.   Chamomile
Listed by Hodgson as a garden escape at Maryport (0.3),

Abbeytown (1.5), Wigton (2.4) and Carlisle (4.5) and by Baker from Colwith Force (3.0) where, half a century later, it was collected by Wilson (1935, YRK). There is a record, also from 1935, from Skinburness (1.5, J. Parkin) and a *Scarce Plants* one for the Lorton square (1.2).

This is a decreasing plant of southern and south-west England and known in the north only as an introduction.

S   (WC)

Anthemis arvensis L.   Corn chamomile
Very rare. The only Survey records are from a sandy track at Aldoth, Aspatria (14.48, CCH, 1981, herb. C.C.H.), the railway embankment at Arnside (46.78s, MB, 1978), the newly-seeded verge of the A590 near Witherslack (44.82s, GH, 1989, LANC) and the Furness coast at Roanhead (20.74, PB, 1989, LANC).

Hodgson records it from Egremont (0.1), Workington (9.2), Silloth (1.5), Borrowdale (2.1) and St John's in the Vale (3.2) and Wilson from near Appleby (6.2). There are also records from Abbeytown (1.5, J. Leitch, 1885, CLE), Woodrow, Wigton (2.4, E.J. Glaister, 1876, CLE) and How Mill (*cf.*5.5, T.S. Johnstone, 1905, CLE). It was later seen at Silloth by J. Parkin in 1935 and at Thwaites, Millom (1.8, M. Cross, 1932, 1949). There are post-1930 *Atlas* dots for the Distington (0.2) and Borrowdale (2.1) squares.

4   WFC

A. cotula L.   Stinking chamomile
Seen during the Survey only on an industrial site at Barrow-in-Furness (18.72, PB, *c.*1980).

Hodgson comments that this species was "completely unknown in Cumberland twenty years ago [1875]; now well established at many stations, and spreading rapidly". He lists it from roadsides, railway banks and rubbish tips along the west coast from Seascale (0.0) to Silloth (1.5) and inland from Flosh-gate, Ullswater (4.2), Penrith (5.3) and Gilsland (6.6,

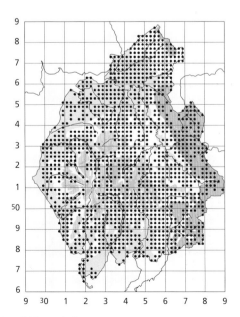

Map 919. *Achillea ptarmica*

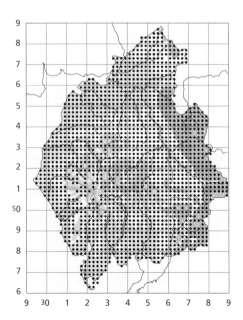

Map 920. *Achillea millefolium*

v.c.67/70). Earlier Baker had recorded it only from near Skelwith Force (3.0) and Wilson gives only Windermere (4.9) and Melkinthorpe (5.2). There are further Cumberland records from Isel, Cockermouth (1.3, M. Cross, 1908), Melmerby (6.3, W.W. Mason, 1917), Buttermere (1.1, G.A.K. Hervey, 1921), Skinburness (1.5, J. Parkin, 1935) and Thwaites, Millom (1.8, M. Cross, 1949). Surprisingly, the *Atlas* shows a further 17 post-1930 squares.

1   (W)F(C)

*A. tinctoria L.   Yellow chamomile
Seen only at Eskmeals (08.92, A. Warburton, 1970s), Melkinthorpe (54.24, CEW, 1990, LANC), on waste ground at Little Salkeld (56.36, RWMC, 1992, LANC) and at Hodbarrow (16.78, PB, *c*.1990).

It is listed by Hodgson from Workington (9.2), Flimby (0.3) and near Silloth docks (1.5, J. Leitch, 1891, CLE). There are other 19th century records from Papcastle (1.3, R. Wood, 1878, 1879, 1908), East Cote, Silloth (1.5, E.J. Glaister, 1879, CLE) and later ones from waste ground by the railway at Carlisle (4.5, C.W. Muirhead, 1943, CLE) and by the River Eden at Great Salkeld (5.3, G.A.K. Hervey, 1956).

4   WC

Chrysanthemum segetum L.   Corn marigold
Fairly frequent as an arable weed on the Solway Plain, becoming rather scarce up the Eden valley, also scattered along the coastal areas of the south and south-west where it may be decreasing. There are also records from a sewage works and a nursery garden.
Alt. limit: 200 m, near Renwick (58.40)
Map 921   145   WFC(W/Y)

*C. coronarium L.
A garden escape found by Hodgson on Flimby beach (0.3) in 1884.

(C)

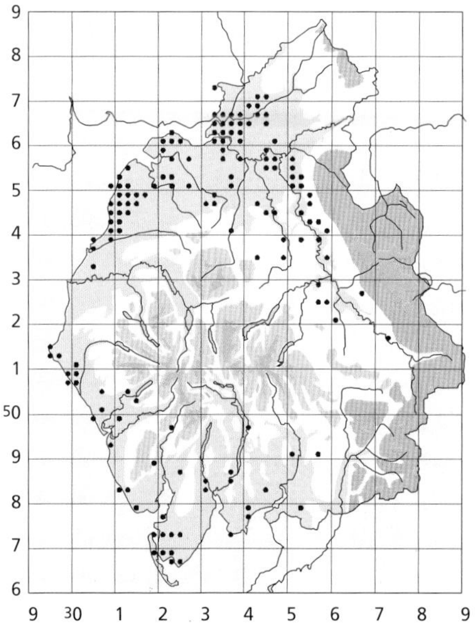

Leucanthemum vulgare Lam. Ox-eye daisy; Dog daisy
Very common in the lowlands occurring in a wide range of grasslands except the most acid and heavily grazed. It is very much a feature of railwaysides, woodland margins, scrub and roadside verges. As C.D. Pigott has pointed out, it thrives on areas where the subsoil has been disturbed and exposed and in such situations, as for example by newly-constructed motorways, it may quickly become dominant.

It is extraordinary that Hodgson says that it "was almost unknown in Cumberland at the beginning of the present [19th] century". The first record appears to be that of Nicolson's in the 17th century from Great Salkeld (Whittaker 1981).
Alt. limit: *c*.590m, roadside at Yad Moss (76.34)
Map 922   1262   LWFCY

*L. lacustre (Brot.) Samp. x **L. maximum** (Ramond) DC.   Shasta daisy
L. x **superbum** (Bergmans ex J. Ingram) Kent
(*Chrysanthemum maximum* auct.)
An uncommon garden escape first noted at Seascale (02.00, A.J. Richards) in 1973. It was first seen in Furness at Mansriggs, Ulverston (28.80, JA, *Watsonia* (1981) 13:339) in 1980 and in Westmorland near Kendal (54.94, CEW) in 1987. Most of the records are from roadside verges.
Alt. limit: 245 m, Newbiggin, Penrith (44.28)
Map 923   22   WFC

Matricaria recutita L.   Wild chamomile
Common around Carlisle and on the Solway Plain, elsewhere uncommon or scarce and often only as a casual. It occurs chiefly as an arable and farmyard weed but also on disturbed and waste land. There is no obvious reason why this southern species should be so common in the north yet scarce in Furness.
Alt. limit: 365 m, Kirkstone (40.06)
Map 924   221   WFCY

Map 921. *Chrysanthemum segetum*

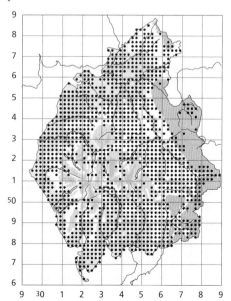

Map 922. *Leucanthemum vulgare*

***M. discoidea** DC.   Pineapple weed; Rayless mayweed
(*M. matricarioides* (Less.) Porter)
Very common throughout the lowlands as a weed of
disturbed, waste and arable land, and particularly
frequent on open trampled ground such as paths, green
lanes and field entrances. This North American species,
first recorded in Britain in 1871, was not discovered in
the county until 1913 when it was seen by W.H.
Pearsall at Barrow docks (2.6s, *BEC Rpt* (1914) 3:402).
The following year W.W. Mason noted it at Melmerby (6.3) and 25 years later Wilson described it as
"common almost everywhere".
Alt. limit: 670 m, Great Dun Fell (70.30)
Map 925   1493   LWFCY

Tripleurospermum maritimum (L.) Koch and
T. inodorum (L.) Schultz-Bip.   Sea mayweed
Common in the lowlands, particularly in the coastal

areas and throughout the Eden valley, but surprisingly
scarce in the Lake District. It is a plant of disturbed and
waste land and a weed of arable fields. It is very
characteristic of roadside grit-piles which are said to
have originated mostly from the Maryport area (0.3)
but are now deposited over much of the county.

No particular attempt was made to record the
two species separately during the Survey. Both occur
but *Tripleurospermum maritimum* is typically a plant of
the strandline.
Alt. limit: 470 m, Birkdale, Teesdale (80.26)
Map 926   907   LWFCY

***Senecio fluviatilis** Wallr.   Broad-leaved ragwort
Apart from a few isolated sites, this tall and handsome
species is confined to riverbanks along the middle and
lower stretches of the River Eden, where it is locally
common, the upper Lune near Gaisgill (6.0), the South

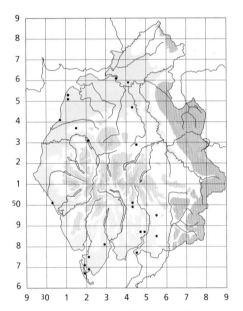

Map 923. *Leucanthemum lacustre* x *L. maximum*

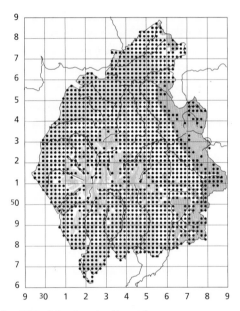

Map 925. *Matricaria discoidea*

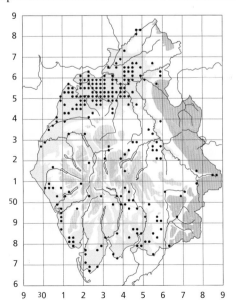

Map 924. *Matricaria recutita*

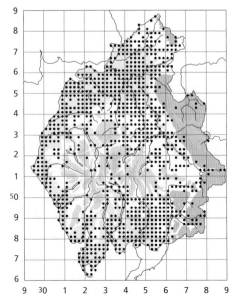

Map 926. *Tripleurospermum maritimum* and *T. inodorum*

Tyne at Alston (7.4) and Preston Patrick (52.82s, CEW, 1994), where there is a large stand near the vicarage, probably Wilson's site. The more isolated sites include an overgrown quarry at Field Broughton (38.80, JA, 1993, LANC) where it occurs in profusion, the only Furness record.

It seems to have disappeared from most of the post-1930 west Cumberland squares shown in the *Atlas*, also from Wilson's Witherslack (4.8s) site. Nevertheless, the Cumbrian populations of this central and southern European species are still substantial and in Britain second only to those of the Somerset levels.
Alt. limit: 190 m, Crosby Garrett (72.08)
Map 927   33   WFC(Y)

[S. paludosus L.   Fen ragwort
(*S. congestus* (R. Br.) DC.)
The 18th century records from Burton Moss (5.7s) are rightly dismissed by Wilson as errors. They probably refer to *Senecio fluviatilis*. *Senecio paludosus* is an extremely rare plant of eastern England.   (W)]

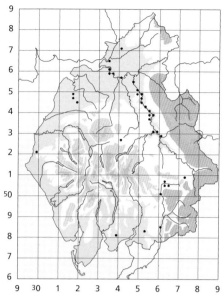

Map 927. *Senecio fluviatilis*

Plate 99. *Senecio jacobaea*, South Walney

S. jacobaea L.   Ragwort
Very common in the lowlands occurring in coarse grassland, on roadside verges and railwaysides, also on waste ground and particularly as a weed of old neglected pastures where the ungrazed stems, unpalatable except to the caterpillars of the cinnabar moth (*Tyria jacobaeae*), are a familiar sight. Although avoided by cattle it is eaten by sheep. Rather surprisingly there are still a number of lowland tetrads in which it has not been found. On the coasts it is an important species of old sand-dunes, its numbers fluctuating from year to year depending on rainfall. Probably nowhere is it more abundant and impressive than at the South End of Walney Island (2.6s) where the dunes, with their rabbit warren and gull colonies, are converted to a sea of yellow in late summer. There are also records from some of the lower crags, for example Dod Crag, Castlerigg Fell (28.20).

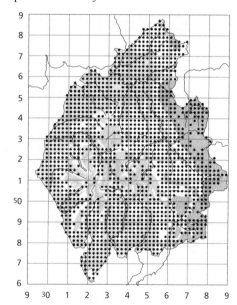

Map 928. *Senecio jacobaea*

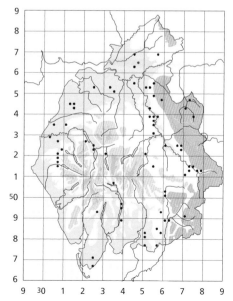

Map 929. *Senecio jacobaea* x *S. aquaticus*

Alt. limit: 560 m, Tynehead Fell (76.34)
Map 928, Plate 99   1468   LWFCY

S. jacobaea x S. aquaticus
S. x ostenfeldii Druce

Not infrequent where the two parents occur together,
as, for example, in damp pastures and on riverbanks. It
is certainly much under-recorded and went entirely
unnoticed by earlier Flora writers. The first record, as
Senecio aquaticus var. *pinnatifolius* Gren. & Godr.,
was from Esthwaite (3.9, G.C. Druce, 1926, *BRC Rpt*
(1927) 8:118). It was first found in Cumberland in
1978 at Ennerdale Bridge (06.16, CCH, CLE, LANC,
Watsonia (1979) 12:357) and in Westmorland near
Low Borrowbridge (60.00, GH, 1970, LANC).

   This hybrid, which is partially fertile and
capable of backcrossing, is characterised by broad
panicles, the lowest branches arising well down the
stem, and the numerous, rather narrow leaf segments.
Alt. limit: *c*.350 m, Leadgate, Alston (70.42)
Map 929   70   WFCY

S. aquaticus Hill   Marsh ragwort

Frequent in the lowlands but surprisingly uncommon in
Furness. It is a characteristic plant of marshy fields,
growing with *Achillea ptarmica* and species of *Juncus*
and *Epilobium*, also of riversides and roadside ditches.
Alt. limit: at least 415 m, Nenthall (76.46)
Map 930   902   LWFCY

S. erucifolius L.   Hoary ragwort

Rare, known only from a few sites on the west coast,
chiefly in the Brigham - Workington area, where it still
grows on the limestone at Broughton Crags (08.30) as
mentioned by Hodgson, on a disused railwayline and on
roadside verges. A number of Survey records from
outside this area have not been substantiated and it
seems likely that several of the *Atlas* records, including
those made by G. Wilson in Furness, are also errors.

   This ragwort reaches its northern limit on the
west coast in Cumbria.
Map 931   11   [(WF)]C

*S. squalidus L.   Oxford ragwort

The dramatic spread of this species following its introduction from southern Europe to Oxford in 1794 has
been documented by Kent. In his paper (1964) dealing
with northern England, he refers to a pre-war record
from Cumberland, probably J. Parkin's from Saltpans,
Maryport (0.3, 1935), and post-war ones from Workington (9.2, R. Walker, 1949) and the railway between
Scotby and Carlisle (4.5, Anon.). His only other record
from Cumbria is from Ramsden dock, Barrow-in-
Furness (2.6s, G. Wilson, 1948, CLE). This specimen
is the curious var. *subinteger* Druce, with unlobed but
sharply serrate leaves. There is a single Survey record
of this from the roadside near Pooley Bridge (46.24,
RWMC, 1990, LANC). The only Westmorland records

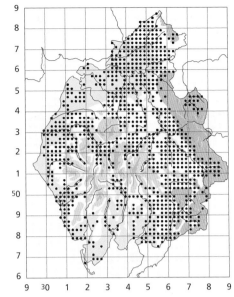

Map 930. *Senecio aquaticus*

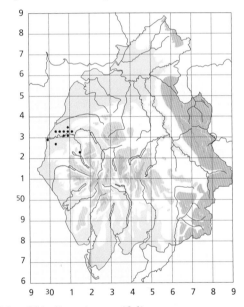

Map 931. *Senecio erucifolius*

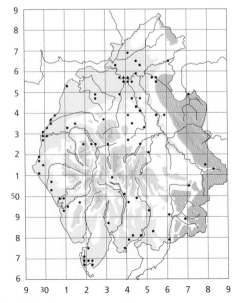

Map 932. *Senecio squalidus*

prior to the Survey are those in the *Atlas* for the Miln-thorpe (4.8s) and Barbon (5.8s) squares.

This ragwort is now quite widespread and particularly well established around the west coast ports, Barrow and Carlisle. Most of the records are from urban sites, where it occurs chiefly on waste ground and tips, also on railway ballast, as at Sellafield (02.02), and there is a thriving population on the Silurian rock exposures of the Lindale by-pass (42.80s). Alt. limit: 200 m, Argill (82.12)

Map 932   82   WFCY

S. vulgaris L.   Groundsel
Extremely common in the lowlands on urban waste sites, on roadside pavements, on tips and as an all too successful weed of cultivated gardens and allotments, less commonly as a weed of arable land.

There are records of radiate plants, var. *hibernicus* Syme, from near Hayton (52.58), the Carlisle area (38.54, 38.56, 40.56), Whitehaven (9.1), Great Langdale (30.06) and Haverthwaite (34.82). Similar plants also occur on coastal sand and shingle at a number of sites on the Furness peninsula and South Walney (18.74, R. Young, 1968, AMDE; 16.72, 20.74, 20.78, PB, LANC), at Haverigg (14.78) and Braystones (00.06), the earliest record being from Sandscale (18.74). These presumably belong to the maritime var. *denticulatus* (Mueller) N. Hylander, being weakly branched, densely arachnoid-hairy, and having fewer than nine capitula (Allen 1967). These are the northernmost records for this variety although it formerly occurred on the Isle of Man.
Alt. limit: 550 m, Moor House, Teesdale (74.32)

Map 933   1309   LWFCY

*S. vernalis Waldst. & Kit.
An eastern European species seen briefly on the newly-seeded verge of the Ravenstonedale by-pass (70.04, GH, 1982, LANC, *Watsonia* (1983) 14:428). The only previous record is Hodgson's from the Derwent Tinplate Works, Workington (9.2) in 1889.

1   W(C)

S. sylvaticus L.   Wood groundsel
Frequent in the lowlands on dry, light acid soils, often occurring in woodland clearings and margins, among gorse and bracken, particularly in cleared areas and following fire, on field banks and around sand and gravel workings.
Alt. limit: *c.*270 m, above Knock (68.28), although Hodgson gives it as 320 m on Robinson (1.1/2.1) and Priest's Crag, Ullswater (42.22).

Map 934   466   WFCY

S. viscosus L.   Sticky groundsel
A frequent lowland plant of open, well-drained ground, chiefly on urban waste sites, along roadside pavements, the central reservation of motorways, railway ballast and on practically every pile of roadside grit, also on

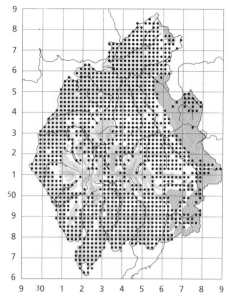

Map 933. *Senecio vulgaris*

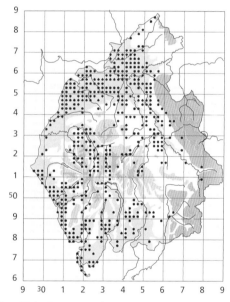

Map 934. *Senecio sylvaticus*

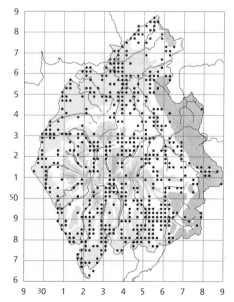

Map 935. *Senecio viscosus*

coastal shingle as at Flimby (02.32).

This ragwort has increased its range considerably in recent years. Wilson listed only four sites. The *Atlas* shows it as occurring in 32 squares, half the number recorded during the Survey.

Alt. limit: at least 430 m, Nenthead (78.42)

Map 935   563   WFCY

Tephroseris integrifolia (L.) Holub
(*Senecio integrifolius* (L.) Clairv.)
Probably now extinct. It was first discovered in the Pennines in 1846 by J. Backhouse on limestone outcrops at "1200 to 1500 ft" above Brough (7.1, E). His two flowering specimens are dated 1880 and were presumably cultivated, the larger being 35 cm tall. It is not impossible that vegetative, sheep-grazed plants still survive. Of the two other localities mentioned by Wilson, "near Winton" (M. Mason) could well refer to the same area and the "Westmorland side of Mickle Fell" (T.J. Foggitt) is presumably somewhere near Little Fell (7.2). There is no evidence that Wilson had seen the plant or indeed of any definite post-1930 record for any Pennine site.

In England this ragwort is typically a plant of short chalk or limestone grassland in the south and south-east. It also occurs in Wales on the sea cliffs of Anglesey. The plants from there are appreciably taller and have been described as subsp. *maritima* (Syme) R. Nordenstam. The Cumbrian plants (BM) bear a closer resemblance to the latter.

S   (W)

****Doronicum pardalianches** L.   Leopard's-bane
A fairly frequent garden escape in the south and in the Eden valley, elsewhere occasional or rare. Most records are of colonies on roadside verges, usually near houses.

The only record from the Furness peninsula is from Swarthmoor (26.76). This is also the only one given by Petty (1896 p.344) who regarded the plant as having escaped from Swarthmoor Hall, citing as evidence the observation of William Fell, aged 11 years, in his *Journeys in Furness in the year 1777* "I went to Swarthmoor Hall, a Place a great Deal of

which is pulled down by that infamous Man Captain Lindow; he has spoiled the Beauty of the Woods surrounding it, by hagging down the trees.... The gardens are overgrown with Thistles, Briars, Thorns, etc., but there are some Garden Flowers growing wild."

This species, considered rare by the earlier Flora writers, is now well established and no doubt still spreading, the Survey having more than doubled the number of *Atlas* squares.

Alt. limit: *c.*280 m, Alston (70.46)

Map 936   97   WFC(W/Y)

***D. pardalianches** x **D. plantagineum** L. x
D. columnae Ten.
D. x **excelsum** (N.E. Br.) Stace
This popular garden plant was discovered during the Survey well naturalised on the roadsides east of Inglewood Bank, Penrith and near Penton station (respectively 52.34, 1984; 42.76n, 1985, both RWMC, LANC, *Watsonia* (1985) 15:40, (1986) 16:193). These, and later collections from Westmorland, were determined by A.C. Leslie, but he tentatively identified a roadside population near Lazonby (52.38, RWMC, 1985, LANC) as the hybrid between *Doronicum plantagineum* and *D. pardalianches* or *D. carpetanum* Boiss.& Reuter ex Willkm. The first Westmorland record was from a copse near Milburn (64.28, CFS & JS, 1986).

Alt. limit: 160 m, Milburn (64.28)

Map 937   8   WC

Tussilago farfara L.   Coltsfoot
Very common on waste ground in villages and towns, on recently disturbed ground, especially on roadsides and tips and preferring heavy, wet soils. In the hills it is a feature of eroding riverside banks and of the more stable shingle beds and islands.

Alt. limit: 650 m, Great Dun Fell (70.32)

Map 938   1464   LWFCY

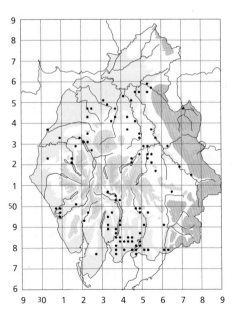

Map 936. *Doronicum pardalianches*

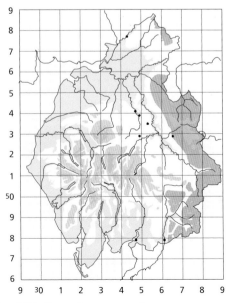

Map 937. *Doronicum* x *excelsum*

Petasites hybridus (L.) P. Gaertner, Meyer & Scherb.

Butterbur

Common in the lowlands where it is essentially a plant of damp riverside woodland and scrub and usually forming dense and conspicuous stands. Like *Tussilago*, it occurs, but not to the same extent, on disturbed ground, particularly in seepage areas on roadside and railwaysides.

Butterbur is dioecious and female plants are largely restricted in Britain to the Midlands and the north Pennines. The Survey records of female plants are mostly from riversides in the north-east but there is an isolated record in the south by the River Kent near Sedgwick (50.86s). There are also earlier records made by C.W. Muirhead from Wetheral Bridge (4.5, 1947, CLE) and the River Caldew at Cummersdale (3.5, 1949, CLE).

Alt. limit: *c*.380 m, near Garrigill (72.40)

Map 939   575   LWFCY

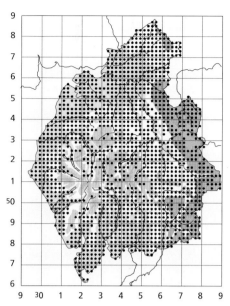

Map 938. *Tussilago farfara*

*P. japonicus (Siebold & Zucc.) Maxim. Giant butterbur
The first record of this very large and strikingly conspicuous plant dates from 1931 when it was collected from outside a garden in Langdale (S. Harmer & R. Clinton-Baker, *BEC Rpt* (1942) 12:283). This is probably the same as Wilson's Clappersgate site (36.02) where it still flourishes on a wet roadside bank. It also grows in the same tetrad by the River Rothay, on a wet, unstable bank above the River Lune in the Casterton woods (62.78s) and, very noticeably, on the roadside verges at Strawberry Bank, Cartmel Fell (40.88s, KAG & FLW, *c*.1974), the first Furness record. These populations probably all originated as garden throw-outs. In the early 1980s material from Clappersgate was unfortunately planted out on roadsides at Crosby Ravensworth (62.14) and near Great Ormside (70.14).

Alt. limit: 190 m, Crosby Ravensworth (62.14) and Great Ormside (70.14)

          6   FW

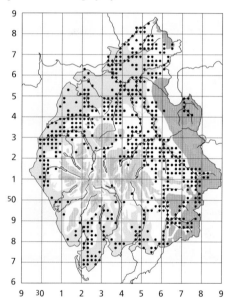

Map 939. *Petasites hybridus* (* female plants)

*P. albus (L.) Gaertner   White butterbur
An uncommon garden escape, usually growing on damp roadside verges and near houses. There are records from shady riversides where, like *Petasites hybridus*, it may dominate considerable areas, as at Ghyll Head, Bowness (38.92) and Gilgarran, Distington (02.22). The first Furness record is probably that from the roadside at Colton (3.8, G. Wilson, 1949, CLE). It has been known for over 40 years at Loweswater (12.20) in the garden of a farm where its leaves were formerly used for wrapping butter.

Alt. limit: *c*.300 m, Alston (72.46)

Map 940   31   WFC

*P. fragrans (Villars) C. Presl   Winter heliotrope
A garden escape which is now frequent in the lowlands of the south and west, growing on roadside verges and banks and in shrubberies and estate grounds. It comes into flower in December but the tender stems are

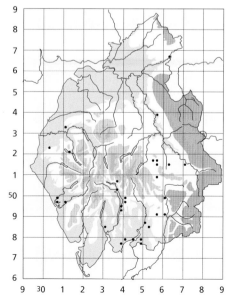

Map 940. *Petasites albus*

often frosted. Remarkably, the only records in the Floras are those given by Petty (1896 p.344) from Aldingham (2.7) and Grange-over-Sands (4.7s). The first Westmorland ones are those shown in the *Atlas* and for Cumberland that from Ravenglass (08.96, A.J. Richards, 1973).

Alt. limit: *c*.170 m, near Dufton (68.24)

Map 941   60   WFCY

*Calendula officinalis L.   Marigold
A rare garden escape seen in 1978 at Newtown, Carlisle (38.56, REG), and later on waste ground at Hesket Newmarket (34.38, REG, 1986), on the beach at Bardsea (30.74s, PB, 1989), by the River Eden near Great Salkeld (54.36, CEW, 1992, LANC) and at Skinburness (12.56, A. Cannell, 1991).

                                          7   FC

*Ambrosia artemisiifolia L.   Ragweed
Seen only at Silloth (10.52, R. Young, 1970s) and with several other casuals on roadside waste during the construction of the Keswick by-pass (26.24, G. Wilson, 1975, LANC).

Although Hodgson makes no reference to this North American species, it is said, according to files in AMDE, to have occurred on the sand-dunes at Silloth (10.52) as early as 1892. In the early 1930s it was reported from Thwaites, Millom (1.8, M. Cross).

                                          2   C

*A. trifida L.
A North American species collected from a corn-mill yard at Whitehaven in 1913 (9.1, H. & G. Adair, YRK).

                                                                            (C)

*Xanthium spinosum L.   Spiny cocklebur
A cosmopolitan weed and native of South America reported by Hodgson as having been seen at Silloth docks (1.5, R. Wood, 1884, CLE) several times during the last half of the 19th century.

                                                                            (C)

*Rudbeckia laciniata L.
A garden escape reported in 1977 by G. Wilson from the north-west side of Derwent Water (24.20) where it is said to have been naturalised for at least a century.

                                          1   C

*Helianthus rigidus (Cass.) Desf. x **H. tuberosus**
H. x laetiflorus Pers.   Perennial sunflower
A popular garden perennial well established on the sand-dunes near Silloth harbour (10.52, M. Armstrong, *c*.1983, LANC, *Watsonia* (1989) 17:477) and later recorded from the roadside between Kelsick and Dundraw (20.50, MP, 1994, LANC).

                                          2   C

*H. tuberosus L.   Jerusalem artichoke
Seen in 1991 on a nettle-covered tip by a laneside between Hartley and Winton (78.08, AAD, LANC).

                                          1   W

*Galinsoga parviflora Cav.   Gallant soldier
A North American weed reported from Thwaites, Millom (1.8, M. Cross) probably early this century.

                                                                            (C)

*G. quadriradiata Ruíz Lopez & Pavón
(*G. ciliata* (Raf.) S.F. Blake)
An uncommon but increasing weed of waste ground and gardens. It was seen in abundance at a nursery garden near Crooklands (52.82s, GH, 1984, LANC, *Watsonia* (1985) 15:401) and probably garden occurrences are mainly from such sources. The first Furness and Cumberland records are from the roadside north of Grange-over-Sands and from Carlisle, where it was widely distributed in the mid-1970s (respectively 38.76s, E.J. Harling, 1978; 4.5, C.W. Muirhead, both *Watsonia* (1979) 12:357).

Alt. limit: 150 m, Brough (78.14)

Map 942   15   WFC

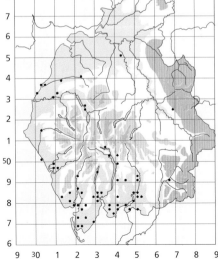

Map 941. *Petasites fragrans*

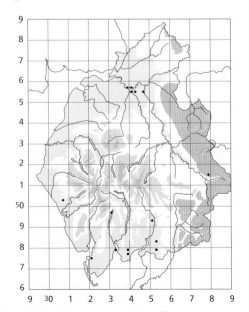

Map 942. *Galinsoga quadriradiata*

*Hemizonia pungens Torrey & A. Gray
A western North American plant seen in 1889 and 1890 at
Silloth (1.5, J. Leitch, CLE) and again in 1920 (C. Waterfall,
BEC Rpt (1921) 6:129).

(C)

*H. kelloggii E. Greene
A western North American weed recorded once from Silloth
(1.5, J. Leitch, 1889, CLE).

(C)

*Madia sativa Molina
An annual species of western North America and Chile listed
by Hodgson as having been collected by him at Floshgate,
Ullswater (4.2, 1882) and by J. Leitch at Silloth (1.5, 1889,
CLE).

(C)

Bidens cernua L.   Nodding bur-marigold
A local plant of the margins of lowland eutrophic
ponds and tarns, chiefly in the Ravenglass area on the
west coast and on the Furness peninsula. The isolated
Eden valley records include ones from Whins Pond,
Edenhall (54.30, RWMC, 1984, LANC), the west side
of Lazonby Fell (50.38, RWMC, 1992, LANC) and a
newly-dug fish-pond at Blencarn (64.30, RWMC,
1984). These are all new sites but the species is
generally in retreat as a result of agricultural drainage.
It appears to have become extinct at its last Westmor-
and site in Dallam Park, Milnthorpe (4.8s, J.A. Wood)
in 1954.

The only Survey records of var. *radiata* DC.
were from near Oulton (22.50) and Whins Pond. There
are a number of old records including one from
Moorhouse, Carlisle (3.5, C.W. Muirhead, 1947, CLE).
Alt. limit: 180 m, Blencarn (64.30)
Map 943   26   (W)FC

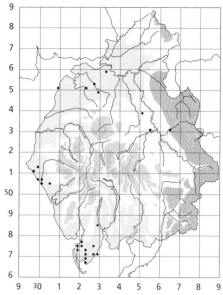

Map 943. *Bidens cernua*

B. tripartita L.   Trifid bur-marigold
Like *Bidens cernua*, this is a local and declining
species. They occur in almost identical habitats and
often grow together, as at Urswick Tarn (26.74). The
first Furness record dates from 1913 when it was
reported by W.H. Pearsall at Dalton (2.7). The only
extant Westmorland site appears to be Whinfell Tarn,
Grayrigg (54.98, TW, 1981), although there is a 1952
record from Rowell, Milnthorpe (4.8s/5.8s, J.A. Wood).
Alt. limit: 125 m, Whinfell Tarn (54.98)
Map 944   22   WFC

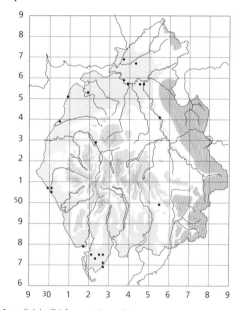

Map 944. *Bidens tripartita*

Eupatorium cannabinum L.   Hemp agrimony
Frequent in the south and west, occasional in the Eden
valley and scarce to rare elsewhere. On the west coast
cliffs it is characteristic of seepage areas and streams.
Inland it is a plant of tall herb communities occurring
in fens, wet ditches and woodland and by slow-flowing
rivers.
Alt. limit: *c.*130 m, Great Ormside (70.16)
Map 945   199   WFCY

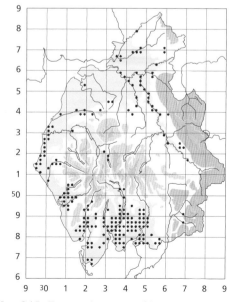

Map 945. *Eupatorium cannabinum*

LILIIFLORAE
(MONOCOTYLEDONES)

BUTOMACEAE

***Butomus umbellatus** L.   Flowering rush
An attractive and not uncommon plant along the middle and lower sections of the River Eden and the Lancaster - Kendal canal but rare elsewhere. It has substantially increased its range this century along the Eden, Hodgson giving only two sites. The first Westmorland record is that in the *Atlas* for the Milnthorpe (5.8s) square, presumably from the canal. The earliest record is that given by Nicolson (Whittaker 1981) in 1690 from the banks of the River Caldew at Carlisle, but it is impossible to say whether the plant is native or introduced in Cumbria; certainly it is capable of effective dispersal along watercourses without the agency of man.
Alt. limit: 100 m, below Eamont Bridge (54.28)
Map 946   43   WC

ALISMATACEAE

[Sagittaria sagittifolia L.   Arrowhead
None of the early Flora writers succeeded in finding this species. Baker and Hodgson refer to old records for Braystones Tarn (0.0) and Baker also mentions an early report of it being "frequent in Furness". This may be the basis of the *Topographical Botany* (1883) reference to v.c.69 mentioned by Wilson. No specimens are known; W.H. Pearsall failed to find it and the records are best treated as doubtful even though Hodgson refers to a drawing of a specimen said to have been collected at Braystones.

    This species occurs in the southern section of the Lancaster - Kendal canal extending to within 10 km of the Cumbrian border.

                            (FC)]

Baldellia ranunculoides (L.) Parl.
                      Lesser water-plantain
A very local plant of ponds and tarn margins, usually in base-poor conditions. Although Baker describes it as "not infrequent", Hodgson considers it rare and doubtless it has been declining throughout this century. It has disappeared from the northern half of the county where it was last seen at Wellington quarry, Aspatria (1.4, J. Parkin) in the 1930s and nearby on Southerfield Moss, Abbeytown (1.4, C.W. Muirhead, CLE) in 1943, and in the east from Whins Pond, Edenhall (5.3) where it was seen by G.A.K. Hervey from 1951 until 1957. In addition to Hodgson's records there is an early one from Monkhill Lough, Burgh-by-Sands (3.5, P. Shepherd, 1884, CLE).

    Baker cites a comment by F. Clowes that *Baldellia* was then, in the mid-19th century, "plentiful in Windermere", the only lake from which it has been recorded. It seems to have survived there almost until the Survey since J. Williamson remembers it from Holm Crag, Wansfell Holme (3.0) between 1942 and 1966. No doubt the intense tourist pressure at this popular spot is responsible for its disappearance from the lake.
Alt. limit: 225 m, Kemp Tarn, Staveley (46.98)
Map 947   10   WFC

Luronium natans (L.) Raf.   Floating water-plantain
Known only from the west side of Derwent Water (24.18, C. Newbould, herb. Eng. Nat.) where a flowering population was found in 1996. This is the only substantiated record although it was listed from there by Baker who also mentions Braystones Tarn (0.0) and Coniston Water. Wilson's only record, attributed to Rev. Dodd, is from Patterdale (4.1) in 1805 but, as pointed out by Q. Kay (*in litt.*), this is derived from a

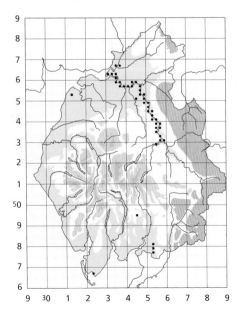

Map 946. *Butomus umbellatus*

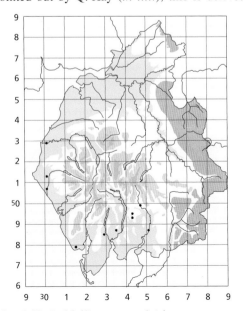

Map 947. *Baldellia ranunculoides*

misreading of the original source (Turner & Dillwyn 1805, p.151), where the record is attributed to Hutchison and cited under *Baldellia ranunculoides*. Although it was reported from Braystones Tarn by J. Parkin in 1935, no specimens are known and it seems likely that the record refers to *Baldellia*.

According to Kay, Wales represents the species' world stronghold following its discovery at several sites in oligotrophic lakes at moderate altitudes. It grows in 1-2.5 m of water, its rosettes forming *Littorella*-like mats although the leaves are flat, slender and tapering.

1 ([W]F)C

Alisma plantago-aquatica L.   Water plantain
Frequent throughout the lowlands by ponds, tarns, and lakes, by rivers and canals and also in dykes. It prefers mesotrophic and eutrophic water and sheltered, ungrazed sites.
Alt. limit: 215 m, Tindale Tarn (60.58)
Map 948   254   WFCY

HYDROCHARITACEAE

[Stratiotes aloides L.   Water soldier
Hodgson cites an old record from Ennerdale Lake (1.1). He himself failed to find it. The occurrence of this predominantly East Anglian species in Ennerdale is highly unlikely and the record is best regarded as an error rather than an introduction as indicated in the *Atlas*, where it is shown in the wrong square (1.2). The *Atlas* record for the Arnside square (4.7s) is in v.c.60.

(C)]

*Elodea canadensis** Michaux   Canadian pondweed
A fairly frequent species of ponds, tarns, rivers, lakes and canals. It may completely dominate the submerged vegetation, as at Blelham Tarn (36.00) where it forms a sward growing to within *c*.15 cm of the surface and where it is no doubt at least partly responsible for the virtual disappearance of the *Potamogeton* flora.

The first record appears to be from Windermere at Newby Bridge where it was noted by E. Hodgson in 1874 (Petty 1898, p.39), some 40 years after the first British record of this American invader. It spread slowly in the south of the county, Wilson giving only five localities, but in Cumberland it quickly became established at the end of the last century around Carlisle, becoming very common in the rivers Eden and Petteril. Its present distribution is probably to some extent the result of increasing eutrophication during recent decades, but since the appearance in 1976 of the more vigorous *Elodea nuttallii* it has declined markedly in certain places, notably in Windermere, Grasmere and in parts of the Lancaster - Kendal canal.
Alt. limit: 215 m, Tindale Tarn (60.58)
Map 949   215   WFC

*E. nuttallii** (Planchon) H. St John
Largely restricted to Bassenthwaite Lake, Grasmere, Rydal Water, Windermere and Coniston Water, with a few outlying sites mainly in canals and rivers. This aggressive American waterweed was first recorded in Britain in 1966 and in Cumbria in 1976 when it was collected at the southern end of Coniston Lake (28.90, C.D. Pigott, 1976 , LANC). The first record for Cumberland was from the north end of Derwent Water in 1977 (24.22, RS, herb. R.S., both records *Watsonia* (1978) 12:176) and for Westmorland from the Lancaster - Kendal canal at Stainton (52.84s, GH, 1977, LANC).

As mentioned under *Elodea canadensis*, *E. nuttallii* has now largely displaced the former in some waterbodies, notably Grasmere and Windermere. Simpson (1984, 1990), who has studied its spread and competitiveness in relation to *E. canadensis*, considers that its success is due in part to its more rapid initial growth during the spring and also to its richer branching. M.C. Lock has recorded shoots 2-3 m long in Bassenthwaite Lake. The map suggests that it is a

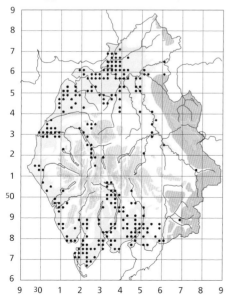

Map 948. *Alisma plantago-aquatica*

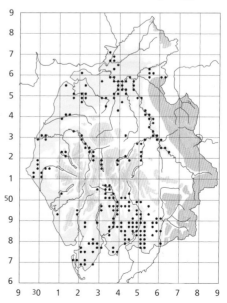

Map 949. *Elodea canadensis*

more lowland species than *E. canadensis* and also less tolerant of flowing water.

The best characters distinguishing this species from *E. canadensis* are the usually much more twisted and often strongly recurved leaves, with narrower tips (0.2-0.8 mm wide, *cf.* 0.7-2.3 mm, measured 0.5 mm from the tip) and with 3-10 rows (*cf.*1 or 2) of marginal hyaline cells.

Map 950   54   WFC

Hydrilla verticillata (L.f.) Royle
The only British station for this delicate *Elodea*-like waterweed was Esthwaite Water (3.9, BM) where it was discovered by W.H. Pearsall and his son in July 1914 and reported the same year by Bennet (1914a & b). It was seen and collected on many occasions over the following 20 years, but its last known sighting was probably in 1941.

Its extinction is generally attributed to increased turbidity as a result of eutrophication by effluent from Hawkshead. It grew in about 2.5 m of water in inorganic clay and associated with *Potamogeton obtusifolius*, *Callitriche hermaphroditica* and *Najas flexilis*.

The only other locality in the British Isles is in west Galway, far removed from the species' nearest station in north-east Germany.

The stems may be distinguished from *Elodea* by their fringed nodal glands and the leaves being in whorls of four to eight.

R   (F)

*Lagarosiphon major (Ridley) Moss
A popular aquarium plant discovered in 1984 filling a field-pond near Helsington (48.88s, KR, 1984; LANC, *Watsonia* (1986) 16:194) and later at Knottallow Tarn, Mansriggs (26.80, M.J.Y. Foley & MP, 1992, LANC). Alt. limit: 210 m, Knothallow Tarn (26.80)

2   WF

Triglochin palustre L.   Marsh arrow-grass
Frequent in mesotrophic and eutrophic marsh communities, in relatively base-rich or calcareous flushes and in the uplands in river shingle. It prefers somewhat open habitats and in the fells is commonly associated with *Juncus articulatus*, *Carex dioica*, *C. hostiana* and *Eleocharis quinqueflora*. It is also frequent in brackish coastal marshes.
Alt. limit: 760 m, Little Dun Fell (70.32)
Map 951   774   WFCY

T. maritimum L.   Sea arrow-grass
Common in all the salt-marshes, occurring mostly in the upper marsh, and growing with *Triglochin palustre* in areas influenced by freshwater seepage, but it is also found at the foot of cliffs and sea-walls.
Map 952   99   WFC

POTAMOGETONACEAE

Most of the records of the rarer pondweeds have been identified by J.E. Dandy or, more recently, by C.D. Preston. Dandy's identifications are in his Index (BM).

Potamogeton natans L.   Broad-leaved pondweed
Widely distributed throughout the lowlands but far less common in the uplands than *Potamogeton polygonifolius* with which it is often confused. The map almost certainly includes a number of records of the latter. *P. natans* has a narrower ecological range occurring in the more mesotrophic and eutrophic lakes and tarns and rarely in ditches.

The most reliable diagnostic characters are the longer stipules (5 cm or more) and fruits (4-5 mm), and the laminar rather than phyllodic submerged leaves. The lamina is usually quite flat, but the oft-quoted

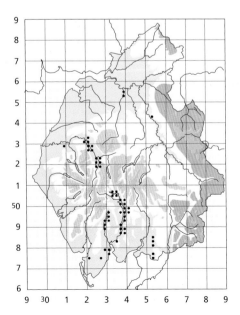

Map 950. *Elodea nuttallii*

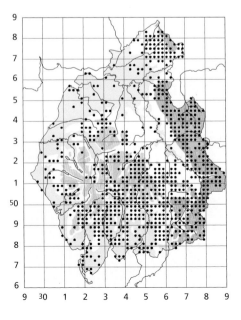

Map 951. *Triglochin palustre*

character of the marked angle between the lamina and petiole is often barely discernible or quite absent.
Alt. limit: 550 m, Moor House, Teesdale (7.3, A. Wilson, 1934, YRK)
Map 953   279   WFC

P. polygonifolius Pourret   Bog pondweed
The commonest species, occurring in tarns, streams, ditches and hillside soakways, also in the lakes in sheltered bays and reedbeds, although here it is often replaced by *Potamogeton natans*. It is the only pond-weed now found in the oligotrophic lakes Wast Water and Buttermere and its frequent association with *Sphagnum* species, *S. recurvum*, for example, in the uplands supports the generally held view that it is a calcifuge, yet it occurs in abundance among tufa in Potts Beck, east of Sunbiggin Tarn (70.08, GH, 1969, LANC), at pH 7.8.

The most useful characters distinguishing this species, from the last are the smallers stipules (2-4 cm) and fruits (1-2.5 mm).
Alt. limit: 705 m, Codale Head (28.08)
Map 954   672   WFCY

P. coloratus Hornem.
The only authenticated record is Baker's from Newton Reigny moss (4.3, 1883, BM, det. J.E. Dandy). It is also listed by Hodgson from Horrock Wood, Ullswater (4.2) and near Dowthwaite Head, Matterdale (3.2), and by Wilson from Barton Fell, Ullswater (4.2, J.C. Varty Smith, 1883). These last three localities seem unlikely. Survey material from Barton Fell (46.22, R.S., 1978, herb. R.S., *Watsonia* (1979) 12:359) and Roudsea Wood (32.82, GH, 1988, LANC) thought to be *Potamogeton coloratus* has been redetermined by C.D. Preston as *P. polygonifolius*.

There is no convincing evidence for the occurrence of this species in the Arnside area (4.7s) although Dandy's Index refers to two specimens from there, one collected by W.P. Hiern (RAMM, 1864) and the other by C. Bailey (1890,

BM). The latter, although labelled Westmorland, was from "the cemetery road, Silverdale" and hence in Lancashire. The same may be true of Hiern's specimen. There is a 1981 collection from a ditch just inside Lancashire (48.78s, S. & M.D.G. Jones, LPL).
S   ([W] C)

P. lucens L.   Shining pondweed
As with the last species, the literature records for this predominantly south-eastern pondweed are few and mostly suspect. Baker lists it from Ponsonby (0.0), Derwent Water and Coniston and Pearsall (1923) from Windermere, Coniston and Blelham Tarn (3.0, 1915, BM, det. J.E. Dandy). The Coniston records may well refer to *Potamogeton* x *zizii*.
(W?F[C])

P. lucens x **P. gramineus**
P. x zizii Koch ex Roth
Extremely rare. The only Survey records are from Coniston Water where it was discovered by R. Stokoe in three tetrads, two at the southern end and one at the north (28.88, 28.90, 30.96, herb. R.S.; LANC). At the southern end the plant grows vigorously in 2-3 m of water, usually in fine, pale silt. The plants are very pale in colour and suffused with a delicate pink when fresh. Of the two parents, they more resemble *Potamogeton lucens*, from which they can be distinguished by their narrower and almost sessile leaves.

The first record for this rather uncommon hybrid dates from 1833 when it was collected in Derwent Water (Anon.). The specimen was seen by J.E. Dandy, but its present whereabouts is unknown. He also identified an anonymous 1851 (LDS) collection from there. A few years later H.S. Fisher collected it in Coniston Water (MANCH) and in 1882 C. Bailey sent material from this lake and also from the north end of Derwent Water (2.2, BM) to A. Bennett (1882) who, unaware of the earlier finds, pronounced the hybrid new to England. The hybrid was last collected from

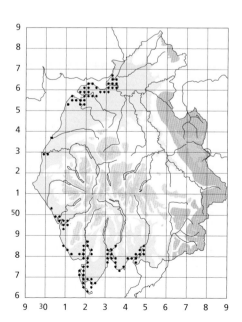

Map 952. *Triglochin maritimum*

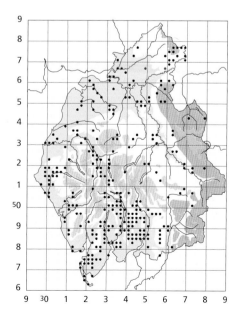

Map 953. *Potamogeton natans*

Derwent Water in 1916 (W.H. Pearsall, BM).

3   F(C)

P. gramineus L.
Scattered in a number of the lakes but mainly restricted to Bassenthwaite and the River Derwent, the middle section of the River Eden and its tributary the Eamont, and Ullswater, where some of the records may refer to *Potamogeton* x *nitens*. Unfortunately none of the Ullswater records is supported by specimens. There are also records from a number of tarns, such as Tindale (60.58) and Overwater (24.34) in the north, and Borwick (44.96) and Knipe (42.94) in the south. In Overwater it occurs with its hybrid with *P. perfoliatus*, *P.* x *nitens*.

Pearsall (1923) commented that it "flourishes" in Loweswater, Coniston Water, Esthwaite and Windermere, but it appears to have disappeared from Loweswater and is very local in the other two lakes. There are no records for Crummock Water subsequent to Hodgson's Flora and the last from Derwent Water is Pearsall's in 1916 (BM). There has been no confirmation of the *Atlas* record from "Troutbeck and Kentmere" (4.0, G. Bell, 1957). The first Furness record is from Esthwaite in 1914 (34.96, W.H. Pearsall, BM).

Potamogeton gramineus usually grows in shallower water than do the other broadleaved species such as *P. perfoliatus* and *P. praelongus*.
Alt. limit: 215 m, Tindale Tarn (60.58)

Map 955   36   WFC

P. gramineus x **P. perfoliatus**
P. x **nitens** Weber
Rare and almost entirely in the north where it has been recorded from Overwater (24.34), where it occurs with *Potamogeton gramineus* by the outflow stream, the nearby Chapelhouse Reservoir (26.34), the River Derwent at Seaton (02.30) and near Isel Bridge (16.32),

the Eamont from Pooley Bridge (46.24) to below Brougham (56.28), and, in the extreme north-east, in the River Irthing near Churnsike Lodge, Butterburn, where it is abundant and very luxuriant (64.76n, 66.76n). The river here forms the county boundary and there is an earlier specimen (H. Milne-Redhead, 1958, BM) attributed to Northumberland. The only records from the south are from the River Leven, where it is exceptionally luxuriant and where it was first seen, new to Furness, by W.H. Pearsall (34.84, 1913, BM, YRK) and Windermere (36.00, A. Darwell, 1996, LANC; both Pearsall (1923)).

The first record for Cumberland was E.S. Todd's from Derwent Water (1907, BM) and for Westmorland Pearsall's from the Westmorland side of Ullswater at Pooley Bridge (4.2, 1919, BM).

Potamogeton x *nitens* is often difficult to distinguished from *P. gramineus*, its main feature being the somewhat amplexicaul leaves on the main stem. As mentioned under that species, it is possible that some of the field records of *P. gramineus*, especially those from Ullswater, may refer to the hybrid.
Alt. limit: 270 m, River Irthing, Churnsike Lodge (64.76n)

Map 956   13   WFC

P. alpinus Balbis   Red pondweed
A fairly frequent species of oligotrophic and mesotrophic tarns in the south-western half of the county but with two outlying sites in the north-east at Spadeadam (62.72n) and Tindale Tarn (60.58). It is uncommon in the lakes, with the exception of the northern half of Windermere, and there are a few records from shallow rivers, such as the upper Winster (40.90, 40.92) and Kirkby Pool (22.86). There are no recent records for Esthwaite, Blelham Tarn or Derwent Water, where it was recorded by Pearsall (1923). Wilson cites two rather surprising records from the limestone country of north Westmorland: Maulds Meaburn (6.1) and near Sun-

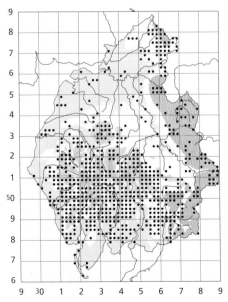

Map 954. *Potamogeton polygonifolius*

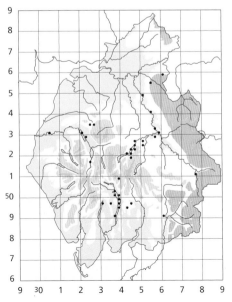

Map 955. *Potamogeton gramineus*

biggin Tarn (6.0), as well as one from the Weel of upper Teesdale. Graham (1988) gives a number of records from the Co. Durham side of the River Tees and Cauldron Snout, but none post-date the completion of the Cow Green Reservoir in 1970. *Potamogeton alpinus* should also be looked for in the extreme north-east as it is known at several sites nearby in Northumberland. Stokoe's (1983) Grasmere (3.0) record is an error.

At the high-level tarns, Hayeswater (42.12) and Angle Tarn, Patterdale (40.14), it grows vigorously in fine silt in 2-3 m of water.
Alt. limit: 490 m, Angle Tarn, Patterdale (40.14)
Map 957   39   WFC

[P. alpinus x P. crispus
P. x olivaceus Baagöe ex G. Fisch.
This rare hybrid is known only from Denmark and Britain where, until recently, it had been recorded only from Scotland and mid-Wales. In 1978 N.T.H. Holmes reported it in the River Eden at Carlisle (40.56, herb. Eng. Nat., *Watsonia* (1979) 12:359) and he subsequently identified material collected by R. Stokoe in 1977 in Small Water, above Haweswater (44.10, herb. R.S., LANC) as the hybrid . This was a surprising find since the site is an upland oligotrophic tarn and one of the hybrid's parents, *Potamogeton crispus*, is characteristic of lowland mesotrophic and eutrophic waters. The site was visited in 1987 by GH and numerous scattered plants were seen in fine silt at about 3 m depth, at the point where the lake bed suddenly deepens. The plants were very distinctive with their narrow, distant leaves. This material has been carefully examined by C.D. Preston who is unconvinced of the original identification although not able to identify the material with certainty. In view of Preston's subsequent discovery of *P. perfoliatus* x *P. crispus* at Holmes' original site at Carlisle, the existence of *P. x olivaceus* in Cumbria should be regarded with some doubt.
                                                        WC]

P. praelongus Wulfen
Very rare. Found during the Survey at five sites in Ullswater: near Glencoyne Bridge (38.18, J.M. Lock,

1978, LANC), east of Aira Point (40.20, M. Wade, 1978, UTLH), Howtown Bay (44.18, GH, 1988, LANC) and Sharrow Bay (44.20, 44.22, RS, 1980, herb. R.S.); in Windermere near Watbarrow Point (36.00, P. Clark & M. Wade, 1981, UTLH); in Hayeswater (42.12, GH, 1982, LANC) and in Talkin Tarn (54.58, RS, 1977, herb. R.S.), where it was apparently first collected by R.C.L. Howitt in 1960 (BM, PLYP). This contrasts with Baker's and Pearsall's (1923) records for Coniston Water, Grasmere and Rydal Water, where, as in Windermere, Baker describes it as "very abundant"; he also gives a record from Angle Tarn, Patterdale (4.1, W. Borrer, 1857, MANCH), where subsequent search has failed to rediscover it. Hodgson cites further records from Overwater (2.3) and from the River Eden at Edenhall (5.3). Finally, Pearsall (1921) lists it from near Dalton-in-Furness (2.7), Esthwaite (3.9) and Derwent Water (2.1/2.2). In this paper he comments on the appreciably broader leaves of the Esthwaite plants. The *Scarce Plants* records for the Maryport (0.3) and Warcop (7.1) squares are errors.

The plants resemble *P. x zizii* in being very luxuriant and growing in deep water, usually 2-3 m. The only flowering material collected during the Survey was from Talkin Tarn.
Alt. limit: 420 m, Hayeswater (42.12), although it previously occurred at 490 m in Angle Tarn, Patterdale (40.14).
Map 958   S 8   W(F)C

P. perfoliatus L.   Perfoliate pondweed
An attractive and very distinctive species which is frequent in Coniston, Bassenthwaite, Windermere and Ullswater, and also in the River Eden. Elsewhere most of the records are from isolated tarns. Surprisingly it is absent from the Cumbrian section of the Lancaster - Kendal canal. Although listed from Crummock Water by Hodgson, there are no subsequent records.

Like *Potamogeton x zizii*, it grows well in deep water: at the north end of Coniston it was seen by

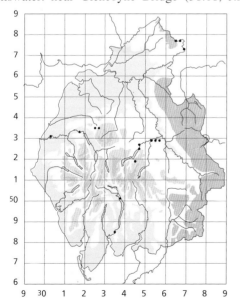

Map 956. *Potamogeton gramineus* x *P. perfoliatus*

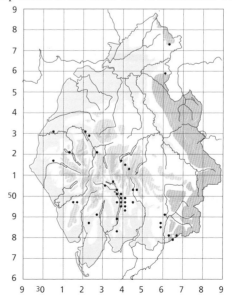

Map 957. *Potamogeton alpinus*

J.M. Lock in 3 m of water and with stems 2 m long.
Alt. limit: 215 m, Tindale Tarn (60.58)
Map 959   62   WFC

P. perfoliatus x **P. crispus**
P. x cooperi (Fryer) Fryer
Found in 1989 in three tetrads along the River Eden at
Carlisle (38.56, 40.56, 42.56, C.D. Preston, CGE;
LANC, *Watsonia* (1993) 19:292). This is the first
definite record of this rather rare hybrid from north-
west England but, as mentioned earlier, it seems quite
likely that the material of *Potamogeton* x *olivaceus*
collected earlier from the same area by N.T.H. Holmes
is *P.* x *cooperi*.

3   C

P. pusillus L.   Lesser pondweed
Local but well scattered, occurring chiefly in eutrophic
lowland tarns especially near the coast. It occurs also in
the Lancaster - Kendal canal and in Esthwaite (36.94)
and Bassenthwaite (20.30, 20.32). Stokoe's (1983)
Windermere record is an error. Early records for this
species are confused since the name was misapplied to
Potamogeton berchtoldii. The first Furness record is
given by Pearsall (1921) who lists it, under *P. panor-
mitanus*, from Esthwaite (1920, K). That for Westmorland was from Lily Bay, Belle Isle, Windermere (3.9,
C. Waterfall, 1895, BM) and for Cumberland from
Derwent Water (2.1/2.2, H. Groves, 1883, BM).

Although very similar to *P. berchtoldii*,
P. pusillus can be recognised by the absence of nodal
glands and its tubular, not convolute, stipules.
Alt. limit: 215 m, Tindale Tarn (60.58)
Map 960   22   WFC

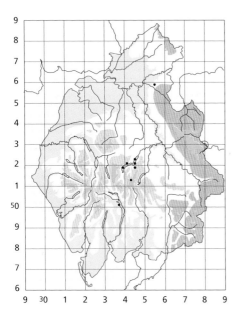

Map 958. *Potamogeton praelongus*

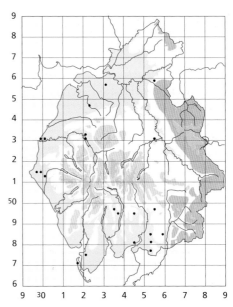

Map 960. *Potamogeton pusillus*

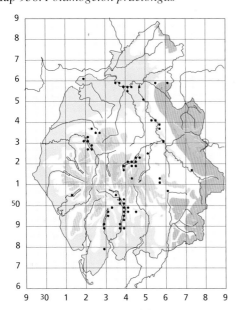

Map 959. *Potamogeton perfoliatus*

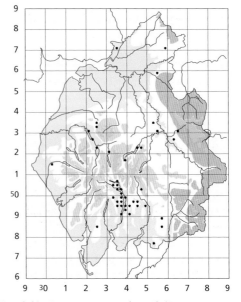

Map 961. *Potamogeton obtusifolius*

P. obtusifolius Mert.& Koch Blunt-leaved pondweed
Frequent in the south-east in tarns and field ponds, also
in Esthwaite (3.9) and the northern half of Windermere. Elsewhere it is rather local and it is absent from
the western lakes, from Coniston to Buttermere. It
occurs sometimes in abundance, as at Eskrigg Tarn
(56.88s), but it apparently avoids calcareous and
running water.

The first Furness record dates from 1914 when it
was found by W.H. Pearsall in Urswick Tarn and Esthwaite (2.7, 3.9, BM). Stokoe's (1983) records for Wast
Water, Ennerdale and Coniston appear to be errors.
Alt. limit: 315 m, Skeggles Water, above Longsleddale
(46.02)
Map 961   44   WFC

P. berchtoldii Fieber
A common pondweed occurring in lakes and tarns and,
like *Potamogeton pusillus*, rarely occurring in running
water. However, unlike that species it is appreciably
more tolerant of oligotrophic waters, being present in
all the major lakes except Buttermere, Ennerdale and
Wast Water, although earlier collected in the last two
by W.H. Pearsall (1921, BM; 1919, OXF). It also
ascends higher. The first Furness record was from
Coniston Water (J. Comber, 1911, BM).

Early records for this species usually appear
under the name *P. pusillus* and deep water Lake
District plants with paler and wider (-2 mm) leaves
were described as subsp. *lacustris* Pearsall & Pearsall f.

Potamogeton berchtoldii is usually readily
identifiable by the pair of conspicuous glistening,
spherical nodal glands and the less obvious convolute
stipules.
Alt. limit: 505 m, Dogber Tarn, Long Fell (78.18)
Map 962   144   WFC

P. crispus L.   Curled pondweed
Widespread in mesotrophic and eutrophic tarns, lakes
and rivers, being particularly frequent in the River
Eden and its tributary the Lowther. It has apparently
disappeared from Crummock Water, Derwent Water,
and Windermere. The first localised Furness record is
from Esthwaite (3.9, Pearsall (1923)).
Alt. limit: 260 m, River Irthing, Butterburn (68.76n)
Map 963   146   WFC

P. pectinatus L.   Fennel pondweed
An uncommon but characteristic species of coastal
pools. These are usually brackish and sometimes
extremely eutrophic. At some sites, such as Cavendish
Dock, Barrow-in-Furness (20.68s) and by Old Park
Wood, Holker (32.76s), it occurs in considerable
quantity, often to the exclusion of other aquatic
macrophytes. This coastal tendency is characteristic of
the species in northern and western Britain, but there
are two recent authenticated inland records in Cumbria:
Thurstonfield Lough (30.56/32.56, RS, 1977, herb.
R.S.) and, in the north-east, Tindale Tarn (60.58,

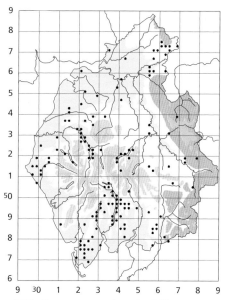
Map 962. *Potamogeton berchtoldii*

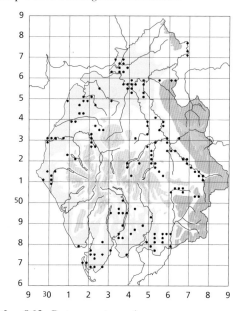
Map 963. *Potamogeton crispus*

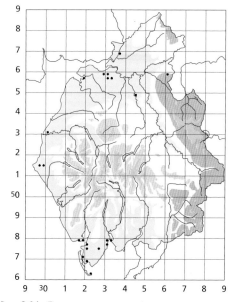
Map 964. *Potamogeton pectinatus*

A. Newton, 1976; herb. R.S.). Baker's record from Bassenthwaite is treated by Hodgson as an error, but there is a surprising Lake District record in J.E. Dandy's Index from Tarn Hows (3.9, W.H. Pearsall, 1913). This and Pearsall's record of the same year from Ormsgill Reservoir, Barrow (1.7, BM) are the first for Furness.

Although superficially similar to other Cumbrian narrow-leaved species, *Potamogeton pectinatus* is distinguished by the stipule being not a separate, tubular sheath around the stem but fused to the base of the leaf forming a leaf-sheath. The divergent branching is usually a good diagnostic feature. In brackish water the plant may be rendered almost rigid by encrusting algae.

Alt. limit: 215 m, Tindale Tarn (60.58)

Map 964   21   FC

Groenlandia densa (L.) Fourr.

       Opposite-leaved pondweed

Common in the upper part of the River Eden and its limestone tributaries the Lowther and Helm Beck but curiously absent from the intervening Hoff and Lyvennet Becks. Outside this area there are single records from the lower Eden (46.56) and the south of the county, Larkrigg pond, Sedgwick (50.86s, CEW, 1986).

       This pondweed has suffered a general contraction of its range this century outside its headquarters in north Westmorland. There is an old record from the River Bela at Milnthorpe (5.8s, W.C. Worsdell, 1906, BM) and Hodgson lists it from Briggle Beck, Langwathby (5.3), a brickfield at Etterby, Carlisle (3.5) and, most remarkably, from the River Ellen below Aspatria (0.3, 1.3). At this last site, in the mill-race at Allerby, he describes it as occurring with *Zannichellia* and other aquatics and forming "in the autumn a mass of floating aquatic vegetation, which in

its downward course becomes a source of trouble to the tenant of the mill".

       There are also early records from the River Eden at Great Salkeld (5.3, H. Britten, 1900-1908), Carlisle (3.5/4.5, J. Parkin, 1935) and, more recently, Warwick Bridge (4.5, C.W. Muirhead, 1956). The *Atlas* dots for the Arnside (4.7s) and Burton-in-Kendal (5.7s) squares refer to Lancashire and that for the Hawkshead square (3.9) is an error.

Alt. limit: 300 m, Hardendale (58.12)

Map 965   21   W(C)

RUPPIACEAE

Ruppia maritima L.   Beaked tasselweed

A rare plant of brackish coastal pools and ditches, although sometimes occurring in abundance. It is encouraging that it still persists at the two sites given by Hodgson: Skinburness (12.54, W. Hodgson, 1890, LIV; LANC) and Workington (00.28), but it has disappeared from Rockcliffe Marsh (3.6), where it was found by H. Duncan in 1949, and from the two 19th century Kent estuary sites (4.7s, 4.8s) given by Wilson. The species clearly has its present headquarters in the Barrow-in-Furness area, where it was originally reported by W.H. Pearsall from South Walney in 1913 (2.6s, *BEC Rpt* (1914) 3:396).

Map 966   13   (W)FC

R. cirrhosa (Petagna) Grande   Spiral tasselweed

Found independently by PB and K. Raistrick in 1991, growing luxuriantly with *Potamogeton pectinatus* in a brackish mine pool near Askam-in-Furness (20.76, PB, LANC, conf. C.D. Preston, *Watsonia* (1992) 19:151).

       This is one of only two extant sites on the west coast of mainland Britain.

       S 1   F

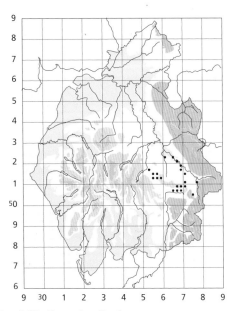

Map 965. *Groenlandia densa*

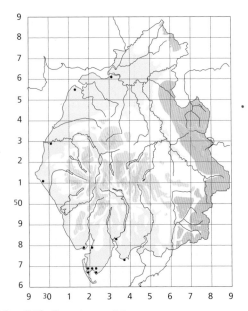

Map 966. *Ruppia maritima*

NAJADACEAE

Najas flexilis (Willd.) Rostkov & W. Schmidt
Known only from Esthwaite Water (34.96, BM) where it was discovered, new to England, by W.H. Pearsall in 1914. The find was immediately reported by Bennett (1914c, 1915). The discovery was presumably only a matter of days before that of *Hydrilla verticillata* with which it grew. Other associated species included *Callitriche hermaphroditica, Potamogeton alpinus* and *P. obtusifolius.* They were growing in peaty water at a depth of almost 2 m.

The *Najas* was said at that time to be abundant but, like *Hydrilla*, and indeed all the associated species, it has decreased considerably over the years, probably as a result of increased eutrophication and turbidity of the lake. It was last seen in 1981 by M. Wade and R. Wright (UTLH).

This is still the only English record for a species which elsewhere in Britain is restricted to the Hebrides, central east Scotland and Kintyre.

S 1   F

ZANNICHELLIACEAE

Zannichellia palustris L.   Horned pondweed
Common and locally abundant in the Eden valley, both in the main river and in tributaries and ponds; elsewhere here and there around the coast, chiefly in brackish pools, but recently found in Windermere.

This is a rare example of a native aquatic plant which has increased markedly, almost explosively yet unnoticed, in Cumbria during the present century. Hodgson gives only five sites, Wilson only one and none of these is from the Eden valley. The *Atlas* shows it in only four post-1930 squares whereas the Survey records are from 29. It is difficult to suggest any satisfactory explanation for this except possibly the general increase in eutrophication. It is strange that it should

have apparently disappeared from its two late-19th century sites in the Lake District: Portinscale, Derwent Water (2.2) and the Goldrill Beck, Ullswater (3.1).

The first Furness record is W.H. Pearsall's from Barrow-in-Furness (2.6/2.7, 1913, *BEC Rpt* (1914) 3:396)
Alt. limit: 215 m, Helbeck Wood (78.16)
Map 967   54   WFC

ZOSTERACEAE

Zostera marina L.   Eelgrass
Hodgson lists two 19th century records from the Cumberland coast, Bootle (0.8) and Maryport (0.3). At both sites it was probably washed up, perhaps originating from the north side of the Solway Firth, rather than growing *in situ.* It was reported from South Walney (2.6s) by the Barrow-in-Furness N.H.S. during the 1950s and by D.S. Ranwell from Cavendish Dock, Barrow (2.7) in 1961. It has not been seen since at either of these Furness sites and it is quite possible that the records refer to the following species.

There appear to be no extant sites for this species in north-west England but it still occurs on the Isle of Man.

S   ([F]C)

Z. angustifolia (Hornem.) Reichenb.
                    Narrow-leaved eelgrass
Extremely rare. Seen during the Survey only from the area around the south end of the Walney Channel where it occurs on both sides of the Roa Island causeway (22.64s, 22.66s, T.C.G. Rich, 1980; LANC), Foulney Island (24.64s, M. Alison, 1977), on Roosecote Sands (20.66s, T.G. Piearce, 1982, LANC) and off Wylock Marsh, South Walney (20.64s, Anon.). It was first reported by W.H. Pearsall from Roa Island and Walney Island in 1913 (*BEC Rpt* (1914) 3:396) and later at these sites by the Barrow-in-Furness N.H.S. during the 1950s and by D. Wood in 1961 (LIV).

Most of the sites are in silty runnels, often between stones and just below the high-water mark of

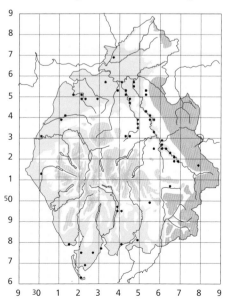

Map 967. *Zannichellia palustris*

Plate 100. *Zostera angustifolia*

between stones and just below the high-water mark of neap tides. The leaves vary very considerably in width, those of the two herbarium specimens referred to ranging from 0.6 to 2.6 mm. The identifications were confirmed by T.G. Tutin.

Plate 100   S 5   F

ARACEAE

*Acorus calamus L.   Sweet flag

Seen during the Survey by Stokoe (*Watsonia* (1978), 12:178; (1979) 12;361) at the following scattered sites: in a fen at Brigham, where it was originally discovered in 1949 (08.30, J.D. Hinde, CLE), by estate ponds at Hutton-in-the-Forest (46.34, 1977), at Holehird, Troutbeck (40.00, 1978), by Esthwaite Water (34.96, 1977, herb. R.S.), and at Eskdale Green (14.00, 1978). There is also a record from near Sowerby Row (38.40, EHR, 1987). It is quite probable that the species has been intentionally introduced at all these localities, which are the most northerly on the west coast.

Although this species is common in the Lancashire section of the Kendal canal, it stops just short of the county boundary.

Alt. limit: 180 m, Hutton-in-the-Forest (46.34)

6   WFC

*Lysichiton americanus Hultén & H. St John

Skunk cabbage

This aroid of the Alaskan - Canadian Pacific coast has been effectively naturalised in a woodland tarn near Witherslack Hall (42.84s) for many years. In 1981 it was found flourishing with *Caltha* and *Chrysosplenium oppositifolium* below a tip of garden rubbish in Holbeck

Gill, Ambleside (38.00, GH, LANC, *Watsonia* (1982) 14:198) and in 1985 apparently self-sown plants were found far from houses by the River Eden and also on an island in the river east of Cotehill (48.52, R. Little; 48.50, FJR). It has since been found at five other sites including one in Furness near Lowick (28.84, J. & R. Clinch, 1987). Most sites are in damp woodland with seepage areas or frequent standing water.

This is a very conspicuous plant early in the year with its large, bright yellow spathes.

Map 968, Plate 101   16   WFC

*Calla palustris L.

Found, new to North-West Yorkshire, v.c.65, by CEW in 1986 growing in a small pond near Sedbergh (66.90, LANC). He subsequently discovered a small, vegetative plant by the Peasey Beck, near Bendrigg, Killington (58.88s, LANC) in 1988. The two sites were between 150 and 200 m from the nearest house.

Alt. limit: 170 m, Bendrigg, Killington (58.88s)

2   WY

Arum maculatum L.   Lords and ladies, Cuckoo pint

Frequent and, like *Orchis mascula*, a good indicator of base-rich woodland soils, not necessarily on limestone, being abundant in the south and in a broad arc along the west side of the upper Eden valley and around the north of the Lake District to the west coast. It is most characteristic of oak - ash woodland, copses and hedgerows. The species becomes far less common immediately to the north across the Solway Firth.

As in the Isle of Man, the vast majority of plants are unspotted.

Alt. limit: 380 m, above Bannerdale (42.14)

Map 969   699   LWFCY

*A. italicum Miller

A colony of this southern arum was found in 1982 in a laneside by Levens Hall (48.84s, MB & R. Dalton).

Plate 101. *Lysichiton americanus*, Witherslack

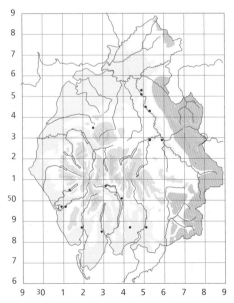

Map 968. *Lysichiton americanus*

It no doubt originated from nearby gardens. A second colony was discovered by PB on the beach at Aldingham (28.70, 1988) with such garden throw-outs as *Petasites fragrans* and *Fallopia sachalinensis*.

2   WF

LEMNACEAE

Lemna gibba L.

Extremely rare although sometimes locally abundant. It was found in 1978 by M. Wigginton in the Lancaster - Kendal canal at Stainton (52.84s, *Watsonia* (1979) 12:361) and in 1982 traced by T.C.G. Rich southwards through four more tetrads to the county boundary (52.76s, 52.78s, 52.80s, 52.82s, LANC). It has also been seen in quantity in the drain at Foulshaw (46.84, CEW, 1992). The only previous record is of a find by N.D. Simpson from Fleswick Bay, St Bees Head (9.1, *BSBI Yr Bk* (1951) p.49) in 1949. This last site was visited by R. Stokoe in 1979 who reported that the plants were "not gibbous but certainly a bit spongy below". No material was collected and the record is best omitted.

These are the most northerly sites in Britain. The material is somewhat atypical in having appreciably less well developed aerenchyma than is found in specimens from the south of England and they can easily be confused with *Lemna minor*.

5   W

L. minor L.   Common duckweed

Common in the lowlands in similar habitats to the last but extending into mesotrophic waters and appreciably more widespread. In small water bodies and occasionally in the Lancaster - Kendal canal it may become so abundant as to completely cover the surface.

Alt. limit: *c*.300 m, Spadeadam (62.74n), but there is a pre-Survey record from Middle Tongue Beck, Cross Fell (6.3) at 420 m.

Map 970   542   WFCY

L. trisulca L.   Ivy-leaved duckweed

A southern species in Britain, and rather scarce in Cumbria, being coastal and almost confined to the south. It occurs in eutrophic pools, tarns, ditches and the Lancaster - Kendal canal and, more rarely, as in Windermere and Esthwaite Water, in sheltered bays of the lakes.

Alt. limit: 130 m, Whinfell Tarn, Grayrigg (54.98)

Map 971   25   WFC

JUNCACEAE

Juncus squarrosus L.   Heath rush

Common to locally dominant on poorly drained acid peat, both on lowland mosses and heaths and on upland blanket peat in the Lake District and the Pennines where it is often associated with *Eriophorum vaginatum* and *Trichophorum cespitosum*.

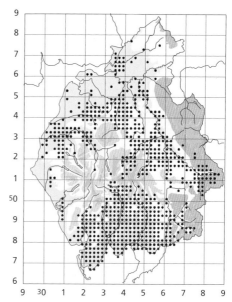

Map 969. *Arum maculatum*

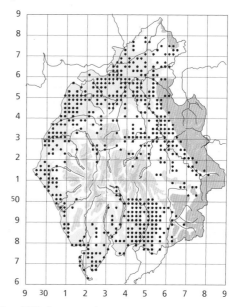

Map 970. *Lemna minor*

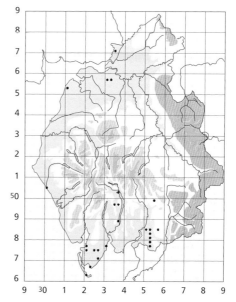

Map 971. *Lemna trisulca*

Alt. limit: 920 m, Helvellyn (34.14, Wilson).
Map 972   1083   WFCY

*****J. tenuis** Willd.

Not uncommon in the lowlands, almost always on or by paths, green lanes, roadsides and rutted forest rides. This North American rush was first recorded in Cumberland in 1924 when it was seen at Nether Wasdale (1.0) and Lady Hall, Millom (1.8) by M. Cobbe and W.H. Pearsall respectively (*BRC Rpt* (1925) 7:599-600) and in Westmorland in 1933 when it was seen at Flakebridge (6.2) and Brackenber Moor (7.1). The earliest Furness records are those shown in the *Atlas*.
Alt. limit: 430 m, by the coal road above Dent station (76.86) and in the Caldbeck Fells (32.34)
Map 973   217   WFCY

J. compressus Jacq.

Very rare, recorded during the Survey only from sandy riverbank sites by the River Eden near Staffield (52.42, CS, 1977; FJR, 1988, herb. F.J.R.), from rocks by the River Eden at Eden Lacy, Great Salkeld (56.38, RWMC, 1991, LANC) and from roadside verges near Crosby Ravensworth (62.14, 62.16, CW, 1984; 60.16, GH, 1994, all LANC), Newby (58.20, CEW, 1992, LANC) and close to Dent railway station (76.86, E. Shorrock, 1981; LANC). The Eden Lacy site is only a short distance downstream from where it was earlier seen on riverside rocks below Nunwick (5.3, G.A.K. Hervey, 1941, 1947).

The *Atlas* records appear largely to be taken from Wilson and Hodgson. Most are coastal, the only inland sites being Kinniside Common, Ennerdale (0.1/1.1) and the roadside near Kings Meaburn (6.2, A. Wilson, 1935, YRK). The first of these is extremely improbable. There appear to be no specimens to substantiate the several coastal records and these are best treated as dubious, especially as four coastal collections submitted to C.A. Stace have all been redetermined as *Juncus gerardii*. These were from the Leven estuary near Cark (3.7s, M.I. Tetley, 1951, CLE) and probably the basis of the *Atlas* dot, Ravenglass - Eskmeals (0.8, C.W. Muirhead, 1950, CLE, PLYP), Anthorn - Kirkbride (2.5, C.W. Muirhead, 1947, CLE) and Silloth (1.5, J. Leitch, 1878, CLE).

Juncus compressus reaches its northern limit on the west coast on the Scottish side of the Solway Firth.
Alt. limit: 370 m, above Dent station (76.86)
       7   W[(F)]CY

J. gerardii Lois.   Salt-marsh rush

Very common along the coast in the upper parts of salt-marshes, in grazed *Puccinellia - Festuca* turf and along the banks of tidal rivers; scarce or absent from exposed shores. It was introduced, presumably unintentionally, *c.*1980 on Hartside (64.40) with other halophytes from the west coast and now occupies about 25 sq. m.
Alt. limit: 575 m, Hartside (64.40) but introduced.
Map 974   99   WFC

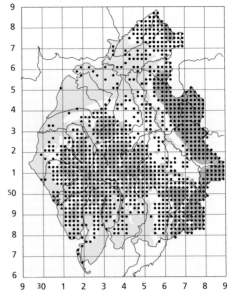

Map 972. *Juncus squarrosus*

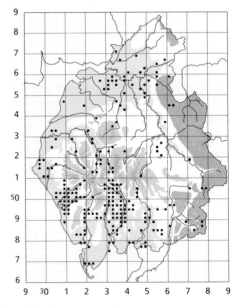

Map 973. *Juncus tenuis*

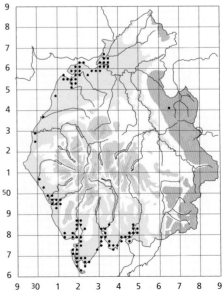

Map 974. *Juncus gerardii*

J. foliosus Desf.

Seen only in a marshy field near St Bees (96.08, CCH, 1978, LANC, det. C.A. Stace, *Watsonia* (1979) 12:360) and by the shore near the south-west end of Wast Water (14.04, CCH, 1979, herb. C.C.H.). The only pre-Survey record is from Borrowdale (2.1, K, det. T. Cope) and based on a specimen collected by A.W. Hill in 1896.

The most obvious distinguishing characters of this predominantly western segregate of the *Juncus bufonius* complex are the wide leaves (>1.5 mm) and the dark line on either side of the midrib of the perianth segments.

$$2 \hspace{5cm} C$$

J. bufonius L. Toad rush

A common and characteristic rush of damp, open communities, especially along rutted tracks and lanes, also on roadside verges, in seasonally wet field entrances and in the upper parts of salt-marshes.
Alt. limit: at least 500 m, Hartside (64.42, 64.44), although Wilson (1956) gives it as 595 m in the Lake District.

Map 975   1406   LWFCY

J. ambiguus Guss.

Seen only at Muncaster salt-marsh (10.96, CCH, 1979, LANC, det. C.A. Stace, *Watsonia* (1981) 13:340), by the saltings on Drigg dunes (06.96, CCH, 1983), on the shore at Askam-in-Furness (20.76, PB, 1990, LANC, det. C.A. Stace) and Arnside (44.78s, MB, 1993).

Cope & Stace (1978) mention an earlier record of this halophytic segregate from Sandscale (18.74, C.A. Stace, 1970, MANCH). Their record for the Arnside square (4.7s) is based on one from Blawith Marsh, Grange-over-Sands, collected by W.M. Hind in 1870 (BM).

This segregate is distinguished by the capsules almost equalling the obtuse inner perianth segments.

$$5 \hspace{5cm} FC$$

J. subnodulosus Schrank

Very rare, seen only at Sandford bog, near Appleby (72.16, CW, 1978, conf. C.A. Stace, *Watsonia* (1979) 12:360), with *Juncus acutiflorus* and *J. articulatus* at Sandscale (18.74, PB, 1993, LANC), in a small marsh near Storth (46.78s, A. Cannell, 1993), and in flushed areas by the shore on both sides of the Leven estuary near Greenodd where it is locally abundant as at Plumpton (32.78s) and Roudsea (32.78s, 32.80, GH, 1986, LANC). Baker records it from this latter area and Hallbank Pond, Aspatria (1.4), and W.H. Pearsall collected it in 1913 from the shore at Roose (2.7, BM, YRK, *BEC Rpt* (1914) 3:499) in 1913.

This generally lowland rush reaches its northern limit on the west coast on the north side of the Solway Firth.
Alt. limit: 140 m, Sandford bog (72.16)

$$7 \hspace{5cm} WF(C)$$

J. alpinoarticulatus Chaix
(*J. alpinus* Villars)

Known only from three tetrads in upper Teesdale where it extends from Co. Durham just into Cumbria near Birkdale Farm (78. 26, 80.26, 80.28), where it was first recorded by W.A. Sledge in 1940. Like another Teesdale speciality, *Kobresia simpliciuscula*, it grows in wet, open, base-rich and often stony flushes.

This rush is restricted in Britain to a few sites in the central Scottish Highlands, the Southern Uplands, Teesdale and the Craven area of Yorkshire.
Alt.: 450 m, Birkdale, upper Teesdale (80.28)

$$S\ 3 \hspace{5cm} W$$

J. articulatus L. Jointed rush

Common, occurring typically in open communities by lakes and ponds and in fellside and upland flushes and seepage areas. It is never dominant to the same extent as *Juncus acutiflorus* and it is usually indicative of some degree of base-enrichment. In the uplands it is a

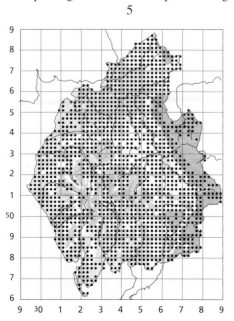

Map 975. *Juncus bufonius*

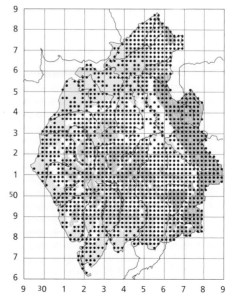

Map 976. *Juncus articulatus*

very characteristic member of base-rich stony flush communities with *Carex hostiana, C. viridula* subsp. *brachyrrhyncha* and *Eleocharis quinqueflora*.
Alt. limit: 810 m, Great Dun Fell (70.30)
Map 976   1427   WFCY

J. articulatus x **J. acutiflorus**
J. x **surrejanus** Druce ex Stace & Lambinon
Very rare but certainly under-recorded. The only Survey records are from five sites in Cumberland: Nethertown (96.08, CCH, 1979, det. C.A. Stace, herb. C.C.H.), Rowrah (06.18, CCH, 1987), Drigg (04.98, GH, 1987, conf. R.H. Roberts) and Coombs Wood, near Armathwaite (50.44, RWMC, 1994, LANC), one in Westmorland at Gillside, Glenridding (36.16, 1980, GH, LANC, conf. C.A. Stace, *Watsonia* (1981) 13:340) and one in Furness at Sandscale (18.84, PB, 1993, LANC), where there are large vigorous clumps. The only previous records are from by Loweswater (1.2, J.D. Hinde, 1949, LANC, PLYP, det. P.W. Richards) and the River Eden at Great Corby (4.5, C.W. Muirhead, 1950, PLYP).

Apart from its low fertility, useful distinguishing characters are the more or less colourless margins of the outer perianth segments and their slightly recurved tips. Blackstock & Roberts (1986) omitted to mention that the capsules may be abortive.
Alt. limit: 215 m, Glenridding (36.16)
Map 977   10   WFC

J. acutiflorus Ehrh.ex Hoffm.
Generally common except in the more intensively agricultural areas of the lower Eden valley, the Solway Plain and around Morecambe Bay; it is also curiously scarce in upper Teesdale and around Alston. It is a plant of ill-drained but not necessarily base-poor soils, often forming dense and extensive stands, both in the lowlands and the uplands, where it is particularly characteristic of valley bottoms.
Alt. limit: at least 600 m, east of Knock Fell (74.30)
Map 978   1289   WFCY

J. bulbosus L.   Bulbous rush
(*J. kochii* F.Schultz)
Very common in the uplands, scattered or even scarce in the lowlands and surprisingly rare in the Alston area (7.4). It occurs almost always in base-poor soils, typically in damp or wet open habitats: by tarns and lakes, in ditches and in small runnels and on damp rutted tracks. As the very distinctive var. *fluitans* (Lam.) Druce it is often dominant in upland rivers, especially in deep dubs, in tarns and in sheltered bays of the lakes.

Plants with six instead of three stamens are often referred to the ill-defined species *Juncus kochii*. It seems likely that, as in Co. Durham (Graham 1988) and the Isle of Man (Allen 1984), such plants are actually appreciably commoner than *J. bulbosus* sensu stricto. Very few records of the latter have been

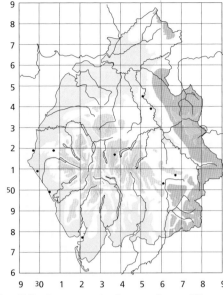

Map 977. *Juncus articulatus* x *J. acutiflorus*

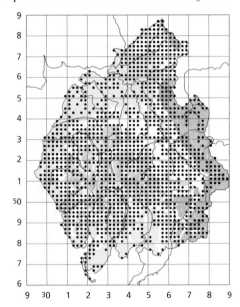

Map 978. *Juncus acutiflorus*

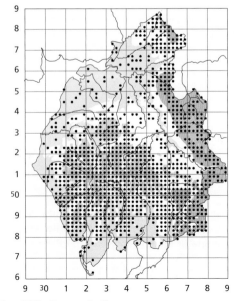

Map 979. *Juncus bulbosus*

sent in and these are all from the coastal lowlands.
Alt. limit: 840 m, Cross Fell (68.34)
Map 979   1080   WFCY

J. triglumis L.   Three-flowered rush
Rather scarce in the Lake District where it occurs
chiefly in the eastern fells, and in the high Pennines. It
is a species of stony, montane flushes and irrigated rock
ledges, rarely, as at Moor House, Teesdale (74.32), on
riverside rocks. Hodgson cites an old unauthenticated
record from Scales Tarn, Saddleback (3.2), but this
seems unlikely.

        South of the Scottish Highlands this rush is
restricted to Cumbria, Teesdale and Snowdonia.
Alt. range: 490 m, Blea Water, Haweswater (44.10) -
855 m, Helvellyn (34.14)
Map 980   41   WC

J. maritimus Lam.   Sea rush
Fairly frequent in the uppermost zone of salt-marshes,
often where there is seepage of fresh water, and
persisting on grazed marshes as small but conspicuous
colonies or isolated clumps. Along the Solway coast it
is surprisingly absent between Bowness and Rockcliffe
Marsh and west of the River Waver.
Map 981   69   WFC

[J. balticus Willd.
A northern rush which reaches its southern limit in Britain on
the Lancashire coast, far distant from its main west coast
sites in the Outer Hebrides. Druce (1932) lists it for v.c.69
and this is repeated by Wilson who adds that it occurs only on
the shores of Furness. The source of this information has
never been traced. In the absence of any herbarium material,
Stace (1972) concluded that the record is best treated as
dubious. The *Atlas* dot is for the Cartmel square (3.7s).
                                        (F)]

J. filiformis L.
Occasional, on stony shores, in wet turf or in marsh
communities by all the major lakes with the exception
of Ennerdale and Wast Water in the west and Ullswater
and Haweswater in the east. It is interesting that it has
been able to withstand the flooding of Thirlmere. There
has been no recent confirmation of the 1946 record
from Dacre Bank (4.2, W. Atkinson, CLE).

        The ecology and distribution of this rush, which
has its British headquarters in the Lake District, has
been discussed in detail by Blackstock (1981).
Alt. limit: 170 m, Thirlmere (32.12)
Map 982   S 25   WFC

J. inflexus L.   Hard rush
Very common in the calcareous meadows and roadside
verges of the upper Eden valley and quite common
throughout much of the northern half of the county
where it prefers the rather heavy clay soils. It is
virtually absent from the siliceous soils of the Lake
District but is locally frequent to the south around

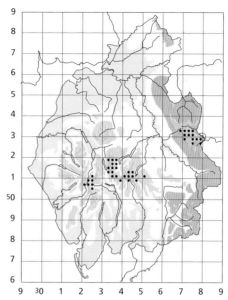

Map 980. *Juncus triglumis*

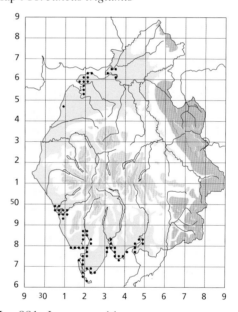

Map 981. *Juncus maritimus*

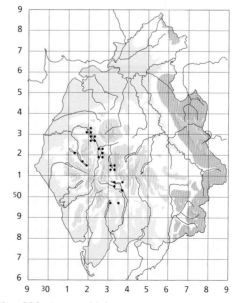

Map 982. *Juncus filiformis*

Morecambe Bay.

Alt. limit: 550 m, Mattergill Sike (7.2, Rawes *et al.* 1969 p.260).

Map 983   695   LWFCY

J. inflexus x **J. effusus**
J. x **diffusus** Hoppe

Very rare but easily overlooked. There are records for eight squares, all additional to that in the *Critical Atlas* for the Crosby Ravensworth (6.1) square and Wilson's for Appleby (6.2, but wrongly mapped as 7.2). The first record for Cumberland is from Lillyhall (00.24, CCH, 1980, herb. C.C.H., det. C.A. Stace, *Watsonia* (1981) 13:340), and for Furness from near the shore at Askam (20.76, PB, 1991, LANC, det. C.A. Stace).

Alt. limit: 305 m, Gilcambon Beck, Greystoke Park (40.32)

Map 984   9   WFC

J. effusus L.   Soft rush

Extremely common, occurring in ill-drained lowland meadows and pastures, often following attempts at reclamation, as well as fellside flushes, in the sykes on blanket peat and abundantly in ditches and around the margins of tarns, also along wet woodland rides and the higher parts of salt-marshes.

Alt. limit: 845 m, Great Dun Fell (70.32)

Map 985   1759   LWFCY

J. conglomeratus L.

Very common but appreciably less so than *Juncus effusus* with which it commonly grows. It is, however, more restricted than that species to base-poor soils and it rarely if ever occurs in ditches, by tarns and in upland flushes. Curiously Eddy & Welch (1967) state that in the Moor House Reserve, Teesdale, it is chiefly on calcareous drift and Allen (1984) says it occurs in the Isle of Man in base-rich bogs.

It is not possible to add anything to the controversy surrounding the degree of hybridisation between these two species. Apparent intermediates could be hybrids or merely one end of the wide range of variation shown by *J. effusus*. According to Stace (1991), hybrids are infrequent, highly fertile and difficult to identify from herbarium material.

Alt. limit: 745 m, Milburn Common (7.3, Wilson 1956)

Map 986   1440   LWFCY

Luzula pilosa (L.) Willd.   Hairy wood-rush

Occasional to frequent in damp woodland on both siliceous and calcareous soils. It is particularly characteristic of rocky wooded gills where it grows with *Vaccinium myrtillus, Luzula sylvatica* and *Melampyrum pratense*, but also on peat in woods bordering the Solway mosses.

Alt. limit: at least 420 m, upper Swindale, Brough (82.18), although Wilson (1956) gives it as 510 m in the Lake District.

Map 987   503   LWFCY

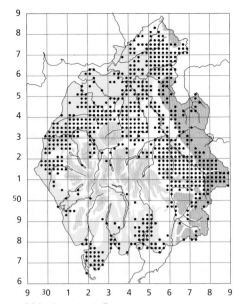

Map 983. *Juncus inflexus*

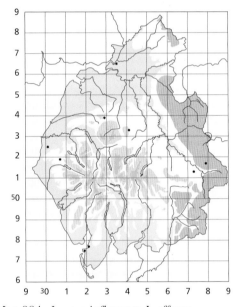

Map 984. *Juncus inflexus* x *J. effusus*

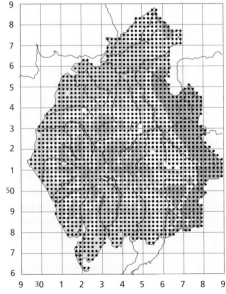

Map 985. *Juncus effusus*

L. sylvatica (Hudson) Gaudin   Great wood-rush
Common except in the more intensively farmed areas
of the Solway Plain and around Morecambe Bay, and
on the limestone uplands of north Westmorland. It is
typically a species of siliceous woodland being an
almost constant and dominant feature of rocky wooded
gills and particularly abundant in the valleys north-east
of Carlisle. It also occurs on sea-cliffs at St Bees (9.1)
and very commonly on rock ledges and by gullies in the
fells.

    Although intolerant of grazing, areas of stunted,
vegetative plants are sometimes encountered high on
the fells as, for example, around the summit of The
Calf at 640 m in the Howgill Fells (66.96), and at much
the same height on the western escarpment of the
Pennines and, somewhat lower, on the fells around the
head of the River Caldew (30.32).
Alt. limit: 855 m, Helvellyn (34.14)
Map 988   785   LWFCY

*L. luzuloides (Lam.) Dandy & Wilm. White wood-rush
A rare alien of usually shady grassland, sometimes
naturalised in estate grounds, as at Appleby Castle
(68.18), and by railwaysides, as on the disused railway
at Troutbeck (32.24) and high up above Dentdale by
Dent station (76.86, 76.88).

    The first record dates from 1923 when it
was found by G. Bacon in woodland at St Bees (9.1,
BEC Rpt (1924) 7:600). A few years later it was seen
by L.C. Kennedy at Grange-in-Borrowdale (2.1, *BEC
Rpt* (1934) 10:545) and later by K. Hollick in the
grounds of Briery Close, Ambleside (3.0, 1950s) and at
Bowness-on-Windermere (4.9, M.I. Tetley, 1951, CLE,
PLYP.

    This widespread Continental species has a very
scattered distribution throughout Britain. It may well
have been introduced for ground-cover.
Alt. limit: 365 m, above Garsdale Head (78.90)
Map 989   11   WCY

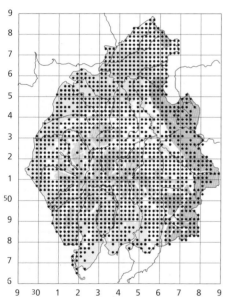

Map 986. *Juncus conglomeratus*

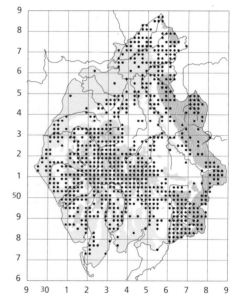

Map 988. *Luzula sylvatica*

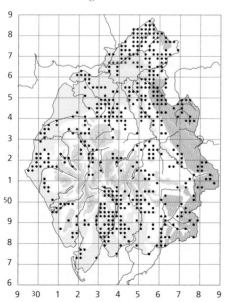

Map 987. *Luzula pilosa*

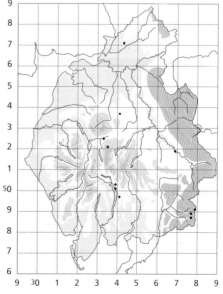

Map 989. *Luzula luzuloides*

L. campestris (L.) DC.   Field wood-rush
Very common in short, usually well-drained grassland, and probably only absent from the highest fells and areas of blanket peat. It also occurs as a weed in lawns. Although sometimes regarded as avoiding base-poor soils and even as rather a lowland species, this is certainly not the case in Cumbria where it is widespread throughout the Lake District.
Alt. limit: 845 m, Great Dun Fell (70.32)
Map 990   1522   LWFCY

L. multiflora (Ehrh.) Lej.   Heath wood-rush
Very common in the uplands, thinning out towards the coast and also, apparently, in the western half of the Lake District. It is a species of base-poor grassland, of lowland heaths and mosses, upland blanket peat and mountain crags.
        Reliable identification of this and the last species often requires careful examination of the anthers, particularly their length, and of the seeds, although *Luzula multiflora* subsp. *congesta* (Thuill.) Arcang. is very distinctive and, as Hodgson observes, often the commoner.
Alt. limit: 915 m, Skiddaw (2.2, Hodgson)
Map 991   1245   WFCY

[L. spicata (L.) DC.
Listed by Baker from Blake Fell, above Lamplugh (1.1), Helvellyn (3.1) and Fairfield (3.0). The occurrence of this species of the Scottish Highlands on these last two mountains is not inconceivable and it is interesting that Wilson gives three sources for the Helvellyn record. Nevertheless, in the absence of any supporting herbarium material these records are best considered suspect.

                                                                (WC)]

CYPERACEAE

Eriophorum angustifolium Honck.
                                        Common cotton-grass
Very common to abundant in the uplands in valley mires and hillside flushes, in the hollows of blanket peat and a vigorous colonist of eroding peat. In the lowlands it is largely restricted to fragmented areas of mossland, wet heath and undrained marshy fields. Although often considered to be calcifuge, it is in fact frequent in calcareous flushes.
Alt. limit: 870 m, Cross Fell (68.34)
Map 992   950   LWFCY

E. latifolium Hoppe
Occasional in rather open vegetation in calcareous or base-rich flushes, chiefly on the Asby - Orton limestones and on the Bewcastle Fells in the northeast, otherwise very local. Some of the sites are surprisingly undistinguished, such as the *Carex rostrata* boggy hollow above Casterton (64.78s) and that on the ridge between Bannisdale and Longsleddale (50.02).
        There are several additional sites where the species has not been recorded since it was seen by DAR

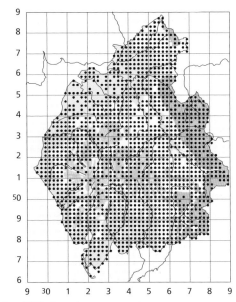

Map 990. *Luzula campestris*

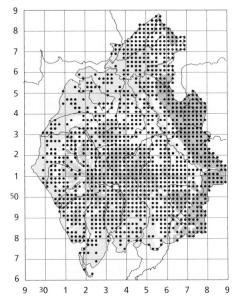

Map 991. *Luzula multiflora*

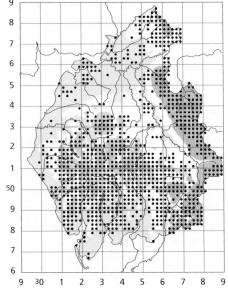

Map 992. *Eriophorum angustifolium*

in the l950s, notably at five sites in the Bewcastle Fells (56.78n, 56.82n, 58.72n, 58.76n, 60.76n), some of which have succumbed to afforestation, Midgeholme Moss, Gilsland (60.66), Cumwhitton Moss (50.50) and Armboth Fell (28.14). It has certainly disappeared from Newton Reigny Moss (46.30, C.W. Muirhead, CLE), where it was last seen in l949, and it is only a matter of time before it disappears from one of the Bewcastle sites (58.76n) as a result of afforestation. However, what is particularly encouraging is the number of records and squares (20) revealed by the Survey compared with the six post-1930 ones in the *Atlas*. Of particular interest is its discovery in quantity in a disused quarry at Kirkland (06.16, P. Ullrich, 1991, LANC).

The only Furness record is from Tottlebank Moss (26.88) where it was found by T.H.L. Adams in l988.
Alt. limit: 520 m, north-west of Yad Moss (76.38)
Map 993   41   WFC

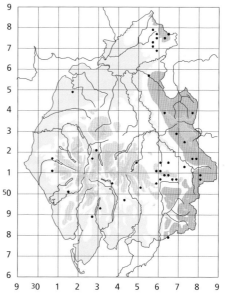

Map 993. *Eriophorum latifolium*

E. vaginatum L.   Hare's-tail cotton-grass
Common to abundant on sheep-grazed upland blanket peat, sometimes, as in the Pennines, co-dominant with *Calluna* over extensive areas. It is uncommon in the lowlands where, like *Eriophorum angustifolium*, it is chiefly a species of the coastal mosslands.
Alt. limit: 890 m, Cross Fell (68.34)
Map 994   720   WFCY

Trichophorum cespitosum (L.) Hartman   Deer grass
Common, like the last species, on upland blanket peat and on lowland mosses and with a substantially similar distribution although appreciably commoner in the western Lake District. Viviparous plants were found in Swindale (50.12, RWMC, 1994, LANC) and they have previously been reported from Scotland and Northumberland.

Swan (1993) has also found plants in Northumberland which appear to match the Continental subsp. *cespitosum*, characterised by a small (1 mm) orbicular opening to the uppermost sheath and spikelets with 10 flowers or fewer.
Alt. limit: 930 m, Helvellyn (3.1, Wilson)
Map 995   658   WFCY

Eleocharis palustris (L.) Roemer & Schultes
                        Common spike-rush
Common and often abundant in a wide range of mainly lowland aquatic habitats: ponds, tarns, lakes and riversides, ditches, wet meadows and swamps, also in coastal dune-slacks and seepage areas in salt-marshes where it maybe confused with *Eleocharis uniglumis*.

The Cumbrian plants probably all belong to the widespread subsp. *vulgaris* Walters.
Alt. limit: c.550 m by the River Tees near Tyne Head (74.34)
Map 996   610   WFCY

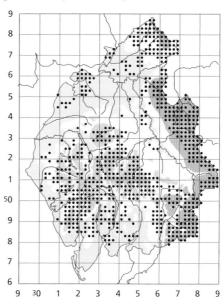

Map 994. *Eriophorum vaginatum*

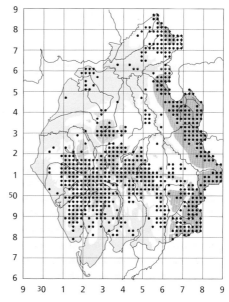

Map 995. *Trichophorum cespitosum*

E. austriaca Hayek
Although collected for the first time in Britain in
Wharfedale in 1947, the specimens were not identified
as this species until 1960. This and subsequent finds
have been described by Walters (1963). It is now
known from several sites in the Craven area of Yorkshire, the North Tyne valley, across the Border in
Selkirkshire, and from the River Irthing above Gilsland.

It was originally found by the River Irthing on
the Northumberland side by G.A. Swan in 1962 and
by FJR on the Cumberland side during the Survey
(68.76n, 1976; 68.72n, GH, 1977, LANC). It is
restricted to the calmer stretches of the river, particularly small, silty backwaters, and the colonies are
quite small and always at least partially submerged. It
is. however, easily overlooked. The distinctive features
include a short rhizome with numerous, clustered
fertile stems, the spikes being somewhat darker and
more broadly conical, compared with the cylindrical to
elliptic ones of *Eleocharis palustris*.

The species has a wide distribution on the
Continent from the mountains of central Europe to
northern Scandinavia.
Alt.: 270 m, River Irthing above Gilsland (68.76n)
R 2   C

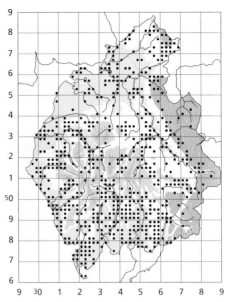

Map 996. *Eleocharis palustris*

E. uniglumis (Link) Schultes
A rather rare coastal species of dune-slacks and salt-
marshes, often growing with *Eleocharis palustris* and
easily confused with it. Field records are notoriously
unreliable and the *Atlas* records for the Great Strickland (5.2) and Tebay (6.0) squares are best treated as
dubious. The only authenticated inland records are
from two seepage areas in the Duddon valley north of
Broughton-in-Furness (20.90, MMG, 1983, LANC), by
Knittleton Tarn (24.86, RS, 1977, herb. R.S.) and by
the upper Irthing (64.68n, 66.70n, FJR, 1991, LANC).
The species should be looked for on the west coast at
Workington, where it was seen at the mouth of the
River Derwent in 1958 (9.2, W. Frame, CLE). The first
record for Cumberland was from Anthorn (1.5, U.K.
Duncan, 1949, CLE) and for Furness from Roudsea
(3.8, JW, 1956).
Alt. limit: 310 m, Broughton-in-Furness (20.90)
Map 997   45   WFC

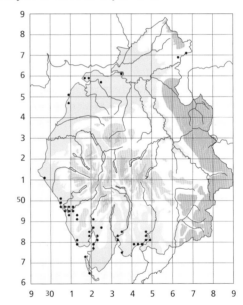

Map 997. *Eleocharis uniglumis*

E. multicaulis (Smith) Desv. Many-stalked spike-rush
A predominantly south-western species with a similar
distribution to *Eleogiton fluitans*. It occurs chiefly in
valley mires in the Lake District, but it extends down to
the coastal mosses by Morecambe Bay and dune-slacks
at Drigg (04.98, 06.96) and near Beckfoot (08.50).

The *Atlas* dot for the Gilsland square (6.6,
v.c.67/70) is probably based on G.A.K. Hervey's record
from Wiley Sike (6.7n). This is possibly a transcription
error since he refers in his diary for 1949 to "creeping
Scirpus". However, it is worth mentioning that this
south-western species occurs on the Co. Durham side of

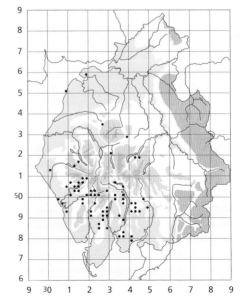

Map 998. *Eleocharis multicaulis*

upper Teesdale (Graham 1988).

The first certain record for Furness is that in the *Atlas* for the Hawkshead square (3.9); Baker's site is in Westmorland.

Alt. limit: possibly 215 m, above Staveley (46.98)

Map 998   68   WFC

E. quinqueflora (F.Hartmann) O.Schwarz

A characteristic species of open, base-rich or calcareous flushes, chiefly in the uplands and commonest in the southeast but also frequent in the Bewcastle Fells in the extreme north-east. It is commonly associated with *Carex lepidocarpa*, *C. hostiana* and *C. dioica*.

On the coast it is frequent in dune-slacks by the Ravenglass estuary (0.9), near Beckfoot (08.50) and at Sandscale by the Duddon (18.74) and it occurs on the upper salt-marsh around Anthorn (1.5, 1.6n) and Bowness-on-Solway (2.6n).

Alt. limit: 825 m, above Red Tarn, Helvellyn (3.1, Wilson)

Map 999   275   WFCY

E. acicularis (L.) Roemer & Schultes

A very rare species of pond, lake and river margins in the north and east. It grows in shallow water and flowers when exposed during the summer. It was seen in a village pond near Abbeytown (18.52, FJR, 1978; LANC, *Watsonia* (1979) 12:361), locally frequent at four sites on the exposed gravel and sandy bed of the River Eden at Carlisle (40.56, 42.56, REG, 1989, LANC) and on an island in the river below Park Broom (42.58, FJR, 1996). It was also refound at Thurstonfield Lough (30.56, 32.56), where it was first noted in l937 (3.5, R.W. Butcher, *BEC Rpt* (1938) 11:512), and in a wet depression by the River Lowther near Knipe (50.18), where it was seen in 1957 by H. Milne Redhead (*Proc. BSBI* (1959) 3:199). Wilson also coll-

ected it from this area (1934, 1937, YRK, *BEC Rpt* (1934) 10:845) but from appreciably higher up, on Knipe Moor, a record erroneously shown in *Scarce Plants* for the Melkinthorpe square (5.2). Hodgson gives additional records from Egremont (0.1), Ennerdale Water (0.1/1.1) and near Skinburness (1.5, E.J. Glaister, 1882, CLE), and there is an undated and unlocalised pre-1930 *Scarce Plants* record for the Castle Carrock square (5.5).

Alt. limit: 185 m, Low Knipe (50.18), although Wilson's site on Knipe Moor was at 330 m.

            S 7   WC

Bolboschoenus maritimus (L.) Palla   Sea club-rush

(*Scirpus maritimus* L.)

Common by the Solway Firth, Duddon and Ravenglass estuaries and in Morecambe Bay; elsewhere scarce but with a single inland occurrence by a brackish spring between Colby and King's Meaburn in the Eden valley (64.18, CW, l981). The coastal sites are tidal riversides and brackish pools and ditches where it sometimes forms dense stands.

Alt. limit: 190 m, near Colby (64.18)

Map 1000   77   WFC

Scirpus sylvaticus L.   Wood club-rush

A local lowland plant, commonest along the Eden valley and absent from the Solway Plain, the south-east and much of the south-west. Most of the records are from riversides, dykes and wet meadows. Where stock have access it is extremely heavily grazed. Prior to the Survey the only Furness record was from Black Beck, Esthwaite (34.96, J. Lund, 1946, herb. FBA) and there is an interesting record from Santon Bridge (1.0, Anon., 1923, CLE).

Alt. limit: 145 m, Great Musgrave (76.12)

Map 1001   60   WFC

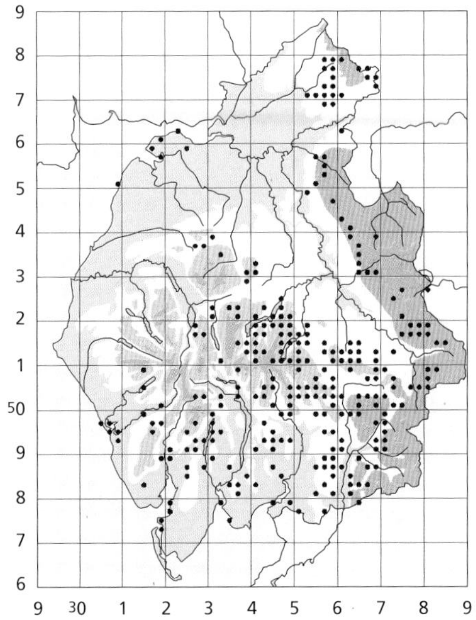

Map 999. *Eleocharis quinqueflora*

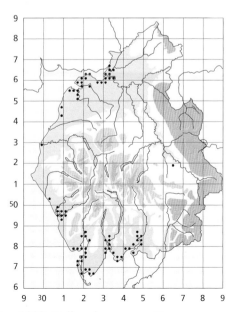

Map 1000. *Bolboschoenus maritimus*

Schoenoplectus lacustris (L.) Palla Common club-rush
Locally frequent in lowland mesotrophic and eutrophic
waters in the west and south and almost absent from
the east, although it occurs in the extreme north-east by
the River Irthing (64.76n, 66.76n). It grows in tarns
and sheltered lakeside bays and by slow-flowing rivers,
usually at some distance from the shore, in deeper
water than *Phragmites*. Coastal records should always
be carefully checked for the following species.
Alt. limit: 405 m, Dock Tarn, Watendlath (26.14)
Map 1002   70   WFC

S. tabernaemontani (C. Gmelin) Palla
Very rare, recorded only from marshes and brackish
pools on the north side of the Kent estuary (4.7s/4.8s)
and at Sandscale (18.74), from where it has since
disappeared. Other Survey records from the Kent
estuary (46.82s, 46.84s, 48.84s), Flookburgh (36.74s),
South Walney (22.62s) and Cavendish Dock, Barrow-

in-Furness (22.68s) have not been confirmed. The plant
may have been misidentified or been lost as a result of
sea defence works.

There are early records from high up the
Duddon estuary (2.8) and in a marsh at Middlebarrow,
Arnside (4.7s). This last site has been destroyed by
quarrying but the record, and the Bassenthwaite Lake
(2.2) one cited by Hodgson, are best regarded as
doubtful since a number of inland records have proved
to be *Schoenoplectus lacustris*. Typical plants are
appreciably smaller than *S. lacustris* and noticeably
glaucous.

Although absent from the south side of the
Solway Firth, it occurs on the Kirkcudbright shore.
Map 1003   6   WF[C]

Isolepis setacea (L.) R. Br.   Bristle club-rush
Fairly frequent in the east, less so in the west but
widespread, easily overlooked and far commoner than

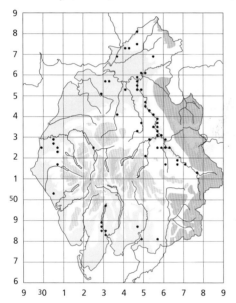

Map 1001. *Scirpus sylvaticus*

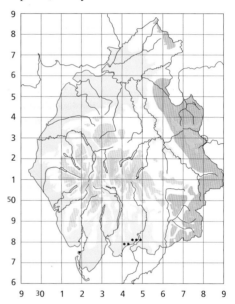

Map 1003. *Schoenoplectus tabernaemontani*

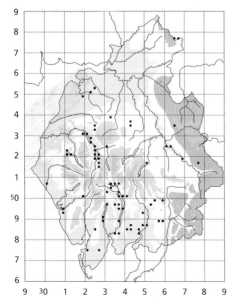

Map 1002. *Schoenoplectus lacustris*

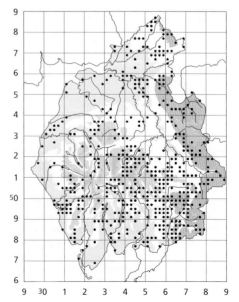

Map 1004. *Isolepis setacea*

is indicated in the *Atlas*. It grows in areas of damp, open soil such as rutted tracks and the eroding banks of becks and rivers. It also occurs in dune-slacks at Sandscale (18.74), with *Carex viridula* subsp *viridula* and *Equisetum variegatum*, in old sand and gravel workings, in seepage areas above the shore, an *Isolepis cernua* habitat, and in the upper salt-marsh turf at Anthorn (18.56).

Alt. limit: 590 m, High Cup Nick (72.26)

Map 1004   480   WFCY

I. cernua (M. Vahl) Roemer & Schultes

Recorded by G. Foggitt and M. Cross in l936 from the coast at Silecroft (*BEC Rpt* (1937) 11:284), where it was rediscovered in 1992 on a bank above the beach by MMG (10.82, LANC, *Watsonia* (1993) 19:293). The population seems likely to disappear as a result of erosion. Although MMG knew of other sites along this coast no details are now available. The species was also found in the l970s by A.J. Richards on the Ravenglass Nature Reserve (0.9) but subsequent searches have proved abortive.

1   C

Eleogiton fluitans (L.) Link   Floating club-rush

A species with a pronounced south-western distribution but with outlying sites in gravel pits at Bowness-on-Solway (20.60n) and Longtown (36.68n), and in plant-ation ponds at Cotehill (48.50). It occurs in base-poor tarns, pools and small becks, and by sheltered shores around the heads of Ennerdale Water and Wast Water. It often forms dense rafts of vegetation and is probably nowhere more abundant than in Lily Mere, Killington (60.90). Two early records from outlying sites are R.W. Butcher's from Thurstonfield Lough (3.5, 1937, *BEC Rpt* (1938) 11:512) and Wilson's from Cliburn Moss (5.2).

Alt. limit: *c.*280 m, Potter Fell, Staveley (50.98)

Map 1005   66   WFC

Blysmus compressus (L.) Panzer ex Link

Nearly confined to a limited area of north Westmorland where it is quite frequent in calcareous flushes and marshes and by becksides, particularly on ground subject to flooding. There are isolated Survey records from north of Sedgwick (50.88, P. Gibson, 1980s, LANC), Hutton Roof (56.76s, CEW, 1989), the River Esk near Netherby (38.72n, CS, 1979, *Watsonia* (1980) 13:147), the White Lyne (44.68, RWMC, 1995, LANC), by the Bailey Water, Kershope (52.80n, GH, 1981, LANC), Greystoke Park (30.38, M. Rawes, 1980s; LANC), flushes in the Caldbeck Fells (28.36, REG, 1980, LANC; 30.38) and, remarkably, on the edge of the salt-marsh near Bowness-on-Solway (20.62, 22.62; LANC), where it was first seen in abundance by DAR in 1957 (CLE, *Watsonia* (1979) 12:361).

Following its widespread decline over much of Britain, this species is now concentrated in the north Pennines, from the Craven area of Yorkshire to south

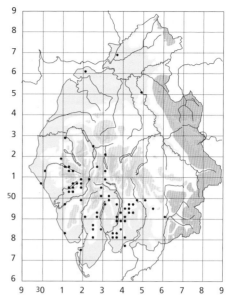

Map 1005. *Eleogiton fluitans*

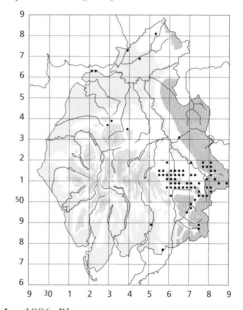

Map 1006. *Blysmus compressus*

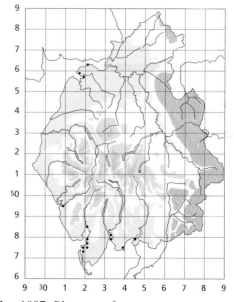

Map 1007. *Blysmus rufus*

Northumberland.
Alt. limit: 455 m, north of Tan Hill (88.08)
Map 1006   71   WCY

B. rufus (Hudson) Link
A northern species known from only a few sites all on
the edge of salt-marshes, often in areas subject to
seepage. At Humphrey Head (38.74s) it grows with
Cochlearia officinalis in a small fragment of *Juncus
gerardii* marsh. Although it has apparently disappeared
from Silloth (1.5, 1932) and Brownrigg, Abbeytown
(1.5, 1936), where it was seen by J. Parkin, and also
from Sellafield (0.0, A. Templeman, 1922, *BEC Rpt*
(1923) 6:751), this species is not as rare in the county
as the *Atlas* suggests.
Map 1007   14   WFC

Schoenus nigricans L.   Black bog-rush
Like *Blysmus compressus*, largely confined to limestone
flushes in north Westmorland and much under-
recorded in the *Atlas*. At some sites, such as Stagmire
Moss, Soulby (72.12) and near Sunbiggin Tarn (66.06),
it is locally abundant, and at Hale Moss, Burton-in-
Kendal (50.76s), it is dominant on the open marl-rich
fen. There are other isolated sites at Newton Reigny
Moss (46.30) and on the coast south of Greenodd
(32.78s), where it occurs with *Eleocharis quinqueflora*
and *Blysmus rufus* in flushes by the shore. Although
absent from the south side of the Solway Firth, it is
frequent in flushes along the Scottish shore.

Baker gives the surprising record "Frequent in
bogs near Kirkby Lonsdale" which he attributes to
I. Hindson. This is not included by Wilson who perhaps
thought it erroneous.
Alt. limit: 330 m, Hardendale (58.10)
Map 1008   20   WFC

Rhynchospora alba (L.) M. Vahl   White beak-sedge
An attractive species which, like *Eleocharis multi-*

caulis, has a pronounced south-western, Lake District
distribution but differs in occurring on raised mosses by
the Solway Firth, at Scaleby (42.62), on Bolton Fell
(48.68), and higher up the Lyne valley and on Butterburn Flow, Spadeadam (66.76n). The species' survival
at some of these last sites is currently threatened by
commercial peat extraction. Wilson's records include
an outlying one from Cliburn Moss (5.2). H. Britten
noted it on Lazonby Fell at the beginning of the century
and it was seen near there (50.40) by DAR in 1958.

Its usual habitat is around the edges of tussocks
in valley and raised mosses, commonly associated with
Sphagnum spp., *Erica tetralix* and *Narthecium ossi-
fragum*.
Alt. limit: 300 m, above Haweswater (46.12)
Map 1009   135   WFC

Cladium mariscus (L.) Pohl   Great fen sedge
Known only from Sunbiggin Tarn (66.06), Cunswick
Tarn (48.92) and Newton Reigny Moss (46.30),
although there are also old records from Sandford Moss
(7.1). At the first two sites it grows by the margin of
highly calcareous tarns, forming in places dense stands.
At Newton Reigny it occurs in tall-herb fen and carr.
The *Atlas* record for the Arnside square (4.7s) is from
Lancashire.
Alt. limit: 255 m, Sunbiggin Tarn (66.06)
                             3   WC

Kobresia simpliciuscula (Wahlenb.) Mackenzie
An arctic-montane species occurring in Britain only in
a small area of the central Scottish Highlands and in
Teesdale where it has its headquarters on Widdybank
Fell but extending into Cumbria near Birkdale Farm
(78.26, 80.26, 80.28).

Its habitat is very similar to that of *Juncus
alpinoarticulatus*, with which it is usually associated,
namely open, calcareous and often stony flushes characterised by *Primula farinosa, Pinguicula vulgaris,*

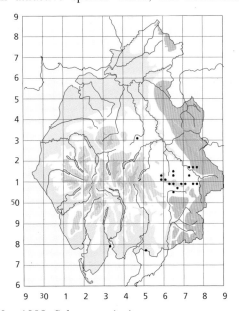

Map 1008. *Schoenus nigricans*

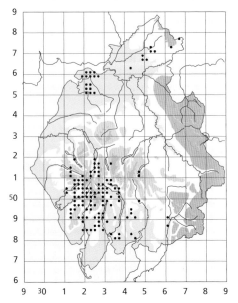

Map 1009. *Rhynchospora alba*

Galium boreale, Carex hostiana and *C. capillaris*.
Alt.: *c*.450 m, Birkdale (80.26)

R 3   W

Carex paniculata L.   Greater tussock sedge
Scattered throughout the lowlands, most frequent in the
Eden valley and occurring in fens and fen carr, often by
tarns, also in dykes and by the Kendal canal. At some
sites, for example Arrow Moss, Silecroft (12.80), the
tussocks are more than a metre high, and by the River
Marron (06.24) there is one said to support a mature
sycamore. The sedge is curiously absent from the
mosses of the Rusland and Duddon valleys in the south
and it has probably disappeared, perhaps through
drainage, from a number of sites on the Solway Plain
where it was known earlier this century.
Alt. limit: 270 m, Tarn Moss, Troutbeck (40.26)
Map 1010   112   WFC

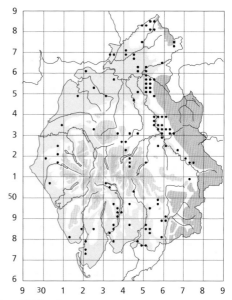

Map 1010. *Carex paniculata*

C. paniculata x **C. remota**
C. x **boenninghauseniana** Weihe
Known only from Roudsea Wood (32.82), where it was
collected by W.H. Pearsall and W.A. Sledge in 1932
(herb. W.A. Sledge) and where it still occurs, from a
wood by the River Ehen near Ennerdale Bridge (06.16),
where it was recorded by Hodgson in 1876 and refound
by CCH in 1979 (LANC, conf. A.O. Chater), from
wooded mossland in the Rusland valley (34.84, GH,
l991, LANC, conf. R.W. David) and from a pondside
near Dalton-in-Furness (20.74, PB, 1994, LANC).

4   FC

C. diandra Schrank   Lesser tussock sedge
A rare species of tarn margins and valley mires and
fens in the east but with an isolated record from
Spadeadam in the far north (58.70n, MG & JP, 1995).
The Furness records must be considered doubtful: it has
never been refound at the extensively botanised
Urswick Tarn (2.7) since it was reported in 1874 and
there is no confirmation of the two *Handbook* records
for Lindal (2.7, G.M. Brown, 1955) and Roudsea Wood
(3.8, Anon., 1950s). It has also not been been refound
at Biglands Bog (24.52, DAR, 1954).

        This sedge has decreased markedly this century
in England and is now as frequent in Cumbria as
anywhere south of the Border.
Alt. limit: 270 m, Spadeadam, (58.70n)
Map 1011   16   W(F)C

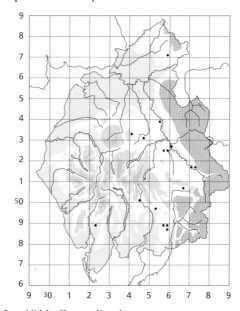

Map 1011. *Carex diandra*

C. otrubae Podp.
Fairly frequent at the upper edge of salt-marshes and by
brackish pools, ditches and coastal meadows, also
by the Kendal canal. From Cumbria and Co. Durham
northwards, this is essentially a coastal sedge.
Authenticated inland records include lanesides near
Crosby Garrett (72.10, KR, 1987; 74.08, R.W. David,
1975), near Greystoke (38.32, EHR & MS, 1982), by
the River Eden above Langwathby (58.30, RWMC,
1978) and by the A590 at Ambleside (36.04, CEW,

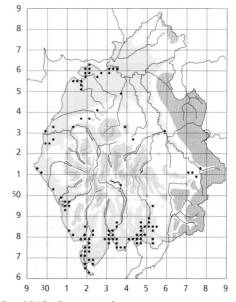

Map 1012. *Carex otrubae*

1989, LANC).
Alt. limit: *c*.210 m, Crosby Garrett (72.10)
Map 1012   103   WFC

C. otrubae x C. remota
C. x pseudoaxillaris K.Richter
Found during the Survey by CCH in 1979, growing with both parents on a damp cliff ledge at St Bees Head (94.10, LANC, det. A.O. Chater). Druce's reference to this rare hybrid in Westmorland is regarded in the *Critical Atlas* as an error .

                              1   C

C. spicata Hudson
Early in the Survey this was considered by R.W. David to be distinctly rare in north-west England. However, an appreciable number of specimens submitted to him proved to be this species. As is evident from the map it is quite frequent in an arc from the upper Eden valley across to the west coast and also in the south towards the Kent estuary. The sites are mainly in moderately base-rich or calcareous grassland, usually roadside verges and lanes.

    Because of past confusion between this and the following species it is impossible to know whether the early Flora records for Cumberland and Westmorland refer to this species, but there is a 1950 Westmorland record from Troutbeck (4.0, M.I. Tetley, CLE).
Alt. limit: *c*.215 m, Crosby Garrett (72.10)
Map 1013   43   WFC(Y)

C. muricata L.
subsp. **lamprocarpa** Celak
Frequent along the middle and lower Eden valley, in the northwest Lake District and in a band across the south of the county, scarce to rare elsewhere. Like *Carex spicata,* it is a sedge of coarse grassland occurring predominantly by roadsides and tracks, sometimes in woodland rides and more tolerant of acid

soils than that species. It occurs, for example, in acid grassland on Whinlatter Pass (18.24).

    It is probable that Hodgson's records of *C. contigua* and Wilson's of *C. muricata* include this species, but the earliest substantiated records are for Westmorland: Rydal (3.0, J. Sidebotham, 1845, CGE, det. R.W. David) and Cumberland: near Great Salkeld (5.3, C.E. Salmon, 1919, BM, det. R.W. David). There appear to be no pre-Survey specimens from Furness.

    Both this species and the previous one are much commoner than is indicated by the *Sedge Handbook*. It is interesting that subsp. *lamprocarpa* is absent from west Yorkshire, Co. Durham and south Northumberland.

    Subsp. *muricata*, characterised by more globose spikes, the lowest somewhat distant, and earlier flowering, in May, occurs only on limestone and should be looked for as it is known from Ribblesdale (v.c.64).
Alt. limit: 310 m, Priest's Crag, Watermillock (42.22)
Map 1014   76   WFCY

C. divulsa Stokes
subsp. **leersii** (Kneucker) Walo Koch
A small colony of this south-eastern sedge was found by PB in 1988 below a roadside hedge near Dalton-in-Furness (20.72, conf. A.O. Chater & R.W. David, LANC, *Watsonia* (1989) 17:483).

    This is the most northerly record on the west coast and the first for north-west England.

                              1   F

C. arenaria L.   Sand sedge
Common around the coast, almost continuously from the Solway Firth to Newbiggin (26.68s) at the entrance to Morecambe Bay. It grows especially well on sand-dunes and in dune-slacks playing an important role in stabilising mobile sand, but it also occurs in a variety of coastal ruderal habitats such as verges, waste ground and car-parks and even on a cinder railway lonning at

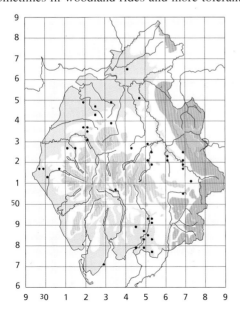

Map 1013. *Carex spicata*

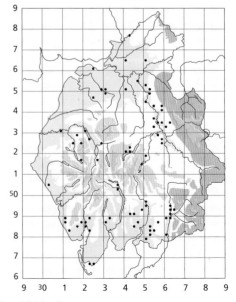

Map 1014. *Carex muricata*

Carlisle (38.56).

Why it should be absent from the Morecambe Bay coast east of Newbiggin is a mystery, although there is an *Atlas* record from the Cartmel square (3.7s) and a 19th century "Westmorland" record cited by Wilson. It recurs just south of the county boundary in Lancashire.

Map 1015   82   (W)FC

C. disticha Hudson
Locally common, especially in the upper Eden valley and the extreme north-east but absent from considerable areas including almost all the Lake District. It is a low-land species of wet meadows, rough grassland and the margins of pools, often forming extensive patches.

Alt. limit: 250 m, Helbeck Wood, Brough (78.16)
Map 1016   160   WFC

C. maritima Gunnerus
A single robust flowering stem of this northern sedge was collected by E.J. Harling in 1971 on the west side of Humphrey Head (38.74s, BM, det. A.C. Jermy, *Watsonia* (1973) 9:278). Despite subsequent searches it has not been seen since and it is possible that the colony was destroyed during the repair of the sea wall a few years later.

At the time, this was the most southerly extant record for the sedge which is largely confined in Britain to north-east Scotland.

S 1†   F

C. remota L.   Remote sedge
Very common in base-rich, wet woodland being particularly common in the south, on the west coast and in the north-east in the wooded valleys of the Gelt, Lyne and Liddel. It commonly grows with *Chrysosplenium oppositifolium, Carex sylvatica, Ajuga reptans* and

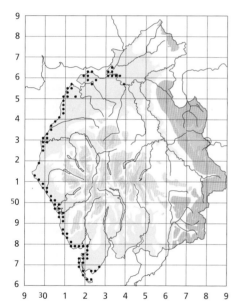

Map 1015. *Carex arenaria*

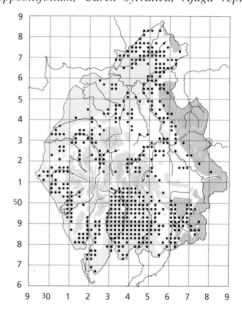

Map 1017. *Carex remota*

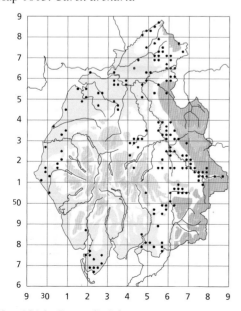

Map 1016. *Carex disticha*

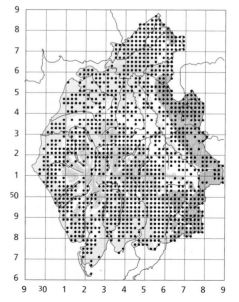

Map 1018. *Carex ovalis*

Stellaria nemorum.
Alt. limit: 270 m, Needlehouse Gill, Sedbergh (72.96)
Map 1017   515   LWFCY

C. ovalis Gooden.   Oval sedge
Very common in acid, base-poor grassland, particularly
in the south-east and north-east. Its habitats range from
lowland heath, roadside verges, woodland margins and
wet marshy meadows to upland hay-meadows and
Calluna - Vaccinium moorland. It can also occur in
moderately base-rich hillside flushes, as below Melmerby High Scar (64.36) and dune-slacks at Sandscale
(18.74).
Alt. limit: 870 m, Cross Fell (68.34)
Map 1018   1191   LWFCY

C. echinata Murray   Star sedge
Extremely common to abundant in the uplands on wet
acid peat and in base-poor flushes; scattered in the low-
lands on mossland and meadows which are at least
seasonally wet.
Alt. limit: 825 m, Little Dun Fell (7.3, Wilson) and
Helvellyn (3.1, Wilson)
Map 1019   1131   LWFCY

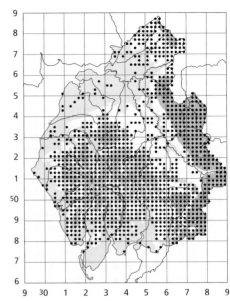

Map 1019. *Carex echinata*

C. dioica L.   Dioecious sedge
A characteristic upland sedge of base-rich flushes in the
east, usually associated with *Carex hostiana, C. viridula* subsp. *brachyrrhyncha* and *Eleocharis quinqueflora*. In the western Lake District it also occurs in
slightly flushed areas of valley mires and on the coast at
Kirksanton (12.78).
        This upland sedge appears to have decreased
dramatically this century in lowland England and is
now largely confined, in England, to the Pennines and
the Lake District.
Alt. limit: 765 m, below Tees Head (68.32)
Map 1020   394   WFCY

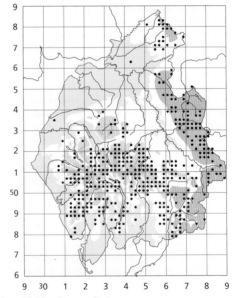

Map 1020. *Carex dioica*

C. elongata L.
Known only from Roudsea Wood (32.80), Rusland
Moss (32.88), by Esthwaite Water (34.96), the head of
Windermere (36.02) and by Derwent Water (24.22,
26.22). At most of these sites it occurs in very wet and
often dense fen carr, usually of *Alnus* and *Salix cinerea*,
and associated with *Carex paniculata, C. remota* and
Dryopteris carthusiana. At Ambleside it also occurs in
the open in a ditch by a field-edge and on alluvium at
the mouth of the River Rothay.
        It was first found in Furness at Roudsea by W.H.
Pearsall and W.A. Sledge in 1932 (herb. W.A.S.).
According to Hodgson, it disappeared from his only
site, Snellings Mire (9.0), last century. The post-1930
Atlas dot for the Keswick square (2.2) was at the time
incorrect being based on Sledge's erroneous record for
Ullock Moss (*Naturalist* (1944) 1944:143).
        This sedge has a very scattered distribution in
Britain. It extends as far north as central Scotland but
has disappeared as a result of drainage and canalisation

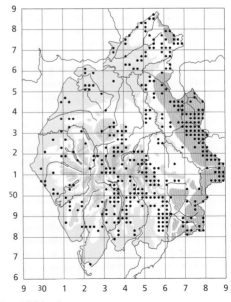

Map 1021. *Carex curta*

C. curta Gooden.   White sedge
Common in the Pennines, well scattered over the rest of
the county though absent from the limestone areas. It
occurs on blanket peat, in upland valley mires and
raised mires, such as Butterburn Flow (66.76n) in the
extreme north-east, and in the lowlands on open or
wooded mossland and small residual areas of acid peat
around tarns. It prefers some water movement and
mineral enrichment: the slight depressions at the head
of the Pennine sikes are often conspicuous by the very
pale colour of this sedge.
Alt. limit: 850 m, Cross Fell (68.34)
Map 1021   390   WFCY

C. hirta L.   Hairy sedge
A not infrequent lowland sedge that is difficult to
categorise ecologically. It occurs chiefly in coarse
grassland, usually roadside verges and railwaysides,
and frequently in areas subject to trampling, also by
woodland rides, pond- and riversides. Although often
occurring in heavy, wet soils, this is certainly not
always the case. It is often the only sedge in intensively
farmed areas.
Alt. range: 470 m, Hartside (66.42)
Map 1022   565   LWFCY

C. lasiocarpa Ehrh.   Slender sedge
Local but well dispersed over much of the centre and
east. It occurs in base-poor valley mires and around
tarns but also, in the Lake District, in fen communities
by Esthwaite Water (3.9), at the head of Windermere
and by Rydal Water (3.0). It probably still occurs at
most of the following sites where it was seen by DAR
in the 1950s: Biglands Bog (24.52, DAR, 1954, CLE),
Wythburn Bog (28.10), Armboth Fell (28.14) and
Shoulthwaite Moss (30.20). Flowering is rare at the
higher sites.
Alt. limit: 440 m, Armboth Fell (28.14, DAR, 1950s)
Map 1023   37   WFC

C. acutiformis Ehrh.   Lesser pond-sedge
Frequent in the east, scarce in the west, predominantly
lowland and absent over large areas. Most of its sites
are in reedswamp communities around tarns, in fen
carr and, especially in the north, in marshy fields where
it often occurs towards the lower side where the fields
abut on to steeper riverside woods. It is frequently
dominant, sometimes covering an appreciable area.
Alt. limit: 370 m, south of Garrigill (74.38)
Map 1024   225   WFCY

C. acutiformis x C. acuta
C. x subgracilis Druce
Wilson's specimen of *Carex acuta* from by the River Lune at
Gaisgill (6.2, 1927 YRK) has been tentatively redetermined
by A.O. Chater as this hybrid.
                                                        (W)

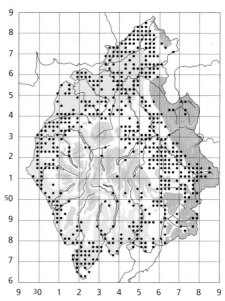

Map 1022. *Carex hirta*

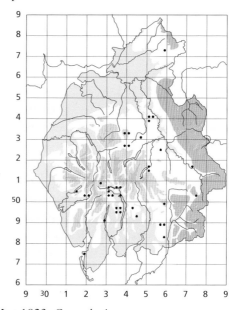

Map 1023. *Carex lasiocarpa*

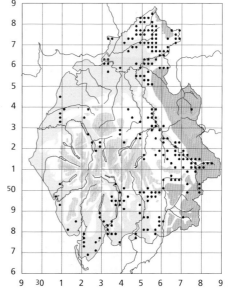

Map 1024. *Carex acutiformis*

C. riparia Curtis   Greater pond-sedge
The only Survey record is from a wet meadow between
Soulby and Warcop (74.12, KR., 1985, LANC,
Watsonia (1986) 16:196). By contrast, there are
numerous early records. Hodgson lists it from Arlecdon
(0.1) and Bolton, near Wigton (2.4), records which
CCH considered very dubious, and Wilson from Kirkby
Lonsdale (6.7s), Burrells, Appleby (6.1), by the River
Leith at Cliburn (5.2), where he saw it himself, and
near Kirkby Stephen (7.0/7.1), where it was seen by
G.C. Druce and H.J. Riddelsdell. There is a 1916
record from Allonby (0.4, W.W. Mason) and a pre-
1930 *Atlas* one from the Great Salkeld square (5.3).
The *Atlas* also has post-1930 dots for the Gilsland
square (6.6, v.c.67/70) and one for 3.7s which is based
on a Humphrey Head record made by the Barrow-in-
Furness N.H.S. during the 1950s, but the sedge has
never been refound there. Finally, G.A.K. Hervey noted
it by the River Leith in 1951.

  All the post-1930 *Atlas* squares have been
searched unsuccessfully by KR and there seems little
doubt that many of the pre-Survey records really refer
to *Carex acutiformis*.

  This is a sedge of lowland Britain with only
scattered localities in the north and west.
Alt.: 150 m, Soulby (74.12)
           1   W[(FC)]

C. pseudocyperus L.
Seen during the Survey at three sites: under alders by
the main stream in Roudsea Wood National Nature
Reserve (32.80), where it was discovered by R.W.
David in 1967, in a small area of fen carr southwest of
Sedbergh (62.90, KR, 1985; LANC) and by a field
pond at Burton-in-Kendal (52.74s, CEW, 1990,
LANC). It also grows near the last site in the pond in
Dalton Park where it was no doubt introduced.

  The record cited by Hodgson from Moorside
Woods, Arlecdon (0.2) is best considered doubtful and

the sedge appears to be extinct at Wilson's localities:
Cliburn Moss (5.2) and Terrybank Tarn (5.8s).
Alt. limit: 100 m, Sedbergh (62.90)
         3   L(W)F(C)Y

C. rostrata Stokes   Bottle sedge
Common in the uplands, rather scattered in the lowlands. It occurs in a wide variety of habitats: in the
hollows and slightly flushed areas of upland mires and
blanket peat, in tarns and in reedswamp communities
by many of the lakes, often forming dense stands, also
in marshy fields and dykes, and by slow-flowing rivers
and the Kendal canal. It grows well in both oligotrophic and eutrophic waters and probably always on a
peaty substrate.
Alt. limit: 820 m, Cross Fell (70.34)
Map 1025   643   LWFCY

C. rostrata x **C. vesicaria**
C. x involuta (Bab.) Syme
Seen during the Survey at Pull Garth, Brathay (36.02,
T. Blackstock, 1977, AMDE, conf. R.W. David), near
the Bowder Stone, Borrowdale (24.16, J. Lister, 1991,
LANC, conf. R.W. David) and by KR in 1991 by ponds
near Edenhall Grange (54.32) and near Wan Fell
(52.34). The only previous record was from by the
River Derwent, north of Keswick (2.2), where it was
seen by C.E. Salmon in 1929 (*BEC Rpt* (1930) 9:141).
Alt. limit: 230 m, Wan Fell (52.34)
         4   FC

C. vesicaria L.   Bladder sedge
Occasional in mesotrophic to eutrophic lowland waters,
growing by tarns and in lake and riverside reed-bed
communities. It often grows with *Carex rostrata*, with
which it is sometimes confused, but even when vegetative the light green, rather than glaucous, leaves
usually distinguish it. It occurs also in fens and in fen
carr, as at Roudsea Wood (32.82) and Calf Close Bay,

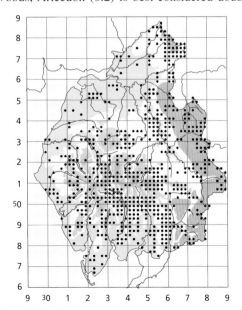

Map 1025. *Carex rostrata*

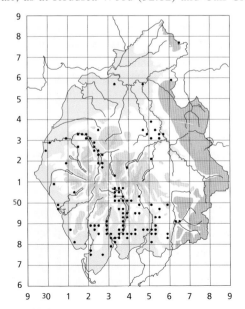

Map 1026. *Carex vesicaria*

Derwent Water (26.20).
Alt. limit: 270 m, Paddaburn, Spadeadam (64.76n)
Map 1026   109   WFC

C. pendula Hudson   Pendulous sedge
Scarce, except in the north-east and around the head of
Morecambe Bay, occurring in heavy, damp and somewhat base-rich woodland, often, like *Carex laevigata*,
by streams or seepage areas. Most of the southern and
western sites are in estate woodland where it has probably been introduced, but in the valley woods of the
north-east, for example along the River Lyne where it
is locally abundant, and on the cliffs at St Bees (9.1),
there is, as Hodgson says, no reason to doubt its native
status.
Alt. limit: 410 m, south-east of Alston (72.44)
Map 1027   89   LWFC

C. sylvatica Hudson   Wood sedge
Very common in the deciduous woods of south Westmorland and Furness and following the more base-rich
soils through the upper Eden valley across to the west
coast, also in the wooded gills from Geltsdale north to
the Border. It tolerates heavy clay soils and is a feature
of rutted and trampled woodland paths.
Alt. limit: at least 450 m, Tyne Head (76.36)
Map 1028   428   LWFCY

C. capillaris L.
A rare sedge of calcareous flushes and damp limestone
grassland on the Asby - Orton limestone, where it is
associated in two places with *Bartsia alpina* and *Polygala amara*, in High Cup valley (74.24, 74.26), in
Teesdale and around Tyne Head (7.3). Despite its rarity
it is dominant over an area of several square metres at
some of its sites. The sedge should be looked for on
Long Fell, Warcop (7.1, J. Chapple, OXF), where it
was collected in 1939.
        The first Cumberland record dates from 1957
when it was found at Tyne Head (76.34, DAR,
Watsonia (1979) 12:361).
        Outside the Scottish Highlands and the Moffat
hills, this arctic-montane species is known in Britain
only from Cumbria, the adjacent Co. Durham part of
upper Teesdale, the Craven area of Yorkshire and
Snowdonia.
Alt. range: 200 m, Waitby (76.08) - 580 m, Tyne Head
(76.34)
Map 1029   S 15   WC

C. strigosa Hudson
An elusive sedge seen during the Survey only in High
Wood, Askam-in-Furness (20.74, 20.76, PB, 1989,
LANC, *Watsonia* (1991) 18:434) where it is locally
abundant growing with *Chrysosplenium oppositifolium*
and *Carex sylvatica*. The only previous records are
those given by Baker for Shaw Woods (*cf*.2.5), now
felled), and by Wilson for Windermere (4.9) and Melkinthorpe (5.2). In 1956 DAR searched unsuccessfully at

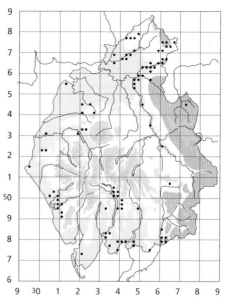

Map 1027. *Carex pendula*

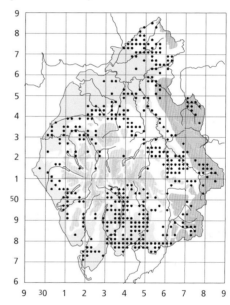

Map 1028. *Carex sylvatica*

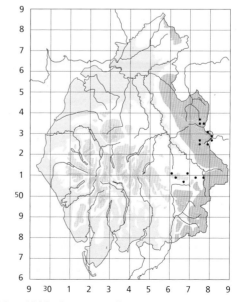

Map 1029. *Carex capillaris*

this last site when it was still a broadleaved wood.

The Furness record in the *Check-list* was based on the apparently misplaced *Atlas* dot for Windermere. The post-1962 record cited in the *Check-list* for Westmorland was a field record from Witherslack woods (42.84s, KAG & FLW, 1966). Although this is a likely site, all attempts to refind the sedge have failed.

Carex strigosa is a sedge of southern and south-eastern England which is not now known north of Cumbria.

2   (W)F(C)

C. flacca Schreber   Glaucous sedge
Very common on the more base-rich and calcareous soils, scarce only in the Lake District, where it is characteristic of basic rock ledges, and in the intensively farmed parts of the Eden valley and the Solway Plain. Like the ecologically similar *Briza media*, it occurs over a wide range of soil moisture: from sandy coastal sites and dry limestone grassland to lowland fens, moist upland roadside verges and base-rich flushes.
Alt. limit: at least 750 m, Fairfield (36.10)
Map 1030   1026   LWFCY

C. panicea L.   Carnation sedge
Very common except in the more intensively farmed parts of the Solway Plain and the Eden valley. It is tolerant of a wide range of soil acidity, occurring in both wet heath and valley mires, often with *Carex viridula* subsp. *oedocarpa*, *Drosera rotundifolia* and *Narthecium ossifragum,* and in calcareous flushes with *C. viridula* subsp. *brachyrrhyncha* and *C. flacca*. It also grows in most types of wet grassland and is particularly luxuriant in ungrazed marshy fields.
Alt. limit: 840 m, north side of Cross Fell (68.34)
Map 1031   1270   LWFCY

C. laevigata Smith   Smooth sedge
Scattered, mainly in the north-east and south-west although the *Atlas* shows several post-1930 records from the Solway Plain and Carlisle area. It grows typically in woodland flushes and seepage areas, particularly in lowland alder woods, and often associated with *Carex sylvatica*, *Ajuga reptans* and *Lysimachia nemorum*. It also occurs in reedswamp at the north end of Bassenthwaite Lake (20.30, 20.32). The tree cover may be quite sparse at higher altitudes.
Alt. limit: 335 m, below Kirksty Brow, Gowbarrow (40.22)
Map 1032   94   LWFC

C. binervis Smith   Green-ribbed sedge
Widespread on relatively well-drained, acid soils, chiefly in the uplands where it commonly occurs in *Nardus* grassland, *Calluna - Vaccinium* moorland and on rock ledges. In the lowlands it is largely restricted to heathy ground, drained and wooded mossland and, at St Bees Head (9.1), sea-cliffs.

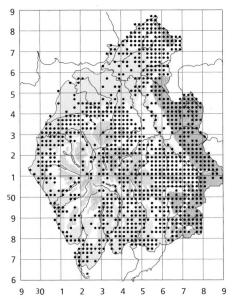

Map 1030. *Carex flacca*

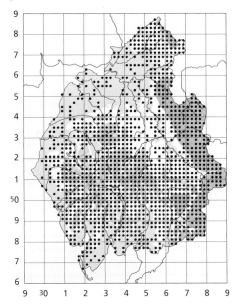

Map 1031. *Carex panicea*

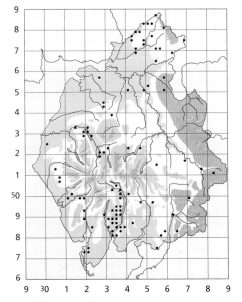

Map 1032. *Carex laevigata*

Its altitudinal range overlaps that of the montane *Carex bigelowii* and within this range there is a tendency for vegetative material of *C. binervis* to be misrecorded.

Alt. limit: 760 m, south of Cross Fell (66.34)

| Map 1033 | 668 | WFCY |

C. distans L.   Distant sedge

Scattered around the margins of salt-marshes, often growing with *Juncus gerardii* and *Carex extensa*, and occasionally occurring among rocks as on the old railway embankment near Arnside (46.78s). The *Atlas* shows inland records from the Hawkshead (3.9), Coniston (2.9), Greystoke (4.3) and Tebay (6.0) squares and Wilson gives others for Borrow Beck (5.0), Barbon (6.8s), Kirkby Stephen (7.0) and Warcop (7.1). This is very puzzling since there are no inland Survey records and the only herbarium sheet seen from an inland locality, Sunbiggin Tarn (6.0, M.I. Tetley, 1950, CLE), has been redetermined by A.O. Chater and R.W. David as a mixture of *C. binervis* and *C. hostiana*.

| Map 1034 | 62 | WFC |

[C. punctata Gaud.

The record from Whitehaven (9.1) cited by Oliver (*Phytologist* (1852) 4:67) was regarded by David (1981a) as an error for *Carex distans*, which does grow there. However, its occurrence in Cumbria is not unlikely since it is known from several sites on the Isle of Man and across the Solway Firth.

C]

C. extensa Gooden.

Occurs at the margins of most of the salt-marshes, often growing with *Carex distans* and *Juncus gerardii*, but frequent only around Morecambe Bay and Walney Island. At St Bees Head (9.1) it grows on rock ledges and in crevices of the sandstone cliffs.

| Map 1035 | 29 | WFC |

C. hostiana DC.   Tawny sedge

A widespread and frequent species of upland base-rich or calcareous flushes, springs, fens and wet meadows, with a generally similar ecology to *Eleocharis quinqueflora*, although rarer at low altitudes and absent from coastal dune-slacks. Other frequent associates include *Carex dioica*, *C. viridula* subsp. *brachyrrhyncha* and *Selaginella selaginoides*.

Alt: 610 m, below Melmerby High Scar (62.38)

| Map 1036 | 463 | WFCY |

C. hostiana x **C. viridula**

C. x fulva Gooden.

Probably quite frequent although only recorded from twelve tetrads, mostly in the south of the county and usually in upland flushes. It has not been possible to decide on herbarium specimens, with any degree of confidence, whether the cross involved *C. viridula* subsp. *brachyrrhycha* or *oedocarpa*, although in the field it is usually evident which is the more likely

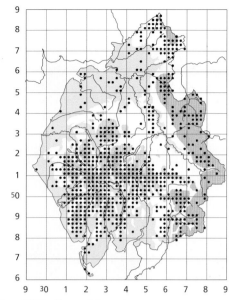

Map 1033. *Carex binervis*

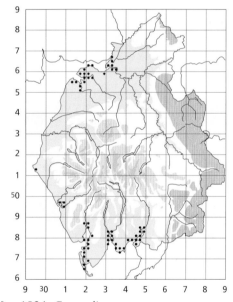

Map 1034. *Carex distans*

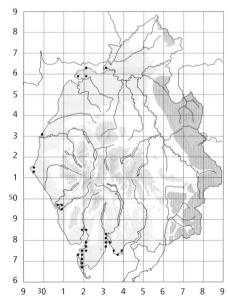

Map 1035. *Carex extensa*

parent.

The Survey record from Lowick Green (28.84, GH, 1975, LANC) is probably the first for Furness. The only previous records are K.M. Hollick's from the fen at Cunswick Tarn (4.9, 1952, DBY), involving subsp. *brachyrrhyncha*, and ones from Bassenthwaite Lake (2.2/2.3, C.E. Salmon, 1929) and Newton Reigny Moss (4.3, A. Wallis, 1929, both *BEC Rpt* (1930) 9:142), involving subsp. *oedocarpa*.

Alt. limit: 440 m, Hesk Fell, Ulpha (16.94)

Map 1037   15   WFC

C. flava L.

This is the largest and rarest member of the *Carex flava* complex and now known in Britain only from Roudsea Wood (32.82) where it occurs on calcareous peat along one of the rides and also in a dry clearing on the limestone itself. It was apparently first discovered here by D. Lumb in 1913 (*BEC Rpt* (1914) 3:504, BM).

A specimen purporting to come from Ennerdale (0.1/1.1) collected in 1836 by J. Dickinson (LIV) undoubtedly belongs to this species but has probably been mislabelled.

This species is more robust than *C. viridula* subsp. *brachyrrhyncha* and has longer utricles (>6 mm) with a longer beak (2-2.5 mm).

Plate 102   R 1   F[C]

C. flava x C. viridula subsp. oedocarpa

A sterile hybrid found by R.W. David and A.C. Jermy in 1975 in Roudsea Wood National Nature Reserve (32.82; LANC) growing by a shaded path on peat and with both parents.

                                  1   F

C. viridula Michaux
subsp. brachyrrhyncha (Celak.) B.Schmid

(*C. lepidocarpa* Tausch)

Frequent in upland base-rich flushes, chiefly in the east and rather rare in most of the Lake District. Common associates include *Carex hostiana*, *C. dioica* and *Eleocharis quinqueflora*.

The flowering stems are generally noticeably different from subsp. *oedocarpa* being stiffly erect, taller and with pale, clustered female spikes, although the utricle/beak ratio tends to be lower than in the lowland fen populations of southern England.

The map is certainly substantially correct although there is likely to have been some confusion between this and the following subspecies, and lowland records, especially from west Cumberland and the Furness peninsula, should be checked.

The first Cumberland and Furness records may well be those shown in the *Atlas*.

Alt. limit: 700 m, Knock Ore Gill (70.30)

Map 1038   278   LWFCY

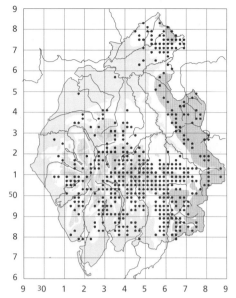

Map 1036. *Carex hostiana*

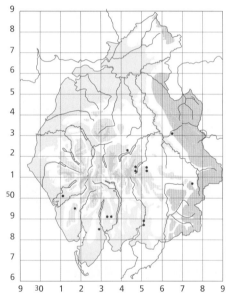

Map 1037. *Carex hostiana* x *C. viridula*

Plate 102. *Carex flava*

subsp. **brachyrrhyncha** x subsp. **oedocarpa**
These two subspecies are said to produce fertile hybrids. Although not recorded during the Survey, they presumably do occur. The only record appears to be one from Buttermere in 1949 (1.1, E.J. Gibbons, det. E. Nelmes, *BSBI Yr Bk* (1951) p.50).

(C)

subsp. **oedocarpa** (Andersson) B. Schmid
(*C. demissa* Hornem.)
Very common in the uplands growing usually in flushes, by the sides of becks and on wet rock ledges. In the lowlands it is characteristic of open peaty ground by ponds and tarns, seepage areas by the coast, damp heaths and the floors of old sand and gravel workings. Contrary to a commonly held view, it is not a strict calcifuge.
It is likely that most of the records for *Carex flava* in the old Floras refer to this subspecies.
Alt. limit: 840 m, Helvellyn (3.1, Wilson 1956)
Map 1039   1101   LWFCY

subsp. **viridula**
(*C. serotina* Mérat)
Seen during the Survey by the Duddon estuary and at Ravenglass, chiefly in dune-slacks where it is often quite abundant and sometimes associated with *Equisetum variegatum*. It also occurs at the upper edge of the salt-marsh at Anthorn (18.56, REG, 1983, LANC). There are inland records from calcareous marl at Hale Moss, Burton-in-Kendal (50.76s, GH, 1970, LANC), presumably the site mentioned by Wilson, damp calcareous ground near Raven's Lodge, Whitbarrow (46.84s, MB, 1980, det. R.W. David), by Bassenthwaite Lake (20.30, KR, 1989; C.E. Salmon, 1919, BM), and in four tetrads by Coniston Water (28.90, 28.92, 30.94, 30.96, KR, 1989, LANC) growing in wet gravel with *Juncus articulatus* and *Littorella*. It was originally reported from there by K.M. Hollick in 1974.
There is also a pre-Survey site from the sandy shore of Thurstonfield Lough (3.5, C.W. Muirhead, 1949, CLE) where it was last seen in 1954 (DAR, CLE) prior to the destruction of the site by the tipping of shingle. A re-examination of several other inland collections in BM and CLE suggests that the material is best referred to subsp. *oedocarpa*. The first Furness record appears to date from 1955 when it was found at Sandscale, Dalton-in-Furness (18.74) by G.M. Brown.
Map 1040   14   WFC

C. pallescens L.   Pale sedge
Fairly frequent and commonest in the northeast. It is a sedge of generally damp meadows, roadside verges, scrub and woodland rides and basic rock ledges.
Alt. limit: 700 m, head of Glencoynedale (34.18, DAR, 1955)
Map 1041   411   WFCY

C. digitata L.   Fingered sedge
Scattered on the low-lying limestones around the Kent

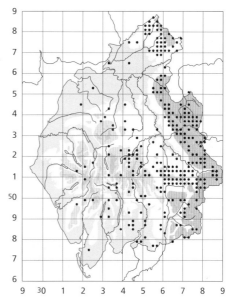

Map 1038. *Carex viridula* subsp. *brachyrrhyncha*

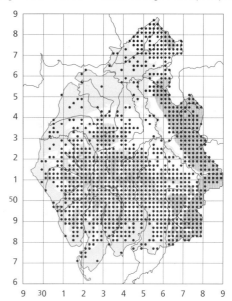

Map 1039. *Carex viridula* subsp. *oedocarpa*

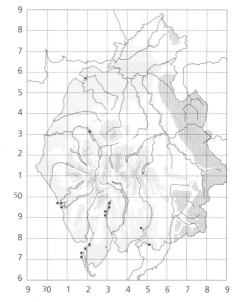

Map 1040. *Carex viridula* subsp. *viridula*

estuary, growing in shallow soil on pavements, rock outcrops and screes, usually with some tree cover. In the Whitbarrow - Witherslack area its range just overlaps with that of the following species with which it is sometimes confused. The references by Baker and Wilson to Hutton Roof refer to the next species. The first Furness record apparently dates from 1946 when it was seen at Roudsea Wood (32.82) by J.A. Whellan.

Carex digitata is readily distinguished from *C. ornithopoda* by the deeper green of its leaves, the deep purple of its leaf-bases and the longer, more distinctly pedicelled spikes.

This sedge reaches its northern limit in Britain in Cumbria and Yorkshire. A full account of its British distribution is given by David (1978a).
Alt. limit: 180 m, Whitbarrow Scar (44.86s)
Map 1042   S 15   WF

C. ornithopoda Willd.   Bird's-foot sedge
Frequent on the limestone around the upper Eden valley, especially the Asby - Orton area, but less common in the south where it is absent from most of the Kent estuary limestone.There are also isolated occurrences in the east on the Stennersceugh Clouds pavement (74.00) and in Cumberland on the silt-covered sandstone banks of the River Eden above Armathwaite (50.44, RWMC, 1986, *Watsonia* (1988) 17:196, Corner & Roberts (1989)), the only auth-enticated Cumberland site and the only one off the limestone. As mentioned above, its distribution over-laps that of *Carex digitata* only on the north side of Morecambe Bay.

It grows in shallow limestone soil on cliff-tops, rock ledges, steep slopes and limestone pavement. It can tolerate light shade, as at Armathwaite, where it grows with *Origanum vulgare* and *Trifolium medium.*

This sedge occurs in only 13 squares in Britain: eleven in Cumbria and two in Derbyshire (David 1980).
Alt. limit: 600 m, Long Fell, Warcop (76.18)
Map 1043   R 34   WC

C. caryophyllea L.   Spring sedge
Common in the east, scarce in the west and almost absent from the Solway Plain. It is a sedge of dry, base-rich or limestone turf and rock ledges, often growing with *Thymus* and *Festuca ovina*. However, it is not restricted to such soils and can be found in well-drained, south-facing *Festuca - Agrostis* fellside grass-lands in the Lake District and the Bewcastle Fells in the north-east. Here it often occurs close to or even intermixed with *Carex pilulifera,* but it prefers the sharper drainage afforded by slight breaks of slope and disruption of the turf. Even when vegetative it can usually be distinguished from that species by its darker, shorter and more leathery leaves. It is very conspicuous in spring as the male spikes open.
Alt. limit: 765 m, Knock Fell (72.30)
Map 1044   792   LWFCY

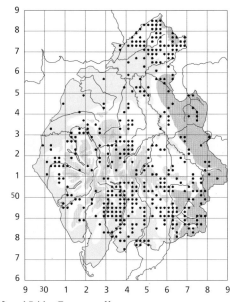

Map 1041. *Carex pallescens*

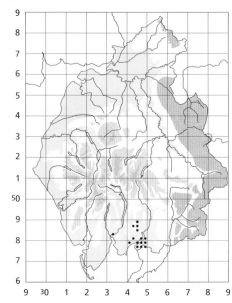

Map 1042. *Carex digitata*

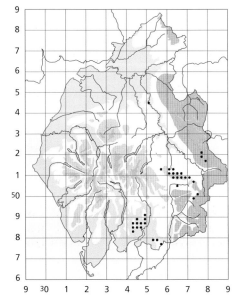

Map 1043. *Carex ornithopoda*

C. ericetorum Pollich

This sedge was thought to be confined in northern England to upper Teesdale until it was discovered, new to Westmorland, on Scout Scar by T.G. Tutin in 1944 (4.9, *BEC Rpt* (1948) 13:316). It was subsequently found to be quite widespread in the short limestone turf by the cliff-edge along most of the scarp. Its range was further extended by its discovery at Arnside Knott (44.76s, Anon., date unknown) and on Orton Scar (6.1) by R.W. David in 1967 (David 1981b). It is now known from several sites around Arnside and on the Asby - Orton limestone.

At the latter area it usually produces only a few flowering spikes, but occasionally there is a good year apparently quite unrelated to grazing pressure. Common associates include *Carex caryophyllea, Thymus, Festuca ovina* and *Galium sterneri*.

This species is restricted in Britain to a narrow belt extending from Cumbria to East Anglia.

Alt. limit: 400 m, Crosby Ravensworth Fell (58.10)

Map 1045   S 10   W

C. pilulifera L.   Pill sedge

Common in the uplands, local and decreasing in the lowlands, occurring in fairly well-drained acid grassland and heath, frequently associated with *Nardus, Calluna* and *Vaccinium*, but known also from dune grassland at Wolsty, Beckfoot (08.50) and coastal pastures at Cardurnock (16.58).

Alt. limit: at least 610 m, Cross Fell (68.34)

Map 1046   659   WFCY

C. limosa L.   Bog sedge

Rare, seen during the Survey on the north-eastern fringes of the Lake District, in the fells north of Kendal, also at Newton Reigny Moss (46.30), Carlatton (52.52) and Biglands Bog (24.52; CLE) and in upper Teesdale. In addition DAR recorded it during the 1950s from Ehenside (98.06, 1955, CLE), Bowness Common (18.58), Armboth Fell (28.14, 1959) and Foul Bog, Bewcastle (58.76n, 1956, CLE). It has probably now disappeared from the lowland sites as a result of drainage and from Foul Bog following afforestation. It probably still persists on Armboth Fell and at Butterburn Flow, Spadeadam (66.76n), where he noted it in 1965. The old record for Cogra Moss, Lamplugh (0.1, A. Templeman, 1922, *BEC Rpt* (1923) 6:752) almost certainly refers to *Carex magellanica*.

This is a sedge of *Sphagnum* mires occurring, often with *C. rostrata*, in areas subject to a very slight amount of mineral enrichment and apparently rather more exacting in its requirements than the ecologically similar *C. magellanica*. Unlike the latter it can grow in the more base-rich mires. Both species occur on Butterburn Flow and at Tarn Moss, Troutbeck (40.26).

Carex limosa is a northern species in Britain and now rather rare outside Scotland.

Alt. limit: 515 m, Teesdale (76.30)

Map 1047   13   WC

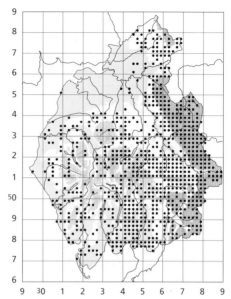

Map 1044. *Carex caryophyllea*

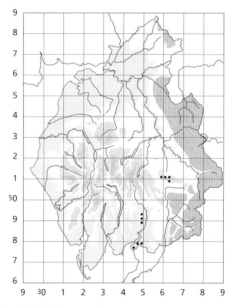

Map 1045. *Carex ericetorum*

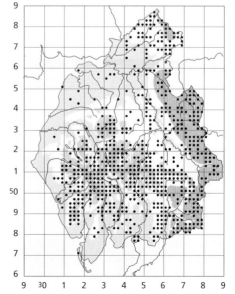

Map 1046. *Carex pilulifera*

C. magellanica Lam.
Rather rare, with scattered sites in the Lake District, in the hills between the rivers Kent and Lune, one locality in the Eden valley (36.50), in Teesdale and near Tyne Head (7.3) but it is well-represented in the extreme north-east. Unfortunately, at least two of its former Teesdale sites have vanished beneath the Cow Green Reservoir (7.3, 8.2).

There are additional records made by DAR between 1956 and 1968 from upper Eskdale (22.04), Greenup Edge (28.10), Foul Bog, Bewcastle (58.76n) and Denton Fell (60.62). The last site has now been afforested and the sedge has probably been lost.

Carex magellanica is a plant of acid *Sphagnum* mires, apparently requiring some water movement and occasionally occurring in the open water of bog pools. In Kershope Forest (5.8n) it persists in residual areas of mire along forest rides but probably will not survive clear-felling.

Occurring from Scotland south to mid-Wales, this sedge is probably most frequent in Cumbria.
Alt. limit: 640 m, Greenup Edge (28.10, DAR, 1960)
Map 1048, Plate 103 S 26   WC

C. atrata L.
A very rare arctic-montane species now known only from base-rich rock ledges in three tetrads on Helvellyn and Fairfield (3.1), and also from a site in the High Street range (4.1, R.J. Birkett, 1974), probably that originally discovered by J. Backhouse. There is also an intriguing record from Piers Gill, Scafell Pike (2.0, W. Frost, 1927, conf. K. Rob).

The only other British localities south of the Scottish Highlands are in the Moffat Hills and Snowdonia.
Alt. range: 650 m, Fairfield (3.1) - 790 m, Helvellyn (3.1)

                              S 4   W(C)

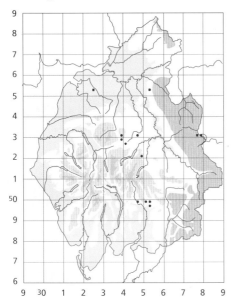

Map 1047. *Carex limosa*

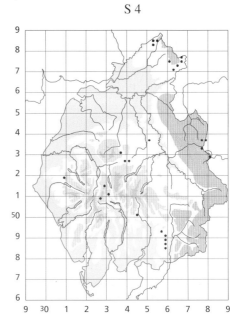

Map 1048. *Carex magellanica*

Plate 103. *Carex magellanica*

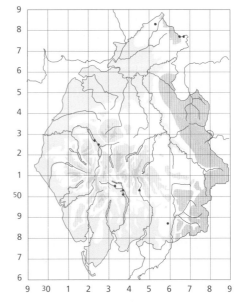

Map. 1049. *Carex aquatilis*

C. aquatilis Wahlenb.
A very rare sedge of rivers and lakesides in the central Lake District and between Derwent Water and Bassenthwaite. It occurs also by Kentmere Tarn (44.02), by a beck in the hills above Killington (58.86s), in Kershope Forest (52.82n) and along county border by the River Irthing near Churnsike Lodge (64.76n, 66.76n).

The first record appears to be Wilson's reference to the site by Blelham Beck on the Furness side of Windermere (36.00) which was probably discovered by W.H. Pearsall. E.C. Wallace's 1942 records from Elterwater and Skelwith (3.0, *BEC Rpt* (1944) 12:508) could refer both to Westmorland and Furness. In Cumberland it was reported by Miss Bacon in 1925 (*BEC Rpt* (1926) 7:902) from between Silloth and Bowness (1.5). This last record is best treated as dubious and the first reliable one is probably that from near Churnsike Lodge, Spadeadam (66.76n, CCH & FJR, 1976; LANC, *Watsonia* (1980) 13:147).

Carex aquatilis has been confused with *C. acuta* and is best distinguished by the longer lower bract, which overtops the male spikes, and the brittle stems.

The only other English sites for this northern sedge are one lower down the Irthing in Northumberland and one by the River Tees in Yorkshire.
Alt. limit: 270 m, near Churnsike Lodge, (66.76n)
Map 1049   10   WFC

C. acuta L.
Known from several places along the River Eden from Great Ormside down to Warwick Bridge (mostly RWMC, LANC), from the River Derwent at Brigham (08.30) and near its mouth at Workington (00.28, RS, 1980, herb. R.S), and from high up the River Irthing near Butterburn (68.72n, GH, 1977, LANC, det. A.O. Chater & A.C. Jermy), only a few kilometres downstream from a *Carex aquatilis* site. The only Survey records from the south are by Terrybank Tarn, west of Barbon (58.82s, R.W. David, 1978) and an estate pond at High Biggins, Kirkby Lonsdale (58.78s, GH, 1984, LANC), where it was probably planted. There is also a pre-Survey record for Beaumont Marsh, near the mouth of the River Eden (3.6, DAR, 1954, CLE).

This sedge has been much confused, chiefly with *C. elata* and *C. acutiformis*, but also with the closely allied *C. aquatilis*. It has not been possible to refind it in three of the west Cumberland *Atlas* squares. The few Furness records cited by Baker and by Petty (1898 p.311) probably refer to *C. acutiformis*. Wilson's specimen from Gaisgill has been tentatively identified by A.O. Chater as the hybrid with *C. acutiformis* and his other cited records from the south: Windermere, Elterwater (3.0) and the Arnside area (4.7s) require confirmation.
Alt. limit: 230 m, Butterburn (68.72n), but formerly at 335 m in Greystoke Park (40.32)
Map 1050   18   W[(F)]C

C. nigra (L.) Reichard   Common sedge
Very common in wetlands, particularly in the uplands, less frequent in the more intensively farmed lowlands but sometimes abundant on the disturbed edges of lowland peat mosses and in 'poor' fen. It occurs over a wide pH range but is commonest in acid and mesotrophic mires, flushes, tarn margins and wet meadows. In blanket peat it occurs around pools and characteristically on slightly sloping ground adjacent to the sikes where it may be dominant.
Alt. limit: 870 m, Cross Fell (68.34)
Map 1051   1318   LWFCY

C. nigra x **C. elata**
C. x turfosa Fries
Several plants of this hybrid were found in 1975 by R.W. David, growing with the parents by Eskrigg Tarn, Killington (56.88s, det. A.C. Jermy). The only previous record is W.H. Pearsall's from Lakeside (3.8, *BEC Rpt* (1914) 3:396) in 1913.

                   1   W(F)

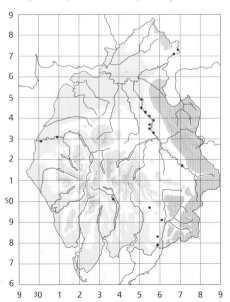

Map 1050. *Carex acuta*

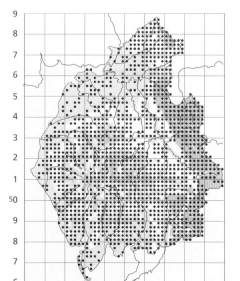

Map 1051. *Carex nigra*

C. elata All.   Tufted sedge

Largely restricted to the south where it occurs by tarns, in fen carr, by the Kendal canal and by river- and lake-sides, particularly from Grasmere and Elterwater south to the Leven estuary. There is an anonymous record from Cliburn (5.2, 1946, K). The post-1930 anonymous *Atlas* record from Wetheral (4.5) almost certainly refers to *Carex acuta*. The *Sedge Handbook* record for the Pooley Bridge square (4.2) is based on a misident-ification and the pre-1950 record for the Allonby square (0.4) seems very dubious.

With its dense tussocks and long, slender, top-heavy flowering stems, this is usually an easy species to recognise.

Alt. limit: 250 m, Sunbiggin Tarn (66.06)

Map 1052   49   WFC

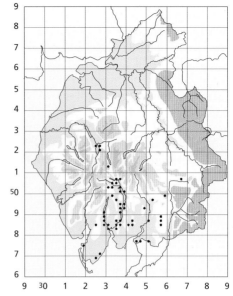

Map 1052. *Carex elata*

C. bigelowii Torrey ex Schwein.   Stiff sedge

Fairly common above 550 m on the high fells of the Lake District and the north Pennines, occasional in the Howgill Fells and on the fells along the Yorkshire border. It grows on the ridges and on rounded summits where it may form a dense sward on the peaty soil, which in many places is being eroded by walkers, and where it is prevented from flowering by the sheep. Vegetative plants and those of *Carex binervis* are sometimes confused but the characteristic orange tips to the old leaves in the latter and the loose, purplish scale leaves of the *C. bigelowii* rhizomes are usually sufficient to distinguish them. By contrast it grows and flowers luxuriantly on ungrazed rock ledges on the higher crags and in the exclosures on Little Dun Fell.

It was first seen in Furness by J.R. Birkett in 1956 on Dow Crag (26.96).

Alt. range: 530 m, south-east of Garrigill (76.38) - 965 m, Scafell Pike (20.06)

Map 1053   126   WFCY

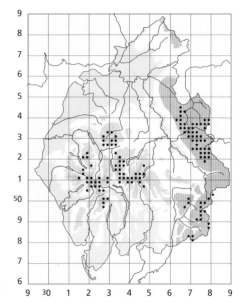

Map 1053. *Carex bigelowii*

C. pauciflora Light.   Few-flowered sedge

A very distinctive, diminutive and easily overlooked sedge of the western and northern Lake District but with a single isolated occurrence on Butterburn Flow (66.76n) in the extreme north-east. The only Furness record is from Wrynose Pass (26.02) where it was dis-covered by R. Mackechnie in 1956 (*Proc. BSBI* (1957) 2:264). It was not refound during the Survey at two sites where it had been seen in the 1950s by DAR: Ashness Fell (28.18) and Birker Moor (16.96).

It usually occurs in slightly flushed areas on or around the margins of hummocks in base-poor mires, often associated with *Sphagnum* spp., *Eriophorum angustifolium*, *E. vaginatum*, *Erica tetralix*, *Molinia*, *Narthecium ossifragum* and *Drosera rotundifolia*.

South of the Scottish Border this Highland sedge is known only from Cumbria and a few sites in North-umberland, east Yorkshire and North Wales.

Alt. limit: *c*.635 m, south of Allen Crags (22.08)

Map 1054   20   FC

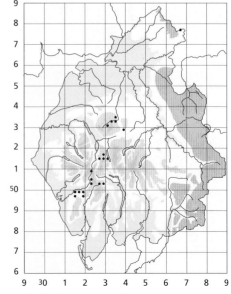

Map 1054. *Carex pauciflora*

C. pulicaris L.   Flea sedge
Very common in the uplands in the eastern half of the county, scattered throughout the western Lake District, scarce in the lowlands and absent over most of the lower Eden valley and the Solway Plain.

       This is typically a sedge of flushed upland grassland and it is particularly characteristic of rock ledges and gillsides, both on and off the limestone.

Alt. limit: 760 m, Middle Tongue, Cross Fell (68.32)

Map 1055   739   WFCY

POACEAE
(*GRAMINEAE*)

*Sinarundinaria anceps (Mitf.) C.S. Chao & Renvoize
A graceful bamboo seen with *Sasa palmata* in woodland near Low Wood, Windermere (38.02, CEW, 1994, LANC, det. D. McClintock).

       1   W

*Sasa palmata (Burb.) Camus
A very conspicuous broad-leaved bamboo noticed at Muncaster (10.96, GH, 1991, LANC), where it is spreading from the estate on to the roadside, by Rusland Pool (32.88, JA, 1987, LANC), in woodland south of Windermere (40.98, CEW, 1994, LANC) and near Low Wood, Windermere (38.02, CEW, 1992), also in Underley Woods, Kirkby Lonsdale (60.80s, T.C.G. Rich, 1991, LANC).

       5   WFC

*Phyllostachys sp.
A single clump was found in estate woodland at Heversham (48.84s, MB, *c*.1980; LANC) but lacking the young culms necessary for identification

       1   W

Nardus stricta L.   Mat grass
Very common in the uplands, often dominating extensive areas giving the fells a bleached appearance throughout the winter, particularly striking in the Howgill Fells. These unpalatable and unproductive *Nardus* grasslands have been derived through overgrazing by sheep from the *Festuca - Agrostis* fellside grasslands and they remain virtually unchanged over extensive periods of time even when grazing ceases. *Nardus* grassland usually gives way at higher altitudes to *Juncus squarrosus - Eriophorum* communities on the poorly-drained plateaux, but persists in *Calluna* and *Vaccinium* moorland and on the drier

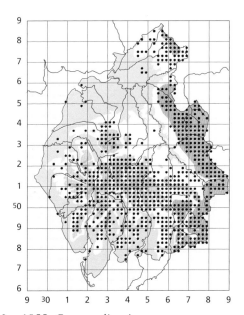

Map 1055. *Carex pulicaris*

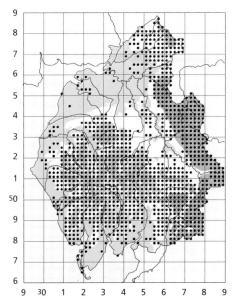

Map 1056. *Nardus stricta*

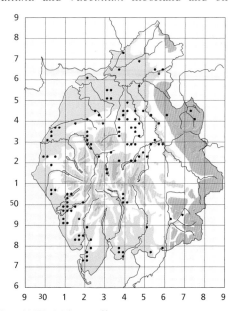

Map 1057. *Milium effusum*

knolls in mires. At high altitudes on the Helvellyn range, the distribution of *Nardus* is thought by RWMC to be related to late snow-lie. On the lower fells it is often replaced by *Pteridium* on the deeper soils. It occurs in lowland heath and rough grassland and in the upper salt-marsh turf along the Solway Firth.
Alt. limit: 945 m, Helvellyn (34.14).
Map 1056   1127   WFCY

[Stipa pennata L.
This species of southern and central Europe is listed by Baker from limestone rocks in Longsleddale (4.0/5.0). This is obviously an error.

(W)]

Milium effusum L.   Wood millet
A southern and eastern species occurring in Cumbria chiefly across the lower Eden valley and more sparsely along the west coast. It is surprisingly scarce around Morecambe Bay. All the sites are from the more base-rich deciduous woodlands.
Alt. limit: at least 380 m, west of Dockray (36.20)
Map 1057   113   WFC

Festuca pratensis Hudson   Meadow fescue
An important fodder grass which is generally common in the lowlands occurring on roadside verges, railway-sides, coarse grassland and wasteland, scrub and woodland margins.
Alt. limit: 575 m, Hartside (64.40)
Map 1058   893   WFCY

F. arundinacea Schreber   Tall fescue
Well distributed in the lowlands and occurring in essentially the same habitats as *Festuca pratensis*, alsoon river gravel and locally abundant on railway-sides on the west coast.
Alt. limit: 430 m, north of Alston (72.48)
Map 1059   414   WFCY

F. gigantea (L.) Villars   Giant fescue
A frequent and prominent member of the ground flora of old, lowland deciduous woodland on the moister soils, although scarce on the Solway Plain and in parts of the lower Eden valley. It is particularly characteristic of woodland paths and shady riversides.
Alt. limit: 370 m, north-west of Nenthead (74.44)
Map 1060   649   LWFCY

F. altissima All.   Wood fescue
An uncommon, graceful and very distinctive grass of wooded rocky ravines and gills, especially by water-falls, chiefly in the Lake District but also in the south-east and extreme north-east. It grows both on and off the limestone, but it does require soils of a moderate base-status and these conditions are met in the old relict areas of deciduous woodland running up many of the Lake District gills. In these woods it is often asso-ciated with *Luzula sylvatica*. Both species are sensitive

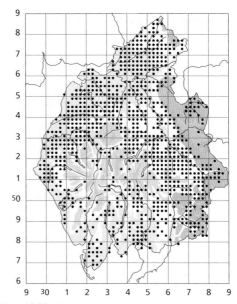

Map 1058. *Festuca pratensis*

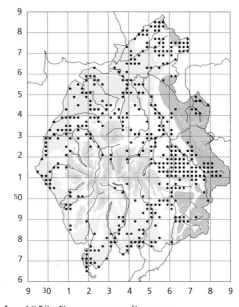

Map 1059. *Festuca arundinacea*

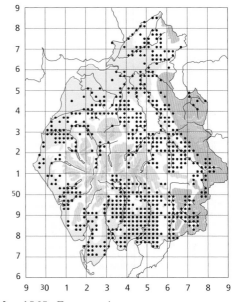

Map 1060. *Festuca gigantea*

to grazing and their large rosettes of persistent broad leaves are prominent on rock ledges and in clefts.

The first Furness record dates from 1917 but is unlocalised (W.H. Pearsall, *BEC Rpt* (1918) 5:135). The grass was not refound during the Survey at Scale Force, Crummock Water (1.1, E.J. Glaister, 1882, CLE), Lodore Falls, Derwent Water (2.1, C.W. Muirhead, 1950, CLE, PLYP) and Melmerby (6.3, W.W. Mason, 1923, *BEC Rpt* (1924) 7:223), nor at the Workington (9.2), Frizington and Egremont (0.1) or Distington (0.2) sites listed by Hodgson.

This grass is commonest in Cumbria, northern and mid-west Scotland.
Alt. limit: 330 m, Guerness Gill, Haweswater (48.02)
Map 1061   70   WFCY

*F. heterophylla Lam.
Known only from deciduous woodland above the River Eden and below Appleby Castle (68.18, REG, 1985, LANC, *Watsonia* (1986) 16:197). As in its Roxburghshire and Selkirkshire localities, it is associated with *Poa chaixii*; both grasses were introduced for ground cover early last century.

This is a very distinctive grass with its slender panicle of small spikelets and long, narrow leaves.
Alt. limit: 100 m, Appleby (68.18)
                         1   W

F. arenaria Osbeck
(*F. juncifolia* Chaub.)
An uncommon coastal segregate of *Festuca rubra* found by the Duddon estuary at Sandscale and Roanhead (18.74, 1992, 20.74, 1996, both PB, LANC) and Askam-in-Furness (20.76, 1991, LANC). It was earlier collected by C.W. Muirhead from Fleswick Bay, St Bees (94.12, 1949, PLYP). The specimens, characterised by the hairy lower surface of the leaves and sheaths, were determined by A.K. Al-Bermani.
                  S 3   F(C)

F. rubra L.   Red fescue
Extremely common in lowland pastures, meadows and amenity grassland, almost ubiquitous on old and seeded roadside verges and as a fodder grass in meadows and pastures. It also occurs in coarse grassland and railwaysides, on the less mobile sand-dunes and dune grassland, on sea-cliffs and sea-walls and in salt-marshes, particularly those which are grazed. In the fells it is largely restricted to damp grassland around flushes, and by becksides and waterfalls.

With such a wide range of habitats it is not surprising that examination of 42 collections submitted to A.K. Al-Bermani and C.A. Stace included all seven subspecies currently recognised in Britain, including two new to England. The material was not a representative sample as half was from coastal sites and a quarter from the Lake District and Pennine hills.

A viviparous plant was found on the rocks by the River Eden at Great Salkeld (56.38, RWMC, 1994, LANC).
Alt. limit: 845, Great Dun Fell (70.32)
Map 1062   1675   LWFCY

subsp. **rubra**
A loosely tufted and probably common subspecies but collected only from a wall at Askam-in-Furness (20.76, PB, 1991, herb. P.B.), from Bridekirk (10.36, AAD, 1994), Aspatria (14.42, AAD, 1987, LANC), and Deepdale, Dentdale (70.82, GH, 1993, LANC).
Alt. limit: 350 m, Deepdale (70.82)
                         4   FCY

subsp. **juncea** (Hackel) K. Richter
A common and strongly coastal taxon recorded from sand-dunes and sea-cliffs from Askam-in-Furness to Skinburness (12.54) but no doubt present along the entire coast. The only inland record is from the old railway track on the Thistle viaduct, north of Longtown (38.72, REG, 1979, LANC).

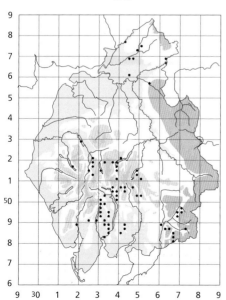
Map 1061. *Festuca altissima*

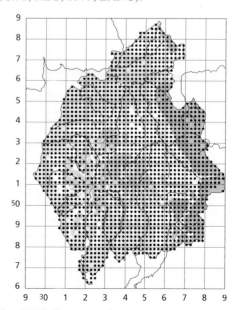
Map 1062. *Festuca rubra*

It is densely tufted, typically forming a deep, springy turf.

12 FC

subsp. **litoralis** (G. Meyer) Auq.
A mat-forming salt-marsh subspecies collected from the Duddon estuary at Sandscale and the marsh at Askam-in-Furness (18.74, 20.76, PB, 1991, LANC).

2 F

subsp. **commutata** Gaudin
Although no doubt very common and frequently sown, the only collections were from Barrow-in-Furness (18.70, 1992; 18.68s, 1992, LANC) and the Dalton-in-Furness by-pass (22.74, PB, 1994, LANC).

This subspecies is unusual in lacking rhizomes and consequently forms dense tufts.

3 F

subsp. **arctica** (Hackel) Govoruchin
Recorded at five rock-ledge sites in the Lake District: Scrubby Crag, Fairfield (36.10, GH, 1981, LANC), the first Westmorland record, Dollywaggon Pike (34.12), Red Screes (38.08), High Street (44.12) and Red Gill, Glaramara (24.10), and four sites in the Pennines: a laneside verge at Ravenstonedale (72.98) and limestone ledges east of Crookburn on the Co. Durham border (78.76), High Cup Nick (74.24) and Cross Fell (66.36, RWMC, 1989, LANC). There are pre-Survey collections from Ardale Head, Cross Fell (C.W. Muirhead, 1952, PLYP) and in the Lake District from near Fleetwith Pike (2.1, C.W. Muirhead, 1952, PLYP).

Although recorded from Scotland and Snowdonia, these are the only known English records (Halliday 1995). All the Cumbrian specimens have the characteristic pale and densely pubescent spikelets.
Alt. limit: 760 m, Red Screes (38.08)

9 WC

Plate 104. *Festuca vivipara*

The lesser grasse-eared Grasse or Grasse upon Grasse. On the Fells betweene Pereth [Penrith] and Kendall, as also on Snowdon.
Mercurius Botanicus, 2:23 (1641),Thomas Johnson

subsp. **scottica** Cunn. ex Al-Bermani
Known only from a Survey collection from Pillar (16.12, GH, 1993, LANC) and an earlier one from the limestone on the Cross Fell side of Crowdundle Beck (6.3, C.W. Muirhead, 1949, PLYP).

This is the first English record for this subspecies which otherwise occurs in central and northern Scotland (Halliday 1995). The long spikelets, often 1 m or more, are particularly characteristic.
Alt.: 730 m, Pillar (16.12)

1

C

subsp. **megastachys** Gaudin
Recorded on roadsides at Aspatria (14.42, AAD, 1987, LANC) and Brampton (66.22, RWMC, 1992, LANC), and on tipped soil by the shore at Askam-in-Furness (20.78, PB, 1992, LANC), the first records for the three vice-counties. It was also seen at Calder Bridge (04.04), Egremont (02.06), at Cocklakes gypsum works (44.50) and by the upper Croglin Water in the Pennines (60.48, RWMC, 1991, LANC), a possible native site.

This is the most robust of the subspecies and is probably quite frequent.
Alt. limit: 305 m, Croglin Water (60.48)

7 WFC

F. ovina L. Sheep's fescue
Extremely common in the uplands, less so in the lowlands. Like the last, this species occurs over a wide ecological range which includes the upland *Festuca - Agrostis* grasslands, the short turf of the coastal and montane limestones, old quarries and well-drained sandy heathy soils around the coast and in the Eden valley.

M.J. Wilkinson has examined a number of Survey specimens and identified most as subsp. *hirtula* (Hackel ex Travis) Wilk., a taxon based on a specimen collected by W.G. Travis at Coniston (2.9, *BEC Rpt* (1914) 3:512) in 1913. All are from Cumberland. The only records of subsp. *ovina*, which is characterised by longer awns and usually glabrous leaf-blades (Stace 1988), are from Belah Scar, Kaber (78.12, GH, 1987, LANC) and Red Gill Beck, Torver (26.94, GH, 1990, LANC.
Alt. limit: 965 m, Scafell Pike (20.06)
Map 1063 1472 LWFCY

F. vivipara (L.) Smith Viviparous sheep's-fescue
A characteristic grass of the Lake District fells and the high Pennines. Although often considered an important component of the higher *Festuca - Agrostis* grasslands, it is probably scarce in that community. Certainly it is much more conspicuous on rock ledges and in rock crevices, both acidic and basic, where it flowers freely. It also occurs along beck and riversides and is not infrequent at low altitudes on rocky shores and islands of the major lakes.
Alt. limit: 965 m, Scafell Pike (20.06)
Map 1064, Plate 104 322 WFCY

F. filiformis Pourret
(*F. tenuifolia* Sibth.)

Six Survey collections (all LANC) have been identified as *Festuca filiformis* by C.A. Stace, M. J. Wilkinson or T. Cope. Most are from Cumberland and were growing in dry, acid grassland, two in the open, the others in woodland where they formed conspicuous bright green tussocks, interspersed with *Anthoxanthum*: near Burththaite Park, Wreay (40.48, GH, 1986, BM), near Gaitsgill (40.44, GH, 1986), Eycott Hill, Berrier (38.30, RWMC, 1987). Flinty Fell, Garrigill (76.42, RWMC, 1987) and Penrith Beacon (52.30, RWMC, 1995, LANC). There is also a Furness collection from dunes at Sandscale (18.74, PB, 1994, LANC).

The earliest records appear to be those of D. Lumb from Kirkby Moor (2.8, 1913, BM, YRK, *BEC Rpt* (1914) 3:512), A.J. Wilmott's from the Cumberland side of Crowdundle Beck, Crossfell (6.3, 1921, BM) and J.E. Lousley's from the Westmorland side of Mickle Fell (7.2/8.2) and by Cauldron Snout, Teesdale (8.2, 1927, *BEC Rpt* (1928) 8:591) and C.W. Muirhead's from Kingmoor, Carlisle (3.5, 1951, BM).

In the light of recent work by Stace and Wilkinson (Stace 1988), records unsupported by specimens are best regarded with suspicion since this grass has been much confused with short-awned specimens of *F. ovina*.
Alt. limit: 550 m, between Garrigill and Nenthead (76.42)

6   [(W)]FC

x **Festulolium loliaceum** (Hudson) P. Fourn.
Festuca pratensis x **Lolium perenne**

Scattered in the lowlands and commonest in the Eden valley, occurring in meadows and old pastures and roadside verges. Both the hybrid and *Lolium perenne* are very variable and some field records may refer to the latter. It is however much commoner than is suggested by the *Critical Atlas*. A reliable character for the hybrid is the presence of a rudimentary inner glume. However, some plants approach the parent species and are presumably backcrosses.

The hybrid with *Festuca arundinacea* almost certainly occurs but has not been recorded separately.

The first Westmorland record was from Witherslack, Grange-over-Sands (4.8s, H.W. Kew, 1925, *BEC Rpt* (1926) 7:905), for Furness from Old Holbeck, Roose (22.68s, GH & MS, 1978, LANC) and for Cumberland those in the *Critical Atlas*.
Alt. limit: 470 m, below Tyne Head (74.36)
Map 1065   102   WFCY

*x **F. braunii** (K. Richter) A. Camus
Festuca pratensis x **Lolium multiflorum**

Recorded from three roadside sites on the Furness peninsula but no doubt overlooked: by Mere Tarn, Scales (26.72, PB, LANC, 1992), at Dalton-in-Furness (22.72, PB, 1989, LANC, *Watsonia* (1993) 19:294) and near Humphrey Head (38.74, E. Chicken, 1995;

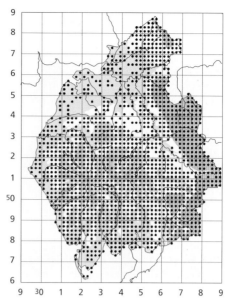

Map 1063. *Festuca ovina*

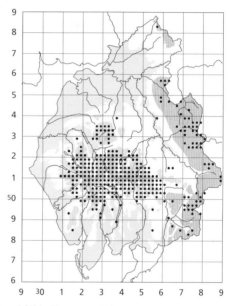

Map 1064. *Festuca vivipara*

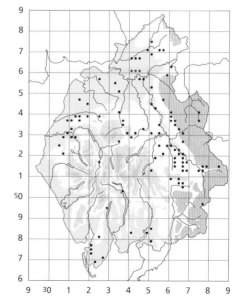

Map 1065. x *Festulolium loliaceum*

LANC).

3   F

[x F. fredericii Cugnac & A. Camus
Festuca rubra x Lolium perenne
There are only two records known of this hybrid, one from Sweden the other from a meadow at Stonethwaite, Borrowdale (2.1, G.A. Hayes, BM, *Proc. BSBI* (1957) 2:381) where it was collected in 1956 and identified by A. Melderis. When visited in 1986 the field had just been rotovated and there was no sign of any likely plant.

  It must be said that the BM specimen is poor and scarcely convincing for the only British record and the record is treated as doubtful by Stace (1991).

(C)]

x **Festulpia hubbardii** Stace & R. Cotton
Festuca rubra x Vulpia fasciculata
Known only from a limited area at Sandscale (18.74, PB, LANC, det. C.A. Stace, *Watsonia* (1995) 20: 435) where a few plants were found in 1994 growing with the parent species.

  This is the northernmost locality on the west coast of Britain for this rare hybrid.

1   F

Lolium perenne L.   Perennial rye-grass
Ubiquitous in anthropogenic grasslands: improved pastures, meadows, leys and amenity grassland, also roadside verges. It is a common and robust weed in towns and villages. Specimens with awned lemmas were seen near Colton (30.86, PB, 1993, LANC).
Alt. limit: 570 m, Alston Moor (72.38)
Map 1066   1517   LWFCY

*L. multiflorum Lam.   Italian rye-grass
Occasional in improved lowland meadows, pastures and leys, also on roadside verges. It is included in most short-term ley seed mixtures.

There appear to be no Furness or Cumberland records prior to the *Atlas*.
Alt. limit: 410 m, above Garrigill (74.40)
Map 1067   321   LWFCY

*L. temulentum L.
Hodgson lists this cornfield weed from several places from Whitehaven to Longtown and Penrith. Baker mentions an additional locality at Kirkby Lonsdale (6.7s) and Wilson comments that it "occurs occasionally in cultivated fields". There are other records from Grinsdale, Carlisle (3.5, T.S. Johnstone, 1901, CLE), Silloth (1.5, E.J. Glaister, 1875, CLE; J. Parkin, 1935), Thwaites, Millom (1.8, M. Cross, 1927), and Askam-in-Furness (2.7, D. Lumb, 1912, YRK, *BEC Rpt* (1913) 3:299). The *Atlas* shows a pre-1930 record from the Barrow-in-Furness square (2.7) and a post-1930 one from the Cartmel peninsula (3.7s). There are no recent records and the species is probably now extinct in the county.

(WFC)

*L. remotum Schrank
This central European grass is known only from the 19th century records given by Hodgson: Workington (9.2), Leegate (1.4, E.J. Glaister, 1877, CLE) and Westward, Wigton (2.4, R. Wood, 1881, CLE), Maryport (0.3) and Silloth (1.5) and a more recent one from near Abbeytown (1.5, C.W. Muirhead, 1951, CLE).

(C)

Vulpia fasciculata (Forsskål) Fritsch
An attractive and local annual grass of sand-dunes on the east side of the Duddon estuary and with an isolated occurrence at South Walney (22.62s). These are the most northerly sites on the west coast.

  The only Cumberland records are from the beach at Seascale (0.0) where it was seen by W.A.P. Sprott in 1934 (*BEC Rpt* (1937) 11:290). It should be looked for on the dunes at Eskmeals (0.9) and Haverigg (1.8). The first Furness record dates from 1939 when it

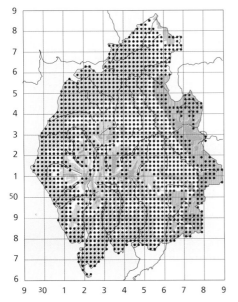

Map 1066. *Lolium perenne*

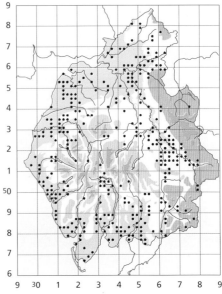

Map 1067. *Lolium multiflorum*

was seen on Walney Island by W.H. Hardaker (*BEC Rpt* (1942) 12:304).

This is a coastal species, generally distributed from East Anglia round to Wales and Cumbria.
Map 1068   7   F(C)

V. bromoides (L.) Gray   Squirreltail fescue
Occasional around the coast and in the Eden valley, rare elsewhere. It occurs in dry, well-drained sites: coastal sands, sand-pits, quarries and wall-tops and is seemingly indifferent to soil pH.

The only pre-Survey records for Furness are those shown in the *Atlas*.
Alt. limit: 245 m, Troutbeck Park (42.06)
Map 1069   77   WFC

V. myuros (L.) C. Gmelin   Rat's-tail fescue
A south-eastern species which is rare and often casual in Cumbria. Eight of the 14 sites are on railway prop-

erty, mostly disused stations and lines. It was extremely common, together with *Poa compressa*, on the old station site at Keswick (26.22), but the site has since been redeveloped. Other sites include Silloth quay (10.52), a farm-track near Seaton (02.28) and a street in Carlisle (40.56). The only post-1930 *Atlas* record is from the Pooley Bridge square (4.2).
Alt. limit: 115 m, Culgaith (58.28)
Map 1070   14   WFC

Cynosurus cristatus L.   Crested dog's-tail
Extremely common, occurring in a wide range of grasslands including some which are quite acid and nutrient-poor. Like *Lolium perenne* and *Poa pratensis*, it is very much a pasture and meadow grass. C.D. Pigott has pointed out that, judging from the prominence of ungrazed culms in close-grazed pastures and upland roadside verges, stock evidently find the culms unpalatable.

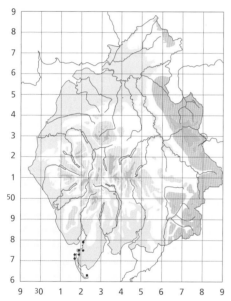

Map 1068. *Vulpia fasciculata*

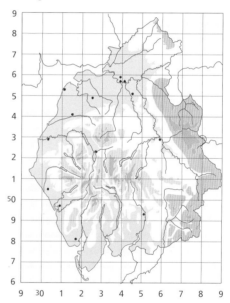

Map 1070. *Vulpia myuros*

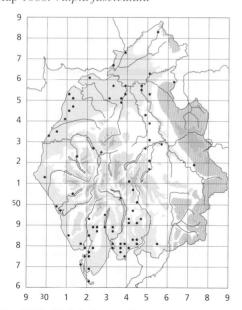

Map 1069. *Vulpia bromoides*

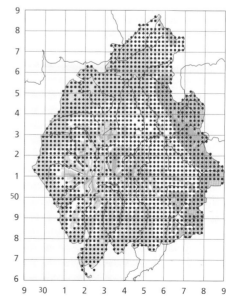

Map 1071. *Cynosurus cristatus*

Viviparous plants were found at Hartside (62.42).

Alt. limit: at least 660 m, Burnhope Seat (78.36)

Map 1071 1615 LWFCY

*C. echinatus L.

The only localities for this southern, weedy alien grass are the three given by Hodgson from Workington (9.2), Penrith (5.2/5.3) and near Silloth (1.5, E.J. Glaister, 1889, CLE), where it was later seen by C.W. Muirhead in 1943 (CLE), and the ones from 'Brackenthwaite' (1.2/2.4, R. Wood, 1878-9), Dalton-in-Furness (2.7, D. Lumb, 1913, *BEC Rpt* (1914) 3:397) and Kirkby Lonsdale (6.7s, C.G. Trapnell, 1923, *BEC Rpt* (1924) 7:222).

(WFC)

Puccinellia maritima (Hudson) Parl.

Common salt-marsh grass

Very common in the upper parts of the saltmarshes, often on bare mud by creeks and along the estuaries. It is most familiar as an important constituent of the extensive areas of grazed salt-marsh turf, growing with *Festuca rubra* which has usually assumed dominance by the time it is cut as highly-prized 'sea-washed turf'.

Map 1072 87 WFC

P. distans (L.) Parl.

Rare, recorded only early in the Survey from brackish or salt-marsh sites on Walney Island and in the Kent estuary and at two sites near Silloth (10.52, 12.56). However, it was subsequently seen on the saline verges of the A6 road on Shap Summit (54.04, 54.06), the M6 slip-road at Penrith (50.28) and by the A66 nearby (52.28) and in some quantity on Stainmore near the Co. Durham border (8.1). It is conspicuous by its late flowering. This species has been known for some time as a colonist of trunk road and motorway verges on the east side of the Pennines.

It was first found in Furness by P. Adam on Tumner Hill Marsh, Walney Island (18.66) in 1979 and in Westmorland in 1970 by E.J. Harling and KAG on Meathop Marsh (42.78s). It has not been refound in the Workington square (9.2) shown in the *Atlas*.

Alt. limit: 430 m, Stainmore (88.12)

Map 1073 22 WFC

P. rupestris (With.) Fern. & Weath.

Recorded by Hodgson from the west Cumberland coast at Workington (9.2), Maryport (0.3) and Silloth (1.5) and from railway ballast inland at Aspatria (1.4). There appear to be no subsequent records from these the most northerly sites on the west coast of Britain.

S (C)

Briza media L. Quaking grass

Very common in base-rich or calcareous grasslands and also, like *Linum catharticum*, widespread in upland flushes. It is absent from much of the Solway Plain, the lower Eden valley and the poorer, more acidic soils of

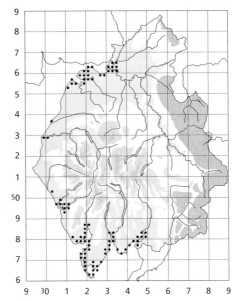

Map 1072. *Puccinellia maritima*

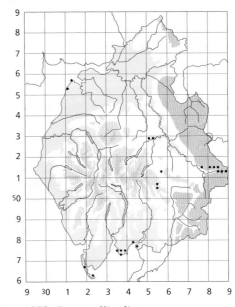

Map 1073. *Puccinellia distans*

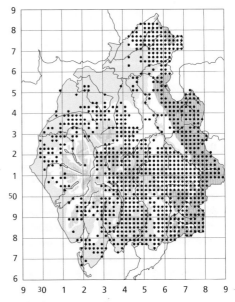

Map 1074. *Briza media*

the western Lake District.
Alt. limit: 720 m, east of Knock Fell (74.28)
Map 1074 962 LWFCY

*B. maxima L.
Known only from an 1884 collection from Silloth (1.5, J. Leitch, CLE).

(C)

Poa annua L. Annual meadow-grass
Extremely common, growing in a wide variety of open or semi-open habitats, especially as an almost ubiquitous weed in towns and villages, by roadsides, waste places and cattle-trampled field entrances. In more natural habitats it grows in rather damp situations such as the margins of shallow pools and tarns and in the fells in areas of damp grassland below small springs or cliffs often associated with *Stellaria uliginosa* and *Montia fontana*, also by fell paths and on bare peat.
Alt. limit: 975 m, Scafell Pike (20.06)
Map 1075 1731 LWFCY

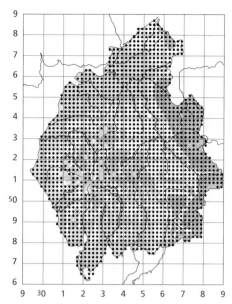

Map 1075. *Poa annua*

P. trivialis L. Rough meadow-grass
Very common but slightly less so than *Poa pratensis*, particularly in the uplands. In contrast to that species it is essentially a grass of damp or wet grassland, and a constant feature of ditchsides and damp roadside verges. It occurs also in meadows and pastures and by the margins of ponds and tarns. In the inhospitable upland peatlands it is restricted, like so many common species, to the sides of roads and tracks although it also occurs with *P. annua, Montia fontana* and *Chrysosplenium oppositifolium* around springs and seepage areas where it is probably frequently under-recorded.

As Hodgson remarks, both this and the last species are valuable agricultural grasses.
Alt. limit: 840 m, Helvellyn (34.14)
Map 1076 1540 LWFCY

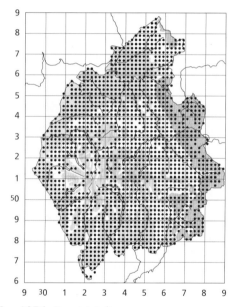

Map 1076. *Poa trivialis*

P. humilis Ehrh. ex Hoffm.
(*P. subcaerulea* Smith)
Common on the Pennines in short limestone grassland, on scars and rock ledges. Most of the records were contributed by A.J. Richards. Its apparent scarcity elsewhere is almost certainly the result of underrecording. It seems, for example, likely that this species rather than *Poa pratensis* sensu stricto is the prevailing plant on the coast; it also occurs on damp, gritty roadside verges in the uplands. There appear to be no preSurvey records for Westmorland.

It is usually recognisable in the field by its rhizomatous habit and single culms.
Alt. limit: at least 600 m, north of Knock Fell (72.32)
Map 1077 76 WFCY

P. pratensis L. Smooth meadow-grass
Very common in dry or fairly well-drained, moderately base-rich or calcareous grassland. Its habitats include meadows and pastures, roadside verges and railway-

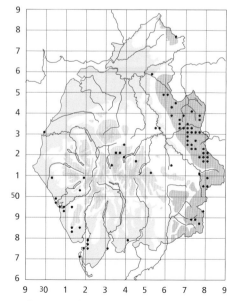

Map 1077. *Poa humilis*

sides, sand-dunes and limestone grassland. In the drier sites and on the limestone many of the map records probably belong to the following segregate.

Alt. limit: at least 620 m, Dufton Fell (70.28), although the record could refer to *Poa humilis*. Wilson's figure of 945 m, presumably from the summit of Helvellyn (32.l4), must refer to *P. annua*. There are no records of *P. pratensis* in this tetrad.

Map 1078　　　　1513　　　　LWFCY

P. angustifolia L.

Seen in 1992 on railway banks at Lazonby, Little Salkeld, Long Marton and Appleby (respectively 54.40, 56.36, 66.24, 68.20, RWMC, LANC, conf. J. Edmondson, *Watsonia* (1993) 19:294). It is probably a relative newcomer to the county.

Sargent *et al.* (l986) have recently given an account of the British distribution of this critical segregate of the *Poa pratensis* complex. They found it to be widespread on the somewhat base-rich soils of the railway sites which they surveyed, often growing in coarse grassland with *Arrhenatherum* and *Achillea millefolium*. Their map shows it as occurring in 16 squares in Cumbria, all but one being on railway sites. There are no previously published records from Cumbria, but in the absence of any grid-references and locality details it is best to regard the 1992 finds as the first substantiated records.

Poa angustifolia differs from *P. pratensis* chiefly in being taller and in having long, inrolled leaves <2 mm wide.

Alt. limit: 145 m, Appleby (68.20)

4　　　　WC

*P. chaixii Villars

Known only from seven sites: in the grounds of Underley School, Kirkby Lonsdale (60.80s, GH, l976, LANC) and Edmond Castle School, Hayton (48.58, FJR, 1980s); by a streamside at Millwood, Dalton-in-

Furness (20.72, PB, 1990, LANC), where it was first recorded in 1912 (D. Lumb, YRK, *BEC Rpt* (1913) 3:296); hotel grounds, Ambleside (36.02, MP, 1995); Skelghyll Woods, Ambleside (38.02, P. Bullard, 1989, herb. P.B.); in and around the churchyard at Skirwith (60.32, LR, 1993, LANC); and in woodland above the River Eden below Appleby Castle (68.18, C.M. Robb, 1969; REG, l985, LANC). At this last site it was associated with *Festuca heterophylla*, another popular ground-cover species during the early l9th century.

Alt. limit: 130 m, Appleby (68.18)

7　　　　WFC

Poa compressa L.　　　　Flattened meadow-grass

Scattered, mostly in the upper Eden valley but probably under-recorded. Most of the records are from man-made habitats, for example the railway platform at Dalton-in-Furness (22.72), the old steelworks site at Barrow-in-Furness (18.68, 1870), the old station site at Keswick (26.22), slate wall-tops near Ambleside (36.04) and Glenridding (38.l6) and a limestone quarry near Morland (60.20). Native sites include the limestone of Hellgill, Mallerstang (78.96) and slate rocks by the River Lune near Sedbergh (62.92).

Alt. limit: 365 m, Mallerstang (78.96)

Map 1079　　　　35　　　　WFCY

P. glauca Vahl　　　　Glaucous meadow-grass

A very rare grass of base-rich gullies and rock ledges in the Lake District, first discovered in Britain on the Helvellyn range (3.l). It was seen during the Survey in five tetrads: on Coniston Old Man (26.96), Helvellyn (32.14, 34.12, 34.14) and Dove Crag (36.10). These were confirmations of earlier records from the three areas made respectively by DAR and R.J. Birkett in 1959, DAR in 1955 and R.J. Birkett in 1960 (*Proc. BSBI* (1963) 5:40).

The grass has not been refound at two of Wilson's sites: Ill Bell (this could be the Kentmere (4.0)

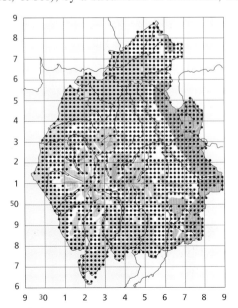

Map 1078. *Poa pratensis*

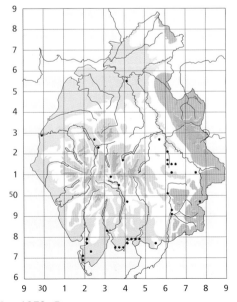

Map 1079. *Poa compressa*

or Haweswater (4.0/4.l) mountain) and rocks at the head of Hartsop Beck (4.l). Hodgson's Cumberland record must be considered doubtful.

Specimens have in the past often been referred to *Poa balfourii* Parnell which is now generally regarded as a shade form of *P. glauca* (Trist l986). However, such plants are easily confused with *P. nemoralis*.

This arctic-montane grass occurs in Britain only in Snowdonia, the Lake District and the Scottish Highlands.
Alt. range: 520 m, Coniston Old Man (26.96) - 850 m, Helvellyn (34.14)

S 5   WFC

P. nemoralis L.   Wood meadow-grass
Generally frequent in the lowlands and quite common in the Eden valley. It usually occurs in rather dry, moderately base-rich deciduous woodland, often in some profusion, but it is not uncommon in the open on wall-tops and dry rock ledges, as on the whinsill rocks of High Cup Nick (74.24) and in the eastern quarries of Coniston Old Man (26.96), where it often escapes detection. Montane plants tend to have more open panicles and larger spikelets (*Poa parnellii* Bab.).
Alt. limit: 670 m , Piers Gill (20.06, DAR, 1954)
Map 1080   422   WFCY

P. alpina L.   Alpine meadow-grass
Extremely rare; known since the mid-l9th century from the eastern cliffs of Helvellyn (3.l), where it still occurs in two tetrads, and on the Pennines near High Cup Nick (7.2), where it grows on two limestone scars.

This is an arctic-montane grass, widely distributed throughout the central Scottish Highlands but with only a few isolated occurrences to the south in Snowdonia, on Ingleborough and in Cumbria. The High Cup Nick populations and that on Ingleborough are unusual in being non-viviparous, although they differ in chromosome number (Lloyd 1964).
Alt. range: 580 m, High Cup Nick (7.2) - 790 m, Helvellyn (3.1)

4   W

Dactylis glomerata L.   Cock's-foot
Extremely common, absent only from upland blanket peat and the higher fells of the Lake District. This valuable fodder grass is essentially a species of pastures and meadows, field edges, roadside verges and waste ground. It is more demanding in its nutrient requirements than either *Poa pratensis* or *P. trivialis*, and this rather than climatic factors is probably the reason for its relative scarcity in the uplands.

A viviparous form noted at Melmerby (6.3) by W.W. Mason in 1922 was seen in three tetrads in that area during the Survey and also in the south-west near Dalton-in-Furness (20.74).
Alt. limit: 550 m, Moor House, Teesdale (74.32)
Map 1081   1518   LWFCY

Catabrosa aquatica (L.) P. Beauv.
Rather rare, restricted to a few localities near Barrow-in-Furness, on the Solway Plain, near Crosby Ravensworth (6.1) and on Warcop Fell and Little Fell (78.18, 78.20). The coastal sites are field ditches, dykes and ponds. It occurs by a spring below Pardshaw Crags (10.24) and on the coast at Askam-in-Furness (20.76) where streams reach the sandy beach. It used to occur in a marshy field at Arnside (46.76s), but this has now been built on,and there is a 1951 record from Humphrey Head (3.7s, C.W. Muirhead, CLE).

In l978 FJR found it in small quantity by flushes on Little Fell (78.20) at 710 m (Roberts & Halliday 1979, LANC). This, like the discovery of *Glyceria maxima* in Sprinkling Tarn, is a remarkable occurrence for a species which Wilson (1956) says is restricted to below 310 m. Surprisingly, no other montane sites have since been discovered.

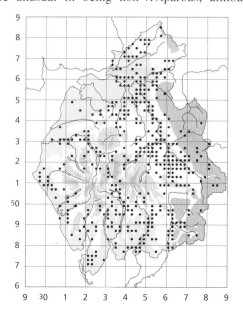

Map 1080. *Poa nemoralis*

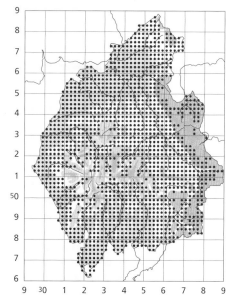

Map 1081. *Dactylis glomerata*

This grass thrives in muddy, cattle-poached pastures. It suffers from being very palatable to stock; it is also very vulnerable to drainage and is probably decreasing.
Alt. limit: 710 m, Little Fell (78.20)
Map 1082   24   WFC

Catapodium rigidum (L.) C.E. Hubb.   Fern-grass
(*Desmazeria rigida* (L.) Tutin)
Fairly common on and near the south coast and extending north to Maryport (02.36) where it reaches its northern limit on the west coast of Britain, although it used to occur at Skinburness (1.5, W.A.P. Sprott, 1935).
        This is an early-flowering annual grass of dry situations, occurring on coastal sands, in sand and gravel pits, in quarries, on walls and on bare soil around limestone outcrops and ledges.
Alt. limit: *c*.145 m, Kendal Fell (50.92)
Map 1083   54   WFC

C. marinum (L.) C.E. Hubb.   Sea fern-grass
(*Desmazeria marina* (L.) Druce)
Rather rare, in open often sandy places along the coast from Arnside (44.76s) north to Maryport (02.36), occuring on sand-dunes, walls, waste ground and pavements.
        There is no recent confirmation of the *Atlas* records from the Whitehaven square (9.1), Silloth (1.5, J. Leitch, 1878, CLE) and Grune Point (1.5, C.W. Muirhead, 1950, CLE). The only Westmorland record was from by the railway at Arnside (R. Dalton, 1977) and the first for Furness was from Holme Island, Grange-over-Sands (Petty 1894 p.333).
Map 1084   31   WFC

Sesleria caerulea (L.) Ard.   Blue moor-grass
(*S. albicans* Kittel ex Schultes)
Common to abundant throughout the limestone hills around Morecambe Bay and the upper Eden valley, also, but sparsely, on limestone outcrops west of Penrith (3.4, 4.3). This early flowering, coarse, vigorous grass is characteristic of the Carboniferous limestone of northern England. It is not particularly palatable to sheep and is often dominant over considerable areas producing a species-poor turf, except where it is disrupted by rock outcrops or by ants' nests. It also grows in open, species-rich communities on rock ledges and in crevices.
        There are no authenticated records from the Lake District despite the statement in Hubbard (1984) that it grows there at 760 m and the fact that it occurs on basic rocks in the central Scottish Highlands.
Alt. limit: 755 m, Knock Ore Gill (70.30)
Map 1085   S 169   LWFCY

Parapholis strigosa (Dumort.) C.E. Hubb.   Hard-grass
A rather uncommon but locally frequent grass of short *Festuca* - *Puccinellia* turf and open, muddy ground in

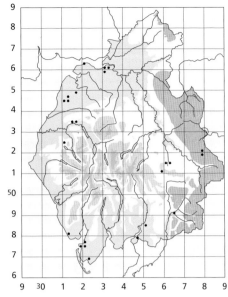
Map 1082. *Catabrosa aquatica*

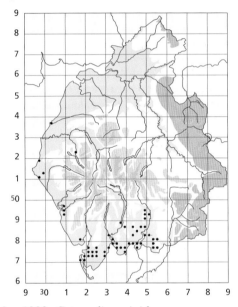
Map 1083. *Catapodium rigidum*

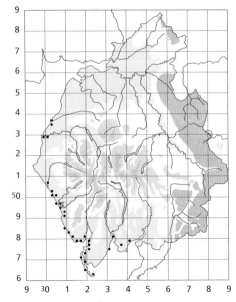
Map 1084. *Catapodium marinum*

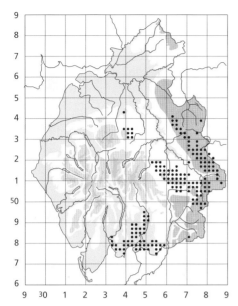

Map 1085. *Sesleria caerulea*

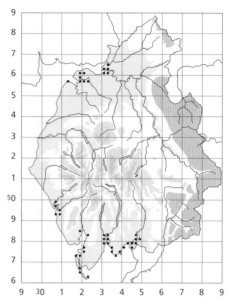

Map 1086. *Parapholis strigosa*

the upper part of salt-marshes and close to the strand-line. There are no Survey records for the post-1930 *Atlas* record from the Allonby square (0.4) or from Grune Point (1.5, 1949, *BSBI Yr Bk* (1949) p. 45).

This grass is probably under-recorded. The populations are said to vary considerably in size from year to year and the grass is very inconspicuous except at anthesis.

Map 1086   44   WFC

Glyceria maxima (Hartman) O. Holmb.
Rather rare; known only from the margins of a few lowland tarns and ponds, mainly in the south and west, but also from the oligotrophic waters of Gurnal Dubbs, on the fells above Staveley (50.98) and even higher in Sprinkling Tarn in the central Lake District (22.08, LANC), its highest site in Britain. At one site near Sedbergh (60.90) it completely fills what was once a field pond and at Monkhill Lough (32.58) it covers many acres. This decorative grass is very palatable to stock and it occurs by a number of estate ponds where it has probably been planted. These records are included in the map.
Alt.limit: 600 m, Sprinkling Tarn, Seathwaite (22.08)
Map 1087, Plate 105   28   WFCY

G. fluitans (L.) R. Br.
Common, often filling ditches, runnels and pools, also in wet meadows and forming a conspicuous raft of vegetation around the more sheltered shores of tarns but avoiding the more oligotrophic waters. It is very palatable to cattle.

The map certainly includes records of the following two species and the hybrid, but nevertheless it is probably still substantially accurate.

This species is usually readily identifiable in the field by its long, narrow panicle which disarticulates as soon as the fruits ripen.

Plate 105. *Glyceria maxima*, Sprinkling Tarn

Alt. limit: 720 m, east of Knock Fell (74.28)
Map 1088   1046   WFCY

G. fluitans x **G. notata**
G. x **pedicellata** F. Towns.
Scattered within the general range of the last species.
There appear to be no published records of this hybrid
prior to those in the *Critical Atlas*.

The panicle branches diverge at a wider angle
than in *Glyceria fluitans* but less than in *G. plicata*.
The anthers fail to dehisce and the empty florets
disarticulate.
Alt. limit: 550 m, Moor House, Teesdale (74.32)
Map 1089   85   WFC

G. declinata Bréb.
Very common in the east and south-west, less so on the
Solway Plain and scarce in the Lake District and
northern Pennines. It is more lowland than *Glyceria*

fluitans and is very characteristic of cattle-trampled
swampy fields and field ponds where it is easily
recognised by its short glaucous leaves.

The only previous Flora writer to mention it was
Wilson who thought it "rare or overlooked?". It is
certainly far commoner than the *Atlas* records suggest.
The only pre-Survey records for Cumberland and
Furness appear to be those shown in the *Atlas*.
Alt. limit: at least 360 m, near Nenthead (76.44)
Map 1090   691   WFCY

G. notata Chevall.
(*G. plicata* (Fries) Fries)
Scattered, apparently commonest in the upper Eden
valley and in west Cumberland but it has probably been
appreciably overlooked and recorded as *Glyceria*
fluitans, which it closely resembles. On the other hand
some field records may be suspect.

The widely-spreading panicle usually disting-

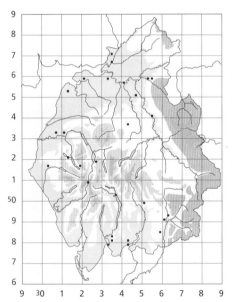

Map 1087. *Glyceria maxima*

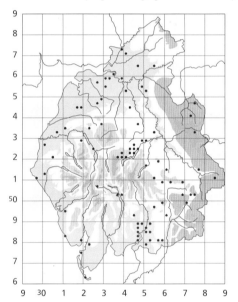

Map 1089. *Glyceria fluitans* x *G. notata*

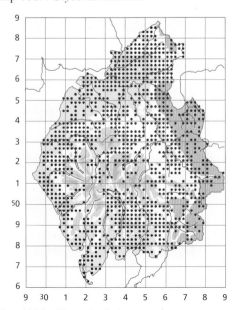

Map 1088. *Glyceria fluitans*

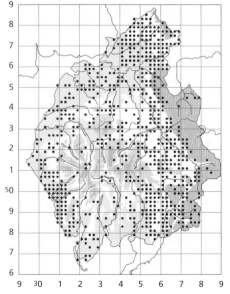

Map 1090. *Glyceria declinata*

uishes this species in the field from *G. fluitans* and *G. declinata*. The light green leaves also separate it from the latter.

Alt. limit: *c*.350 m, near Garrigill (74.42)

Map 1091   181   WFCY

Melica nutans L.   Mountain melick

Fairly frequent on both wooded and treeless limestone pavements and scars around the Kent estuary and in north Westmorland, otherwise scarce and off the limestone restricted to shady, moderately base-rich rock ledges in upland gills. There is an outlying pre-Survey record from a gully on Wasdale Screes (14.02, DAR, 1954, CLE).

Alt. limit: 550 m, Helvellyn (32.18)

Map 1092   83   LWFCY

M. uniflora Retz.   Wood melick

Common in the south and, in the north, around the Caldew valley, elsewhere somewhat sporadic and absent from most of the Solway Plain. It prefers the more base-rich soils and like so many woodland grasses it thrives best by the side of streams and gills and by woodland rides and margins. It is also frequent in shady hedgerows.

Alt. limit: at least 330 m, near Garrigill (74.40)

Map 1093   453   LWFCY

Helictotrichon pubescens (Hudson) Pilger

(*Avenula pubescens* (Hudson) Dumort.)

Frequent in the east, scattered to the north of the Lake District and across to the west coast, otherwise largely absent. Although, like the last species, occurring in limestone grassland, it is also frequent in unimproved meadows and on road- and railwaysides on moderately base-rich, non-calcareous and somewhat moist soils.

Alt. limit: 550 m, Moor House, Teesdale (74.32)

Map 1094   332   LWFCY

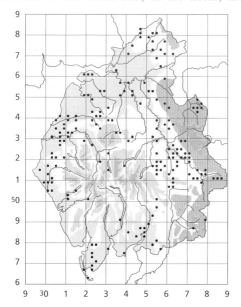

Map 1091. *Glyceria notata*

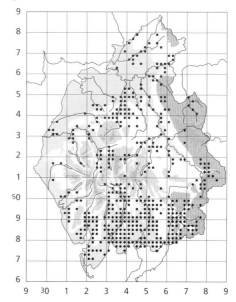

Map 1093. *Melica uniflora*

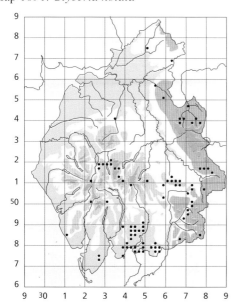

Map 1092. *Melica nutans*

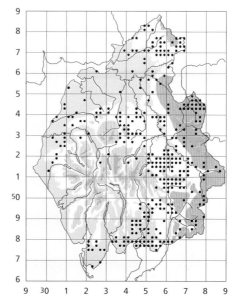

Map 1094. *Helictotrichon pubescens*

H. pratense (L.) Besser
(*Avenula pratensis* (L.) Dumort.)
Fairly frequent on the limestone around the upper Eden valley and the Kent estuary, elsewhere very rare and absent from large areas including the whole of the Lake District. Most of the sites are in limestone grassland, one exception being that on slate rocks by the River Lune near Sedbergh (62.90, 62.92, RWMC, 1988).

The first Furness record dates from 1977 when it was found on Hampsfield Fell, Grange-over-Sands (40.78s, GH, LANC).
Alt. limit: at least 510 m, east of Connypot Beck, Teesdale (82.20)
Map 1095   124   LWFCY

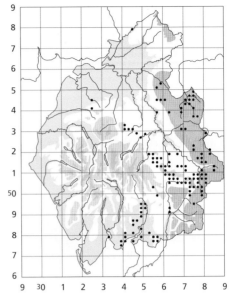

Map 1095. *Helictotrichon pratense*

Arrhenatherum elatius (L.) P. Beauv.
ex J.S. & C. Presl   False oat-grass
A very common grass of unimproved meadows, roadside verges, hedgebanks, railwaysides and waste ground; in the lowlands probably only absent from areas of peat moss and woodland.

Var. *bulbosum* (Willd.) St Amans, which is sometimes a troublesome arable weed in southern England, has been found only at Silloth (1.5, E.J. Glaister, 1877, CLE), Talkin Tarn (5.5, C.W. Muirhead, 1946, CLE), allotments at Whitehaven (98.16, CCH, 1989, LANC) and near Levens Hall (46.84s, CEW, 1991, LANC). This is in marked contrast to the situation in the Isle of Man where Allen (1984) says it "overwhelmingly predominates".
Alt. limit: 500 m, Ardale, Cross Fell (66.34)
Map 1096   1449   LWFCY

*Avena strigosa Schreber
The only records for this uncommon and decreasing cornfield weed are the 19th century ones cited by Hodgson from St Bees (9.1, R. Wood, 1879, CLE) and Workington (9.2), a specimen from Silloth docks (1.5, J. Leitch, 1884, CLE) and Baker's from Kirkby Lonsdale (6.7s), where it was said to be frequent in cornfields.

(WC)

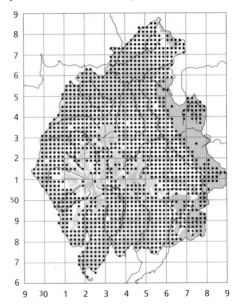

Map 1096. *Arrhenatherum elatius*

*A. fatua L.   Wild oat.
Rather rare, chiefly along the M6 and trunk roads where the dead panicles are conspicuous in late summer. Elsewhere it occurs only sporadically, and rarely in cornfields, its main habitat in southern and eastern England.

It is listed by Hodgson from Workington (9.2) and Maryport (0.3). Baker gives a record from Kirkby Lonsdale (6.7s) and Wilson one from Crosby Ravensworth (6.1). There are also records from Askam-in-Furness (2.7, D. Lumb, 1912, *BEC Rpt* (1913) 3:299), Newton Reigny (4.3, J. Parkin, 1935) and Gilsland (6.6, G.A.K. Hervey, 1943, v.c.67/70).
Alt. limit: 440 m, above Swindale, Brough (80.18)
Map 1097   43   W(F)C(Y)

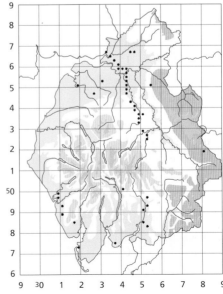

Map 1097. *Avena fatua*

*A. sterilis L.

A southern European species once seen in an oatfield at Askam-in-Furness (2.7, D. Lumb, 1912, *BEC Rpt* (1913) 3:299).

(F)

*A. sativa L.   Oat

Occasionally persisting after cultivation, often with *Hordeum distichon*, and also around farms but generally ignored during the Survey.

6   WFC

Trisetum flavescens (L.) P. Beauv.   Yellow oat-grass

Common in the upper Eden valley, frequent to scattered elsewhere, particularly along the south-west coast. This attractive lowland grass occurs on roadside verges and in unimproved meadows and pastures, generally on the rather better-drained and less acid soils.

Alt. limit: 550 m, Moor House, Teesdale (74.32)

Map 1098   710   WFCY

Koeleria macrantha (Ledeb.) Schultes

Common in short limestone turf and on rock ledges around the upper Eden valley, in the south from Kendal down to the Kent estuary, also in open sandy soil or dry grassland around the coast.

Alt. limit: at least 510 m, Helbeck Fell (82.20)

Map 1099   171   LWFCY

Deschampsia cespitosa (L.) P. Beauv. Tufted hair-grass subsp. **cespitosa**

Extremely common in a wide range of habitats subject to a periodically high water-table. In the lowlands it forms conspicuous tussocks in wet meadows, on roadside verges, in deciduous woodland and along wet, compacted woodland rides. At higher altitudes it grows mainly by drainage ditches and the margins of mires but also on apparently rather dry roadside banks and verges. Of its agricultural value, Hodgson commented that "This is a worthless grass which nothing short of absolute hunger can induce cattle or horses to eat".

Viviparous plants were seen in in the Lake District by Ley & Linton (1906) in 1905 on Crinkle Crags (22.08) and later by DAR in Skew Gill, Great End (2.0, 1959, LANC), by C.W. Muirhead, DAR and CCH in Piers Gill, Scafell Pike (20.06, respectively 1952, PLYP; 1954, LANC; 1976, herb. C.C.H.), where they grew with non-viviparous individuals, and on Kirk Fell (18.10, GH, 1989, LANC). There is also a record from Loo Gill, Hartside (62.42, LR, 1993, LANC) and from a lowland site at Wetheral (46.54, RWMC, 1995, LANC). These probably represent tetraploid populations of subsp. *cespitosa* and may be the only English records. According to H.A. McAllister (1989 *in litt.*), both diploid and tetraploid plants of this subspecies are likely to occur in Cumbria.

Alt. limit: 855 m, Helvellyn (34.14)

Map 1100   1632   LWFCY

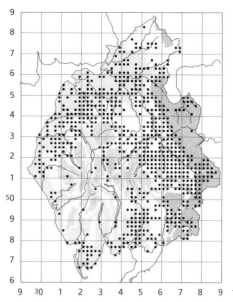

Map 1098. *Trisetum flavescens*

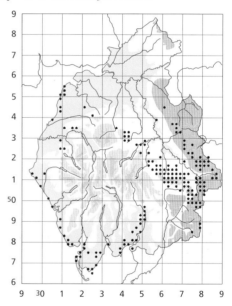

Map 1099. *Koeleria macrantha*

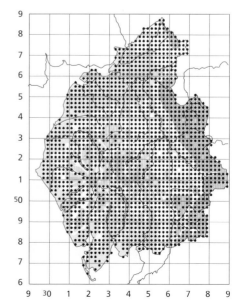

Map 1100. *Deschampsia cespitosa*

subsp. **parviflora** (Thuill.) Dumort.
The only records of this shade-loving diploid are from near Welton (34.44, J. Clarke & A. Franks, 1993, LANC, conf. R. Payne), the west side of Thirlmere and above the River Eden above Armathwaite (30.16, 50.44, RWMC, 1994, LANC, det. P.J. Trist) and a pre-Survey record from Penton Linns (4.7n, N.D. Simpson, 1949, *BSBI Yr Bk* (1951) p.50.

    This subspecies has shorter spikelets (2.5-3 (-3.5) mm) and shorter hairs at the base of the lemmas.
Alt. limit: 190 m, Thirlmere (30.16)

               3   C

D. flexuosa (L.) Trin.   Wavy hair-grass
Very common in the uplands as a constant and sometimes major component of acid grassland, and in *Calluna* and *Vaccinium* moorland. It is scattered in the lowlands on areas of heath and moorland, on acid and often rather dry hedgerow banks and in acid woodlands, usually of birch or oak. It is sometimes dominant in wooded riverside ravines.
Alt. limit: 960 m, Scafell Pike (20.06)
Map 1101   1249   LWFCY

Holcus lanatus L.   Yorkshire fog
Extremely common on virtually all types of lowland grassland, also but less frequently in somewhat open, rather dry woodland. In the uplands it occurs in hay-meadows and on roadside verges, by rocky becksides and waterfalls and on rock ledges.
Alt. limit: at least 650 m, Cross Fell (68.34)
Map 1102   1711   LWFCY

H. mollis L.   Creeping soft-grass
Common, particularly on the dry sandstone soils along the east side of the Eden valley and in the extreme north. It is much more of a woodland plant than the last, occurring and sometimes becoming dominant in rather dry woods of birch and oak, also in the more open coniferous woods and in hedgebanks. It frequently occurs under bracken, notably on the open fellsides.
Alt. limit: 550 m, Moor House, Teesdale (76.32) and Kentmere (44.08)
Map 1103   1381   LWFCY

Aira caryophyllea L.   Silver hair-grass
Fairly frequent on dry, sandy soils along the coast; rather sparse inland where some of its sites are on railway ballast, a common habitat across the Pennines in Durham (Graham 1988). Interestingly, it occurs on a limestone scarp at the head of Dentdale immediately below the railway viaduct (76.84). Although calcifuge like *Aira praecox*, it appears less tolerant of heavy rainfall and the map shows a noticeable clustering of inland records in the Eden valley. This is an attractive grass, notwithstanding Hodgson's dismissal of it as "a worthless species".

    No information is available on the occurrence, or otherwise, of subsp. *multiculmis* (Dumort.) Bonnier

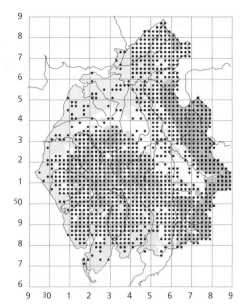

Map 1101. *Deschampsia flexuosa*

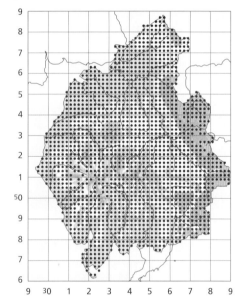

Map 1102. *Holcus lanatus*

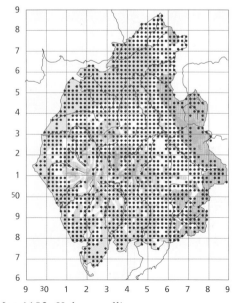

Map 1103. *Holcus mollis*

& Layens which is taller and has a narrower panicle.
Alt. limit: 320 m, above Milburn (68.30)
Map 1104   219   WFCY

A. praecox L.   Early hair-grass
Very common in the Lake District, in the fells to the
east and also along the coast; elsewhere rather
scattered. It grows commonly in areas of thin bare
ground around siliceous rock outcrops, often with
Sedum anglicum and *Polytrichum piliferum*, also on
sandy banks, coastal cliffs and in small pockets of
humus on the top of siliceous walls, where it is very
conspicuous by its precocious flowering.
Alt. limit: 600 m, Kidsty Howes, Haweswater (44.12)
Map 1105   867   LWFCY

[Hierochloe odorata (L.) P. Beauv.   Holy grass
Reported by Hodgson as having appeared in a garden at
Workington (9.2). It is known in Britain only from a few
stations in Scotland, three of the *Atlas* records being on the
north side of the Solway Firth.

                                              (C)]

Anthoxanthum odoratum L.   Sweet vernal grass
Rather unexpectedly this is the most widespread species
having been recorded in every tetrad in Cumbria except
one on Rockcliffe Marsh (30.60). It thrives over a wide
range of soil pH, being ubiquitous in unimproved grasslands and on verges but also occurring among *Calluna*
and *Vaccinium* on the upland peats, and in open woodland in the lowlands.
Alt.limit: 950 m, Helvellyn (34.14)
Map 1106   1780   LWFCY

Phalaris arundinacea L.   Reed canary-grass
Very common in the lowlands occurring in fens, in
reedswamp by tarns and lakes, in ditches and by river
and canalsides where it often forms a conspicuous

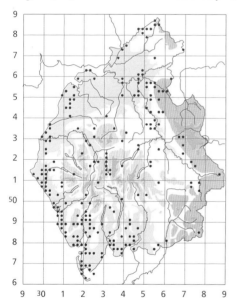

Map 1104. *Aira caryophyllea*

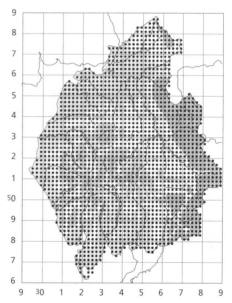

Map 1106. *Anthoxanthum odoratum*

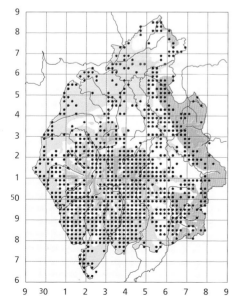

Map 1105. *Aira praecox*

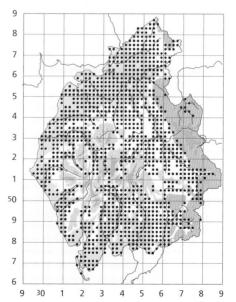

Map 1107. *Phalaris arundinacea*

fringing margin. It is also occasionally found in wet woodland and even in such apparently well-drained sites as the central reservation of the M6 motorway.

All the early Flora writers record the variegated garden escape var. *picta* L. and this was seen seven times during the Survey.

Alt. limit: *c*.370 m, below Hartside (72.42)

Map 1107   1099   LWFCY

*P. canariensis L.   Canary grass

Fairly widely, if sparsely, distributed, occurring as a casual chiefly by roadsides, on waste land and in gardens, where it probably originates as bird seed. It is recorded from only five squares in the *Atlas* and the recent increase in records is probably attributable to the increased use of bird seed.

The first Furness records are those shown in the *Atlas* for the Broughton-in-Furness (2.8) and South Walney (2.6s) squares, both dating from the 1950s.

Alt. limit: 430 m, Nenthead (78.42)

Map 1108   27   WFC

Agrostis capillaris L.   Common bent

Extremely common. It occurs in most types of grassland, particularly the *Festuca - Agrostis* grasslands of the uplands, in heaths and moorland and in the more acid and open woodlands. It can, however, occur over a wide range of soil pH. On the wetter soils it tends to be replaced by *Agrostis canina*.

Alt. limit: 940 m, Helvellyn (34.14)

Map 1109   1687   LWFCY

A. gigantea Roth   Black bent

Fairly frequent as an arable weed in the Eden valley, usually in barley fields, and conspicuous by its tall culms and large, delicate, spreading panicle. It has also a number of scattered localities, mostly along the south coast, where it occurs by roadsides and on waste

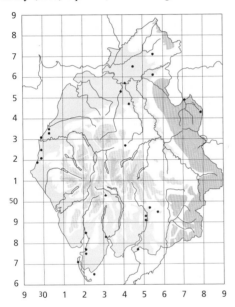

Map 1108. *Phalaris canariensis*

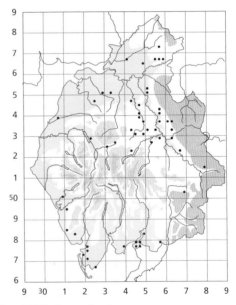

Map 1110. *Agrostis gigantea*

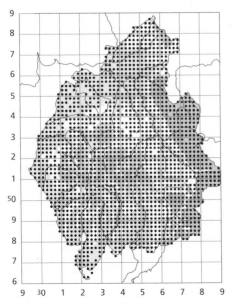

Map 1109. *Agrostis capillaris*

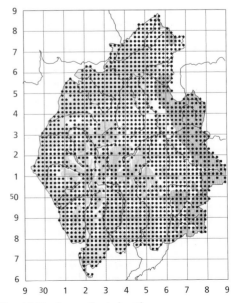

Map 1111. *Agrostis stolonifera*

ground.
Alt. limit: 215 m, Brough (78.14)
Map 1110 54 WFC

A. stolonifera L. Creeping bent
Very common, particularly in the lowlands where it
occurs in meadows and pastures, on roadside verges, in
allotments and waste places and on rubbish dumps. It
requires a higher nutrient status than either of the last
two species and Hodgson comments on the invigorating
effect of domestic sewage on plants on Maryport beach
(02.36).

It is also characteristic of wetter habitats being
often abundant and semi-floating on the muddy edges
of ponds, and in the fells it is a feature of rock ledges by
waterfalls.
Alt. limit: at least 650 m, Cross Fell (68.34)
Map 1111 1578 LWFCY

A. canina L. and **A. vinealis** Schreber Brown bent
Frequent but chiefly in the uplands. *Agrostis canina* is
predominantly a grass of mires, wet moorland and
ditches where it is conspicuous by its very fine, bright
green leaves and late flowering.

Plants occurring in rather dry heath and moorland, where it is appreciably less common, may
represent the related tetraploid species *A. vinealis*. This
can only really be distinguished morphologically by the
absence of stolons and presence of rhizomes. Recorders
have been asked to look for such plants but only three
convincing specimens have been forthcoming: Long
Moss, Plumpton (50.36, RWMC, 1984, LANC), slate
waste near Hall Dunnerdale (20.94, PB, 1992, LANC)
and wet rocks below Helvellyn (34.14, GH, 1992,
LANC). This is the reverse of the situation in Co.
Durham where Graham (1988) states that *A. vinealis* is
frequent while *A. canina* is only occasional, but "not
recognised by most of our recorders".
Alt. limit: 845 m, Little Dun Fell (70.32)
Map 1112 945 WFCY

Calamagrostis epigejos (L.) Roth
Occasional by Morecambe Bay, occurring in rough
grassland and scrub, on the limestone and by woodland
paths and lanesides. In the Asby - Orton area it grows
in the grikes of limestone pavements, and, by contrast,
in damp dune-slacks by the Duddon and Ravenglass
estuaries. There are no Survey records for the post-1930
Atlas records from the Kirkbride (2.5) and Seathwaite
(Duddon, 2.9) squares; the latter seems a most unlikely
locality.
Alt. limit: 370 m, above Great Asby (66.08)
Map 1113 25 WFC

C. canescens (Wigg.) Roth
Very rare, known only from fen woodland in Roudsea
Wood (32.82), around the head of Windermere (36.00,
36.02), Elterwater (32.04) and in Flakebridge Wood,
Appleby (68.22), and fringing lakeside vegetation at

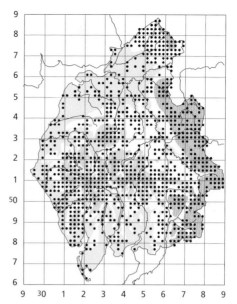

Map 1112. *Agrostis canina*

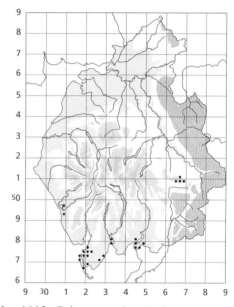

Map 1113. *Calamagrostis epigejos*

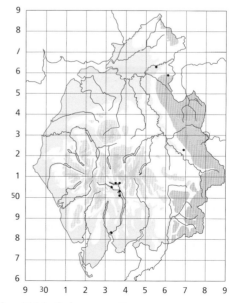

Map 1114. *Calamagrostis canescens*

Rydal Water (34.06), the lake at Naworth (54.60, 54.62) and Tindale Tarn (60.58). At Rydal it grows behind the *Carex elata* fringe, at Naworth with *Phragmites* and *Filipendula*, and at Tindale in a *Carex* swamp.

The post-1930 *Atlas* record from the Hawkshead square (3.9) probably refers to the next species. The earliest records for Furness are the pre-1930 *Atlas* one for the Goldmire Wood, Dalton-in-Furness (2.7, D. Lumb, 1917, YRK, *BEC Rpt* (1918) 5:256), an area now much ravaged by quarrying, and Wilson's reference to the Blelham Beck site (36.00). The first Cumberland record was from Monkhill Lough (3.5, 1935, J. Parkin).

This is an eastern grass which in Cumbria approaches its north-western limit.

Alt. limit: 215 m, Tindale Tarn (60.58)

Map 1114   9   WFC

C. purpurea (Trin.) Trin.

Known only from the fen at the north end of Esthwaite Water (34.96; LANC) where it is locally plentiful in willow carr. T.G. Tutin was long aware that the Esthwaite plant differed from both *Calamagrostis epigejos* and *C. canescens* and about 1960, or earlier, was suggesting that it might be *C. purpurea*, a grass of northern, central and eastern Europe. Unaware of this, O.M. Stewart started to examine some deviating Scottish *Calamagrostis* populations in 1979 and later the Esthwaite plant and came to the same conclusion (Stewart 1988, 1991). In addition to Esthwaite, the species is now known from six sites in Scotland.

Calamagrostis purpurea is apomictic and a high polyploid, possibly derived from a hybrid between the two previous species. It is best distinguished from *C. epigejos* and *C. canescens* by the taller culms, with 5-8 nodes, often glabrous leaves, although not in the Esthwaite plant, slightly larger ligules (-14 mm), wider (1-1.1 mm), lanceolate-acuminate glumes, 5-veined lemmas and empty, indehiscent anthers. The British and Scandinavian plants belong to subsp. *phragmitoides* (Hartman) Tzvelev.

Fig. 20   R 1   F

Ammophila arenaria (L.) Link   Marram grass

Frequent around the coast but, like Leymus, scarce in the inner parts of the Solway Firth and Morecambe Bay. It is restricted to sandy coasts and, as is well known, it is the single most important species in the stabilisation of sand-dunes and has been planted at a number of sites to encourage dune regeneration following damage and blow-outs.

In an interesting historical aside, Hodgson mentions a special Act of Parliament during the reign of George II "imposing penalties for cutting or otherwise damaging the bent. From the provisions of this Act the people of Cumberland were specifically exempted.".

Map 1115   66   WFC

*Gastridium ventricosum (Gouan) Schinz & Thell.

This distinctive grass is native only in the southern half of England and Wales where it has decreased considerably this century. The only Cumbrian record is Hodgson's from the Derwent Tinplate Works, Workington (9.2, J. Hodgson, 1889, CLE).

(C)

*Lagurus ovatus L.   Hare's-tail

Seen only once during the Survey on a pavement at Arnside (44.78s, MB, 1991). This decorative southern grass is listed by Hodgson from a ballast tip at Silloth (1.5, E.J. Glaister, 1878, CLE); the *Atlas* record for the Wigton square (2.4) may refer to this locality.

1   W(C)

*Apera spica-venti (L.) P.Beauv.

A predominantly south-eastern grass seen during the Survey by a lane at Great Salkeld (54.36, GH, 1983, BM, LANC) and by a new roundabout at Dalton-in-Furness (24.74, PB, 1993, LANC). Hodgson recorded it from the usual 19th century alien sites at Workington (9.2), Maryport (0.3) and Silloth (1.5, E.J. Glaister, 1877, CLE; J. Leitch, 1889, CLE) and it was seen on Walney Island (1.6, 2.6s) by the Barrow-in-Furness N.H.S. during the 1950s.

S 2   FC

*Polypogon monspeliensis (L.) Desf.

The only localised Cumbrian records are those given by Hodgson from Floshgate, Ullswater (4.2), Workington (9.2), Maryport (0.3) and Silloth (1.5, J. Leitch, 1879, CLE), and from an oatfield at Askam-in-Furness (2.7, D. Lumb, 1912, *BEC Rpt* (1913) 3:299). There is also a pre-1930 *Atlas* record from the Keswick square (2.2).

This grass is native in Britain only around the coast from Norfolk to Dorset.

S   (C)

Alopecurus pratensis L.   Meadow foxtail

A common grass in the lowlands and with a very similar distribution and ecology to *Phleum pratense*. Like that species, it is of considerable importance as a fodder grass. It is at its most conspicuous in late spring in haymeadows and along roadside verges being one of the earliest perennial grasses to flower.

Alt. limit: 845 m, Great Dun Fell (70.32)

Map 1116   1282   LWFCY

A. geniculatus L.   Marsh foxtail

Common and chiefly in the lowlands where it is an almost constant feature of seasonally wet ground: wet rutted tracks, wet depressions in fields and the margins of ponds, tarns and shallow ditches.

Alt. limit: 845 m, Great Dun Fell (70.32)

Map 1117   1288   LWFCY

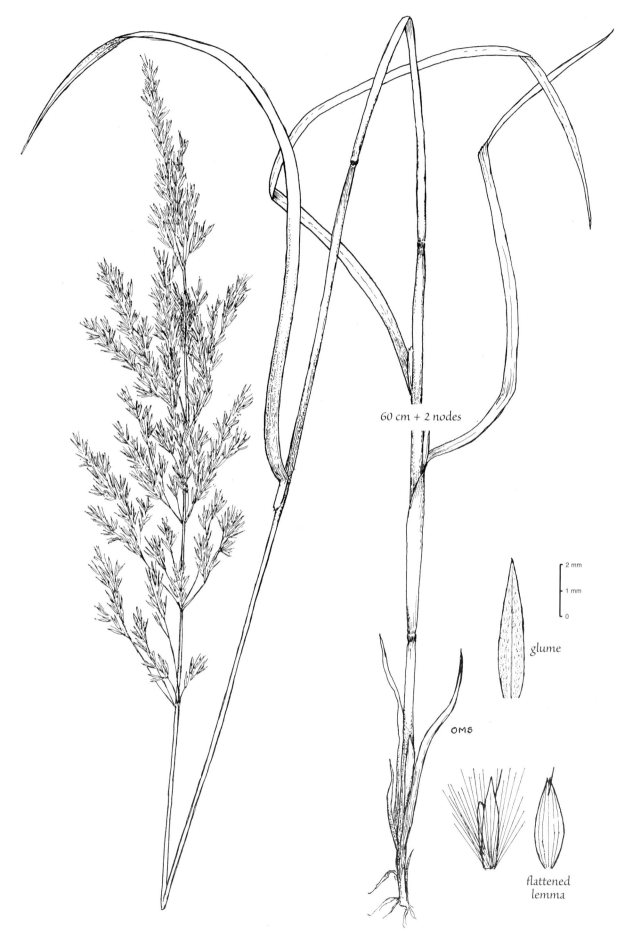

60 cm + 2 nodes

2 mm
1 mm
0

glume

OMS

*flattened
lemma*

Fig. 20. *Calamagrostis purpurea* subsp. *phragmitoides*

A. borealis Trin.   Alpine foxtail
(*A. alpinus* Sm.)
Extremely rare, known from only four tetrads between Cross Fell (68.34) and Knock Fell (70.28, 72.30). It was first recorded in 1959 by DAR on Cross Fell and by A. Eddy on Little Dun Fell (70.32) (Ratcliffe & Eddy 1960). There is an additional pre-Survey record from Ardale Head (66.34, DAR, 1962), on the west side of Cross Fell, but attempts to refind it there have failed.

The favoured habitat is a cold, bryophyte-rich, slightly sloping, stony and very wet flush. Associated species include *Sedum villosum*, *Epilobium anagall-idifolium*, *E. alsinifolium* and *Montia fontana*. The largest population consists of about 200 plants.

Apart from these records and a problematical site beneath the Co. Durham side of the Cow Green Reservoir, this circumpolar high-arctic species is known in Britain only from a few sites in the Cheviot, Moffat and Tweedsmuir hills, the Ochil Hills and the Scottish Highlands.
Alt. range: 770 m, Little Dun Fell (70.32) - 820 m, Cross Fell (68.34)

S 4   WC

A. myosuroides Hudson   Black grass
Very rare, seen by a farm gateway on the north side of Muncaster Fell (10.98, AAD, 1982, LANC, *Watsonia* (1983) 13:434), by a laneside at Irton Road Station (14.98, MMG, 1991, LANC), a farmyard near The Green, Millom (18.84, CEW, 1996, LANC), in a garden at High Hesket (46.42, MS, 1984, LANC) and in nearby barley fields (44.42, 46.40, 46.44, GH, 1985, LANC). It may prove to be not particularly rare in arable fields in the Eden valley.

This arable weed of southern and southeast England is listed by Hodgson from Workington goods station (9.2) and the ballast tips at Maryport (0.3) and Silloth (1.5, E.J. Glaister, 1877, CLE). Refuting claims to its native status in Cumberland, he comments "nor...would its acquisition be desirable". The only other records are from the 1950s, when it was seen by the Barrow-in-Furness N.H.S. in the mid-Walney (1.6) and Rampside (2.6s) squares, and the post-1930 *Atlas* dot for the Matterdale square (3.2).
Alt. limit: 150 m, Calthwaite (46.40)

7   (WF)C

Phleum pratense L.   Timothy; Cat's tail
An extremely common lowland grass of roadside verges, amenity grassland, meadows and pastures, and commonly included in seed mixtures.

Small specimens may be confused in the field with *Phleum bertolonii*, which is probably much less common, but they can usually be identified by the length of the lemmas and awns (Wigginton & Graham 1981). In view of this confusion the map combines the field records for both segregates, although it probably presents a fairly accurate picture of *P. pratense* sensu stricto. Viviparous plants were seen at Melmerby (6.3)

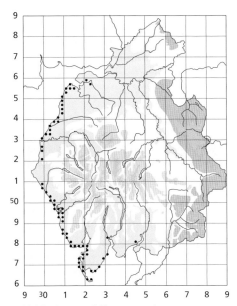

Map 1115. *Ammophila arenaria*

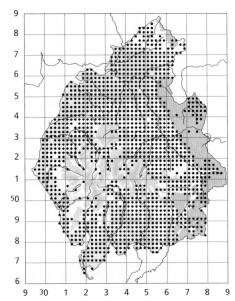

Map 1116. *Alopecurus pratensis*

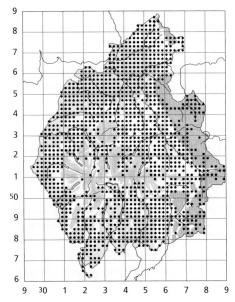

Map 1117. *Alopecurus geniculatus*

by W.W. Mason in 1922.
Alt. limit: 845 m, Great Dun Fell (70.32)
Map 1118   1281   LWFCY

P. bertolonii DC.   Smaller cat's-tail
During the Survey numerous records of this species
were sent in, particularly from the Eden valley. Because
it was thought that recorders might be confusing it in
the field with *Phleum pratense*, a sample of 39 collections was examined in 1988 by CCH who found only
three specimens of *P. bertolonii*, using the characters
given by Wigginton & Graham (1981) but particularly
lemma and awn lengths. This may not be a fair
reflection of the frequency of this species as a sample of
eight 'small Phleums' submitted by CEW in 1988 from
south-east Westmorland included only one *P. pratense*.
According to T. Cope (1988 *in litt.*), robust agricultural
strains of *P. bertolonii* are morphologically indistinguishable from small *P. pratense* and the only
reliable distinction is the chromosome number, the
former being diploid and the latter tetraploid. Any
attempt at present to map the distribution of
P. bertolonii would be misleading.

In its general distribution and ecology *P. bertolonii* is probably very similar to the preceding species
but perhaps less demanding in its soil requirements,
occurring also in drier and more disturbed habitats.
See Map 1118   WFC

P. alpinum L.
Extremely rare, a few plants surviving at two sites on
the Pennines near Cross Fell (6.3, 7.3) where it grows
in limestone turf, associated at one site with *Festuca
rubra*, *Thymus polytrichus* and *Minuartia verna*.

This attractive montane and low-arctic grass
was first discovered by G. Adair in 1913 (*BEC Rpt*
(1914) 3:397) growing on cliff ledges on the east side
of Helvellyn (3.1, R.H. Williamson, undated, CLE).
G. Moule was of the opinion that Adair also saw it on

the Cumberland side of the mountain a short distance
below the summit. It is not known when it was last seen
on Helvellyn. In 1923 Wilson reported it from the
Cumberland side of Crowdundle Beck, Cross Fell (*BEC
Rpt* (1924) 7:220) where it occured on unstable
limestone ledges. In his Flora he mentions its discovery
the following year by H. Britten on the Westmorland
side. The Pennine populations suffered initially from
gross over-collecting, and also from rock-falls, the
Cumberland population being last seen in 1976. The
grass, however, still occurs on the Westmorland side of
Crowdundle Beck and a second population was
discovered nearby by DAR in 1987. These are the only
extant localities in Britain south of the Scottish
Highlands.
Alt.: *c.*745 m, Crowdundle Beck (6.3) but earlier at
*c.*900 m on Helvellyn (3.1, DAR, 1955).

2   WC†

P. arenarium L.   Sand cat's-tail
A fairly frequent winter annual of sandy shores and
sand-dunes on Walney Island, the Duddon and Ravenglass estuaries and the Solway Firth where it reaches its
northern limit on the west coast. There are also post-
1930 *Atlas* records for the mid-Walney (1.6) and Gosforth (0.0) squares.
Map 1119   27   FC

*Bromus arvensis L.
Hodgson gives records from Gilgarran (0.2) and Workington
(9.2); there is one from Silloth (1.5, J. Leitch, 1890,CLE) and
later ones from Thwaites,near Millom (2.8, M. Cross, 1907),
Seascale (0.0, R.H. Williamson, 1927, CLE), Buttermere (1.1,
G.A.K. Hervey, 1922) and Gilsland (6.6, G.A.K. Hervey,
1945, v.c.67/70). The *Atlas* shows post-1930 records for the
Seascale (0.0) and Millom (1.8) squares and there are also
field records made during the 1950s by the Barrow-in-Furness
N.H.S. from the Broughton-in-Furness square (2.8).

This southern species has declined considerably in

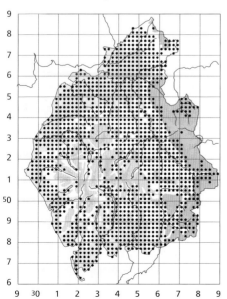

Map 1118. *Phleum pratense* and *P. bertolonii*

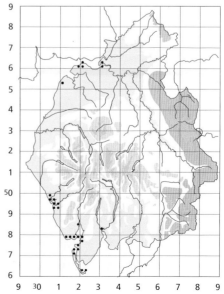

Map 1119. *Phleum arenarium*

Britain during the present century.

(FC)

*B. commutatus Schrader
Seen only once during the Survey, in a hayfield on the west side of Ambleside (36.02, GH, 1977, LANC). It was recorded by Hodgson from Maryport (0.3), Silloth (1.5), Ullswater (4.2) and Greystoke (4.3), and by Baker from Clifton (5.2). Only the last of these is shown in the *Atlas* together with records from the Buttermere (l.l) and Penrith (5.3) squares.

1   W(C)

B. racemosus L.
The single Survey record is from a roadside verge near Cockermouth (12.26, CCH, l981, herb. C.C.H., det. P.M. Smith, *Watsonia* (1982) 14:199).

Although Hodgson cites records from Workington (9.2) and Maryport (0.3), he omits others from St Bees (9.1, W. Dickinson) and Silloth (1.5, J. Leitch, 1890). The only other published record is the post-1930 *Atlas* dot for the Hawkshead square (3.9).

1   (F)C

B. hordeaceus L.   Soft brome
subsp. **hordeaceus**
Common throughout the lowlands, chiefly in meadows, pastures and roadside verges, also in open sites and on waste ground, especially towards the coast. It is a frequent constituent of agricultural seed mixtures. Luxuriant meadow plants may be misidentified as *Bromus commutatus* or *B. racemosus*, while stunted coastal ones may resemble the following subspecies.
Alt. limit: 490 m, below Tyne Head (76.36)
Map 1120   1194   LWFCY

subsp. **thominei** (Hardouin) Braun-Blanquet
The only authenticated records are from the Duddon estuary sand-dunes at Haverigg (l4.78, CCH, l981,

herb. C.C.H., *Watsonia* (1982) 14:199), Sandscale (18.74, PB, 1980s) and North Walney (16.70, GH, l982, LANC, *Watsonia* (l983) 14:433).

As pointed out by Smith (l968), *B. thominei* sensu Tutin, and mapped as such in the *Critical Atlas*, actually refers to hybrid derivatives of *B. hordeaceus* and *B. lepidus* (*B. x pseudothominei* P.M. Smith) and these have not yet been reliably recorded from Cumbria. The true *B. thominei*, now generally relegated to a subspecies of *B. hordeaceus*, is a plant of coastal dunes in western Europe.

3   FC

[*B. x pseudothominei* P.M. Smith
B. hordeaceus x *B. lepidus*
(*B. thominei* auct.)
The *Critical Atlas* shows a number of Cumbrian records of *Bromus thominei* which, as mentioned under the last species, probably belong to *B. pseudothominei*, a complex of hybrid segregates of *B. hordeaceus* and *B. lepidus*. In view of the confusion between these taxa and the fact that none of the records has been authenticated by P.M. Smith, it is best to regard the occurrence in Cumbria of *B. pseudothominei* as not proven.

WFC]

B. lepidus O. Holmb.
Recorded during the Survey only from Cocklakes, Carlisle (44.50, FJR, l981, LANC, *Watsonia* (1982) 14:192), the roadside east of Ireby (26.38, MS & B. Bain, 1990, LANC) and at two sites at Sandscale and Dalton-in-Furness (20.74, PB, 1993, LANC), the last three collections being determined by P.J. Trist. The only previous records are W.W. Mason's from Melmerby (6.3, *BEC Rpt* (1930) 9:146, and regarded by the editor as doubtful) and the three post-l930 *Atlas* dots for the Cartmel (3.7s), Brampton (5.6) and Spadeadam (6.7n) squares.

3   FC

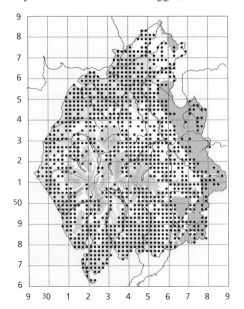

Map 1120. *Bromus hordeaceus*

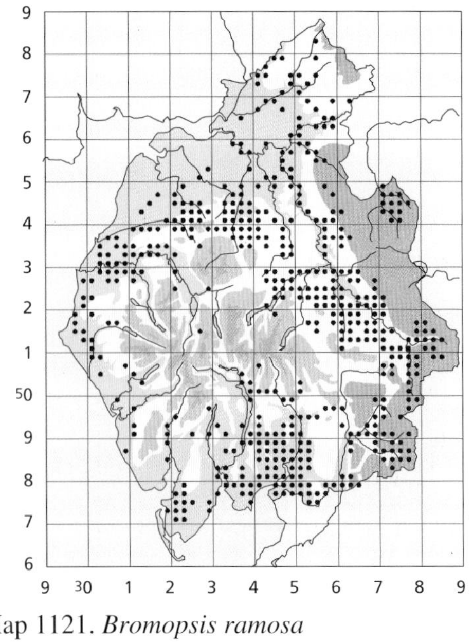

Map 1121. *Bromopsis ramosa*

*B. secalinus L.

Hodgson cites late l9th century records from Workington (9.2), Maryport (0.3) and Tallentire (l.3) and there is also one from Silloth (l.5, J. Leitch, 1882, CLE, det. P.M. Smith). There is a single post-l930 *Atlas* dot for the Brampton square (5.6).

(C)

*B. japonicus Thunb. ex Murray

Material of this eastern and central European species in CLE and collected at Silloth (l.5) by E.J. Glaister in 1877 and by J. Leitch in l882 has been identified by P.M. Smith (*Watsonia* (l986) 16:197) as this species.

(C)

Bromopsis ramosa (Hudson) Holub

(*Bromus ramosus* Hudson)   Hairy brome

Common in damp deciduous woods and copses in the lowlands, chiefly in the south, the Eden valley and to the north of the Lake District. Like *Festuca gigantea*, this is a characteristic grass of woodland margins and path-sides and is an indicator of at least moderately base-rich or calcareous soils.

Alt. limit: 420 m, south of Garrigill (74.40)

Map 1121   492   LWFCY

B. erecta (Hudson) Fourr.   Upright brome

(*Bromus erectus* Hudson)

Very rare, known only from a few sites by the Kent estuary and near Kendal, also in the upper Eden valley where one site (68.16) may be the same as the Appleby record given by Wilson. This and three others are by or close to railway lines; those near Kendal are in a limestone quarry and on a roadside verge, and the only 'natural' site is that on the west side of Arnside Knott (44.76s).

It seems very likely that these records, all the west Cumberland ones given by Hodgson, and G.A.K. Hervey's record from Great Salkeld (5.3, 1956) represent introductions of this predominantly south-

eastern grass.

Alt. limit: 200 m, near Crosby Garrett (70.12)

Map 1122   12   WF(C)

*B. inermis (Leysser) Holub

(*Bromus inermis* Leysser)

Found at the Haverigg amusement ground in l977 (14.78, JW, LANC, *Watsonia* (l980) 13:148) and two years later on the Spadeadam range (60.72n, GH, *Watsonia* (1981) 13:342), where it grew luxuriantly by the edge of a beck and close to a road. More recently it was seen on the roadside between Kaber and Barras (80.10, JMA, 1993, LANC, *Watsonia* (1995) 20:300).

This is a species of eastern and central Europe but with a wide distribution further west as a casual. It may be recognised by the spikelets, which widen above, and the shortly-awned (-1.5 mm) or awnless lemmas.

Alt. limit: 265 m, Spadeadam (60.72n)

3   WC

*Anisantha diandra (Roth) Tutin ex Tzvelev

(*Bromus diandrus* Roth)

A handsome, long-awned Mediterranean species discovered in l981 in the grounds of a factory at Cocklakes, Carlisle (44.50, FJR, LANCS, *Watsonia* (1982) 14:199).

1   C

*A. rigida (Roth) N. Hylander

(*Bromus rigidus* Roth)

A Continental species recorded by Hodgson only from Maryport (0.3) and Silloth (l.5) during the 1890s.

(C)

A. sterilis (L.) Nevski   Barren brome-grass

(*Bromus sterilis* L.)

A southern grass and locally frequent around the coast and in the lower half of the Eden valley, occurring on dry, open ground, chiefly by roadsides, hedgebanks and pavements, by farms and on waste ground. It is most

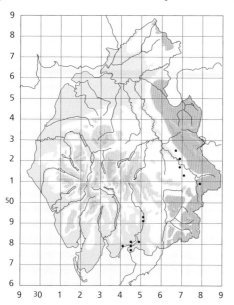

Map 1122. *Bromopsis erecta*

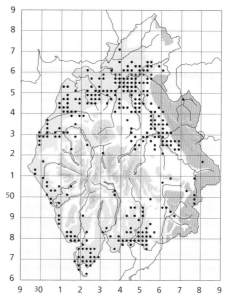

1123. *Anisantha sterilis*

frequent in towns and villages.
Alt. limit: 340 m, Garrigill (74.40)
Map 1123   301   WFCY

*A. tectorum (L.) Nevski
(*Bromus tectorum* L.)
A widespread Continental species said by Hodgson to have
been seen at Silloth (1.5) several times towards the end of the
19th century.

                         R   (C)

*A. madritensis (L.) Nevski
(*Bromus madritensis* L.)
A widespread grass of southern and western Europe found at
Silloth (1.5, CLE) by E.J. Glaister in 1877 and J. Leitch in
1882.

                         R   (C)

*Ceratochloa carinata (Hook. & Arn.) Tutin
(*Bromus willdenowii* Kunth)
Seen on a laneside verge near a farm at Sellafield in
1992 (02.04, MMG; LANC). The only other records for
this rare casual are from Silloth (1.5, J. Leitch, 1882,
CLE) and South American ballast at Maryport (0.3),
cited by Hodgson under *Bromus schraderi*.

                         1   C

Brachypodium pinnatum (L.) P. Beauv.   Tor grass
A rare grass, occurring chiefly in north Westmorland
where it occurs on railwaysides, also on limestone
pavement and in Bampton cemetery (52.18). The only
Furness records are a Survey record from semi-open
sand-dunes at Sandscale, Dalton-in-Furness (18.74, PB,
1994, LANC) and the unlocalised and dubious one in
the *Atlas* from the Newby Bridge square (3.8). The first
record for Westmorland was from the railway at
Appleby (6.2, J.H. Clarke, 1956) and for Cumberland
from the railway at Duncowfold, Cotehill (48.50, FJR,
1979, LANC, *Watsonia* (1980) 13:148).

        This is a south-eastern grass of chalk and lime-
stone which on the east coast extends as far north as
Northumberland. One might well doubt its native status
in Cumbria were it not for its occurrence on the north
Westmorland limestone pavements.
Alt. limit: at least 300 m, above Great Asby (66.10)
Map 1124   18   WFC

B. sylvaticum (Hudson) P. Beauv.   False brome
Common to very common in deciduous woodland at
relatively low altitudes but avoiding the more acid soils
and absent from much of the intensively farmed Solway
Plain. It is often the commonest woodland grass,
frequently growing with *Festuca gigantea*, *Bromopsis
ramosa* and *Carex sylvatica*, but unlike those species it
is more light tolerant, persisting for some time in the
open after tree-felling, as well as occurring in hedgerows.
Alt.limit: 365 m, Helbeck Wood, Brough (78.16)
Map 1125   570   LWFCY

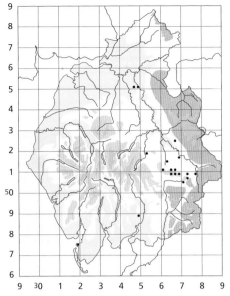

Map 1124. *Brachypodium pinnatum*

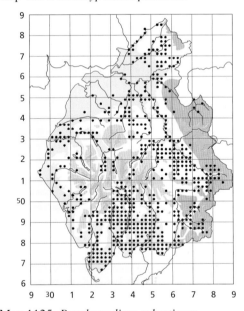

Map 1125. *Brachypodium sylvaticum*

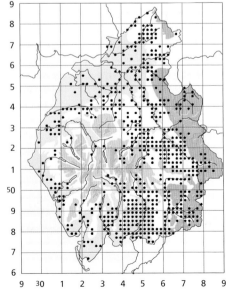

Map 1126. *Elymus caninus*

Elymus caninus (L.) L.

Frequent to common in deciduous woodland and shady hedgerows chiefly in the east. It is especially frequent on moist, base-rich or limestone soils.

Alt. limit: at least 400 m, Cross Gill, Alston Moor (72.38)

Map 1126   580   LWFCY

Elytrigia repens (L.) Desv. ex Nevski   Couch grass
(*Elymus repens* (L.) Gould)

A common lowland grass of hedgerows and waste ground, especially near the coast, and also still a persistent weed of arable land. Along the coast it can grow in much the same habitats as *Elytrichia atherica*, at the upper edge of the salt-marshes and by sea-walls, and may sometimes be mistaken for that species, particularly as coastal plants are often strongly glaucous. Long-awned plants (*Agropyron repens* var. *aristatum* Baumg.) are not infrequent.

Alt. limit: at least 430 m, Nenthead (78.42)

Map 1127   1162   LWFCY

E. repens x **E. atherica**

E. x oliveri (Druce) Kerguelen ex Carreras Mart. and
E. atherica (Link) Kerguelen ex Carreras Mart.
(*Elymus pycnanthus* (Godron) Melderis)

                    Sea couch-grass

Fairly common at the upper edge of salt-marshes and by sea-walls in the south, rather less so on the west coast and becoming progressively rarer northwards, reaching its northern limit in Britain on the Scottish side of the Solway Firth. It is difficult to know how to interpret Hodgson's records of the maritime couch grasses but it seems likely that his 'plentiful' *Triticum acutum* belongs here. The *Atlas* surprisingly shows it as occurring in only four squares.

     Material from 14 sites from Morecambe Bay to the Solway has been examined by T. Cope. All were male sterile and in his opinion they are best regarded as the hybrid. Whether *E. atherica* actually occurs is therefore open to doubt. The only definite pre-Survey records of the hybrid are from Silloth (1.5, G.C. Druce, 1926, *BEC Rpt* (1927) 8:143), Eskmeals (0.9, C.W. Muirhead, 1950, CLE, det. C.E. Hubbard) and Humphrey Head (3.7s, W.L. Smith, 1910, LANC, det. T. Cope).

Map 1128   69   WFC

E. repens x **E. juncea**

E. x laxa (Fries) Kerguelen

A fairly distinctive hybrid recorded only from Beckfoot (08.48, GH, 1978), between Silloth and Skinburness (12.54, GH, 1978) and south of Allonby (06.40, AAD, 1993; all LANC, conf. T. Cope).

3   C

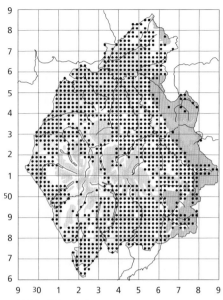

Map 1127. *Elytrigia repens*

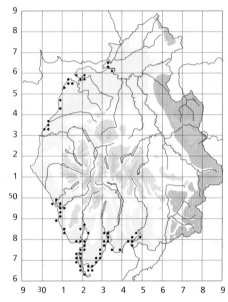

Map 1128. *Elytrigia repens* x *E. atherica/E.atherica*

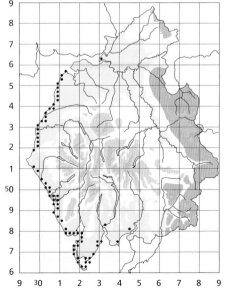

Map 1129. *Elytrigia juncea*

E. atherica x **E. juncea**
E. x obtusiuscula (Lange) N. Hylander
Both Hubbard (1984) and Melderis (1975) refer to the occurrence of this hybrid in Cumberland and it may well be widespread but overlooked. As mentioned above, Hodgson's *Triticum acutum* probably refers to *Elytrigia atherica* or *E.* x *oliveri*.

(C)

E. juncea (L.) Nevski   Sand couch-grass
(*Elymus farctus* (Viv.) Runem. ex Meld.)
Fairly frequent along the coast from Walney Island to Skinburness (12.56) but only rarely in the inner parts of the Solway Firth and Morecambe Bay. It grows on or just above the strandline, nearly always in sand and it plays an important role in stabilising and initiating dune formation.

The only Westmorland record is an unsubstantiated one from the north side of the Kent estuary (44.80s, Anon., 1970s).
Map 1129   70   [W]FC

Leymus arenarius (L.) Hochst.   Lyme grass
Well distributed along the coast except for the inner parts of Morecambe Bay and less frequent than *Ammophila*. It occurs on sandy shores, higher up the beach than the last species and often near the foot of small cliffs; on the sand-dunes it is characteristic of the more seaward, younger dunes. Although Wilson mentions it from two sites by the Kent estuary, there are no recent Westmorland records. The *Atlas* shows it as an introduction in the Warcop square (7.1).
Map 1130   49   (W)FC

Hordelymus europaeus (L.) Jessen   Wood barley
Seen by the Swindale Beck, above Brough (80.16, LH, 1967, LANC, *Watsonia* (1974) 10:213), nearby by the Augill Beck (82.14, JMA, 1989), and in Middlebarrow

Wood, Arnside (46.76s, P. Jepson, 1971) *Watsonia* (1973) 9:278). The only other record is Baker's from Grange-over-Sands (4.8s) The *Scarce Plants* record for the upper Teesdale square (7.3) is an error.

At Swindale the grass grows in *Ulmus - Fraxinus - Corylus* woodland and associated with *Galium odoratum, Mercurialis, Sanicula* and *Paris*, and at Arnside with *Paris* under *Fraxinus, Corylus* and *Tilia cordata*. All three sites are in limestone woodland and are close to the species' northern limit in Britain.
Alt. limit: 440 m, Swindale, Brough (80.16)
                    S 3   W(F)

*Hordeum distichon L.   Two-rowed barley
A casual recorded only from Brathay, Ambleside (36.02, R. Young, 1969, AMDE), where it appeared in stable rubbish, as a crop relic near Brough (78.14, LH, 1970s), near Catlowdy (44.74n, MG, 1993) and on waste ground at Silloth (10.52, R. Young, 1970, AMDE).

These appear to be the only records but as an impermanent relic of cultivation it is probably generally ignored.
                    4   WFC

H. murinum L.   Wall barley
A southern species, here close to its northern limit on the west coast and not surprisingly commonest on the Furness peninsula. It is essentially a weedy, often casual species of waste ground and the sides of roads and pavements. It is a conspicuous feature of Barrow-in-Furness and the Workington - Maryport area.

The first Westmorland records are those shown in the *Atlas* for the Arnside (4.7s, v.c.60/69) and Crosby Ravensworth (6.1) squares.
Alt. limit: 450 m, Kirkstone Pass car-park (40.08)
Map 1131   61   WFC

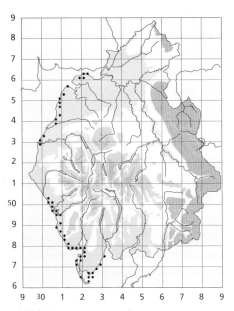

Map 1130. *Leymus arenarius*

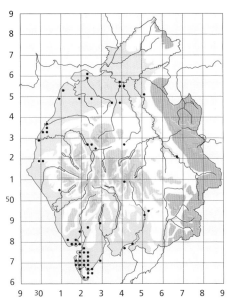

Map 1131. *Hordeum murinum*

*H. jubatum L.

Scattered along the M6 motorway, usually at the very edges of the verges, central reservation and slip roads, and by the A66 on Stainmore close to the Durham border (86.12). These are highly saline sites and the distribution and spread along these roads where it was first noted in l978 between Tebay and Hardendale (56.l8, GH, *Watsonia* (l980) 13:148) is similar to that in north-east England. There are also Survey records from the Ravenstonedale (70.04) and Brampton (54.60) by-passes, a roadside grit-pile near Berrier (38.30) and reclaimed industrial sites at Barrow-in-Furness (18.70) and Workington (98.28, 98.30, the Silloth record (*Watsonia* (l98l) 13:343) being an error for Siddick). At only a few of these sites can the grass be said to be firmly established and it seems to have been declining over the last five to ten years. The first records for Furness are from Sandscale (1.7) and Dalton-in-Furness (2.7, D. Lumb, 1913, *BEC Rpt* (1914) 3:397, 403), and for Cumberland from Silloth (1.5, J. Leitch, 1890, CLE). There is also a later record from White-haven (9.1, M. Cobbe, *BEC Rpt* (1925) 7:605).

With its iridescent panicles this is one of our most beautiful and conspicuous grasses when in flower from mid-July onwards, when it easily qualifies as a "70 m.p.h. plant".
Alt. limit: 370 m, Stainmore (86.12)
Map 1132, Plate 106   22   WFC

*H. secalinum Schreber

A grass of southern and south-east England known only from Hodgson's l9th century records from Workington (9.2) and Maryport (0.3).

(C)

*H. marinum Hudson   Sea barley

The only records for this southern grass are those given by Hodgson: Coulderton (9.0) and Silloth (l.5), and a later one from Grinsdale, Carlisle (3.5, T.S. Johnstone, l901, CLE).

S   (C)

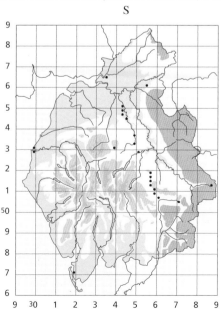

Map 1132. *Hordeum jubatum*

*Triticum aestivum L.   Bread wheat

Recorded by PB in 1992 from a roadside near Carlisle (42.58), the car-park at Sandscale (20.74) and the old steelworks at Barrow-in-Furness (18.70). It no doubt has a wide distribution as a casual but has generally not been recorded.   4   FC

Danthonia decumbens (L.) DC.   Heath grass

Common in the uplands on relatively well-drained soils, often growing with *Calluna* and *Vaccinium*, also on rock ledges, in *Festuca - Agrostis* grassland, and, on the limestones, on slightly leached soil or drift. In the lowlands it occurs, though rather sparingly, on coastal heaths, in rough grassland and on roadside verges.
Alt. limit: 550 m, Moor House, Teesdale (74.32)
Map 1133   872   WFCY

Molinia caerulea (L.) Moench   Purple moor-grass

Common and sometimes dominant in wet moorland, fens and carr and particularly in coastal and valley mosses and birchwoods, scarce on the limestones and in the intensively farmed areas and absent from the higher fells and most of the upper Tees basin.

No attempt was made during the Survey to distinguish the two subspecies. Specimens in LANC from Hale Moss (50.76s, GH, 1991) and Finglandrigg Moss (26.56, GH, 1991) both appear to be the more robust subsp. *arundinacea* (Schrank) K. Richter which may well be the commoner on the lowland marshes, fens and mosses.

Plate 106. *Hordeum jubatum* by the M6, Penrith

Alt. limit: 550 m, Moor House, Teesdale (74.32) but given as 610 m by Wilson.

Map 1134   1074   LWFCY

Phragmites australis (Cav.) Trin. ex Steudel

Common reed

Common in brackish dykes and the less exposed estuaries around the Solway Firth and Morecambe Bay, also in fens and fringing protected shorelines of mesotrophic to eutrophic lowland tarns and lakes in the Lake District, elsewhere rather scattered and absent from the high ground. Both coastal and lakeside habitats are vulnerable to the pressures of boating and development.

Alt. limit: *c*.260 m, Sunbiggin (64.08)

Map 1135   328   WFCY

*Cynodon dactylon (L.) Pers.   Bermuda grass

A southern grass found only once during the Survey growing by the footpath from Carleton to Brougham Castle (52.28, LR, 1978-1980). The only other record is Hodgson's, under *Finnichia umbellata*, from the Derwent Tinplate Works, Workington (9.2) where he discovered it in 1889.

R 1   C

Spartina anglica C.E. Hubb.   Common cord-grass

Frequent in almost all the saltmarshes in Morecambe Bay and covering extensive areas especially off Rampside (2.6s) where, according to Whiteside (1987), it was first noted in 1949. This is the first record for Morecambe Bay and it is suggested that the grass had spread there from the Wyre estuary to the south. In 1960 it was seen in the Leven estuary at Plumpton (3.7s) and by l974 it had reached Arnside (4.7s).

The first Cumberland record was from the Ravenglass estuary (0.9) where it was seen by T.G. Tutin in l975 (*Watsonia* (l978) 12:180). Although it is well established on the Scottish side of the Solway Firth, the only record from the south side is of a few colonists on Newton Marsh (16.54, F. Mawbry, 1996).

With its vigorous rhizomes and the effective dispersal of detached pieces, this is an aggressive colonist of bare mud, sand and especially silt. Its spread is a source of concern to the residents around the Bay as they watch the clumps extending, merging into a sward, spreading seawards across the sand and shorewards over the salt-marsh.

It seems likely that most, if not all the plants are *Spartina anglica*, the fertile amphidiploid, rather than the more southern, sterile, hybrid parent *S.* x *townsendii* Groves & J.Groves.

Map 1136   42   WFC

*Panicum capillare L.

A North American grass listed by Hodgson from the Derwent Tinplate Works, Workington (9.2, J. Hodgson, 1892).

(C)

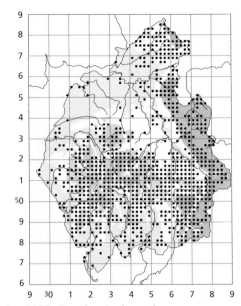

Map 1133. *Danthonia decumbens*

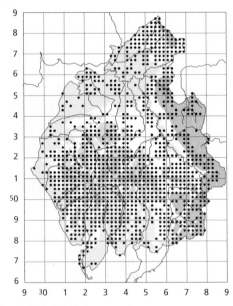

Map 1134. *Molinia caerulea*

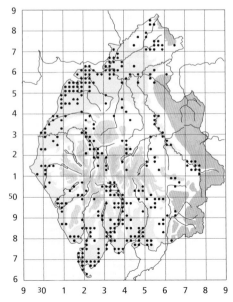

Map 1135. *Phragmites australis*

*P. miliaceum L.   Common millet

Seen in 1992 by PB on a newly-seeded roadside near Carlisle (42.58, LANC) and by a pavement in Kendal (50.92, A. Boucher, 1996). This one-time cereal grass of southern and central Europe was found by Hodgson in 1893 at the same Workington site (9.2) as *Panicum capillare*.

            2   C

*Echinochloa crus-galli (L.) P. Beauv.   Cockspur

The only records for this distinctive and robust grass of southern Europe are Hodgson's own from Workington (9.2, 1889-1892), one from Durran Hill, Carlisle (3.5, R. Dixon, 1942, CLE), and the two *Atlas* records from Walney Island (1.7, 2.6s) contributed by the Barrow-in-Furness N.H.S. during the 1950s.

                                  (FC)

*Setaria pumila (Poiret) Roemer & Schultes

Seen only once, in 1995, growing by the pavement in Bowness-on-Windermere (40.96, A. Boucher; LANC) It was listed by Hodgson, under *Setaria glauca*, from the Derwent Tinplate Works, Workington (9.2) where it persisted from 1889 until 1892, and by Thomson (1902), under *Panicum glaucum*, from near Carlisle where it was collected in 1901 by T.S. Johnstone. It is a native of southern Europe.

            1   W(C)

*S. verticillata (L.) P. Beauv.

A southern European grass known only from Hodgson's 1891 record from the Derwent Tinplate Works, Workington (9.2).

                                    (C)

*S. viridis (L.) P. Beauv.

The only Survey records are from a recently landscaped picnic area at Eden Bridge, Lazonby (54.40, REG, 1984, LANC), where it was growing with other casuals including *Amaranthus retroflexus* and *Erucastrum gallicum*, by the roadside near Ghyll Head, Storrs (38.92, CEW, 1989, LANC) and in a potato field at Smithfield (44.64, MG & JP, 1995).

    This grass was found by Hodgson at Workington (9.2, 1889, CLE) and by J. Leitch at Silloth (1.5, 1893, CLE), where it was seen again by T.S. Johnstone (CLE) in 1901. There is also an 1882 record from Kirkbride (2.5, J. Wood, CLE). The *Atlas* has post-1930 dots for the Workington square and also for North Walney (1.7), where it was seen by the Barrow-in-Furness N.H.S. during the 1950s.

            3   FC

SPARGANIACEAE

Sparganium erectum L.   Branched bur-reed

Common throughout the lowlands in mesotrophic and eutrophic waters forming an often prominent fringe of vegetation to rivers, lakes, tarns and the Kendal canal. It is also frequent in dykes and ditches.

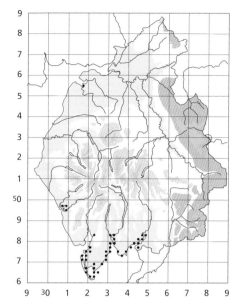

Map 1136. *Spartina anglica*

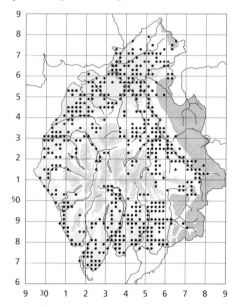

Map 1137. *Sparganium erectum*

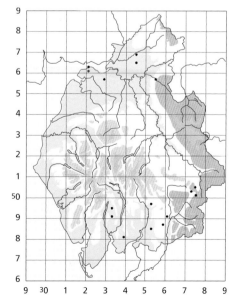

Map 1138. *Sparganium erectum* subsp. *microcarpum*

Of the 28 Survey collections 16 were of the commoner subsp. *microcarpum* (Neumann) Domin and ten of the rarer subsp. *neglectum* (Beeby) Schinz & Thell., with narrowly ellipsoid fruits. This is similar to the situation in Northumberland (Swan 1993). Two, from Field Broughton (38.80) and Rickerby, Carlisle (40.56), with mostly abortive fruits are probably subsp. *oocarpum*. There is an early record of subsp. *neglectum* from near Dalston (3.5, D. Graham, 1949, PLYP).
Alt. limit: *c*.270 m, near Ravenstonedale (74.02) and Paddaburn, Spadeadam (64.76n)

Maps 1137-1139   523   WFCY

S. emersum Rehmann   Unbranched bur-reed
Rather scarce, much less frequent than the last although occurring in similar habitats but apparently preferring, less mesotrophic waters. Plants with only floating and submerged leaves are frequent and probably belong here rather than to *Sparganium*

erectum.
Alt. limit: 280 m, Paddaburn (62.76n) and at least that on Harter Fell, Ravenstonedale (72.00)

Map 1140   83   WFC

S. emersum x **S. angustifolium**
S. x **diversifolia** Graebner
A stand of this extremely rare hybrid was found in Stickle Tarn, Langdale (28.06, GH, 1995, LANC) and identified by C.D.K. Cook. The only other English record is from Northumberland (Swan 1993).
Alt. limit: 470 m, Stickle Tarn (28.06)

      1   W

S. angustifolium Michaux   Floating bur-reed
Frequent in the Lake District in sheltered bays of upland oligotrophic tarns, rarely in the lakes. These are the only localities in England for this north-western species. It usually occurs in fine mud or silt in 1-2 m of

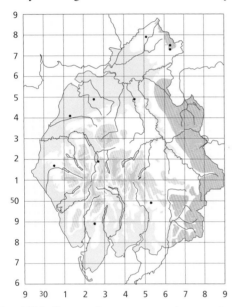

Map 1139. *Sparganium erectum* subsp. *neglectum*

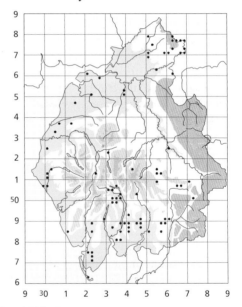

Map 1140. *Sparganium emersum*

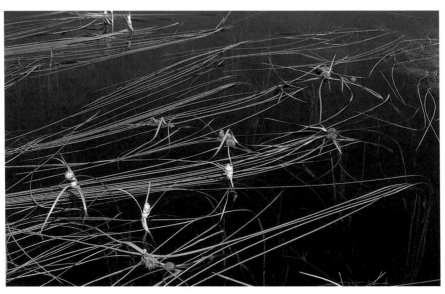

Plate 107. *Sparganium angustifolium*, Sprinkling Tarn

water, the floating leaves forming a raft on the surface. Flowering is rather infrequent.

Alt. range: 60 m, Wast Water (14.02) - 710 m, Sergeant Man (28.08)

Map 1141, Plate 107   52   WFC

S. natans L.   Least bur-reed

(*S. minimum* Wallr.)

Scattered in mesotrophic to oligotrophic lakes and peaty tarns, chiefly in the lowlands and in the south. In Lily Mere, near Sedbergh (60.90), it is co-dominant with *Eleogiton fluitans* over a considerable area.

As is evident from the the early Floras, it has been confused both with the last species and non-flowering *Sparganium emersum*. The same errors may account for a number of dubious records in the *Atlas*, for example that from the Spadeadam square (6.7n). That for square 3.5 is based on a specimen from a pond near Dalston (D. Graham, 1949, CLE, PLYP). Wilson's own record of it growing abundantly at 440 m in a tributary of the Black Beck, Teesdale (7.2) must be an error: searches have failed to reveal any *Sparganium* nor are there any *Atlas* records for that square.

This species, which is commoner in Cumbria than anywhere else south of the Border, flowers only in shallow water. Unlike *S. angustifolium*, its leaves fail to reach the surface when growing in deep water. The only definite record of both occurring together is from Rydal Water (34.06).

Alt. limit: *c*.300 m, Ravenstonedale Moor (68.06)

Map 1142   32   WFC

TYPHACEAE

Typha latifolia L.   Bulrush

Fairly frequent in lowland mesotrophic and eutrophic tarns and lakes and by rivers, forming an important component of the fringing vegetation, frequently associated with *Sparganium erectum* and *Phragmites*. It occurs also in dykes, swamps and gravel pits.

Alt. limit: 445 m, Alston (68.42)

Map 1143   246   WFCY

T. angustifolia L.   Lesser bulrush

Rare, known only from Rydal Water and the head of Windermere (3.0) and a few lowland tarns in the south, also one in the north at Naworth (56.60, 56.62) and one in the west near Cleator Moor (02.14). Unfortunately the substantial colony at this last site has succumbed to land-fill operations. Like *Typha latifolia*, it occurs in reedbeds and fringing vegetation and at some sites, notably Knittleton Tarn, Grizebeck (24.84), it is the dominant plant. The first Furness record was from Urswick Tarn in 1968 (26.74, KAG & FLW).

Foley & Porter (1993) found that at eight of the sites up to a quarter had the female part of the spike divided into two more or less equal parts.

Alt. limit: 155 m, Knittleton Tarn, Grizebeck (24.84).

Map 1144   14   WFC

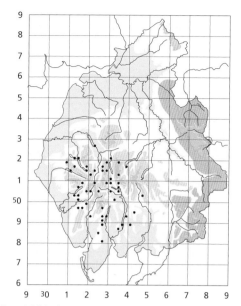

Map 1141. *Sparganium angustifolium*

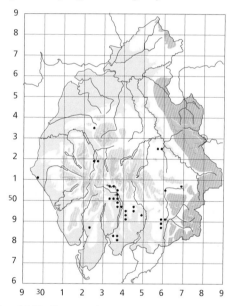

Map 1142. *Sparganium natans*

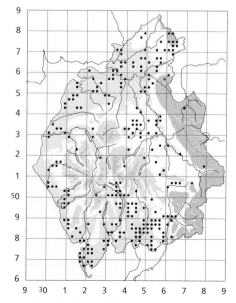

Map 1143. *Typha latifolia*

LILIACEAE

Narthecium ossifragum (L.) Hudson   Bog asphodel
A common plant of lowland raised mires, of upland blanket peat and, especially in the Lake District, of valley mires, sometimes within *Myrica gale*. It also occurs in somewhat open and rocky hillside flushes, characterised also by *Drosera rotundifolia*. If ungrazed, it often flowers in profusion and is particularly luxuriant in ungrazed woodland mires.
Alt. limit: 825 m, Cross Fell (68.34)
Map 1145   782   WFCY

*Hemerocallis fulva (L.) L.   Orange day-lily
A well-established colony far from houses was found in a lane between Pattenfoot and Boltongate in 1990 (22.42, REG).

1   C

*H. lilioasphodelus L.   Yellow day-lily
A garden throw-out known only from roadside verges south of Orton Moss (34.54, REG, 1989, LANC, *Watsonia* (1991) 18:432), south-east of Carlisle (42.50, FJR, 1991), and in the south at Arnside (44.76s, CEW, 1995, LANC).

3   WC

Colchicum autumnale L.   Meadow saffron
Recorded from the sand-dunes at Eskmeals (08.92, MMG, 1986), woodland near Carleton Hall (10.96, MMG, 1989), a roadside bank near Abbeytown (14.48, G. Armstrong, 1989, LANC), by the River Eden in Wetheral Woods (46.52, FJR, 1981), by Fisher Tarn, Kendal (54.92, CEW, 1989) and in a meadow at Newby Bridge (36.86), where it has been known for nearly two centuries.

There are post-1930 *Atlas* records for Broughton Mills (2.9, G. Wilson, 1950s) and the Hawkshead (3.9) and Milnthorpe (5.8s) squares. Other early records include Baker's from Blennerhasset (1.4), north of Kendal (5.9) and Kirkby Lonsdale (6.7s), and Hodgson's from Silloth (1.5) and Tirril (5.2), for which the *Atlas* dot is wrongly placed. Wilson gives a further record from Asby (6.2) and G.A.K. Hervey recorded it from Gilsland (6.6, 1943, v.c.67/70) and Keswick (2.2, 1954).

While some of these sites no doubt represent introductions, others, such as that at Newby Bridge, probably represent native plants here at their northern limit in Britain.
Alt. limit: 195 m, Fisher Tarn (54.92)
6   WFC

Gagea lutea (L.) Ker Gawler Yellow star-of-Bethlehem
Rather rare, usually occurring by shady riversides, particularly by the rivers Kent, Eden and Eamont, growing in areas of silt deposition and often associated with *Chrysosplenium alternifolium* and *Saxifraga granulata*. There are also records from limestone woodland

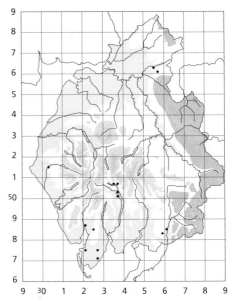
Map 1144. *Typha angustifolia*

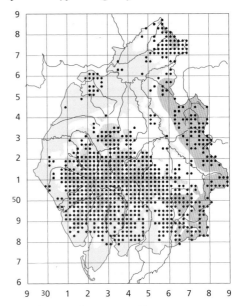
Map 1145. *Narthecium ossifragum*

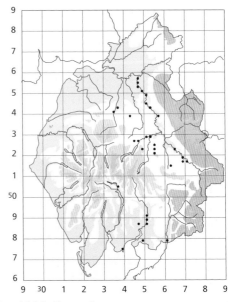
Map 1146. *Gagea lutea*

at Dalemain (46.26) and former woodland at Kirkby Lonsdale (60.78s). It is particularly vulnerable in these small woodlands and was lost at a site near Brigham (0.2) in 1949 and one near Appleby (68.l8) as a result of felling.

Alt. limit: *c*.140 m, Great Ormside (70.16)

Map 1146   32   WFC

*Erythronium dens-canis L.   Dog's-tooth violet
A garden escape recorded by G.A.K. Hervey from the banks of the River Eden at Great Salkeld (5.3, 1948-1958).

(C)

*Fritillaria meleagris L.   Fritillary, Snake's head
A rare garden escape which was not seen during the Survey. It was listed by Hodgson as introduced in estate grounds near Waverton (2.4, E.J. Glaister, 1877, CLE) and at Old Church, Ullswater (4.2). Baker gives an additional record from between Troutbeck and Ambleside (3.0) and Wilson two more from Burton-in-Kendal (5.7s) and near Kendal (5.8s/5.9). The only other localities are a meadow at Brigsteer (4.8s, Anon., 1940s) and by the River Calder near Sella Park (0.0, D. Gillies, 1950, LANC).

(WC)

*Lilium martagon L.   Martagon lily
A conspicuous and not infrequently naturalised species of damp, base-rich or calcareous woodland and shady riversides from south Westmorland north to Warwick Bridge (4.5). It is particularly fine in the woods along the River Caldew above and below Sebergham (3.4) and also in Wetheral Woods by the River Eden (46.52, 46.54). Hodgson gives a record from by the River Duddon, but this probably refers to *Lilium pyrenaicum*.

Alt. limit: *c*.200 m, Grayrigg (56.96)

Map 1147   64   WFCY

*L. pyrenaicum Gouan   Pyrenean lily
An occasional garden escape occurring, like *Lilium martagon*, in damp, riverside woodland as, for example, at Duddon Bridge (18.88) but also as a garden outcast on roadside banks and verges. There are apparently no published pre-Survey records although, as mentioned above, Hodgson's record of *L. martagon* from the Duddon probably belongs here. The first certain Cumberland record is from Cockermouth (10.30, RS, 1974) and for Westmorland and Furness those from Arnside (46.76s, MB, 1976) and the Furness side of Duddon Bridge (18.88, M. Stewart, 1975).

Alt. limit: *c*.170 m, Morland (58.22)

Map 1148   17   WFC

Convallaria majalis L.   Lily-of-the-valley
Present in the limestone woods around Morecambe Bay, in the grikes of the limestone pavements in the Asby - Orton area and on lightly wooded limestone hillsides by the upper Eden valley. By contrast it also grows in a fen at Brathay (36.02), on acid, non-calcareous soils such as on islands in Rydal Water (34.06) and Windermere (38.96), where C.D. Pigott recorded a

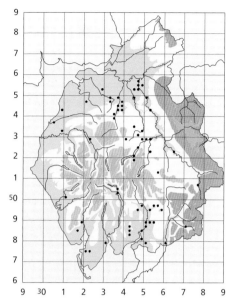

Map 1147. *Lilium martagon*

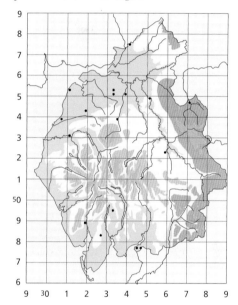

Map 1148. *Lilium pyrenaicum*

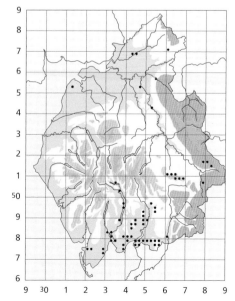

Map 1149. *Convallaria majalis*

pH of 4.0, and by the River Lune at Kirkby Lonsdale (60.80s) and the River Eden at Great Corby (46.52). Records from the north of the county probably represent naturalised populations.
Alt. limit: 335 m, Helbeck Wood, Brough (78.16)
Map 1149   61   LWFC

Polygonatum multiflorum (L.) All.   Solomon's seal
and **P. multiflorum** x **P. odoratum**
P. x **hybridum** Brügger
These two plants were not reliably distinguished during the Survey. It seems likely that, as Wilson suggests, *Polygonatum multiflorum* is native only on the wooded limestones of south Westmorland. Elsewhere the records probably represent naturalised populations of garden origin of both the species and the hybrid. The latter, the commoner garden plant, may actually be less frequently naturalised than *P. multiflorum*. Most sites are shaded roadside verges, but at Eskmeals (06.92, 08.92) the plants are on the open dunes.
Alt. limit: 245 m, Helbeck Wood, Brough (78.16)
Map 1150   113   LWFC

P. odoratum (Miller) Druce   Angular Solomon's-seal
Occasional in the grikes of the limestone pavements around the Kent estuary and with two isolated occurrences on the pavements in the Asby - Orton area, the most northerly native records of this scarce British plant. It is mainly restricted to the grikes although occasionally occurring in the surrounding grassland. There is an isolated locality on a roadside verge near Armathwaite (52.44, M. Rawes, 1990; LANC) where it has been known for at least 20 years. There are pre-1930 *Atlas* records, presumably also of introduced plants, for the Ambleside (3.0), Barbon (6.8s) and Melkinthorpe (5.2) squares.
Alt. limit: 355 m, above Great Asby (64.08)
Map 1151   S 15   WFC

Paris quadrifolia L.   Herb Paris
Not uncommon on moderately base-rich or calcareous soils in woodland in south Westmorland, around the upper Eden valley and along its western edge to Wigton (2.4), elsewhere rather rare. In the Asby - Orton area it occurs sparingly on the grikes of the high, treeless limestone pavements at about 360 m, its highest altitude in Britain. It is less sensitive to waterlogging than *Mercurialis* and occurs, for example, in willow carr near Field Broughton (38.80).
Alt. limit: 360 m, Great Asby Scar (64.08) and above Garrigill (74.40)
Map 1152   99   LWFCY

*****Ornithogalum pyrenaicum** L.   Spiked star-of-Bethlehem
The only records, both from the mid-1960s, are from the vicinity of the lighthouse on South Walney (2.6s, M. Braithwaite) and Sandscale (1.7, H. Kellet).
                         S   (F)

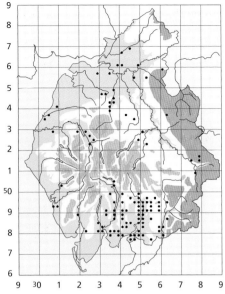

Map 1150. *Polygonatum multiflorum and P.* x *hybridum*

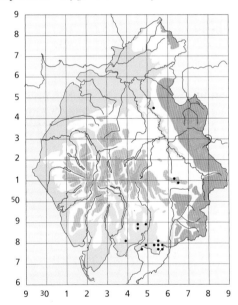

Map 1151. *Polygonatum odoratum*

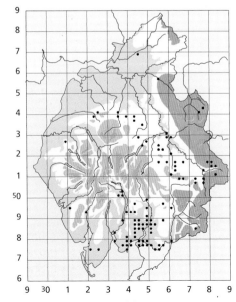

Map 1152. *Paris quadrifolia*

*O. angustifolium Boreau   Star-of-Bethlehem
(*O. umbellatum* auct.)
A fairly widespread and increasing garden escape occurring usually in the short grassland of roadside verges and by paths. There are also records from lawns and cemeteries. Hodgson gives only four sites and Wilson three, while Baker refers only to "orchards in Furness".
Alt. limit: 185 m, Great Asby (68.12)
Map 1153   82   WFC

*O. nutans L.
A native of the south-eastern Balkans listed, somewhat sceptically, by Hodgson from Isel Woods, near Cockermouth (1.3). The only subsequent records are from Great Salkeld (5.3, H. Britten, 1908) and near Gosforth (0.0, M. Cross, 1940).
                                     (C)

*Scilla siberica Haw.   Siberian squill
Found well established on a wooded bank by the River Caldew north of Sebergham (34.42, MP, 1992, det. B. Mathew, LANC) but not far from a house.
                 1   C

*S. lilio-hyacinthus L.   Pyrenean squill
A single plant has been seen by RWMC on a roadside verge near Johnby, Greystoke since 1988 (42.32, LANC, det. B. Mathew).
                 1   C

Hyacinthoides non-scripta (L.) Chouard ex Rothm.
                                  Bluebell
Very common in the lowlands in hedges and particularly in deciduous woodland where it often dominates the ground flora. It prefers damp but not waterlogged soils with a moderately high nutrient status. Nevertheless, it is strangely absent from many apparently suitable woods. It occurs in the open on the cliffs

between St Bees Head and Harrington (9.1), also occasionaly in enclosed allotments in the Lake District and elsewhere, as around Tebay and Ravenstonedale, but the striking blue carpets on some Lake District fellsides usually indicate areas of former woodland.
Alt. limit: 550 m, Rydal Head (36.10)
Map 1154   1168   LWFCY

*H. non-scripta x **H. hispanica** (Miller) Rothm.
An occasional garden escape on roadsides, waste ground and tips, usually near houses. All the records of cultivated bluebells are included here since all ten samples from Furness and south Westmorland submitted during the Survey to K.W. Page proved to be the hybrid and not, as originally supposed, *Hyacinthoides hispanica*. There are no reports of the latter or the hybrid prior to the Survey and it is impossible to decide which are the first vice-county records.
      The most important diagnostic characters of the hybrid are the moderately spreading but hardly reflexed perianth lobes, the usually blue anthers and the insertion of the filaments at two levels just below the middle of the perianth segments.
Alt. limit: *c.*180 m, Great Asby (68.12)
Map 1155   62   LWFC

*Chionodoxa forbesii Baker   Glory-of-the snow
A popular garden plant seen by RWMC on roadside verges east of Greystoke (46.32, 1984), near Salkeld Dykes (54.34, 1992) and north-east of Penrith (52.32, 1984, all LANC, det.B. Mathew).
Alt. limit: 230 m, north-east of Penrith (52.32)
                                       C

*Muscari armeniacum Leichtlin ex Baker
                                 Grape hyacinth
Seen during the Survey on several roadside verges, in a plantation near Burneside (48.96, CEW, 1988) and at Catlands quarry, Sandale (24.40, MP, 1991, *Watsonia*

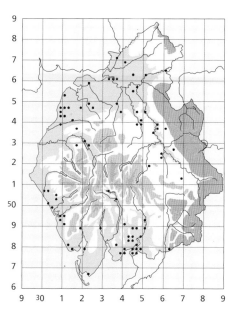

Map 1153. *Ornithogalum angustifolium*

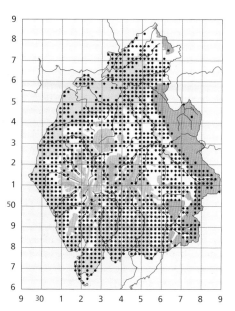

Map 1154. *Hyacinthoides non-scripta*

(1992) 19:151, under *Muscari neglectum*). It is also not infrequent in the Arnside area (4.7s) but always near houses.

This popular garden plant was first recorded *c.*1840 by a Miss Burton from "fields about Windermere" (3.9/4.9, KDL) and later listed by Hodgson as almost certainly introduced at Low Mill, Egremont (0.0). The only pre-Survey records this century are G. Wilson's from South Walney (2.6s) and Sandscale (1.7) in 1954, although these probably both refer to *M. comosum*.

Alt. limit: 425 m, south of Alston (72.42)

Map 1156   14   W[(F)]C

*__M. comosum__ (L.) Miller   Tassel hyacinth
A very rare garden escape persisting in open sandy ground near the sea, as at Silloth (10.52), Eskmeals (06.92, 08.92), Sandscale (18.74) and South Walney (20.62s, 22.62s). It also appeared briefly, under bracken, at Crosthwaite (42.92, H. Caldwell, 1978). It has been known at Silloth since at least 1932 when it was found by M. Forster. The first Furness record is that from Foulney Island (2.6s, R.J. Elliott, 1954).

Map 1157   7   WFC

AMARYLLIDACEAE

*__Allium schoenoprasum__ L.   Chives
Very rare. A thriving colony was found on the roadside south of Cumwhinton (44.50, C. Smith, 1974). This probably originated as a garden throw-out and the same is also true of a former roadside colony near Arnside Tower (44.76s, L. Livermore, *c.*1974). There are later records from Swindale (50.12, CEW, 1988, LANC), where it was growing on wet rocks above the road , a laneside east of Heversham (48.84s, MB, 1990), a slag-bank at Askam-in-Furness (20.76, PB, 1993, LANC) and the car-park by the Wolsty dunes, Beckfoot (08.50, REG, 1992).

It is listed by Hodgson from the Roman Wall near Gilsland (6.6), although this is probably on the whin sill in Northumberland, by Baker from "Chivey Sike", Cartmel Fell (4.8s) and by Wilson from the limestone at Rus Mickle, Whitbarrow (4.8s, A. Wilson, 1939, YRK), probably the same site and where it was collected by T. Gough about 1840 (KDL). There are also collections by E.J. Glaister from Parkgate and Woodrow (2.4, 1876; 1875, CLE).

Alt. limit: 290 m, Swindale (50.12)

     S 6   WFC

*__A. moly__ L.
A native of south-west France and central Spain recorded by Wilson as having been found in a field at Kirkby Lonsdale (6.7s, J.B. Foggitt, 1934, YRK) where it was presumably an escape from cultivation.

           (W)

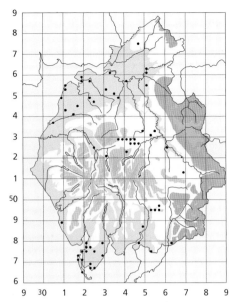

Map 1155. *Hyacinthoides non-scripta* x *H. hispanica*

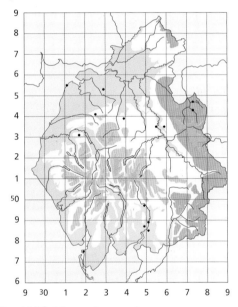

Map 1156. *Muscari armeniacum*

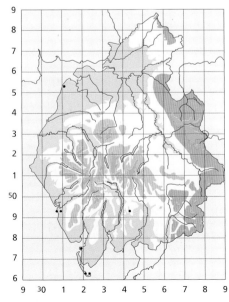

Map 1157. *Muscari comosum*

[A. triquetrum L.

Hodgson records this from Lamplugh (0.2), Salta (0.4) and Clifton (5.2), but he regarded them all as errors. There is a post-1930 *Atlas* record for the Cartmel square (3.7s), but this is perhaps best regarded as doubtful.

(FC)]

*A. paradoxum (M. Bieb.) Don

Found during the Survey on a roadside near Blenner-hasset (16.40, CCH, 1986, LANC, *Watsonia* (1988) 17:195), on laneside banks with *Allium ursinum* along the west side of the River Rothay above Ambleside (36.04, A. Franks, 1989, LANC, *Watsonia* (1991) 18:433), at Chapels, Kirkby-in-Furness (24.82, M. Miller, 1992) and by a track at Blaithwaite, Wigton (2.4, MP, 1994). This last site is probably that where it was seen by J. Parkin in 1949 (CLE, PLYP), the only previous record.

This species is probably increasing and might prove as invasive as elsewhere in the country.

6   WFC

A. ursinum L.   Wild garlic; Ramsons

Common in the lowlands except on the intensively farmed and largely woodland-free Solway Plain. It occurs chiefly in damp and often base-enriched wood-lands, below ash and elm on river alluvium and often in valley bottoms under alder, and also in hillside oak-ash woods where it frequently and very noticeably dom-inates the spring flora over extensive areas. It some-times occurs in the open, in the grikes of limestone pavement and in areas of former woodland where it can be surprisingly persistent, despite its palatability to cattle and sheep.

Alt. limit: *c*.450 m, Tynehead (76.36)

Map 1158   827   LWFCY

A. oleraceum L.   Field garlic

Generally rather rare but locally frequent in the Eden valley. There are also records from the mid-1930s from Abbey Bridge and Wedholme (1.5, 2.5, J. Parkin) and it was last seen at the edge of Wedholme Flow in 1946 (J.S. Muirhead, CLE, PLYP, *BEC Rpt* (1948) 13:311).

Most of the records are from roadside verges but it is also known from a streamside near Curthwaite (32.48) and sandy alluvium by the River Eden near Holmwrangle (50.48), where it was growing with *Allium scorodoprasum*. Like that species it is most readily spotted in the spring before the leaves are overtopped by grasses or cut. Most if not all the records are of var. *complanatum* Fries with many flowers and bulbils.

There has been some confusion in the past between this species and *A. scorodoprasum*, from which it is readily distinguished when vegetative by its narrow, thin leaves which develop rapidly and initially overtop the surrounding grasses; it also flowers later. The records for House Holm (Norfolk) Island, Ullswater (38.18) cited by Wilson and Hodgson must, as the latter suggests, refer to *A. scorodoprasum* which does occur there.

Alt. limit: 230 m, Waitby (74.06)

Map 1159   40   WC

*A. carinatum L.   Keeled garlic

The only record for this species is from near Inglewood Bank, Penrith (52.34, RWMC, 1984, LANC, *Watsonia* (1985) 15:403), where there is a conspicuous and flour-ishing colony on both sides of the road.

Alt.: 185 m, Inglewood Bank, Penrith (52.34)

1   C

A. scorodoprasum L.   Sand leek

Absent over large areas but not uncommon between the rivers Crake and Kent in the south, along the west coast and down the Eden valley to Rockcliffe Marsh (3.6). Handley (1898) gives an outlying site by the River Rawthey near Sedbergh (6.9) and there is an *Atlas*

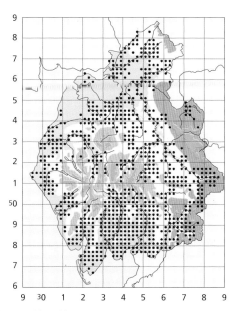

Map 1158. *Allium ursinum*

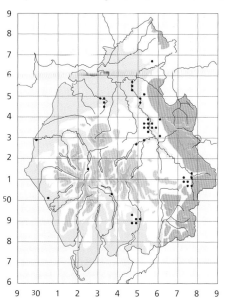

Map 1159. *Allium oleraceum*

record for the Dent square (7.8). Although found chiefly along roadside verges and on railway banks, it also occurs at the upper limit of salt-marshes and on the strand-line, on river banks, occasionally in copses and, in Ullswater, on a rocky island (38.18).

This species is largely confined in Britain to a band extending from the River Humber to the Solway Firth.

Alt. limit: 300 m, Helbeck Wood, Brough (78.16)

Map 1160   148   WFC(Y)

A. vineale L.   Wild onion, Crow garlic
Scattered down the Eden valley and by the Solway Firth, also in the south around Morecambe Bay, elsewhere very rare and now absent from most of Hodgson's localities in west Cumberland. Most Survey records are from well-drained roadside verges, grassland and waste ground.

Practically all the plants belong to var. *compactum* (Thuill.) Boreau with bulbils only but the predominantly floriferous var. *vineale* has been reported from Grune Point (12.56, REG).

Alt. limit: *c.*280 m, North Stainmore (82.14)

Map 1161   75   WFC

***Nectaroscordum siculum** (Ucria) Lindley
An extremely rare garden escape found established on a roadside verge near Dalston (36.48, EEM, 1994, LANC).

                    1   C

***Leucojum aestivum** L.   Summer snowflake
Seen during the Survey by the roadside near Newby Bridge (36.86, L. Livermore, 1970s), the first definite Furness record, and near Witherslack (42.86s, E. Chicken, 1979, herb. E.C.), also by the River Mint west of Grayrigg (54.96, CEW, 1986, LANC) and in a marshy field by Derwent Water (24.22, A. Cannell, 1992; LANC). There are also two records from Arnside

(44.78s, 46.78s, MB) but these are close to gardens.

Wilson's only record for this garden escape is from an island in the River Kent near Sedgwick (5.8s) where it was seen in 1796. There appear to be no more records until 1910 when it was collected at "Studley", in the Lyth valley (4.8s, Anon., LANC) and in the 1940s when it was seen by N.M. Williamson by the River Rothay at Ambleside (36.02) and where it persisted into the 1960s. The only Cumberland records are from Nunwick Wood, Great Salkeld (5.3, G.A.K. Hervey, 1947-1958).

Alt. limit: 120 m, Grayrigg (54.96)

          R 6   WFC

***L. vernum** L.   Spring snowflake
A garden escape known only from a site south of Keswick (26.18, A. Franks, 1988), where there are two colonies in mixed oak woodland, and the roadside verge at the entrance to Rose Castle (36.46, MP, 1994). The first site is probably the basis of the pre-1930 *Atlas* record for the Keswick square (2.2).

          R 2   C

***Galanthus nivalis** L.   Snowdrop
A frequent lowland garden escape occurring on damp roadside verges, in hedgerows and copses and by riversides but usually not far from houses. It is sufficiently common, at least in south Westmorland, for it to be considered as possibly native, but it is evidently much commoner here than in Wilson's day and it has increased similarly in Cumberland since Hodgson's Flora.

All snowdrop records are included here but they probably include some of the other closely related garden species, such as *Galanthus elwesii* Hook. f., and possibly also hybrids. Practically all naturalised Cumbrian populations are single-flowered.

Alt. limit: 370 m, Nenthall (74.44)

Map 1162   534   LWFCY

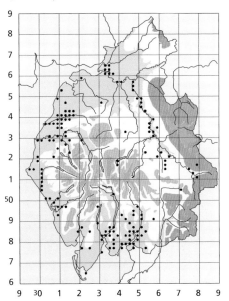

Map 1160. *Allium scorodoprasum*

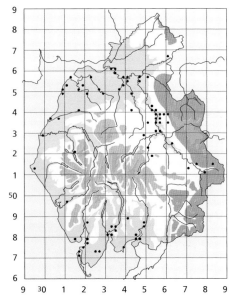

Map 1161. *Allium vineale*

Narcissus

Semi-naturalised garden daffodils, probably mostly cultivars of *Narcissus poeticus* and *N. pseudonarcissus*, but also of their hybrid, *N.* x *incomparabilis* Miller, and of *N.* x *medioluteus*, are being widely planted for their amenity value. They are often far from houses, for example along roadside verges, on roundabouts and in the grounds of large estates. They have also long been planted on roadsides near houses and they appear with other garden throw-outs on waste-tips. Because most colonies have been so obviously planted no systematic attempt was made during the Survey to record any daffodils other than apparently wild *N. pseudonarcissus*.

*N. tazetta L. x N. poeticus L.
N. x **medioluteus** Miller
(*N. biflorus* Curtis)
Listed by Baker from fields above High Lodore (2.1) and below Torver (2.9), and by Petty (1896 p.324, 1898 p.44) from Hawkfield, Urswick and Old Hall, Ulverston (2.7).

(FC)

*N. poeticus L.   Pheasant's-eye daffodil
Reported by Petty (1898 p.44) as thoroughly naturalised on a streambank near Ashslack Hall (24.84) and also in a meadow by the River Duddon (1.8/1.9).

(FF/C)

N. pseudonarcissus L.   Wild daffodil
Common in the southern Lake District and south to Morecambe Bay, also in lower Wasdale, the lower Calder valley and around Bassenthwaite Lake, elsewhere rather scarce, even "Beside the lake, beneath the trees" of Ullswater. It occurs in valley and lakeside oak woods and in oak-ash woodland by Morecambe Bay. It flowers rather poorly in mature woodland but responds dramatically, although briefly, to coppicing and thinning. Like *Hyacinthoides non-scripta*, it is often very conspicuous in open grassland sites of former woodland.

Although undoubtedly native in the south of the county, the wild daffodil has been widely planted, for example in churchyards, and Hodgson refers to frequent plants in "orchards and near farmhouses". The *Atlas* treats all the Scottish records as introductions.
Alt. limit: at least 200 m, lower Bannisdale (52.00)
Map 1163   249   LWFC

*Asparagus officinalis L.
subsp.**officinalis**   Garden asparagus
Seen on coastal limestone rocks at Grange-over-Sands (40.76s, A. Gibson, 1990), presumably the site referred to by Baker, and also on the opposite side of the estuary on the beach at Arnside (44.78s, C. Nicholas, 1977). These are the northernmost records on the west coast.

2   FW

When we were in the woods beyond Gowbarrow Park we saw a few daffodils close to the water-side. We fancied that the lake had floated the seeds ashore, and that the little colony had so sprung up. But as we went along there were more and yet more; and at last, under the boughs of the trees, we saw that there was a long belt of them along the shore, about the breadth of a country turnpike road. I never saw daffodils so beautiful. They grew among the mossy stones about and about them; some rested their heads upon these stones as on a pillow for weariness; and the rest tossed and reeled and danced, and seemed as if they verily laughed with the wind, that blew upon them over the lake; they looked so gay, ever glancing, ever changing.

The Grasmere Journal, April 15th, 1802, Dorothy Wordsworth

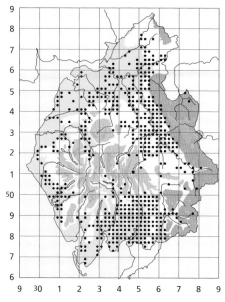

Map 1162. *Galanthus nivalis*

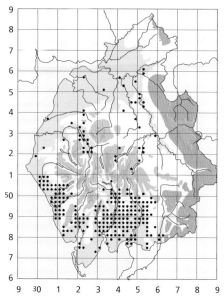

Map 1163. *Narcissus pseudonarcissus*

***Ruscus aculeatus** L.   Butcher's broom
Naturalised in a few places, chiefly in the south around the Kent estuary, usually occurring in or near estates and with a preference for hedgebanks. Nicolson (Whittaker 1981) noted it near Millom Castle (1.8) in 1690. It has probably been long-established in Furness and Westmorland although the first records appear to date from the Survey; these are from Grange-over-Sands (42.78s, FLW) and Arnside (44.78s, MB) respectively.
Map 1164   12   WFC

***Sisyrinchium montanum** E.Greene   Blue-eyed grass
(*S. bermudiana* auct.)
Seen during the Survey on the dunes at Silloth (10.52, Anon.), Ramsden Dock, Barrow-in-Furness (20.66s, B. Fisher, 1991), the island in Fisher Tarn, Kendal (54.92, CEW, 1987) and at Meathop quarry (42.78s, O.M. Stewart, 1987).
      This North American species was listed by Hodgson as naturalised in estate grounds at Gilgarron, near Distington (0.2, not 9.2 as shown in the *Atlas*). In 1946 it was seen by a path south of High Wray (36.98, J. Lund, FBA). Ten years later it was noted as having been deliberately sown on the limestone hills west of Kendal (4.9/5.9, E. Hyde, *Proc. BSBI* (1957) 2:379) and the following year it was recorded from the railway embankment at Griseburn, west of Soulby (7.1) by G.A.K. Hervey.
Alt.limit: 195 m, Fisher Tarn (54.92)
                    4   WFC

***S. californicum** (Ker Gawler) Dryander
                    Yellow-eyed grass
A North American species found in 1983 at two grassland sites on North Walney (16.70, 16.72, P. Kirkland).
                    2   F

Iris pseudacorus L.   Yellow iris
A common lowland species, particularly towards the coast, and a conspicuous feature of tarn and lakeside marshes, of canal and riversides, marshy fields and ditches, also of wet woods and fen carr. It is occasionally planted.
Alt. limit: 480 m, Nenthead (78.44). A.J. Richards (Rawes 1981) recorded it from at least 530 m by Troutbeck, upper Teesdale (7.3) but attempts to refind it have failed..
Map 1165   631   WFCY

***I. versicolor** L.   Purple iris
A North American iris known with certainty only from Aira Point, Ullswater (40.18, J.R. Ellis, 1968, AMDE; LANC), and the edge of Derwent Water near Lingholm (24.22, ES, 1985, LANC, det. B. Mathew (*Watsonia* (1992) 19:152). Ellis (1975) also mentions a site by Esthwaite Water (3.9) but no specimens have been located and it has not been refound.
      A.J. Richards (1986 *in litt.*) found a population with deep amethyst flowers by the River Rothay at Brathay (36.02) which may be this species. *Iris versicolor* is a most beautiful plant, the flowers being much deeper in colour, though smaller, than the following hybrid.
Alt. limit: 150 m, Aira Point, Ullswater (40.18)
Plate 108   R 2   (F)C

***I. versicolor** x **I. virginiana** L.
***I. x robusta** E.S. Anderson
Known since 1965, but probably long-established, from wet meadows, lake and riverside sites at Brathay and across the River Rothay on the Ambleside bank (36.02, K.M. Hollick; LANC). It has been reported by A.J. Richards (1986 *in litt.*) from Pull Wyke, in the same tetrad, and further south on the north-west side of Blelham Tarn (36.00).
      This sterile hybrid is more vigorous than *Iris versicolor* and has larger, paler flowers.
Plate 109   2   WFC

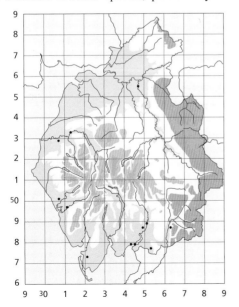

Map 1164. *Ruscus aculeatus*

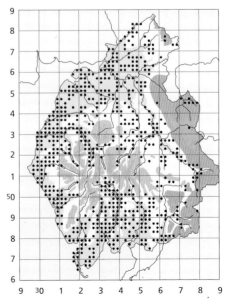

Map 1165. *Iris pseudacorus*

*I. foetidissima L.   Stinking iris
Found on the verge of the Heversham by-pass (48.82s, CEW, 1990, LANC, *Watsonia* (1991) 18:433) and in woodland at Blawith, Grange-over-Sands, (40.78s, G. Huse, early 1980s). Hodgson recorded it from St Bees (9.1), Egremont (0.1) and Aspatria (1.4), and it is mapped in the *Atlas* for the Pooley Bridge square (4.2).

This species is native only in southern Britain and the above records are close to its northern limit.

2   WF(C)

*Crocus vernus (L.) Hill   Spring crocus
(*C. purpureus* Weston)
Very rare. It was noted by G. Wilson in the 1950s as naturalised at Dendron (24.70), Pennington (26.76) and Blawith (28.88). These sites were checked in 1986 by W. Kydd who found the crocuses established on roadside verges and a nearby field, sometimes in abundance and in every case at no great distance from populations in nearby churchyards. At the same time it was found in a valley meadow west of Dalton-in-Furness (22.72, 22.74, PB, 1986, 1987, LANC). In Cumberland it occurs by the River Eden at the Swifts, Carlisle (40.56, 1977-1979), growing with *Petasites hybridus*, and below Armathwaite (48.48, 1993, both REG), also by the River Caldew north of Sebergham (36.42, MP, 1992) and on the roadside west of Renwick (58.42, LR, 1996). It is quite possible that not all these records of purple crocuses belong to this species.

There are earlier records given by Baker from by the River Mint, Kendal (5.9), from Ulverston and Swarthmoor Hall (2.8), by Petty (1898 p.43) from several other Furness sites and by Hodgson from Calvo, near Silloth (1.5). During the same period W.H. Youdale collected it from Daltonleigh, Cockermouth (1.3, LIV) in 1894.

Alt. limit: 245 m, Renwick (58.42)

Map 1166   R 13   (W)FC

*C. nudiflorus Smith   Autumn crocus
The only Survey records are of roadside plants near Rose Castle (36.46, EEM, 1992, *Watsonia* (1993) 19:295) and a colony which persisted in a neglected orchard near Arnside (44.78s, MB) until 1975 when the site was 'improved'.

Earlier records of this garden escape are Wilson's from Arnside (4.7s), probably dating from the 19th century, and the post-1930 *Atlas* dot for the Sebergham square (3.4). The reference to Furness in the *Check-list* is an error.

1   WC

*C. flavus Watson   Yellow crocus
A garden escape found in a hedgerow near Deanscales (08.26, NB, 1986, LANC, *Watsonia* (1988) 17:195), established on roadside verges near Wigton and Great Orton (26.48, 32.52, REG, 1992), and at Mawbray (08.46, Anon., 1990s).

4   C

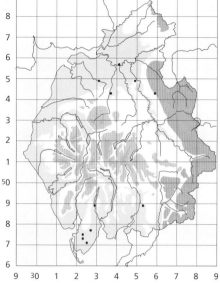

Map 1166. *Crocus vernus*

Plate 108. *Iris versicolor*

Plate 109. *Iris vericolor* x *I. virginiana*

***Crocosmia paniculata** (Klatt) Goldblatt
Found in 1992 well established in Muncaster Park (08.94, MMG, LANC, *Watsonia* (1993) 19:295), and by A. Cannell by the old railway at Harrington (24.98, 1994) and at Raisbeck (64.06, 1993).

3   FC

*__C. aurea__ (Hook.) Planchon x **C. pottsii**   Montbretia (Macnab ex Baker) N.E. Br.
C. x crocosmiflora (Lemoine ex Burb. & Dean) N.E. Br.
Common towards the west coast and not infrequent in the south and on the Solway Plain but otherwise scarce or absent. Most of the sites are on waste ground and roadside verges and ditches, common dumping ground for garden rubbish, but in the west there is an increasing tendency for it to become established in damp woodland margins, on river banks and around the edges of mossland.

Surprisingly, the only pre-Survey records for this hybrid are the six squares shown in the *Atlas*.
Alt. limit: 270 m, Greystoke Moor (40.28)
Map 1167   194   WFC

DIOSCOREACEAE

Tamus communis L.   Black bryony
Common in a remarkably tightly circumscribed lowland area in the south but even there absent along the coast from Millom to Bootle and at the southern end of the Furness peninsula. Handley (1898) gives an outlying record from near Sedbergh (6.9) but it has not been refound there and Wilson's record from Melkinthorpe (5.2) is probably an error for *Bryonia cretica*.

It is a characteristic climber of old hedges and woodland margins being especially evident when the rapidly elongating 'croziers' of the young shoots project above the hedge-tops in mid-May. It is particularly

abundant on the limestone but is by no means confined to it.

Black bryony is here at its northern, climatically determined limit on the west coast. Although these peripheral populations produce plenty of fertile seed, the experimental sowing of considerable numbers of seeds below hedges failed to produce any seedlings (Pigott 1992b).
Alt. limit: 130 m, Garnett Bridge, Longsleddale (52.98)
Map 1168   210   LWFC(Y)

ORCHIDACEAE

Cypripedium calceolus L.   Lady's slipper
Extinct. Baker gives four records for this much sought-for orchid: Legburthwaite (3.1), Blennerhasset (1.4), Whitbarrow (4.8s) and the "north-west of High Furness". Of these, only that from Whitbarrow seems likely to be correct. There is an 1860 specimen from Scout Scar (4.9, ex herb. Linton, BM) and Wilson added a possible 20th century record from the Arnside area (4.7s, v.c.60/69). There is also an *Atlas* dot for the Dentdale (7.8) square based on a 19th century vegetative specimen and best disregarded..

There is no doubt that, as a result of over-collecting, this orchid is now extinct in the county and it has proved impossible to discover, even approximately, when it was last seen.

With the exception of a single site in Derbyshire, this species was restricted in Britain to a belt across northern England, from Cumbria to north-east Yorkshire. It now survives only at a single locality in Yorkshire and at a site in Lancashire where it might just conceivably be native.

R   (WFC[Y])

[Cephalanthera damasonium (Miller) Druce
                                       White helleborine
The records for Gosforth (0.0) and Askham (5.2) given by Baker and Wilson must, as they suggest, refer to *Cephalanthera longifolia*.

WC]

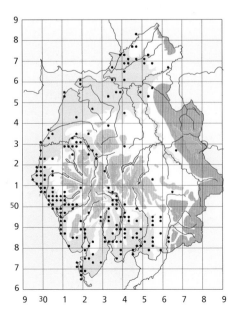

Map 1167. *Crocosmia aurea* x *C. pottsii*

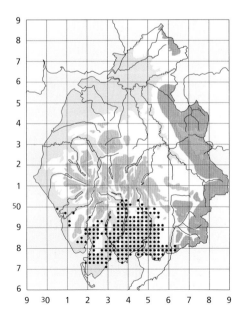

Map 1168. *Tamus communis*

Cephalanthera longifolia (L.) Fritsch
                    Narrow-leaved helleborine
Extremely rare, surviving at only three sites, two in
north Cumberland (5.5) and the other in north Westmorland (8.1). All are in woodland. The populations
are small and at one site the plants are in danger of
being shaded out by brambles. It is possible that the
species still survives in the south, for example in the
Whitbarrow limestone woodlands.

There are several sites referred to in the old
Floras. Baker lists it from Cartmel (3.7s), Yewbarrow
and Whitbarrow (4.8s), Barrowfield Wood (4.9) and
the Lowther Woods (5.2). To these Hodgson added
records from estate grounds at Workington (9.2), where
it may have been introduced, Abbey Holm (1.5) and
Talkin (5.5), and to these should be added Baker's
Gosforth (0.0) record for *Cephalanthera damasonium*
and one from Eden Brows, Armathwaite (4.4/5.4, M.S.
Slater, 1879, NMW). Wilson gives further records from
Melkinthorpe Wood (5.2) and Swindale Beck (8.1, E.S
& C.E. Salmon, 1892, BM), where it was seen,
presumably by him, in 1914. These last two sites are
Wilson's only records from the present century.
However, there is in his herbarium a specimen
collected on the west side of Arnside Knott (4.7s, YRK)
by H.L. Overy in 1938, too late for inclusion in his
Flora. She had been taken to the site having noticed a
vase of the orchids in a house at Arnside. Sadly,
searches of the area by Miss Overy and others in
subsequent years were unsuccessful.

The Cumbrian sites are the only extant ones in
northern England.

S 3   W(F)C

Epipactis palustris (L.) Crantz   Marsh helleborine
Known only from dune-slacks on both sides of the
Duddon estuary and on limestone between Orton and
Kirkby Stephen. The dune-slack populations are large
and vigorous and associated species include *Anagallis*

tenella, Salix repens and *Hydrocotyle vulgaris*. On the
limestone the plants grow in semi-open damp grassland.

In addition to the above sites, Hodgson records it
from Isel Woods near Cockermouth (1.3) and Newton
Reigny moss (4.3), where it persisted at least until
*c.*1950, Baker gives a Coniston locality (2.9) and
Wilson mentions several sites west of Kendal including
Cunswick Tarn (4.9) where it was last seen in 1952.
There are also records from early this century from
near Keswick (2.2, Miss Reynolds) and Melmerby (6.3,
W.W. Mason) and a later one from Blaithwaite station,
Wigton (2.4, J. Parkin, 1935).
Alt. limit: 250 m, Orton (62.08)
Map 1169   10   WFC

E. atrorubens (Hoffm.) Besser   Dark-red helleborine
Fairly frequent on the limestone screes and pavements
around the Kent estuary where it usually grows in some
shade. It also occurs very locally on limestone scars and
steep slopes by the upper Eden valley above Brough
(78.16) and Hartley (78.06), and it was recently
discovered at Hodbarrow (16.78, R. Ellwood, 1995).
Alt. limit: 400 m, above Helbeck Wood, Brough
(78.16)
Map 1170   22   WFC

E. helleborine (L.) Crantz   Broad-leaved helleborine
Still fairly frequent in the rolling country between the
Lake District and the Solway Plain, also around the
Kent estuary; elsewhere scarce or absent although, as
indicated by the *Atlas*, formerly much more widespread. Its decrease may perhaps be partly attributed to
the indiscriminate felling of small, species-rich woods
and the conversion of broad-leaved woodlands to
conifers, although it sometimes persists along rides. It
occurs in usually damp, deciduous woodland, woodland margins, scrub and wooded limestone pavement
on calcareous or moderately base-rich soils.

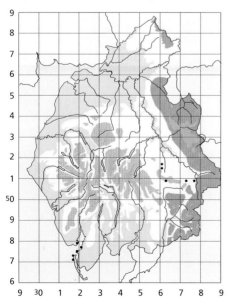

Map 1169. *Epipactis palustris*

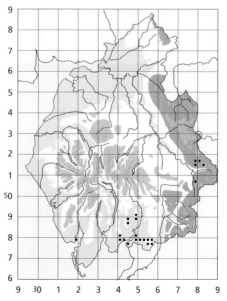

Map 1170. *Epipactis atrorubens*

Alt. limit: *c*.300 m, Augill Beck, North Stainmore (82.14)

Map 1171   132   WFCY

E. leptochila (Godfery) Godfery
var. **leptochila**

Although known for some time from heavy metal contaminated riverside gravels by the South Tyne in Northumberland, it was not until 1993 that a few plants were found above Alston (7.4, LR, LANC, det. A.J. Richards) growing under birch trees.

1   C

var. **dunensis** Stephenson & T.A. Stephenson
(*E. dunensis* (Stephenson & T.A. Stephenson) Godfery

Dune helleborine

Known only from dune-slacks at Sandscale by the Duddon estuary (18.74, K) where the population is well in excess of 1000. It was apparently first recorded by D.E. Allen in 1951 (BM). It is easily confused with the following species and is best distinguished by its patent flowers. R.J. Cooper (1996 *in litt.*) points out that the leaves are more than twice as long as broad, the epichile is seldom longer than broad, the capsules are oblong rather than pyriform, and the leaf margin is minutely crenulate rather than minutely serrulate. It appears to favour the higher slopes of the dunes

This variety is known from the coasts of Anglesey, Lancashire, Cumbria and Northumberland and a number of inland sites, for example on waste tips in the Glasgow area. There are also records elsewhere of plants intermediate between var. *dunensis* and typical *Epipactis leptochila*.

Plate 110   S 1   F

E. phyllanthes G.E. Smith

Although not a coastal species like *Epipactis leptochila* var. *dunensis*, this species is restricted in Cumbria to dune-slacks at Sandscale (18.74), where the population exceeds 1000, and North Walney (16.72), where there are extremely few plants. The species is here at its northern limit on the west coast of Britain. Flowering is irregular.

Characters distinguishing it from *E. leptochila* var. *dunensis* are given under that species. It appears to have been first discovered in Cumbria at Sandscale in 1951 (I.W. Evans & A.W. Westrup, CLE), although it seems most unlikely that W.H. Pearsall was unaware of the populations of these two orchids.

Plate 111   S 2   F

Neottia nidus-avis (L.) Rich.   Bird's-nest orchid

Rare, known only from a few, well-scattered lowland sites in base-rich woodland, chiefly in the north and east. It appears to be declining, particularly in the south, there being no Survey records from four Furness squares shown in the Atlas.

Alt. limit: 250 m, Swindale, Brough (80.16)

Map 1172   25   WFC(Y)

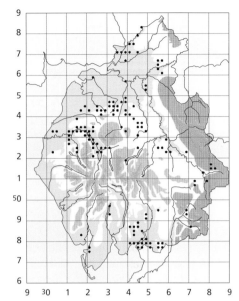

Map 1171. *Epipactis helleborine*

Plate 110. *Epipactis leptochila* var. *dunensis*

Plate 111. *Epipactis phyllanthes*

Listera ovata (L.) R. Br.   Common twayblade
Still a fairly common and even locally abundant lowland orchid of moderately base-rich or calcareous deciduous woodland, roadside verges and old meadows. The map picks out well the crescent of base-rich soil to the north of the Lake District. It occurs also in railway cuttings, dune-slacks, limestone grassland and the grikes of limestone pavements, also in flushed marshy grassland and occasionally, as on the Solway Plain, on peaty and sandy soils.
Alt. limit: 550 m, Moor House (74.32)
Map 1173   326   WFCY

L. cordata (L.) R. Br.   Lesser twayblade
Scarce and inconspicuous, mostly in high-level *Calluna* moorland in the Lake District and the north Pennines. It also occurs under *Calluna* on drift soil over limestone on Helsington Barrows (48.90), on two lowland mosses in the Eden valley and in a pine plantation on the Solway Plain. The plants usually grow in a carpet of moss: *Sphagnum* or *Pleurozium schreberi* and *Hylocomium splendens*.

Sadly, it seems to have disappeared from all the south Westmorland localities mentioned by Wilson, also from Arnside Knott (4.7s, Anon., 1908, LANC), but on the credit side are several new Lake District records, mostly contributed by E.G. Hall. The species is now probably commoner in Cumbria than anywhere in Britain south of the Scottish Highlands.
Alt. limit: at least 600 m, Alston Moor (72.36)
Map 1174   58   WCY

Spiranthes spiralis (L.) Chevall.
Autumn lady's-tresses
Very rare, known only with certainty from eight tetrads on the coastal limestone of Morecambe Bay where it grows in short *Festuca - Thymus* turf, sometimes associated with *Orchis morio*. In addition there is an unsubstantiated report from Heversham Head (5.8s) in the 1970s and an *Atlas* record based on a Barrow-in-Furness N.H.S. record of the early 1950s from near Greenodd (3.8). Both areas have been unsuccessfully searched in the 1980s by M.J.Y. Foley.

The populations, the most northerly in Britain, are mostly small but the largest produced at least 380 spikes in 1985, a particularly good year.
Map 1175   8   WF

Goodyera repens (L.) R. Br.   Creeping lady's-tresses
Scattered but locally abundant in several pinewoods from Cliburn Moss (56.24) north to Warwick-on-Eden (44.56). The woods are either long-established on lowland mosses or relatively recent plantations in the lowlands or on the sandstone ridge southeast of Penrith; most include birch. The future of the plantation populations is uncertain: Stoneraise Wood, Durdar (40.50) was felled in 1946 and another, Warwick Moor Wood (44.56) in 1987. It was introduced at this last site by E. Blezard in 1931 from a

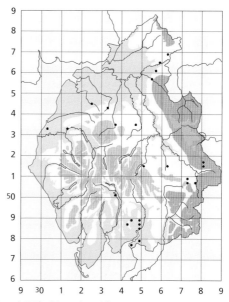

Map 1172. *Neottia nidus-avis*

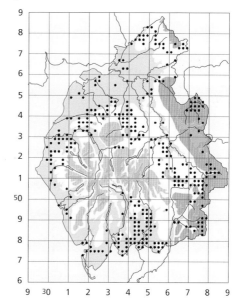

Map 1173. *Listera ovata*

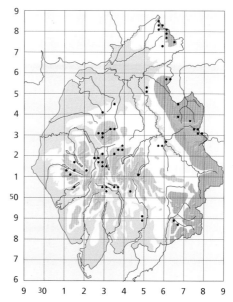

Map 1174. *Listera cordata*

colony threatened by felling. An account of these Cumberland sites is given by Blezard (1946). The orchid frequently grows amongst *Calluna* or *Vaccinium* and in at least two with *Listera cordata*. It also occurs in bare pine litter and even in *Sphagnum*, as at Cumwhitton Moss (50.50, 50.52), the most native-looking of the sites.

The first Cumberland record dates from 1877 when it was found by G. Brook in a pinewood near Armathwaite (4.4/5.4, *Naturalist* (1877) 3:26) and where it was collected shortly afterwards by F.A. Lees (1879, BM, *BRC Rpt* (1880) p.79). Perhaps significantly, there is a note with the specimen to the effect that it was "probably introduced with firs" and in the *Report* a comment that "..in the Borders it is spreading with the plantations".

Surprisingly, it was not seen in Westmorland until 1978 when it was recorded by D.J. Clarke in Whinfell Forest (58.26, CLE, *Watsonia* (1981) 13:340)

The general picture in Cumbria is similar to that in the other English counties from which the orchid has been recorded, and it seems quite likely that it has been unwittingly introduced in at least some sites from the pinewoods of north-east Scotland.

Alt. limit: 220 m, Whinfell Forest (56.26)

Map 1176   S 8   WC

Hammarbya paludosa (L.) Kuntze   Bog orchid
Rare but no doubt overlooked. It was found during the Survey in ten tetrads, nine in the Lake District and one in the Sedbergh area. Although this species is declining nationally, it still occurs in Cumbria in the two post-1930 *Atlas* squares, Borrowdale (2.1) and Carrock Fell (3.3), and it was also seen at single sites in the following squares: Broughton-in-Furness (2.8), Hawkshead (3.9), Scafell (2.0), Shap (5.1) and Sedbergh (6.9), and three in the Ambleside square (3.0). Hodgson gives records from Wast Water (1.0) and Grasmoor (1.1/1.2); the *Scarce Plants* record for the Langwathby

square (5.3) is an error.

This is a plant of lowland mires, often growing by small runnels, and associated species include *Eriophorum angustifolium*, *Juncus squarrosus* and *Drosera rotundifolia*. The pseudobulbs are frequently only half-buried in *Sphagnum*, particularly *S. auriculatum*, or bare peat and they are very sensitive to disturbance. Most populations are small but two usually exceed 100 spikes.

This orchid has declined nationally, particularly in England, where it is now largely restricted to the New Forest, Purbeck and Cumbria.

Alt. limit: 365m, Carrock Fell (3.3)

                              S 14   WFCY

Corallorhiza trifida Chatel.   Coralroot
Known only from dune-slacks at Sandscale (18.74) on the Duddon estuary, on the opposite side of the Walney Channel on North Walney (16.72) and on the Eskmeals firing range (08.92, A.B. Warburton, 1976, *Watsonia* (1977) 11:402). The only other Cumberland record dates from 1920 when it was seen by Postgate in a damp wood at Blaithwaite, Wigton (2.4, *BEC Rpt* (1921) 6:147). The orchid was seen at Sandscale in the 1950s and this is possibly the first Furness record.

The Eskmeals and Walney populations are small, but at Sandscale at least 3,000 plants were counted in 1991 in five slacks, making this the largest English population. Associated species include *Salix repens*, *Anagallis tenella* and *Hydrocotyle vulgare*.

This is a northern orchid, being commonest in the pinewoods of north-east Scotland.

                              3   FC

Platanthera chlorantha (Custer) Reichb.
                         Greater butterfly-orchid
Generally scarce or infrequent, although not uncommon between Kendal and the Kent estuary and in the Sebergham area (3.4). It is typically a plant of open

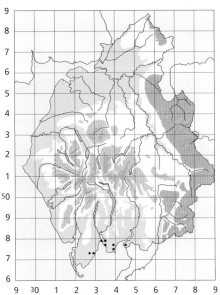

Map 1175. *Spiranthes spiralis*

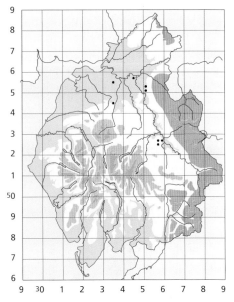

Map 1176. *Goodyera repens*

woodland, hay-meadows and railwaysides on base-rich or calcareous soils. It also occurs in dune-slacks on the west coast south of Silloth (0.4). The only authenticated record of this and the following species occurring together is from Hoff Lunn, Crosby Ravensworth (64.16) where they were seen by CW in a heathy pasture in the early 1970s prior to agricultural reclamation. Both orchids have also declined considerably as a result of the 'improvement' of hay-meadows.
Alt. limit: *c.*350 m, Helbeck Wood, Brough (78.16)
Map 1177   94   WFCY

P. chlorantha x **P. bifolia**
P. x hybrida Bruegger
A very rare hybrid collected in 1926 by H. Britten from a drove road near Skirwith (6.3, conf. G.C. Druce, CLE, *BEC Rpt* (1927) 8:136). Hunt (1975) has suggested that records of this hybrid probably refer to abnormal forms of the parents.
(C)

P. bifolia (L.) Rich.   Lesser butterfly-orchid
Apparently less widely distributed than the last species but often occurring in some quantity. Its overall distribution is generally similar but its ecological range is somewhat wider. Although occurring in base-rich grassland sites, it is also found in damp, base-poor heathy soils, particularly on acid drift over limestone, and in *Sphagnum,* as at Biglands Bog (24.52).
Alt. limit: *c.*280 m, near Ravenstonedale (74.02)
Map 1178   55   WFCY

Anacamptis pyramidalis (L.) Rich.   Pyramidal orchid
Very rare but probably increasing on the west coast. There are small populations on dunes at North and South Walney (16.70, 16.72, 22.62s) and Eskmeals (08.94), in a limestone quarry near Whitehaven (00.12) and on a roadside verge near Workington (02.26). The last two populations are the survivors of five transplanted by R. Stokoe in 1974 from a threatened

industrial site at Workington. There is a large population on the dunes at Haverigg (14.78) and in 1980 one, estimated in 1991 to be at least 2500 plants, was discovered by E.H. Shackleton on a disused industrial site at Maryport (02.36, *Watsonia* (1981) 13:341). In 1988 another was found in coastal grassland between Flimby and Siddick (00.32, W. Byers, 1988) and this appears to be of a similar size. These populations are extraordinary when seen in the context of a handful of small, mostly declining populations in the north of England. The Maryport site has recently been saved from a redevelopment scheme.

The orchid has gone from all the squares shown in the *Atlas* with the exception of North Walney. In the east it has disappeared from Melmerby (6.3, W.W. Mason, 1915) and Melkinthorpe (5.2, G.A.K. Hervey, 1951), in the south from Roudsea (3.8) and in the north from a field near Lamonby, Greystoke (4.3, EHR, *c.*1960), now 'improved', and from Tallentire Hill (1.3, 1939), and Sebergham (3.4, 1960s). As a result of development and dumping it has been lost during the period of the Survey from sand-dunes at Sellafield (0.0), the industrial site at Workington (9.2) and a quarry at Stainton, Dalton-in-Furness (2.7). This is a sad catalogue of extinctions for this most attractive southern orchid.
Map 1179   9   (W)FC

Pseudorchis albida (L.) Á. & D.Löve
Small white orchid
Very rare, known now only from a single site in each of four tetrads in the Sedbergh (6.8s, 6.9, 7.8) area, and the Spadeadam (6.7n), Borrowdale (2.1), Matterdale (3.2) and Crosby Ravensworth (6.1) squares. As a result of agricultural 'improvement', it disappeared *c.*1980 from a meadow at Hardendale (5.1). This attractive orchid has decreased markedly this century, largely for a similar reason. The early Floras and

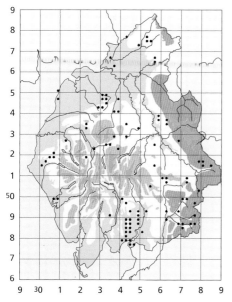

Map 1177. *Platanthera chlorantha*

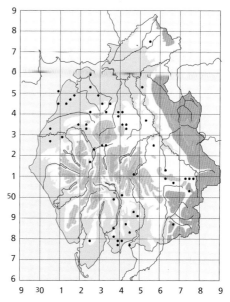

Map 1178. *Platanthera bifolia*

herbarium specimens indicate that it occurred in 23 well-scattered sites, from the valleys of the Lune to the Duddon, and from Kendal (4.9) to Bewcastle (5.7n). Of these, only three occur in the same squares as the present sites; the remaining 20 have been lost. Most of the records were from upland becksides, hay-meadows, open grassland or banks and this is still true, with the exception of one on a cliff-ledge in the central Lake District discovered by C.W. Muirhead in 1951. Here the orchid grows with *Sedum rosea* and *Galium boreale*.

Apart from this last colony, all the populations are small, consisting of only a few individuals, usually less than five, and further losses appear inevitable. Despite this decline, the national situation is such that, as with *Hammarbya* and *Listera cordata*, Cumbria has more extant sites than any county south of the Scottish Highlands.

Alt. limit: 425 m, Borrowdale square (2.1), but seen earlier at 500 m on Alston Moor (7.3, DAR, 1955).

9   W(F)CY

Gymnadenia conopsea (L.) R. Br.   Fragrant orchid
This orchid of base-rich and calcareous grassland, quarries and railwaysides appears to be also on the decline, particularly in the west. Until recently, it used to be abundant on the roadside verges between Lamonby and Sowerby Row (3.3/4.3) and there is only one site left in the Tallentire area (1.3) where it was once common. It is commonest and even locally frequent on the north Pennine limestone though probably never now, as recorded by Hodgson, "sufficiently abundant to diffuse their fragrance to some distance".
Alt. limit: *c*.440m, south of Garrigill (74.38)
Map 1180   142   WFCY

There has been no systematic recording of the three subspecies recognised by Stace (1991). MP con-

siders that all occur and he has provided the following comments.

subsp. **conopsea**
The commonest subspecies and the earliest to flower. The flowers have a labellum which is scarcely wider than long.

subsp. **densiflora** (Wahlenb.) Camus, Bergon & A. Camus
Common, often growing with the preceding but flowering later. The labellum is much wider than long.

subsp. **borealis** (Druce) F. Rose
Apparently rare, having been seen only at Spadeadam (66.72n) and Watendlath (26.16). At both sites it was flowering simultaneously with subsp. *conopsea.* The labellum is only weakly lobed.

Gymnadenia conopsea x Dactylorhiza fuchsii
x **Dactylogymnadenia st-quintinii** (Godfery) J. Duvign.
(x *D. cookei* (Heslop-Harrison) Soó)
The only records for this rare intergeneric hybrid are H. Britton's from Dufton (6.2, 1926, CLE) and Asby (6.1, 1925, CLE). Both collections were identified by G.C. Druce.
                                              (W)

Coeloglossum viride (L.) Hartman   Frog orchid
Not infrequent in the upper Eden valley, otherwise rare and absent from large areas. Most of the sites are in limestone turf, in old limestone quarries and on relatively base-rich, moist roadside verges and railwaysides. Exceptions include the populations in the Irthing gorge above Gilsland (6.7n) and one on thin drift overlying sandstone on Whinny Fell, Talkin (56.56). It has declined rapidly in recent decades, and is now virtually absent from west Cumberland and the middle Eden valley. The reasons for this are not clear but certainly, as with *Gymnadenia conopsea*, agricult-

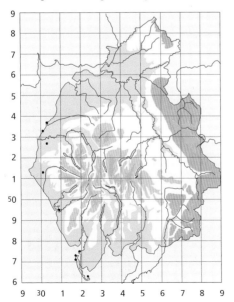

Map 1179. *Anacamptis pyramidalis*

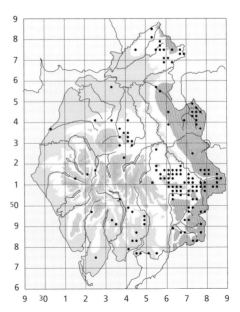

Map 1180. *Gymnadenia conopsea*

ural 'improvement', for example of the limestone meadows around Tallentire (2.3), and disturbance of old roadside verges have been important factors.

The *Atlas* record for the North Walney square (1.7) was contributed by G. Wilson in 1956. It is curious that it has never been refound there despite the attention that the area receives.

This orchid has a particularly long flowering season, from May until September. There are roadside population of exceptionally robust individuals above Orton (60.08) and above Ravenstonedale (74.02).

Alt. limit: *c*.400 m, near Nenthead (76.44)

Map 1181   42   WFCY

Dactylorhiza fuchsii (Druce) Soó

Common spotted orchid

Frequent and widespread on base-rich or calcareous soils and occurring in a wide range of mainly lowland habitats: industrial waste sites, roadside verges, limestone grassland and hay-meadows, marshy fields and fens and damp woodland.

Wilson records an albino plant from near Melkinthorpe (5.2) as var. *okellyi* Druce, but most apparently albino plants are in fact very faintly coloured. This predominantly Irish taxon is quite widespread in the Isle of Man.

Alt. limit: 530 m, east of Garrigill (76.40)

Map 1182   814   WFCY

D. fuchsii x **D. maculata**

D. x **transiens** (Druce) Soó

Recorded only from Swindale, Brough (80.16, LH, 1970s), Threaplands, Sleagill (58.16, CW, 1978) and Lowick Green (28.84, A. Parker, 1973) but probably overlooked.

Alt. limit: at least 280 m, Swindale (80.16)

       3   WF

D. fuchsii x **D. incarnata**

D. x **kerneriorum** (Soó) Soó

Seen during the Survey at Sandscale (18.74, A. Parker, 1977) and Hale Moss, Burton-in-Kendal (50.76s, GH, 1984, conf. P.H. Hunt). At the latter site *Dactylorhiza incarnata* is represented only by subsp. *pulchella*. There is also the record given by Wilson from Asby (6.1) where it was collected by H. Britten and determined by G.C. Druce.

       2   WF

D. fuchsii x **D. purpurella**

D. x **venusta** (Stephenson & T.A. Stephenson) Soó

This hybrid, and hybrid swarms, are not uncommon where the parents occur together although previously known only from H. Britten's 1926 records from Patterdale (3.1/4.1) and Skirwith (6.3, CLE, *BEC Rpt* (1927) 8:135).

Alt. limit: at least 360 m, near Nenthead (76.44)

Map 1183   40   WFC

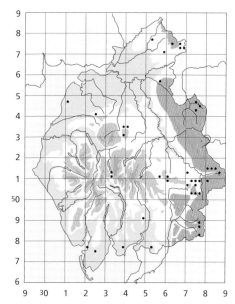

Map 1181. *Coeloglossum viride*

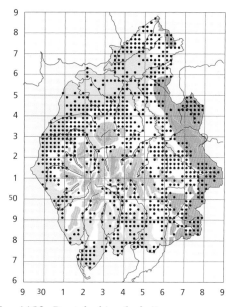

Map 1182. *Dactylorhiza fuchsii*

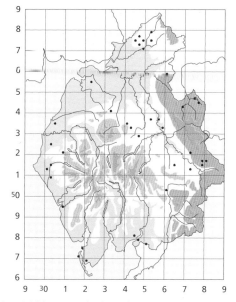

Map 1183. *Dactylorhiza fuchsii* x *D. purpurella*

D. maculata (L.) Soó   Heath spotted-orchid
A fairly frequent orchid of base-poor soils and particularly characteristic of lowland mosses and valley mires, commonly associated with *Sphagnum* spp., *Erica tetralix* and *Drosera rotundifolia*. It also occurs in damp meadows, especially in the uplands.

Hodgson and Wilson both comment on occasional albino plants but, as with *Dactylorhiza fuchsii*, such plants are probably not entirely devoid of pigment.
Alt. limit: at least 610 m, Knock Fell (72.30)
Map 1184   413   WFCY

D. maculata x **D. purpurella**
D. x formosa (Stephenson & T.A. Stephenson) Soó
Found during the Survey at Hardendale (56.14, M.J.Y. Foley, 1986), Wet Sleddale (52.10, I.R. Bonner, 1979), Bampton (50.18, CW, 1981), Butterburn, Spadeadam and near Langwathby (66.74n, 56.32, both A. Parker, 1977), a specimen from the last site being verified by V.S. Summerhayes.

In addition, Wilson gives a 1937 record from Middleton Common, Barbon (6.8s) and there are herbarium specimens from Skirwith (6.3, H. Britten 1926, CLE), Keswick (2.2, Mrs Robinson, 1847, BM) and Salta Moss, near Allonby (0.4, C.W. Muirhead, 1951, PLYP, det. V.S. Summerhayes) and a specimen from Red Moss, Abbeytown (1.5, E.J. Glaister, 1876, CLE) may belong here.
Alt. limit: *c*.240 m, Butterburn (66.74n)
                         5   WC

D. incarnata (L.) Soó   Early marsh-orchid
A local and decreasing species of dune-slacks, fens and calcareous flushes, chiefly in the south and east but with isolated sites in the extreme north-east. Although now present in 17 squares, there are no recent records for seven of the nine shown in the *Atlas*. Most records are of the pale-flowered subsp. *incarnata*.

The purple-flowered subsp. *pulchella* (Druce) Soó occurs at Combe Scar, Dentdale (66.86s), Hale Moss, Burton-in-Kendal (50.76s), Cunswick Tarn, Underbarrow (48.92), Skelsmergh Tarn (52.96), Sunbiggin Tarn (66.06) and Butterburn, Spadeadam (66.74n), and formerly at Salta Moss, near Allonby (0.4, C.W. Muirhead, 1951, PLYP, det. V.S. Summerhayes

The smaller, deep crimson-flowered subsp. *coccinea* (Pugsley) Soó is the rarest of the three, occurring only in dune-slacks by the Duddon estuary: Haverigg (14.78), North Walney (16.70) and Sandscale (18.74). There is also an *Atlas* record for the Gilsland square (6.6, v.c.67/70). According to Allen (1984), it flowers later than subsp. *incarnata*. The latter grows with subsp. *coccinea* at the last two dune-slack sites and at most of the subsp. *pulchella* sites.
Alt. limit (subsp. *incarnata*): *c*.260 m, Matterdale (40.22)
Map 1185   33   WFCY

D. incarnata x **D. purpurella**
D. x latirella (P. Hall) Soó
An extremely rare hybrid known only from three Survey records: Urswick Tarn (26.74, GH, 1984, det. P.F. Hunt), Sandscale (18.74, M.J.Y. Foley, 1985, 1990) and Sunbiggin Tarn (66.06, M.J.Y. Foley, 1986). At the last site there were hybrids with both *Dactylorhiza incarnata* subsp. *incarnata* and subsp. *pulchella*.
                         3   WF

[D. praetermissa (Druce) Soó   Southern marsh-orchid
As mentioned above, it seems likely that all Wilson's records of this southern species refer to *Dactylorhiza purpurella*. The northernmost records of *D. praetermissa* are on the Lancashire coast.
                                         W]

Map 1184. *Dactylorhiza maculata*

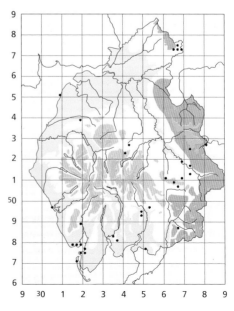

Map 1185. *Dactylorhiza incarnata*

D. purpurella (Stephenson & T.A. Stephenson) Soó

Northern marsh-orchid

A frequent and beautiful northern orchid of dune-slacks, fens, wet meadows, calcareous flushes, quarries and roadside verges. It is commonest in the east and very rare in the Lake District. Probably all Wilson's records of *Orchis praetermissa* belong here.

Like *Dactylorhiza incarnata*, it has decreased appreciably in recent decades as a result of agricultural 'improvement'.

Alt. limit: at least 450 m, south of Nenthead (76.40) and on Hartside (66.42)

Map 1186   293   WFCY

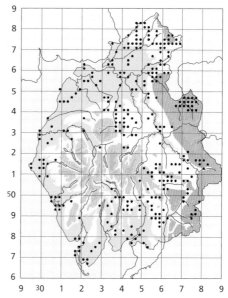

Map 1186. *Dactylorhiza purpurella*

Orchis mascula (L.) L.   Early purple orchid

Frequent on the more base-rich and calcareous soils to the south and north of the Lake District and extending eastwards to the limestone of north Westmorland and the upper Eden valley. It is scarce in the north-east, on the Solway Plain and, surprisingly, virtually absent from the Lune valley. Although characteristically a species of rich, deciduous woodland, it is frequent on areas of treeless limestone pavement and in the Lake District it grows high up on rock ledges on crags and in several of the gills.

Alt. limit: 660 m, Glencoyne Head, Helvellyn (34.18)

Map 1187   406   LWFCY

O. mascula x **O. morio**

O. x **morioides** Brand

Two spikes of this extremely rare hybrid were seen by M.J.Y. Foley in 1985 and 1986 growing with the parents in limestone grassland near Arnside (44.78s, *Watsonia* (1986) 16:195). The hybrid has not been seen since and its disappearance probably reflects a change in management of the field.

Plate 112   1   W

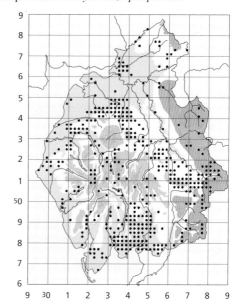

Map 1187. *Orchis mascula*

O. morio L.   Green-winged orchid

A nationally declining southern orchid, now quite rare in Cumbria where it is restricted to 19 populations in 14 tetrads, occurring in limestone grassland by the Kent estuary and by old limestone quarries near Ulverston (28.72, 28.76). However, a recently discovered population was in sandy grassland at Rampside (24.66). There are old records from higher up the Kent estuary and also ones from Gosforth (0.0) and Tallentire (1.3, J. Parkin, 1935). Handley (1898) gives a record from lower Dentdale (6.8s) but this seems very unlikely.

The populations are mostly small and, like those of *Spiranthes spiralis*, vulnerable to changing grazing regimes and scrub invasion. One site by the Kent was partly destroyed during the Survey by the dumping of dredged mud. However, there are two populations which produce several hundred spikes.

Map 1188   15   WF(C[Y])

Plate 112. *Orchis mascula* x *O.morio*

O. ustulata L.   Burnt orchid
Apparently now extinct not having been seen at its last
site at Lamonby (4.3, EHR) since the early 1970s. Like
Cephalanthera longifolia, this species has suffered a
major decline this century, both nationally and within
Cumbria (Foley 1987, 1992), but in this instance the
cause is primarily the agricultural 'improvement' of
limestone grassland and riverside meadows rather than
collecting.

Baker gives records from Blindcrake and the
Tallentire area (1.3), where it persisted until at least
1933 when it was seen at Bridekirk, probably by Mrs G.
Yeomans, and also from Raughton Head (3.4),
Stainton, Carlisle (3.5), Arnside Knott (4.7s) and
Warcop (7.1). There is also an early record from by the
River Eden at Wetheral (4.5, S.A. Naylor, 1877, BM).
To these localities Hodgson adds Edenhall (5.3) and
Crosby (4.5). The orchid was later seen at Crosby, also
at Newby Grange, by Miss Hodgson in 1903, by the
River Eden at Wetheral (4.5, S.A. Naylor, 1877, BM).
To these localities Hodgson adds Edenhall (5.3) and
Crosby (4.5). The orchid was later seen at Crosby, also
at Newby Grange, by Miss Hodgson in 1903, by the
River Eden at Grinsdale, Carlisle by D. Stewart in 1926
and in the middle Eden valley at Armathwaite bridge
(5.4, F. Simpson, 1940), in riverside meadows near
Great Salkeld (5.3) by H. Britten in 1903, and by
G.A.K. Hervey from 1947-1956, at a site which is now
a plantation, and also at Langwathby (5.3, J.R. Parker,
1932), although this could be the same site. It was also
seen by E. Blezard by the River Eden at Carlisle (4.5)
in the 1930s. Wilson gives a record from the turn of the
century from Drybeck (6.1) and a 1934 one from
Warcop. The orchid was known from a number of
riverside meadow sites in this last area, between
Appleby and Brough, but became extinct about 1970. In
the Arnside area (4.7s) it was seen at Hazelslack by
G. Higgins in 1956 (KDL) and at New Barns and Black
Dyke in the 1930s. Handley (1898) includes this

orchid in his Sedbergh list but with reservation and no
localities.

S 1†   (WC[Y])

Ophrys insectifera L.   Fly orchid
Locally frequent in ash woodland, scrub and open
grassland on the limestone around Morecambe Bay and
on steep limestone slopes by the upper Eden valley.
These latter sites, curiously omitted from the Atlas,
represent the species' northern limit in Britain.
Extensive disturbance in Park Wood, Dalton-in-Furness
(2.7), the basis of the *Atlas* record, resulted in its
extinction there about 1970. It is still present in
Roudsea Wood (32.82), a site which Petty (1898 p.42)
said was "guarded more strictly" since "Dozens of
people with buckets (and sometimes trowels too)
invaded the woods.". Like *Ophrys apifera*, it is very
sporadic in its appearance.
Alt. limit: 390 m, Helbeck Wood, Brough (78.16)
Map 1189   29   WF

O. apifera Hudson   Bee orchid
Occasional around Morecambe Bay and the Duddon
estuary but in the west, between Cockermouth and
Egremont, it is now reduced to only two sites. Most of
the extant sites are artificial, usually old limestone
quarries, but there are also records from a slag bank
and a wetland urban waste area. Such sites are very
vulnerable and waste tipping is responsible for the
disappearance of the orchid in at least some of its
former west Cumberland sites. It occurs also in the
Duddon dune-slacks.

Wilson's sites also include two inland records
from this century from the Ambleside area (3.0),
Hodgson mentions its occurrence near Aspatria (1.4),
Skiddaw (2.2) and Caldbeck (3.3), and W.H. Youdale
collected it south-west of Cockermouth (0.2, 1907,
LIV); it was also seen near Cockermouth in 1933 by
Mrs G. Yeomans. These last two records are probably

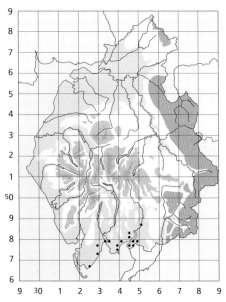

Map 1188. *Orchis morio*

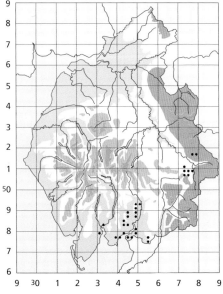

Map 1189 *Ophrys insectifera*

from the quarry at Brigham (08.28) where it was seen by J. Parkin in 1935, J.D. Hinde in 1946 and also during the Survey; unfortunately the site was destroyed in the late 1980s.

Surprising is the absence of any Furness records in Baker's and Petty's accounts and none can be traced earlier than the *Atlas* records of G. Wilson dating from the 1950s.

Most of the populations are small, but as with many orchids the number of flowering plants varies appreciably from year to year. More than 1,000 spikes were counted by M.J.Y. Foley at one of the west coast sites in 1985, quite remarkable for a species here at its northern limit. Plants with a straight, elongate, acute apex to the labellum and a more random pattern of markings (*cf.* var. *trollii* Hegetschw.) have been seen by Foley at Hodbarrow (16.78).

Map 1190   20   WFC

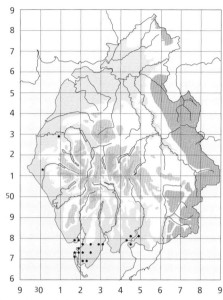

Map 1190. *Ophrys apifera*

Plate 113. December sunset across the Duddon estuary

This is the shore, the line dividing
The dry land from the waters, Europe
From the Atlantic; this is the mark
That God laid down on the third day.
Twice a year the high tide sliding,
Unwrapping like a roll of oilcloth, reaches
The curb of the mud, leaving a dark
Swipe of grease, a scaled-out hay

Of wrack and grass and gutterweed. Then
For full three hundred tides the bare
Turf is unwatered except for rain;
Blown wool is dry as baccy; tins
Glint in the sedge with not a sight of man
For two miles round to drop them there.
But once in spring, and once again
In autumn, here's where the sea begins.

On Duddon Marsh, Norman Nicholson

Appendices

Appendix 1: Gazetteer

Only places referred to in the introductory sections are listed here. Grid references are given as tetrads

Abbeytown 35/16.50
Abbot Moss 35/50.42
Aglionby 35/44.56
Allonby 35/08.42
Alston 35/70.46
Ambleside 35/36.04
Angle Tarn 35/24.06
Anthorn 35/18.58
Appleby 35/68.20
Argill 35/82.12
Armathwaite 35/50.46
Armboth Fells 35/28.16
Arnside 34/44.78
Ash Gill 35/76.40
Askam-in-Furness 34/20.76
Askham Hall 35/50.22
Aspatria 35/1.4
Auchertree Fell 35/26.36
Augill 35/80.14
Bampton 35/52.18
Bannerdale 35/34.28
Barfield Tarn 34/10.86
Baronwood 35/50.42
Bassenthwaite Lake 35/2.3
Baugh Fell 34/7.9
Biglands Bog 35/24.52
Binsey 35/22.34
Birk Fell 35/40.18
Black Combe 34/1.8
Black Dub 35/52.52
Black Dyke Moss 35/40.24
Black Force 34/64.98
Black Moss 35/02.10
Blea Tarn 35/28.04
Blelham Tarn 35/36.00
Blencarn 35/62.30
Blencathra 35/32.26
Blencowe 35/44.32
Blindcrake 35/14.34
Bolton Fell 35/48.68
Bolton Fell Moss 35/48.68
Bootle 34/10.88
Borrow Beck 35/58.00
Borrowdale 35/2.1
Bothel 35/18.38
Bowfell 35/24.06
Bowness-on-Solway 35/22.62
Bowscale Tarn 35/32.30
Braithwaite Moss 35/22.24
Brampton 35/52.60
Brandreth 35/20.10
Braystones 35/00.04
Braystones Tarn 35/00.06
Brigsteer Moss 34/46.88
Brocklebank 35/28.44
Broomhill Moss 35/48.66
Brothers Water 35/40.12
Brough 35/78.14
Bullgill 35/08.38
Burgh 35/32.58

Burnhope Seat 35/78.36
Burnmoor Tarn 35/18.04
Burnsbeck Moss 34/58.86
Burton-in-Kendal 34/52.76
Butterburn 35/66.74
Buttermere 35/18.14
Caldbeck 35/32.38
Cam Spout 35/20.04
Cardew Mires 35/34.50
Carrock Fell 35/34.32
Cash Force 35/70.38
Casterton 34/62.78
Castle Carrock 35/54.54
Catlands Fell 35/24.40
Cautley Crag 34/66.96
Cautley Spout 34/68.96
Christianbury Crag 35/56.82
Cleator Moor 35/02.14
Cliburn Moss 35/56.24
Clints Crags 35/16.34
Cockermouth 35/12.30
Cold Fell 35/60.54
Coniston Old Man 34/26.96
Cotehill 35/46.50
Crag Hill 34/68.82
Crosby Ravensworth Fell 35/62.14
Cross Fell 35/68.34
Culgaith 35/60.28
Cumwhitton Moss 35/50.52
Cunswick Scar 34/48.94
Dalegarth 34/16.98
Dalston 35/36.50
Dalton-in-Furness 34/2.7
Dentdale 34/7.8
Dollywaggon Pike 35/34.12
Dovedale 35/38.10
Dove Crag 35/36.10
Dow Crag 34/20.86
Drigg Nature Reserve 34/06.96
Duddon estuary 34/1.7
Dufton 35/68.24
Dufton Pike 35/68.26
Dukerdale 35/80.16
Dunmail Raise 35/32.10
Eden Lacy 35/56.38
Egremont 35/00.10
Ehenside Tarn 34/00.06
Elterwater 35/32.04
Ennerdale Water 35/10.14
Eskmeals 34/08.92
Esthwaite Water 34/34.96
Eycott Hill 35/38.28
Fairfield 35/34.10
Farleton Fell 34/54.80
Faugh Moss 35/50.54
Faulds Brow 35/28.40
Fell End Clouds 34/72.98
Finglandrigg 35/20.56
Flakebridge Wood 35/70.22
Foulney 34/2.6

Foulshaw Moss 34/46.82
Frizington 35/02.16
Garsdale 34/7.9
Gillercomb 35/20.10
Gilsland 35/62.66
Glencoyne 35/38.18
Gowk Bank 35/66.74
Grasmere 35/32.06
Grasmoor 35/16.20
Great Asby 35/68.12
Great Dun Fell 35/70.32
Great Gable 35/20.10
Great Mell Fell 35/38.24
Great Ormside 35/70.16
Great Orton 35/32.54
Great Salkeld 35/54.36
Great Strickland 35/54.22
Greystoke 35/42.30
Greystoke Park 35/40.32
Grinsdale 35/36.58
Grune Point 35/14.56
Hale Moss 34/50.76
Hallin Fell 35/42.18
Halsenna Moor 35/06.00
Haltcliff Common 35/34.34
Harrington 25/98.24
Hartside 35/68.42
Haverigg 34/16.78
Haweswater 35/4.1
Hayton 35/50.56
Helbeck 35/78/16
Helvellyn 35/32.14
Hethersgill 35/46.66
High Cup Nick 35/74.26
High Pike 35/30.34
High Street 35/44.10
Hillbeck 35/78.14
Hobcarton Crag 35/18.22
Hodbarrow 34/18.78
Holme 34/52.78
Honeybee Wood 34/48.90
Honister 35/22.12
Honister Crag 35/20.14
Howgill Fells 34/9.6
Howk 35/30.38
Humphrey Head 34/38.72
Hutton Roof 34/54.78
Johnby Moor 35/42.32
Kentmere 35/44.04
Kershope 35/5.8
Keswick 35/26.22
Kingmoor 35/38.58
Kinkry Hill 35/50.74
Kirkbride 35/22.56
Kirkby Moor 34/24.82
Kirkby Stephen 35/76.06
Kirklinton 35/42.66
Kirksanton 34/14.80
Knock Fell 35/72.30
Knock Ore Gill 35/70.30

Knock Pike 35/68.28
Lamonby 35/40.34
Langdale Combe 35/26.08
Langwathby 35/56.32
Lazonby Fell 35/50.38
Leck 34/64.76
Little Dun Fell 35/70.32
Little Fell 35/78.22
Little Mell Fell 35/42.24
Lodge Hags 35/80.02
Lodore 35/26.18
Long Fell 35/76.18
Longsleddale 35/4.0
Longtown 35/38.68
Lorton Vale 35/14.24
Low Church Moss 35/00.06
Low Row 35/58.62
Low Tarn 35/16.08
Low Wood 35/40.12
Ludderburn Moss 34/40.90
Mallerstang Edge 35/78.00
Martin Tarn 35/24.50
Maryport 35/02.36
Matterdale 35/20.34
Mayburgh 35/50.28
Meathop Moss 34/44.80
Melbreak 35/14.18
Melkinthorpe 35/54.24
Melmerby 35/60.36
Mickle Fell 35/80.24
Middlebarrow 34/46.76
Middleton Fell 34/64.86
Millom 34/16.80
Mockerkin 35/08.22
Mollen Wood 35/56.70
Monkhill Lough 35/32.58
Moor House 35/74.22
Moorthwaite Moss 35/50.50
Muncaster 34/10.96
Mungrisdale 35/36.30
Naddle Forest35/50.14
Nenthead 35/78.42
Nether Wasdale 35/12.04
Newbiggin 35/46.28
Newton Arlosh 35/18.54
Newton Reigny Moss 35/46.30
North Walney 34/1.7
Orton 35/62.08
Orton Moss 35/3.5
Orton Scar 34/62.10
Overwater 35/24.34
Pavey Ark 35/28.08
Penrith 35/5.3
Penruddock 35/40.26
Penton Linns 35/42.76

Piers Gill 35/20.08
Pillar 35/16.12
Podgill 35/78.06
Pooley Bridge 35/46.24
Potrigg 35/58.12
Rampside 34/24.66
Raughten Head 35/38.44
Ravenglass 34/08.96
Ravenstonedale 35/7.0
Rawthey valley 34/6.9
Red Tarn (Helvellyn) 35/34.14
Red Tarn (Wrynose) 35/26.02
Riggindale 35/44.12
Rise Hill 35/7.8
Robinson 35/20.16
Rockcliffe 35/34.60
Roudsea 34/3.8
Roughtongill 35/30.34
Rusland 34/32.88
Rusland Moss 34/32.88
Rydal Mount 35/36.06
Salta Moss 35/08.44
Sandford Mire 35/72.16
Sandscale Haws 34/18.74
Sandside 34/48.80
Scafell 35/2.0
Scaleby Mire 35/42.62
Scales Tarn 34/32.28
Scales Wood 35/14.16
Scordale 35/74.20
Scout Scar 34/48.90
Seascale 35/02.00
Seathwaite 35/22.12
Seathwaite Tarn 34/24.94
Sebergham 35/34.40
Sedbergh 35/92.64
Sellafield 35/02.02
Sellafield Tarn 35/02.02
Shap 35/56.14
Shap Wells 35/56.08
Sides Wood 35/10.14
Sighty Crag 35/60.80
Silecroft 34/12.82
Silloth 35/10.52
Silver Tarn 25/98.06
Skelsmergh 34/52.96
Skiddaw 35/26.28
Skinburness 35/12.54
Smardale Gill 35/72.08
Snellings Mire 25/98.08
Solway Moss 35/34.68
South Walney 34/2.7
Sowerby Row 35/38.40
Spadeadam Waste 35/6.7
St Bees Head 25/94.10

Staffield 35/54.42
Staveley 34/46.98
Stenkrith 35/76.06
Stock Ghyll 35/36.04
Sunbiggin Tarn 35/66.06
Swarth Fell 35/74.96
Swarthmoor 34/26.76
Swindale 35/80.14
Swirrel Edge 35/34.14
Tailbridge Hill 35/80.04
Talkin Tarn 35/54.58
Tarn Moss 35/40.26
Tarns Dub 35/10.46
Tebay 35/60.04
Temple Sowerby 35/60.26
The Ings 35/26.22
Thirlmere 35/3.1
Threlkeld 35/30.24
Thurstonfield Lough 35/30.56
Thwaites 34/16.84
Tindale Fell 35/60.56
Tindale Tarn 35/60.58
Todhills Moss 35/36.62
Udford Moss 35/56.30
Uldale 35/24.36
Ulverston 34/2.7
Unity 35/52.58
Waitby 35/74.06
Walby Moor 35/46.60
Walla Crag 35/26.20
Walton Moss 35/50.66
Walton Wood 35/54.64
Wan Fell 35/52.36
Warcop 35/74.14
Wast Water 35/1.0
Watendlath 35/28.14
Watermillock 35/44.22
Wedholme Flow 35/2.5
Westward 35/26.44
Wetheral 35/46.54
Wetherlam 35/28.00
Whernside 34/7.8
Whinfell Forest 35/56.26
Whins Pond 35/54.30
Whitbarrow Scar 34/44.86
Whitbeck Moss 34/10.84
Whitehaven 25/9.1
Whiteside 35/16.20
Wigton 35/24.48
Wild Boar Fell 34/76.98
"Williamsons Moss" 34/08.90
Winster valley 34/4.8
Witherslack 34/42.84
Wreay 35/42.48
Wythburn 35/32.12

Appendix 2: Native species and hybrids not seen during the Survey (1970 - 1996) but previously recorded from Westmorland (W), Furness (F) or Cumberland (C)

(Excluding species of *Hieracium*, *Rubus* and *Taraxacum*)

The numbers refer to the period of the last record: 1 - ⇒1900; 2 - ⇒1930; 3 - ⇒1969. The absence of an entry does not necessarily indicate that a species still occurs in that area; it may never have been recorded. The list does not include species which appear to have become extinct during the period of the Survey: *Mertensia maritima* (Furness, v.c.69), *Verbena officinalis* (Cumberland) and *Carex maritima* (Furness, v.c.69, Cumbria).

	W	F	v.c.69	C	Cumbria
Agrostemma githago		3		3	
Alchemilla monticola				3	3
Alopecurus myosuroides	3	3	3		
Anacamptis pyramidalis	3				
Anagallis minima		1			
Anthemis cotula	3			3	
Apium graveolens	3				
Arabis glabra				2	
Arctostaphylos uva-ursi	3		3		
Asplenium ruta-muraria x A. trichomanes	1		1		1
A. septentrionale		2			
A. septentrionale x A. trichomanes	3		3	3	3
A. trichomanes x Phyllitis scolopendrium	1		1		1
Astragalus glycyphyllos	2	1	2		
A. danicus	3			3	3
Bidens cernua	3				
Brassica nigra	3	3	3		
B. oleracea	3		3		3
Bromopsis erecta				3	
Bromus racemosus		3	3		
Carduus crispus x C. nutans				1	1
Carduus nutans		3			
Carex acuta x C. acutiformis	2		2		2
C. arenaria	1				
C. atrata				2	
C. diandra		1			
C. elata x C. nigra		2			
C. pseudocyperus	2			1	
C. strigosa	3			1	
Carum verticillatum				3	3
Cephalanthera longifolia		1			
Cerastium alpinum				1	
C. arvense	3				
Ceratophyllum submersum				3	
Clinopodium ascendens	2	1	2	3	3
Crataegus laevigata		2			
Crepis mollis		2			
Cuscuta epithymum	3		3	1	3
Cypripedium calceolus	2	1	2	1	2
Cystopteris montana	1		1		1
Dactylorhiza fuchsii x Gymnadenia conopsea	2		2		2
Dianthus deltoides	1				
Drosera longifolia	2				
D. longifolia x D. rotundifolia				3	
Dryopteris carthusiana x D. dilatata	1				
D. carthusiana x D. filix-mas	1		1		1
D. filix-mas x D. oreades	3				
D. oreades		3			
D. submontana				3	
Epilobium alsinifolium x E. palustre	1		1	2	
E. montanum x E.parviflorum	3		3		3
E. obscurum x E. parviflorum	3		3		3
E. palustre x E. parviflorum				3	3
Epipactis palustris				3	
Equisetum pratense	1		1		1
Euphorbia exigua				3	
Euphrasia arctica x E. officinalis				3	3
E. frigida		3			
E. frigida x E. scottica				3	
E. micrantha	3	2	3		
E. ostenfeldii	3		3		
Fumaria purpurea	2			3	

Galeopsis angustifolia		3		3	
Gentiana pneumonanthe	2			1	
Gnaphalium sylvaticum		3			
Groenlandia densa				2	
Hordelymus europaeus	1				
Hydrilla verticillata		3	3		3
Juncus subnodulosus				2	
Lathyrus japonicus				2	2
Leymus arenarius	3				
Linum bienne				2	2
Lithospermum officinale				3	
Lycopodium annotinum				1	
Meum athamanticum				1	
Moenchia erecta		1		1	
Myosurus minimus	3		3	1	3
Oenanthe. aquatica	1		1		
O. fistulosa	3	1	3		
Ononis reclinata				2	2
Orchis morio				3	
O. ustulata	3		3	3	3
Orobanche rapum-genistae				3	
Papaver argemone	3				
Phleum alpinum				3	
Picris hieracioides	2	3	3	3	3
Pimpinella major				3	
Polygonum maritimum				1	1
Potamogeton coloratus				1	1
P. gramineus x P. lucens				2	
P. lucens		2	2		2
P. praelongus		2			
Primula farinosa		3			
Pseudorchis albida		1			
Puccinellia rupestris				1	1
Pyrola media	1		1	3	3
Pyrola rotundifolia	3				
Radiola linoides	3				
Ranunculus arvensis	3	3	3		
R. fluitans x R. peltatus	3		3		3
R. hederaceus x R. peltatus		3	3		3
Rhinanthus angustifolius				1	1
Ribes alpinum		1			
Rosa arvensis x R. canina				2	2
R. canina x R. pimpinellifolia				3	3
R. canina x R. tomentosa	2		2	2	2
R. mollis x R. pimpinellifolia	2		2		2
R. rubiginosa x R. sherardii	2				2
R. sherardii x R. tomentosa	3		3		3
Rumex crispus x R. obtusifolius	1	1	1		
R. hydrolapathum	3			3	
R. maritimus		3	3	1	3
Ruppia maritima	1				
Salix aurita x S. phylicifolia	3		3		
S. aurita x S. caprea	3		3		
S. aurita x S. repens	3		3		
S aurita x S. viminalis	3		3		3
S. caprea x S. myrsinifolia	3		3		
Scandix pecten-veneris	3	3	3	3	3
Tephroseris integrifolia	2		2		2
Thalictrum flavum				3	
Thelypteris palustris		3			
Torilis nodosa				3	
Valerianella dentata	3	3	3	3	3
V. rimosa	3		3	3	3
Veronica anagallis-aquatica x V. catenata	3		3		3
Vicia lutea		3	3	3	3
V. tetrasperma	3				
Viola arvensis x V. tricolor		2	2		2
V. lutea		1			
Vulpia fasciculata				3	
Wahlenbergia hederacea	2	2	2		2
Woodsia ilvensis	1		1		1
Zostera marina				1	1

Appendix 3: Native species and hybrids recorded for the first time in Westmorland (W), Furness (F) or Cumberland (C) during the Survey (1970-1996)

(Excluding species of *Hieracium*, *Rubus* and *Taraxacum*)

	W	F	v.c.69	C	Cumbria
Alchemilla glaucescens	*		*		*
Atriplex longipes	*		*		*
A. longipes x A. prostrata		*	*		*
Brachypodium pinnatum				*	
Callitriche obtusangula		*		*	
C. platycarpa		*			
Carex aquatilis				*	
C. divulsa		*	*		*
C. elata x C. nigra	*				
C. flava x C. viridula		*	*		*
C. hostiana x C. viridula		*		*	
C. maritima†		*	*		*
C. ornithopoda				*	
C. otrubae x C. remota				*	*
C. rostrata x C. vesicaria		*	*		
C. strigosa		*			
Centaurium erythraea x C. littorale				*	
Cochlearia danica	*				
Crepis praemorsa	*		*		*
Dactylorhiza fuchsii x D. incarnata		*			
D. fuchsii x D. maculata	*	*	*		*
D. fuchsii x D. purpurella		*			
D. incarnata x D. purpurella	*	*	*		*
Dryopteris affinis x D. filix-mas		*	*	*	*
D. carthusiana x D. dilatata		*		*	
D. dilatata x D. expansa				*	*
D. filix-mas x D. oreades		*			
Elatine hexandra	*		*		
Eleocharis austriaca				*	*
Elytrichia juncea	*				
Epilobium alsinifolium x E. montanum				*	*
E. montanum x E. obscurum				*	*
E. roseum		*	*		
Epipactis atrorubens				*	
Equisetum arvense x E. fluviatile	*	*	*		
E. hyemale x E. variegatum				*	*
Eriophorum latifolium		*			
Euphrasia arctica x E. confusa	*	*	*		
E. arctica x E. micrantha				*	*
E. arctica x E. nemorosa		*	*	*	*
E. confusa x E. officinalis	*	*	*		
E. confusa x E. scottica	*	*	*	*	*
E. confusa x E. tetraquetra		*	*		
E. frigida x E. scottica	*				
E. nemorosa x E. scottica	*	*	*	*	*
E. nemorosa x E. tetraquetra		*	*	*	*
Festuca arenaria		*	*		*
x Festulolium loliaceum		*			
x Festulpia hubbardii		*	*		*
Geum rivale x G. urbanum		*			
Goodyera repens	*		*		
Helictotrichon pratense		*			
Hypericum maculatum x H. perforatum	*	*	*	*	*
Hypochaeris glabra		*	*		
Isoetes echinospora		*	*	*	*
Juncus acutiflorus x J. articulatus	*	*	*		
J. ambiguus				*	
J. effusus x J. inflexus		*		*	
Lathyrus aphaca	*				
Lemna gibba	*		*		*
Limonium britannicum	*				
Myosotis laxa x M. scorpioides				*	*
Nuphar lutea x N. pumila				*	*
Orchis mascula x O. morio	*		*		*

† since lost

Poa angustifolia	*		*	*	*
Polypodium cambricum				*	
P. interjectum x P. vulgare				*	
Polystichum aculeatum x P. setiferum	*		*	*	*
Potamogeton crispus x P. perfoliatus				*	*
Potentilla anglica x P. erecta	*				
P. anglica/erecta x P. reptans				*	
Puccinellia distans	*	*	*		
Ranunculus aquatilis x R. fluitans	*		*	*	*
Rosa caesia x R. mollis				*	*
R. caesia x R. sherardii				*	*
R. micrantha	*		*		*
R. mollis x R. sherardii	*		*		*
R. pimpinellifolia x R. sherardii	*		*		*
Rumex crispus x R. obtusifolius				*	
R. longifolius x R. obtusifolius				*	*
Ruppia cirrhosa		*	*		*
Salicornia fragilis		*	*	*	*
Salix aurita x S. repens				*	
S. caprea x S. cinerea				*	
S. caprea x S. cinerea x S. viminalis				*	*
S. cinerea x S. phylicifolia		*			
S. repens x S. viminalis		*	*		*
S. triandra x S. viminalis	*		*		
Senecio jacobaea x S. aquaticus	*			*	
Silene dioica x S. latifolia		*		*	
Spartina anglica	*			*	
Stellaria pallida		*	*	*	*
Thymus pulegioides				*	*
Trifolium striatum	*				
Ulmus glabra x U. minor		*	*		*
Viola canina x V. riviniana	*	*	*	*	*
V. hirta x V. odorata	*		*		*

Appendix 4: Partial Tetrads

The following peripheral tetrads have only a small area in Cumbria or above high water. Their records have therefore been reallocated to one of the adjacent tetrads as shown below.

25/92.14 to	94.14	20.80	22.80	35/00.02	02.02	42.78	42.76
94.10	96.10	20.82	22.82	06.42	08.42	46.82	48.82
96.08	98.08	22.60	22.62	06.44	08.44	50.84	50.82
96.20	98.20	24.62	24.64	06.46	08.46	56.88	56.86
96.22	98.22	32.74	30.74	14.56	12.56	62.52	60.52
34/06.86	08.86	32.76	32.78	14.58	16.58	64.46	64.44
06.90	08.90	34.72	34.74	16.56 north	16.58	64.58	62.58
10.80	12.80	38.72	38.74	south	16.54	64.68	62.68
12.76	14.78	42.76	44.76	16.60	18.60	66.46	66.44
14.76	14.78	44.74	44/76	24.62	24.64	78.34	76.34
16.66	18.66	56.74	54.74	26.60	24.60	80.00	78.00
16.76	16.78	58.76 west	56.76	28.60	28.58	82.04	80.04
18.62	18.64	north	58.78	30.64	32.64	84.04	84.06
20.60	20.62	60.76	60.78	32.70	32.68	86.04	86.06
20.64	18.64	62.76	62.78	38.74	40.74		
20.66	20.68	68.80	66.80	40.76	40.74		

Appendix 5: Cumbrian higher plant Red data book species

Alchemilla minima Y
Alchemilla glaucescens W
Aster linosyris F
Bartsia alpina W
Calamagrostis purpurea F
Carex flava F
Carex ornithopoda WFC
Crepis praemorsa W
Eleocharis austriaca C
Euphrasia rivularis WC
*Gentiana verna WC
Hypochaeris maculata F
Kobresia simpliciuscula W

Ledum palustre subsp. groenlandicum C
*Lychnis alpina C
Myosotis alpestris W
Polygala amarella W
Potentilla fruticosa WC
Ranunculus reptans WC
*Saxifraga hirculus WC
Sorbus lancastriensis WF
*Trichomanes speciosum WC
Veronica spicata subsp. hybrida WF
Viola rupestris W
*Woodsia ilvensis C

Extinct
Alchemilla monticola
Cypripedium calceolus

* protected by law

Appendix 6: Addenda

p.116

LAURACEAE

***Laurus nobilis** L.   Laurel
A single tree was found in neglected former estate
woodland at Heversham in 1995 (48.84s, GH, LANC).

1   W

p.162

***Dianthus barbatus** L   Sweet William
There is a further record from a quarry west of Kendal
(50.92, CEW, 1997, LANC).

4   WFC

p.181

***Hibiscus trionum** L.
Seen briefly by the Keswick by-pass during its construction (26.24, G. Wilson, 1975, LANC).

1   C

p.287

***Cotoneaster rotundifolius** Lindley
Found by the roadside south of Uldale in 1995 (22.54,
MP, 1995, det. J. Fryer).

This species differs from *Cotoneaster microphyllus* in being more erect (-2 m) and having generally
longer leaves (-2 cm) and larger flowers (1-1.3 cm
across), which are always solitary.

1   C

p.302

Trifolium micranthum Viv.
Seen at Grange-over-Sands in an overgrown putting
green (40.76s, W. Nelson, 1997, LANC), the second
confirmed record.

2   [W]F[(C)]

p.307

***Eleagnus commutata** Bernh.
This North American shrub was originally planted in
the mid-1970s on a derelict site north of Workington
(98.30). The resulting population of vigorous suckers
and seedlings now covers an area of up to 5 hectares
(AAD, 1995, LANC, det. E. Clement (Dudman 1997)).

1   C

p.310

Epilobium parviflorum x **E. ciliatum**
Found at Askam-in-Furness in 1996 (20.76, PB,
LANC, det. T.D. Pennington), the only record.

1   F

E. montanum x **E. alsinifolium**
A specimen from Cautley Spout (6.9, 1887, BM), collected and identified as this by F.A. Lees, has been
redetermined by T.D. Pennington as *Epilobium alsinifolium*. Sledge (1945) mentions a collection from
Uldale (7.9), redetermined by G.M. Ash as *E. montanum*, and one by J.F. Pickard in 1902, also from
Cautley and determined by Lees and C.E. Salmon as
the hybrid. It seems best to regard this too as doubtful.

1   C[(Y)]

p.411

***Cephalaria gigantea** (Ledeb.) Bobrov.
Found in 1997 on a roadside verge east of Kirkby
Thore (64.26, RWMC, LANC), the first Westmorland
record.
Alt. limit: 120 m, Kirkby Thore (64.26)

2   WF

References

ALLEN, D.E. (1967). Taxonomy and nomenclature of the radiate variants of *Senecio vulgaris* L. *Watsonia*, **6**:280-282.

ALLEN, D.E. (1984). *Flora of the Isle of Man*. Douglas.

ALSTON, A.H.G. (1940). Notes on the supposed hybrids in the genus *Asplenium* found in Britain. *Proceedings of the Linnean Society of London*, **152**: 132-144.

ATKINSON, T.C., BRIFFA, K.R. & COOPE, G.R. (1987). Seasonal temperatures in Britain during the past 22,000 years, reconstructed using beetle remains. *Nature*, **325**: 587-593.

BACKHOUSE, J., fil. (1844). Notes of a botanical ramble in Yorkshire etc. in the summer of 1844. *Phytologist*, **1**:1065-1069

BACKHOUSE, J., fil. (1846). Note of a few plants growing on Helvellyn or in its vicinity. *Phytologist*, **2**:422-423.

BACKHOUSE, J., fil. (1847). Account of a few days ramble among the mountains of Cumberland and Westmorland. *Phytologist*, **2**:1044-1047.

BAKER, H.G. (1956). *Geranium purpureum* Vill. and *G. robertianum* L. in the British flora -II. *Geranium robertianum. Watsonia*, **3**:270-279.

BAKER, J.G. (1885). *A Flora of the English Lake District*. London.

BALFOUR, J.H. (1854). Notice of a botanical trip to Helvellyn, with pupils, in July 1853. *Phytologist*, **5**:26-29.

BANGERTER, E.B. (1966). Further notes on *Veronica peregrina* L. *Proceedings of the botanical Society of the British Isles*, **6**:215-220.

BARBER, K.E. (1981). Peat stratigraphy and climatic change. In Balkema, A.A., *A palaeoecological test of the theory of cyclic peat bog regeneration*. Rotterdam.

BENNET, A. (1914a). *Naias flexilis* Rost. & Schmidt; a new record for England. *Naturalist*, **1914**:274.

BENNET, A. (1914b). *Hydrilla verticillata* in England. *Naturalist*, **1914**:289.

BENNET, A. (1914c). *Hydrilla verticllata* Casp. in England. *Journal of Botany, London*, **52**:256-258

BIRKS, H.J.B. (1982). Mid-Flandrian forest history of Roudsea Wood National Nature Reserve, Cumbria. *New Phytologist*, **90**:339-354.

BIRKS, H.J.B. (1989). Holocene isochrone maps and patterns of tree-spreading in the British Isles. *Journal of Biogeography*, **16**: 503-540.

BLACKSTOCK, T.H. (1981). The distribution of *Juncus filiformis* L. in Britain. *Watsonia*, **13**:209-214.

BLACKSTOCK, T.H. & ROBERTS, R.H. (1986). Observations on the morphology and fertility of *Juncus* x *surrejanus* Druce ex Stace & Lambinon in north-western Wales. *Watsonia*, **16**:55-63.

BLEZARD, D. (1946). Creeping Lady's tresses in Cumberland. *Transactions of the Carlisle Natural History Society*, **7**:70-74.

BOLTON, J. (1862). On a deposit with insects, leaves, etc. near Ulverston. *Proceedings of the Geological Society of London*, **18**:274-277.

BONSALL, C. (1980). The coastal factor in the Mesolithic settlements of north-west England. *Veröffentlichungen des Museums För Ur- und Frühgeschichte Potsdam*, **14/15**:451-472.

BORRER, W. (1846). Notices of north of England plants. *Phytologist*, **2**:424-437.

BREWIS, A., BOWMAN, P. & ROSE, F. (1996). *The Flora of Hampshire*. Colchester.

BROWN, R. (1990). The genus *Utricularia* in Cumbria. Unpublished undergraduate dissertation. University of Lancaster.

CLAPHAM, A.R., TUTIN, T.G. & WARBURG, E.F. (1952). *Flora of the British Isles*. Cambridge.

CONOLLY, A.P. (1977). The distribution and history in the British Isles of some alien species of *Polygonum* and *Reynoutria. Watsonia*, **11**:291-311.

COOK, C.D.K. (1975). *Ranunculus* subgenus *Batrachium*, in Stace, C.A., ed., *Hybridization and the flora of the British Isles*, pp.125-129. London.

COOPE, G.R. & PENNINGTON, W. (1977). The Windermere interstadial of the late-Devensian. *Philosophical Transactions of the Royal Society*, **B 280**:337-339.

COPE, T.A. & STACE, C.A. (1978). The *Juncus bufonius* L. aggregate in western Europe. *Watsonia*, **12**:113-128.

CORNER, R.W.M. (1989). Observations on inland populations of *Viola canina* L. in south-eastern Scotland and north-western England. *Watsonia*, **17**:351-352.

CORNER, R.W.M. (1990). *Cardamine amara* L.: its occurrence in montane habitats in Britain. *Watsonia*, **18**:200-201.

CORNER, R.W.M. & ROBERTS, F.J. (1989). *Carex ornithopoda* Willd. in Cumberland. *Watsonia*, **17**:437-438.

CRABBE, J.A., JERMY, A.C. & WALKER, S. (1970). The distribution of *Dryopteris assimilis* S. Walker in Britain. *Watsonia*, **8**:3-15.

CRACKLES, F.E. (1990). *Hypericum* x *desetangsii* Lamotte nm. *desetangsii* in Yorkshire, with special reference to its spread along railways. *Watsonia*, **18**:63-67.

DARBYSHIRE, R.D. (1874). Notes on discoveries in Ehenside Tarn, Cumberland. *Archaeologia*, **44**:273-292.

DAVID, R.W. (1978a). The distribution of *Carex digitata* L. in Britain. *Watsonia*, **12**:47-48.

DAVID, R.W. (1978b). The distribution of *Carex elongata* L. in the British Isles. *Watsonia*, **12**:158-160.

DAVID, R.W. (1980). The distribution of *Carex ornithopoda* Willd. in Britain. *Watsonia*, **13**:53-54.

DAVID, R.W. (1981a). The distribution of *Carex punctata* Gaud. in Britain, Ireland and Isle of Man. *Watsonia*, **13**:318-321.

DAVID, R.W. (1981b). The distribution of *Carex ericetorum* Poll. in Britain. *Watsonia*, **13**:225-226.

DAWSON, H.J. & INGROUILLE, M.J. (1995). A biometric survey of *Limonium vulgare* Miller and *L. humile* in the British Isles. *Watsonia*, **20**:239-254.

DICKINSON, W. (1973). The development of the raised bog complex near Rusland in the Furness district of north Lancashire. *Journal of Ecology*, **61**:871-886.

DICKINSON, W. (1975). Recurrence surfaces in Rusland Moss, Cumbria (formerly North Lancashire). *Journal of Ecology*, **63**:913-935.

DILLWYN, L.W. & TURNER, D. (1805). *The botanist's guide through England and Wales*. London.

DRUCE, G.C. (1932). *The comital Flora of the British Isles*. Arbroath.

DUDMAN, A.A. (1997). *Eleagnus commutata* at Workington Oldside (v.c.70). *B.S.B.I. News*, **74**:47-48.

DUDMAN, A.A. & RICHARDS, A.J. (1997). *Dandelions of Great Britain and Ireland.* B.S.B.I. Handbook no. 9. London.

DUNCAN, J.E. & ROBSON, R.W. (1977). *Pennine flowers.* Clapham.EDDY, A. & WELCH, D. (1967). A species-list of flowering plants and ferns for the Moor House National Nature Reserve in Upper Teesdale. *Proceedings of the Botanical Society of the British Isles*, **6**:325-336.

EDDY, A., WELCH, D. & RAWES, M. (1969). The vegetation of Moor House National Nature Reserve in the northern Pennines, England. *Vegetatio*, **16**:239-284.

EDEES, E.S. & NEWTON, A. (1988). *Brambles of the British Isles.* London.

ELLIS, J.R. (1975). *Iris*, in STACE, C.A., ed., *Hybridization and the flora of the British Isles*, pp.472-473. London.

FOLEY, M.J.Y. (1987). The current distribution and abundance of *Orchis ustulata* L. in northern England. *Watsonia*, **16**:409-415.

FOLEY, M.J.Y. (1992). The current distribution and abundance of *Orchis ustulata* L. (Orchidaceae) in the British Isles - an updated summary. *Watsonia*, **19**:121-126.

FOLEY, M.J.Y. & PORTER, M.S. (1993). *Typha angustifolia* L. in Cumbria. *B.S.B.I.News*, **63**:49.

FRYER, G. (1991). *A natural history of the lakes, tarns and streams of the English Lake District.* Ambleside.

GAME, A.G. (1983). Taxonomic studies of montane *Erigeron* species in Europe, and of the *Sorbus rupicola* (Syme) Hedl. complex in north-western England. Unpublished Ph.D. thesis, University of Lancaster.

GIBSON, G.S. (1846). Botanical notes for 1845. *Phytologist*, 2:373-377.

GILL, J.J.B., McALLISTER, H.A. & FEARN, G.M. (1978). Cytotaxonomic studies on the *Cochlearia officinalis* L. group from inland stations in Britain. *Watsonia*, **12**:15-21.

GORNALL, R.J. (1987). Notes on a hybrid spearwort *Ranunculus flammula* L. x *R. reptans* L. *Watsonia*, **16**: 383-388.

GRAHAM, G.G. (1988). *The flora and vegetation of County Durham.* Durham.

GRAHAM, G.G. & PRIMAVESI, A.L. (1993). *Roses of Great Britain and Ireland.* B.S.B.I. Handbook no. 7. London

HALLIDAY, G. (1978). *Flowering plants and ferns of Cumbria.* Centre for North-West Regional Studies. University of Lancaster, Occasional paper no.4. Lancaster.

HALLIDAY, G. (1990). *Crepis praemorsa* (L.) Tausch, new to western Europe. *Watsonia*, **18**:85-87.

HALLIDAY, G. (1995). Two subspecies of *Festuca rubra* L. new to England. *Watsonia*, **20**:412.

HANDLEY, J. (1898). *Catalogue of plants growing in the Sedbergh District.* Leeds.

HARDY, J. (1881). Report of meetings for 1880. *History of the Berwickshire Naturalists' Club*, **9**:214-290.

HATCHER, P. & ALEXANDER, K.N.A. (1994). The status and conservation of the netted carpet *Eustroma reticulata* (Denis & Schiffermüller, 1775) (Lepidoptera: Geometridae), a threatened moth species in Britain. *Biological Conservation*, **67**:41-47.

HAY, T. Physiographical notes on the Ullswater area. *The Geographical Journal*, **90**: 426-445.

HEADLAM, C., ed. (1939). *The three northern counties of England.* Gateshead.

HENRY, J. (1897). New plant records for Lake-Lancashire. *Naturalist*, **1897**:127.

HERVEY, G.A.K. & BARNES, J.A.G. eds. (1970). *Natural history of the Lake District.* London.

HODGSON, E. (1862). On a deposit containing Diatomaceae, leaves, etc., in the iron-ore mines near Ulverston. *Quarterly Journal of the Geological Society of London*, **19**:19-31.

HODGSON, E. (1874). North or Lake Lancashire; a sketch of its botany, geology and physical geography. *Journal of Botany (London)*, **3**:268-277, 296-305.

HODGSON, W. (1898). *Flora of Cumberland.* Carlisle.

HOFMANN, H., EDMONDSON, J.R. & HALLIDAY, G. (1990). The Cumbrian herbarium of W.H. Youdale. *Watsonia*, **18**:204-206.

HOLLINGWORTH, S.E. (1934). Some solifluction phenomena in the northern part of the Lake District. *Proceedings of the Geological Association*, **15**:167.

HUBBARD, C.E. (1984). *Grasses*, 3rd ed. Harmondsworth.

HUDSON, W. (1778). *Flora Anglica*, 2nd ed. London.

HUNT, P.F. (1975). *Platanthera*, in Stace, C.A., ed., *Hybridization and the flora of the British Isles*, p.488. London.

HUTCHINSON, W. (1794). *The history of the county of Cumberland, and some places adjacent,....* Carlisle.

HUTCHINSON, W. (1823). The history and antiquities of the county palatine of Durham, 2nd ed. Durham.

JACKSON, F. (1965). In British Pteridological Society Meetings 1965. *British Fern Gazette*, **9**:241.

JERMY, A.C., ARNOLD, H.R., FARRELL, L. & PERRING, F.H. (1978). *Atlas of ferns of the British Isles.* London.

JERMY, A.C., CHATER, A.O. & DAVID, R.W. (1982). *Sedges of the British Isles*, 2nd ed. B.S.B.I. Handbook no. 1. London.

JOBLING, J. (1990). *Poplars for wood production and amenity.* Forestry Commission Bulletin no. 92. London.

KAY, Q.O.N. (1995). *The conservation of scarce and declining plant species in lowland Wales.* Countryside Council for Wales, Science Report no. 110.

KELLY, P.G. & PERRY, K.A. (1990). *Wildlife habitat in Cumbria.* Nature Conservancy Council, Research and survey in nature conservation, no. 30. Peterborough.

KENDALL, J.D. (1881). Interglacial deposits of west Cumberland and north Lancashire. *Quarterly Journal of the Geological Society of London*, **37**:29-39.

KENT, D.H. (1964). *Senecio squalidus* L. in the British Isles, 6. Northern England. *Proceedings of the Botanical Society of the British Isles*, **5**:217-220.

KENT, D.H. (1992). *List of vascular plants of the British Isles.* London.

LAWSON, A. (1992). A comparative study involving the morphology and ecology of two *Myriophyllum* species within the county of Cumbria. Unpublished undergraduate dissertation, University of Lancaster.

LEES, F.A. (1888). *The flora of West Yorkshire.* London.

LEES, F.A. (1900). Spring's pageant in Westmorland and Lancashire. *Naturalist*, **1900**:277-284.

LEES, F.A. (1938). The vegetation of Yorkshire and supplement to the Floras of the county, ed. Cheetham, C.A. *Naturalist*, **1938**:101-108.

LESLIE, A.C. (1978). *Fragaria moschata* Duchesne and *Fragaria vesca* L. *B.S.B.I. News*, **18**:27.

LEY, A. & LINTON, W.R. (1906). Some plants of the English Lake District. *Journal of Botany (London)*, **44**:171-173.

LIVERMORE, L.A. & LIVERMORE, P.D. (1987). *The flowering plants and ferns of north Lancashire.* Lancaster.

LIVERMORE, L.A. & LIVERMORE, P.D. (1988). *Azolla filiculoides* in the Lancaster canal. *Pteridologist*, **1**: 214-216.

LLOYD, P.S. (1964). *Poa alpina. Proceedings of the Botanical Society of the British Isles*, **5**:232-233.

LOUSLEY, J.E. & KENT, D.H. (1981). *Docks and knotweeds of the British Isles*. B.S.B.I. Handbook no. 3. London.

LOVIS, J.D. (1975). *Asplenium*, in STACE, C.A., ed., *Hybridization and the flora of the British Isles*, pp.106-111. London.

McALLISTER, H.A. & RUTHERFORD, A. (1990). *Hedera helix* L. and *H. hibernica* (Kirkner) Bean (Araliaceae) in the British Isles. *Watsonia*, **18**:7-15.

MANLEY, G. (1952). *Climate and the British scene*. London.

MANLEY, G. (1973). Climate, in Pearsall, W.H. & Pennington, W., *The Lake District*, pp.106-120. London.

MANTON, I. (1950). Problems of cytology and evolution in the Pteridophyta. Cambridge.

MATTHEWS, J.R. (1955). *Origin and distribution of the British flora*. London.

MEIKLE, R.D. (1975). *Salix*, in Stace, C.A. ed., *Hybridization and the flora of the British Isles*, pp.304-338. London.

MEIKLE, R.D. (1984). *Willows and poplars of Great Britain and Ireland*. B.S.B.I. Handbook no. 4. London.

MELDERIS, A. (1975). *Agropyron*, in Stace, C.A. ed., *Hybridization and the flora of the British Isles*, pp.568-570. London.

METEOROLOGICAL OFFICE (1952). *Climatological Atlas of the British Isles*. London.

MOGFORD, J. (1974). Flower colour polymorphism in *Cirsium palustre. Heredity*, **33**:241-256.

MOSELEY, F. ed. (1978). *The geology of the Lake District*. Yorkshire Geological Society Occasional publication no.3. Leeds.

NEW, J.K. (1958). A population study of *Spergula arvensis*. 1. *Annals of Botany*, n.s. **22**:457-477.

OLDFIELD, F. (1960). Studies in the post-glacial history of British vegetation: Lowland Lonsdale. *New Phytologist*, **59**: 192-217.

OLDFIELD, F. (1965). Pollen analysis and man's role in the ecological history of the south-east Lake District. *Geografiska Annaler*, **45(1)**:23-40.

OLDFIELD, F. & STATHAM, D.C. (1963). Pollen-analytical data from Urswick Tarn and Ellerside Moss, North Lancashire. *New Phytologist*, **62**:53-66.

PAGE, C.N. (1988). *Ferns*. London.

PAGE, C.N. (1989). Three subspecies of bracken, *Pteridium aquililum* (L.) Kuhn, in Britain. *Watsonia*, **17**:429-434.

PARKER, D.M. (1986). Dactyloid saxifrages. *Watsonia*, **16**:223-224.

PAUL, A.M. (1987). The status of *Ophioglossum azoricum* (Ophioglossaceae: Pteridophyta) in the British Isles. *The Fern Gazette*, **13**:173-187.

PEARSALL, W.H. & PEARSALL, W.H., fil. (1921). *Potamogeton* in the English Lakes, 1. *Journal of Botany (London)*. **59**:160-164.

PEARSALL, W.H. & PEARSALL, W.H., fil. (1923). *Potamogeton* in the English Lakes, 2. *Journal of Botany (London)*. **61**:1-7

PEARSALL, W.H., fil. (1950). *Mountains and moorlands*. London.

PEARSALL, W.H., fil. & PENNINGTON, W. (1973). *The Lake District*. London.

PENNINGTON, W. (1947). Studies of the postglacial history of British vegetation 7. Lake sediments: Pollen diagrams from the bottom deposits of the North Basin of Windermere. *Philosophical Transactions of the Royal Society*, **B 233**:137-175.

PENNINGTON, W. (1964). Pollen analyses from the deposits of six upland tarns in the Lake District. *Philosophical Transactions of the Royal Society*, **B 248**:205-244.

PENNINGTON, W. (1965). The interpretation of some postglacial vegetation diversities at different Lake District sites. *Proceedings of the Royal Society*, **B 161**:310-323.

PENNINGTON, W. (1970). Vegetation history in the north-west of England: a regional synthesis. In Walker, D. & West, R.G., eds., *Studies in the vegetational history of the British Isles*, pp.41-79. Cambridge.

PENNINGTON, W. (1975). The effect of Neolithic man on the environment in north-west England: the use of absolute pollen diagrams. In Evans, J.G., Limbrey, S. & Cleere, H., eds, *The effect of man on the landscape: The Highland Zone.* Council for British Archaeological Research Report no. 11, pp.74-86.

PENNINGTON, W. (1977). The late-glacial flora and vegetation of Britain. *Philosophical Transactions of the Royal Society*, **B 280**:247-271.

PENNINGTON, W. (1979). The origin of pollen in lake sediments: an enclosed lake compared with one receiving inflow streams. *New Phytologist*, **82**:189-213.

PENNINGTON, W. (1984). Long-term natural acidification of upland sites in Cumbria: evidence from post-glacial lake sediments. In *Annual Report of the Freshwater Biological Association*, pp.28-46.

PENNINGTON, W. (1991). Palaeolimnology in the English Lakes - some questions and answers over fifty years. *Hydrobiologia*, **214**:9-24.

PERRING, F.H. (1973). Mistletoe, in Green, P.S., ed., *Plants wild and cultivated*, pp. 139-145. London.

PERRING, F.H. & SELL, P.D. (1968). *Critical supplement to the Atlas of the British flora*. London.

PERRING, F.H. & WALTERS, S.M. (1962). *Atlas of the flora of the British Isles*. London.

PETTY, S.L. (1894-1900). The constituents of the North Lancashire flora, 1597(?)-1893. *Naturalist*, **1894**:117-126; 149-156, 243-250,289-296; **1895**:53-60, 133-140, 261-268; **1896**:101-108,197-204, 341-348; **1897**:85-92, 229-236, 309-316, 325-332; **1898**:37-44, 149-156, 309-316, 321-333; **1900**:333-335.

PETTY, S.L. (1903). North Lancashire botanical notes in 1902. *Naturalist*, **1902**:84-86.

PIGOTT, C.D. (1992a). The clones of common lime (*Tilia* x *vulgaris* Hayne) planted in England during the seventeenth and eighteenth centuries. *New Phytologist*, **122**:487-493.

PIGOTT, C.D. (1992b). Are the distributions of species set by failure to set seed? In Marshall, C. & Grace, J. eds, *Fruit and seed production*, pp.203-216. Cambridge.

PIGOTT, C.D. & HUNTLEY, J.P. (1978). Factors controlling the distribution of *Tilia cordata* at the northern limits of its geographical range. 1. Distribution in north-west England. *New Phytologist*, **81**:429-441.

PIGOTT, C.D. & HUNTLEY, J.P. (1980). Factors controlling the distribution of *Tilia cordata* at the northern limits of its geographical range. 2. History in north-west England. *New Phytologist*, **84**:145-164.

PRESTON, C.D. & CROFT, J.M. (1997). *Aquatic plants in Britain and Ireland*. Colchester.

PROCTER, J. & JOHNSTON, W.R. (1977). *Lychnis alpina* L. in Britain. *Watsonia*, **11**:199-204.

PROCTOR, M.C.F. (1965). The distinguishing characters and geographical distributions of *Ulex minor* and *U. gallii*. *Watsonia*, **6**:177-187.

PUGSLEY, H.W. (1930). A revision of the British Euphrasiae. *Journal of the Linnean Society of London (Botany)*, **48**:467-544.

PUGSLEY, H.W. (1948) A prodromus of the British Hieracia. *Journal of the Linnean Society of London (Botany)*, **54**.

RATCLIFFE, D.A. (1960). The mountain flora of Lakeland. *Proceedings of the Botanical Society of the British Isles*, **4**:1-25.

RATCLIFFE, D. A. & EDDY, E. (1960). *Alopecurus alpinus* Sm. in Britain. *Proceedings of the Botanical Society of the British Isles*, **3**:389-391.

RAVEN, P.H. (1963). *Circaea* in the British Isles. *Watsonia*, **5**:262-272.

RAY, J. (1670). *Catalogus Plantarum Angliae*. London.

RAY, J. (1688). *Historia Plantarum*, **2**. London.

RAY, J. (1696). *Synopsis methodica Stirpium Britannicarum*, 2nd ed. London.

RICH, T.C.G. (1991). *Crucifers of Great Britain and Ireland*. B.S.B.I. Handbook no. 6. London.

RICH, T.C.G. & BAECKER, M. (1986). The distribution of *Sorbus lancastriensis* E.F. Warburg. *Watsonia*, **16**:83-85.

RICH, T.C.G. & RICH, M.D.B. (1988). *Plant crib*. London.

RICHARDS, A.J. (1972). The *Taraxacum* flora of the British Isles. *Watsonia* **9**, Suppl.

RICKARD, M.H. (1972). The distribution of *Woodsia ilvensis* and *W. alpina* in Britain. *The British Fern Gazette*, **10**:269-280.

ROBERTS, F.J. & HALLIDAY, G. (1979). The altitudinal range of *Catabrosa aquatica* (L.) Beauv. *Watsonia*, **12**:342-343.

RODWELL, J.S. ed. (1991-1996). *British plant communities*, **1-4**. Cambridge.

ROGERS, W.M. (1907). Plants of the English Lake District. *Journal of Botany (London)*, **45**:8-12.

ROWLAND, S.P. (1938). *Dryopteris rigida* Underwood in Cumberland. *The North Western Naturalist*, **13**:231-232.

RUMSEY, F.J., Thompson, P. & SHEFFIELD, E. (1993). *Isoetes echinospora* (Isoetaceae: Pteridophyta) in northern England. *British Fern Gazette*, 14:215-221.

SALMON, C.E. (1926). A new *Myosotis* from Britain. *Journal of Botany (London)*, **64**:289-295.

SARGENT, C., MOUNTFORD, O. & GREEN, D. (1986). The distribution of *Poa angustifolia* L. in Britain. *Watsonia*, **16**:31-36.

SCHROEDER, F.-G. (1970). Exotic *Amelanchier* species naturalised in Europe and their occurrence in Great Britain. *Watsonia*, **8**:155-162.

SELL, P.D. (1986). The genus *Cicerbita* Wallr. in the British Isles. *Watsonia*, **16**:121-129.

SELL, P.D. (1988) *Pilosella*, in Rich, T.C.G. & Rich, M.D.B., *Plant crib*. London.

SELL, P.D. (1989). The *Sorbus latifolia* (Lam.) Pers. aggregrate in the British Isles. *Watsonia*, 17:385-399.

SILVERSIDE, A.J. (1991). The identity of *Euphrasia officinalis* L. and its nomenclatural implications. *Watsonia*, **18**:343-350.

SILVERSIDE, A.J. (1994). *Euphrasia rostkoviana* Hayne subsp. *rostkoviana*, in Stewart, A., Pearman, D.A. & Preston, C.D., *Scarce plants in Britain*. Peterborough.

SIMPSON, D.A. (1984). A short history of the introduction and spread of *Elodea* Michx in the British Isles. *Watsonia*, **15**:1-9.

SIMPSON, D.A. (1990). Displacement of *Elodea canadensis* Michx by *Elodea nuttallii* (Planch.) H. St-John in the British Isles. *Watsonia*, **18**:173-177.

SISSONS, J.B. (1980). The Loch Lomond Advance in the Lake District, northern England. *Transactions of the Royal Society of Edinburgh: Earth Sciences*, **71**:13-27.

SLEDGE, W.A. (1945). Report of Botanical Records Committee, in The Yorkshire Naturalists' Union eighty-third Annual Report. *Naturalist*, **1945**:19-38.

SMITH, A.G. (1958). Two lacustrine deposits in the south of the English Lake District. *New Phytologist*, **57**:363-386.

SMITH, A.G. (1959). The mires of south-western Westmorland: stratigraphy and pollen analysis. *New Phytologist*, **58**:105-127.

SMITH, P.M. (1968). The *Bromus mollis* aggregate in Britain. *Watsonia*, **6**:327-344.

SPROTT, W.A.P. (1936). New light on a *Pentandra* willow. *Journal of Botany (London)*, **74**:230-233.

STACE, C.A. (1972). The history and occurrence in Britain of hybrids in *Juncus* subgenus Genuinii. *Watsonia*, **9**:1-11.

STACE, C.A. (1975), *Epilobium*, in Stace, C.A., ed., *Hybridization and the flora of the British Isles*, pp.246-265. London.

STACE, C.A. (1988). *Festuca*, in Rich, T.C.G. & Rich, M.D.B., *Plant crib*, p.130. London.

STACE, C.A. (1991). *New Flora of the British Isles*. Cambridge.

STACE, C.A. & INGROUILLE, M. (1986). The *Limonium binervosum* aggregate (Plumbaginaceae) in the British Isles. *Journal of the Linnean Society (Botany)*, **92**:177-217.

STANSFIELD, F.W. (1919). Editorial notes. *British Fern Gazette*, n.s., **4**:1-4.

STEWART, A., PEARMAN, D.A. & PRESTON, C.D. (1994). *Scarce plants in Britain*. Peterborough.

STEWART, O. (1988). *Calamagrostis purpurea* (Trin.) Trin. from Scotland and other Scottish records. *B.S.B.I. News*, **49**:51.

STEWART, O. (1991). A Scottish miscellany. *B.S.B.I. News*, **57**:55.

STEWART, O. (1990). Flowering plants of Kirkcudbrightshire. *The Transactions of the Dumfriesshire and Galloway Antiquarian Society*, **65**.

STOKOE, R. (1962). *The birds of the Lake counties*. Carlisle.

STOKOE, R. (1978). *Isoetes echinospora* Durieu new to northern England. *Watsonia*, **12**:51-52.

STOKOE, R. (1983). *Aquatic macrophytes in the tarns and lakes of Cumbria*. Freshwater Biological Association,

THOMSON, W. (1902). Botanical finds near Carlisle in 1901. *Naturalist*, **1902**:129.

THOMPSON, D.B.A. & BADDELEY, J.A. (1991). Some effects of acidic deposition on montane *Racomitrium lanuginosum* heaths. In Woodwin, S.J. & Farmer, A., *The effects of acid deposition on nature conservation in Great Britain*, pp.17-28. Peterborough.

TRIST, P.J.O. (1986). A reconsideration of the taxonomic status of *Poa balfourii* Parnell (Gramineae). *Watsonia*, **16**:37-42.

TROTMAN, M.J. (1995). Irish lousewort (*Pedicularis sylvatica* subsp. *hibernica*) new to England. *B.S.B.I. News*, **68**:16.

TURNER, J. (1984). Pollen diagrams from Cross Fell and their implication for former tree-lines. In Haworth, E.Y. & Lund, J.W.G., *Lake sediments and environmental history*, pp. 317-357. Leicester.

TYLER, B.J. (1988). From clouds to lakes. *The North Western Naturalist*, **1988**:1-7.

VALENTINE, D.H. & HARVEY, M.J. (1961). *Viola rupestris* Schmidt in Britain. *Proceedings of the Botanical Society of the British Isles*, **4**:129-135.

WALKER, D. (1965). The post-glacial period in the Langdale Fells, English Lake District. *New Phytologist*, **64**:488-510.

WALKER, D. (1966). The late Quaternary history of the Cumberland lowland. *Philosophical Transactions of the Royal Society*, **B 251**:1-210.

WALTERS, S.M. (1963). *Eleocharis austriaca* Hayek, a species new to the British Isles. *Watsonia*, **5**:329-335.

WARBURG, E.F. (1962). *Ribes*, in Clapham, A.R., Tutin, T.G. & Warburg, E.F., *Flora of the British Isles*, 2nd ed. Cambridge.

WEBSTER, S.D. (1988). *Ranunculus penicillatus* (Dumort.) Bab. in Great Britain and Ireland. *Watsonia*, **17**:1-22.

WHITTAKER, E.J. (1981). *A seventeenth century flora of Cumbria*. Gateshead.

WHITTAKER, E.J. (1986). *Thomas Lawson*. York.

WIGGINTON, M.J. & GRAHAM, G.G. (1981). *Guide to the identification of some of the more difficult vascular plant species*. Nature Conservancy Council England Field Unit, Occasional paper no.1. Banbury.

WILLIAMSON, S. (1989). *Betula pubescens* and *B. pendula* in Cumbria. Unpublished undergraduate dissertation, University of Lancaster.

WILSON, A. (1938). *The Flora of Westmorland*. Arbroath.

WILSON, A. (1956). The altitudinal range of British Plants. *The North Western Naturalist*, Supplement.

YEO, P.F. (1975). *Euphrasia*, in Stace, C.A., ed., *Hybridization and the flora of the British Isles*, pp.373-381. London.

YOUNG, D.P. (1958). *Oxalis* in the British Isles. *Watsonia*, **4**:51-69.

Index

The Latin names of taxa recorded during the Survey are in bold type. English names and non-Survey taxa are in normal type, synonyms are in italics and dubious records in square brackets. Page numbers include references in the introductory sections. The main entry in the systematic accounts is in bold, the suffixes **f** and **p** refer to figures and plates respectively and **m** to the map where it is on a different page to the main entry.